OXFORD LOGIC GUIDES: 38

OXFORD LOGIC GUIDES

1. Jane Bridge: *Beginning model theory: the completeness theorem and some consequences*
2. Michael Dummett: *Elements of intuitionism*
3. A.S. Troelstra: *Choice sequences: a chapter of intuitionistic mathematics*
4. J.L. Bell: *Boolean-valued models and independence proofs in set theory* (1st edition)
5. Krister Seberberg: *Classical propositional operators: an exercise in the foundation of logic*
6. G.C. Smith: *The Boole–De Morgan correspondence 1842–1864*
7. Alec Fisher: *Formal number theory and computability: a work book*
8. Anand Pillay: *An introduction to stability theory*
9. H.E. Rose: *Subrecursion: functions and hierarchies*
10. Michael Hallett: *Cantorian set theory and limitation of size*
11. R. Mansfield and G. Weitkamp: *Recursive aspects of descriptive set theory*
12. J.L. Bell: *Boolean-valued models and independence proofs in set theory* (2nd edition)
13. Melvin Fitting: *Computability theory: semantics and logic programming*
14. J.L. Bell: *Toposes and local set theories: an introduction*
15. R. Kaye: *Models of Peano arithmetic*
16. J. Chapman and F. Rowbottom: *Relative category theory and geometric morphisms: a logical approach*
17. Stewart Shapiro: *Foundations without foundationalism*
18. John P. Cleave: *A study of logics*
19. R.M. Smullyan: *Gödel's incompleteness theorems*
20. T.E. Forster: *Set theory with a universal set: exploring an untyped universe*
21. C. McLarty: *Elementary categories, elementary toposes*
22. R.M. Smullyan: *Recursion theory for metamathematics*
23. Peter Clote and Jan Krajíček: *Arithmetic, proof theory, and computational complexity*
24. A. Tarski: *Introduction to logic and to the methodology of deductive sciences*
25. G. Malinowski: *Many valued logics*
26. Alexandre Borovik and Ali Nesin: *Groups of finite Morley rank*
27. R.M. Smullyan: *Diagonalization and self-reference*
28. Dov M. Gabbay, Ian Hodkinson, and Mark Reynolds: *Temporal logic: mathematical foundations and computational aspects: Volume 1*
29. Saharon Shelah: *Cardinal arithmetic*
30. Erik Sandewall: *Features and fluents: Volume I: a systematic approach to the representation of knowledge about dynamical systems*
31. T.E. Forster: *Set theory with a universal set: exploring an untyped universe* (2nd edition)
32. Anand Pillay: *Geometric stability theory*
33. Dov M. Gabbay: *Labelled deductive systems*
34. Raymond M. Smullyan and Melvin Fitting: *Set theory and the continuum problem*
35. Alexander Chagrov and Michael Zakharyaschev: *Modal logic*
36. G. Sambin and J. Smith: *Twenty five years of constructive type theory*
37. Maria Manzano: *Model theory*
38. Dov M. Gabbay: *Fibring logics*

Fibring Logics

DOV M. GABBAY

King's College, London

CLARENDON PRESS • OXFORD

1999

Oxford University Press, Great Clarendon Street, Oxford OX2 6DP
Oxford New York
Athens Auckland Bangkok Bogota Buenos Aires Calcutta
Cape Town Chennai Dar es Salaam Delhi Florence Hong Kong Istanbul
Karachi Kuala Lumpur Madrid Melbourne Mexico City Mumbai
Nairobi Paris São Paolo Singapore Taipei Tokyo Toronto Warsaw
and associated companies in
Berlin Ibadan

Oxford is a registered trade mark of Oxford University Press

Published in the United States by
Oxford University Press Inc., New York

A catalogue record for this book is available from the British Library

Library of Congress Cataloging in Publication Data
(Data available)

ISBN 0 19 850381 4 (Hbk)

Typeset using LaTeX
Printed in Great Britain by
Biddles Ltd, Guildford & King's Lynn

To Matthias Rivlin
who appreciated learning and always helped

PREFACE

I am happy to present my book on fibring logics and systems. It is based on my research in the period 1990–1996, mainly during my EPSRC Senior Research Fellowship. The book is based to a large extent on expanded versions of my papers on fibring in that period.

In preparing the book I had to consider whether to let the various topics and chapters be slightly independent (at the cost of slight repetition of the basic definitions) or to concentrate all basic concepts at the beginning. I decided to opt for the former. The reason is mainly that fibring is used in many diverse research areas and the different chapters and topics will be of interest to several different and possibly disjoint communities of readers. It seems reasonable to make it easy for the reader to get into the fibring relevant to his interests. Furthermore, the fibring definitions and concepts are never exactly the same across different application areas, and so further clarification would have been necessary anyway.

Chapter 13 is based on a joint paper with M. Finger.

Chapter 14 is based on a joint paper with M. D'Agostino.

Chapter 15 is based on a joint paper with M. D'Agostino and A. Russo.

Chapter 16 is based on a joint paper with V. Shehtman.

Chapters 17 and 18 are based on joint papers with B. Beckert.

Chapter 19 is based on a joint paper with G. Govenatori.

I am fortunate and grateful to have such wonderful colleagues, and thank them for helping revise the respective chapters.

Throughout the book I use the following conventions for notation:

$\sim, \wedge, \vee, \rightarrow$	classical connectives
$\neg, \wedge, \vee, \rightharpoonup$	intuitionistic connectives
$\twoheadrightarrow$	special implication
$\&$	the and symbol (in programs)
$(x), (Ex)$	quantifiers over existing elements
$(\forall x), (\exists x)$	quantifiers over all possible elements
$\mathbb{L}, \mathbb{N}$	languages
$\mathbf{L}, \mathbf{K}, \mathbf{S4}$	logical systems
$\mathcal{L}, \mathcal{K}, \mathcal{M}$	semantics (families of models)
$\mathbf{m}, \mathbf{n}$	models
h	assignment to predicates
g	assignments to constants and function symbols
$\mathcal{A}$	algebra of labels
A, B, C	formulas
α, β, γ	formulas or sequences or whatever is special in the local context
t, s	possible worlds; labels

$\mathbf{F}$	fibring function
$\mathbb{J}$	jump operator
$\mathbb{C}, \mathbb{K}, \mathbb{R}$	special sets of whatever is in the local context
$\mathfrak{M}$	automaton
$\vdash, \mid\!\sim$	consequence relations
$\models$	semantic satisfaction
$\prec, \leq$	orderings
$\mathcal{S}^u_a$	substitution operator
$>$	conditional
$a : A$	declarative unit (label: formula)
$\mathsf{T}A : a$	The formula A is to be true at the world (label) a. Similarly $\mathsf{F}A : a$.

I thank my administrator, Mrs Jane Spurr, for producing yet another of my manuscripts with her exceptional efficiency and dedication.

D. M. G.
London
March 1998

CONTENTS

1

AN OVERVIEW OF FIBRED SEMANTICS AND THE COMBINATION OF LOGICS

1.1 Introduction

The purpose of this chapter is to present an intuitive explanation of the author's fibred semantics methodology for combining logics and systems, and to give a brief overview of the main results, ideas and open problems in the book.

The problem of combining logics and systems is central to modern logic, both pure and applied. The need to combine logics arises both from applications and from within logic itself as a discipline. As logic is being used more and more to formalise field problems in philosophy, language, artificial intelligence, logic programming and computer science, the kind of logics required become more and more complex. An increasing number of features from the application area need to be formally represented. These features are highly mutually interactive and a formal study of their combined nature becomes necessary. A methodology for combining systems has now become essential for most logic applications.

A simple example will illustrate the complexity of the problem we are facing. There exist well known and well studied logical systems and theories of belief and knowledge. Such systems use notation such as $B\varphi$ (φ is believed) and $K\varphi$ (φ is known). The formal theory of such notions has been studied in the past 30 years. Similarly there are well known systems of temporal logics, with operators such as $G\varphi$ (φ will always be true) and $H\varphi$ (φ has always been true). Such systems were developed in connection with the study of language and some problems in the philosophy of time. With the rise of artificial intelligence as a consumer of logic, various theories and models of reactive agents were put forward. Such theories involve a cycle of temporally evolving and changing repositories of knowledge and beliefs, as well as theories of actions for agents making use of their changing knowledge and beliefs.

Clearly any logical system modelling reactive agents should be a combined system of logics of knowledge, belief, time and modal logics (of actions).

It is therefore clear that a good understanding and practical methodology for combining logics can provide the consumer of logic (e.g. a robotics designer) with tools which make him more competent in designing his own system. Such a methodology can help decompose the problem of designing a complex system into developing components (logics) and combining them.

Furthermore, such a methodology might allow the designer to use existing components (which may be very well understood and/or efficiently implemented) and put them together into a desired useful combined system, at a relatively acceptable intellectual and practical cost.

In fact, a good combining methodology allows the study of the complex application area in terms of some of its pure components, yielding both theoretical and practical benefits.

A good methodology of combining systems should study *transfer theorems*. Such theorems prove that if the components satisfy property $\mathcal{P}$, then we are assured that the combined system satisfies (a possibly slightly different) property $\mathcal{P}'$. Moreover, transfer theorems can also help validate existing complex systems by verifying properties of their components.

It turns out that our fibring methodology not only helps realise many of the above expectations, but also seems to bring out unexpected theoretical properties of pure logic itself, showing new connections between existing themes of logic.

The following is a list of the main areas where the fibring methodology makes a serious contribution.

- Straightforward combination of logics such as multimodal logics or modal intuitionistic logics, or modal substructural logics.
- Allowing one logic to act as a metalevel language for talking about another logic (e.g. transition systems).
- Combining a metalevel language with its object level language. For example the problem of bringing a consequence relation into the object language as a conditional, or allowing one part of a language to talk about another.
- Allowing for a single language to be syntactically ambivalent, and regarding the various ambivalent 'aspects' as a combined system. For example allowing classical logic formulas to apply to themselves (e.g. $\varphi(\varphi)$).
- Modelling the temporal behaviour of a system by fibring it with a temporal logic.
- Making a system fuzzy by fibring it with a fuzzy logic.

The power of the concept of fibring is also emphasised by some other recent research activities that arose independently from our own work. In [Pfalzgraf, 1991; Pfalzgraf and Stokkermans, 1994] the introduction of *logical fibrings* has been strongly motivated by the classical notion of fibre bundles and sheaves. Concrete applications aim at modelling logical control of cooperating robots (agents) scenarios. In [Baader and Schulz, 1994; Baader and Schulz, 1995a], another variant of fibring is used for combining solution domains and constraint solvers for symbolic constraints.

1.1.1 *Combining pure logical systems*

Problems of combining two logics arise both in pure logical theory and in practical applications. We give several typical examples

1. *Modal intuitionistic logics*
   The logic of modality $\Box$ was originally philosophically motivated. The same is true for intuitionistic logic. Intuitionistic logic arose as the logic for constructive mathematics. It was natural therefore, for philosophers and pure logicians, to try and study the nature of modal necessity $\Box$ and possibility $\Diamond$ from a constructive point of view. Such studies gave rise to systems involving both intuitionistic implication $\Rightarrow$ and the modal operators $\Box$ and $\Diamond$. In the past 30 years

many modal intuitionistic systems were put forward, by philosophically motivated considerations. These systems were formulated directly for the mixed language and were not presented as combined systems. Various methods were used in the presentation, some semantical, some theoretical, some straightforward and some rather roundabout and highly individual. Among the early philosophically motivated systems by famous logicians were Fitch [1948], Bull [1965] and Ono [1977; 1987].

Later, as modal logic had serious applications in computer science, a new wave of interest in modal intuitionistic logic arose, by virtue of its applicability. The modality could stand for possible algorithmic options and the intuitionistic proof theory for the constructive nature of the execution. Sample papers in this category are Wijesekera [1990] and A. K. Simpson [1993].

After a while logicians got fascinated by modal intuitionistic logics for formal technical reasons. Many papers were written without any special philosophical or field applications motivation. Among them are [Blackburn and de Rijke, 1995; Font, 1986, Fischer-Servi, 1977; Fischer-Servi, 1984; Fischer-Servi, 1980; Došen, 1985; Suzuki, 1990; Božić and Došen, 1984, Ewald, 1986; Ono, 1987; Suzuki, 1988].

Among the technical problems involved in the modal intuitionistic logic area are:

- proof theory for the system. Hilbert formulation, Gentzen formulation, tableaux formulation, etc.;
- Kripke type semantics for the system;
- decision procedures;
- finite model property;
- comparison with other systems of modal intuitionistic logics, in terms of expressive power and relative interpretability.

We shall see that our fibring methodology can yield many of the above systems and results automatically and in bulk. Our main paper in this area is [Gabbay, 1995b], which includes a detailed discussion of the main systems in the literature. In Chapters 3–5 we present the main theorems of this paper.

2. *Syntactical modality*

   Here we mean that a modal operator may arise naturally in another logic by a syntactic definition. For example in Girard linear logic, which is independently and very deeply motivated, some modalities (the exclamation !) arise naturally by definition. Studying their properties is of course the study of a combined system. Another example is the modality $\Box A = \text{def}(A \Rightarrow A) \Rightarrow A$ in Anderson and Belnap's *Entailment and Relevant Logics* [Anderson and Belnap Jr, 1975]. The above does turn out to be some sort of modal-relevant system. See also the paper of Meyer and Mares [1993].

   Modalities can also be obtained as by-products of translations and interpretations. We mention two well-known examples.

   (a) the strict implication of modal **S4** for boxed formulas is intuitionistic implication, thus giving rise to a combined modal intuitionistic logic.

(b) the semantics translation of modal logic into classical predicate logic gives rise to a modal intuitionistic logic when classical predicate logic is weakened into intuitionistic predicate logic.

3. *Multimodal logics*
These arise from applications where several modal operators interact. The application may be a philosophical study of some notions and its paradoxes, such as systems of obligation and permission, see Jones and Pörn, [1985; 1986] or a direct formalisation of some practical application area, e.g. logical omniscience, see [Fagin *et al.*, 1990].

A large very useful area of multimodal logic is dynamic logic. In fact, some applied logicians believe that multimodal logic is *the logic* for computer science applications.

The bulk of the literature on combined systems is mainly concerned with the presentation of individual combined systems. There are a few papers on methodological questions and transfer theorems. Besides my own papers on fibring methodology, there are some theoretical studies of multiple modalities as well as some algebraic studies of transfer properties. See [Fischer-Servi, 1977; Fitting, 1969; Finger and Gabbay, 1996; Kracht and Wolter, 1991; Goranko and Passy, 1992]. Again, I am happy to say that the fibred semantics methodology gives bulk results in multimodal logics. See [Gabbay, 1995b].

1.1.2 *Combining meta-level with the object level*

This is a new application of the fibred methodology.

Consider a consequence relation of the form $\Delta \mid\!\sim B$ between wffs of a logic. Assume it is a non-monotonic consequence relation on classical language satisfying Reflexivity ($\Delta \mid\!\sim A$ if $A \in \Delta$); Restricted monotonicity ($\Delta \mid\!\sim A$ and $\Delta \mid\!\sim B$ imply $\Delta, A \mid\!\sim B$) and Cut ($\Delta \mid\!\sim A$ and $\Delta, A \mid\!\sim B$ imply $\Delta \mid\!\sim B$). See [Gabbay, 1985a; Kraus *et al.*, 1990]. We would like to bring $\mid\!\sim$ into the language as a connective $A > B$ and find a conservative extension $\mid\!\sim^*$ of $\mid\!\sim$ such that the following holds

- $\varnothing \mid\!\sim^* A > B$ iff $A \mid\!\sim B$.

The above means that $>$ represents $\mid\!\sim$ in the language itself.

The fibred semantics methodology allows us to do this in a methodological way.

Delgrande has already observed [Delgrande, 1988] the formal similarities between non-monotonic consequence relations $\mid\!\sim$ and the conditional $>$. Of course $>$ is an object level connective and $\mid\!\sim$ is a metalevel consequence. Our fibred methodology turns $\mid\!\sim$ into a conditional. In our paper [Gabbay, 1995c] we indeed get the traditional semantics for the conditional out of the fibred semantics of the construction that brings $\mid\!\sim$ into the object level.

1.1.3 *Self-fibring of predicate logics*

This is another surprising application of the fibred methodology. It allows us to give solid and meaningful semantics to expressions of the form $A(t, \varphi(f(\alpha)))$ where $A(x, y)$

is a formula, $\alpha, \varphi(x)$ are formulas and $f(x)$ a function symbol. As you can see, we have substituted formulas within other formulas and other function symbols.

Such expressions and languages are widely used in logic, philosophy and linguistics and have extensive and important applications. We mention a few typical examples:

- Logics of non-denoting singular terms
- In the logical analysis of language, we try and analyse statements like 'John said that Mary believed that he did not love her'
  Here we use the predicates, *say* (t, φ) and *believe* (t, φ).
- In logic programming there is a widespread use of metapredicates such as
  Demo(A, B) and
  Y if *Not*(X).
  Ordinary clauses for the transitive closure of a relation R are written as

$$(cl(Z))(X, Y) \text{ if } Z(X, Y)$$
$$(cl(Z))(X, Y) \text{ if } Z(X, V) \wedge ((cl(Z))(V, X))$$

  or expressions like

$$\forall x. \text{ bel(John, friend (John, } (x)) \rightarrow \text{ exists } (y, \text{ loves } (x, y))),$$

- Tarski truth predicate

$$\mathbf{T}(\varphi) \leftrightarrow \varphi$$

  and the diagonalisation function symbol

$$d(\varphi) =\sim \varphi(\varphi).$$

We make sense of the above expressions using fibred semantics.

1.1.4 *Temporalising a system*

This example is very practical. Given a system $\mathcal{S}$, we can describe it in some logic $\mathbf{L}$. We now let the system vary in time. We can get snapshots $\mathcal{S}_t$ of it at different times t. What would be a suitable logic to describe the evolving and changing system?

The answer is the combination of $\mathbf{L}$ with a suitable temporal logic of our choice. The fibred semantics methodology will tell us what is the semantics of the new combined logic and the availability of transfer theorems will tell us what properties to expect of the combined system, given what we know of the components. This is done in, e.g. [Finger and Gabbay, 1992] and in Chapter 13.

1.1.5 *Making your logic fuzzy*

This is the most satisfactory application of fibred semantics. Fuzzy logic has been controversial among pure logicians for a long time. It lacked respectability. People had the impression that we can take any system and make it fuzzy in an ad hoc way, by taking any $\{0, 1\}$ function we can find and turning it into a fuzzy $[0, 1]$ function. There is no methodology and no logic.

Take modal logic for example. Its models have the form (S, R, a, h) where S is the set of possible worlds, $a \in S$, R is a crisp $\{0, 1\}$ relation and h is the assignment, giving for each $t \in S$ and atomic q a crisp $\{0, 1\}$ value $h(t, q) \in \{0, 1\}$.

How can we make modal logic fuzzy?

We can turn h into a $[0, 1]$ function. We can turn R into a $[0, 1]$ function. We can even make fuzzy which world is the actual world. This seems arbitrary. See Fitting [1991; 1992; 1994] for examples of where it is done. Can we make sense of it?

The answer is yes. The very same fibring methodology used to combine intuitionistic and modal logics and used to combine two modalities can be used here to combine modal logic with Łukasiewicz infinite valued logic.

Depending on how we fibre we can get the above (or all) different options of making modal logic fuzzy.

1.1.6 *Combining proof systems*

Fibring can be done on the proof theoretic level. Given two proof systems Π_1 and Π_2 for $\mathbf{L}_1$ and $\mathbf{L}_2$, respectively, how do we combine them into a proof system Π for the combined language $\mathbf{L}$? This problem is still under investigation, however for some particular cases, such as tableaux systems, it can be done in a fairly general way. See Chapters 17–20.

1.2 The Idea of Fibring

The basic problem of combining systems can be formally understood as follows:

Combining systems problem

We are given two systems $\mathcal{S}_1$ and $\mathcal{S}_2$ in languages $\mathbb{L}_1$ and $\mathbb{L}_2$. We combine the languages to form $\mathbb{L}$. We ask the following:

1. How can we define a system $\mathcal{S}$ for the combined language, which is a conservative extension of each $\mathcal{S}_1$ and $\mathcal{S}_2$? This is not an easy problem, because $\mathcal{S}_1$ and $\mathcal{S}_2$ may be presented to us in two completely different and incompatible ways. For example even though $\mathcal{S}_1$ and $\mathcal{S}_2$ may be two modal logics with modalities $\Box_1$ and $\Box_2$, the system for $\Box_1$ may be defined via some algebraic semantics while the system for $\Box_2$ may be a tableaux system.
2. How many options for $\mathcal{S}$ are there and how do they relate to each other?
3. If we want $\mathcal{S}_1$ and $\mathcal{S}_2$ to interact, how do we do that? Can we develop a methodology for interaction and have a set of prearranged and well understood 'interactive axioms'?
4. Suppose we know that both $\mathcal{S}_1$ and $\mathcal{S}_2$ satisfy some property $\mathcal{P}$. Can we prove transfer theorems which ensure that the combined $\mathcal{S}$ satisfies property $\mathcal{P}'$ (a variation of $\mathcal{P}$)?
5. Suppose we are given a system $\mathcal{S}$ directly defined in a language $\mathbb{L}$. Assume $\mathbb{L}$ is a syntactic combination of $\mathbb{L}_1$ and $\mathbb{L}_2$. Can we decompose $\mathcal{S}$ into $\mathcal{S}_1$ and $\mathcal{S}_2$, where $\mathcal{S}_i$ are some projections onto the sublanguages $\mathbb{L}_i$, and then can we reconstruct $\mathcal{S}$ back as some combination of $\mathcal{S}_1$ and $\mathcal{S}_2$ with possibly additional interaction

axioms? If $\mathcal{S}_i$ happens to be well understood systems, we would have certainly 'simplified' $\mathcal{S}$.

The fibring methodology we are about to describe allows one to combine systems through their semantics and is a very successful framework for answering the above questions.

1.2.1 *Appreciation of the difficulties involved in combining systems*

The problem of combining systems can be pretty difficult in practice, not only because the two systems to be combined may be presented in two completely different ways, but also because that even when they are presented in the same way, it is not clear how to combine them. The next examples will illustrate.

Example 1.1 (Two modal logics). Let $\Box_1$ and $\Box_2$ be two modalities. We define two systems as follows:

Let $\mathcal{S}_1$ be the system obtained by adding to classical propositional logic the modality $\Box_1$ with the following axioms and rules:

1. $\Box_1 A$, where A is a substitution instance of a truth functional tautology;
2. $\Box_1(A \to B) \to (\Box_1 A \to \Box_1 B)$
3. $\Box_1(\Box_1(A \to B) \to (\Box_1 A \to \Box_1 B))$
4. $\Box_1 A \to \Box_1\Box_1 A$
5. $\Box_1(\Box_1 A \to \Box_1\Box_1 A)$
6. $\Box_1 A \to A$
7. Modus ponens: $\vdash A; \vdash A \to B$ imply $\vdash B$.

Note that we do not require necessitation. So although $\Box_1 A \to A$ is a theorem $\Box_1(\Box_1 A \to A)$ is not provable.

Let $\mathcal{S}_2$ be the system obtained by adding $\Box_2$ to classical propositional logic. The theorems of $\mathcal{S}_2$ are defined semantically as follows.

Let E^2 be the Euclidean plane with the usual topology. An assignment h is a function giving to each atom q, a set of points $h(q) \subseteq E^2$.

h can be extended to all wffs of $\mathcal{S}_2$ as follows

$$
\begin{aligned}
h(\neg A) &= E^2 - h(a) \\
h(A \wedge B) &= h(A) \cap h(B) \\
h(A \vee B) &= h(A) \cup h(B) \\
h(A \to B) &= (E^2 - h(A)) \cup h(B) \\
h(\Box_2 A) &= \text{Topological interior of } h(A).
\end{aligned}
$$

Let $\vDash_2 A$ mean that for all h, $h(A) = E^2$.

Problem: Combine $\mathcal{S}_1$ and $\mathcal{S}_2$ into a system $\mathcal{S}$ which is the smallest logical system for the combined language which is a conservative extension of both $\mathcal{S}_1$ and $\mathcal{S}_2$.

The two systems are presented in totally different ways. How are we going to combine them?

We shall fibre (combine) them in the next subsection by observing that the $\Box_2$ modality is really **S4** modality, complete for the class of Kripke models (S, R, a, h)

where R is reflexive and transitive and that the $\Box_1$ modality is complete for all Kripke models (S, R, a, h) such that R is transitive and aRa holds. Without this extra knowledge we cannot combine them. With this extra knowledge we can apply our methodology and combine them, as we shall see in the next subsection.

Even when the two systems are presented in the same way, it is not clear what to do and how to combine them. In the next example, we try to combine classical implication with intuitionistic implication. The two systems will be presented in the same way, with classical implication having one more axiom than intuitionistic implication. We cannot be more 'compatible' than that, can we?

We shall combine them in a straightforward way by taking the union of the axioms, taking each axiom for its own language. The next example will show that we still get into trouble!

See [Crocco *et al.*, 1995] and [Fariñas del Cerro and Herzig, 1996] on how to do it correctly.

Example 1.2 (Fibring counterexample). Let $\mathbb{L}_1$ be the language with $\wedge_1$ and $\twoheadrightarrow_1$ and $\mathbb{L}_2$ with $\wedge_2$ and $\twoheadrightarrow_2$. $\wedge_i$ is intended to be conjunction and $\twoheadrightarrow_i$ is intended to be implication. Let $\mid\!\sim_1$ and $\mid\!\sim_2$ be the respective consequence relations for $\mathbb{L}_1$ and $\mathbb{L}_2$, with $\mid\!\sim_1$ intending to define the $\wedge, \rightarrow$ fragment of intuitionistic logic and $\mid\!\sim_2$ intended to define the $\wedge, \rightarrow$ fragment of classical logic. Both $\mid\!\sim_i$ can be defined as the smallest consequence relations satisfying Reflexivity, Monotonicity and Cut (i.e. $X \mid\!\sim A$ and $X \mid\!\sim A \twoheadrightarrow B$ implies $X \mid\!\sim B$) and closed under the additional rules below. $\mid\!\sim_2$, which is supposed to define classical implication can be assumed to be closed also under Peirce's rule $(A \twoheadrightarrow_2 B) \twoheadrightarrow_2 A \mid\!\sim_2 A$.

- $A \wedge_i B \mid\!\sim_i C$ iff $A \mid\!\sim_i B \twoheadrightarrow_i C$
- $A \wedge_i B \vdash_i A$
- $A \wedge_i B \vdash_i B$
- $\dfrac{A \mid\!\sim_i B;\ A \mid\!\sim_i C,\ B \wedge_i C \mid\!\sim_i D}{A \mid\!\sim_i D}$.

Let us now combine the languages and let $\mid\!\sim$ be the smallest consequence relation in the combined language closed under both rules, for the respective languages. We show that $\twoheadrightarrow_1$ and $\twoheadrightarrow_2$ become equal.

Since $A \twoheadrightarrow_i B \mid\!\sim_i A \twoheadrightarrow_i B$ we get $(A \twoheadrightarrow_i B) \wedge_i A \mid\!\sim_i B$.

We also have:

1. $(A \twoheadrightarrow_1 B) \wedge_2 A \mid\!\sim A$
2. $(A \twoheadrightarrow_1 B) \wedge_2 A \mid\!\sim A \twoheadrightarrow_1 B$.

Hence from (1) and (2) we get

3. $(A \twoheadrightarrow_1 B) \wedge_2 A \mid\!\sim B$

and hence

$$(A \twoheadrightarrow_1 B) \mid\!\sim (A \twoheadrightarrow_2 B).$$

The above is impossible since $\twoheadrightarrow_1$ can be classical implication and $\twoheadrightarrow_2$ intuitionistic implication, and we know from **S4** Kripke model semantics that they do not collapse when combined..

1.2.2 *The basic idea of fibring*

This subsection explains how fibring works. We do it in two stages. First we take a well-known logic, modal logic, and show how it works in that case, and then, in the next section, we give a general schematic definition.

Let $\mathbb{L}_1$ and $\mathbb{L}_2$ be two modal languages with $\Box_1$ and $\Box_2$ built up from the respective sets Q_1 and Q_2 of atoms. It is convenient to assume that $\mathbb{L}_1$ and $\mathbb{L}_2$ share (are built up from) the same set Q of atomic propositions. Consider two logics $\mid\!\sim_1$ in $\mathbb{L}_1$ and $\mid\!\sim_2$ in $\mathbb{L}_2$. These are our *systems* $\mathcal{S}_1$ of $\mathbb{L}_1$ and $\mathcal{S}_2$ of $\mathbb{L}_2$. To be specific let $\mathcal{S}_1$ be modal logic $\mathbf{K}_1$ for $\Box_1$ and let $\mathcal{S}_2$ be modal logic **S4** for $\Box_2$, as described in Example 1.1. We have to say how these logics are *presented* to us. To simplify matters let us assume that the logics are presented to us via classes $\mathcal{K}_1$ and $\mathcal{K}_2$ respectively, of Kripke models for which the logics are sound and complete. Indeed for our fibring methodology to work there is no need to assume the classes are frame classes or normal or anything special. Let $\mathcal{K}_1$ be the class of all Kripke models of the form (S, R, a, h), where R is transitive and aRa holds. Let $\mathcal{K}_2$ be the class of all models where R is reflexive and transitive. To distinguish between the classes, we write $\mathbf{m}^1 = (S^1, R^1, a^1, h^1)$ for a model in $\mathcal{K}_1$ and $\mathbf{m}^2 = (S^2, R^2, a^2, h^2)$ for a model in $\mathcal{K}_2$.

We can now have two notions of fibring. The most general notion is when $Q_1 \cap Q_2 = \varnothing$, i.e. the languages have no atoms in common. This case is referred to as *disjoint fibring* (or algebraic fibring, because this is how we naturally fibre algebras, see Section 16.3). The more common case for logics is (ordinary) fibring, where $Q_1 = Q_2 = Q$. From now on we deal with ordinary fibring, unless otherwise stated.

To explain in principle how fibring works, consider a mixed wff of the form $\alpha = \Diamond_1\Box_2 q$, q is an atom of $\mathbb{L}_2$.

We say this wff α is in the language $\mathbb{L}_{(1,2)}$, namely we have outer connectives of $\mathbb{L}_1$ and inside there are connectives of $\mathbb{L}_2$. We shall define later the language $\mathbb{L}_{(x_1,\ldots,x_n)}$ in general, where $(x_1, \ldots, x_n)$ is an alternating sequence of numbers in $\{1, 2\}$.

We now motivate the definition of fibred models for formulas of $\mathbb{L}_{(1,2)}$, by looking at our example α.

Let us consider α as a formula of $\mathcal{S}_1$ (since its outer connective is $\Diamond_1$). From the point of view of $\mathbb{L}_1$ this formula has the form $\Diamond_1 p$, where $p = \Box_2 q$ is *atomic*, since $\mathcal{S}_1$ does not recognise $\Box_2$.

To give a model for $\Diamond_1 p$ we take any $\mathcal{S}_1$ model. Take for example $\mathbf{m}^1 = (S^1, R^1, a^1, h^1)$ and check whether $a^1 \vDash_1 \Diamond_1 p$.

For this to hold we need to check whether for some $t \in S^1$ such that $a^1 R^1 t$, we have $t \vDash_1 p$, i.e. whether $t \vDash_1 \Box_2 q$. Since $\Box_2$ is not in the language of $\mathcal{S}_1$, we do not know how to evaluate it! We observe, however, that we need no more than a truth value for $t \vDash_1 \Box_2 q$. We need an answer, yes or no.

The basic idea of fibring is to associate, with each $t \in S^1$, a model $\mathbf{m}_t^2 = (S_t^2, R_t^2, a_t^2, h_t^2)$ of $\mathcal{S}_2$, and get our answer by evaluating at the associated model[1] i.e. we have

(*) $t \vDash_1 \Box_2 q$ iff $a_t^2 \vDash_2 \Box_2 q$.

Of course $\Box_2 q$ can be evaluated at $\mathbf{m}_t^2$ because it is in the right language.

Let $\mathbf{F}^1$ be the fibring function, (namely the function formally associating the model $\mathbf{m}_t^2$ with t), i.e. let $\mathbf{F}^1(t) = \mathbf{m}_t^2$.

We can now say that our fibred semantics for the combined language $\mathbb{L}_{(1,2)}$ has models of the form $(S^1, R^1, a^1, h^1, \mathbf{F}^1)$, where $\mathbf{F}^1$ is as above and we use (*) above in our evaluation. Figure 1.1 below describes the model schematically:

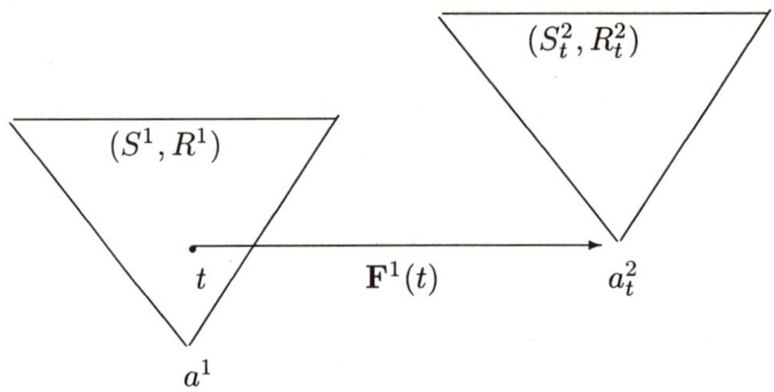

FIG. 1.1.

In principle fibred models for formulas with nestings of modalities (e.g. $\Diamond_1\Box_2\Diamond_1\Diamond_1 q$) can be defined inductively by iterating the process of fibring.

We shall see later that these models can be (just for the case of modal logic) greatly simplified, using various technical devices, which will lead us to the notion of *SFM* models.

Before we do that let us discuss the general case. The above idea of fibring can be schematically generalised, and the rest of this section is going to give an intuitive description of general fibring.

1.2.3 *General fibring*

Since the fibring methodology is applicable in diverse and unconnected areas of logic, the best approach is to give a very general definition of fibring and then show how the fibred semantics can be simplified for each case (area) it is used. The general definition will highlight the basic general assumptions needed to execute the fibring methodology.

Let $\mathbb{L}_1$ and $\mathbb{L}_2$ be two languages, and let $\mathcal{S}_1$ and $\mathcal{S}_2$ be two systems (usually logics, but they could be any systems, for example theorem provers or programming environments). To be specific we advise the reader to think of three typical examples:

- combining two modal logics $\mathcal{S}_1$ and $\mathcal{S}_2$

[1] In the general case we associate a family of models $\mathbf{m}_t^{2,i}$, $i = 1, 2, 3, \ldots$

- $\mathcal{S}_1$ a logic programming language
  $\mathcal{S}_2$ a C language
  For example, we want to allow C code to be put into PROLOG code[2]
- Combining two partially ordered systems, with equality $=$ and ordering $\leq$.

We assume the systems satisfy the following assumptions:

Basic assumption for general fibring

1. SYNTACTICAL ASSUMPTION
   The expressions E^i of the language $\mathbb{L}_i$ is built up using $\mathbb{L}_i$-constructors (connectives) from a set of atomic units Q_i. It is sometimes convenient to assume that $Q_1 = Q_2 = Q$, i.e. the languages share the atoms. We schematically write $E^i = E^i(q_1^i, \ldots, q_n^i)$ to indicate that E^i is built up from the atoms $q_1^i, \ldots, q_n^i \in Q_i$. Let $\mathbb{C}^i$ be the set of constructors of $\mathcal{S}_i$. We assume $\mathbb{C}^1 \cap \mathbb{C}^2 = \varnothing$.
2. SEMANTICAL ASSUMPTION
   The system $\mathcal{S}_i$ is 'characterised' (in case of logics $\mathcal{S}_i$ this means completeness) by a class $\mathcal{K}_i$ of models $\{\mathbf{m}_1^i, \mathbf{m}_2^i, \ldots\}$. Each model $\mathbf{m}_n^i$ is built up from a set S_n^i of basic semantic components $t \in S_n^i$. In many cases the model $\mathbf{m}_n^i$ comes with a distinguished point $a_n^i \in S_n^i$ (the actual world in Kripke models). This distinguished point plays a role in the semantic evaluation in the model.
   For technical reasons it is convenient to assume that for $m \neq n, S_n^i \cap S_m^i = \varnothing$. Let $S^i = \bigcup_n S_n^i$. We also assume that $S^1 \cap S^2 = \varnothing$. This can always be achieved by renaming.[3]
3. THE EVALUATION FUNCTION
   An evaluation function val^i is available which can give values ($\top$ or $\bot$ or solutions or any other output) to the following:
   (a) $val^i(\mathbf{m}_n^i, E^i)$.
   This is the value of a general expression E^i of $\mathcal{S}_i$, evaluated at a model $\mathbf{m}_n^i$. In many cases (e.g. modal logic) $val^i(\mathbf{m}_n^i, E^i)$ is defined to be $val^i(a_n^i, E^i)$ where a_n^i is the distinguished point of $\mathbf{m}_n^i$.
   (b) $val^i(t, E^i), t \in S_n^i$.
   The value of a general expression at a unit $t \in S_n^i$.
   (c) The value $val^i(\mathbf{m}_n^i, E^i(q_1, \ldots, q_k))$ and $val^i(t, E^i)$ is functionally reducible through some possibly algorithmic, inductive process to the values $\{val^i(t, q_j)\}$ where $t \in S^i$ and q_j are the atoms appearing in E. These latter values are arbitrary within an acceptable family of assignments. Let us write this as $\{val^i(t, q)\} \in Acceptable^i$.
   Note that there may be some metapredicates involved in the formulation of *Acceptable*, relating $val^i(t, q)$ with $val^i(s, q)$. The most common (and in

[2]Our fibring methodology has been widely applied in the logic context. I believe it can also be applied for general systems as well. The details for programming systems have not yet been worked out. I am asking the reader to think of the PROLOG-C example because it can help understand the logic case studies.

[3]Note that the nature of this assumption is not clear for the case of programming languages. See the previous footnote.

fact most general) is to impose an ordering $\prec$ on S^i and some relation on *val* (also denoted by $\prec$) and require something like

- $t \prec s$ implies $val^i(t,q) \prec val^i(s,q)$.

4. DEFINITION OF THE FIBRED LANGUAGE

   We now define the fibred languages $\mathbb{L}_{(x_1,\ldots,x_n)}$, as follows:

   - Let $\mathbb{L}_{(1)}$ be $\mathbb{L}_1$ and $\mathbb{L}_{(2)}$ be $\mathbb{L}_2$.
   - Let $\bar{y} = (1, y_1, \ldots, y_k)$. Let $\mathbb{L}_{(2)*\bar{y}}$ be defined as the family of all expressions of the form $\alpha \in \mathbb{L}_{(y_1,\ldots,y_k)}$ or $\alpha = E^2(q_1/A_1, \ldots, q_n/A_n)$ where $E^2(q_1, \ldots, q_n) \in \mathbb{L}_2$ and $A_1, \ldots, A_n$ are in $\mathbb{L}_{\bar{y}}$, and q_j/A_j indicate the substitution of A_j for q_j in E^2.
     In other words, $\mathbb{L}_{(2,1,y_1,\ldots,y_k)}$ is the set of all expressions with outer constructor from $\mathbb{L}_2$ and with no more than $k+2$ nested alternations of constructors from $\mathbb{L}_1$ and $\mathbb{L}_2$.
   - We similarly define $\mathbb{L}_{(1,2,y_1,\ldots,y_k)}$.
   - Let $\mathbb{L}_\infty = \bigcup_{\bar{y}} \mathbb{L}_{\bar{y}}$.

5. THE FIBRING FUNCTION **F**

   We are now ready for our last assumption for fibring. We require that the values $val^2(\mathbf{m}_n^2, E^2)$, for models $\mathbf{m}_n^2$ of $\mathcal{S}_2$ and E^2 of $\mathbb{L}_2$ are acceptable values for the function $val^1(t,q)$, for $t \in S^1$ and q atomic. In symbols:

   - For all $\mathbf{m}_n^2$ and E^2, $\{val^2(\mathbf{m}_n^2, E^2)\} \in Acceptable^1$.

   Let us see what this means. Take for example $\mathbb{L}_1$ as intuitionistic logic and take a Kripke model for intuitionistic logic, $(S, \leq, a, h)$. The assignments satisfy persistence:

   $$t \leq s \text{ and } h(t,q) = 1 \text{ imply } h(s,q) = 1.$$

   So if we fibre models $\mathbf{m}_t$ to t and $\mathbf{m}_s$ to s then for any formula α of $\mathbb{L}_2$ (the other language, which can be for example, modal logic) we must have $\mathbf{m}_t \vDash_2 \alpha$ implies $\mathbf{m}_s \vDash_2 \alpha$. This has to be ensured, so we cannot fibre arbitrarily in the intuitionistic case. In general the nature of the allowed fibring function **F** has to be worked out for each application with possibly some correctness theorems involved.

   Consider the case of the PROLOG and C languages.

   If we encounter a C code α in the middle of a PROLOG program, then $val^C(\alpha)$ will probably be a numerical value. This value must be usable in the PROLOG program, which only recognises predicates. So α may be an equation or constraints verified in C, etc.

   Under the above assumptions we can define the family of fibred models for $\mathbb{L}_{(1,2)}$. These have the form $(\mathbf{m}_n^1, \mathbf{F}_n^{(1,2)})$, where $\mathbf{F}_n^{(1,2)}$ is a fibring function on S_n^1, assigning for each $t \in S_n^1$ a model $\mathbf{m}_t^2 \in \mathcal{K}_2$.

   The basic evaluation clause for $t \in S^1$ is

   - $val^{(1,2)}(t, E^2) = val^2(\mathbf{m}_t^2, E^2)$

6. SIMPLIFYING THE FIBRING FUNCTION.

   The fibring function can be simplified, for the following reasons:

(a) Since $S^1_n, n = 1, 2, \ldots$ are all disjoint, the domain of $\mathbf{F}^{(1,2)}_n$ can be $S^1 = \bigcup_n S^1_n$.

(b) Since for each t, $\mathbf{F}^{(1,2)}_n(t)$ is a model $\mathbf{m}^2_t$, this model must have an index n_t, i.e. $\mathbf{m}^2_t = \mathbf{m}^2_{n_t}$.

It is sufficient to regard $\mathbf{F}^{(1,2)}_n$ as giving either the numerical value n_t or some $a \in S^2_{n_t}$. Since the models $S^2_1, S^2_2, \ldots$ are pairwise disjoint, any $a_t \in S^2$ will characterise a model $\mathbf{m}^2_{n_t}$ uniquely. In fact we mentioned in item 2 above that the models $\mathbf{m}^i_n$ can come with a distinguished element $a^i_n \in S^i_n$. We can let $\mathbf{F}^{(1,2)}(t)$ be $a^2_{n_t} \in S^2$.

(c) Since we can always duplicate names of models, we can always assume that

(*) $t \neq s, t, s, \in S^1$ imply $n_t \neq n_s$.

Thus we can view the fibring function $\mathbf{F}^{(1,2)}$ as a one-to-one function from S^1 into S^2 satisfying (*) above.

(d) Since the models $\mathbf{m}^1_n$, $n = 1, 2, \ldots$ can themselves be duplicated, we can assume that if the $t \in S^1_k$ and $s \in S^1_m$ and $k < m$ then the fibred models $\mathbf{F}^{(1,2)}(t) = a^2_{n_t}$ and $\mathbf{F}^{(1,2)}(s) = a^2_{n_s}$ satisfy $n_t < n_s$.

(e) Furthermore a function $\mathbf{F}^{(1,2)}$ can be similarly defined from S^2 to S^1 for the fibred language $\mathbb{L}_{(2,1)}$. The two functions can be joined as one function

$$\mathbf{F} : S^1 \cup S^2 \mapsto S^1 \cup S^2$$

satisfying the following

- If $t \in S^1$ then $\mathbf{F}(t) = a^2_{n_t} \in S^2$.
- If $t \in S^2$ then $\mathbf{F}(t) = a^1_{n_t} \in S^1$.
- $\mathbf{F}$ is one-to-one and satisfies the condition in (d).

We shall see that for each particular fibring application, astonishing simplifications are possible, showing connections between fibring and other known concepts in logic.

7. GENERAL DEFINITION OF FIBRING.

We can now define the fibred model for the mixed language $\mathbb{L}_\infty = \bigcup_{\bar{x}} \mathbb{L}_{\bar{x}}$ as follows:

(a) A fibred model for $\mathbb{L}$ has the form $\mathbf{n} = (\mathbf{m}^i_n, \mathbf{F})$ where $\mathbf{F}$ is the simplified fibring function as in (6e) above. Note that although $\mathbf{F}$ was defined in (6e) for the case of $\mathbb{L}_{(1,2)}$ and $\mathbb{L}_{(2,1)}$, the form it was given in (6e) allows it to be used for any $\mathbb{L}_{\bar{x}}$.

(b) We now define *val*$(\mathbf{n}, E)$ as follows.

We can assume that $E \in L_{(x_1,\ldots,x_n)}$. E has the form $E = E^i(q_j/A_j)$ where $A_j \in \mathbb{L}_{(x_1,\ldots,x_n)}$. We know that the *val*$^i(\mathbf{m}^i_n, E^i)$ depends directly on

$$V = \{val^i(t, q_j) \mid t \in S^1, i = 1, 2, \ldots\}.$$

Let $f_{E^i}(V)$ symbolise this dependency. Therefore if we replace $val^i(t, q_j)$ by the new $V' = \{val(t, A_j)\}$, which we can assume we know how to get, then the same dependency will give us

$$val(\mathbf{n}, E) = f_{E^1}(V').$$

8. THE NOTION OF DOVETAILING.
   We saw in item 6 above that the function $\mathbf{F}$ can be viewed as a function giving, for each $t \in S^1 \cup S^2$, an element $\mathbf{F}(t) \in S^1 \cup S^2$ such that if $t \in S^i$ then $\mathbf{F}(t) \in S^j, i \neq j$.
   Consider the case where the languages $\mathbb{L}_1$ and $\mathbb{L}_2$ share the same set of atoms Q. We can compare the values $val^i(t, q)$ and $val^j(\mathbf{F}(t), q)$ for atom q. These two values need *not* be identical. If we require from our fibring function $\mathbf{F}$ that for each $t \in S^i, \mathbf{F}(t) \in S^j$ and each $q \in Q$ we have
   † $val^i(t, q) = val^j(\mathbf{F}(t), q)$
   Then this fibring case is referred to as *dovetailing*. We shall see later that dovetailing allows for serious simplifications in the combined system.

1.2.4 *Case study: modal logic fibring and dovetailing*

We take this opportunity to show how the simplifications can be done in the case of fibring modal logics.

In the modal logic case each model involved has the form $\mathbf{m}^i_n = (S^i_n, R^i_n, a^i_n, h^i_n)$.

Since we assumed that all S^i_n are pairwise disjoint, we can put both semantics together in one big set of possible worlds W:

$$W = \bigcup_{(i,n)} S^i_n.$$

The relations R^i_n can all be unified under a single relation $R \subseteq W^2$:

$$R = \bigcup_{(i,n)} R^i_n.$$

This can be done without loss of generality. For any two elements $t, s \in W, tRs$ can hold only if they are both in the same S^i_n and only if $tR^i_n s$ holds. Thus $R^i_n = R \upharpoonright (S^i_n)^2$. Can we retrieve S^i_n out of W? The answer is yes if we record the actual worlds a^i_n. We can assume, from modal logic considerations, that the model (S^i_n, R^i_n, a^i_n) has the property that each $t \in S^i_n$ is accessible from a^i_n via a chain of elements, i.e.

$$\forall t \exists k \exists t_1, \ldots, t_k (a^i_n R^i_n t_1 \wedge \ldots t_{k-1} R^i_n t_k \wedge t_k R^i_n t).$$

With this assumption S^i_n is retrievable from a^i_n using R^i_n. So if we let $W_i = \{a^i_n \mid n = 1, 2, \ldots\}$ and $W_a = W_1 \cup W_2$ (a for 'actual') be the subsets of W for all actual worlds, then knowing W_a, we can get back our models. For each $x \in W_a$ let $S^x = \{t \mid \exists k \exists t_1, \ldots, t_k (x R t_1 \wedge \ldots \wedge t_k R_t)\}$. So for $x = a^i_n, S^x = S^i_n$.

We can now simplify the fibring function $\mathbf{F}$. For each $t, \mathbf{F}(t)$ is a model $\mathbf{m}^i_n$. Since this model is characterised by a^i_n, we can let $\mathbf{F}(t)$ be a^i_n. Thus

$$\mathbf{F} : W \mapsto W_a.$$

$\mathbf{F}$ satisfies the same properties as before, namely it is one to one and always switching semantics, (i.e. for $t \in S^1, \mathbf{F}(t) \in W_2$ and for $t \in S^2, \mathbf{F}(t) \in W_1$).

The above considerations make $\mathbf{F}$ a jump function which corresponds to a unary operator $\mathbb{J}A$ of the form:

- $t \vDash \mathbb{J}A$ iff $\mathbf{F}(t) \vDash A$

We thus conclude that fibring two modalities is reduced to one modality with a jump connective $\mathbb{J}$.

We can now define the notion of *SFM* model (simplified fibred models) for multi-modal logic, with modalities $\{\Box_i \mid i \in I\}$.

Definition 1.3 (SFM-model). An SFM-model has the form $(W, W_i, W_a, R, w_0, h, \mathbf{F}_i)$, $i \in I$, where

1. W is a set of worlds for some $i_0, w_0 \in W_{i_0}$.
2. $W_i \subseteq W, i \in I$ are pairwise disjoint and non-empty, with $W_a = \bigcup_i W_i$.
3. For $t, s \in W_i$ let $S^t = \{x \mid \exists n \;\; tR^n x\}$.
   Then
   - $t \neq s \to S^t \cap S^s = \varnothing$.
   - $W = \bigcup_{t \in W_a} S^t$.
4. $\mathbf{F}_i$ is a function satisfying
   - $x \in S^t$ and $t \in W_i \to \mathbf{F}_i(x) = x$
   - $x \in S^t$ and $t \notin W_i \to \mathbf{F}_i(x) \in W_i$
   - $x \neq y \to \mathbf{F}_i(x) \neq \mathbf{F}_i(y)$.
5. For each $t \in W_i$ the model $\mathbf{m}_t = (S^t, R \upharpoonright S^t \times S^t, t, h \upharpoonright S^t)$ is in the semantics $\mathcal{K}_i$ of $\mathbf{L}_i$.
   In case $\mathbf{L}_i$ is complete for the models $\{\mathbf{m}_t \mid t \in W_i\}$ we say the model is a *universal model.*
6. We can also assume that $\mathbf{F}$ *generates* W as follows.
   Let $W^0 = S^{w_0}$
   $W^{n+1} = W^n \cup \bigcup_{y \in W^n, i \text{ arbitrary}} S^{\mathbf{F}_i(y)}$
   Then $W = \bigcup_n W^n$.

Condition (6) can be assumed because evaluation at a model is evaluation at w_0. Any point not in $\bigcup_n W^n$ is not reachable from w_0 by embedded modalities so cannot affect truth values. See Figure 1.2.

The figure shows a model which has possible worlds $W = S^t \cup S^s \cup \ldots$ $t \in W_j$ means t is an actual world of a model of $\Box_j$. $\mathbf{F}_i(x) = s$ means the model with actual world s (of $\Box_i$) is fibred at the point x.

When dovetailing, the point s is identified with the point x. To distinguish between xRz in S^t and xRy in S^o, we write $xR_j z$ and $xR_i y$.

We now take a new look at the fibring function $\mathbf{F}$. We can simplify the fibred semantics even further. Let us introduce new unary connectives of the form $\mathbb{J}_i A$ ($\mathbb{J}_i$ for a 'jump' operator) and modal operators $\Box$ and $\Diamond$ with the table:

- $t \vDash \mathbb{J}_i A$ iff $\mathbf{F}_i(t) \vDash A$.
- $t \vDash \Diamond A$ iff for some $s(tRs$ and $s \vDash A)$.
- $t \vDash \Box A$ iff for all $s(tRs \to s \vDash A)$.

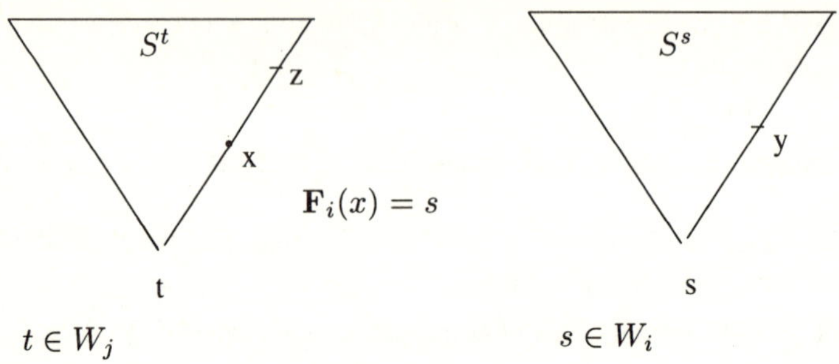

FIG. 1.2.

According to this definition we can let $\Box_i = \mathbb{J}_i\Box, \Diamond_i = \mathbb{J}_i\Diamond$. We are referring to $\mathbb{J}_i$ as 'jump' operators because their truth table 'jumps' the evaluation from a world t to the world $\mathbf{F}_i(t)$.

Thus fibring several modalities together is like adding several unary jump operators to the modalities of the fibred semantics.

Conversely, assume we are given a modal logic with modalities $\Box$ and $\Diamond$ and several $\mathbb{J}_i$ and a class $\mathcal{K}$ of SFM models of the form $(W, W_i, R, w_0, h, \mathbf{F}_i)$, as defined above.

We can define modal semantic classes $\mathcal{K}_i$ by letting $\mathcal{K}_i$ be the set of all models of the form $(S^t, R \upharpoonright S^t \times S^t, t, h \upharpoonright S^t), t \in W_i$.

In the case of a finite number of modalities, one fibring function $\mathbf{F}$ and one jump operator $\mathbb{J}$ are sufficient. $\Box_1$ can be interpreted as $\Box$ and $\Box_k$ as $\mathbb{J}^k\Box$. $\mathbf{F}$ can be chosen to take any point x into $\mathbf{F}(x)$, an actual world of the next semantics in cyclical order.

Note that the jump operators are slightly more expressive. For q atomic, $\mathbb{J}q$ changes the index of evaluation of q from t to $\mathbf{F}(t)$. This cannot be done using modalities.

Example 1.4 (Dovetailing modal Kripke semantics). Continuing the previous example, the case of dovetailing is when we require that for all atomic q

$$h(t, q) = h(\mathbf{F}_i(t), q)$$

for all i. This means the actual world of the model fibred at t can be identified with t. We must be careful however. For each t, we must first introduce the notation tR_iy iff (definition) $\mathbf{F}_i(t)Ry$. Then we can identify the point $\mathbf{F}_i(t)$ with t and the dovetailed model can be equivalently represented as (S, R_i, a, h) with

$$t \vDash \Box_i A \text{ iff } \forall y(tR_iy \to y \vDash A)$$
$$t \vDash \Diamond_i A \text{ iff } \exists y(tR_iy \wedge y \vDash A).$$

The function $\mathbf{F}$ is no longer needed, since we identified t with $\mathbf{F}_i(t)$, so the dovetailed model reduces to (W, R_i, w_o, h).

It is interesting to see the role of the jump operators in the simplified dovetailed semantics.

We have in the fibred model $(W, R, w_0, h, \mathbf{F}_i)$

$$\begin{aligned} t \vDash \mathbb{J}_i \Box A \;\; &\text{iff} \;\; \mathbf{F}_i(t) \vDash \Box A \\ &\text{iff} \;\; \forall x(\mathbf{F}_i(t) R x \text{ implies } x \vDash A) \end{aligned}$$

This becomes

$$t \vDash \mathbb{J}_i \Box A \text{ iff } \forall x(t R_i x \text{ implies } x \vDash A).$$

$\mathbb{J}_i$ becomes a mode shifting operator, changing the mode of evaluation, namely of which R_i to use.

We can consider the two clauses for a satisfaction $\vDash_i$ dependent on mode i.

- $t \vDash_i \Box A$ iff $\forall y(t R_i y$ implies $y \vDash_i A)$.
- $t \vDash_i \mathbb{J}_j A$ iff $t \vDash_j A$.

Example 1.5 (Evaluation mode). Our models need not have more than one binary relation. We can do mode shifting in the semantic evaluation in a traditional Kripke model of the form (S, R, a, h), with R an arbitrary binary relation. Consider two modes

- $t \vDash_1 \Box A$ iff for all x such that tRx we have $x \vDash A$.
- $t \vDash_2 \Box A$ iff for all x such that $t = x \vee tRx$ we have $x \vDash A$.
- $t \vDash_1 \mathbb{J} A$ iff $t \vDash_2 A$.
- $t \vDash_2 \mathbb{J} A$ iff $t \vDash_1 A$.

In general the mode can tell us which points x of S are accessible to t via some general formula $\Psi(t, x)$, associated with the mode

- $t \vDash_{\text{mode}} \Box A$ iff for all x such that $\Psi(t, x)$ we have $x \vDash A$.

The $\mathbb{J}$ operator shifts modes.

1.2.5 *Step by step scenario for fibring two logics*

In view of the previous discussion, we need to describe the steps we take in fibring two arbitrary logics $\mathcal{S}_1$ and $\mathcal{S}_2$.

Step 1: Check the form in which these two logics are presented and see whether they can be recast into the format of Subsection 1.2.2.

Step 2: If worst comes to worst, extract the consequence relations $\mid\!\sim_1$ and $\mid\!\sim_2$ of these two logics.

Step 3: Using theorems of the next section, give $\mid\!\sim_1, \mid\!\sim_2$ basic relational semantics $\mathcal{K}_1$ and $\mathcal{K}_2$.

Step 4: Using the definitions of the next section, fibre the two logics.

Step 5: We can assume, one way or another, that we fibred the two systems, using some semantics.

Depending on what the logics are (e.g. modal logics, intuitionistic logics, fuzzy logics, etc.), use general theorems available in the following respective sections to assert transfer of decidability, finite model property, recursive axiomatisation, etc.

2

LOGICS AND THEIR SEMANTICS

Our considerations in the general discussion of combining logics in the previous chapter indicates that we may be presented with two logics that are completely incompatible in their style and formulation.

In such a case we cannot combine them directly. Our strategy in this case is to do the following:

Step 1: Extract from the given presentation of the logics two respective consequence relations $\mid\!\sim_1$ and $\mid\!\sim_2$ for them.

Step 2: Use the method of this section (as presented below) to give general basic point relational semantics for $\mid\!\sim_1$ and $\mid\!\sim_2$.

Step 3: Fibre the basic point relational semantics to obtain the combined system $\mid\!\sim$.

Note that we offer here a technical solution. The basic point relational semantics may not mean much for the applications or motivations of these logics. We merely need it to bring the logic into some common background so that they can be combined. We can then use the semantics to prove transfer theorems and formulate an axiom system for the combined logic and possibly some confirmed properties of this system (e.g. decidability).

The purpose of this chapter is then to explain the following points.

- What is a logic and what is a basic relational semantics?
- Under what conditions on $\mid\!\sim$ can we give it basic point relational semantics?
- How do we fibre two such semantics?
- Transfer properties for the above.

Let us begin.

2.1 Fibring Basic Relational Semantics

Definition 2.1 (Basic logic).

1. A propositional language is a tuple $\mathbb{L} = (Q, \mathbb{C})$ where Q is a set of atomic propositions and $\mathbb{C}$ is a set of connectives. It may be convenient to include the connectives $\bot$ and $\top$.
2. The set of well formed formulas of $\mathbb{L}$, $\text{WFF}_{\mathbb{L}}$, is defined inductively in the usual manner.
3. Let $\mathbb{L}_1 \cup \mathbb{L}_2$ denote the language $\mathbb{L} = (Q_1 \cup Q_2, \mathbb{C}_1 \cup \mathbb{C}_2)$.

4. A *basic logic* on a language $\mathbb{L}$ is a binary relation between wffs of $\mathbb{L}$ of the form $A \mid\!\sim B$ satisfying the following conditions:[4]
   - Identity
   $$A \mid\!\sim A$$
   - Logical equivalence
   Let $A \equiv B$ abbreviate '$A \mid\!\sim B$ and $B \mid\!\sim A$'. Then for all A, B, C
   $$\frac{A \equiv B}{(C \mid\!\sim A \text{ iff } C \mid\!\sim B)}$$
   and
   $$\frac{A \equiv B}{A \mid\!\sim C \text{ iff } B \mid\!\sim C}.$$
   - Transitivity
   $$\frac{A \mid\!\sim B;\ B \mid\!\sim C}{A \mid\!\sim C}.$$
   - If $\top$ and $\bot$ are present we have
   $$\bot \mid\!\sim A \mid\!\sim \top.$$

Definition 2.2 (Strengthening). Let $\mathbb{L}$ be a language, let $\sharp(A_1, \ldots, A_n)$ be a connective in the language and $\mid\!\sim$ be a basic logic for $\mathbb{L}$.

1. We say that $\sharp$ satisfies upward (respectively, downward) strengthening in coordinate i, relative to $\mid\!\sim$, iff the following holds:
$$\frac{X \mid\!\sim Y \,(\text{respectively } Y \mid\!\sim X)}{\sharp(A_1, \ldots, A_{i-1}, X, A_{i+1}, \ldots, A_n) \mid\!\sim \sharp(A_1, \ldots, A_{i-1}, Y, A_{i+1}, \ldots, A_n)}$$

[4]The general notion of a *logic* allows for relations between sets of wffs $\Delta \mid\!\sim \Gamma$, and even lists or other structures. Most common is the form $\Delta \mid\!\sim A$, where Δ is a set of wffs including the empty set $\varnothing$. In such a case we require $\mid\!\sim$ to satisfy:
- Reflexivity
  $\Delta \mid\!\sim A$, if $A \in \Delta$
- Restricted monotonicity
$$\frac{\Delta \mid\!\sim A;\ \Delta \mid\!\sim B}{\Delta, A \mid\!\sim B}$$
- Cut
$$\frac{\Delta, A \mid\!\sim B;\ \Delta \mid\!\sim A}{\Delta \mid\!\sim B}$$

2. We say $\mid\!\sim$ satisfies strengthening iff each connective $\sharp$ satisfies either upward or downward (or both) strengthening in each coordinate.
3. A logic is said to be *isotonic* (I-Logic) iff it satisfies strengthening.

Lemma 2.3. *Let $\mathbb{L}$ be a language and let $\mid\!\sim$ be an I-logic in $\mathbb{L}$. Let $C(q)$ be a wff of $\mathbb{L}$ with atomic q and let $C(q/X)$ denote the substitution of the wff X for q in C. Then the following holds*

$$\frac{A \equiv B}{C(q/A) \equiv C(q/B)}.$$

Proof By structural induction on C, using strengthening. □

Definition 2.4 (Rules). Let $\mathbb{L}$ be a language and let $\mid\!\sim$ be a logic.

1. A logic rule ρ has the form

$$\frac{A_i \mid\!\sim B_i, i = 1, \ldots, m}{A \mid\!\sim B}.$$

2. $\mid\!\sim$ is said to satisfy the rule ρ iff whenever for all i, $A_i(q_j/C_j) \mid\!\sim B(q_j/C_j)$ holds then $A(q_j/C_j) \mid\!\sim B(q_j/C_j)$ holds.
3. Let $\mathbb{R}$ be a set of rules. Denote by $\mid\!\sim_{\mathbb{R}}$ the smallest consequence relation satisfying all the rules in $\mathbb{R}$.

Lemma 2.5. *Let $\mathbb{L}$ be a language and $\mid\!\sim$ be an I-logic on $\mathbb{L}$. Let $\sharp$ be a connective such that $\sharp$ satisfies both upward and downward strengthening in its first coordinate. Then $\mid\!\sim$ is independent of its first coordinate.*

Proof We will prove the lemma for the case when either $\top$ or $\bot$ are present in the language. Since $\bot \mid\!\sim A \mid\!\sim \top$ for arbitrary A, we get $\sharp(\top) \equiv \sharp(\bot) \equiv \sharp(A)$ and hence $\sharp(A)$ is independent of A. If neither $\bot$ nor $\top$ are present the proof is more complex, and we are not going to address it here. See Gabbay [1994a; 1996b]. □

Definition 2.6.

1. Let $\mathbb{L}$ be a language and let $\mid\!\sim$ be an I-logic. Let $\sharp$ be a connective. We can assume that each coordinate in $\sharp$ satisfies uniquely either downward or upward strengthening. We can therefore present $\sharp$ in the form $(A_1, \ldots, A_r) \twoheadrightarrow_\sharp (B_1, \ldots, B_s)$ where A_i are the wffs in the downward coordinates and B_j are the wffs in the upward coordinates.
2. From now on we present all I-logics using connectives of the form $\{\twoheadrightarrow_1, \ldots, \twoheadrightarrow_k\}$, where for each i, the arities of the connective $\twoheadrightarrow_i$ are written as (r_i, s_i).
3. Let $\mathbb{L}$ be a language with connectives $\mathbb{C} = \{\twoheadrightarrow_i \mid i = 1, \ldots, k\}$ with arities (r_i, s_i), respectively. We denote by $\mid\!\sim_{I(\mathbb{C})}$ the smallest I-logic (i.e. smallest consequence relation) with the connectives $\mathbb{C}$.

Definition 2.7 (Basic point relational structures). Let $\mathbb{L}$ be a language and $\mid\!\sim$ an I-logic for $\mathbb{L}$. Assume its connectives are $\{\twoheadrightarrow_1, \ldots, \twoheadrightarrow_k\}$ with arities $m_i = r_i + s_i, i = 1, \ldots, k$. We present the connectives in the form $(A_1, \ldots, A_{r_i}) \twoheadrightarrow_i (B_1, \ldots, B_{s_i})$.

1. A basic relational model for this language has the form

$$\mathbf{m} = (S, \prec, a, R_1, \ldots, R_k, h)$$

where S is a non-empty set, $a \in S$, $\prec$ is a reflexive and antisymmetric and transitive relation on S. We assume that for some $t_S \in S$ we have $\forall x \in S(t_S \prec x)$. Further, for $i = 1, \ldots, k$, R_i is an $m_i + 1$ place relation on S satisfying the following conditions.

1.1. $\forall x_1, \ldots, x_{r_i}, y_1, \ldots, y_{s_i} \exists! t$
$[R_i(t, x_1, \ldots, y_1, \ldots) \wedge \forall u(R_i(u, x_1, \ldots, y_1, \ldots) \Leftrightarrow t \prec u)]$.
We can denote this unique t by $t = f_i(x_1, \ldots, x_{r_i}, y_1, \ldots, y_{s_i})$, and call the function f_i the function associated with R_i. By 1.2 below, we have that f_i is monotonic upwards in $y_1, \ldots, y_{s_i}$ and downwards in $x_1, \ldots, x_{r_i}$ (relative to $\prec$).

1.2. $\forall x_1, \ldots, x_{r_i}, y_1, \ldots, y_{s_i}, x'_1, \ldots, x'_{r_i}, y_1, \ldots, y'_{s_i}$
$[R_i(t, x_1, \ldots, y_1, \ldots) \wedge \bigwedge_j x_j \prec x'_j \wedge \bigwedge_j y'_j \prec y_j \Rightarrow$
$R_i(t, x'_1, \ldots, y'_1, \ldots s)]$.

h is an assignment to the atoms. For atomic q, there exists a $t_q \in S$ such that $h(q) = \{x \mid t_q \prec x\}$.

2. Satisfaction is defined as follows:
   - $t \vDash q$ iff $t \in h(q)$, q atomic.
   - $t \vDash (A_1, \ldots, A_{r_i}) \twoheadrightarrow_i (B_1, \ldots, B_{s_i})$ iff for all $x_j \vDash A_j, j = 1, \ldots, r_i$ there exist $y_j \vDash B_j, j = 1, \ldots, s_i$ such that $R_i(t, x_1, \ldots, x_{r_i}, y_1, \ldots, y_{s_i})$.
   - $\mathbf{m} \vDash B$ iff $a \vDash B$.
   - Let $\mathcal{K}$ be a class of models. We define $A \vDash_{\mathcal{K}} B$ iff for all $\mathbf{m} \in \mathcal{K}, \mathbf{m} \vDash A$ implies $\mathbf{m} \vDash B$.

   We let $\vDash_{\mathcal{K}} B$ iff for all $\mathbf{m} \in \mathcal{K}, \vDash_{\mathbf{m}} B$.
3. Note that there is another way of defining satisfaction in a model, which we denote by $\vDash^*$. Let $\hat{A} = \{t \mid t \vDash A\}$. Define $A \vDash^*_{\mathbf{m}} B$ iff for all $\prec$ minimal points $x \in \hat{A}$, we have $x \vDash_{\mathbf{m}} B$.
Similarly $A \vDash^*_{\mathcal{K}} B$ is defined. It is easy to see that if $\mathcal{K}$ is a class of models such that if $(S, \prec, a, R_i, h) \in \mathcal{K}$ then for all $x \in S$ we have also $(S, \prec, x, R_i, h)$ is in $\mathcal{K}$ then $A \vDash^*_{\mathcal{K}} B$ iff $A \vDash_{\mathcal{K}} B$.
4. Consider the first-order language of classical logic with binary relation $\prec$ and relations $R_1, \ldots, R_k$ and possibly function symbols $f_1, \ldots, f_k$. Then the class of basic relational structures can be taken as the class of all models of some first-order theory β, comprised of all the conditions in (1) above.

Lemma 2.8. *Let* $\mathbf{m}$ *be a basic relational structure. Let* $\hat{B}$ *be defined as* $\hat{B} = \{t \mid t \vDash B\}$. *Then there exists a* $t_B \in S$ *such that*

$$\hat{B} = \{t \mid t_B \prec t\}.$$

Since $\prec$ is transitive, we have $A \vDash_{\mathbf{m}} B$ iff $t_B \prec t_A$.

Proof By definition, for q atomic we have:

$$h(q) = \{t \mid t_q \prec t\} = \hat{q},$$

where t_q is in S.

We show by induction that for any B there exists a t_B such that $\hat{B} = \{t \mid t_B \prec t\}$. We examine the sample inductive case of a binary $B \twoheadrightarrow C$. Let $R(t, x, y)$ be its relation.

Using the induction hypothesis we assume that t_B and t_C exist for $\hat{B}$ and $\hat{C}$ and that:

$$\widehat{B \twoheadrightarrow C} = \{x \mid \forall y(t_B \prec y \rightarrow \exists z(t_C \prec z \wedge R(x, y, z)))\}.$$

We are looking for $t_{B \twoheadrightarrow C}$. By condition (1.2) on R, there exists a t_0 such that $R(t_0, t_B, t_C)$ holds and for every u, $R(u, t_B, t_C)$ holds if $t_0 \prec u$ (i.e. $t_0 = f(t_B, t_C)$). We claim t_0 is $t_{B \twoheadrightarrow C}$.

Let $t_0 \prec u$. Show $u \in h(B \twoheadrightarrow C)$. We know that $R(u, t_B, t_C)$ holds. Let $t_B \prec y$, and choose $z = t_C$. We need to show $R(u, y, t_C)$.

By condition (1.1) since $t_B \prec y$ and $t_C \prec t_C$ we get $R(u, y, t_C)$.

Assume $u \in \widehat{B \twoheadrightarrow C}$ and show $t_0 \prec u$. Choose $y = t_B$. Then for some z such that $t_C \prec z$ we have $R(u, t_B, z)$ and by (1) we have $R(u, t_B, t_C)$ and we have $t_0 \prec u$.

This concludes the proof of the lemma. □

Remark 2.9. The proof of the previous lemma shows the basic point relational structures can be equivalently presented in the form

$$\mathbf{m} = (S, \prec, a, f_1, \ldots, f_k, h)$$

where each $f_i(x_1, \ldots, x_{r_i}, y_1, \ldots, y_{s_i})$ is monotonic upwards in the y coordinates and downward in the x coordinates. The part without the h is called the *frame* of the model and h, of course, is the assignment.

The inductive truth definition is the following:

- $t \vDash (A_1, \ldots, A_{r_i}) \twoheadrightarrow_i (B_1, \ldots, B_{s_i})$ iff for all $x_j \vDash A_j, j = 1, \ldots, r_i$, there exist $y_j \vDash B_j, j = 1, \ldots, s_i$ such that $f_i(x_1, \ldots, x_{r_i}, y_1, \ldots, y_{s_i}) \prec t$.

The next definition and lemma show how to calculate t_B for any wff B in such a model.

Definition 2.10. Let $\mathbf{m}$ be a basic point relational model for a language $\mathbb{L}$.

With each formula $C(q_1, \ldots, q_n)$ with atoms $q_1, \ldots, q_n$ we can define a function $f_C(x_1, \ldots, x_n)$ on the domain of $\mathbf{m}$ as follows

- $f_{q_i}(x_i) = x_i$
- $f_C(x_1, \ldots, x_n) = f_i(f_{A_1}, \ldots, f_{A_{r_i}}, f_{B_1}, \ldots, f_{B_{s_i}})$ where $C = (A_1, \ldots, A_{r_i}) \twoheadrightarrow_i (B_1, \ldots, B_{s_i})$.

Lemma 2.11. *Let* **m** *be a basic point relational model and let* $C(q_1, \ldots, q_n)$ *be a wff built up from the atoms* $q_1, \ldots, q_n$. *Assume that* $h(q_i) = \{t \mid x_i \prec t\}$. *Then*

$$\hat{C} = \{t \mid f_C(x_1, \ldots, x_n) \prec t\}.$$

Note that $\hat{\top}$ *is the smallest element of* $(S, \prec)$ *and* $\hat{\bot}$ *is the largest.*

Proof Follows from the proof of the previous lemma. □

Theorem 2.12 (Strong completeness). *Let* $\mathbb{L}$ *be a language and* $\mid\!\sim$ *be an I-logic. Let* $\mathbb{R}$ *be a set of rules such that* $\mid\!\sim \, = \, \mid\!\sim_{\mathbb{R}}$.

Then there exists a class $\mathcal{K}$ *of basic point relational structures such that* $\mid\!\sim \, = \models_{\mathcal{K}}$, *and such that for each rule* ρ *of* $\mathbb{R}$ *of the form:*

$$\frac{A_i(q_1, \ldots, q_m) \mid\!\sim B_i(q_1, \ldots, q_m)}{A(q_1, \ldots, q_m) \mid\!\sim B(q_1, \ldots, q_m)}$$

the models satisfy in classical logic the following formula β_ρ:

$$\begin{array}{l} \forall x_1, \ldots x_m [\bigwedge_i f_{B_i}(x_1, \ldots, x_m) \prec f_{A_i}(x_1, \ldots, x_n) \rightarrow \\ \quad f_B(x_1, \ldots, x_n) \prec f_A(x_1, \ldots, x_m)]. \end{array}$$

Proof

1. Soundness
   This can be verified.
2. Completeness
   Let S be the set of equivalence classes of $\mathrm{WFF}_{\mathbb{L}}$ over $\equiv$.
   Let $A_{/\equiv} \prec B_{/\equiv}$ hold iff
   $$B \mid\!\sim A.$$
   For each connective $\twoheadrightarrow_i$ of $\mathbb{L}$, let
   $$\begin{array}{l} f_i(A_{1/\equiv}, \ldots, A_{r_i/\equiv}, B_{1/\equiv}, \ldots, B_{s_i/\equiv}) = \\ \quad ((A_1, \ldots, A_{r_i}) \twoheadrightarrow_i (B_1, \ldots, B_{s_i}))_{/\equiv}. \end{array}$$
   Let $R_i(t, x_1, \ldots, x_{r_i}, y_1, \ldots, y_{s_i})$ hold iff $f_i(x_1, \ldots, x_{r_i}, y_1, \ldots, y_{s_i}) \prec t$.
   Let $h(q) = \{t \mid q_{/\equiv} \prec t\}$.
   It is easy to show by induction that
   - $\hat{A} = \{B_{/\equiv} \mid B \mid\!\sim A\} = \{t \mid A_{/\equiv} \prec t\}$.
   - $f_C(q_{1/\equiv}, \ldots, q_{n/\equiv}) = C(q_1, \ldots, q_n)_{/\equiv}$.
   - $A \models B$ iff $A \mid\!\sim B$ iff $f_B \prec f_A$
   - Each formula β_ρ, for $\rho \in \mathbb{R}$ holds in the model because $\mid\!\sim$ satisfies ρ.

   The above shows that the canonical model we have defined is the desired model. Assume $A \mid\!\not\sim B$, then a countermodel would be $(S, \prec, A_{/\equiv}, R_i, h)$. □

We shall fibre the basic relational semantics for the logics.

Remark 2.13 (Fibring of canonical models). We are now going to be more specific about the classes of models $\mathcal{K}_1$ and $\mathcal{K}_2$ which we use. The completeness proof in Theorem 2.12 shows that the semantics of the logic $\mid\!\sim_i$ is obtained by taking one canonical frame, namely $\mathbf{m}^i_\alpha = (S^i, \prec^i, \alpha_{/\equiv}, f^i_1, \ldots, f^i_{k_i}, h^i)$, (where $S^i = \mathrm{WFF}_{\mathbb{L}_i/\equiv}$ and $A_{/\equiv} \prec^i B_{/\equiv}$ is $B\mid\!\sim_i A$ etc.) and letting α run over all wffs. Whenever $A\not\mid\!\sim B$ we let $\alpha = A$ and the model $\mathbf{m}^i_\alpha \vDash A$ but $\mathbf{m}^i_\alpha \nvDash B$. Thus we can let $\mathcal{K}_i = \{\mathbf{m}^i_\alpha \mid \alpha \in \mathrm{WFF}_{\mathbb{L}_i}\}$. These models, however, are not pairwise disjoint because $S^i_\alpha = \mathrm{WFF}_{\mathbb{L}_i/\equiv}$. To obtain pairwise disjoint models we can best base our WFF on *new atoms*.

Let

$$\mathbb{L}_1 = (Q, \mathbb{C}^1)$$
$$\mathbb{L}_2 = (Q, \mathbb{C}^2).$$

Assume $Q = \{q_1, q_2, q_3, \ldots\}$. Let us make more copies of the atoms. Let

$$Q^{(i,n)} = \{q^{i,n}_1, q^{i,n}_2, \ldots, \}.$$

Consider the language

$$\mathbb{L}_{i,n} = (Q^{(i,n)}, \mathbb{C}^i).$$

Each formula $\alpha(q^*_1, \ldots, q^*_m) \in \mathrm{WFF}_{\mathbb{L}_i}$ built up from $q^*_1, \ldots, q^*_m \in Q$ has its 'isomorphic' counterpart $\alpha^n \in \mathrm{WFF}_{\mathbb{L}_{i,n}}$ obtained by the simultaneous substitution of any atom $q_r \in Q$ by its corresponding $q^{i,n}_r \in Q^{(i,n)}$. Let $\{\alpha^i_1, \alpha^i_2, \alpha^i_3, \ldots\}$ be an enumeration of $\mathrm{WFF}_{\mathbb{L}_i}$.

We can now regard the semantics $\mathcal{K}_i$ as having the models

$$\mathbf{M}^i_{\alpha^i_n} = (\mathrm{WFF}_{\mathbb{L}_{i,n}/\equiv}, \prec^i_n, \alpha^{i,n}_n, f^i_{r,n}, h^i_n).$$

Now all models in $\mathcal{K}_i$ are pairwise disjoint.

A fibring function $\mathbf{F}$ should associate with each $\beta^{i,n}_{n/\equiv} \in \mathrm{WFF}_{\mathbb{L}_{i,n}}$ an actual world of a model of the other language, i.e. a formula $\alpha^{j,m}_m, j \neq i$. The persistence requirement can be observed by the following trick.

Let $\mathbf{f}$ and $\mathbf{g}$ be two functions $\mathbf{f} : \mathrm{WFF}_{\mathbb{L}_1} \mapsto \mathrm{WFF}_{\mathbb{L}_2}$ and $\mathbf{g} : \mathrm{WFF}_{\mathbb{L}_2} \mapsto \mathrm{WFF}_{\mathbb{L}_1}$ such that

$$\alpha\mid\!\sim_1\beta \text{ implies } \mathbf{f}(\alpha)\mid\!\sim_2\mathbf{f}(\beta)$$
$$\gamma\mid\!\sim_2\delta \text{ implies } \mathbf{g}(\gamma)\mid\!\sim_1\mathbf{g}(\delta).$$

Now define a fibring function $\mathbf{F}$ as follows:

$$\mathbf{F}(\beta^{i,n}_n) = \alpha^{j,m}_m, m > n$$

satisfying the following

- if $i = 1$ then $\alpha^{2,m}_m = \mathbf{f}(\beta^{1,n}_n)$
- if $i = 2$ then $\alpha^{1,n}_m = \mathbf{g}(\beta^{2,n}_n)$.

The properties of **f** and **g** will ensure that **F** satisfies persistence.

Theorem 2.14. *Let $\mid\!\sim_1$ and $\mid\!\sim_2$ be two I-logics in $\mathbb{L}_1$ and $\mathbb{L}_2$ respectively. Let $\vDash$ be the fibred consequence relation arising from the fibred semantics for the combined language as defined in Remark 2.13. Then $\vDash$ is the smallest I-logic in the combined language which is a conservative extension of each pure component.*

Proof Let $\mid\!\sim$ be the smallest I-logic in the combined language conservatively containing both $\mid\!\sim_1$ and $\mid\!\sim_2$. Such a logic exists as the intersection of all conservative extensions, $\vDash$ being one of them. Note that being an extension does not mean that $\mid\!\sim$ is closed under the rules defining $\mid\!\sim_1$ or $\mid\!\sim_2$, but only that $A\mid\!\sim_i B$ implies $A'\mid\!\sim B'$ where A', B' are substitution instances (in the combined language) of A and B respectively. We show $\mid\!\sim\, =\vDash$. We know that $\mid\!\sim\, \subseteq\vDash$, because $\mid\!\sim$ is the smallest conservative extension. We need to show $A_0\not\mid\!\sim B_0$ implies $A_0 \not\vDash B_0$. Let $\mathbf{m}_{A_0}$ be a canonical model of $\mid\!\sim$. We know that the canonical models as constructed in Theorem 2.12 have the form

$$\mathbf{m}_\alpha = (\mathrm{WFF}_{/\equiv}, \prec, \alpha_{/\equiv}, f_i, g_j, h)$$

where $A_{/\equiv} \prec B_{/\equiv}$ iff $B\mid\!\sim A$, and f_i, g_j are the functions associated with the connectives f_i for $\mathbb{L}_1$ and g_j for $\mathbb{L}_2$.

We will turn $\mathbf{m}_\alpha$ into a fibred model and this will show that $A_0 \not\vDash B_0$.

Each node $\beta_{/\equiv}$ in the model is either atomic or has the form $(C_1, \ldots, C_k) \twoheadrightarrow (D_1, \ldots, D_m)$ where $\twoheadrightarrow$ is a connective in the language $\mathbb{L}_i, i = 1$ or $i = 2$.

$\mathbf{m}_\alpha$ can be viewed as a model of $\mathbb{L}_i$, where any β with main connective $\twoheadrightarrow$ not from $\mathbb{L}_i$ is regarded as atomic. We display this point of view by writing $\mathbf{m}^i_\alpha$. Thus $\mathbf{m}^i_\alpha$ is the same model $\mathbf{m}_\alpha$ but where any wff with main connective from $\mathbb{L}_j, j \neq i$, is regarded as atomic. We can now define a fibring function $\mathbf{F}$ on $\mathbf{m}^i_\alpha$. Take any $\beta_{/\equiv}$ in the model $\mathbf{m}^i_\alpha$.

We can let $\mathbf{F}(\beta) = \mathbf{m}^j_\beta$ now regarded as a model of the $\mathbb{L}_j$ language, $j \neq i$. Thus $\mathbf{F}$ fibres, with each node $\beta_{/\equiv}$, a model of the other language. The assignment h^i_α in $\mathbf{m}^i_\alpha$ is defined as $h^i_\alpha(q) = \{t \mid q \prec t\}$.

We know that satisfaction in the canonical model is equal to $\mid\!\sim$ i.e. B holds at $A_{/\equiv}$ iff $A\mid\!\sim B$. We now show that satisfaction in the fibred model is also the same i.e. $A_{/\equiv} \vDash_{i,\beta} B$ iff $A\mid\!\sim B$.

The proof is by induction.

1. Case when q atomic
   $A_{/\equiv} \vDash_{i,\alpha} q$ iff $A_{/\equiv} \in h^i_\alpha(q)$ iff $A\mid\!\sim q$.
2. Case when the main connective of B is in the language $\mathbb{L}_i$
   $A_{/\equiv} \vDash_{i,\alpha} (C_m) \twoheadrightarrow (D_n)$ iff $\forall x_m \vDash_{i,\alpha} C_m\, \exists y_n \vDash_{i,\alpha} D_n$ s.t. $f_{\twoheadrightarrow}(x_m, y_n) \prec t$. By the induction hypothesis $x_m \vDash_{i,\alpha} C_m$ is the same $x_m\mid\!\sim C_m$ and $y_n \vDash_{i,\alpha} D_n$ is the same as $y_n\mid\!\sim D_n$ and hence we continue, iff $A\mid\!\sim B$.
3. Case when B has the main connective in the language $\mathbb{L}_j \neq \mathbb{L}_i$
   $A_{/\equiv} \vDash_{i,\alpha} B$ iff $\mathbf{F}(A_{/\equiv}) \vDash_{j,A_{/\equiv}} B$ iff, by the previous case, $A\mid\!\sim B$.
   Thus the model $\mathbf{m}_{A_0} \not\vDash B_0$.

This completes the proof of the theorem. □

2.2 Fibring Preferential Semantics

The last section dealt with basic relational structures. This kind of semantics is not the most general one. This section will show how to fibre two very general semantics.

Definition 2.15 (Non-monotonic logic).

1. A (propositional) *language* is a tuple $\mathbb{L} = (Q, \mathbb{C})$, where Q is a set of atomic propositions and $\mathbb{C}$ is a set of connectives. It may be convenient to include $\top, \bot$ as connectives.
2. The set of well-formed formulas of $\mathbb{L}$, $\mathrm{WFF}_{\mathbb{L}}$ is defined in the usual manner. Let $\mathbb{L}_1 \cup \mathbb{L}_2$ be defined as the language $\mathbb{L}_{1,2} = (Q_1 \cup Q_2, \mathbb{C}_1 \cup \mathbb{C}_2)$.
3. A (non-monotonic) *logic* on the language $\mathbb{L}$ is defined to be any relation $\mid\!\sim\, \subseteq \mathrm{WFF}^2_{\mathbb{L}}$, satisfying the following conditions, where $A \equiv B$ abbreviates ($A \mid\!\sim B$ and $B \mid\!\sim A$):
   - Identity $A \mid\!\sim A$
   - Left logical equivalence

   $$\frac{A \equiv B, A \mid\!\sim C}{B \mid\!\sim C}$$

   - $\bot \mid\!\sim B \mid\!\sim \top$, when $\top, \bot$ are in the language.
4. By right logical equivalence we mean the rule

   $$\frac{A \equiv B; C \mid\!\sim A}{C \mid\!\sim B}.$$

Definition 2.16. Let $\mathbb{L}$ be a language.

1. A Scott model for $\mathbb{L}$ is any function $\mathbf{s} : \mathrm{WFF}_{\mathbb{L}} \mapsto \{0, 1\}$.
2. A GKLM (Generalised Kraus–Lehmann–Magidor) model for $\mathbb{L}$ has the form $\mathbf{m} = (S, \prec, \mathbf{l})$, where S is a non-empty set (of states), $\prec$ is a binary relation on S and $\mathbf{l}$ is a function associating with each $t \in S$ a set of Scott models $\mathbf{l}(t)$ of $\mathbb{L}$. The model is required to satisfy the smoothness condition defined later in Definition 2.17.
3. A Scott semantics is any set $\mathcal{K}$ of Scott models.
4. A GKLM semantics is any set $\mathbb{M}$ of GKLM models.

The concept of the GKLM model should be compared with the Kraus–Lehmann–Magidor cumulative semantics [1990]. From our point of view, if the function $\mathbf{l}(t)$ gives a set of classical models, then we get the cumulative semantics for the language $\mathbb{L}$ of classical propositional logic. In the general case $\mathbb{L}$ is arbitrary and $\mathbf{l}(t)$ gives Scott models for $\mathbb{L}$. We shall see later that classes of Scott models yield monotonic consequence relations and, when we do the GKLM semantics using the Scott models, we are actually producing a non-monotonic logic *based* on the original monotonic one. This is only incidental to our purpose, which is to get a convenient semantics and fibre logics, but it does show that we can deal with general non-monotonic logics. Compare with Section 11.2, Definition 11.2.

Definition 2.17. Let $\mathbf{m} = (S, \prec, \mathbf{l})$ be a GKLM model. Let $t \in S$ be a state and A be a wff. We define the satisfaction relation $t \vDash A$ in $\mathbf{m}$ as follows:

1. $t \vDash A$ iff for all $\mathbf{s} \in \mathbf{l}(t)$ we have $\mathbf{s}(A) = 1$.
2. Let $S' \subseteq S$. We say that t is $\prec$ minimal in S' iff for all $t' \in S'$ such that $t' \prec t$ we have $t' = t$.
   We say S' is smooth if for every $t \in S'$, either t is minimal in S' or for some $s \in S'$, $s \preceq t$ and s is minimal in S'.
   We say S' is strongly smooth if there exists a $t_0 \in S'$ such that $S' = \{x \in S \mid t_0 \preceq x\}$.
3. Let $\hat{A} = \{x \in S \mid x \vDash A\}$. We say $\mathbf{m}$ is smooth iff for all A, $\hat{A}$ is smooth. We say $\mathbf{m}$ is strongly smooth iff for all A, $\hat{A}$ is strongly smooth.
   It may be convenient to require that S itself is smooth (respectively strongly smooth). In case we have *truth* $\top$ in the language, we can then have $\hat{\top} = S$.
4. We say $A \vDash_{\mathbf{m}} B$ iff for all t minimal in $\hat{A} = \{x \in S \mid x \vDash A\}$, we have $t \vDash B$. We say $\varnothing \vDash_{\mathbf{m}} B$ iff for all t minimal in S we have $t \vDash B$. If we have a symbol for *truth*, $\top$, then $\varnothing \vDash B$ is the same as $\top \vDash_{\mathbf{m}} B$.
5. Let $\mathbb{K}$ be a class of GKLM models. We say $A \vDash_{\mathbb{K}} B$ iff $A \vDash_{\mathbf{m}} B$ for all $\mathbf{m} \in \mathbb{K}$. Similarly $\vDash_{\mathbb{K}} B$.

Lemma 2.18. *Let $\mathbb{K}$ be a class of smooth models. Then $\vDash_{\mathbb{K}}$ as a binary relation on* WFF *satisfies the left logical equivalence condition. If all models in $\mathbb{K}$ are strongly smooth then $\vDash_{\mathbb{K}}$ also satisfies the right logical equivalence.*

Proof

1. Assume $\mathbb{K}$ is smooth. Assume $A \vDash_{\mathbb{K}} B$ and $B \vDash_{\mathbb{K}} A$. Let $\mathbf{m} = (S, \prec, \mathbf{l})$ be in $\mathbb{K}$. We show that $\hat{A}$ and $\hat{B}$ have the same minimal points. Let $t \in \hat{A}$ be minimal. Then $t \in \hat{B}$. We show t is minimal in $\hat{B}$. Otherwise, by smoothness, for some s, such that $s \neq t$ and $s \prec t$ we have that s is minimal in $\hat{B}$. Hence $s \in \hat{A}$, contradicting the minimality of t in $\hat{A}$. Thus we have shown that all minimal points in $\hat{A}$ are minimal in $\hat{B}$. By symmetry the converse holds.
   Assume now that for some arbitrary C, $A \vDash_{\mathbf{m}} C$. This means that for any minimal $t \in \hat{A}$, $t \vDash_{\mathbf{m}} C$. Hence for every minimal point t in $\hat{B}$, $t \vDash_{\mathbf{m}} C$ and so $B \vDash_{\mathbf{m}} C$.
2. Assume $\mathbb{K}$ is strongly smooth. Let $\mathbf{m} \in \mathbb{K}$ as above. Assume $A \vDash_{\mathbf{m}} B$ and $B \vDash_{\mathbf{m}} A$ hold. Then $\hat{A}, \hat{B}$ have the same minimal point t_0, and hence $\hat{A} = \hat{B}$. Assume, for arbitrary C with minimal point $t_c \in \hat{C}$ that $C \vDash_{\mathbf{m}} A$. We want to show $C \vDash_{\mathbf{m}} B$. We know that $t_c \vDash A$ hence either $t_c = t_0$, in which case $t_c \vDash B$ or $t_0 \prec t_c$. In the latter case $t_c \in \hat{B}$ and so $t_c \vDash B$. □

Theorem 2.19 (Completeness).

1. *Let $\mid\!\sim$ be a non-monotonic logic. Then there exists a semantics $\mathbb{K}$ such that $\mid\!\sim \; = \vDash_{\mathbb{K}}$.*

2. *If* $\mid\!\sim$ *satisfies right logical equivalence then* $\mathbb{K}$ *can be assumed to be strongly smooth semantics.*

Proof Soundness has already been verified. We show completeness. We construct a canonical model $\mathbf{m}_{\mid\sim} = (S, \prec, \mathbf{l})$.

Definition 2.20 (Canonical model for a non-monotonic logic). Let $\mid\!\sim$ be a non-monotonic logic in a language $\mathbb{L}$.

1. Let A be a wff of $\mathbb{L}$. Define a function $\mathbf{s}_A : \mathrm{WFF}_{\mathbb{L}} \mapsto \{0, 1\}$ as follows:

$$\mathbf{s}_A(X) = 1 \text{ iff } A\mid\!\sim X.$$

   Let $\mathcal{K}_{\mid\sim}$ be $\{\mathbf{s}_A \mid A \in \mathrm{WFF}_{\mathbb{L}}\}$. Note that $A \equiv B$ implies $\mathbf{s}_A = \mathbf{s}_B$, and that if $\mid\!\sim$ satisfies right logical equivalence, then $A \equiv B$ implies for all $\mathbf{s} \in \mathcal{K}_{\mid\sim}, \mathbf{s}(A) = \mathbf{s}(B)$.
   We now define $\mathbf{m}_{\mid\sim}$.
2. Let $S = \mathrm{WFF}_{\mathbb{L}/\equiv}$. The elements of S are equivalence classes modulo $\equiv$, denoted by $A_{/\equiv}$.
3. To define $\prec$ on S, let $A_{/\equiv} \prec B_{/\equiv}$ hold iff there exists an $X \equiv A$ such that $B\mid\!\sim X$.
   Note that if $\mid\!\sim$ satisfies right logical equivalence the relation $\prec$ can be identically defined as $A_{/\equiv} \prec B_{/\equiv}$ iff $B\mid\!\sim A$.
4. Let $\mathbf{l}(A_{/\equiv})$ be $\{\mathbf{s}_A\}$.

Lemma 2.21.

1. $\mathbf{m}_{\mid\sim}$ *is smooth.*
2. $\mathbf{m}_{\mid\sim}$ *is strongly smooth if* $\mid\!\sim$ *satisfies right logical equivalence.*

Proof

1. First we note that $A_{/\equiv} \in \hat{A}$. This holds since $A\mid\!\sim A$, i.e. $\mathbf{s}_A(A) = 1$.
   We now show that $\hat{A}$ is smooth. Let $B_{/\equiv} \in \hat{A}$. We seek a minimal point below or equal to $B_{/\equiv}$. We shall show that this minimal point is $A_{/\equiv}$.
   Since $B_{/\equiv} \in \hat{A}$, we know by definition that $B\mid\!\sim A$. This implies that $A_{/\equiv} \prec B_{/\equiv}$.
   We now show that $A_{/\equiv}$ is minimal. Let $C_{/\equiv} \prec A_{/\equiv}$ be in $\hat{A}$. Then there exists $C_0 \equiv C$ such that $A\mid\!\sim C_0$. Since we also have $C_{/\equiv} \in \hat{A}$ we get that $C_0\mid\!\sim A$. Hence $C_{/\equiv} = A_{/\equiv}$.
   This shows that $A_{/\equiv}$ is minimal in $\hat{A}$ below $B/_{\equiv}$.
2. We now show that $\mathbf{m}_{\mid\sim}$ is strongly smooth when $\mid\!\sim$ also satisfies right logical equivalence. Under this assumption we need to show for any $B_{/\equiv}$ that $B_{/\equiv} \in \hat{A}$ iff $A_{/\equiv} \prec B_{/\equiv}$.
   Assume $A_{/\equiv} \prec B_{/\equiv}$ and show $B_{/\equiv} \in \hat{A}$. Since $A_{/\equiv} \prec B_{/\equiv}$, then by definition, there exists $A_0 \equiv A$ such that $B\mid\!\sim A_0$. Because of right equivalence, we have $B\mid\!\sim A$. Hence by definition $B/_{\equiv} \in \hat{A}$.

Assume $B_{/\equiv} \in \hat{A}$ and show $A_{/\equiv} \prec B/_{\equiv}$. By definition we get $B \mid\!\sim A$. Hence by definition of $\prec$ we get $A_{/\equiv} \prec B_{/\equiv}$. □

We have thus shown that the model $\mathbf{m}_{\mid\sim}$ satisfies the respective semantical properties depending on the respective properties of $\mid\!\sim$. We now show completeness, i.e. if $A \not\mid\!\sim B$ then $A \not\models_{\mathbf{m}} B$.

Lemma 2.22. *$A \mid\!\sim B$ iff $A_{/\equiv} \in \hat{B}$.*

Proof Assume $A \not\mid\!\sim B$. Then $\mathbf{s}_A(B) = 0$. Since $\mathbf{s}_A \in \mathbf{l}(A_{/\equiv})$ we get by definition that $A_{/\equiv} \not\models B$, and since $A_{/\equiv}$ is a minimal point we get $A \not\models_{\mathbf{m}} B$.

Assume $A \mid\!\sim B$. Let $C_{/\equiv} \in \hat{A}$ be a minimal point. We want to show $C_{/\equiv} \models B$, i.e. $C \mid\!\sim B$. Since $C_{/\equiv} \in \hat{A}$, we have $C \mid\!\sim A$. But the latter means $A_{/\equiv} \prec C_{/\equiv}$. Since $C_{/\equiv}$ is a minimal point we must have either $C \equiv A$, in which case $C \mid\!\sim B$ or $C_{/\equiv} \prec A_{/\equiv}$ which again implies $C \equiv A$ and hence $C \mid\!\sim B$. □

This concludes the proof of the completeness Theorem 2.19. □

Remark 2.23.

1. If $\mid\!\sim$ satisfies right logical equivalence, then we have in the canonical model that
$$B_{/\equiv} \models A \text{ iff } B \mid\!\sim A$$
2. Note that if $\mid\!\sim$ is transitive, then it also satisfies right logical equivalence, and is complete for the semantics which is strongly smooth and $\prec$ is transitive.
3. A logic $\mid\!\sim$ is a relation between pairs of wffs. Since we have a semantical interpretation for $\mid\!\sim$, we can use it to extend in a conservative way the relation $\mid\!\sim$ to sets of wffs Δ in the antecedent, as follows:
   - $\{A_1, \ldots, A_n\} \mid\!\sim B$ iff for all models $\mathbf{m} = (S, \prec, \mathbf{l})$ and all minimal points $t \in \hat{A}_1 \cap \ldots \cap \hat{A}_n$ we have $t \models B$. It is easy to prove the following.
   - Restricted monotonicity:
   $$\frac{\Delta \mid\!\sim A; \Delta \mid\!\sim B}{\Delta, A \mid\!\sim B}.$$
   - A version of Cut
   $$\frac{\Delta, A \mid\!\sim B; \Delta \mid\!\sim A}{\Delta \mid\!\sim B}.$$

   We can also define
   - $\varnothing \mid\!\sim C$ iff for all models $\mathbf{m} = (S, \prec, \mathbf{l})$ and all points t minimal in S we have $t \models C$.
   To allow for this definition we need to assume that S itself is smooth (strongly smooth, resp.). The above definition will be equivalent to the following, in view of the completeness theorem relative to the canonical model $\mathbf{m}_{\mid\sim}$:
   $$\varnothing \mid\!\sim C \text{ iff } \forall A[\forall B(B \mid\!\sim A \Rightarrow A \mid\!\sim B) \Rightarrow A \mid\!\sim C].$$
   One can assume we have $\top$ in the language and define $\varnothing \mid\!\sim C$ as $\top \mid\!\sim C$.
   In the canonical model $\top_{/\equiv}$ is the minimum of S.

4. Given a GKLM model $\mathbf{m} = (S, \prec, \mathbf{l})$, we can use it to define a Scott model $\mathbf{s_m}$ as follows:
$$\mathbf{s_m}(A) = 1 \text{ iff } \vDash_{\mathbf{m}} A.$$
Any Scott model $\mathbf{s}$ can be seen as a derived model of the form $\mathbf{s_m}$, for a suitable $\mathbf{m}$.
We can let
$$\mathbf{m} = (\{\mathbf{s}\}, \{(\mathbf{s}, \mathbf{s})\}, \{(\mathbf{s}, \mathbf{s})\})$$
i.e. a one point GKLM model with the point being $\mathbf{s}$ itself.

Definition 2.24. Let $\mid\!\sim$ be a non-monotonic logic.

1. We define the associated Scott semantics $\mathcal{K}_{\mid\sim}$ to be
$$\mathcal{K}_{mid\sim} = \{\mathbf{s}_A \mid A \in WFF_{\mathbb{L}}\}$$
where $\mathbf{s}_A$ is the function defined by
$$\mathbf{s}_A(X) = 1 \text{ iff } A \mid\!\sim X.$$
2. Let $\mathcal{K}$ be a Scott semantics. We define the companion GKLM semantics to $\mathcal{K}$ denoted by $\mathbb{K}_{\mathcal{K}}$ to be the class of all GKLM models $(S, \prec, \mathbf{l})$ with $\mathbf{l}$ having a range of subsets of $\mathcal{K}$.
In the case of $\mathcal{K} = \mathcal{K}_{\mid\sim}$ for some logic $\mid\!\sim$, we denote the companion semantics by $\mathbb{K}_{\mid\sim}$.
3. Let $\mathcal{K}$ be a class of Scott models. Define a relation $\vDash_{\mathcal{K}}$ by $A \vDash_{\mathcal{K}} B$ iff for all $\mathbf{s} \in \mathcal{K}$ such that $\mathbf{s}(A) = 1$ we have also $\mathbf{s}(B) = 1$.
4. Let $\mid\!\sim$ be a logic. Let $\mathcal{K}$ be $\mathcal{K}_{\mid\sim}$. Then we denote $\vDash_{\mathcal{K}_{\mid\sim}}$ by $\vDash_{\mid\sim}$.
5. Let $\mathbb{K}$ be a class of GKLM models of the form $\mathbf{m} = (S, \prec, \mathbf{l}) \in \mathbb{K}$. We observed in item 4 of Remark 2.23 that we can derive for each $\mathbf{m}$ a Scott model $\mathbf{s_m}$. Let $\mathcal{K}_{\mathbb{K}}$ denote the class of all such models.

We can therefore define the companion semantics of $\mathbb{K}$ to itself, denoted by $\mathbb{K}_{\mathbb{K}}$ as the class of all GLKM models $\mathbf{n} = (S^{\mathbf{n}}, \prec^{\mathbf{n}}, \mathbf{l}^{\mathbf{n}})$, with $\mathbf{l}^{\mathbf{n}}$ having range from $\mathcal{K}_{\mathbb{K}}$.

The advantage of this companion semantics is that the range of $\mathbf{l}^{\mathbf{n}}$ is really a GLKM model 'used' or 'viewed' as a Scott model. In the next section, when we come to fibre and combine logics, we may choose to 'view' these models sometimes as Scott and sometimes as GKLM models.

Lemma 2.25.

1. *Let $\mathcal{K}$ be a Scott semantics. Then $\vDash_{\mathcal{K}}$ is a non-monotonic logic, satisfying Transitivity.*
2. *Let $\mid\!\sim$ be a non-monotonic logic. Then $\vDash_{\mid\sim} \subseteq \mid\!\sim$.*
3. *$\mid\!\sim$ is transitive iff $\vDash_{\mid\sim} = \mid\!\sim$.*

Proof

1. Is simple to show.

2. Note that $A \vDash_{\mid\sim} B$ iff for all $\mathbf{s} \in \mathcal{K}_{\mid\sim}, \mathbf{s}(A) = 1$ implies $\mathbf{s}(B) = 1$ iff for all wffs X ($X {\mid\sim} A$ implies $X {\mid\sim} B$).
   Thus if $A \vDash_{\mid\sim} B$, choose $X = A$ in the quantifier in item 2 above, and get $A {\mid\sim} B$.
3. If $\mid\sim$ satisfies transitivity then $\mid\sim\ \subseteq \vDash_{\mid\sim}$. In fact the converse also holds: if $\mid\sim\ \subseteq \vDash_{\mid\sim}$ then $\mid\sim$ is transitive. □

Theorem 2.26. *Let* $\mid\sim$ *be a non-monotonic logic. Then* $\vDash_{\mid\sim}$ *is the associated transitive logic of* $\mid\sim$. *Let* $\mathbb{K}$ *be the companion semantics to* $\mathcal{K}_{\mid\sim}$. *Then* $\mid\sim\ = \vDash_{\mathbb{K}}$.

In other words, every non-monotonic logic is complete for the companion semantics to its transitive associated logic.

Proof Follows from the proof of the completeness theorem, Theorem 2.19. The canonical model constructed is exactly in the companion semantics to its associated transitive logic. □

Another way of looking at the previous result is that all logics can be obtained by preferential semantics from the semantics of some (monotonic) transitive logic. It is therefore sufficient to analyse and study transitive logics.

We are now ready for a general semantical method for fibring logics. Consider two languages $\mathbb{L}_1$ and $\mathbb{L}_2$, based on the same set of atomic propositions Q, but with completely different sets of connectives $\mathbb{C}_1$ and $\mathbb{C}_2$ (with $\mathbb{C}_1 \cap \mathbb{C}_2 = \varnothing$). We consider the language $\mathbb{L}_{1,2} = (Q, \mathbb{C}_1 \cup \mathbb{C}_2)$. Our first step is to start with two Scott semantics $\mathcal{K}_1$ and $\mathcal{K}_2$ for $\mathbb{L}_1$ and $\mathbb{L}_2$ respectively and try to 'fibre' them together into some semantics $\mathcal{K}_{1,2}$ for $\mathbb{L}_{1.2}$. Since any transitive logic $\mid\sim_i$ is complete for a Scott semantics $\mathcal{K}_i$, we might hope for a fibring construction yielding $\mathcal{K}_{1,2}$ such that the logic $\mid\sim_{1.2}$ corresponding to $\mathcal{K}_{1,2}$ is the smallest conservative extension of $\mid\sim_1$ and $\mid\sim_2$. It all depends on how satisfactory the 'fibring' of the two semantics is.

To explain our intuitive idea, prior to giving formal definitions, let us assume $\mathbb{C}_1 = \{\Box\}$ and $\mathbb{C}_2 = \{\rightarrow\}$ This would be, for example, the case of combining modality $\Box$ with some substructural implication $\rightarrow$. The language $\mathbb{L}_{1,2}$ requires Scott models $\mathbf{s}$ capable of giving $\{0, 1\}$ values to mixed formulas of the form $B \vDash (q_2 \rightarrow \Box q_1)$.

Consider the formula $A_2 = q_2 \rightarrow q$. This is a pure $\mathbb{L}_2$ formula with atoms q and q_2. The formula R is obtained from A_2 by substituting the pure $\mathbb{L}_1$ formula $A_1 = \Box q_1$ for the atom q.

Somehow we want to define a model $\mathbf{s}$ for $\mathbb{L}_{1,2}$ capable of giving value $\mathbf{s}(B)$ which is reasonable, using a construction which can be generalised for arbitrary B of $\mathbb{L}_{1,2}$.

Let $\hat{A}_i$ be $\{\mathbf{s} \in \mathcal{K}_i \mid \mathbf{s}(A_i) = 1\}$. We can get a value $\mathbf{s}(A_2)$ for any $\mathbf{s} \in \mathcal{K}_2$, but A_2 has q in it and we really want a value for the case of q being A_1. However A_1 is from a different language $\mathbb{L}_1$ and a different semantics $\mathcal{K}_1$. There is no connection between the two languages, unless we create such a connection. We therefore have no choice but to establish such a connection. Let $\mathbf{F}_{2,1}, \mathbf{F}_{1,2}$ be (fibring) functions, making the following connection:

$$\mathbf{F}_{i,j} : \mathcal{K}_i \mapsto \mathcal{K}_j \quad i \neq j$$

giving for each $\mathbf{s} \in \mathcal{K}_i$ another model $\mathbf{s}' = \mathbf{F}_{i,j}(\mathbf{s}) \in \mathcal{K}_j$.

Consider the following subset of $\mathcal{K}_2$.

$$\mathcal{S}_{\Box q_1} = \{\mathbf{s} \in \mathcal{K}_2 \mid \mathbf{F}_{2,1}(\mathbf{s})(\Box q_1) = 1\}.$$

Assume we are lucky and there exists an atom k such that $\hat{k} = \mathcal{S}_{\Box q_1}$. Then from the point of view of $\mathbb{L}_2$, $A_1 = \Box q_1$ is just like an atom k. Thus using $\mathbf{F}_{2,1}$ we fibred for each $\mathbf{s} \in \mathcal{K}_2$ a model of $\mathcal{K}_1$. We considered $A_1 = \Box q_1$ as if it were atomic. To give it a value at $\mathbf{s}$ we consider $\mathbf{F}_{2,1}(\mathbf{s})$ and take that as the value. The atom k has the same truth set as the set of values we get using $\mathbf{F}$ and so we give $\mathbf{s}(B)$ the same value as $\mathbf{s}(q_2 \to k)$.

What we did can be explained in a diagram as follows:

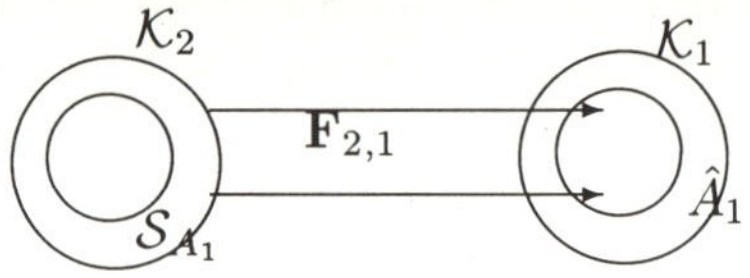

The function $\mathbf{F}_{2,1}$ allows us to identify a set $\mathcal{S}_{A_1}$ which just happens to be equal to $\hat{k}$. We can thus reasonably say that

$$\mathbf{s}(B) = \mathbf{s}(q_2 \to k).$$

Definition 2.27. Let $\mathcal{K}_1, \mathcal{K}_2$ be two Scott semantics for languages $\mathbb{L}_1$ and $\mathbb{L}_2$, respectively.

1. A subset $\mathcal{S} \subseteq \mathcal{K}_i$ is said to be a truth set iff there exists a formula A of the language such that
$$\mathcal{S} = \hat{A} = \{\mathbf{s} \in \mathcal{K}_i \mid \mathbf{s}(A) = 1\}.$$
2. Let $\mathbf{F}_{1,2}$ and $\mathbf{F}_{2,1}$ be two functions of the form
$$\mathbf{F}_{i,j} : \mathcal{K}_i \mapsto \mathcal{K}_j \quad i \neq j.$$
The above functions are said to be *Fibring functions* iff the following holds:
   - for every truth set $\mathcal{S} \subseteq \mathcal{K}_j$, the inverse image set $\{\mathbf{s} \in \mathcal{K}_i \mid \mathbf{F}_{i,j}(\mathbf{s}) \in \mathcal{S}\}$ is also a truth set in $\mathcal{K}_i$, of some formula which we conveniently denote by $k_{\mathcal{S}}$.
3. Let $\mathbb{L}_{1.2}$ be the fibred language of $\mathbb{L}_1$ and $\mathbb{L}_2$. We shall extend the models $\mathbf{s}_i \in \mathcal{K}_i$ to unique models $\mathbf{s}_i^{\mathbf{F}}$ of $\mathbb{L}_{1,2}$ using a fibring pair of functions $\mathbf{F}_{1,2}$ and $\mathbf{F}_{2,1}$. Thus $\mathbf{F}$ will allow us to create a new class of $\mathbb{L}_{1,2}$ models
$$\mathcal{K}_{1,2}^{\mathbf{F}} = \{\mathbf{s}^{\mathbf{F}} \mid \mathbf{s} \in \mathcal{K}_1\} \cup \{\mathbf{s}^{\mathbf{F}} \mid \mathbf{s} \in \mathcal{K}_2\}$$
capable of giving values $\mathbf{s}^{\mathbf{F}}(B)$ for arbitrary $B \in WFF_{\mathbb{L}_{1,2}}$
The definition is by induction on the structure of the fibred formulas, the induction being on the maximal number of nested changes of embedding of connectives from $\mathbb{L}_i$ in $\mathbb{L}_j$, for $i \neq j$.

Case 0
The fibred formula B is pure.

Subcase 0.1
B is pure formula of $\mathbb{L}_i$.
Let $\mathbf{s}_i^{\mathbf{F}}(B) = \mathbf{s}_i(B)$

Subcase 0.2
B is pure $\mathbb{L}_j, j \neq i$. Let $\mathbf{s}_i^{\mathbf{F}}(B) = \mathbf{s}_i(k_{\hat{B}})$, where

$$\hat{k}_{\hat{B}} = \{\mathbf{s} \in \mathcal{K}_i \mid \mathbf{F}_{i,j}(\mathbf{s})(B) = 1\}.$$

Note that at any rate there exists a formula C_B of $\mathbb{L}_i$ such that $\mathbf{s}_i(C_B) = \mathbf{s}_i^{\mathbf{F}}(B)$, and $\hat{C}_B = \{\mathbf{s} \in \mathcal{K}_i \mid \mathbf{F}_{i,j}(\mathbf{s})^{\mathbf{F}}(B) = 1\}$.

Case $n+1$
Subcase $(n+1)$*.1*

The main connective of B is from $\mathbb{L}_i$. Then there exists a formula $B_0(q_1, \ldots, q_n)$ of $\mathbb{L}_i$ and mixed formulas $B_1, \ldots, B_n$ of lower complexity such that

- $B = B_0(q_1/B_1, \ldots, q_n/B_n)$ where p/A indicates the substitution of wff A for the atom p.
- By the induction hypothesis for each $t_i \in \mathcal{K}_i, \mathbf{F}_{i,j}(t_i)^{\mathbf{F}}(B_m)$ is uniquely defined for $m = 1, \ldots, n$ and there exist $\mathbb{L}_j$ wffs D_{B_m} such that $\mathbf{F}_{i,j}(t_i)$ $(D_{B_m}) = \mathbf{F}_{i,j}(t_i)^{\mathbf{F}}(B_m)$. Consider $\hat{D}_{B_m}$. By the properties of the fibring there exist k_m in $\mathbb{L}_i$ such that

$$\hat{k}_m = \{t_i \mid \mathbf{F}_{i,j}(t_i)(D_{B_m}) = 1\}.$$

We now let

$$\mathbf{s}_i^{\mathbf{F}}(B) = \mathbf{s}_i(B_0(q_1/k_1, \ldots, q_n/k_n)).$$

Subcase $(n+1)$*.2*
The main connective of B is from $\mathbb{L}_j, j \neq i$. In this case the definition of $\mathbf{s}_i^{\mathbf{F}}(B)$ will be similar to subcase 0.1, relying on the inductive hypothesis of subcase $(n+1).1$ which we have just defined.

4. Let $\mathcal{K}_{1,2} = \bigcup_{\mathbf{F}} \mathcal{K}_{1,2}^{\mathbf{F}}$ i.e. $\mathcal{K}_{1,2}$ models are all possible models from all possible fibrings.

Lemma 2.28. *$\mid\!\sim_{1,2}$ is a conservative extension of $\mid\!\sim_1$ and $\mid\!\sim_2$.*

Proof Clear from the definition, as $\mathbf{s}_i^{\mathbf{F}} \upharpoonright WFF_{\mathbb{L}_i} = \mathbf{s}_i$. □

The above general fibring definition has a computational disadvantage. We have to ensure that the fibring functions satisfy the conditions in item 2 of Definition 2.27. It is much easier to define $\mathbf{F}$ for some (inductive) basic case and be confident that the conditions are automatically satisfied for all cases.

We can achieve that, however, only for special cases.

Definition 2.29. Let $\mathbf{n}$ be a GKLM model. Then $\mathbf{n}$ has the form $\mathbf{n} = (S^{\mathbf{n}}, \prec^{\mathbf{n}}, \mathbf{l}^{\mathbf{n}})$, where for each $t \in S^{\mathbf{n}}$, $\mathbf{l}^{\mathbf{n}}(t)$ is a family of Scott models. Let $\mathbf{m}$ be a basic relational model of the form $\mathbf{m} = (S^{\mathbf{m}}, \prec^{\mathbf{m}}, R_i^{\mathbf{m}}, h^{\mathbf{m}})$. We can regard $\mathbf{m}$ as a Scott model $\mathbf{s_m}$ where

$$\mathbf{s_m}(A) = 1 \text{ iff } \vDash_{\mathbf{m}} A.$$

We say $\mathbf{n}$ is a RKLM structure (a relational GKLM structure) iff it has the form $\mathbf{n}$ above with $\mathbf{l}^{\mathbf{n}}(t)$ a family of basic relational structures $\mathbf{m}$ regarded as Scott models $\mathbf{s_m}$ as explained above.

We now take a slightly different presentation of (relational) RKLM structures which will enable us to fibre them more easily.

Let $\mathbf{n} = (S, \prec, \mathbf{l})$ be such a structure. Here S is a set of states, $\prec$ is the preferential relation, and $\mathbf{l}$ is a function associating with each $s \in S$ a non-empty set $\mathbf{l}(s)$ of basic relational models. Our first inessential change is to write $\mathbf{n}$ as

$$\mathbf{n} = (W^{\mathbf{n}}, S^{\mathbf{n}}, \prec^{\mathbf{n}}, \mathbf{l}^{\mathbf{n}}, \mathbf{m}^{\mathbf{n}})$$

where $W^{\mathbf{n}}$ is a set of labels, and $\mathbf{l}^{\mathbf{n}}(s)$, for $s \in S^{\mathbf{n}}$, is a non-empty subset of $W^{\mathbf{n}}$. $\mathbf{m}^{\mathbf{n}}$ is a function associating with each $u \in W^{\mathbf{n}}$ a basic relational model $\mathbf{m}^{\mathbf{n}}(u)$ of the form $\mathbf{m}^{\mathbf{n}}(u) = (S^{\mathbf{n},u}, \prec^{\mathbf{n},u}, R_i^{\mathbf{n},u}, h^{\mathbf{n},u})$. The above is essentially the same semantics as before, except that now we are using the intermediary labels $W^{\mathbf{n}}$. This device allows us to attach other structures of possibly other languages or of different kinds to any $u \in W^{\mathbf{n}}$. So if we need a truth value for any kind of wff A at the label u, we can evaluate using a suitable structure associated with u.

The GKLM semantics contains essentially two semantic evaluation functions. First, the evaluation $\mathbf{m}^{\mathbf{n}}(u) \vDash A$, which is basic relational and second the evaluation $s \vDash A$, which is GKLM. We have

$$s \vDash A \text{ iff for all } u \in \mathbf{l}^{\mathbf{n}}(s), \mathbf{m}^{\mathbf{n}}(u) \vDash A.$$

We will distinguish between the two by writing the name of the model as an index for one and a function notation for the other.

Thus we write

$$s \vDash_{\mathbf{n}} A \text{ iff for all } u \in \mathbf{l}^{\mathbf{n}}(s), \mathbf{m}^{\mathbf{n}}(u)(A) = 1.$$

Let $\mathbb{L}$ be a language and $\mid\!\sim$ a logic in $\mathbb{L}$ and $\mathbb{K}$ its semantics. Assume $\mathbb{K}$ is a class of RKLM models. Let $>$ be a new binary connective.

Let us see what happens when we allow $>$ into the language.
Let $\mathbb{L}^* = \mathbb{L} \cup \{>\}$. Let A, B be two wffs of $\mathbb{L}$. Assume that they have the form

$$A = A[x_1, \ldots, x_n, y_1, \ldots, y_m]$$
$$B = B[x_1, \ldots, x_n, y_1, \ldots, y_m]$$

where the display of atoms means that the formula is built out of xs and ys.

Let P_i, Q_i be arbitrary formulas of $\mathbb{L}$ and consider

$$A^* = A[y_i/P_i > Q_i]$$
$$B^* = B[y_i/P_i > Q_i]$$

obtained, as the above notation suggests, by substituting $P_i > Q_i$ for y_i.

A^*, B^* have now the general form of a wff in $\mathbb{L}^*$.

Suppose that we want to check whether $A^* \mid\!\sim_{\mathbf{n}} B^*$. **n** is not a model for the language $\mathbb{L}^*$ but let us evaluate the above anyway. We need to check whether in all minimal states $s \vDash A^*$, we also have $s \vDash B^*$.

We have a problem here. To decide whether s is minimal in $\{t \mid t \vDash A^*\}$, we have to obtain a value for the semantic function $t \vDash A^*$. The recipe for this is to check the value of $\mathbf{m^n}(u)(A^*)$ for each $u \in \mathbf{l^n}(t)$. $\mathbf{m^n}(u)$ is a basic relational model of $\mathbb{L}$, capable of giving values to wffs of $\mathbb{L}$, with $h^{\mathbf{n},u}$ an assignment to the atoms x_i, y_j, but it has no value for $P_i > Q_i$ which appear in A^*. If we have values for the above wffs, for each u, then we can obtain a value for $s \vDash A^*$.

How we proceed at this point depends on our intentions for the meaning of the connective $P > Q$. There are several possibilities.

1. $>$ is a connective from another language. In this case we need for each $x \in S^{\mathbf{n},u}$, a model $\mathbf{F}^{\mathbf{n},u}(x)$ in the language of $>$ to enable us to evaluate whether $\vDash_{\mathbf{F}^{\mathbf{n},u}(x)} P > Q$ or not, i.e. we check $\vDash P > Q$ and if it holds we let $x \in h^{\mathbf{n},u}(P > Q)$.
2. $>$ may be a formalisation of another consequence relation $\mid\!\sim$. In which case we need to attach another RKLM model $\mathbf{F}^{\mathbf{n},u}(x)$ and we check whether $P \vDash_{\mathbf{F}^{\mathbf{n},u}(x)} Q$ or not.

No matter what our intention is, we need to attach a suitable model of the suitable semantics at the points $x \in S^{\mathbf{n},u}$. We denote this model by $\mathbf{F}^{\mathbf{n},u}(x)$. Note that the model $\mathbf{m}^{\mathbf{n},u}$ may have a restriction on the assignment h of the form $h^{\mathbf{n},u}(q) = \{t \mid t_{\mathbf{n},u} \prec^{\mathbf{n},u} t\}$. The function $\mathbf{F}^{\mathbf{n},u}$ must be chosen in such a way that it respects this restriction.

$\mathbf{F^n}$ is a fibring function. For each $\mathbf{n}$, each $s \in S^{\mathbf{n}}$, and each $u \in \mathbf{l^n}(s)$, and each $x \in S^{\mathbf{n},u}$ it associates a model $\mathbf{n}' = \mathbf{F}^{\mathbf{n},u}(x)$.

Consider now the function $g^{\mathbf{n}}(u)$ defined by

$$t \in g^{\mathbf{n}}(u)(y_i) \text{ iff } P_i > Q_i \text{ holds at } \mathbf{F}^{\mathbf{n},u}(t).$$

Thus the assignment $h^{\mathbf{n}}(u)$, giving $h^{\mathbf{n},u}(x_i), i = 1, \ldots, n$, to x_i and the value $g^{\mathbf{n}}(u)(y_i), i = 1, \ldots, m$, to y_i, can give a value to $t \vDash A^*$. We can now obtain a value for $A^* \vDash_{\mathbf{n}} B^*$.

Therefore we are led to the following intuitive definition

Definition 2.30 (Graded fibred semantics).

1. A graded fibred structure of level 0 has the form $\mathbf{n} = (W^{\mathbf{n}}, S^{\mathbf{n}}, \prec^{\mathbf{n}}, \mathbf{l^n}, \mathbf{m^n})$ as discussed above.
2. A graded fibred structure of level $n + 1$ has the form $\mathbf{n} = (W^{\mathbf{n}}, S^{\mathbf{n}}, \prec^{\mathbf{n}}, \mathbf{l^n}, \mathbf{m^n}, \mathbf{F^n})$, where $\mathbf{F^n}$ is a function which, for each $s \in S^{\mathbf{n}}$ and $u \in \mathbf{l^n}(s)$, and $t \in S^{\mathbf{n},u}$ gives a graded fibred structure $\mathbf{F}^{\mathbf{n},u}$ of level n. There are no restrictions on $\mathbf{F}^{\mathbf{n},u}$.

3. We can now define satisfaction of a formula of $\mathbb{L}^*$ in a graded fibred structure. We assume that the level of the structure is no less than the maximal number of nested occurrences of '>' in the formula. We define three semantic valuations, $s \vDash A, A \vDash_{\mathbf{n}} B$, and $u \vDash A$, for a label u and state s:
   - $A \vDash_{\mathbf{n}} B$ means that the structure $\mathbf{n}$ validates $A {\mid\!\sim} B$.
   - $s \vDash A$ means that A holds in s of the structure.
   - $u \vDash A$ means A holds at the label u.

   We also have that:
   - $A \vDash_{\mathbf{n}} B$ iff $s \vDash B$ for all s that are minimal in $\hat{A}$.
   - $s \vDash A$ iff for all $u \in \mathbf{l}(s), u \vDash A$.

The evaluation of $\vDash_{\mathbf{m}^{\mathbf{n}}} A$ follows by induction as in Definition 2.7. The case of $P > Q$ is as follows

- $t \vDash P > Q$ iff $\mathbf{F}^{\mathbf{n},u}(t) \vDash P > Q$.

Remark 2.31. As we remarked before, the evaluation of $\mathbf{F}^{\mathbf{n},u}(t) \vDash P > Q$ in the previous definition depends on what meaning we give the new connective '>'. If $>$ is just a connective from another language, then $\mathbf{F}^{\mathbf{n},u}(t) = \mathbf{n}'$ is just another basic relational structure for the language of $>$ and $\mathbf{n}' \vDash P > Q$ is the same as $t_{S^{\mathbf{n}'}} \vDash P > Q$, as in Definition 2.7.

If $P > Q$ is supposed to mean $P {\mid\!\sim} Q$, for another consequence relation ${\mid\!\sim}$, then $\mathbf{F}^{\mathbf{n},u}(t) = \mathbf{n}'$ is a basic relational structure for ${\mid\!\sim}$ and we evaluate $\mathbf{n}' \vDash P > Q$ as $P \vDash_{\mathbf{n}'} Q$.

It is interesting to note that ${\mid\!\sim}$ may be the original consequence relation of the class of structures under study. In other words, we start with ${\mid\!\sim}$ complete for a class $\mathbb{K}$ of RKLM structures and add $>$ to the language with the intended meaning of

$$\varnothing {\mid\!\sim} A > B \text{ iff } A {\mid\!\sim} B.$$

This means that we are bringing ${\mid\!\sim}$ into the language as a connective $>$. In this case we fibre $\mathbf{F}^{\mathbf{n},u}(t) = \mathbf{n}'$ another structure from $\mathbb{K}$. This kind of fibring was studied in Chapter 11 of this book. See also [Gabbay, 1995c]. It gives rise to conditional logics and explains the formal similarities between non-monotonic consequence and conditional logic.

2.3 Completeness Theorems—A Discussion

We now turn to the problem of completeness. Can we axiomatise the logic (consequence relation) corresponding to $\mathcal{K}_{1,2}$ in terms of ${\mid\!\sim}_1$ and ${\mid\!\sim}_2$ (of $\mathcal{K}_1$ and $\mathcal{K}_2$)?

This turns out to be not so simple and can vary from logic to logic. In Chapters 3–5 below (see also [Gabbay, 1995b]) we axiomatise a variety of modal and intuitionistic combinations but the general theory is illusive. Let us demonstrate the difficulties involved.

Let $\mathcal{K}_1$ and $\mathcal{K}_2$ be two Scott semantics in $\mathbb{L}_1$ and $\mathbb{L}_2$ respectively and let ${\mid\!\sim}_1$ and ${\mid\!\sim}_2$ be the respective logics. Define the fibred semantics $\mathcal{K}_{1,2}$ as in Definition 2.27. Let us see whether we can axiomatise $\mathcal{K}_{1,2}$. Let ${\mid\!\sim}$ be the logic in $\mathbb{L}_{1,2}$ complete for $\mathcal{K}_{1,2}$. Can we represent it in terms of ${\mid\!\sim}_1$ and ${\mid\!\sim}_2$?

Consider the rules

$$\frac{A|\!\sim_i B}{A|\!\sim B}\; i = 1, 2.$$

Can we obtain all consequences of $|\!\sim$ by substituting in consequences of $|\!\sim_1$ and $|\!\sim_2$?

Clearly this rule is sound. Can we show completeness?

Assume $A|\!\not\sim B$. We want some fibring functions $\mathbf{F}_{1,2}$ and $\mathbf{F}_{2,1}$ and some $\mathbf{s}^{\mathbf{F}} \in \mathcal{K}_{1,2}$ such that $\mathbf{s}^{\mathbf{F}}(A) = 1$ and $\mathbf{s}^{\mathbf{F}}(B) = 0$.

Let us look into the inductive structure of A and B and illustrate our ideas.

Both A and B are in the mixed language and suppose they are of complexity m of nested changes of different language connectives. This means that A and B are constructed as follows:

- Let $C_1^0, \ldots, C_{m_0}^0$ be wffs of the pure language $\mathbb{L}_1$.
- Let $D_1^1(y_i^1), \ldots, D_{n_1}^1(y_i^1)$ be wffs of the pure language $\mathbb{L}_2$, with atoms which include $y^1, \ldots, y_{m_0}^1$.
- In general let

  $$C_1^r(x_1^r, \ldots, x_{n_r}^r), \ldots, C_{m_r}^r(x_1^r, \ldots, x_{n_r}^r)$$

  be pure $\mathbb{L}_1$ wffs and

  $$D_1^r(y_1^r, \ldots, y_{m_{r-1}}^r), \ldots, D_{n_r}^r(y_1^r, \ldots, y_{m_{r-1}}^r)$$

  be pure $\mathbb{L}_2$ wffs.

The above pure wffs will be successively substituted in each other to yield A and B. This is done as follows:

- Let

$$\begin{aligned} F_j^1 &= D_j^1(y_i^1/E_i^0) \\ E_j^0 &= C_j^0 \\ F_j^r &= D_j^r(y_i^r/F_i^r) \\ E_j^r &= C_j^r(x_i^r/F_j^r) \end{aligned}$$

Then there exists A_0, B_0 with atoms q_j in the pure language $\mathbb{L}_1$ (or $\mathbb{L}_2$ depending on whether m is even or odd) such that:

- $A = A_0(q_i/F_i^m)$
- $B = B_0(q_i/F_i^m)$.

For example A may be

$$A = A_0(q_1/D_1^1(x_1/C_1^0, x_2/C_2^0), q_2/D_2^1(x_1/C_1^0, x_2/C_2^0))$$

$B = B_0$ (same substitution).

Here $m = 1, m_0 = 2, m_1 = 2$; C_1^0, C_2^0 are $\mathbb{L}_1$ pure; D_1^1, D_2^1 are $\mathbb{L}_2$ pure; A_0, B_0 are $\mathbb{L}_1$ pure; and $F_i^1 = D_i^1(x_1/C_1^0, x_2/C_2^0)$.

If $A|\!\not\sim B$ then $A_0|\!\not\sim_1 B_0$. Hence for some $\mathbf{s}_1 \in \mathcal{K}_1$, $\mathbf{s}_1(A_0) = 1$. $\mathbf{s}_1(B_0) = 0$. We are looking for an $\mathbf{F}$ and $\mathbf{s}^{\mathbf{F}}$ such that $\mathbf{s}^{\mathbf{F}}(A) = 1, \mathbf{s}^{\mathbf{F}}(B) = 0$.

From now on we continue with the particular case of $m = 1$, so that we do not obtain too many indices. Consider $\mathbf{s}_1$ above. This Scott model gives $\mathbf{s}_1(A_0) = 1$ and $\mathbf{s}_1(B_0) = 0$. This is our countermodel not for $A \not\mid\!\sim B$ but for $A_0 \not\mid\!\sim B_0$. We are looking for $\mathbf{F}$ such that $\mathbf{s}_1^{\mathbf{F}}(A) = 1$ and $\mathbf{s}_1^{\mathbf{F}}(B) = 0$. A and B are mixed formulas so by Definition 2.27 case $(n+1).1$ $\mathbf{s}_1^{\mathbf{F}}(A)$ and $\mathbf{s}_1^{\mathbf{F}}(B)$ are defined inductively in terms of $\mathbf{s}^{\mathbf{F}}(F_1^1)$ and $\mathbf{s}_1^{\mathbf{F}}(F_2^1)$. By the inductive step of Case $(n+1).1$ there exist wffs α_1, α_2 of the pure $\mathbb{L}_2$ language such that for all $t \in \mathcal{K}_1$,

$$\mathbf{F}_{1,2}(t)(\alpha_i) = \mathbf{F}_{1,2}(t)^{\mathbf{F}}(F_i^1).$$

If we can engineer $\mathbf{F}_{1,2}$ and $\mathbf{F}_{2,1}$ such that

$$\hat{q}_i = \{t \mid \mathbf{F}_{1,2}(t)(\alpha_i) = 1\} = \hat{k}_{\hat{\alpha}_i}$$

then we get our $\mathcal{K}_{1,2}$ countermodel.

Let us work our way up from the basic substitutions and see what happens.

$\hat{A}_0, \hat{B}_0, \hat{C}_1^0$ and $\hat{C}_2^0$ are subsets of $\mathcal{K}_1$, since these formulas are $\mathbb{L}_1$ pure. They form a Boolean Venn diagram. Suppose we define $\mathbf{F}_{2,1}$ in such a way that the sets $\hat{k}_{\hat{A}_0}, \hat{k}_{\hat{B}_0}, \hat{k}_{\hat{C}_1^0}, \hat{k}_{\hat{C}_2^0}$ stand in the same Boolean relationships and assume (or hope!) that any subsets of $\mathcal{K}_2$ standing in this relationship can be chosen, so we have this freedom which we may exercise later.

Consider now

$$\alpha_1 = D_1^1(k_{\hat{C}_1^0}, k_{\hat{C}_2^0}), \alpha_2 = D_2^1(k_{\hat{C}_1^0}, k_{\hat{C}_2^0}).$$

These are the α_1, α_2 mentioned above.

If they stand to each other in the same Boolean relationship as $\hat{q}_i$ stand among themselves, then maybe we can arrange that $\hat{q}_i = \hat{k}_{\hat{\alpha}_i}$ and get a countermodel.

We do not know if we can ensure all of that because of the intervening D_1^1 and D_2^1. We therefore need:

1. Assume $\mathbb{L}_1, \mathbb{L}_2$ contain the Boolean connectives so that we can express axioms to ensure completeness.
2. Propose suitable axioms and assumptions on $\mid\!\sim_1, \mid\!\sim_2$ and $\mid\!\sim$ that will allow us to have a completeness proof.
   These problems are difficult and are yet open. Some completeness proofs are given later in this book. For discussions and further references see the concluding chapter in this book and [Blackburn and de Rijke, 1995].

3

COMBINING MODAL LOGICS

3.1 Introducing Modal Fibring

The most successful case of fibring is the case of modal logic. The reason for the ease in which it can be done is that the modality operators allow us to speak directly about the possible worlds, where the local fibring is done.

We start with a motivating example for modal and intuitionistic fibring.

Example 3.1 (Motivating example for fibred semantics for modal logic). This example introduces the idea of fibred semantics. Imagine we want to form a combined logic with the intuitionistic arrow $\Rightarrow$ and the **K** modality $\Box$. We know the Kripke semantics for $\Rightarrow$ and the Kripke semantics for $\Box$. We also know the syntax of the combined logic, namely all wffs built up from the atoms using any of $\{\Rightarrow, \Box\}$ as connectives. What we do not have is any axioms or semantics for a combined system **L**.

Let us proceed nevertheless and take an arbitrary formula of the combined language, say $B = (q \Rightarrow \Box(p \Rightarrow r))$. We do not have models for the combined language, but we do notice that the main connective of B is $\Rightarrow$ and that $B = (q \Rightarrow A)$ with $A = \Box(p \Rightarrow r)$. We do have intuitionistic models for $\Rightarrow$ so let us take such a model of the form $\mathbf{m} = (S, \leq, a, h)$. S is the set of possible worlds, $\leq$ is a reflexive and transitive relation on S, $a \in S$ is the actual world and h is an assignment to the atoms, satisfying

$$t \in h(q) \text{ and } t \leq s \text{ imply } s \in h(q),$$

where $h(q) \subseteq S$.

We can evaluate, or try to evaluate, B in $\mathbf{m}$, since its main connective is $\Rightarrow$. For B to hold in $\mathbf{m}$ we must have that it holds in the actual world a. Thus we continue:
$a \vDash_{\mathbf{m}} B$ iff $a \vDash_{\mathbf{m}} (q \Rightarrow A)$ iff for all t such that $a \leq t$ and $t \vDash_{\mathbf{m}} q$ we have $t \vDash_{\mathbf{m}} A$.

We know how to evaluate $t \vDash_{\mathbf{m}} q$, because q is atomic:

$$t \vDash_{\mathbf{m}} q \text{ (def) } t \in h(q).$$

The problem is to evaluate $t \vDash_{\mathbf{m}} A$, which is $t \vDash_{\mathbf{m}} \Box(p \Rightarrow r)$. $\Box$ is not in the intuitionistic language and so we do not have a recursive evaluation clause for it. Nor is A atomic so we cannot use the assignment h. So what shall we do?

We notice that in order to complete the evaluation of $a \vDash_{\mathbf{m}} B$ all we need is to have an answer, for each $t \in S$ such that $a \leq t$, to the question of whether $t \vDash_{\mathbf{m}} A$. Any answer will do! It is at this stage that we can introduce the basic idea of fibring! We notice that $A = \Box(p \Rightarrow r)$ is a formula beginning with the modality $\Box$. So if we take any Kripke model for $\Box$ (and we do have a semantics for $\Box$) then we can evaluate A

there and get a value, and this value we give to $t \vDash_{\mathbf{m}} A$. So let us associate with each $t \in S$, a modal Kripke model

$$\mathbf{n}_t = \mathbf{F}(\mathbf{m}, t) = (T^t, R^t, a^t, h^t)$$

where $\mathbf{n}_t$ depends on $\mathbf{m}$ and on $t \in S$ and where T^t is the set of possible worlds of $\mathbf{n}_t$, R^t is the **K** accessibility relation, a^t is the actual world and h^t is the assignment.

We stipulate:

$$t \vDash_{\mathbf{m}} \Box A \text{ iff (definition) } \mathbf{n}_t \vDash \Box A$$

The latter is the same as $a^t \vDash_{\mathbf{n}_t} \Box A$.

Since $\mathbf{n}_t$ is a modal model, a value can be found.

We continue the evaluation: $a^t \vDash \Box A$ iff for all $x \in T^t, a^t R^t x$ implies $x \vDash A$.

If A is in the modal language we can continue to evaluate. If A is atomic we get our value from h^t. What if A contains the intuitionistic $\Rightarrow$, i.e. $A = (p \Rightarrow r)$? Then we need to evaluate $x \vDash_{\mathbf{n}_t} (p \Rightarrow r)$.

Since $\mathbf{n}_t$ is a modal model, we do not know how to evaluate the above. However, all we need is a value. If we associate with x an intuitionistic model

$$\mathbf{m}_x = \mathbf{F}(\mathbf{n}_t, x) = (S^x, \leq^x, 0^x, h^x)$$

then we can continue our evaluation. $\mathbf{m}_x$ is functionally dependent on x and $\mathbf{n}_t$ through the function $\mathbf{F}$. We can represent the dependence as $\mathbf{F}(t, x)$.

We stipulate:

$$x \vDash_{\mathbf{n}_t} p \Rightarrow r \text{ iff (definition) } \mathbf{m}_x \vDash (p \Rightarrow r)$$

which is evaluated at the actual world of $\mathbf{m}_x$.

Thus we continue:

$$\mathbf{m}_x \vDash p \Rightarrow r \text{ iff } 0^x \vDash (p \Rightarrow r) \text{ iff for all } y \in S^x; 0^x \leq y \text{ and } y \vDash p \text{ implies } y \vDash r.$$

This process can be continued as many times as we need. To carry out the evaluation of arbitrary mixed formulas of $\mathbf{L}$, we need Kripke models of each logic and functions associating models of the other logic to each possible world, ad infinitum, as illustrated in Figure 3.1.

These fibred models are not properly defined at the moment but are intuitively defined by the needs of combining logics. Note that we cannot put arbitrary modal models $\mathbf{n}_t$, for points $t \in S$, because they must satisfy the intuitionistic persistence conditions

$$t \leq s \text{ and } \mathbf{n}_t \vDash \Box A \text{ imply } \mathbf{n}_s \vDash \Box A$$

for any $\Box A$ of the modal language.

We need to secure the above condition (for arbitrary A). Let

$$\mathbf{n}_t = (T^t, R^t, a^t, h^t)$$
$$\mathbf{n}_s = (T^s, R^s, a^s, h^s)$$

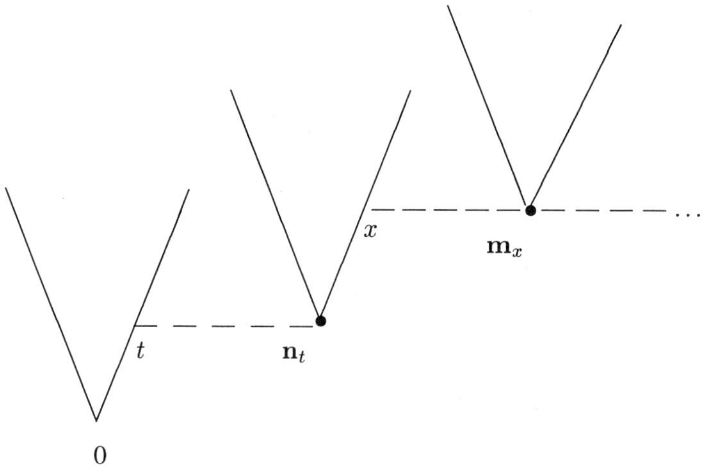

FIG. 3.1.

We can assume without loss of generality that our Kripke models $\mathbf{n}_t$ are pairwise disjoint and that each model $\mathbf{n} = (T, R, a, h)$ satisfies the condition:

$$x \in T \text{ iff } (\exists m \geq 0) aR^m x$$

where xR^0y iff $x = y$ and $xR^{m+1}y$ iff $\exists z(xRz \wedge zR^m y)$.

We will see in Section 3.3 that without loss of generality we can conservatively ensure that condition (*) holds for arbitrary A by asking that we have:

(**) $t \leq s$ implies $T^t \supseteq T^s$ and $R^t \supseteq R^s$, and $h^s \supseteq h^t \restriction T^s$.

To summarise, we needed to have associated (*or fibred*) with each intuitionistic world a modal model and with each modal world an intuitionistic model. When we were forced to evaluate a formula D in a world belonging to the other semantics, we continued the evaluation in the associated model.

The basic fibring can be done by a two place function $\mathbf{F}$. If $\mathbf{m}$ is a model and t is a possible world in $\mathbf{m}$ then $\mathbf{F}(\mathbf{m}, t)$ is a Kripke model of the other semantics. We have $t \vDash_{\mathbf{m}} A$ iff $\mathbf{F}(\mathbf{m}, t) \vDash A$, iff $a^{\mathbf{F}(\mathbf{m},t)} \vDash A$ where A is a formula with main connective not in the logic of the model $\mathbf{m}$ and $a^{\mathbf{F}(\mathbf{m},t)}$ is the actual world of the model $\mathbf{F}(\mathbf{m}, t)$.

The model $\mathbf{F}(\mathbf{m}, t)$ can be presented in many forms. If the semantics of the two logics involved are very similar, the model $\mathbf{F}(\mathbf{m}, t)$ may be a slight modification of the model $\mathbf{m}$ itself.

It may even be the case that $\mathbf{F}(\mathbf{m}, t)$ is $\mathbf{m}$ itself with a change in evaluation procedures. For example fibring a classical model to an intuitionistic model can be achieved by a change of the evaluation procedure.

The intuitive idea of the fibred semantics can also be understood in terms of classical model theory. Suppose we are dealing with classical models of a binary relation R and unary relation Q. These models have the form (S, R, Q) where S is a set and $R \subseteq S \times S$ and $Q \subseteq S$. Where do R and Q come from? For the purpose of the model theory of binary and unary relations, we do not care where R and Q come from. All we need is a subset of $S \times S$ and a subset of S. However, we could get R and Q in a more elaborate way. We could, for example, map S onto a group G using $f : S \to G$ and let xRy hold in S iff $f(x) \cdot f(y) = f(y) \cdot f(x)$ in G and let $x \in Q$ hold if for example $f(x)^2 = 1$. Our way of looking at this is to say that we are fibring group structure onto S.

In computer science terms this can be looked upon as opening a window or what is called 'delegation' in object oriented programming. To compute whether xRy holds, we open a window and place $f(x), f(y)$ in the window (i.e. go to the associated group) and then compute something in the window (e.g. commutativity) and then come back. The group G itself may be fibred further, for example, into a field F. We need a function $g : G \to F$ with some translation of the group operations into field operations. For example $x \cdot y = z$ holds in G iff $g(x) \cdot g(y) = g(z)$ holds in F. (This means that G is viewed as a multiplicative group of a field.)

The above is not exactly fibring, it is more like representation. Fibring occurs when we double back into S. For example we can map G into S using a mapping $h : G \to S$ and require that for some formula $\Psi(u_1, u_2, u_3)$ of S the following holds: $x \cdot y = z$ in G iff $\Psi(h(x), h(y), h(z))$ holds in S.

3.1.1 *Basic definitions and examples*

The next series of definitions and examples will formally present the concept of fibring of this chapter. Compare with Section 2.1.

Definition 3.2 (Logics and semantics).

1. A propositional logical language $\mathbb{L}$ comprises an infinite set of atoms and a set of connectives $\{\sharp_1, \ldots, \sharp_k\}$. The connective $\sharp_j$ has n_j places.
2. The wffs of $\mathbb{L}$ are defined by induction
   - An atom q is a wff.
   - If $A_1, \ldots, A_{n_j}$ are wffs then $\sharp_j(A_1, \ldots, A_{n_j})$ is a wff with main connective $\sharp_j$.
3. A consequence relation $\mid\!\sim$ in the language $\mathbb{L}$ is a binary relation on wffs of $\mathbb{L}$ of the form $A \mid\!\sim B$ satisfying:
   - Identity
   $$A \mid\!\sim A.$$
   - Logical equivalence
   Let $A \equiv B$ abbreviate '$A \mid\!\sim B$ and $B \mid\!\sim A$'. Then for all A, B, C
   $$\frac{A \equiv B}{(C \mid\!\sim A \text{ iff } C \mid\!\sim B)}$$
   and
   $$\frac{A \equiv B}{A \mid\!\sim C \text{ iff } B \mid\!\sim C}.$$

- Transitivity

$$A \mid\!\sim B \text{ and } B \mid\!\sim C \text{ imply } A \mid\!\sim C.$$

4. A classical monadic logic language $\mathbb{CL}$ for a possible world semantics for a logic $\mathbf{L}$ with connectives $\{\sharp_1, \ldots, \sharp_k\}$ has the non-logical predicates $\varphi_1, \ldots, \varphi_k$, $R_1, \ldots, R_k$ where R_i is an $n_i + 1$ place predicate (corresponding to the n_i place connective $\sharp_i$) and φ_i is an m_i place predicate supporting (or connected to) the connective $\sharp_i$, $(m_i \leq n_i + 1)$.
5. An interpretation for the connectives $\sharp_i$ is a classical first-order formula $\Psi_i(t, Q_1, \ldots, Q_{n_i})$ of the language $\mathbb{CL}$ where t is the only free element variable in Ψ_i and Q_j are monadic variables (subset variables).[5]
6. A possible world structure for the language $\mathbb{L}$ in the language $\mathbb{CL}$ with interpretation $\{\Psi_i\}$ is any classical structure $\mathbf{m} = (S^{\mathbf{m}}, \varphi_j^{\mathbf{m}}, R_j^{\mathbf{m}}, \mathbf{e}^{\mathbf{m}}, h^{\mathbf{m}})$, where $\mathbf{e} \in S, \varphi_j$ and $R_j, \ldots$ are relations as in (4) above, and h is an assignment giving a subset $h(q) \subseteq S$, to each atom q of $\mathbb{L}$.
7. Given a structure $\mathbf{m}$, the function h can be extended to an arbitrary wff A of $\mathbb{L}$ by structural induction as follows

$$h(\sharp_j(A_1, \ldots, A_{n_j})) = \{t \mid \mathbf{m} \vDash \Psi_j(t, h(A_1), \ldots, h(A_{n_j}))\}.$$

   We also write $t \vDash_h A$ when $t \in h(A)$.
   We are abusing notation here. '$\vDash$' is classical satisfiability. $\Psi_j(t, Q_1, \ldots, Q_{n_j})$ is a formula with free variable t and free subset variables $Q_1, \ldots, Q_{n_j}$. By substituting $h(A_i)$ for Q_i we get a wff which we can evaluate at $\mathbf{m}$ as a classical model.
8. $\mathbf{m}$ is said to satisfy $A, \mathbf{m} \vDash A$, if $\mathbf{e} \in h(A)$.
9. We say that a class of model, $\mathcal{K}$, characterises a consequence relation $\mid\!\sim$ (or $\mid\!\sim$ is complete for $\mathcal{K}$) if we have for all A, B:

$$A \mid\!\sim B \text{ iff for all } \mathbf{m} \in \mathcal{K} \text{ (if } \mathbf{m} \vDash A \text{ then } \mathbf{m} \vDash B).$$

Definition 3.3 (Fibred semantics). Let $\mathbb{L}_i$ be logic languages based on the same set of atomic propositions, with connectives $\{\sharp_1^i, \ldots, \sharp_{k_i}^i\}$, each of n_j^i places and $\mathbb{CL}_i$ be the corresponding monadic classical logic languages with interpretations Ψ_j^i for the respective connectives. Let $\mathbb{L}$ be a language containing (and based on) the disjoint union of the connectives of $\mathbb{L}_i$. Let $\mathbb{CL}$ be a monadic classical language based on the disjoint union of the predicates of $\mathbb{CL}_i$. We define a special class of structures for $\mathbb{CL}$, obtained by 'fibring' the structures of $\mathbb{CL}_i$.

Let $\mathcal{K}_i$ be a class of structures for $\mathbb{L}_i$ of the form $\mathbf{m} = (S^{\mathbf{m}}, \varphi_j^{\mathbf{m}}, R_j^{\mathbf{m}}, \mathbf{e}^{\mathbf{m}}, h^{\mathbf{m}})$. We can assume that all sets of possible worlds $S^{\mathbf{m}}$ of all structures in $\cup_i \mathcal{K}_i$ are all pairwise disjoint. We now construct the fibred structures for $\mathbb{L}$.

Let $W = \bigcup_{\mathbf{m} \in \cup_i \mathcal{K}_i} S^{\mathbf{m}}$. Since all $S^{\mathbf{m}}$ are pairwise disjoint, for each $w \in W$ there exists a unique $\mathbf{m}, i$ such that $w \in S^{\mathbf{m}}$ and $\mathbf{m} \in \mathcal{K}_i$. Let $\tau(w)$ be the function giving

[5] This clause is one dimensional. Fibring can be done in any dimension.

this unique i. Let $\mathbf{a}(w) = \mathbf{e^m}$, i.e. $\mathbf{a}$ gives the actual world of the model $\mathbf{m}$ to which w belongs. Let $\mathbf{F}(i, w), \mathbf{F} : I \times W \mapsto W$ be an arbitrary function giving for each i and each w a value $w' = \mathbf{e^m}, \mathbf{m} \in \mathcal{K}_i$. The function $\mathbf{F}$ is called a fibring function, associating with each possible world w and each semantic class $\mathcal{K}_i$ a structure $\mathbf{m} \in \mathcal{K}_i$. We also write $\mathbf{F}_i(w)$ when convenient. Let h be an assignment into W defined by

$$w \in h(q) \text{ iff } w \in h^{\mathbf{m}}(q), \text{ for the unique } \mathbf{m} \text{ such that } w \in S^{\mathbf{m}}.$$

We call any $\mathbf{n} = (W, \mathbf{a}, w_0, h, \mathbf{F}, \tau)$ a fibred structure, where $w_0 \in W$ is the 'actual' world of the structure.

Let $\mathcal{K}$ be the class of all fibred structures, over the languages and semantics as above. Let A be a wff of $\mathbb{L}$. We define the notion of satisfaction as follows:

- $w \vDash_h q$ iff $w \in h(q)$, for q atomic.
- $w \vDash_h \sharp^i_j(A_1, \ldots, A_{n^i_j})$ iff
    1. $\tau(w) = i$ and for the unique $\mathbf{m} \in \mathcal{K}_i$, such that $w \in S^{\mathbf{m}}$ we have $\mathbf{m} \vDash \Psi^i_j(h(A_1) \cap S^{\mathbf{m}}, \ldots, h(A_{n^i_j}) \cap S^{\mathbf{m}})$, or
    2. $\tau(w) \neq i$ and $\mathbf{e^m} = \mathbf{F}(i, j)$ and $\mathbf{m} \vDash \sharp^i_j(A_1, \ldots, A_{n^i_j})$.
- We say $\mathbf{n} \vDash A$ iff $w_0 \vDash A$.
- We say $\mathcal{K} \vDash A$ iff for every $\mathbf{n}$ in $\mathcal{K}$, $\mathbf{n} \vDash A$.

Fibring the semantics of a modal logic to itself can be considered as a special case of two-dimensional modal logics.

Example 3.4 (Fibring modal Kripke semantics). Let us see what the fibred semantics looks like in the case of Kripke models for modal logics.

Let $\mathbb{L}_i$ be a modal language with connectives $\Box_i, \Diamond_i$ whose semantics involves models of the form (S, R, a, h) where S is the set of possible worlds, $a \in S$ is the actual world and $R \subseteq S^2$ is the accessibility relation. h is the assignment function,[6] which we present for convenience as a binary function, giving a value $h(t, q) \in \{0, 1\}$ for any $t \in S$ and atomic q. Assume the usual truth definitions for the $\Box$ and $\Diamond$ of the language, namely

- $t \vDash \Box A$ iff $\forall s(tRs \rightarrow s \vDash A)$
- $t \vDash \Diamond A$ iff $\exists s(tRs \wedge s \vDash A)$
- satisfaction in the model is defined as satisfaction at a.

Assume $\mathbb{L}_i$ is a language with either a modality $\Box_i$ or $\Diamond_i$ or both. Let $\mathcal{K}_i$ be a set of Kripke models which defines a logic $\vdash_i$ for this language. For example $\Box_1$ may be **S4** modality, $\Box_2$ may be the Löb modality and $\Box_3$ may be some non-normal modal extension of **K4**. In general, the logic $\vdash_i$ is defined by the class of models $\mathcal{K}_i$. $\mathcal{K}_i$ need not be a frame class or anything in particular, just a class of models of the form (S, R, a, h).

We can assume that the models satisfy the following condition:

[6]Note that we can consider two options for fibring. The disjoint fibring where the sets Q_i of $\mathbb{L}_i$ are all pairwise disjoint and the ordinary option, where they are all equal. In the case of disjoint fibring if we try to evaluate $h(t, q)$ for q not in the language of the model at t, we will know to go to the appropriate fibred model.

- $S = \{x \mid \exists n \ \ aR^n x\}$.

This assumption does not affect satisfaction in models because points not accessible from a by any power R^n of R do not affect truth values at a.

We can assume that all sets of possible worlds in any $\mathcal{K}_i$ are all pairwise disjoint, and that there are infinitely many isomorphic (but disjoint) copies of each model in $\mathcal{K}_i$. We use the notation $\mathbf{m}$ for a model and present it as $\mathbf{m} = (S^{\mathbf{m}}, R^{\mathbf{m}}, a^{\mathbf{m}}, h^{\mathbf{m}})$ and write $\mathbf{m} \in \mathcal{K}_i$, when the model $\mathbf{m}$ is in the semantics $\mathcal{K}_i$. Thus our assumption boils down to

- $\mathbf{m} \neq \mathbf{n} \Rightarrow S^{\mathbf{m}} \cap S^{\mathbf{n}} = \varnothing$.

In fact a model can be identified by its actual world, i.e.

- $\mathbf{m} = \mathbf{n}$ iff $a^{\mathbf{m}} = a^{\mathbf{n}}$.

We can now present the fibred semantics arising from Kripke models, in a much simplified form. Let

- $W = \bigcup_{\mathbf{m} \in \cup_i \mathcal{K}_i} S^{\mathbf{m}}$
- $W_i = \{a^{\mathbf{m}} \mid \mathbf{m} \in \mathcal{K}_i\}$
- $W_a = \bigcup_i W_i$
- $R = \bigcup_{\mathbf{m} \in \cup_i \mathcal{K}_i} R^{\mathbf{m}}$
- $h(t, q) = h^{\mathbf{m}}(t, q)$, for the unique $\mathbf{m}$ such that $t \in S^{\mathbf{m}}$.[7]

Note that $R \restriction S^{\mathbf{m}} = R^{\mathbf{m}}$ so that $R^{\mathbf{m}}$ are all retrievable from R. In fact $S^{\mathbf{m}}$ is retrievable from R and $a^{\mathbf{m}}$ as follows

$$S^{\mathbf{m}} = \{x \mid \exists n \ \ a^{\mathbf{m}} R^n x\}.$$

Notice that the master model (W, W_i, W_a, R, h), $W_a \subseteq W$ satisfies the following properties

1. $\forall x \in W \exists y \in W_a \exists n (y R^n x)$.
2. For any $t, s \in W_a$ let $S^t = \{x \mid tR^n x, \text{ for some } n\}$; then $t \neq s \rightarrow S^t \cap S^s = \varnothing$.

The semantics $\mathcal{K}_i$ is retrievable from the above master model by letting

$$\mathcal{K}_i = \{(S^t, R \restriction S^t \times S^t, t, h \restriction S^t) \mid t \in W_i\}.$$

We can thus view the fibring function $\mathbf{F}$ as a function giving for each i and each $w \in W$ another point (actual world) in W_i as follows:

$$\mathbf{F}_i(w) = \begin{cases} w & \text{if } w \in S^{\mathbf{m}} \text{ and } \mathbf{m} \in \mathcal{K}_i \\ \text{a value in } W_i, & \text{otherwise.} \end{cases}$$

The model $(W, W_i, W_a, R, w_0, h, \mathbf{F})$ with $w_0 \in W_a$ is referred to as a *master simplified fibred semantics model*. We have

$$t \vDash \Box_i A \text{ iff } \begin{cases} t \in \mathbf{m} \text{ and } \mathbf{m} \in \mathcal{K}_i \\ \text{and } \forall s \{tRs \rightarrow s \vDash A\} \\ \text{or} \\ t \in \mathbf{m} \text{ and } \mathbf{m} \notin \mathcal{K}_i \\ \text{and } \mathbf{F}_i(t) \vDash \Box_i A. \end{cases}$$

[7]For disjoint fibring $h(t, q)$ will not be defined for q not in the appropriate language of $\mathbf{m}$.

It is obvious from the definition that the truth definition of $t \vDash \Box A$ is compatible with the truth condition:

$$t \vDash \Box_i A \text{ iff } \mathbf{F}_i(t) \vDash \Box_i A.$$

Similarly we define

$$t \vDash \Diamond_i A \text{ iff } \begin{cases} t \in \mathbf{m} \text{ and } \mathbf{m} \in \mathcal{K}_i \\ \text{and } \exists s\{tRs \wedge s \vDash A\} \\ \text{or} \\ t \in \mathbf{m} \text{ and } \mathbf{m} \notin \mathcal{K}_i \\ \text{and } \mathbf{F}_i(t) \vDash \Diamond_i A. \end{cases}$$

We say the model satisfies A iff $w_0 \vDash A$.

There is a further simplifying assumption on $\mathbf{F}$ which we can require. Let $x, y, x \neq y$, be two points in W. We know there exist unique t, s such that $x \in S^t$ and $y \in S^s$. Consider $\mathbf{F}_i(x)$ and $\mathbf{F}_i(y)$.

If $t \in W_i$ then $\mathbf{F}_i(x) = x$.
If $s \in W_i$ then $\mathbf{F}_i(y) = y$.
If $t \notin W_i$ then $\mathbf{F}_i(x) \in W_i$.
If $s \notin W_i$ then $\mathbf{F}_i(y) \in W_i$.

The above allows for the possibility that $\mathbf{F}_i(x) = \mathbf{F}_i(y)$, in case $t, s \notin W_i$. We want to exclude this possibility, so we add the further assumption that

- $x \neq y \to \mathbf{F}_i(x) \neq \mathbf{F}_i(y)$.

This assumption can be made without loss of generality because we assumed that each Kripke model in each $\mathcal{K}_i$ has infinitely many isomorphic but pairwise disjoint copies of itself in $\mathcal{K}_i$.

We can now give an independent definition of what it means to be a *simplified fibred master model* (SFM model) of a family of logics $\mathbf{L}_i, i \in I$, as follows.

Definition 3.5 (SFM model). An SFM model has the form $(W, W_i, W_a, R, w_0, h, \mathbf{F}_i)$, where

1. W is a set of worlds for some $i_0, w_0 \in W_{i_0}$.
2. $W_i \subseteq W, i \in I$, are pairwise disjoint and non-empty, with $W_a = \bigcup_i W_i$.
3. For $t, s \in W_i$ let $S^t = \{x \mid \exists n \;\; tR^n x\}$.
   Then
   - $t \neq s \to S^t \cap S^s = \varnothing$.
   - $W = \bigcup_{t \in W_a} S^t$.
4. $\mathbf{F}_i$ is a function satisfying
   - $x \in S^t$ and $t \in W_i \to \mathbf{F}_i(x) = x$
   - $x \in S^t$ and $t \notin W_i \to \mathbf{F}_i(x) \in W_i$
   - $x \neq y \to \mathbf{F}_i(x) \neq \mathbf{F}_i(y)$.
5. For each $t \in W_i$ the model $\mathbf{m}_t = (S^t, R \restriction S^t \times S^t, t, h \restriction S^t)$ is in the semantics $\mathcal{K}_i$ of $\mathbf{L}_i$.
   In case $\mathbf{L}_i$ is complete for the models $\{\mathbf{m}_t \mid t \in W_i\}$ we say the model is a *universal model*.

6. We can also assume that $\mathbf{F}$ *generates* W as follows.
   Let $W^0 = S^{w_0}$;
   $W^{n+1} = W^n \cup \bigcup_{y \in W^n, i \text{ arbitrary}} S^{\mathbf{F}_i(y)}$
   Then $W = \bigcup_n W^n$.

Condition 6 can be assumed because evaluation at a model is evaluation at w_0. Any point not in $\bigcup_n W^n$ is not reachable from w_0 by embedded modalities so cannot affect truth values.

We now take a new look at the fibring function $\mathbf{F}$. We can simplify the fibred semantics even further. Let us introduce new unary connectives of the form $\mathbb{J}_i A$ ($\mathbb{J}_i$ for a 'jump' operator) and modal operators $\Box$ and $\Diamond$ with the table:

- $t \vDash \mathbb{J}_i A$ iff $\mathbf{F}_i(t) \vDash A$.
- $t \vDash \Diamond A$ iff for some $s(tRs$ and $s \vDash A)$.
- $t \vDash \Box A$ iff for all $s(tRs \to s \vDash A)$.

According to this definition we can let $\Box_i = \mathbb{J}_i \Box$, $\Diamond_i = \mathbb{J}_i \Diamond$. We are referring to $\mathbb{J}_i$ as 'jump' operators because their truth table 'jumps' the evaluation from a world t to the world $\mathbf{F}(i, t)$.

Thus fibring several modalities together is like adding several unary jump operators to the modalities of the fibred semantics.

Conversely, assume we are given a modal logic with modalities $\Box$ and $\Diamond$ and several $\mathbb{J}_i$ and a class $\mathcal{K}$ of SFM models of the form $(W, W_i, R, w_0, h, \mathbf{F}_i)$, as defined above.

We can define modal semantic classes $\mathcal{K}_i$ by letting $\mathcal{K}_i$ be the set of all models of the form $(S^t, R \upharpoonright S^t \times S^t, t, h \upharpoonright S^t), t \in W_i$.

Note that the jump operators are slightly more expressive. For q atomic $\mathbb{J}q$ changes the index of evaluation of q from t to $\mathbf{F}_i(t)$. This cannot be done using modalities.

Example 3.6 (Dovetailing modal Kripke semantics). Continuing the previous example, the case of dovetailing is when we require that for all atomic q

$$h(t, q) = h(\mathbf{F}_i(t), q)$$

for all i. This means the actual world of the model fibred at t can be identified with t. We must be careful however. For each t, we must first introduce the notation $tR_i^t y$ iff (definition) $\mathbf{F}_i(t)R^t y$. Then we can identify the point $\mathbf{F}_i(t)$ with t and the dovetailed model can be equivalently represented as (S, R_i^t, a, h) with

$$\begin{aligned} t \vDash \Box_i A &\text{ iff } \forall y(tR_i^t y \to y \vDash A) \\ t \vDash \Diamond_i A &\text{ iff } \exists y(tR_i^t y \wedge y \vDash A). \end{aligned}$$

The function $\mathbf{F}$ is no longer needed, since we identified t with $\mathbf{F}_i(t)$.

Dovetailing looks particularly simple if we consider the simplified fibred semantics model $(W, R, w_0, h, \mathbf{F})$ of Example 1.4. Then the corresponding relations R_i are not dependent on t.

It is interesting to see the role of the jump operators in the simplified dovetailed semantics.

We have in the fibred model $(W, R, w_0, h, \mathbf{F}_i)$

$$\begin{aligned} t \vDash \mathbb{J}_i \Box A \ &\text{iff} \ \mathbf{F}_i(t) \vDash \Box A \\ &\text{iff} \ \forall x (\mathbf{F}_i(t) R x \text{ implies } x \vDash A). \end{aligned}$$

This becomes

$$t \vDash \mathbb{J}_i \Box A \text{ iff } \forall x (t R_i x \text{ implies } x \vDash A).$$

$\mathbb{J}_i$ becomes a mode shifting operator, changing the mode of evaluation, namely of which R_i to use.

We can consider the two clauses for a satisfaction $\vDash_i$ dependent on mode i.

- $t \vDash_i \Box A$ iff $\forall y (t R_i y$ implies $y \vDash_i A)$.
- $t \vDash_i \mathbb{J}_j A$ iff $t \vDash_j A$.

3.1.2 *Ways of combining*

This section discusses relationships between the logics $\mathbf{L}_i$ of $\mathbb{L}_i$ and the fibred logic $\mathbf{L}$ in the language $\mathbb{L}$, for various conditions on $\mathbf{F}$ and h. Let $\delta(\mathbf{F}, h)$ be a condition on the combination (i.e. on $\mathbf{F}$ and h). We denote by $\mathbf{CL}^\delta$ the class of fibred structures obtained and by $\mathbf{L}^\delta$ the resulting logic (consequence relation). For some δ's, $\mathbf{L}^\delta$ is a conservative extension of each $\mathbf{L}_i$. For other δ's, we get a specific combined logic. The general problem to study is: given a δ and $\mathbf{L}_i$, can we characterise $\mathbf{L}^\delta$? If we have syntactical axiomatisations of $\mathbf{L}_i$, can we put them together to obtain an axiomatisation of $\mathbf{L}^\delta$? Can we identify known combined logics from the literature and present them as the result of some special conditions δ on the fibring?

Our method is to show how we can fibre logics $\mathbf{L}_i$ into a logic $\mathbf{L}^F$ through the most general fibring construction. Once we get semantics and axiomatisation for this general logic $\mathbf{L}^F$, any combined logic existing in the literature can, in principle, be given a semantics which is related to the $\mathbf{L}^F$ semantics. The situation is similar to that of modal logic $\mathbf{K}$. Any modal logic stronger than $\mathbf{K}$ can, in principle, be given semantics related to that of $\mathbf{K}$. So, by studying the most general combination $\mathbf{L}^F$ of logics $\mathbf{L}_i$, any existing axiomatic combined logic might, in principle, be given semantics. Conversely, we can try and axiomatise any restriction on the $\mathbf{L}^F$ semantics.[8]

For example, if we restrict our choice of fibred structures and require that for any atomic q and any w and i

$$w \vDash q \text{ iff } \mathbf{F}(i, w) \vDash q$$

we get a construction which we call dovetailing. Does this condition correspond to an axiom on $\mathbf{L}$? We shall see that in modal logic this requirement is very natural.

Before we start fibring modal logics and proving completeness theorems, we need to clarify several points. To explain our problems, let us choose a very simple language. Let $\mathbb{L}$ be the language with propositional atoms $\{q_1, q_2, \ldots\}$ and classical conjunction $\{\wedge\}$ only. Let $\mathbb{L}_1$ and $\mathbb{L}_2$ be two copies of this conjunction denoted by $\wedge_1$ and $\wedge_2$. We

[8] We agree that different modal logics can be based on the same semantics and that a given modal logic may have different semantics. However, once we have any reasonable semantics for a logic, we can hope to find some semantics for its extensions.

would like to form a combined language $\mathbb{L}_{1,2}$ with both conjunctions in it such that each conjunction behaves like classical conjunction in its own language fragment.

The first decision we have to make is whether we share the atoms in $\mathbb{L}_{1,2}$. In other words, do we formulate $\mathbb{L}_{1,2}$ with two sets of atomic propositions $\{q_1^1, q_2^1, \ldots\}$ and $\{q_1^2, q_2^2, \ldots\}$ (this process is called *disjoint* or *general fibring*), one coming from each language, or do we say we have only one set of atoms $\{q_1, q_2, \ldots, \}$ and two connectives $\wedge_1$ and $\wedge_2$ (called *ordinary fibring*).

Experience from combined logics in the literature suggests that the atoms must be shared and only the connectives are combined. This however will present a slight technical problem as we shall now explain. It is not a logical problem, just a slightly awkward technicality.

We want, of course, the combined system to be a conservative extension of each component. Since $p \wedge_2 q \vdash p$ holds in $\mathbb{L}_2$, for p, q atomic, it must also hold in $\mathbb{L}_{1,2}$. Consider a typical fibred model of $\mathbb{L}_{1,2}$. Let t be a point of $\mathbb{L}_1$ (it is a classical model) and let us evaluate $t \vDash p \wedge_2 q$ and $t \vDash p$.

The first cannot be evaluated at t since the main connective is $\wedge_2$. We therefore go to a fibred model $s = \mathbf{F}_{1,2}(t)$ and evaluate $s \vDash p \wedge_2 q$.

However, since p is in the common language, when we evaluate $t \vDash p$ we need not go to the fibred models. Thus $t \vDash (p \wedge_2 q)$ iff $s \vDash p \wedge_2 q$ iff $s \vDash p$ and $s \vDash q$ and this does not imply $t \vDash p$. Thus $p \wedge_2 q \vdash p$ is not valid in the fibred semantics for 'technical reasons'.

This would not have happened if p was considered an atom of $\mathbb{L}_2$ only because in this case the evaluation of $t \vDash p$ would have required us to evaluate $s \vDash p$.

Since we do want to keep atomic propositions in the common language we must overcome this difficulty. This problem does not arise in the case of dovetailing. This is because in a dovetailed model the fibred model must agree with the current world on the values of atoms.

In modal logic ordinary fibring, any theorem of the form $A \to \alpha$, where A is a Boolean combination of atoms while α is such that all atoms occur within a scope of a modality, will not necessarily be a theorem of the fibred logic. Thus the reflexivity axiom $\Box q \to q$ will not necessarily survive the fibring process.

The completeness theorem, Theorem 3.10, for fibred logic will show exactly what survives and how.

Example 3.7 (Sample problem). Devise a logic with the classical connective '$\wedge$', the intuitionistic implication '$\Rightarrow$', and the **S4** modality '$\Box$'. In other words, we want a logic **L** with connectives $\{\wedge, \Rightarrow, \Box\}$ where these connectives behave in **L** respectively as they behave in their original logics. (Actually '$\wedge$' is the same in all three logics.)

To solve the above problem, we adopt an intuitive way of thinking. Given two logics, $\mathbf{L}_1$ and $\mathbf{L}_2$, how do we combine them? Never mind the question of whether we get a (minimal) conservative combination; we just want to put them together in some natural way.

Well, whether we can do anything depends on how the logics are presented to us and what we know about them. Here are some possibilities:

1. If $\mathbf{L}_1$ and $\mathbf{L}_2$ are presented in the same methodology, say proof theoretically as Gentzen systems, or say as Hilbert systems, then we can let $\mathbf{L}$ be the union of the languages and axioms and rules and allow substitution in $\mathbf{L}$ in both sets of axioms and rules. Thus if $\mathbf{L}_1$ is a modal **K4** for $\Box_1$ and $\mathbf{L}_2$ is modal **K4** for $\Box_2$ and the common language is classical logic then $\mathbf{L}$ will have both modalities with **K4** axioms and rules for each. There is no interaction between the two modalities. In fact, if we add the interactive axioms

$$A \to \Box_2 \Diamond_1 A$$
$$A \to \Box_1 \Diamond_2 A$$

   we get the temporal logic $\mathbf{K}_t$ (with $\Box_1$ future and $\Box_2$ past).
2. It may be that $\mathbf{L}_1$ is presented as a Hilbert system and $\mathbf{L}_2$ as a Gentzen system. How do we combine them?
3. Worse still, suppose $\mathbf{L}_1$ is presented proof theoretically and $\mathbf{L}_2$ semantically. How do we combine them?
4. If both $\mathbf{L}_1$ and $\mathbf{L}_2$ are faithfully translatable into a third logic $\mathbf{M}$ then we can combine the two logics in $\mathbf{M}$ and translate back. Let τ_1 and τ_2 translate (respectively) $\mathbf{L}_1$ and $\mathbf{L}_2$ into $\mathbf{M}$. Assume that the translation is done by translating each connective $\sharp$ of $\mathbf{L}_i$ of the form $\sharp(x_1, \ldots, x_n)$ into a formula $\varphi_\sharp(x_1, \ldots, x_n) = \tau_i(\sharp)(x_1, \ldots, x_n)$ of $\mathbf{M}$.
   Thus any formula of $\mathbf{L}_{1,2}$ can be translated into $\mathbf{M}$ by applying recursively either τ_1 or τ_2 depending on the connective. Let τ be this 'combined' translation. Thus
   - $\tau(\sharp(A_1, \ldots, A_n)) = \tau_i(\sharp)(\tau(A_1), \ldots, \tau(A_n)), \sharp \in \mathbb{L}_i$
   - $\tau(x) = x$, for x atomic.

   Define $\mathbf{L}_{1,2}$ by

$$\mathbf{L}_{1,2} \vdash A \text{ iff (def) } \mathbf{M} \vdash \tau(A).$$

   For example both $\Box_1$ and $\Box_2$ are translatable into classical logic via transitive possible world relations R_1 and R_2. Combining them in $\mathbf{M}$ gives a translation into (essentially Kripke semantics) $\mathbf{M}$ using both relations. In semantical terms we get Kripke semantics of the form (S, R_1, R_2). In this semantics $t \vDash \Box_i A$ iff for all s, tR_is implies $s \vDash A$. Do we get the same system as when we combine their Hilbert style formulation? In this case the answer is yes.

Going back to the example with $\{\wedge, \Box\}$ of modal logic **S4** and $\{\Rightarrow\}$ of intuitionistic logic, we note that there are several ways of combining them. Are they identical?

- Combine through the translation into classical logic via their respective Kripke semantics.
- Translate modal logic via the Kripke semantics into classical logic $\mathbf{M}$. Consider the translation as a syntactical translation into $\mathbf{M}$ as a Hilbert system. Weaken $\mathbf{M}$ into $\mathbf{I}$ (intuitionistic predicate logic). Let $\mathbf{L}$ of $\{\wedge, \Box, \Rightarrow\}$ be defined via the translation into $\mathbf{I}$.
- Combine $\wedge, \Box$ and $\Rightarrow$ via the translation of intuitionistic logic into modal **S4**.

- Combine $\{\wedge, \Box\}$ and $\{\Rightarrow\}$ as Hilbert systems. Take the union of languages and all instances of axioms and rules involving the connectives. Alternatively, take a natural deduction or Gentzen formulation of the two systems and combine the two proof theories.
- Combine the two systems by fibring their semantics as proposed in this chapter.

In fact, we develop several case studies of combining logics:

- combining two modalities;
- intuitionistic modal logic.

The first case study, combining two modalities, is the simplest. This should be straightforward, as we are combining two logics of the same kind which, in general, have no special requirement on the assignments to the atoms.

The second case combines two slightly different logics; namely modal logic and intuitionistic logic. The novelty in this case is that intuitionistic logic has a restriction on the assignment to atoms which has to be satisfied. This slight complication has to be addressed.

We will introduce several methods of combining logics. Each method will have its own name. So for example we shall define what it means to 'fibre' two logics or to 'dovetail' two logics or to 'fuzzle' (make fuzzy) a logic by another or to 'join' two logics. Our purpose now is to give an intuitive idea of what these methodologies are. We explain the concepts by taking a simple example. Suppose we want to combine two modal logics $\mathbf{L}_1$ and $\mathbf{L}_2$. Let $\mathcal{K}_1, \mathcal{K}_2$ be the respective Kripke semantics of the logic. Let $\mathbf{m}$ be a model in $\mathcal{K}_1$ and let t be a possible world of $\mathbf{m}$. The semantic construction which combines the logics associates a model $\mathbf{n}$ with t. The different methodologies of combination differ in the kind of model $\mathbf{n}$ that we use.

1. For *fibring* logics, we require that $\mathbf{n}$ be any model in $\mathcal{K}_2$.
2. For *dovetailing*, we require that $\mathbf{n}$ be a model of $\mathcal{K}_2$ such that for any *atomic* q

$$t \vDash q \text{ iff } \mathbf{n} \vdash q$$

   (i.e. the fibred model must agree with the values t gives to atoms).
3. For *joining* we form a multi-dimensional system out of the participating logics.

This chapter deals with combinations of modal logics. The problem in its general form is how to construct a multimodal logic containing several unary modalities, each coming with its own system of specific axioms. For example, how to combine a necessity $\Box A$ with belief BA and knowledge KA systems, or how to combine necessity $\Box A$ with temporal connectives GA and HA and the like.

There are many such combined systems in the literature but no methodology for generating them. Each combined system was presented and motivated by a different author for different reasons and different applications.

The method presented in this section for modal logics is as follows. Most known modal systems already have good semantics. Thus each logic $\mathbf{L}_i$ can be identified via a class of Kripke structures $\mathcal{K}_i$, which characterises it. Thus we will give a methodology for fibring (combining) classes of Kripke structures and study its properties. We shall then investigate what other modes of syntactical combinations are available. If each $\mathbf{L}_i$ is axiomatisable and $\mathbf{L}$ is the logic of the semantic combination of the semantics of $\mathcal{K}_i$ of $\mathbf{L}_i$, can we axiomatise $\mathbf{L}$ by some combination of the individual axiomatisations of $\mathbf{L}_i$? We continue to study other features such as decidability, finite-model property, etc.

We begin by combining arbitrary extensions of modal **K**. It makes no difference to our methodology whether these extensions are normal or not. We find out how to axiomatise the combination in terms of the theorems (axiomatisations) of the combined components. We see that in certain cases, the combined axiomatisation is very simple.

The fibring construction is very simple intuitively but unfortunately very awkward to write on paper.

The rest of the chapter is structured as follows.

We begin with a short discussion and present a canonical fibring construction. Definition 3.8 gives the general definition of fibred Kripke semantics for modal logics. Its idea is essentially to replace the notion of a possible world as a 'unit' by another Kripke structure. Armed with this definition, we can fibre arbitrary modal logics through their semantics. We need a completeness theorem for the combined logic, and this is done in Theorem 3.10. The proof is involved notationally, though simple conceptually. It does require an intermediate construction for an inductive argument to go through. This is called 'grafting' and is defined in Definition 3.9. Examples 3.11 and 3.12 show how two modal logics, which are quite typical and arbitrary, can be combined and axiomatised, using the previous completeness theorem. Theorem 3.13 gives more properties of the construction.

The second part of this chapter is concerned with a more specific construction, called dovetailing. Here the combination of the logics is done in a tighter manner, though the combined logic is still a conservative extension of each of its components. Again, Definitions 3.14 and 3.15 describe the notion and Theorem 3.16 gives the completeness theorem. Properties of dovetailing and some examples are given following the completeness theorem. We especially look for conditions on the logics which make dovetailing the same as the previous notion of fibring.

The next series of definitions and theorems have to do with fibring fragments of modal logics. It is important as a preliminary to introducing modality into other systems such as intuitionistic logic or relevance or linear logics. These other logics do not have all the classical connectives and so they 'take in' only the modality. For this reason we must develop techniques of combining pure modalities. We may have both modalities $\Box$ and $\Diamond$ in modal intuitionistic logic or modal relevance logic and because of lack of negation, one is not reducible to the other, and in fact, one may not have all the dual properties of the other. Thus, for all practical purposes we have here a fragment with combined pure modalities. Example 3.25 introduces the problems involved and Theorem 3.26 is the completeness theorem for this case.

We conclude with Theorem 3.27 about the transfer of decidability.

3.2 Fibring Modalities

Let us now start the business of this section. Imagine we want to combine two modalities $\Box_1$ and $\Box_2$. We start with the simple case where both $\Box_i$ are **K** modalities, i.e. they each satisfy the axioms below.

1. All truth functional tautologies.
2. Modus ponens
$$\frac{\vdash A; \vdash A \to B}{\vdash B}.$$
3. Necessitation
$$\frac{\vdash A}{\vdash \Box A}.$$
4. Axiom schema
   (a) $\Box(A \to B) \to (\Box A \to \Box B)$.

The above is a Hilbert system presentation. Such presentations are standard in modal logics. One may be tempted to take the combined system with both modalities and the **K** axioms and rules for each of them and let this be, together with substitution, the desired fibred system $\mathbf{K}_{1,2}$. This procedure, however, depends on the proof formulation of the components. **K** may be formulated, for example, without necessitation, as a system with an infinite number of axioms, namely all **K** theorems. If $\Box_1$ and $\Box_2$ are fibred under this new Hilbert formulation we get a combined system with less theorems because necessitation is not available. We cannot derive $\Box_1\Box_2(A \to A)$ for example unless we put in some additional rules. The lesson we learn from the above example is that we should not just formally combine proof systems but that we need some semantical interpretation, an intended meaning, to guide us in doing the fibring. This intuition can also help us with the choice of axioms for the resulting combined system.

Turning now to combining modal logics, we approach the problem semantically. The logic **K** is complete for the Kripke semantics with possible world structures of the form (S, R, a, h) where $R \subseteq S \times S$, $a \in S$, is the actual world and h is the assignment to the atoms ($h(q) \subseteq S, q$ atomic). To distinguish between the semantics for $\Box_1$ and $\Box_2$, we write (S^1, R^1, a^1, h^1) and (S^2, R^2, a^2, h^2) respectively. We now study in detail how to fibre several modalities. We have already defined a clean notion of fibred modal models in Example 3.4. We give here a more general definition.

Definition 3.8 (Fibring of propositional modal logics).

1. Let $\mathbf{L}_i, i \in I$, be a family of modal logics with $\Box_i, i \in I$, respectively. Let $\mathcal{K}_i$ be a class of models for which $\mathbf{L}_i$ is complete.
   A fibred model for the logic $\mathbf{L}_I^F$ has the form $(W, S, R, \mathbf{a}, w_0, h, \mathbf{F}, \tau)$ where W is a set of labels. S is a function giving for each w a set of possible worlds (labels) $S^w \subseteq W$. R is a function giving for each w, a relation $R^w \subseteq S^w \times S^w$. $\mathbf{a}$ is a function giving the actual world $\mathbf{a}^w$ of the model labelled by w.
   h is the assignment function $h^w(q) \subseteq S^w$, for each atomic q. τ is the semantical identifying function $\tau : W \to I$. $\tau(w) = i$ means that the model $(S^w, R^w, \mathbf{a}^w, h^w)$ is a model in $\mathcal{K}_i$. $\mathbf{F}$ is the fibring function, for each w and $t \in S^w$ and $i \in I$,

such that $\tau(w) \neq i$, $w' = \mathbf{F}(i, w, t) \in W$ and $\tau(w') = i$. $w_0 \in W$ is the distinguished label.

In the case of $\tau(w) = i$ we do not need to fibre. However, we can still define the function $\mathbf{F}$. We can let $\mathbf{F}(i, w, t)$ give us the part of the model w truncated at t. In other words, if w is the model $(S, R, w, h) \in \mathcal{K}_i$ and $t \in S$ then $\mathbf{F}(i, w, t)$ is essentially $(s, R, t, h \restriction S_t)$, where $S_t = \{s \in S \mid \exists n \geq 0 t R^n s\}$. Formally we define $\mathbf{F}(i, w, t) = w'$ and $\mathbf{a}^{w'} = t$ and $S^{w'} = \{s \in S^w \mid ((\exists n \geq 0) t (R^w)^n s)\}$ and $R^{w'} = R^w \cap (S^{w'} \times S^{w'})$ and $h^{w'} = h^w \restriction S^{w'}$. The model $\mathbf{F}(i, w, t)$, for the case $\tau(w) = i$ may not be in the semantics $\mathcal{K}_i$, as is the case for example for modal logics which are not normal, violating the requirement on the fibring function $\mathbf{F}$. Thus to have a strictly correct definition, we must assume that $\mathcal{K}_i$ is such that all truncated models obtained from models in $\mathcal{K}_i$ are also in $\mathcal{K}_i$.

2. Satisfaction is defined as follows:
   - $(w, t) \vDash q$ iff $t \in h^w(q)$, for q atomic and $t \in S^w$.
   - $(w, t) \vDash \Box_i A$ iff $\tau(w) = i$ and for all $s(t R^w s \rightarrow s \vDash A)$ or $\tau(w) = j \neq i$ and $(w', t') \vDash \Box_i A$ where $w' = \mathbf{F}(i, w, t)$ and $t' = \mathbf{a}^{w'}$.
   - Note that because of our requirement on $\mathbf{F}(i, w, t)$ for the case of $\tau(w) = i$, the second clause above is identical to the first clause for this case.
3. We say A is valid if for all models of $\mathbf{L}_I^F$, with actual world $w_0 \in W$ we have $(w_0, \mathbf{a}^{w_0}) \vDash A$.
4. Note that according to the above definition $\mathbf{L}_I^F$ is not guaranteed to be a conservative extension of each of its components.

   Let $\Box_1$ and $\Box_2$ be two $\mathbf{T}$ modalities. Assume the actual world w_0 is from a model of $\mathbf{L}_2$. Then when evaluating $(w_0, \mathbf{a}^{w_0}) \vDash \Box_2 q \rightarrow q$, $\Box_2 q$ will be evaluated at $\mathbf{F}(2, w_0, \mathbf{a}^{w_0})$ while q will be evaluated at w_0. This happens because q is in the common language and so the truth definition does not see the need to evaluate it in the fibred model.

   One way around the problem is to require that for the actual world we do fibre and not dovetail, i.e. that $\mathbf{F}(i, w_0, a^{w_0})$ always agrees with $(w_0, \mathbf{a}^{w_0})$ on atoms.

   In the case of dovetailing, we have agreement on atoms at all fibred points, but for general fibring, only for the actual world.

   This will make, in our example, $\Box_1 q \rightarrow q$ a theorem of $\mathbf{L}_I^F$ but not $\Box_2(\Box_1 q \rightarrow q)$. This does not sort out the problem altogether, because we have $\vdash \top \leftrightarrow \Box_1 q \rightarrow q$ and $\vdash \Box_2 \top$ but $\nvdash \Box_2(\Box_1 q \rightarrow q)$.

Definition 3.9 (Fibring and grafting models). We would like to define certain constructions with fibred structures.

1. Let $\mathbf{L}_i, i \in I$, be logics. Let $\mathbf{m}_0 = (S_0, R_0, a_0, h_0)$ be a model of one of the logics, say $\mathbf{L}_{i_0}$. For each $t \in S_0$ and each $i \neq i_0$ let $\mathbf{m}_{i,t} = (W^{i,t}, S^{i,t}, R^{i,t}, \mathbf{a}^{i,t}, w_0^{i,t}, h^{i,t}, \mathbf{F}^{i,t}, \tau^{i,t})$ be fibred structures with $\tau^{i,t}(w_0^{i,t}) = i$.

   We want to fibre $\mathbf{m}_{i,t}, i \neq i_0, t \in S_0$ onto $\mathbf{m}_0$ as follows.

First assume S_0 and $W^{i,t}$ are all pairwise disjoint sets. Also assume that $\mathbf{m}_{i,t}$ is named by its actual world $w_0^{i,t}$. We define the fibred model $\mathbf{m} = (W, S, R, \mathbf{a}, a_0, h, \mathbf{F}, \tau,)$ as follows.

- Let $W = S_0 \cup \bigcup_{i \neq i_o, t \in S_0} W^{i,t}$.
- Let S^w, for $w \in W$ be

$$\begin{aligned} S^w &= (S^{i,t})^w \text{ for } w \in W^{i,t} \\ &= S_{o,w} \text{ for } w \in S_0 \\ &\quad \text{where } S_{o,w} = \{t \in S_0 \mid (\exists n \geq 0) w R_0^n t\}. \end{aligned}$$

- Let R^w, for $w \in W$ be

$$\begin{aligned} R^w &= (R^{i,t})^w \text{ for } w \in W^{i,t} \\ R^w &= R_0 \restriction S_{0,w} \text{ for } w \in S_0. \end{aligned}$$

- Let h^w, for $w \in W$ be

$$\begin{aligned} h^w &= (h^{i,t})^w, \text{ for } w \in W^{i,t} \\ h^w &= h_0 \restriction S_{0,w} \text{ for } w \in S_0. \end{aligned}$$

- Let $\mathbf{a}$ be defined as follows

$$\begin{aligned} \mathbf{a}^w &= (\mathbf{a}^{i,t})^w, \text{ for } w \in W^{i,t} \\ \mathbf{a}^w &= w \text{ for } w \in S_0. \end{aligned}$$

- Let τ be defined as follows

$$\begin{aligned} \tau(w) &= \tau^{i,t}(w), \text{ for } w \in W^{i,t} \\ \tau(w) &= w \text{ for } w \in S_0. \end{aligned}$$

- Let $\mathbf{F}(j, w, s)$, for $s \in S^w$ be:
  $\mathbf{F}(j, w, s) = \mathbf{F}^{i,t}(j, w, s)$ for $w \in W^{i,t}$
  $\mathbf{F}(j, w, s) = w_0^{j,s}$ for $w \in S_0$ and $j \neq i_0$
  $\mathbf{F}(i_0, w, s) = s$ for $w \in S_0$.

The following holds in the new fibred model.

- For any formula A of $\mathbf{L}_{i_0}$ and any $w \in S_0$ we have
  $w \vDash A$ in $\mathbf{m}_0$ iff $w \vDash A$ in $\mathbf{m}$.
- For any formula $\Box_i B$ of $\mathbf{L}_I^F, i \neq i_0$ and any $w \in S_0$ we have
  $w \vDash \Box_i B$ in $\mathbf{m}$ iff $\mathbf{m}_{i,w} \vDash \Box_i B$.

2. Let $\mathbf{m}$ be a fibred model with distinguished label w_0, with $\tau(w_0) = i_0$ and fibring function $\mathbf{F}^0$. Let $\mathbf{m}_i$ for $i \neq i_0$ be fibred models with distinguished labels w_i and fibring functions $\mathbf{F}^i$. We want to *graft* $\mathbf{m}_i$ onto $\mathbf{m}$ at w_0 by replacing $\mathbf{F}^0(i, w_0, w_0)$ by the model $\mathbf{m}_i$.

$$\begin{aligned} \text{Let } \mathbf{m} &= (W^0, S^0, R^0, \mathbf{a}^0, w_0, h^0, \mathbf{F}^0, \tau^0) \\ \mathbf{m}_i &= (W^i, S^i, R^i, a^i, w_i, h^i, \mathbf{F}^i, \tau^i). \end{aligned}$$

We can assume all w^i, W^0 are pairwise disjoint and that they are named by their actual worlds, i.e. $\mathbf{m} = w_0$ and $\mathbf{m}_i = w_i$.

Let $\mathbf{m}_{i,t}$, for $t \in S^{o,w_0}, t \neq w_0$ be $\mathbf{F}^0(i, w_0, t)$. Let $\mathbf{m}_{i,w_0} = \mathbf{m}_i = w_i$.
Then the grafting of $\mathbf{m}_i$ onto $\mathbf{m}$ is defined to be the same as the fibring of $\mathbf{m}_{i,t}$ onto $(S^{0,w_0}, R^{0,w_0}, w_0, h^{0,w_0})$.

Theorem 3.10 (Completeness theorem for the fibred logic $\mathbf{L}_I^F$). *Let $\mathbf{L}_i, i \in I$, be modal logics in the respective languages $\mathbb{L}_i$ with classes of structures $\mathcal{K}_i$ and set of theorems $\mathbf{T}_i$, (i.e. $\mathbf{T}_i = \{A \text{ of } \mathbb{L}_i \mid A \text{ is valid in all } \mathcal{K}_i \text{ models }\}$). We consider two options for fibring. The general option where the logics $\mathbf{L}_i$ are assumed to have pairwise disjoint sets of atomic propositions (referred to as disjoint fibring) and the ordinary fibring where all $\mathbf{L}_i$ share the same set of atoms.*

Let $\mathbf{T}_I^F$ be the following set of wffs of $\mathbf{L}_I^F$.

1. (a) $\mathbf{T}_i \subseteq \mathbf{T}_I^F$.
   (b) *For the case of ordinary fibring of n logics, we need axioms (1) and (2) below. If we are fibring an infinite number of logics, (1) is not needed.*
   (1) If A is a Boolean combination of atoms and α_i is in the ith fibred language $i = 1, \ldots, n$, then $A \to \alpha_i \in \mathbf{T}_i$ implies $A \to \bigvee_i \alpha_i \in \mathbf{T}_I^F$ where in α_i every atom is in the scope of a modality.
   (2) If $A(x_j) \in \mathbf{T}_i$ then $A(x_j/\Box_{\sigma(j)}\alpha_j) \in \mathbf{T}_I^F$, for any $\Box_{\sigma(j)}\alpha_j$ in the language $\mathbf{L}_{\sigma(j)}, j = 1, 2, \ldots$.
2. **Modal fibring rule:**[9]
   If $\Box_i$ is the modality of $\mathbf{L}_i$ and $\Box_j$ of $\mathbf{L}_j, i, j$ arbitrary $i \neq j$ and $C = \bigwedge_{k=1}^n \Box_i A_k \to \bigvee_{k=1}^m \Box_i B_k \in \mathbf{T}_I^F$ then for all d, $\Box_j^d C \in \mathbf{T}_I^F$.
3. *$\mathbf{T}_I^F$ is the smallest set closed under (1), (2), modus ponens and substitution.*

Then $\mathbf{T}_I^F$ is the set of all wffs of $\mathbf{L}_I^F$ valid in all the fibred structures of $\mathbf{L}_I^F$.[10]

Proof

1. Soundness
   The case of soundness is easy to verify. We check the case of axiom 1(b1). Given any model, the actual world must be in one of the participating languages, say $\mathbb{L}_1$. If A holds in the world and $\bigvee \alpha_i$ does not hold, then we get a countermodel for $A \to \alpha_1$.
2. Completeness
   To show completeness we use induction on the number n of nested different lan-

[9]The meaning of the modal fibring rule will be apparent from the proof below. Intuitively, it has to do with substitutions of wffs of one language into a formula of the other language. If the substituted wffs are related (proof theoretically) we want to propagate this relation into the other language.

There are formal similarities between the modal fibring rule and necessitation. Consider a formula C built up only from 'atoms' of the form $\Box_i B_k$. Then our special rule of necessitation says that from $\vdash C$ we can deduce $\vdash \Box_j^d C$ for any modality $\Box_j$ *other than* $\Box_i$.

[10]Note that for languages that are disjoint, axiom (1b) implies axiom (1a). We can pretend that the atoms q_j^i of the language $\mathbb{L}_i$ are modal expressions of the form $I_i q_j$, based on a common vocabulary $\{q_j\} = Q$. I_i is the identity modality $I_i x = x$. Let $\alpha(q) \in \mathbf{T}_i, q \in Q_i$. We can view it as $\alpha(I_i q) \in \mathbf{T}_i, q \in Q$, i.e. as an axiom where all atoms are within the scope of a modality. Thus, using axiom 1(b2) we get $\alpha \in \mathbf{T}_I^F$.

guage modalities in the formula, to show that every consistent formula B with at most n different nestings has a model.

Case $n = 1$: disjoint fibring

B is a formula with only one nested different modality. This means B is a Boolean combination of wffs of $\mathbf{L}_i$ for several $i \in I$. We can assume B is a prenex normal form and we present it as

$$B = \bigvee_k \bigwedge_j B_{k,j}$$

where $B_{k,j} = A_{k,j} \wedge \Box_{\sigma(j)} C_{k,j} \wedge \bigwedge_r \Diamond_{\sigma(j)} D^r_{k,j}$ and where $A_{k,j}$ and $D^r_{k,j}$ are conjunctions of atomic literals of the language of $\mathbb{L}_{\sigma(j)}$ and $C_{k,j}$ is a disjunction of such literals.

We want to construct a fibred model for B. It is sufficient to construct for at least one index k, $\mathbf{L}_{\sigma(j)}$ models for each $B_{k,j}, j = 1, 2, \ldots$, because we can then graft them together. If this is not possible then for each k there exists a j such that $\mathbf{L}_{\sigma(j)} \vdash\sim B_{k,j}$. Hence by axiom (1a), $\mathbf{L}^F_I \vdash\sim B_{k,j}$ and so $\mathbf{L}^F_I \vdash\sim B$, a contradiction.

Case $n = 1$: ordinary (shared) fibring

In this case B has the form

$$B = \bigvee_k A_k \wedge \bigwedge_j B_{k,j}$$

where $B_{k,j} = \Box_{\sigma(j)} C_{k,j} \wedge \bigwedge_r \Diamond_{\sigma(j)} D^r_{k,j}$ and A_k is a full conjunction of literals in the common atomic language, i.e. $\vdash \bigvee_k A_k$ in classical logic and $C_{k,j}$ and $D^r_{k,j}$ are as before.

We try and graft an $\mathbf{L}^F_I$ model for at least one disjunct $A_k \wedge \bigwedge_j B_{k,j}$. Let us examine what can happen.

Subcase 1

There is a fibred logic $\mathbf{L}_m$ which has no wff appearing in this disjunct. (This case will always hold if we are fibring an infinite number of logics.) In this case we try to fibre a logic whose actual world is from the logic $\mathbf{L}_m$. A_k can be made true there and we need models for $B_{k,j}, j = 1, \ldots$. This will fail only if for some j $\mathbf{L}_{\sigma(j)} \vdash\sim B_{k,j}$.

Subcase 2

The fibred logics are all represented in the disjunct. In this case, choose a j and consider $A_k \wedge B_{k,j}$. If this is $\mathbf{L}_{\sigma(j)}$ consistent, and has a model $\mathbf{m}_j$ then if we can find $\mathbf{L}_{\sigma(r)}$ models $\mathbf{m}_r$ for $B_{k,r}$, $r \neq j$, then we will be able to graft a model for the disjunct. This is so since the actual world of $\mathbf{m}_j$ will validate $A_k \wedge B_{k,j}$ while the other $B_{k,r}$ will be evaluated at the fibred models $\mathbf{m}_r$. A_k need not be valid in $\mathbf{m}_r, r \neq j$. So the actual world of the fibred model will be the actual world of $\mathbf{m}_j$, and that is where A_k holds.

Summarising both subcases we see that we fail to give a model to the kth disjunct if one of the following holds:

$(k, 1)$ All the fibred logics are represented in the disjunct and for all j.
$\vdash_{\sigma(j)} A_k \rightarrow\sim B_{k,j}$

$(k, 2)$ For some $r \vdash_{\sigma(r)}\sim B_{k,r}$.

Consider an arbitrary disjunct with index k. If case $(k,1)$ holds then as we have seen $\mathbf{L}_{\sigma(j)} \vdash A_k \to \sim B_{k,j}, j = 1, \ldots$.

If $(k,2)$ holds then for some r, $\mathbf{L}_{\sigma(r)} \vdash \sim B_{k,r}$, and by axiom 1(2b) we have $\mathbf{L}_I^F \vdash \sim B_{k,r}$.

Hence B is equivalent to

$$\bigvee_{k_1} A_{k_1} \wedge \bigwedge B_{k_1,j}$$

where k_1 ranges over the disjunct where case $(k,1)$ holds. By axiom 1(b1) we have $\mathbf{L}_I^F \vdash A_{k_1} \to \bigvee \sim B_{k_1,j}$ contradicting the consistency of B.

Case $n > 1$

We can assume B is obtained from formulas of the form

- $B_i(\overline{x}^1, \overline{x}^2, \ldots \overline{x}^k, \overline{y})$,
- $\Box_{\sigma(1)} A_1^1(\overline{y}), \ldots, \Box_{\sigma(1)} A_{n(1)}^1(\overline{y})$,

  $\vdots$
- $\Box_{\sigma(k)} A_1^k(\overline{y}), \ldots, \Box_{\sigma(k)} A_{n(k)}^k(\overline{y})$.

Here $\Box_{\sigma(j)}$ is the modality of $\mathbf{L}_{\sigma(j)}$ and $\overline{x}^j$ is the sequence of variables $x_1^j, \ldots, x_{n(j)}^j$ and B_i is a formula of the logic $\mathbf{L}_i$ and B is obtained from B_i by the substitution Θ where $\Theta(x_j^r) = \Box_{\sigma(r)} A_j^r$ and $\Theta(y_i) = y_i$.

Thus

$$B = B_i(x_j^r / \Box_{\sigma(r)} A_j^r, \overline{y}), r = 1, \ldots, k; j \in I.$$

Subcase a: In B_i all of x_j^r are within the scope of the modality $\Box_i$.

In this subcase, by the induction hypothesis, each $\Box_{\sigma(r)} A_j^r$ has a lower number of nested modalities than B. We assume B is $\mathbf{L}_I^F$ consistent. Hence B_i is $\mathbf{L}_i$ consistent. (Consistency follows for disjoint fibring from axiom 1(a) and for ordinary fibring from axiom 1(b2).) We want to show that B has a fibred model. Our strategy is to consider all $\mathbf{L}_i$ models of B_i and try to augment at least one of them into a fibred model of B.

Let $\mathbf{m} = (S^{\mathbf{m}}, R^{\mathbf{m}}, a^{\mathbf{m}}, h^{\mathbf{m}})$ be any model of B_i. Since B is $B_i\Theta$, we might be able to augment $\mathbf{m}$ into a fibred model of B if we can realise the assignment $h^{\mathbf{m}}$ by fibred models of $\Theta(x_r^j)$. To be more specific, let $t \in S^{\mathbf{m}}$ be any possible world in the model $\mathbf{m}$. Consider the assignment $h^{\mathbf{m}}(t, x_r^j)$, for any atomic proposition x_r^j of B_i, of the sequence $\overline{x}^j$. It is more convenient, for the clarity of the proof, to write $h(t,q) = 1$ instead of $t \in h(q)$. $\Theta(x_r^j)$ is a formula $\Box_{\sigma(j)} A_r^j$ of the logic $\mathbf{L}_{\tau(j)}$. In the fibred semantics, any such $\Box_{\sigma(j)} A_r^j$ is evaluated at a fibred model $\mathbf{n}_j$. We want to have $\mathbf{n}_j \vDash \Box_{\sigma(j)} A_r^j$ iff $h^{\mathbf{m}}(t, x_r^j) = 1$. We can choose the model $\mathbf{n}_j$. The question is whether we can realise the truth values $h^{\mathbf{m}}(t, x_r^j)$ in a suitable model $\mathbf{n}_j$, for all the atoms of $\overline{x}^j$. If we can do that for all j, then we can build a fibred model for B. Of course we cannot realise the values $h^{\mathbf{m}}(t, x_r^j)$, i.e. if we fibre an arbitrary model $\mathbf{n}_t'$ at the point t, we might change the value of $B = B_1\Theta$ at $a^{\mathbf{m}}$.

We must consider two possibilities:

- t is a point such that the values of the atoms x_r^j at t do not affect the truth value of $B_1(x_r^j, \overline{y})$ at $a^{\mathbf{m}}$. In this case we don't mind what model $\mathbf{n}_t'$ we fibre at t, because even though $h^{\mathbf{m}}(t, x_r^j)$ is not the same as the value of $\Theta(x_r^j)$ at $\mathbf{n}'_t$, the values of $B = B_1\Theta$ at $a^{\mathbf{m}}$ will not change.

- t is a point such that the values of x_r^j at t do affect the value of $B_1(x_r^j, \overline{y})$ at $a^{\mathbf{m}}$. In this case the model $\mathbf{n}_t$ must realise these values.

Our first observation is that the points t we need worry about are the points at distance less than $d+1$ from $a^{\mathbf{m}}$ (i.e. $a^{\mathbf{m}}(R^{\mathbf{m}})^k t$ holds for $k \leq d+1$), where d is the maximal depth of nested modalities in B_1. This can be proved by induction. For $d = 0$ there are no modalities and so $a^{\mathbf{m}}$ is all that matters. For $\Box E$ evaluated at $a^{\mathbf{m}}$, the values that matter are values of E at points s such that $a^{\mathbf{m}}R^{\mathbf{m}}s$. By the induction hypothesis the value of E at s is not affected by values of atoms at points at distance more than d from s.

We thus conclude that we need realise all values $h^{\mathbf{m}}(t, x_r^j)$, for all points t at a distance less than $d+1$ from $a^{\mathbf{m}}$.

How can we fail to do that? Let us check! Let

$$D_t^j = \bigwedge_{s=1}^{n(j)} [x_s^j]^{\varepsilon(j,s)}$$

$$C_t^j = \bigwedge_{s=1}^{n(j)} [\Box_{\sigma(j)} A_s^j]^{\varepsilon(j,s)} = D_t^j\Theta.$$

where $\varepsilon(j,s) = h^{\mathbf{m}}(t, x_s^j)$. We use the notation: $[\varphi]^\varepsilon = \varphi$ if $\varepsilon = 1$ and $\sim \varphi$ if $\varepsilon = 0$. If the above C_t^j is consistent, then by the induction hypothesis a fibred model would exist. If C_t^j is inconsistent, then no fibred model exists. C_t^j is a conjunction of formulas of the form $\pm\Box_{\sigma(j)} A_s^j$. Since the models for the different modalities do not interact, if all C_t^j are consistent then the conjunction $\wedge_j C_t^j$ is consistent. Assume now that for every model $\mathbf{m} \in \mathcal{K}_i$, there exists a j and a $t \in S^{\mathbf{m}}$ at distance $d(t)$ (with $d(t) \leq d+1$) from the actual world $a^{\mathbf{m}}$ (i.e. $a^{\mathbf{m}}(R^{\mathbf{m}})^{d(t)}t$) such that C_t^j is inconsistent. Since the number of possible different C_t^j and $d(t)$ is finite, this means that for some finite number of wffs $D_1, \ldots, D_m$ built up as conjunctions of atoms and their negations from among $\overline{x}^j$, we have that $\{D_s\Theta\}$ covers all the C_t^j and we have $\vdash\sim D_s\Theta$, $s = 1, \ldots m$. Therefore we have that

$$\vdash_{\mathbf{L}_i} (\bigwedge_{s=1}^{m} \Box_i^{d(s)} \sim D_s) \to\sim B_i$$

since in each model $\mathbf{m}$ of B_i one of the D_i is satisfied.

Hence

$$\vdash_{\mathbf{L}_I^F} (\bigwedge_{s=1}^{m} \Box_i^{d(s)} \sim D_s\Theta) \to\sim B_i\Theta$$

but $B_i\Theta$ is B and $\vdash_{L_I}\sim D_s\Theta$. Hence (by axiom (2) of $\mathbf{L}_I$)

$$\vdash_{\mathbf{L}_I} \Box_i^{d(s)} \sim D_r\Theta$$

and hence

$$\vdash_{\mathbf{L}_I} \sim B$$

which contradicts the assumptions.

We therefore conclude that there exists a model $\mathbf{m}$ of B_i for which all C_t^j, for all j and all $t \in S^m$ such that $d(t) \leq d+1$ are consistent. By the induction hypothesis there exist fibred models $\mathbf{W}_t^j$ for each C_t^j. In this fibred model we have for each x_r^j and $t \in S^m$

$$h^{\mathbf{m}}(t, x_r^j) = \text{ value of } \Box_{\sigma(j)} A_r^j \text{ in } \mathbf{W}_t^j.$$

For points t such that $d(t) > d+1$, let $\mathbf{W}_t^j$ be an arbitrary fibred model. These models will not necessarily realise the values of $h^{\mathbf{m}}(t, x_r^j)$ but we know that these values do not affect the value of B_1 at $a^{\mathbf{m}}$. Now that we have $\mathbf{W}_t^j$ for all $t \in S^{\mathbf{m}}$, we can assume that $\mathbf{W}_t^j$ are all pairwise disjoint. We can fibre $\mathbf{m}$ with $\mathbf{W}_t^j$ and get a new fibred model for $B = B_i\Theta$, as we do in Definition 3.9.

Subcase b:
We now examine the subcase where the conditions of *Subcase a* are not necessarily satisfied. Let B be an arbitrary formula. Certainly there exists a classical Boolean combination $\Psi(\overline{x}_1, \ldots, \overline{x}_n, \overline{y})$, where $\overline{x}_i = (x_1^i, \ldots, x_{k(i)}^i)$ and $\overline{y} = (y_1, \ldots, y_k)$ and a substitution Θ such that $B = \Psi\Theta$, where $\Theta x_r^j = \Box_{\sigma(j)} A_r^j$ and $\Theta y_i = y_i$ where for disjoint fibring y_i is in $\mathbf{L}_{\sigma(i)}$ and for shared ordinary fibring y_i is a common atom. Since we are trying to find a model for Ψ, we can assume that Ψ is in a disjunctive normal form $\Psi = \bigvee_s \alpha_s$.

Let α be such a disjunct:

$$\alpha_s = \bigwedge_{r,j} [x_r^j]^{\varepsilon(j,r,s)} \wedge \bigwedge_r y_r^{\varepsilon(r,s)}$$

where $\varepsilon(j, r, s)$ is the sign of x_r^j in α_s and similarly $\varepsilon(r, s)$ is the sign of y_r in α_s.

At least for one such α_s, $\alpha_s\Theta$ must be $\mathbf{L}_I$ consistent otherwise we will have $\vdash_{\mathbf{L}_I} \sim \Psi\Theta$.

Let this disjunct be α. (We omit reference to s.) Consider an arbitrary i and consider the formula

$$E_{\sigma(i)} = \bigwedge_{r=1}^{k(i)} [\Box_{\sigma(i)} A_r^i]^{\varepsilon(i,r)} \wedge \bigwedge_{y_r \text{ in } \mathbf{L}_{\sigma(i)}} y_r.$$

This is a consistent formula which has the form

$$E_{\sigma(i)} = B_{\sigma(i)}(\overline{z}_1, \ldots, \overline{z}_m, \overline{u})\Theta_1$$

with substitution Θ_1, where $\Theta_1(\overline{z}_r) = \Box_{\sigma(i)} A_r^i$, $\Theta_1(\overline{u}) = \overline{u}$ and this case satisfies the conditions of *Subcase a* above. It therefore has a fibred model $\mathbf{W}_i$.

The models $\mathbf{W}_i$ can now be fibred into a model of $\alpha\Theta$ and hence of $\Psi\Theta$, as we have done in Definition 3.9.

This completes the induction and the proof of the theorem. □

Example 3.11. Consider two logics $\mathbf{K}_1$ and $\mathbf{K}_2$ with two $\mathbf{K}$ modalities $\Box_1$ and $\Box_2$. Consider the fibred combination $\mathbf{K}^F_{1,2}$. By the previous theorem it can be axiomatised by taking all theorems of $\mathbf{K}_1$ and $\mathbf{K}_2$ together with the following modal fibring rule:

$$\frac{\vdash \bigwedge_k \Box_i A_k \to \bigvee_k \Box_i B_k}{\vdash \Box_j[\bigwedge_k \Box_i A_k \to \bigvee_k \Box_i B_k]}$$

where $i, j \in \{1, 2\}, i \neq j$.

Note that necessitation is not available. This example investigates under what conditions necessitation is admissible.

The logic $\mathbf{K}$ has the following special properties:

1. $\vdash \Box\top$
2. $\vdash \Box(A \wedge B) \leftrightarrow \Box A \wedge \Box B$
3. The disjunction property

$$\frac{\vdash \Box A \to \Box B \vee \Box C}{\vdash (\Box A \to \Box B) \text{ or } \vdash (\Box A \to \Box C)}.$$

The above three properties can be used to prove that the modal fibring rule is equivalent to necessitation.

First the rule can be reduced to

$$\frac{\vdash \Box_i A \to \Box_i B}{\vdash \Box_j(\Box_i A \to \Box_i B)}$$

using (2) and (3) above.

Second, since $(\top \to A) \leftrightarrow A$ we get from (1) that $(\Box\top \to \Box A) \leftrightarrow \Box A$.

This can be used to derive

$$\frac{\vdash \Box_i A}{\vdash \Box_j \Box_i A}.$$

Since necessitation is admissible in each component we get that

$$\frac{\vdash A}{\vdash \Box_i A}$$

is admissible in the combination $\mathbf{K}_{1,2}$.

The key property which we have used is the disjunction property. This holds for $\mathbf{K}$, $\mathbf{K4}$, and some other system and for all of these systems the above reduction of the modal fibring rule to necessitation stands.

Example 3.12. This example deals with non-normal logics. The example we give is typical. Consider the following axiomatisation of a modal logic $\mathbf{L}$.

0. All substitution instances of truth functional tautologies.

1. $\Box A$ if A is a truth functional tautology.
2. $\Box(A \wedge B) \leftrightarrow (\Box A \wedge \Box B)$
3. $\Box(\Box(A \wedge B) \leftrightarrow (\Box A \wedge \Box B))$
4. $\Box A \rightarrow \Box\Box A$
5. $\Box(\Box A \rightarrow \Box\Box A)$
6. $\Box A \rightarrow A$.

We allow the rule of modus ponens but not necessitation.

This system can derive all theorems of **K4** together with $\Box A \rightarrow A$ and all their modus ponens consequences. We cannot derive $\Box(\Box A \rightarrow A)$ because we do not have necessitation. This logic is complete for all frames of the form (S, R, a) where R is transitive and aRa holds. We have:

$$\vdash_{\mathbf{L}} A \text{ iff } a \vDash A \text{ in all } (S, R, a, h).$$

It is interesting to see what happens if we fibre $\mathbf{L}$ with itself. This is a case where axiom (1b) of Theorem 3.10 applies. We get two $\mathbf{L}$ modalities $\Box_1$ and $\Box_2$ together with all $\mathbf{L}$ theorems (1)–(5) above for each $\Box_1$ and $\Box_2$ and the modal fibring rule. Axiom (6) above gives rise (using 1(b1)) to the fibred axiom $\Diamond_1 q \vee \Diamond_2 q$. Unlike the previous example the fibring rule cannot be simplified.

Theorem 3.13. *Assume $\mathbf{L}_i, i \in I$, has the finite model property. Then so does $\mathbf{L}_I^F$. Assume $\mathbf{L}_i, i \in I$, are recursively axiomatisable. Then so is $\mathbf{L}_I^F$.*

Proof

1. **Finite model property.** To prove this claim we use the fibred semantics formulation of Example 3.4. Assume $\mathbf{L}_i$ is complete for the class $\mathcal{K}_i$ of models. Assume the models have the form (S, R, a, h), where R is a binary relation on S and $a \in S$ is the actual world. The semantics $\mathcal{K}_i$ is not necessarily frame or anything in particular, just a collection of models.
   As in the case of Example 3.4 we can assume the fibred structures have the form $(W, R, W_0, w_0, h, \mathbf{F})$. Assume now that each $\mathcal{K}_i$ is a class of finite models. Then clearly by the construction of the fibred models described in Example 3.4, for each t the set S^t is finite. We are now ready to show that the logic arising from the fibred semantics also has the finite model property.
   We can prove by induction on the number of nested different occurrences of $\Box_i$ within the scope of $\Box_j$ $i \neq j$, that if A has a fibred model it has a finite fibred model.
   We can assume for simplicity that the number of different fibred modalities is finite. This is allowed since each formula A has a finite number of modalities in it. Let $\Box_1, \ldots, \Box_n$ be the different modalities in A. Let $\mathbf{n}$ be a finite model of the fibred language built as follows. Choose (S_i, R_i, a_i, h_i), finite models of the modality $\Box_i$. Assume $S_i \cap S_j = \varnothing$ for $i \neq j$. Let

$$\mathbf{n}_k = (\bigcup_i S_i, \bigcup_i R_i, a_k, \bigcup_i h_i) = (S^{\mathbf{n}}, R^{\mathbf{n}}, a_k, h^{\mathbf{n}}).$$

This becomes a fibred model if we add the fibring function

$$\mathbf{F}^{\mathbf{n}}(i,t) = \begin{cases} t & \text{if } t \in S_i \\ a_i & \text{otherwise.} \end{cases}$$

$\mathbf{n}_k$ will be used as dummies in building our finite fibred models for consistent formulas A. These models have no other significance. We now prove the finite model property by structural induction on A.

1.1 **Basic case**. A is in the pure language $\mathbb{L}_i$. In this case if A is consistent it has a fibred model $\mathbf{m} = (W, R, W_0, w_0, h, \mathbf{F})$ such that $w_o \vDash A$. Since A is in the pure language $\mathbb{L}_i$, we have that $\mathbf{M}_0 = (S_0, R_0, w_0, h_0)$ where $S_0 = S^{w_0}, R_0 = R \restriction S^{w_0}$, and $h_0 = h \restriction S^{w_0}$ is a finite model of A. We can assume $S^{w_0} \cap S^{\mathbf{n}} = \varnothing$. To turn it into a finite fibred model let

$$\begin{aligned} S^A &= S_0 \cup S^{\mathbf{n}} \\ R_A &= R_0 \cup R^{\mathbf{n}} \\ h^A &= h_0 \cup h^{\mathbf{n}} \\ \mathbf{F}^A(k,t) &= \begin{cases} t \text{ if } k = i, t \in S^{w_0} \\ a_k \text{ if } k \neq i, t \in S^{w_0} \\ \mathbf{F}^{\mathbf{n}}(k,t) \text{ otherwise.} \end{cases} \end{aligned}$$

The model (S^A, R^A, w_0, h^A) is a finite model of A.

1.2 **Induction case.** Let $A = A_0(x_1, \ldots, x_n, y_1/A_1, \ldots, y_m/A_m)$ where $A_0(x_1, \ldots, x_n, y_1, \ldots, y_m)$ is a pure $\mathbb{L}_i$ formula with atoms $\{x_i, y_i\}$ and A_j are fibred formulas of lower rank (for which the induction hypothesis holds). We can assume that $A_j = \Box_{\sigma(j)} C_j$.
A is obtained by the simultaneous substitution of A_j for y_j in A_0.
Assume A is consistent. Let $\mathbf{m}$ be a fibred model of A. Let $\mathbf{m}_0$ be the model based on S^{w_0} as defined before.
Define a new assignment h_1 by letting

$$h_1(t,x) = \begin{cases} h_0(t,x) & \text{if } x = x_i, i = 1, \ldots, n \\ 1 & \text{if } x = y_j, j = 1, \ldots, m \text{ and } t \vDash A_j \\ 0 & \text{if } x = y_j, j = 1, \ldots, m \text{ and } t \nvDash A_j. \end{cases}$$

Then $\mathbf{m}_1 = (S_0, R_0, w_0, h_1)$ is a model of A_0.
Consider the formula A_j. If $t \vDash A_j$ then A_j is consistent and so by the induction hypothesis, there exists a finite fibred model $\mathbf{n}_j^+ = (S_j^+, R^+, a_j^+, h_j^+, \mathbf{F}_j^+)$ of A_j. Otherwise $\sim A_j$ is consistent and there exists a finite fibred model $\mathbf{n}_j^-$ of $\sim A_j$. We can assume that all $S_j^{\pm}$ are all pairwise disjoint.
Define a model $\mathbf{n}$ by letting

$$\begin{aligned} S^{\mathbf{n}} &= S_0 \cup \textstyle\bigcup_j S_j^+ \cup \bigcup_j S_j^- \\ R^{\mathbf{n}} &= R_0 \cup \textstyle\bigcup_j R_j^+ \cup \bigcup_j R_j^- \\ h^{\mathbf{n}} &= h_1 \cup \textstyle\bigcup_j h_j^+ \cup \bigcup_j h_j^-. \end{aligned}$$

Define $\mathbf{F^n}$ as follows:

$$\mathbf{F^n}(j,t) = \begin{cases} + \text{ if } + \in S_0 \text{ and } j = i \\ a_j^+ \text{ if } t \in S_0, j \neq i \text{ and } t \vDash A_j \\ a_j^- \text{ if } t \in S_0, j \neq i \text{ and } t \nvDash A_j \\ \mathbf{F}^{\pm}(j,t) \text{ if } t \in S_j^{\pm}. \end{cases}$$

We claim that in the model $\mathbf{n} = (S^{\mathbf{n}}, R^{\mathbf{n}}, w_0, h^{\mathbf{n}}, \mathbf{F^n})$ we have $w_0 \vDash A$.

2. **Recursive axiomatisability.** The proof of this claim follows from Theorem 3.10. □

3.3 Dovetailing Modal Logics

Dovetailing arises in many applications where the fibred model at t has the world t itself as its actual world. A major example is multimodal logics for action. In this example, assume we are at a given state t of a system. There are several possible actions α one can take. With each α we associate an accessibility relation R_α, where $tR_\alpha s$ means that s can non-deterministically arise after action α is applied to state s. We associate the modality $\Box_\alpha A$ to read: A holds in all states arising from the application of α. Such a multi-modal logic, with modalities $\{\Box_\alpha\}$, is dovetailed, not fibred.

Definition 3.14 (Dovetailing of propositional modal logics).

1. Let $\mathbf{L}_i, i \in I$, be modal logics as in the previous definition, with $\mathcal{K}_i$ the class of models for $\mathbf{L}_i$. Let $\mathbf{L}_I^D$ (the dovetailing combination of $\mathbf{L}_i, i \in I$) be defined semantically through the class of all (dovetailed) models of the form $(W, \mathbf{R}, a, h)$, where W is a set of worlds, $a \in W$, h is an assignment into W and for each $i \in I, \mathbf{R}(i) \subseteq W \times W$. We require that for each i $(W, \mathbf{R}(i), a, h)$ is a model in $\mathcal{K}_i$.[11] Wc further require the following:

   Let $t \in W$ be such that there exists $n_1, \ldots, n_k$ and $i_1, \ldots, i_k$ such that $a\mathbf{R}^{n_1}(i_1) \circ \mathbf{R}^{n_2}(i_2) \ldots \circ \mathbf{R}^{n_k}(i_k)t$ holds. Let $i \neq i_k$ and let $W^t = \{y \in W \mid \text{for some } n \geq 0, t\mathbf{R}^n(i)y\}$. Then the model $(W^t, \mathbf{R}(i) \upharpoonright W^t \times W^t, t, h \upharpoonright W^t)$ is in $\mathcal{K}_i$.
2. We define the notion of $w \vDash A$ by induction.
   - $w \vDash q$ if $w \in h(q)$ for q atomic.
   - $w \vDash \Box_i A$ if for all $y \in W$, such that $w\mathbf{R}(i)y$ we have $y \vDash A$.
   - $\vDash A$ iff for all models and actual worlds $a \vDash A$.

Definition 3.15 (Dovetailing models). Let $\mathbf{L}_i, i \in I$, be logics. Let $\mathbf{m}_0 = (S_0, R_0, a_0, h_0)$ be a model in the logic $\mathbf{L}_{i_0}$. For each $i \neq i_0$ and $t \in S_0$, let

$$\mathbf{m}_{i,t} = (S^{i,t}, \mathbf{R}^{i,t}, a^{i,t}, h^{i,t})$$

be a dovetailed model. We want to dovetail the models $\mathbf{m}_{i,t}$ onto $\mathbf{m}_0$. We can assume that S_0 and $S^{i,t}$ are all pairwise disjoint. We define a model $\mathbf{m} = (S, \mathbf{R}, a_0, h)$ as follows.

[11] To be more precise, the following model is in $\mathcal{K}_i$: $(\{x \mid \exists n a\mathbf{R}^n(i)x\}, \mathbf{R}(i), a, h)$.

- Let $S = S_0 \cup \bigcup_{\substack{i \neq i_0 \\ t \in S^0}} S^{i,t}$
- Let $\mathbf{R}$ be defined as follows:
  $x\mathbf{R}(j)y$ iff $x \in S^{i,t}$ and $x\mathbf{R}^{i,t}(j)y$
  or $x \in S_0$ and $j \neq i_0$ and $a^{j,x}\mathbf{R}^{j,x}(j)y$
  or $x \in S_0$ and $j = i_0$ and xR_0y.
- Let $h = h_0 \bigcup_{\substack{i \neq i_o \\ t \in S_0}} h^{i,t}$.

What we have basically grafted is the situation in Figure 3.2.

The following holds at the new dovetailed model:

- For any formula A of $\mathbf{L}_{i_0}$, and any $w \in S_0$ we have
  $w \vDash A$ in $\mathbf{m}_0$ iff $w \vDash A$ in $\mathbf{m}$.
- For any formula $\Box_i B$ of $\mathbf{L}_I^D$, $i \neq i_0$ and any $w \in S_0$ we have
  $w \vDash \Box_i B$ in $\mathbf{m}$ iff $w \vDash \Box_i B$ in $\mathbf{m}_{i,w}$.

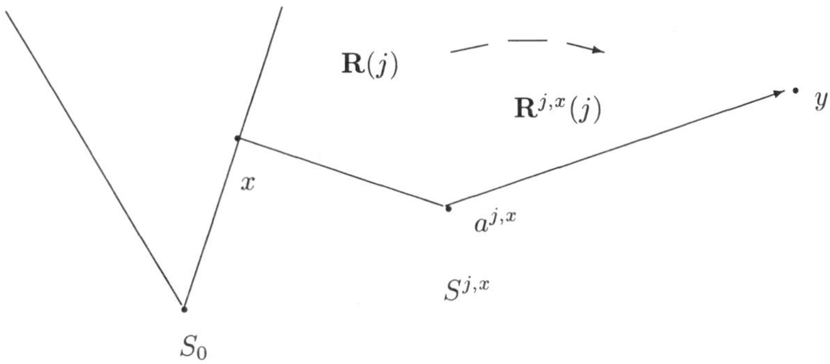

FIG. 3.2.

Theorem 3.16 (Completeness theorem for the dovetailed logic $\mathbf{L}_I^D$). *Let $\mathbf{L}_i, i \in I$, be modal logics with semantical classes of structures $\mathcal{K}_i$ and set of theorems $\mathbf{T}_i$. Let $\mathbf{T}_I^D$ be the following set of wffs of $\mathbf{L}_I^D$.*

1. $\mathbf{T}_i \subseteq \mathbf{T}_I^D$
2. **Modal dovetailing rule**:[12]
   If $\Box_i$ is the modality of $\mathbf{L}_i$ and $\Box_j$ of $\mathbf{L}_j, i, j$ arbitrary $i \neq j$ and

$$C = \bigwedge_{k=1}^{n} \Box_i A_k \wedge \bigwedge_{k=1}^{m} \Diamond_i \sim B_k \to \bigvee_{k=1}^{r} q_k \in \mathbf{T}_I^D$$

[12]The modal dovetailing rule is really a necessitation rule. It says that if C is a wff built up from 'atomic' units of the form $\Box_i A_s$ and ordinary atoms q_k and $\Box_j$ is a modality *different* from $\Box_i$, then a limited necessitation rule holds: $\vdash C$ implies $\vdash \Box_j^d C$ for any natural number d.

then for all d $\square_j^d C \in \mathbf{T}_I^D$, *where* q_k *are atoms or their negations, and* $q_1, \ldots, q_r$ *list all the atoms or their negations appearing in any* A_k *or* $B_k, k = 1, 2, \ldots$.

3. $\mathbf{T}_I^D$ *is the smallest set closed under (1), (2) modus ponens and substitution.*

Then $\mathbf{T}_I^D$ *is the set of all wffs of* $\mathbf{L}_I^D$ *valid in all the dovetailed structures of* $\mathbf{L}_I^D$.

Proof

1. The case of soundness is easy to verify.
2. To prove completeness we proceed along the lines of the proof of the previous theorem, the completeness theorem for $\mathbf{L}_I^F$.

The case to examine is *Subcase a* of case $n > 1$. We have a model $\mathbf{m} = (S^{\mathbf{m}}, R^{\mathbf{m}}, a^{\mathbf{m}}, h^{\mathbf{m}})$ of $\mathbf{L}_i$ of the formula B_i and we want to augment it into a dovetailed model of B.

For this purpose we have to realise the assignment of $h^{\mathbf{m}}(t, x_r^j)$, for points t such that $d(t) \leq d + 1$ for all x_r^j by a fibred model for the substitution $\Theta(x_r^j) = \square_{\sigma(j)} A_r^j$ and in addition the model must realise all the values $h^{\mathbf{m}}(t, y_i)$, for the atoms y_i because the model will be dovetailed at t. Thus the formulas C_t^j and D_t^j to be considered in our case are the following:

$$D_t^j = \bigwedge_{s=1}^{n(j)} [x_s^i]^{\varepsilon(j,s)} \wedge \bigwedge_r [y_r]^{\varepsilon(r)}$$

$$C_t^j = \bigwedge_{s=1}^{n(j)} [\square_{\sigma(j)} A_s^j]^{\varepsilon(j,s)} \wedge \bigwedge_r [y_r]^{\varepsilon(r)}$$

where

$$\varepsilon(j, s) = h^{\mathbf{m}}(t, x_s^j)$$
$$\varepsilon(r) = h^{\mathbf{m}}(t, y_r)$$

and

$$C_t^j = D_t^j \Theta.$$

Just like the previous proof, if for every $\mathbf{m}$ there is a $t \in S^{\mathbf{m}}$ at distance $d(t)$ and a j for which C_t^j is inconsistent, then for some finite $D_1, \ldots, D_m$ we have

$$\vdash_{\mathbf{L}_i} (\bigwedge_{s=1}^{m} \square_i^{d(s)} \sim D_s) \rightarrow \sim B_i.$$

Hence

$$\vdash_{\mathbf{L}_I^D} (\bigwedge_{s=1}^{m} \square_i^{d(s)} \sim D_s \Theta) \rightarrow \sim B.$$

But from the inconsistency of C_t^j for all j, t we get

$$\vdash \sim D_s \Theta.$$

The modal dovetailing rule allows us to conclude

$$\vdash_{\mathbf{L}_I^D} \Box_i^{d(s)} \sim D_s\Theta$$

and so we get

$$\vdash_{\mathbf{L}_I^D} \sim B,$$

a contradiction.

We can thus assume that for some model $\mathbf{m}$ of B_i, all the formulas C_t^j are realisable by a model $\mathbf{m}_t^j = (S^{j,t}, \mathbf{R}^{j,t}, a^{j,t}, h^{j,t})$.

We can now dovetail these models with $\mathbf{m}$ into a model of B, as done in Definition 3.15. Note that for each t and atom y, $h^{\mathbf{m}}(t, y) = h^{j,t}(a^{j,t}, y)$.

Subcase b of $n > 1$ is also proved similarly to the fibred case. We assume there exists a classical Boolean combination $\Psi(\overline{x}_1, \ldots, \overline{x}_n, \overline{y})$ and a Θ such that $B = \Psi\Theta$, where $\Theta(x_r^j) = \Box_{\sigma(j)} A_r^j$, and $\Theta(y_i) = y_i$. Write Ψ in a disjunctive normal form and let α be any disjunct. At least one such α must be $\mathbf{L}_I^D$ consistent. Assume α is such a disjunct.

Consider the subformula of α.

$$E_{\sigma(i)} = \bigwedge_{r=1}^{k(i)} [\Box_{\sigma(i)} A_r^i]^{\varepsilon(i,r)} \wedge \bigwedge_r y_r^{\varepsilon(r)}.$$

This formula is consistent and has the form

$$E_{\sigma(i)} = B_{\sigma(i)}(\overline{z}_1, \ldots, \overline{z}_m, \overline{u})\Theta_1$$

with substitution Θ_1 falling under *Subcase a*. It therefore has a dovetailed model $\mathbf{m}_i$.

The models all agree on the value to $\overline{y}$ and can all be dovetailed into a single model of α and hence of $B = \Psi\Theta$ as in Definition 3.15. □

Theorem 3.17. *Assume $\mathbf{L}_i, i \in I$ are all extensions of* **K** *formulated using traditional Hilbert axioms and the rule of necessitation. Then $\mathbf{L}_I^D$ (the dovetailing of $\mathbf{L}_i$) can be axiomatised by taking the union of the axioms and the rules of necessitation for each modality $\Box_i$ of each $\mathbf{L}_i$*

Proof Let $\mathbf{L}_I^D$ be the modal dovetailed system and let $\mathbf{L}_{\text{union}}$ be the system obtained by taking the union of axioms of each $\mathbf{L}_i$ and necessitation for each $\Box_i$. Clearly $\mathbf{L}_I^D \subseteq \mathbf{L}_{\text{union}}$. To show the other direction it is sufficient to prove the following rule:

$$\vdash C \text{ implies } \vdash \Box_j C$$

for arbitrary C of the form

$$C = \bigwedge_s \Box_{\sigma(s)} A_s \wedge \bigwedge_r \Diamond_{\sigma(r)} B_r \to \bigvee_k q_k.$$

This is sufficient because every wff is a conjunction of formulas of the above form. We note that we already have necessitation for the case where the modalities in C are all equal to $\Box_i \neq \Box_j$. If $\mathbf{L}_j$ itself satisfies necessitation then we do not need the restriction $\Box_i \neq \Box_j$.

We need to show necessitation for the general case where more than one modality is involved in C. This we now proceed to do. Using the classical tautology

$$(x \wedge y \to z) \leftrightarrow ((x \to z) \vee (y \to z))$$

we can write C as $\bigvee_i C_i$ where each C_i involves only a single modality $\Box_i$ and has the form

$$C_i = \bigwedge_s \Box_i A^i_s \wedge \bigwedge_r \Diamond_i B^i_r \to \bigvee_k q_k$$

where $\bigvee_k q_k$ is the same in all C_i.

If $\vdash \bigvee_i C_i$ then for some $i \vdash C_i$. For otherwise we can graft together all the countermodels of each C_i to get a countermodel for C, since they all agree on the truth values to the atoms $\{q_k\}$.

We can therefore deduce $\vdash \Box C_i$ and hence $\vdash \Box C$.

This concludes the proof of the theorem. □

Theorem 3.18. *If* $\mathbf{L}_i, i \in Im$ *admit necessitation and satisfy the disjunction property, then* $\mathbf{L}^F_I = \mathbf{L}^D_I$.

Proof Follows from the considerations of Example 3.11 and the previous theorem. □

Theorem 3.19. *If* $\mathbf{L}_i, i \in I$, *all have the finite model property (are finitely axiomatisable), so is* $\mathbf{L}^D_I$.

Proof From the completeness theorem.

1. If all participating models are finite then the dovetailed semantics (as described in Example 3.6) will contain models which may not be finite but are made (fibred) from finite models. Any particular formula A of the fibred logic when evaluated at a fibred model requires only a finite depth of fibring relative to the actual world and the model can be changed to a finite fibred model. See the proof of Theorem 3.13.
2. The transfer of finite axiomatisability is immediate from the completeness theorem. □

We want to study under what conditions fibring is identical with dovetailing. The condition on the semantics is that it be *forward looking*.

Definition 3.20.

1. A model (S, R, a) is *forward looking* iff for any assignment h, any t and any h' such that $h \restriction S_t = h' \restriction S_t$ and any wff $A(\Box B_1, \ldots, \Box B_k)$ we have $t \vDash_h A$ iff $t \vDash_{h'} A$, where $S_t = \{s \mid \exists n, tR^n s, n \geq 1\}$.
2. $\mathcal{K}_i$ is said to be *forward looking* if all of its structures are forward looking.
3. A logic is *forward looking* if it is complete for a forward looking semantics.

Example 3.21.

1. If R is reflexive then it is not forward looking.

2. Modal **K** and **K4** are forward looking.

Theorem 3.22. *Let $\mathbf{L}_i, i \in I$, be logics with semantics $\mathcal{K}_i$. Let A be a wff of $\mathbf{L}_I^F$ which has a fibred model. Then A has a dovetailed-like model. The dovetailed-like model may not satisfy the condition that $(S, \mathbf{R}((i), a, h) \in \mathcal{K}_i)$. This condition can be guaranteed when each $\mathcal{K}_i$ is forward looking, in which case we get a proper dovetailed model.*

Proof We show that any fibred model $(W, S, R, \mathbf{a}, h, \mathbf{F}, \tau)$ can be turned into an equivalent dovetailed model $(S, \mathbf{R}(i), a, h^D)$. Note that the dovetailed model relies on the particular function h^D, and that $(S, \mathbf{R}(i), a)$, for any fixed i may *not* be in the class $\mathcal{K}_i$.

Turning to the definition of $(S, \mathbf{R}_i, a, h^D)$, let S be W. We may assume that for any $x \in W$ there exists a unique w such that $x \in S^w$. Let $x, y \in W$. We let $x\mathbf{R}(i)y$ hold iff either

1. for some w, we have that $x, y \in S^w$ and $\tau(w) = i$, and $xR^w y$;
2. $x \in S^w$ and $y \in S^{w'}$ and $w' = \mathbf{F}(i, w, x)$ and $\mathbf{a}^{w'} R^{w'} y$.
   Let $h^D(q) = \cup_{w \in W} h^w(q)$.

Notice that in clause (2) we took the model $(S^{w'}, R^{w'}, \mathbf{a}^{w'})$ and effectively replaced it by the model $(S^{w'}, R^{w'}, x)$, as we dovetailed the possible worlds of $S^{w'}$ (except $\mathbf{a}^{w'}$) onto x. When the semantics of the logic of $\tau(w')$ is forward looking, this new model can also be assumed to be in the semantics and hence the dovetailed model is proper, according to Definition 3.14. □

Lemma 3.23. *For any (w, y) and A*

$$(w, y) \vDash_F A \text{ iff } y \vDash_D A$$

where $\vDash_F$ is satisfaction in the fibred model and $\vDash_D$ is satisfaction in the dovetailed model.

Proof By induction. Note that since $\forall y \exists! w (y \in S^w)$ holds, we can write $y \vDash_F A$ in place of $(w, y) \vDash_F A$.

The lemma holds for atomic q by definition. $y \vDash_F \Box_i A$ iff $\mathbf{a}^{w'} \vDash_F \Box_i A$, where $w' = \mathbf{F}(i, w, y)$ iff $\forall x [\mathbf{a}^{w'} R^{w'} x \to x \vDash_F A]$ iff $\forall x [y R(i) x \to x \vDash_F A]$ iff $y \vDash_D \Box_i A$, by the induction hypothesis.

The lemma just proved that $(S, \mathbf{R}(i), a, h^D)$ is an equivalent model to $(W, S, R, \mathbf{a}, h, \mathbf{F}, \tau)$.

This equivalence holds for any semantics. We have not used in the proof of the lemma the assumption that the semantics we have fibred is forward looking. We need this assumption to show that under any assignment h^* (other than h^D) the structure $(S, \mathbf{R}(i), a, h^*)$ satisfies only $\mathbf{L}_I^F$ theorems. □

Example 3.24. We fibre a **T** modality $\Box_1$ with a **K** modality $\Box_2$, to show the difference between dovetailing and fibring.

Let q be atomic. Consider

$$\Box_2 \Box_1 q \wedge \Diamond_2 \sim q.$$

In the fibred semantics the above formula has a model. We can have a pair (w, y), such that $wR^w y, (w, y) \vDash \sim q$ and in the fibred model $w' = \mathbf{F}(w, w, y)$ $\mathbf{a}^{w'} \vDash \Box_1 q$. It is true that in w' by reflexivity $\mathbf{a}^{w'} \vDash q$ but this leads to no contradiction. In the dovetailed model, however, we will need to have at some y such that wR_2y $y \vDash\sim q \wedge \Box_1 q$, which is impossible.

3.4 Fibring and Dovetailing Modal Fragments

Example 3.25 (Fibring modal fragments). The previous theorems dealt with fibring and dovetailing full modal logics, with modalities $\Box_i$ and all the Boolean connectives. We found theorems telling us how to fibre the semantics $\mathcal{K}_i$ of $\mathbf{L}_i$ and at the same time how to generate the theorems of $\mathbf{L}_i, i \in I$. The proof also shows that any semantics $\mathcal{K}_i$ for $\mathbf{L}_i$ will do for the purpose of fibring. If $\mathbf{L}_i$ is complete for $\mathcal{K}_i$ and for another semantics $\mathcal{K}'_i$, then fibring $\mathcal{K}_i, i \in I$, or $\mathcal{K}'_i, i \in I$, yields the same resulting logic and the same set of theorems for $\mathbf{L}^F_I$ or $\mathbf{L}^D_I$ respectively. We refer to this property as semantical independence of the fibring process.

This independence does not hold if we are dealing with modal fragments . Consider for example two modal fragments $\mathbf{L}_1$ and $\mathbf{L}_2$, where $\mathbf{L}_1$ is the fragment of modal $\mathbf{K}$ with $\{\Diamond_1\}$ only while $\mathbf{L}_2$ is the fragment of $\mathbf{K}$ with $\{\Box_2, \wedge\}$. Each fragment is complete for the class $\mathcal{K}$ of all $\mathbf{K}$ Kripke models with arbitrary binary relations R. What do we mean by complete? Here we must use the consequence relation. Let $A_1, \ldots, A_n \vDash_{\mathcal{K}} B$ mean that in every Kripke model $(S, R, , a, h) \in \mathcal{K}$. If $a \vDash A_i$, for all i then $a \vDash B$. Completeness for the fragment with $\Diamond_1$ means that for all A_i, B in the fragment

$$\{A_i\} \vdash_{\mathbf{K}} B \text{ iff } \{A_i\} \vDash_{\mathcal{K}} B$$

where $\vdash_{\mathbf{K}}$ is provability in some syntactical formulation of the full $\mathbf{K}$. If we fibre $\mathbf{L}_1$ and $\mathbf{L}_2$ according to (i.e. using) the $\mathbf{K}$ Kripke semantics, we get what is expected i.e. we get the $\{\Diamond_1, \{\Box_2, \wedge\}\}$ fragment of the full fibring of two full versions of $\mathbf{K}$ namely $\{\{\Diamond_1, \wedge, \sim\}, \{\Box_2, \wedge, \sim\}\}$. However because $\mathbf{L}_1$ with $\Diamond_1$ is a very weak language, all we can express in it is sequents of the form

$$\Diamond_1^{n_i} q_i \vdash_{\mathbf{L}_1} \Diamond_1^{m_j} p_j$$

where q_i, p_j are atomic.

The above will hold in $\mathbf{K}$ iff for some i and j, $q_i = p_j$ and $n_i = m_j$.

Another semantics for $\mathbf{L}_1$ is the class of all linear Kripke models of the form $(\mathcal{N}, <_1, 0, h)$ where $\mathcal{N}$ is the set of natural numbers and $x <_1 y$ iff $y = x + 1$. This semantics gives $\Diamond_1 q$ the reading 'q holds the next day'. The $\mathbf{L}_1$ language cannot distinguish between linear and branching Kripke models.

If we fibre $\mathbf{L}_1$ and $\mathbf{L}_2$ using the natural numbers Kripke semantics for $\mathbf{L}_1$ we get the new theorem

(*) $\Diamond_1\Box_2 q_1 \wedge \Diamond_1\Box_2 q_2 \vdash_{\mathbf{L}_{1,2}} \Diamond_1\Box_2(q_1 \wedge q_2)$.

This theorem does not hold if we fibre $\mathbf{L}_1$ and $\mathbf{L}_2$ using the ordinary branching $\mathbf{K}$ semantics for $\mathbf{L}_1$. The reason that we can get (*) is that conjunction becomes available to $\mathbf{L}_1$ through the fibring, and $\Diamond_1$ with $\wedge$ can distinguish between the different semantics chosen for $\mathbf{L}_1$, while $\mathbf{L}_1$ on its own without $\wedge$ cannot tell the difference.

It is instructive to see why the inductive argument of the fibring completeness theorem works for the full language, in the case of the natural numbers semantics.

Following our strategy we try to show that $\Diamond_1\Box_2 q_1, \Diamond_1\Box_2 q_2 \nvdash \Diamond_1\Box_2(q_1 \wedge q_2)$ in the fibred semantics.

This form is a substitution instance of the form

$$\Diamond_1 x, \Diamond_1 y \nvdash \Diamond_1 z$$

using the substitution Θ, with $\Theta(x) = \Box_2 q_1, \Theta(y) = \Box_2 q_2$ and $\Theta(z) = \Box_2(q_1 \wedge q_2)$.

A natural numbers counter model $(\mathcal{N}, <_1, 0, h)$ can be obtained with $h(1, x) = h(1, y) = 1$ and $h(1, z) = 0$.

The realisation of h for the substitution Θ will give us a fibred countermodel for the original consequence form. We need to fibre a model yielding

$$h(1, \Box_2 q_1) = h(1, \Box_2 q_q) = 1$$

and

$$h(1, \Box_2(q_1 \wedge q_2)) = 0.$$

This is not possible because in $\mathbf{L}_2$

(**) $\Box_2 q_1 \wedge \Box_2 q_2 \vdash_{\mathbf{L}_2} \Box_2(q_1 \wedge q_2)$.

We now want to use this fact, (**), to show that (*) must hold.

When the full Boolean connectives are available to us in $\mathbf{L}_1$ we can write the form:

$$\Box_1(x \wedge y \to z), \Diamond_1 x, \Diamond_1 y \vdash \Diamond_1 z;$$

the addition of $\Delta_1 = \Box_1(x \wedge y \to z)$ excludes the above countermodel. If we find such a wff D_r for each unrealisable model, we get

(***) $\{D_r\}, \Diamond_1 x, \Diamond_1 y \vdash \Diamond_1, z$

and after substitution we get

$$\{D_r\Theta\}, \Diamond_1\Box_2 q_1, \Diamond_1\Box_2 q_2 \vdash \Diamond_1\Box_2(q_1 \wedge q_2).$$

Since $\vdash_{\mathbf{L}_{1,2}} D_r\Theta$, we get $\Diamond_1\Box_2 q_1, \Diamond_1\Box_2 q_2 \vdash_{\mathbf{L}_{1,2}} \Diamond_1\Box_2(q_1 \wedge q_2)$, thus showing (*) is provable.

Without Boolean operations in $\mathbf{L}_1$ we cannot write D_r. We may try to write the Gentzen Rule:

$$\frac{x, y \vdash_{\mathbf{L}_1} z}{\Diamond_1 x, \Diamond_1 y \vdash_{\mathbf{L}_1} \Diamond_1 z}$$

but the meaning of the above as a rule involves validity in all models and is not what we want.

The above discussion shows that fibring weak modal fragments may depend on the semantics chosen for the fragments, the reason being that we may have a fragment $\mathbf{L}$ which is syntactically too weak to distinguish between two different semantics $\mathcal{K}_1$ and $\mathcal{K}_2$ but when fibred with another logic $\mathbf{M}$, the fibred result can make the distinction.

Thus we will have no way of predicting the set of theorems of the fibred result simply by looking at the axiomatic formulation of **L** and **M**. The question still remains of what happens in cases where all the logics involved, i.e. **L** and **M**, are so weak that fibring does *not* help distinguish between the choice of semantics? Can we get a completeness theorem and thus establish the semantical independence of the fibring process? The answer is yes under certain conditions.

The weakest (least expressive) modal logic is the *pure modality* logic with just $\Box$ and $\Diamond$ and atoms and no other connectives. The wffs of such a logic have the form $\overline{M}x$, where $\overline{M}$ is a string of modalities and x atomic. Any syntactical fibring of logics with pure modalities, $\mathbf{L}_i, i \in I$, will yield similar mixed wffs $\overline{M}x$, where x is atomic and $\overline{M}$ a string of modalities from some $\mathbf{L}_i, i \in J, J \subseteq I, J$ finite. Such fragments are of independent interest because when fibring modalities into other logics such as modal intuitionistic logic, one tends to fibre only the pure modal fragment. Such fragments have also arisen in modal formulations of terminological logics. A completeness theorem for fibring pure modalities will therefore be very useful.

Theorem 3.26 (Completeness for fibring and dovetailing pure modalities). *Let $\mathbf{L}_i, i \in I$, be modal logics in a language with connectives $\Box_i$ and/or $\Diamond_i$ together with atomic propositional variables. We assume for convenience that $\top, \bot$ are available.*

Let $\mathcal{K}_i$ be semantics (class of Kripke models) for which $\mathbf{L}_i$ is complete. Let $\mathbf{L}_I^F$ be the fibring (resp. let $\mathbf{L}_I^D$ be the dovetailing) of $\mathbf{L}_i, i \in I$. Let $\mid\!\sim_i$ be the consequence relation of $\mathbf{L}_i$. Let $\mid\!\sim$ be the consequence relation of $\mathbf{L}_I^F$ (resp. $\mathbf{L}_I^D$) defined by the following rules.

1. *Any substitution instance of $\Delta_i \mid\!\sim_i \Gamma_i$, where $\Delta_i \mid\!\sim \Gamma_i$ is in the language of $\mathbf{L}_i$.*
2. *Pure modalities fibring (resp. dovetailing) rules*

$$\frac{A_1, \ldots, A_n \mid\!\sim C_1, \ldots, C_m}{\Box A_1, \ldots, \Box A_n, \Diamond B \mid\!\sim \Diamond C_1, \ldots, \Diamond C_m}.$$

   In case of fibring, A_i and C_j must begin with a modality.

$$\frac{A_1, \ldots, A_n \mid\!\sim C, B_1, \ldots, B_m}{\Box A_1, \ldots, \Box A_n, \mid\!\sim \Box C, \Diamond B_1, \ldots, \Diamond B_m}.$$

   In case of fibring, A_i, B_j and C must begin with a modality.

3. *$\mid\!\sim$ is the smallest consequence relation (monotonic, reflexive and transitive) closed under the above rules.*

The following holds.

Let Δ, Γ be two finite sets of wffs of $\mathbf{L}_I^F$ (resp. $\mathbf{L}_I^D$). If $\Delta \not\mid\!\sim \Gamma$, then there exists a fibred model $\mathbf{m}$ of $\mathbf{L}_I^F$ (resp $\mathbf{L}_I^D$) such that

- *$A \in \Delta$ implies $\mathbf{m} \vDash A$*
- *$A \in \Gamma$ implies $\mathbf{m} \nvDash A$.*

Proof The proof is by induction on the modal complexity of wffs in $\Delta \cup \Gamma$. Each formula of the language has the form $\overline{M}^i y_i$, with $\overline{M}^i$ a string of modalities and y_i atomic. Let $l(i)$ be the length of the string. The induction is on the number $n = \max_i l(i)$.

The inductive assumption on finite sets of wffs Δ, Γ of max $l(i) \leq n$ is:
Completeness
$\Delta \not\mid\!\sim \Gamma$ implies for some fibred model $\mathbf{m}$, $\mathbf{m} \vDash$ all of Δ and $\mathbf{m} \nvDash$ any $B \in \Gamma$.

Case $n = 0$
In this case all wffs of $\Delta \cup \Gamma$ are atomic. We therefore have that $\Delta \mid\!\sim \Gamma$ iff $\Delta \cap \Gamma \neq \varnothing$; clearly both inductive assumptions hold.

Case $n \geq 1$
In this case the elements of $\Delta \cup \Gamma$ have the form $\alpha_i = N_i \overline{M}^i y_i$, where N_i is the top front modality, which may not exist if α_i is atomic. $\overline{M}^i$ is the rest of the string, which may be empty. For a given $\Delta \not\mid\!\sim \Gamma$, we need to show a fibred (resp. dovetailed) model $\mathbf{m}$ as desired. We first observe that if $q_1, \ldots, q_k$ are all the atoms involved in $\Delta \cup \Gamma$, then we can assume that they are distributed among $\Delta \cup \Gamma$. In fact, if Ω is any set of formulas, we can assume that $\Omega \subseteq \Delta \cup \Gamma$ and $\Delta \not\mid\!\sim \Gamma$. This follows from the cut rule, because if for all partitions of Ω into Ω_1 and Ω_2 we have $\Delta \cup \Omega_1 \mid\!\sim \Gamma \cup \Omega_2$, then $\Delta \mid\!\sim \Gamma$ by cut.

Secondly we observe that we can assume that all modalities N_i for all i belong to the same logic, say $\mathbf{L}_{\Box,\Diamond}$. The reason we can do that is that we can decompose Δ into $\Delta_1 \cup \Delta_2 \cup \ldots$ and Γ into $\Gamma_1 \cup \Gamma_2 \cup \ldots$ such that $\Delta_1 \not\mid\!\sim \Gamma_1, \Delta_2 \not\mid\!\sim \Gamma_2$ etc., and Δ_j, Γ_j are maximal in containing all the atoms and all the formulas of the same top modalities N_i of the same logic. If we find countermodels $\mathbf{m}_j$ for $\Delta_j \not\mid\!\sim \Gamma_j$, then we can put them together as done in Definition 3.9. Since Δ_i, Γ_i contain all atoms, the case of dovetailing is also covered, i.e. the models can be dovetailed.

We therefore assume that all N_i are of the same logic, $i = 1, 2, \ldots$.

We therefore need to find a model $\mathbf{m}$ of the logic $\mathbf{L}$ such that

- $\mathbf{m} \vDash A$ for $A \in \Delta$
- $\mathbf{m} \nvDash A$ for $A \in \Gamma$.

The third step in our analysis is to ask whether we can find a countermodel $\mathbf{m}$ in which $\Diamond\top$ holds. If such an $\mathbf{m}$ exists then we can assume $\Diamond\top \in \Delta$. If such a model does not exist then all fibred Kripke models which satisfy Δ and falsify Γ have only one possible world, the actual world, which is not accessible to anything.

In this latter case $\Delta \mid\!\sim \Box\bot$ and Γ contains only wffs $\Diamond A$. We have thus found a very special countermodel for $\Delta \not\mid\!\sim \Gamma$.

We have now finished our initial analysis of Δ, Γ.

Let us assume that

$$\Delta = \{\Box D_1, \ldots, \Box D_{r_1}, \Diamond E_1, \ldots, \Diamond E_{r_2}\} \cup \Omega_1$$

and

$$\Gamma = \{\Box D'_1, \ldots, \Box D'_{r'_1}, \Diamond E'_1, \ldots, \Diamond E'_{r'_2}\} \cup \Omega_2$$

and that $E_1 = \top$. $\Omega_1 \cup \Omega_2$ cover all basic atoms appearing in wffs of $\Delta \cup \Gamma$.

We can assume that D_j, E_j, D'_j, E'_j are wffs of the form $\overline{M}^i_1 y_i$, where $\overline{M}^i_1$ are the tails of the strings of modalities in $\Delta \cup \Gamma$.

We need a fibred (resp. dovetailed) Kripke model $\mathbf{m} = (W, S, R, \mathbf{a}, h, \mathbf{F}, \tau, w_0)$ such that $\tau(w_0) = \mathbf{L}$ and $(S^{w_0}, R^{w_0}, w_0, h^{w_0}) \in \mathcal{K}$ such that for all $s \in S^{w_0}, w_0 R^{w_0} s \to (s \vDash D_j$ and $s \nvDash E'_j)$ and for some t_j and $s_j, w_0 R^{w_0} t_j$ and $w_0 R^{w_0} s_j$ and $t \vDash E_j$ and $s_j \nvDash D'_j$.

If for all j we have that both (*1) and (*2) below hold, then by the induction hypothesis we can find appropriate countermodels and put them together in the obvious way, and complete the induction step.

(*1) $\{D_1, \ldots, D_{r_1}, E_j\} \not\vdash \{E'_1, \ldots, E'_{r'_2}\}$

(*2) $\{D_1, \ldots, D_{r_1}\} \not\vdash \{E'_1, \ldots, E'_{r'_2}, D'_j\}$

Otherwise either

(1**) $\Box D_1, \ldots, \Box D_{r_1}, \Diamond E_j \vdash \Diamond E'_1, \ldots, \Diamond D'_{r'_2}$

or

(2**) $\Box D_1, \ldots, \Box D_{r_1} \vdash \Diamond E'_1, \ldots, \Diamond E'_{r'_2}, \Box D'_j$.

In either case this contradicts $\Delta \not\vdash \Gamma$.

This completes the inductive step and the proof of the theorem. □

3.5 Decidability

We now examine conditions under which if $\mathbf{L}_i$ are all decidable then so are $\mathbf{L}_I^F$ and $\mathbf{L}_I^D$. Unlike the preservation theorems of the finite model property and finite axiomatisability, the case of decidability is more tricky. If each $\mathbf{L}_i$ has its own totally different decision procedure, we will have a 'fibred', not so simple, decision procedure for $\mathbf{L}_I^D$ and $\mathbf{L}_I^F$. If they are all logics of the same type with the same kind of decision procedure we may be able to simplify decidability. We can simplify decidability if the procedure is tableaux like. The argument goes as follows. Take any formula A of the mixed language. A is of the form $B_i(D_1, \ldots, D_k)$ where $B_i(x_1, \ldots, x_k)$ is in the logic $\mathbf{L}_i$ and $D_1, \ldots, D_k$ are in some other of the fibring logics. Apply the tableaux method characteristic for the logic $\mathbf{L}_i$, regarding $D_1, \ldots, D_k$ as atomic. The tableaux will require assignment of values to the 'atoms' $D_1, \ldots, D_k$. When such values are needed, apply the tableaux methods of the logic $\mathbf{L}_j$ of the external connectives of each D_j. This can be done because D_j itself is of the form $D_j^*(C_j^1, \ldots, C_j^{m_j})$ where $D_j^*(x_1, \ldots, x_{m_j})$ is in the logic $\mathbf{L}_j$ and C_j^r are in some other logic. The process will terminate because the complexity goes down.

In the general case, we need a general theorem that 'fibres' the individual decision procedures of the participating logics. This we now describe.

Theorem 3.27. *Assume $\mathbf{L}_i, i \in I$, are all decidable. Then so are $\mathbf{L}_I^F$ and $\mathbf{L}_I^D$.*

Proof The proof is by structural induction on the wff B.

Basic case 1

The wff B is in the logic $\mathbf{L}_i$. Then we can use the decision procedure of $\mathbf{L}_i$, since $\mathbf{L}_I$ is a conservative extension of $\mathbf{L}_i$.

Basic case 2

The formula B is a Boolean combination of pure formulas $C_s^j, s = 1, \ldots, n(j)$, where

C_s^j is in the language $\mathbf{L}_{\sigma(j)}$, and of the atoms $y_1, \ldots, y_k$. For $i \neq j$ we have $\mathbf{L}_{\sigma)i)} \neq \mathbf{L}_{\sigma(j)}$. We can assume B is written in a conjunctive normal form, where each conjunct has the form

$$D = \bigvee_r B_r \vee E$$

where E contains a disjunction of literals (atoms or their negations) and each B_r is a disjunction of pure formulas (or their negations) for the language $\mathbf{L}_{\sigma(r)}$.

Clearly to decide whether B is a theorem it is sufficient to decide whether each conjunct D is a theorem.

We claim that D is a theorem iff at least one of $B_r \vee E$ is a theorem, for the dovetailed case and iff one of B_r or E is a theorem, for the fibring case. For if we have a pure $\mathbf{L}_{\sigma(r)}$ countermodel for each $B_r \vee E$, (resp. each B_r and E), they can be dovetailed (resp. fibred) into a countermodel of D.

Since the theoremhoods of $B_r \vee E$, B_r and E are decidable so is that of B.

Inductive case 1
B has the form

$$B_i(x_s^j / C_s^j, y_1, \ldots, y_k)$$

where $B_i(x_s^j, y_r)$ is a pure wff of $\mathbf{L}_i$ and C_s^j are pure wffs of $\mathbf{L}_{\sigma(j)}$, and $y_1, \ldots, y_k$ are atomic. Let $C_j = \bigvee_s C_s^j$ and $D_j = \bigvee_s x_s^j$. Consider all wffs of the form

$$\Box^{d_1} D_{j_1} \wedge \ldots \wedge \Box^{d_m} D_{j_m} \to B_i$$

where $d_1, \ldots, d_m$ are less than the maximal number of nested modalities in B_i. If any such wff is a theorem we check whether $C_{j_1}, \ldots, C_{j_m}$ are theorems. If they are then B is a theorem. Otherwise B is not a theorem.

We need to show that our decision procedure is complete. We need to show that if B is a theorem then our decision procedure says it is a theorem and vice versa. One direction is clear. To show the other direction, assume B is a theorem. If B_i is a theorem of $\mathbf{L}_i$, then our decision procedure will recognise B as a theorem. If B_i is not a theorem of $\mathbf{L}_i$, consider $\sim B$ and consider any model $\mathbf{m}$ of $\sim B_i$. We try to fibre models of $\pm C_s^j$ to realise the assignment $h^{\mathbf{m}}$ of $\mathbf{m}$ and get a model of $\sim B$. This is the same reasoning as in the proof of Theorems 3.10 and 3.16. Since B is a theorem, we will not succeed and thus by the same kind of reasoning we find $\sim D_1, \ldots, \sim D_m$ and $d_1, \ldots, d_m$, such that

$$\vdash_{\mathbf{L}_i} \bigwedge_j \Box_i^{d_j} D_j \to B_i$$

and such that $\vdash_{\mathbf{L}_i} C_j, j - 1, \ldots, m$.

We thus get that B is a recognised theorem by the decision procedures.

Induction case 2
B is a Boolean combination of formulas for which induction case 1 applies. In this case we proceed as in the proof of basic case 2.

This completes the proof of the theorem. □

4

INTUITIONISTIC MODAL LOGICS

4.1 Introduction

This chapter deals with fibring modality into intuitionistic logic. The key difference between fibring modality into modal logic (to form a bi-modal logic) and fibring modality into intuitionistic logic is that in the latter case we are fibring into a logic whose semantics requires a condition of persistence. The semantics contains a reflexive relation $\leq$ (usually transitive, see Chapter 2) and the following condition must be satisfied:

(*) $t \vDash A$ and $t \leq s$ implies $s \vDash A$.

Since any modal formula, such as $\Box q$ or $\Diamond q$ is considered atomic from the intuitionistic point of view, we need to have (*) holding for $\Box q$ and for $\Diamond q$.

In practical terms this means that we are not free to fibre any modal model we want at t and s but the models must be related.

Let $\mathbf{m}_t = (S_t, R_t, a_t, h_t)$ be the model fibred at t and similarly $\mathbf{m}_s$. Assume $t \vDash \Box q$. This means that $\forall y[a_t R_t y \to y \vDash q]$. Similarly for $s \vDash \Box q$.

Since q is atomic, the simplest way of ensuring condition (*) for $\Box$ is to require:[13]

($\Box$) $t \leq s \wedge a_s R_s y \to a_t R_t y$

for $\Box$ and

($\Diamond$) $t \leq s \wedge a_t R_t y \to a_s R_s y$

for $\Diamond$.

Thus the best way of fibring both $\Box$ and $\Diamond$ to intuitionistic logic without having the fibring degenerate, is to consider them as independent logics involving two accessibility relations $R_t^{\Box}$ and $R_t^{\Diamond}$, satisfying ($\Box$) and ($\Diamond$), respectively.

The above considerations apply to any logic which has a persistence condition in its semantics. These include Łukasiewicz infinite valued logic (see the considerations in Section 12.3) and substructural logics (see Chapters 14–15). The general algebraic considerations for fibring when persistence is present will be addressed in Section 20.3.

We are now ready to address in detail the fibring of modality to intuitionistic logic.

[13] In the case of dovetailing, we have $t = a_t$ and $s = a_s$ and the conditions can be written as

- $t \leq s \wedge sRy \to \exists z(tRz \wedge z \leq y)$
- $t \leq s \wedge tRy \to \exists z(sRz \wedge y \leq z)$.

4.2 Fibring Modality into Intuitionistic Logic

This section studies the combination of pure modality $\Box$ and/or $\Diamond$ into intuitionistic logic. It begins with a general discussion of what is involved leading to Definition 4.1—fibred semantics for intuitionistic modal logic. This is the most general combination. The completeness theorem is proved in Theorem 4.3. The proof is rather involved, mainly because classical negation is not available, and we need to deal with negative information in a roundabout way. However, the effort is worthwhile, as Example 4.7 shows.

A lot of effort needs to be spent to ensure that the persistency condition

$$(*) \quad t \vDash A \text{ and } t \leq s \text{ implies } s \vDash A$$

holds in the fibred model. The above condition is the main difference between fibring of modalities alone and fibring of modalities with intuitionistic and intermediate logics.

The semantics for general logics, investigated in [Gabbay, 1995b] also requires persistence, so our efforts here are worthwhile.

Definition 4.1.

1. **Fibred semantics for intuitionistic modal logic**
   Let $\mathbf{L}_i, i \in I$, be a family of logics. Each $\mathbf{L}_i$ is either a fragment of modal logic with either a modality $\Box_i$ or a modality $\Diamond_i$ or both (we may not have negation in the fragment) or a fragment of some extension of intuitionistic logic (i.e. an intermediate logic). A fibred model for the logic $\mathbf{L}_I^F$ has the form $(W, S, R, \mathbf{a}, w_0, h, \mathbf{F}, \tau)$ as in Definition 3.8. We add the following additional fibring conditions.
   If $\tau(w) = i$ and $\mathbf{L}_i$ is an intuitionistic fragment and $\mathbf{L}_j$ is a modal or another intuitionistic fragment and $t, s \in S^w$ and $tR^w s$ and $t' = \mathbf{F}(j, w, t)$ and $s' = \mathbf{F}(j, w, s)$ (i.e. t' and s' are the fibred models at t and s respectively) then (**) and [**] hold whenever $\Box$ and/or $\Diamond$ are the modalities respectively present at $\mathbf{L}_j$, or (**) holds if $\mathbf{L}_j$ is intuitionistic:
   (**) $S^{t'} \supseteq S^{s'}, R^{t'} \supseteq R^{s'}, h^{s'} = h^{t'} \upharpoonright S^{s'}$
   [**] $S^{t'} \subseteq S^{s'}, R^{t'} \subseteq R^{s'}, h^{t'} = h^{s'} \upharpoonright S^{t'}$.
2. **Dovetailed semantics for intuitionistic modal logic**
   Let $\mathbf{L}_i$ be as before. A dovetailed model for the logic $\mathbf{L}_I^D$ has the form $(W, \mathbf{R}, a, h)$, where W is a set of worlds, $a \in W$, h is an assignment into W and for each $i \in I, \mathbf{R}(i) \subseteq W \times W$ is the accessibility relation. For each $i \in I$, and $t \in W$, $(W, \mathbf{R}(i), a, h)$ is a model of the logic $\mathbf{L}_i$ (modal or intuitionistic). We also require for any intuitionistic $\mathbf{L}_i$ and any $\mathbf{L}_j$ that
   (a) If $\mathbf{L}_i$ is intuitionistic and $\mathbf{L}_j$ is intuitionistic or with only modal $\Box$ we require that $t\mathbf{R}(i)s \wedge s\mathbf{R}(j)t'$ imply $t\mathbf{R}(j)t'$.
   (b) If $\mathbf{L}_i$ is intuitionistic and $\mathbf{L}_j$ is $\Diamond$ modal then $t\mathbf{R}(i)s \wedge t\mathbf{R}(j)t'$ imply $sR(j)t'$.
   (c) If $\mathbf{L}_j$ is $\Box, \Diamond$ modal then both (a), (b) hold.
   (d) We also require the following:
   Let $t \in W$ be such that there exist $n_1, \ldots, n_k$ and $i_1 \ldots, i_k$ such that $a\mathbf{R}^{n_1}(i_1) \circ \mathbf{R}^{n_2}(i_2) \circ \ldots \circ \mathbf{R}^{n_k}(i_k)t$ holds. Assume $i \neq i_k$ and let

$$W^t = \{y \in W \mid t\mathbf{R}^m(i)y \text{ holds}\}.$$

Then the model $(W^t, \mathbf{R}(i) \restriction W^t, t, h \restriction W^t)$ is in $\mathcal{K}_i$.

These conditions are the ones mentioned in 2(a)–(c) of Definition 4.1. Condition 2(d) of Definition 4.1 is also satisfied automatically.

Example 4.2. We illustrate Definition 4.1 by constructing a fibred model and a dovetailed model for three logics $\mathbf{L}_1, \mathbf{L}_2$ and $\mathbf{L}_3$. $\mathbf{L}_1$ contains $\Rightarrow$ and is an intermediate logic, $\mathbf{L}_2$ is a modal logic with $\Box$ only and $\mathbf{L}_3$ a modal logic with $\Diamond$ only.

Let $\mathcal{K}_i$ be an enumerated class of Kripke models for which the logics are respectively complete. We denote the structures by

$$\mathbf{m}_n^i = (S_{i,n}, R_{i,n}, a_{i,n}, h_{i,n}) \in \mathcal{K}_i.$$

Since $\mathbf{L}_1$ is an intermediate logic, $R_{1,n}$ is assumed to be reflexive and transitive and the following holds for all $t, s \in S_{i,n}$ and any wff A

(*) $tR_{i,n}s$ and $t \vDash A$ imply $s \vDash A$.

We also have $A \vdash_i B$ iff for all $n, a_{i,n} \vDash A$ implies $a_{i,n} \vDash B$.

We can assume the following additional properties without loss of generality (compare Example 3.4).

- $x \in S_{i,n}$ iff for some $m, a_{i,n}R_{i,n}^m x$.
- $(i,m) \neq (k,n)$ implies $S_{i,n} \cap S_{k,n} = \varnothing$.
- For any finite conjunction A of literals (i.e. atoms or their negations) there exist infinitely many n such that $a_{i,n} \vDash A$.

We now form a fibred model out of the above. Let

- $W = \bigcup_{i,n} S_{i,n}$
- $W_i = \{a_{i,n} \mid n = 1, 2, \ldots\}$
- $W_a = \bigcup_i W_i$
- $R = \bigcup_{i,n} R_{i,n}$
- $h(t,q) = h_{i,n}(q)$, for the unique (i,n) such that $t \in S_{i,n}$.

Note that our assumptions imply that for each $x \in W$ there exist unique i and n such that $x \in S_{i,n}$. Let $(\tau_1(x), \tau_2(x))$ pick up the unique (i,n).

Define $\mathbf{F}_i(x)$ as follows:

$$\mathbf{F}_i(x) = \begin{cases} x \text{ if } \tau_1(x) = i \\ a_{i,n} \text{ if } \tau_1(x) \neq i, \text{ where } a_{i,n} \text{ is arbitrary.} \end{cases}$$

We can assume that if $\mathbf{F}_i(x) \neq x$ and $\mathbf{F}_i(y) \neq y$ then $\mathbf{F}_i(x) \neq \mathbf{F}_i(y)$. This means that when we actually fibre models to different points we always choose pairwise disjoint models.

We now have a prefibred model $(W, W_i, R, h, \mathbf{F}_i)$ which is almost the desired fibred model, except for one problem. Persistence (*) must hold in the fibred model, not only for ordinary atoms but also for the 'new atoms', i.e. wffs of the form $\Box Q$ and $\Diamond Q$. Since for $t, s \in S_{1,n}, \mathbf{F}_1(t), \mathbf{F}_2(s)$ are arbitrary , (*) need not necessarily hold. (*) cannot be ensured without either imposing additional assumptions on $\mathbf{F}_i$ or possibly doing additional construction in the model. In practice, we opt for the additional construction. We let **F** be arbitrary but we make all the worlds of $\mathbf{F}_2(s)$ accessible to the actual world of $\mathbf{F}_2(t)$. Thus when $\Box A$ holds at $\mathbf{F}_2(t)$, it will hold at $\mathbf{F}_2(s)$. For $\mathbf{F}_3$ it is the other way round; we make all the worlds of $\mathbf{F}_3(t)$ accessible to the actual worlds of $\mathbf{F}_3(s)$.

The construction to follow is important not only as an example illustrating how we build a fibred model but also the specific construction we offer can be used to prove that if the finite model property holds for each individual logic then it also holds in the fibred and dovetailed logics.

For this reason it is worth our while to present the construction in more detail. Consider the situation described in Figure 4.1. Our starting point is the model $\mathbf{m}^1_n$, with actual world $a_{1,n}$ and two future worlds $a_{1,n}R_{1,n}t$ and $tR_{1,n}s$. Consider the fibring function $\mathbf{F}_i(i = 2 \text{ or } i = 3)$. Let $a_{i,k(i)} = \mathbf{F}_i(t)$ and $a_{i,m(i)} = \mathbf{F}_i(s)$. This means that at the point t we fibre the model $\mathbf{m}^i_{k(i)}$ and at point s we fibre the model $\mathbf{m}^i_{m(i)}$. We want to have for any wff Q that $a_{i,k(i)} \vDash \Box Q$ implies $a_{i,m(i)} \vDash \Box Q$ for $i = 2$ and $a_{i,k(i)} \vDash \Diamond Q$ implies $a_{i,m(i)} \vDash \Diamond Q$ for $i = 3$.

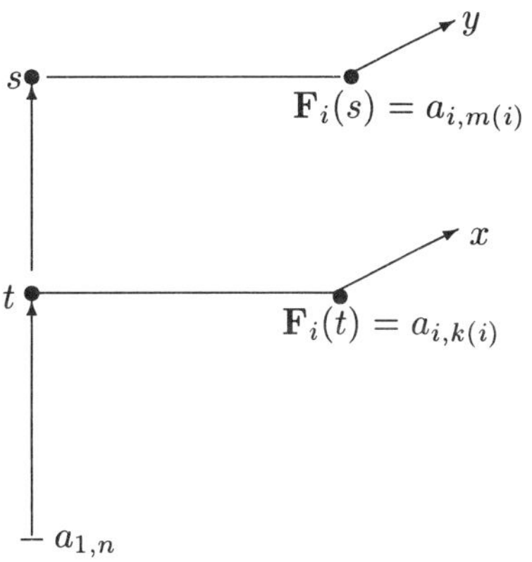

FIG. 4.1.

This cannot be ensured if $\mathbf{F}_i$ gives arbitrary values to t and s.

One way to overcome this problem and at the same time allow $\mathbf{F}_i$ to be arbitrary is to change the evaluation procedure. Instead of the standard fibred evaluation

$$t \vDash \Box Q \text{ iff(def) } \mathbf{F}_2(t) \vDash \Box Q$$
$$t \vDash \Diamond Q \text{ iff(def) } \mathbf{F}_3(t) \vDash \Diamond Q$$

we can say[14]

(†) $t \vDash \Box Q$ iff for all s such that $tR_{1,n}s, \mathbf{F}_2(s) \vDash \Box A$

[†] $t \vDash \Diamond Q$ iff for all s such that $sR_{1,n}t, s \vDash Q$.

In algorithmic terms this means

- $t \vDash \Box Q$ iff for all y such that $y \in S_2^t$, we have $y \vDash Q$.
- $t \vDash \Diamond Q$ iff for some y such that $y \in S_3^t$ we have $y \vDash Q$.

  where

$$y \in S_2^t \text{ iff for some } s, tR_{1,n}s \text{ and } sR_{\tau(\mathbf{F}_2(s))}y$$

$$y \in S_3^t \text{ iff for some } s, sR_{1,n}t \text{ and } sR_{\tau(\mathbf{F}_3(s))}y.$$

  If we let

$$\mathbf{F}_2(t)R_2^t y \text{ iff } y \in S_2^t$$
$$\mathbf{F}_3(t)R_3^t y \text{ iff } y \in S_3^t$$

  we get

- $t \vDash \Box Q$ iff for all y such that $\mathbf{F}_2(t)R_2^t y$ we have $y \vDash Q$.

  $t \vDash \Diamond Q$ iff for some $y, \mathbf{F}_3(t)R_3^t y, y \vDash Q$.

To turn the above into a good model we must allow R_2^t (resp. R_3^t) to include $R_{\tau(\mathbf{F}_2(s))}$ (resp. $R_{\tau(\mathbf{F}_3(s))}$) for all s such that $tR_{1,n}s$ (resp $sR_{1,n}t$).

The definitions (†) and [†] above will be useful in our considerations of the finite model property.

Let us now define the fibred model formally. Assume we are given $\mathbf{m}_n^1$, with the fibring functions $\mathbf{F}_2, \mathbf{F}_3$, defined on $S_{1,n}$, yielding for each $t \in S_{1,n}$ the fibred models $\mathbf{m}_{\tau_2(\mathbf{F}_2(t))}^{\tau_2(\mathbf{F}_2(t))}$ and $\mathbf{m}_{\tau_1(\mathbf{F}_3(t))}^{\tau_1(\mathbf{F}_3(t))}$. We want to define new functions $\mathbf{F}'_i(t)$ yielding proper fibred models. This we do simultaneously for all $t \in S_{1,n}$ by letting $\mathbf{F}'_i(t) = (S_i^t, R_i^t, a_{\tau(\mathbf{F}_i(t))}, h_i^t)$ where

$$S_2^t = \bigcup_{\{s \mid tR_{1,n}s\}} \{x \mid \mathbf{F}_2(s)R_{\tau(s)}x\}$$

$$S_3^t = \bigcup_{\{s \mid sR_{1,n}t\}} \{x \mid \mathbf{F}_3(s)R_{\tau(s)}x\}.$$

We have

$$tR_{1,n}s \text{ implies } S_2^t \supseteq S_2^s \text{ and } S_3^t \subseteq S_3^s.$$

Let

[14] If $\vDash$ is the old evaluation in intuitionistic models, define a new evaluation $\vDash_*$ by

$$t \vDash_* A \text{ iff } \forall s(tR_{1,n}s \to s \vDash A).$$

In the intuitionistic models $\mathbf{m}_n^1$ for the $\mathbf{L}_1$ language we have $\vDash_* = \vDash$. When we evaluate for $\mathbf{L}_2$ or $\mathbf{L}_3$ we get (†) and [†].

$$R_2^t = (\{\mathbf{F}_2(t)\} \times S_2^t) \cup \bigcup_{\{s|tR_{1,n}s\}} R_{\tau(s)}$$

$$R_3^t = (\{\mathbf{F}_3(t)\} \times S_2^t) \cup \bigcup_{\{s|sR_{1,n}t\}} R_{\tau(s)}.$$

Let

$$h_2^t = \bigcup_{\{s|tR_{1,n}s\}} h_{2,s}$$

$$h_3^t = \bigcup_{\{s|sR_{1,n}t\}} h_{3,s}.$$

First note the following properties of $\mathbf{F}$ and $\mathbf{F}'$

- $\bigcup_{t\in S_{1,n}} S_i^t = \bigcup_{t\in S_{1,n}} S_{\tau(\mathbf{F}_i(t))} = U_{1,n}$
- Let $R_i = \bigcup_{t\in S_{1,n}} R_i^t$. Then $R_i^t = R_i \upharpoonright S_i^t \times S_i^t$.

The above means that the fibred models $\{\mathbf{F}_i'(t \mid t \in S_{1,n}\}$ use the same stock of possible worlds that $\{\mathbf{F}_i(t) \mid t \in S_{1,n}\}$ use. Note also that $U_{1,n}$ are pairwise disjoint and that $\mathbf{F}_i, i = 1, 2, 3, \ldots$ are defined on any $x \in U_{1,n}$.

Second note that if $t \neq s \in S_{1,n}$ then $\mathbf{F}_i(t) \neq \mathbf{F}_i(s), i = 2, 3, \ldots$. Furthermore, we assumed that our fibring functions $\mathbf{F}_i$ *always* fibre *new* models. This means that for all other $x \in S_{1,r}, r \neq n$ we have $\mathbf{F}_i(t) \neq \mathbf{F}_j(x)$.

This second point is important. The old functions $\mathbf{F}_i$ fibre models $\mathbf{F}_i(t)$ to all $t \in S_{1,n}$. These models were replaced simultaneously by the new models $\mathbf{F}_i'(t), t \in S_{1,n}$, using the stock $U_{1,n}$ of possible worlds of the old models $\mathbf{F}_i(t)$. Since these old models are fibred *nowhere else*, our construction is unique to each index $(1, n)$ and the constructions for different indices do not relate to each other and involve completely different worlds and models.

For $t \in S_{2,n}$ or $S_{3,n}$ let $\mathbf{F}_i'(t) = \mathbf{F}_i(t)$.

We can assume that the modal models

$$(S_i^t, R_i^t, a_{\tau(\mathbf{F}_i(t))}, h_i^t)$$

are already members of $\mathcal{K}_i$ (since we can close $\mathcal{K}_i$ under the construction) and hence replacing $\mathbf{F}_i$ with $\mathbf{F}_i'$ gives us the desired fibred model.

We conclude the fibred part of our example by an observation. We started with the fibred system $\{\mathbf{F}_i(t) \mid t \in S_{1,n}\}$ and constructed the fibred system $\{\mathbf{F}_i'(t) \mid t \in S_{1,n}\}$. This is a different system of models. Truth values may change. However the following does hold:

- If A is a wff such that for all $t, s, t \vDash A$ and $tR_{1,n}s$ imply $s \vDash A$, in the model with the system $\mathbf{F}_i$, then its truth value does not change when $\mathbf{F}_i$ is replaced by $\mathbf{F}_i'$. In other words
- $t \vDash A$ in the fibred model with $\mathbf{F}_i$ iff $t \vDash A$ in the fibred model with $\mathbf{F}_i'$.

We now examine the dovetailed version of the above semantics. Dovetailing is obtained by assuming that the actual world $a_{\tau(\mathbf{F}_i(t))}$ of $\mathbf{F}_i(t)$ and the original world t give the same truth values, i.e. for all q

$$h(t,q) = h(a_{\tau(\mathbf{F}_i(t))}, q).$$

We can proceed to simplify the presentation of the semantics along the lines of Example 3.6.

First take the case that $t \in S_{i,n}$ and consider $t, s \in S_{1,n}$ such that $tR_{1,n}s$ holds. See Figure 4.1. Our assumption is that for all q:

- $h_{1,n}(t,q) = h_{i,k(i)}(a_{i,k(i)}, q)$
- $h_{1,n}(s,q) = h_{i,m(i)}(a_{i,m(i)}, q)$
- If $h_{1,n}(t,q) = 1$ then $h_{1,n}(s,q) = 1$
- $a_{2,m(2)}R_2y \to a_{2,k(2)}R_2y$
- $a_{3,k(3)}R_3y \to a_{3,m(3)}R_3y$.

If we identify the point t with $a_{i,k(i)}$ and s with $a_{i,m(i)}$, we get that the following holds:

- $tR_{1,n}s \wedge sR_2y \to tR_2y$
- $tR_{1,n}s \wedge tR_3y \to sR_3y$.

These conditions are the ones mentioned in 2(a)–(c) of Definition 4.1. Condition 2(d) of Definition 4.1 is also satisfied automatically.

We are now ready to state a completeness theorem for the fibring and dovetailing of modalities with intuitionistic connectives. We want a general fibring of an intuitionistic or intermediate logic connective $\Rightarrow$ with one or several modalities $\Box_1$ and/or $\Diamond_2$. Thus for example we would like to fibre the $\Rightarrow$ of Dummett's **LC** with the **K** modalities $\Box$ and $\Diamond$. The result of the fibring, as well as the axiomatisation, is very much dependent on the semantics we take for the modalities and on the additional connectives which happen to be present, such as $\wedge, \vee$ and $\neg$. Thus no general theorem can be easily given. We will therefore choose one type of modality, **K** modality, and only $\Rightarrow$ and do the fibring and dovetailing for that. The methodology will be clear from the proofs and we shall later indicate what happens if the modalities come from a stronger (possibly non-normal) extension of **K**.

The reader will notice throughout that the proofs use properties of intuitionistic logic in an essential way and one needs to think what one can do in a general case.

We now fibre several intermediate logics $\Rightarrow_i$ with several **K** modalities $\Box_j, \Diamond_j$. Since we might not have $\neg, \wedge$ and $\vee$, we need to formulate the logics to be fibred using the notion of consequence relation. We choose the notion $\Delta \mid\!\sim \Gamma$ where both Δ and Γ are sets of wffs. Our proof works for the notion $\Delta \mid\!\sim B$, i.e. where Γ is a single formula. We therefore have to assume that the case of $\Delta \mid\!\sim \Gamma$ can be reduced to that of $\Delta \mid\!\sim B$. If the logic has disjunction then we can take B to be $\bigvee \Gamma$. Otherwise we an assume that $\mid\!\sim$ is *disjunctive*, namely that $\Delta \mid\!\sim \Gamma$ iff for some $B \in \Gamma, \Delta \mid\!\sim B$.

This condition is not too restrictive. The implicational fragment of intuitionistic logic is disjunctive, i.e. if Δ, Γ are in the pure implicational fragment then $\Delta \mid\!\sim \bigvee \Gamma$ iff for some $B \in \Gamma, \Delta \mid\!\sim B$.

Dummett's **LC** does not satisfy the above. Its implicational fragment can be axiomatised by the additional axiom

$$((A \Rightarrow B) \Rightarrow C) \Rightarrow (((B \Rightarrow A) \Rightarrow C) \Rightarrow C).$$

Thus the conditions of Theorem 4.3 below do not apply to it. This does not mean, however, that we cannot prove the result for this case by some other means.

Theorem 4.3 (Completeness theorem for fibred (resp. dovetailed) intuitionistic modal logics).

1. *Let $\mathbf{L}_i$ be a logic with either an intermediate intuitionistic implication $\Rightarrow_i$ (and possibly $\wedge, \vee$ and $\bot$) or with one or both of the $\mathbf{L}_i$ modalities $\Box_i, \Diamond_i$. Let $\mathcal{K}_i$ be the semantics for $\mathbf{L}_i$. Let $\mid\!\sim_i$ be the consequence relation of $\mathbf{L}_i$. Then $\{A_1, \ldots, A_n\} \mid\!\sim_i \{B_1, \ldots, B_m\}$ means that in every Kripke model of $\mathbf{L}_i$ (modal or intuitionistic) of the form (S, R, a, h) whenever for all $j, a \vDash A_j$ then for some $k, a \vDash B_k$. We regard the consequence relation as a relation between finite sets of formulas. The completeness theorem characterises the consequence relation of $\mathbf{L}_I^F$ (denoted by $\mid\!\sim$) in terms of the consequence relations of $\mathbf{L}_i$ (denoted by $\mid\!\sim_i$). We make the following assumptions.*
   (*) *For intuitionistic $\mathbf{L}_i$, we assume that either disjunction is available or that it is disjunctive.*
2. *Let $\mid\!\sim$ be the smallest consequence relation containing $\mid\!\sim_i$ for $i \in I$, closed under substitution (and of course under reflexivity, monotonicity and cut) and closed under the following rules, where $\Rightarrow$ is a connective of an arbitrary $\mathbf{L}_{i_1}$ and $\Box, \Diamond$ of an arbitrary $\mathbf{L}_{i_2}$.*
   (a)
   $$\frac{A_1, \ldots, A_n \mid\!\sim C_1, \ldots, C_m}{\Box A_1, \ldots, \Box A_n, \Diamond B \mid\!\sim \Diamond C_1, \ldots, \Diamond C_m}$$
   for the case of fibring A_i and C_j must begin with $\Box, \Diamond$ or $\Rightarrow$;
   $$\frac{A_1, \ldots, A_n \mid\!\sim C, B_1, \ldots, B_m}{\Box A_1, \ldots, \Box A_n \mid\!\sim \Box C, \Diamond B_1, \ldots, \Diamond B_m}$$
   for the case of fibring A_i, B_j and C must begin with $\Box, \Diamond$ or $\Rightarrow$.
   (b) *Deduction theorem for each $\Rightarrow_i$.*

Then $\mid\!\sim$ is the consequence relation of all the fibred models of $\mathbf{L}_I^F$ (resp. dovetailed models of $\mathbf{L}_I^D$).

Proof

1. Soundness is easy to verify.

2. Completeness can be proved along similar lines to the proof of Theorem 3.10.

In the case of modal logics, completeness was proved by induction on the number of nested modalities of different logics. Thus one would expect that we use induction in the modal intuitionistic case on the number of nested different connectives of the different logics. Thus for a formula of the form $\Box_2(A \Rightarrow_1 \Box_2\Diamond_3 B) \Rightarrow_4 B$ one may expect the inductive nesting number to be 4. This does not work, however, because the modal logics with $\Box$ and $\Diamond$ only do not have enough expressive power to carry the inductive case step. It is more convenient to count only nestings of intuitionistic $\Rightarrow$, because we can use the expressive power of the $\Rightarrow$ to carry the induction through. Thus the above formula falls under the inductive case for $n = 2$. The nestings are $A \Rightarrow_1 \Box_2\Diamond_3 B$ and the full $\Box_2(A \Rightarrow_1 \Box_2\Diamond_3 B) \Rightarrow_4 B$. Modalities do not count and they join the outer $\Rightarrow$ governing them. The above will be analysed as:

Case $n = 1$
Substitution into

$$B_1(x, y) = [x \Rightarrow_1 \Box_2\Diamond_3 y]$$

by

$$\Theta_1(x) = A, \Theta_2(y) = B.$$

Case $n = 2$
Substitution into

$$B_2(x, y) = [\Box_2 x \Rightarrow_4 y]$$

by

$$\begin{aligned} \Theta_1(x) &= A \Rightarrow_1 \Box_2\Diamond_3 B \\ \Theta_2(y) &= B. \end{aligned}$$

Case n is described inductively as composed of wffs $B_n(\overline{x}, \overline{y})$, whose outer connective is some intuitionistic $\Rightarrow_i$ of the logic $\mathbf{L}_i$ and within it some modalities and atoms from $\mathbf{L}_j, j \in J$ (abbreviated $\mathbf{L}_J$). A substitution $\Theta_n(x_i)$ gives wffs of case $n - 1$.
□

Let $\Delta \not\vdash \Gamma, \Delta, \Gamma$ sets of wffs of $\mathbf{L}_I^F$ (resp. $\mathbf{L}_I^D$). We show by induction on the maximal number n of nested intuitionistic connectives (of different logics) in $\Delta \cup \Gamma$ that there exists a fibred (resp. dovetailed) model in which all members of Δ hold and all members of Γ do not hold.

Case n = 0
This case means that we have only connectives of the form $\Box$ and $\Diamond$ of various modal logics $\mathbf{L}_j, j \in J$. Let $\mathbf{L}_J$ be the logic with modalities of $\mathbf{L}_j, j \in J$. We need a lemma:

Lemma 4.4. *Given the assumptions of Theorem 4.3, let* $A_1 \ldots A_n \not\vdash B_1, \ldots, B_k$ *in* $\mathbf{L}_J$. *Then for some fibred (resp. dovetailed) model* $\mathbf{m}$ *of* $\mathbf{L}_J^F, \mathbf{m} \vDash A_i$ *for all* i *and* $\mathbf{m} \not\vDash B_j$ *for some* j.

Proof From the theorem on fibring (resp. dovetailing) pure modalities in the previous section, there exists a fibred modal model. The intuitionistic fibred part can be added arbitrarily. □

Case $n = 1$, involving only one $\Rightarrow$.
We prove that every Δ, Γ s.t. $\Delta \not\mid\!\sim \Gamma$ has a fibred (resp. dovetailed) countermodel. Our strategy is to regard all modally prefixed atoms in $\Delta \cup \Gamma$ as new atoms in a purely intuitionistic form $\Delta' \not\mid\!\sim \Gamma'$. We can find an intuitionistic countermodel for that form and realise the assignments to the new atoms by fibring (resp. dovetailing) modal models. This would give us a fibred countermodel for $\Delta \not\mid\!\sim \Gamma$. We now go through the details.

Let (S, R, a, h) be any intuitionistic countermodel for $\Delta' \not\mid\!\sim \Gamma'$ (we assumed that Δ, Γ are in a language of a single $\Rightarrow$) and let $\{A_1, \ldots, A_m, B_{m+1}, \ldots, B_k\}$ be the new modal atoms. Thus

$$\Delta' = \{\alpha_i(\overline{x}, \overline{y})\}, \Gamma' = \{\beta_j(\overline{x}, \overline{y})\}$$

where $\overline{x} = (x_1, \ldots, x_k), \overline{y} = (y_1, \ldots, y_k)$ and

$$\Theta(x_1) = A_1, \ldots, \Theta(x_m) = A_m, \Theta(x_{m+1}) = B_{m+1}, \ldots, \Theta(x_k) = B_k.$$

m is an arbitrary division to be chosen later. Let $t \in S$ and assume $h(t, x_i) = 1, i \leq m(t)$, and $h(t, x_i) = 0, m(t) < i \leq k$. If we can realise, for arbitrary t (m depends on t) $\{A_i\} \not\mid\!\sim \{B_j\}$ then we shall show later that we get a fibred (resp. dovetailed) countermodel for case $n = 1$. Otherwise $\{A_i\} \mid\!\sim \{B_j\}$. By the property (*) assumed in Theorem 4.3, $\{A_i\} \mid\!\sim B$ where B is either $\bigvee B_j$ or B_{j_0} for some j_0 depending on the logic involved. Let z be $x_{m+1} \vee \ldots \vee x_k$ or let z be x_{j_0}. Let $\delta_t = x_1 \Rightarrow_i (x_2 \Rightarrow_i \ldots \Rightarrow_i z) \ldots)$. There is only a finite number of possibilities for δ_t. Hence

$$\{\delta_t\} \cup \Delta' \mid\!\sim_{\mathbf{L}_i} \Gamma'$$

so

$$(\delta_t \Theta), \Delta' \Theta \mid\!\sim \Gamma' \Theta.$$

Since $\mid\!\sim \delta_t \Theta$ and $\Delta = \Delta' \Theta, \Gamma = \Gamma' \Theta$ we get $\Delta \mid\!\sim \Gamma$, a contradiction.

Thus for some intuitionistic countermodel (S, R, a, h) of $\Delta' \not\mid\!\sim \Gamma'$, all assignments at any t are realisable by fibred models $\mathbf{m}_t$. These models satisfy the persistence condition, $i = 1, \ldots, k$:

- $\mathbf{m}_t \vDash \Theta(x_i)$ and tRs implies $\mathbf{m}_s \vDash \Theta(x_i)$.

These models can be grafted onto (S, R, a, h) as in Definition 3.9. This grafted model can be turned to satisfy the requirements (**) and [**] of Definition 4.1 by the construction method described in Example 4.2.

This completes the proof for case $n = 1$, single $\Rightarrow$.

Case $n = 1$ *several* $\Rightarrow_r, r \in J$.
In this case assume $\Delta' = \{\alpha_i^r(\overline{x}, \overline{y})\}, \Gamma' = \{\beta_j^r(\overline{x}, m\overline{y})\}$, where r runs over the logics $\mathbf{L}_r, r \in J$. α^r and β^r need not necessarily contain $\Rightarrow_r$; they may be atoms from $\overline{y}$ or from $\overline{x}$, in which case we say they are atomic. $\overline{x}$ are the atoms for which Θ substitutes modalities and $\overline{y}$ are the atoms for which no modal wffs are substituted.

If $\Delta' \not\mid\!\sim \Gamma'$ then $\Delta'_r \not\mid\!\sim_r \Gamma'_r$ for all r, where Δ'_r, Γ'_r are as follows:

$$\Delta'_f = \{\alpha_i^s \mid s = r \text{ or } \alpha_i^s \text{ is atomic}\}$$

in case we are dovetailing

$$\Delta'_r = \{\alpha_i^s \mid s = r \text{ and } \alpha_i^s \text{ is } \textit{not} \text{ atomi c from } \overline{y}\}$$

in case we are fibring.

Similarly for the respective definition of Γ'_r.

We prove this case by finding countermodels for each $\Delta'_r \not\vdash_r \Gamma'_r$ and grafting them together.

Case $n > 1$

We assume Δ, Γ are obtained from a Δ_i, Γ_i of a pure intuitionistic modal component logic $\mathbf{L}_J$ as in case $n = 1$ via a substitution Θ. In other words

$$\Delta_i = \{B_i^r \mid r = 1, \ldots, k\}$$
$$\Gamma_i = \{B_i^r \mid r = k + 1 \ldots m\}$$

and each B_i^r has the form

$$B_i^r(\overline{x}^r, \overline{y}^r).$$

Let $\Theta_0(x_r) = \overline{M}^r z_r$, and let $\Theta_1(z_r) = A_r \Rightarrow_j B_r$. Let $\Theta = \Theta_0\Theta_1$ (Θ_0 first then Θ_1). Thus $\Theta(x_r) = \overline{M}^r(A_r \Rightarrow_j B_r)$, where $\overline{M}^r$ is a string of modalities from $\mathbf{L}_J$ and $A_r \Rightarrow_j B_r$ are from a logic $\mathbf{L}_j$.

Subcase a

We assume that each x variable is under the scope of the $\Rightarrow_i$ connective. Since $\Delta_i\Theta \not\vdash \Gamma_i\Theta$ we can assume $\Delta_i\Theta_0 \not\vdash_{\mathbf{L}_J} \Gamma_i\Theta_0$ and hence there are $\mathbf{L}_J$ models $\mathbf{m} = (S^{\mathbf{m}}, R^{\mathbf{m}}, a^{\mathbf{m}}, h^{\mathbf{m}})$ in which $\Delta_i\Theta_0$ holds and $\Gamma_i\Theta_0$ does not hold as shown in case $n = 1$. Reasoning very much like the case of Theorem 3.10 we want to show that there exists at least one $\mathbf{m}$ in which we can realise the assignment $h^{\mathbf{m}}(t, x_s^{r,j})$ for all t by a fibred model for $\Theta(x_s^{r,j})$, of the logic $\mathbf{L}_j$. Assume this is not possible. Then by the induction hypothesis for every $\mathbf{m}$, there exists a j and t such that $\Phi_{j,t}\Theta \vdash \Psi_{j,t}\Theta$, where

$$\Phi_{j,t} = \{x_s^{r,j} \mid h^{\mathbf{m}}(t, x_s^{r,j}) = 1, r, s \text{ varying}\}$$
$$\vdash \Psi_{j,t} = \{x_s^{r,j} \mid h^{\mathbf{m}}(t, x_s^{r,j}) = 0, r, s \text{ varying}\}.$$

Our aim is to reach a contradiction by using the fact that for every $\mathbf{m}$ there exists a j and $t \in S^{\mathbf{m}}$ such that $\Phi_{j,t} \mid\!\sim \Psi_{j,t}$. The contradiction is to show that $\Delta \mid\!\sim \Gamma$ contrary to assumptions. To achieve this end we need to extract some formulas $\delta_1, \delta_2, \ldots$ so that $\Delta_i\Theta_0, \delta_1\Theta_0, \delta_2\Theta_0, \ldots \mid\!\sim \Gamma_i\Theta_0$ and $\mid\!\sim \delta_1\Theta, \mid\!\sim \delta_2\Theta, \ldots$ and hence by cut, $\Delta = \Delta_i\Theta \mid\!\sim \Gamma = \Gamma_i\Theta$. We need a small lemma to the effect that we can assume that $\Psi_{j,t}$ contains (or is equivalent in $\mathbf{L}_j$ to) only one element $\beta_{j,t}$. Using this fact we can write

$$\delta_{j,t} = \varphi_1 \Rightarrow_j (\varphi_2 \Rightarrow_j \ldots \Rightarrow (\varphi_{k(j,t)} \Rightarrow_j \beta_{j,t} \ldots))$$

where

$$\Phi_{j,t} = \{\varphi_1, \ldots, \varphi_{k(j,t)}\}.$$

Since there is only a finite number of different possibilities for $\delta_{j,t}$, for different models $\mathbf{m}$, we get by case $n = 1$ that for some $\delta_1, \delta_2 \ldots, \Delta_i\Theta_0, \delta_1\Theta_0, \delta_2\Theta_0, \ldots \mid\!\sim \Gamma_i\Theta_0$.

Hence

$$\Delta_i\Theta, \delta_1\Theta, \ldots, \vdash \Gamma_i\Theta$$

and since $\mid\!\sim \delta_r\Theta$ we get $\Delta \mid\!\sim \Gamma$, a contradiction.

We now prove the lemma.

Lemma 4.5. *We can assume that $\Psi_{j,t}$ contains only one element.*

Proof Note that all the elements of $\Phi_{j,t}\Theta$ and $\Psi_{j,t}\Theta$ are of the form $(A \Rightarrow B)$. If $\mathbf{L}_j$ contains no disjunction in the language we get from 4.4 and the induction hypothesis that for some $x \in \Psi_{j,t}$ we have $\Phi_{j,t}\Theta\mid\!\sim\Theta(x)$. If $\mathbf{L}_j$ does contain disjunction then clearly $\Phi_{j,t}\Theta\mid\!\sim \bigvee \Psi_{j,t}\Theta$.

In either case we get a wff $\beta_{j,t}$ such that $\Phi_{j,t}\Theta\mid\!\sim\beta_{j,t}\Theta$. We can thus assume that $\Psi_{j,t}$ contains only one element. □

We have thus concluded the proof of case $n > 1$, subcase (a).

Subcase b

We now examine the subcase where the conditions of subcase (a) are not necessarily satisfied. We thus have $\Delta\not\mid\!\sim\Gamma$ and Δ, Γ are obtained by substitution into $\Delta_i\not\mid\!\sim_{\mathbf{L}_J}\Gamma_i$ and the formulas in $\Delta_i \cup \Gamma_i$ are of the form $B_i^r(\overline{M}, \overline{x}, \overline{y})$ as in subcase (a).

We do not have in this case that each variable x is under the scope (in $B_i^r(\ldots x \ldots)$) of the intuitionistic connective $\Rightarrow_j$. This means that $\Delta_i \cup \Gamma_i$ is comprised of several formulas whose outer connectives are different intuitionistic ones $\Rightarrow_{j_1}, \Rightarrow_{j_2}$. Since these formulas are evaluated in different fibred models, we can apply subcase (a) to them separately. The following lemma explains how:

Lemma 4.6. *Let $j \neq j'$ and assume*

$$A \Rightarrow_j B, A' \Rightarrow_{j'} B'\mid\!\sim C \Rightarrow_j D, C' \Rightarrow_{j'} D'.$$

Then either

$$A \Rightarrow_j B\mid\!\sim C \Rightarrow_j D$$

or

$$A' \Rightarrow_{j'} B'\mid\!\sim C' \Rightarrow_{j'} D'.$$

Example 4.7 (Some dovetailed modal intuitionistic systems). Note that Theorem 4.3 allows us to fibre or dovetail any intermediate logic with any modal logic provided the intermediate logic has disjunction in it or is disjunctive. The theorem gives us the axiom system for the combined logic. Especially for the case of dovetailing, the method is very powerful. Consider any modal logic where axioms are reduction of modalities, for example

- $\Box A \vdash A$
- $\Box A \vdash \Box\Box A$
- $\Diamond A \vdash \Box\Diamond A$
- $\Diamond\Box A \vdash \Box\Diamond A$
- $\Diamond^5\Box^6 A \vdash \Diamond^{18}\Box^{11}\Diamond^{12} A$

- $\Box\Diamond A \wedge \Diamond\Box B \vdash \Diamond\Diamond(A \wedge B)$.

Some of these axioms you would recognise and some I have just invented. It does not matter. In the combined logic any $\bigwedge_i M_i A_i \vdash MA$ becomes a theorem $\bigwedge_i M_i A_i \Rightarrow MA$, where M_i, M are strings of modalities. We can thus immediately have semantics and axioms for the dovetailed system of any intermediate logic with disjunction and any major modal logic (e.g. **K**, **T**, **B**, **S4**, **S5**, etc.).

To be more specific, the above theorem allows us to fibre or dovetail the following:

1. The intermediate logic **KC** obtained by adding the schema $\neg A \vee \neg\neg A$ to intuitionistic logic with **S4** modality $\Box$ and $\Diamond$.
2. Dummett's **LC** (obtained by adding the schema $(A \Rightarrow B) \vee (B \Rightarrow A)$ with the **K** modality $\Diamond$).

The axioms of the combined system are obtained from the completeness theorem. For example, dovetailing the systems in (1) yields:

1. **KC** axioms with modus ponens
2. $\Box A \wedge \Box(A \Rightarrow B) \Rightarrow \Box B$
   $\Box(A \Rightarrow B) \wedge \Diamond A \Rightarrow \Diamond B$
   $\Diamond\Diamond A \Rightarrow \Diamond A$
   $\Box A \vee \Box\Box A$
   $\Box A \Rightarrow \Diamond A \vee \Box B$
   $\Diamond\neg A \vee \Box\neg\neg A$
   $\Diamond\neg\neg A \vee \Box\neg A$
   $\Box A \Rightarrow A$
   $A \Rightarrow \Diamond A$
3. $\dfrac{\vdash A}{\vdash \Box A}$.

Further note that the dovetailing method of adding modality to any intermediate logic is systematic. We check all modalities reductions of the form $\bigwedge_i M_i A_i \vdash MA$ and turn them into an axiom

$$M_1 A_1 \Rightarrow (\ldots \Rightarrow (M_n A_n \Rightarrow MA)\ldots).$$

There is nothing arbitrary here.

Another way to obtain intuitionistic modal logic is to take the modal logic formulated axiomatically and change the underlying logic from classical to intuitionistic. Thus modal **K**, when axiomatised by

- $\Box(A \to B) \to (\Box A \to \Box B)$
- $\dfrac{\vdash A}{\vdash \Box A}$

yields by chance, when '$\to$' is changed to '$\Rightarrow$', the same system as dovetailing. but if we use an equivalent axiom

- $\Box(\neg A \vee B) \vee \neg\Box A \vee \Box B$

we get a different modal intuitionistic system.

Remark 4.8. In the above examples we assumed the fibring is done by a function $\mathbf{F}(w,t) = z$. Thus the value of a new connective (e.g. $\Box$) is evaluated at a single model labelled by z. It is more general to associate with (w,t), not a single model z but a family of such models in which case $\mathbf{F}$ becomes a set function or a relation $\mathbf{R}$. We thus get in $\mathbf{m}$:

$$t \vDash \sharp(A_i) \text{ iff for all } \mathbf{n} \text{ such that } (\mathbf{m}, t)\mathbf{Rn}, \text{ we have } \mathbf{n} \vDash \sharp(A_i).$$

The advantage of such an approach is that it is more general. Imagine some system where classical disjunction is a new connective. Then we do not necessarily want to have

$$\frac{A \mathrel{|\!\sim} C, B \mathrel{|\!\sim} C}{A \vee B \mathrel{|\!\sim} C}.$$

Theorem 4.9. *Let* $\mathbf{L}_1, 1 \in I$, *be an intermediate logic and let* $\mathbf{L}_i, i \neq 1, i \in I$, *be modal logics. Let* $\mathbf{L}_I^D$ *and* $\mathbf{L}_I^F$ *be the dovetailing, resp. fibring of the logics. Then if* $\mathbf{L}_i, i \in I$, *are all decidable, so are* $\mathbf{L}_I^D$ *and* $\mathbf{L}_I^F$.

Proof Assume $\mathbf{L}_i$ are all decidable. We show that $\mathbf{L}_I^D$ nd $\mathbf{L}_I^F$ are also decidable. The proof is by structural indiction on the number of nested implications in the expression $A_i \mathrel{|\!\sim} ?B_j$.

Case $n = 0$

In this case the wffs A_i, B_j are pure modalities and the decision procedure is that of fibred or dovetailed modal logics. In this case $A_i \mathrel{|\!\sim} B_j$ iff $\vdash \bigwedge_i A_i \rightarrow \bigvee_j B_j$ in the modal logic.

Case $n > 0$.

In this case we have intuitionistic formulas $\{\alpha_i(\overline{x}, \overline{y})\}, \{\beta_j(\overline{x}, \overline{y})\}$ with $A_1, \ldots, A_m$, $B_{m+1}, \ldots, B_k$ modal formulas (without $\Rightarrow$) such that the main decision problem is whether

$$\{\alpha_i(\overline{x}, \overline{y})\Theta\} \mathrel{|\!\sim} ?\{\beta_j(\overline{x}, \overline{y})\Theta\}$$

where $\Theta(x_1) = A_1, \ldots, \Theta(x_n) - A_m$, $\Theta(x_{m+1}) = B_{m+1}, \ldots, \Theta(x_k) = B_k$.

We now describe an algorithm for the main decision problem. Let $\delta_{r,m}$ be a formula

$$x_1 \Rightarrow (x_2 \Rightarrow \ldots \Rightarrow (x_m \Rightarrow z) \ldots)$$

where $z = x_r$, in case disjunction is not in the language, with $m < r \leq k$ and where $z = x_{m+1} \vee \ldots \vee x_k$ if disjunction is available.

Consider the following decision problem, which we call the first (r, m) problem:

$$\delta_{r,m}; \alpha_i(\overline{x}, \overline{y}) \mathrel{|\!\sim} ?\beta_j(x, y).$$

If for all r, m the answer is negative, then the answer to the main decision problem is also negative. Let r, m be such that the answer is positive. Then we check an additional decision problem, which we call the second (r, m) problem:

$$A_1, \ldots, A_n \mathrel{|\!\sim} ?B$$

where $B = B_r$ or $B = B_{m+1} \vee \ldots \vee B_k$ depending on the language.

If the answer to this problem is also yes, then we say that the answer to the main problem is positive.

If for no (r, m) do we have that both the first and second (r, m) problems are positive, then the answer to the main problem is negative.

We now have to show that $\Delta \mid\!\sim \Gamma$ iff the answer to the decision problem is positive. Clearly if the answer is positive then $\Delta \mid\!\sim \Gamma$.

Assume $\Delta \mid\!\sim \Gamma$. We try to construct a fibred (or dovetailed) countermodel for $\Delta \not\mid\!\sim \Gamma$. This construction should fail. The considerations of the proof of Theorem 4.3 will show that the decision procedure must succeed. □

Theorem 4.10. *In the notation of Theorem 4.9, if* $\mathbf{L}_i, i \in I$, *all have the finite model property, so do* $\mathbf{L}_I^D$ *and* $\mathbf{L}_I^F$.

Proof Similar to the proof of the modal case. We rely on the special fibred/dovetailed models of Example 4.2. □

5

DISCUSSION AND COMPARISON WITH THE LITERATURE

The previous chapters presented a general methodology for combining logics. There was an emphasis on semantical methods, especially for the cases of intuitionistic and modal logics. To see where our work stands, we must compare it with existing published works in the area. Combined systems of modal and intuitionistic logics have been around for a long time, and the landscape is a jungle. As shown by the examples below, authors simply defined what they wanted for good or bad reasons and proceeded to investigate the systems they presented. There was no clear methodology, and no coordinate system of milestones for comparison. We aim in this chapter to show how the existing literature fits into our methodology. We proceed by summarising some typical existing papers and discussing their systems from our point of view.

The conclusion we reach is the following:

- The dovetailing methods allow the identification of a coordinate system of pure dovetailed logics of the form $\mathbf{L}_1 \otimes \mathbf{L}_2$ where $\mathbf{L}_1, \mathbf{L}_2$ range over well-known logics (see Example 4.7) e.g. $\mathbf{KC} \otimes \mathbf{S4}$, etc. The semantics for these dovetailed 'milestone' logics has the form (S, R_1, R_2, h) where R_1 and R_2 satisfy the conditions of $\mathbf{L}_1, \mathbf{L}_2$ respectively, together with some simple interaction conditions. Additional interaction conditions on R_1, R_2 give rise to meaningful additional logics.
- The above yields enough well-understood combined logics to allow us to put some order and understanding in the chaotic landscape of existing published combined logics.

The examples and the discussions below abundantly support our claims.

5.1 Intuitionistic Modal Logics

Such logics are abundant in the literature. The first syntactical attempt to introduce modality into intuitionistic logic was by Fitch [1948]. Many others followed. The motivation was mainly philosophical in search for intuitionistic constructive modality, as both intuitionistic logic and modal logic have solid and independent philosophical reasons. Recently with increased successful applications of these logics in computer science, many combined systems of the intuitionistic $\Rightarrow$ (representing constructive formula as type implication) and modal $\Box$ (representing actions) were defined. We begin with such an example.

Example 5.1 (The system of Wijesekera 1990).

1. **Motivation**
   The system arises in connection with dynamic logic and computer science applications, involving actions, partial information and states of knowledge. The

worlds in the semantics represent states of partial knowledge, the partial order represents increase in information and the accessibility relation represents change of state. The author claims that

'This formalism covers all computer science aplications of which we are aware.'

2. **Language and semantics**
   This system is presented in [Wijesekera, 1990]. We deal with the propositional part since we have not yet dealt with quantified versions. The language contains the intuitionistic connectives $\wedge, \vee, \Rightarrow, \neg$ and the modal connectives $\Box$ and $\Diamond$.
   The structures have the form $(S, \leq, R, h)$ where S is a non-empty set of worlds, $\leq$ is a partial order on S and R is any binary relation on S. h is the assignment satisfying for all atomic q the hereditary condition
   (*) $t \in h(q)$ and $t \leq s$ imply $s \in h(q)$.
   Satisfaction is defined as follows, for $s \in S$ and wff A:
   - $s \vDash q$ iff $s \in h(q)$, for q atomic
   - $s \vDash A \vee B$ iff $s \vDash A$ or $s \vDash B$
   - $s \vDash A \wedge B$ iff $s \vDash A$ and $s \vDash B$
   - $s \vDash A \Rightarrow B$ iff for all s', $s \leq s'$ and $s \vDash A$ imply $s' \vDash B$
   - $s \vDash \neg A$ iff for all s', $s \leq s'$ implies *not* $s \vDash A$
   - $s \vDash \Box A$ iff for all s' and s'' such that $s \leq s'$ and $s'Rs''$ we have that $s'' \vDash A$
   - $s \vDash \Diamond A$ iff for all s' such that $s \leq s'$ there exists an s'' such that $s'Rs''$ and $s'' \vDash A$
   - $A \vdash B$ iff for all structures and all s, if $s \vDash A$ then $s \vDash B$.

   Note that the hereditary condition (*) holds for all wffs in this definition.
3. **Hilbert axiomatic formulation**
   The author proposes the following Hilbert system for his logic, as follows:

   Axioms:
   (a) axioms for intuitionistic logic
   (b) $\Box(A \Rightarrow B) \Rightarrow (\Box A \Rightarrow \Box B)$
   (c) $\Box(A \Rightarrow B) \Rightarrow (\Diamond A \Rightarrow \Diamond B)$
   (d) $\Box A \wedge \Diamond(A \Rightarrow B) \Rightarrow \Diamond B$
   (e) $\Diamond(A \wedge \neg A) \Rightarrow A \wedge \neg A$

   Rules of inference:
   (a)
   $$\frac{\vdash A; \vdash A \Rightarrow B}{\vdash B}$$
   (b)
   $$\frac{\vdash A}{\vdash \Box A}.$$

   A completeness theorem is proved.
4. **Discussion of the Wijesekera system in our methodology**
   To see how this system fits in with our methodology we examine the modal part

and the intuitionistic part separately. The modal part is **K** and the intuitionistic part is intuitionistic logic itself under the usual Kripke semantics.

Let us use our methodology and *dovetail* these two systems. Dovetailing means that Definition 4.1 applies here. Thus the semantic structures of the dovetailed logic have the form $(S, \leq, R_\Box, R_\Diamond)$ or $(S, \leq, R_{\Box,\Diamond})$ depending on whether we consider $\{\Box, \Diamond\}$ as coming from the same logic or as modalities from distinct copies of (the same) logic.

According to Definition 4.1, item 2, the following must hold, as part of the dovetailing process:

(DV1) $t \leq s \wedge sR_\Box t' \to tR_\Box t'$

(DV2) $t \leq s \wedge tR_\Diamond t' \to sR_\Diamond t'$.

If we dovetail with $\Box$, $\Diamond$, then it must satisfy both conditions.

We now claim that every dovetailed structure $(S, \leq, R, h)$ is also a Wijesekera structure and the two semantic evaluations, $t \vDash_W A$ (denoting satisfaction in Wijesekera structures) and $t \vDash_D A$ (denoting satisfaction in the dovetailed structure) in the structure coincide.

We prove this by structural induction. The important cases are $\Diamond A$ and $\Box A$.

- Assume $t \vDash_D \Box A$. Then for all s such that tRs we have $s \vDash_D A$ and hence $s \vDash_W A$. Let s be such that for some $x, t \leq x$ and xRs. Then by condition (1) we have tRs and hence $s \vDash_D A$. This proves that $t \vDash_W \Box A$.
- Assume $t \vDash_W \Box A$. Then for all x and s such that $t \leq x$ and xRs we have $s \vDash_W A$. Choose $x = t$ and use the induction hypothesis to get $t \vDash_D \Box A$.
- Assume $t \vDash_D \Diamond A$. Then for some s, tRs and $s \vDash_D A$. By the induction hypothesis, $s \vDash_W A$. By condition (2) on R and $\leq$, we get that for all x such that $t \leq x$, we have xRs. Hence $t \vDash_W \Diamond A$.
- Assume $t \vDash_W \Diamond A$. Then for all x such that $t \leq x$, there exists an s such that xRs and $s \vDash_W A$. Choose $x = t$ and use the induction hypothesis to get an s such that tRs and $s \vDash_D A$. Hence $t \vDash_D \Diamond A$.

The above means that if $A \nvDash_D B$ then $A \nvDash_W B$. This can be verified directly by showing that all the Wijesekera axioms are theorems of the dovetailed system.

Let us examine the axioms proposed. Since in intuitionistic logic

$$A, A \to B \vdash B$$

we can use rule (2a) of Theorem 4.3 and get

$$\begin{array}{c}\Box A, \Diamond(A \Rightarrow B) \vdash \Diamond B \\ \Diamond A, \Box(A \Rightarrow B) \vdash \Diamond B \\ \Box A, \Box(A \Rightarrow B) \vdash \Box B.\end{array}$$

Using the deduction theorem we get the first three axioms of Wijesekera.

The fourth axiom is basically

$$\Diamond\bot \Rightarrow \bot.$$

Semantically this means that for all t

$$t \not\models \Diamond\bot.$$

This certainly holds in any modal semantics. Now since $\Diamond\bot \vdash \bot$ holds in the modal logic it holds in the combined logic and by rule (2b) of theorem 4.3 we get $\vdash \Diamond\bot \Rightarrow \bot$.

Conversely, the dovetailed system is the smallest logic which can prove the rules of 4.1. It is obvious that the Wijesekera system can do that.

5. **Conclusion**
   The Wijesekera system coincides with the dovetailed system of intuitionistic $\Rightarrow$ with the **K** modalities $\{\Box, \Diamond\}$.
   It is hence also decidable and has the finite model property by Theorems 4.9 and 4.10.

Example 5.2 (The system of F. B. Fitch).

1. **Motivation**
   Fitch [1948] presents axiomatically an intuitionistic modal logic **M** with $\{\neg, \wedge, \vee, \Rightarrow, \Box, \Diamond\}$. He discusses its formal properties but does not motivate it. We shall present here the propositional part.
2. **Hilbert formulation**
   The following are the axioms of **M**.
   (a) Axioms for the intuitionistic propositional calculus
   (b) $\Box A \Rightarrow A$
   (c) $\Box(A \Rightarrow B) \Rightarrow (\Box A \Rightarrow \Box B)$
   (d) $A \Rightarrow \Diamond A$
   (e) $\Box(A \Rightarrow B) \Rightarrow (\Diamond A \Rightarrow \Diamond B)$.
3. **Discussion in our methodology**
   Of course we do not have semantics for the system and must rely on the axioms alone. I would take as a starting point the dovetailing of intuitionistic logic with $\{\Box, \Diamond\}$ of modal logic **T**. Using Theorem 4.3 we will immediately get the axioms of Fitch above, plus the two axioms

$$\Box A \Rightarrow A \text{ from } \Box A \vdash A$$
$$A \Rightarrow \Diamond A \text{ from } A \vdash \Diamond A.$$

The dovetailing yields two extra axioms. The first is

$$\Box A \wedge \Diamond(A \Rightarrow B) \Rightarrow \Diamond B.$$

This is provable in Fitch because $\vdash A \Rightarrow ((A \Rightarrow B) \Rightarrow B)$ and hence $\vdash \Box A \Rightarrow \Diamond((A \Rightarrow B) \Rightarrow B)$ and hence $\vdash \Box A \Rightarrow (\Diamond(A \Rightarrow B) \Rightarrow \Diamond B)$. The second group of dovetailing axioms has the form $M_1 A \Rightarrow M_2 A$, where M_1 and M_2 are modalities. For example $\Diamond\bot \vdash \bot$. If these are provable in the system of Fitch, then it is the same as dovetailing intuitionistic logic with modal **T**. Otherwise it is weaker. I do not know the answer.

4. **Conclusion**
   Fitch's system is weaker or equal to the system resulting from dovetailing the intuitionistic $\Rightarrow$ with the **T** modality $\Box$. If certain modal reductions are provable in Fitch's system then there is equality. In such a case the system is decidable and has the finite model property by Theorems 4.9 and 4.10.

Example 5.3 (The system of A. N. Prior and R. A. Bull).

1. **Motivation and presentation**
   In [1957] A. N. Prior introduced an intuitionistic modal logic, **MIPC**, as a Hilbert system by adding to intuitionistic propositional logic the following rules for $\Box$ and $\Diamond$.
   (a)
   $$\frac{\vdash A \Rightarrow B}{\vdash \Box A \Rightarrow B}$$
   (b)
   $$\frac{\vdash A \Rightarrow B}{\vdash A \Rightarrow \Box B}$$
   provided every occurrence of a propositional variable in A is within a scope of a modal connective
   (c)
   $$\frac{\vdash A \Rightarrow B}{\vdash A \Rightarrow \Diamond B}$$
   (d)
   $$\frac{\vdash A \Rightarrow B}{\vdash \Diamond A \Rightarrow B}$$
   provided every occurrence of a propositional variable in B is within a scope of a modal connective.
2. **Semantics through translation**
   The above system can be translated into monadic intuitionistic predicate logic by taking the translation $A \mapsto [A](x)$, x a variable as follows:
   - $[q](x) = Q(x)$, for q atomic and Q a corresponding monadic predicate
   - $[A \wedge B](x) = [A](x) \wedge [B](x)$
   - $[A \vee B](x) = [A](x) \vee [B(x)]$
   - $[A \Rightarrow B](x) = [A](x) \Rightarrow [B](x)$
   - $[\neg A](x) = \neg([A](x)]$
   - $[\Box A](x) = \forall y([A](y))$
   - $[\Diamond A](x) = \exists y([A](y))$.

   Bull proves that A is a theorem of the intuitionistic modal logic iff $[A]$ is a theorem of monadic intuitionistic predicate logic.
   Bull also considers the additional axiom D
   $$D : \Box(\Box A \vee B) \Rightarrow (\Box A \vee \Box B).$$

3. **Discussion of Bull's system**
   This system follows the methodology of translating one logic into another and combining the two logics in the translation. This method was presented in Example 3.7, item 4. Bull's logic is extensively studied by Fischer-Servi and so we can discuss it there; see the discussion at the end of Example 5.5, which will let us conclude the following.
4. **Conclusion**
   Bull's logic is the result of dovetailing intuitionistic $\Rightarrow$ with **S5** modality $\Box$.
   The system is therefore decidable and has the finite model property by Theorems 4.9 and 4.10.

Example 5.4 (The approach of Kit Fine). In a lecture in 1987, Kit Fine proposed obtaining intuitionistic modal logic via the standard translation into classical logic, by weakening classical logic. If the standard translation of the metapredicate $t \vDash \Box A$ is $||\Box A||_t$ then $||\Box A||_t = \forall s(tRs \rightarrow ||A||_s)$. Thus any modal wff can be inductively translated into intuitionistic logic as follows:

- $||q||_x = Q(x)$, q atomic, Q unary
- $||A \wedge B||_x = ||A||_x \wedge ||B||_x$
- $||A \vee B||_x - ||A||_x \vee ||B||_x$
- $||A \rightarrow B||_x = ||A||_x \Rightarrow ||B||_x$
- $||\Box A||_x = \forall y(xRy \Rightarrow ||A||_y)$
- $||\Diamond A||_x = \exists y(xRy \wedge ||A||_y)$
- $||\bot||_x = \bot$.

Note that when a wff A of modal logic is translated into a wff of intuitionistic logic, we must pay attention to the particular way the translation is done. Classically equivalent translations are not intuitionistically the same.

If we look at the Kripke semantics for intuitionistic predicate logic, we get semantics for intuitionistic modality.

It is not clear whether the system obtained by translating modal logic into classical logic and then weakening it to be intuitionistic logic yields the same result as dovetailing.

Example 5.5 (The systems of G. Fischer–Servi).

1. **Motivation**
   The author, in [Fischer-Servi, 1977], tries to answer the following formal question:

   > 'Can we find a general criterion that will give us "the" intuitionistic analogue of some of the most usual modal systems? The problem as stated is of a technical nature and therefore the philosophical issues relating to the plausibility of an intuitionistic logic of modality will be ignored in this context.'

   The author notes correctly that adding the axioms of modality to those of intuitionistic logic faces the problem that equivalent modal formulations, when added to intuitionistic logic, give rise to different new systems.

The author proposes to make use of known translations of intuitionistic logic into modal logic and proposes to obtain intuitionistic modal logic as a subsystem of bimodal logic.

2. **Semantic translation of the system**
Let $\Box_1$ be $\mathbf{L}_1$ modality satisfying the Hilbert axioms for the logic $\mathbf{L}_1$. We can let $\mathbf{L}_1$ be **S4** in which case the axioms are:
   - $\Box_1 A \to A$
   - $\Box_1(A \to B) \to (\Box_1 A \to \Box_1 B)$
   - $\Box_1 A \to \Box_1\Box_1 A$
   - $\vdash A$ implies $\vdash \Box_1 A$.

   Let $\Box_2$ be $\mathbf{L}_2$ modality satisfying the Hilbert axioms for $\mathbf{L}_2$. We can choose $\mathbf{L}_2$ to be **S5** in which case the axioms are:
   - $\Box_2 A \to A$
   - $\Box_2(A \to B) \to (\Box_2 A \to \Box_2 B)$
   - $\Diamond_2 A \to \Box_2\Diamond_2 A$
   - $\vdash A$ implies $\vdash \Box_2 A$.

   Let $(\mathbf{L}_1 \otimes \mathbf{L}_2)$ (in our case we get $(\mathbf{S4} \otimes \mathbf{S5})$) be the bimodal logic with both connectives $\Box_1$ and $\Box_2$ and the above axioms for any wff of the combined language together with the following connecting axioms $\mathbf{C}$:
   - $\Box_1\Box_2 A \to \Box_1\Box_1 A$.

   For the case of $\mathbf{L}_2 = \mathbf{S5}$, the author, G. Fischer-Servi, also takes:
   - $\Diamond_2\Box_1 A \to \Box_1\Diamond_2 A$.

   The intuitionistic modal logic $\{\Rightarrow, \Box\}$ is obtained by translating $\Rightarrow$ into $\mathbf{L}_1$ and $\Box$ into $\mathbf{L}_2$ as follows:
   - $\tau(q) = \Box_1 q$, q atomic
   - $\tau(A \wedge B) = \tau(A) \wedge \tau(B)$
   - $\tau(A \vee B) = \tau(A) \vee \tau(B)$
   - $\tau(\neg A) = \Box_1 \sim \tau(A)$
   - $\tau(A \Rightarrow B) = \Box_1(\tau(A) \to \tau(B))$
   - $\tau(\Diamond A) = \Diamond_2 \tau(A)$
   - $\tau(\Box A) = \Box_1\Box_2 \tau(A)$.

   For each $\mathbf{L}_1, \mathbf{L}_2$ and each set of connecting (interaction) axioms $\mathbf{C}$ we can get an intuitionistic modal system $\mathbf{L}$ by letting

$$\mathbf{L} \vdash A \text{ iff (definition) } (\mathbf{L}_1 \otimes \mathbf{L}_2, \mathbf{C}) \vdash \tau(A).$$

   In another paper [Fischer-Servi, 1980], the author provides semantics for these intuitionistic modal logics. The semantics has the following form:
   - A *double modal structure* (for $\mathbf{L}_1$ and $\mathbf{L}_2$ modalities) has the form (S, R_1, R_2) where R_1, R_2 are binary relations on S (S is non-empty) satisfying the accessibility requirements of $\mathbf{L}_1, \mathbf{L}_2$ resp., and the following connecting conditions:

   (a) $R_1 \circ R_2 \subseteq R_2 \circ R_1$

(b) $mR_1x \wedge mR_2y \to \exists u(xR_2u \wedge yR_1u)$

In case $\mathbf{L}_1$ is **S4** we have R_1 is reflexive and transitive and in case $\mathbf{L}_2$ is **S5** we have R_2 is an equivalence relation.
Under an assignment h, satisfaction is defined as follows for the modalities:

- $t \vDash \Box_1 A$ iff for all s, tR_1s implies $s \vDash A$
- $t \vDash \Box_2 A$ iff for all s, tR_2s implies $s \vDash A$.

The τ-interpretation of the intuitionistic modal system $\{\Rightarrow, \Box\}$ into the bimodal logic $(\mathbf{L}_1 \otimes \mathbf{L}_2)$ induces a semantical interpretation for it.
These induced intuitionistic modal structures have the form (S, R_1, R_2, h) where (S, R_1, R_2) is as before (a double modal structure with R_1 reflexive and transitive and the assignment h satisfies the hereditary condition:
$(*)$ $\quad t \in h(q) \wedge tR_1s$ implies $s \in h(q)$, for each atomic q.
Satisfaction can be defined as follows:

- $t \vDash q$ iff $t \in h(q), q$ atomic
- $t \vDash \neg A$ iff for all s such that tR_1s we have *not* $s \vDash A$
- $t \vDash A \vee B$ iff $t \vDash A$ or $t \vDash B$
- $t \vDash A \wedge B$ iff $t \vDash A$ and $t \vDash B$
- $t \vDash A \Rightarrow B$ iff for all s such that tR_1s and $s \vDash A$ we have $s \vDash B$
- $t \vDash \Diamond A$ iff for some s such that tR_2s we have $s \vDash A$
- $t \vDash \Box A$ for all s such that there exists a t' with tR_1t' and $t'R_2s$ we have $s \vDash A$
- $A \vdash B$ iff for all structures and all t, if $t \vDash A$ then $t \vDash B$.

3. **Axiomatisation**

In another paper, [Fischer-Servi, 1984], the author provides Hilbert axioms for some of the above systems as follows:

(a) axioms for intuitionistic logic
(b) $\Diamond(A \vee B) \Rightarrow \Diamond A \vee \Diamond B$
(c) $\Box A \wedge \Box B \Rightarrow \Box(A \wedge B)$
(d) $\neg\Diamond(A \wedge \neg A)$
(e) $\Diamond(A \Rightarrow B) \Rightarrow (\Box A \Rightarrow \Diamond B)$
(f) $(\Diamond A \Rightarrow \Box B) \Rightarrow \Box(A \Rightarrow B)$
(g) From A and $A \Rightarrow B$ infer B
(h) From $A \Rightarrow B$ infer $\Diamond A \to \Diamond B$
(i) From $A \Rightarrow B$ infer $\Box A \Rightarrow \Box B$.

The above system is supposed to introduce the **K** modality $\Box$ into intuitionistic logic and corresponds to the τ-translation into $(\mathbf{S4} \otimes \mathbf{K})$, i.e. $\mathbf{L}_1 = \mathbf{S4}, \mathbf{L}_2 = \mathbf{K}$.
If we strengthen $\mathbf{L}_2$ to be **T**, we need to add

(j) $\Box A \Rightarrow A$
(k) $A \Rightarrow \Diamond A$.

If we add the transitivity axiom to $\mathbf{L}_2$, the corresponding axioms are

(l) $\Box A \Rightarrow \Box\Box A$
(m) $\Diamond\Diamond A \Rightarrow \Diamond A$

and symmetry in $\mathbf{L}_2$ corresponds to

(n) $\Diamond\Box A \Rightarrow A$

(o) $A \Rightarrow \Box\Diamond A$.

If $\mathbf{L}_2$ is **S5**, we need to add

(p) $\Diamond A \Rightarrow \Box\Diamond A$

(q) $\Diamond\Box A \Rightarrow \Box A$

The author proves completeness.

4. **Discussion of the Fischer–Servi approach**

This approach is a combination of dovetailing two modalities, $\Box_1$ and $\Box_2$ with some interaction axioms and then translating intuitionistic logic into one modality ($\Box_1$) to obtain intuitionistic modal logic via the translation. Although Fischer-Servi seems to have done a neat and thorough job, in the light of our methodology, some aspects of it seem arbitrary.

First why go through the bimodal logic route? We can dovetail directly intuitionistic logic with any modality. In fact looking at the Fischer-Servi axioms for intuitionisitc modal **K**, given in (3) above, we recognise all the dovetailing axioms (or Wijesekera axioms) with the exception of (b) and (f).

(b) is valid in the dovetailed semantics and is therefore provable. (f) is an additional axiom which probably arises from the interaction conditions between the modalities which Fischer-Servi postulated, and can be considered in our dovetailing approach as an interaction axiom. In fact, it is a very natural one. It looks like the axiom corresponding to the following condition on our dovetailed semantics:

(FS) $\quad tRs \wedge s \leq x \to tRx.$

This is a natural condition to ask for. Compare it to the dovetailing requirement (DV1) below which $\leq$ and R satisfy:

(DV1) $\quad t \leq s \wedge sRx \to tRx.$

The additional modalities axioms, (j)–(q) naturally yield dovetailing of stronger modal logics, as discussed in Example 4.7,

To summarise, how do we present the Fischer-Servi intuitionistic modal logics in our framework? Very simply. We dovetail intuitionistic logic with either **K**, **T**, **S4** or **S5**. For each one of these systems we investigate the axiom

(g) $\quad (\Diamond A \Rightarrow \Box B) \Rightarrow \Box(A \Rightarrow B)$

which probably corresponds to the condition[15]

(FS) $\quad tRs \wedge s \leq x \to tRx.$

That is all!

Note that the case of dovetailing intuitionistic logic with **S5** and adding this axiom gives us Bull's logic, as Fischer-Servi has shown. Thus we know what Bull's logic looks like in our aproach, thanks to Fischer-Servi.

The cases of bimodal logics of Fischer-Servi also dovetail two modalities with natural aditional interaction axioms, namely single direction commutativity $\Box_1 \Box_2 \to \Box_2\Box_1$.

[15] I did not prove completeness but it seems likely to be true.

Note in passing that bidirectional commutativity ($\Box_1\Box_2 = \Box_2\Box_1$) yields two-dimensional modal logic.

5. **Conclusion**

   The Fischer-Servi logics are probably the dovetailing of intuitionistic $\Rightarrow$ with modality $\Box$ of various systems plus an interaction axiom.

Example 5.6 (The systems of H. Ono).

1. **Logics obtained from translation**

   Ono [1987] considers formally the possibilities offered by the translation of Bull of intuitionistic modal logic into the monadic intuitionistic predicate logic. He proposes two systems $\mathbf{M_H}$ and $\mathbf{M_D}$ defined as follows:

   $\mathbf{M_H}$ is obtained from propositional intuitionistic logic by adding the following axioms for the modalities $\Box$ and $\Diamond$:

   (a) $\Box A \Rightarrow A$
   (b) $A \Rightarrow \Diamond A$
   (c) $\Box A \Rightarrow \Box\Box A$
   (d) $\Diamond\Diamond A \Rightarrow \Diamond A$
   (e) $\Box(A \Rightarrow B) \Rightarrow (\Box A \Rightarrow \Box B)$
   (f) $\Box(A \Rightarrow B) \Rightarrow (\Diamond A \Rightarrow \Diamond B)$
   (g) $\Diamond(A \vee B) \Rightarrow (\Diamond A \vee \Diamond B)$
   (h) $\Diamond A \Rightarrow \Box\Diamond A$
   (i) $\Diamond\Box A \Rightarrow \Box A$
   (j) $(\Box A \Rightarrow \Box B) \Rightarrow \Box(\Box A \Rightarrow \Box B)$
   (k) The rules are modus ponens, substitution and necessitation.

   This system is the same as the Prior–Bull system of Example 5.3 and the theorem there holds.

   One can take any quantificational axiom X and add it to the monadic intuitionistic logic, giving rise to an intermediate logic $\mathbf{X}$. Through the translation $A \mapsto [A]$ we get a stronger modal logic $\mathbf{M_X}$. We can ask whether the system can also be obtained by adding some additional modal axiom Y to the axioms of $\mathbf{M_H}$, so that the following holds:

   $$\mathbf{M_H} + Y \vdash A \text{ iff monadic predicate } \mathbf{X} \vdash [A].$$

   Ono answers yes for the case Y = the axioms D of Bull, see Example 5.3, and X being the axiom CD of constant domains.

   $$CD = \forall x(A(x) \vee B) \Rightarrow (\forall x A(x) \vee B), x \text{ not in } B.$$

2. **Axiomatic approach**

   In his paper [1977], Ono considers several intuitionistic modal logics presented axiomatically, containing $\Box$ only. The basic system, which he calls $\mathbf{L_0}$, is obtained by adding the following three axiom schema to those of intuitionistic logic,

(a) $\Box A \Rightarrow A$
(b) $\Box A \Rightarrow \Box\Box A$
(c) $\Box(A \Rightarrow B) \Rightarrow (\Box A \Rightarrow \Box B)$

together with the rule of necessitation. Clearly if the law of excluded middle is added to intuitionistic logic, then $\mathbf{L}_0$ becomes **S4**.
Ono considers the following additional axioms:

$$A_1 : \neg\Box A \Rightarrow \Box\neg\Box A$$
$$A_2 : (\Box A \Rightarrow \Box B) \Rightarrow \Box(\Box A \Rightarrow \Box B)$$
$$A_3 : \Box(\Box A \vee B) \Rightarrow (\Box A \vee \Box B)$$
$$A_4 : \Box A \vee \Box\neg\Box A.$$

It is easy to see that classically, $\mathbf{S4} + A_i = \mathbf{S5}$, for each i.
Note also that $\mathbf{L}_0 + A_2$ is the $\Box$ fragment of **MIPC** of Bull, see Example 5.3.
Ono proposes semantics for these logics. His I-structures have the form $(S, \leq, R, h)$ with $\leq$ a partial order on S and R a reflexive and transitive relation on S. The following holds:
(Ono): $x \leq y$ implies xRy.
Assignment h must satisfy:

- $t \in h(q)$ and $t \leq s$ imply $s \in h(q)$
  Satisfaction is defined as usual
- $t \vDash q$ iff $t \in h(q)$ for q atomic
- $t \vDash A \wedge B$ iff $t \vDash A$ and $t \vDash B$
- $t \vDash A \vee B$ iff $t \vDash A$ or $t \vDash B$
- $t \vDash \neg A$ iff for any s such that $t \leq s$, *not* $s \vDash A$
- $t \vDash A \Rightarrow B$ iff for all s such that $t \leq s$ and $s \vDash A$ we have $s \vDash B$
- $t \vDash \Box A$ iff for all s such that tRs we have $s \vDash A$.

The above semantics is sound for $\mathbf{L}_0$.
We now add conditions on I-structures:
(0) Type 0 structures have no additional conditions.
(1) Type 1 structures satisfy

$$xRy \rightarrow \exists y'(x \leq y' \wedge yRy').$$

(2) Type 2 structures satisfy

$$xRy \rightarrow \exists y'(x \leq y' \wedge yRy' \wedge y'Ry).$$

(3) Type 3 structures satisfy

$$xRy \rightarrow \exists x'(xRx' \wedge x'Rx \wedge x' \leq y).$$

(3,j) Type (3, j) structures satisfy both condition (3) and condition $(j = 1, 2)$
(4) Type 4 structures satisfy the condition that R is symmetric.

Note that the semantics obtained by Bull from the translation of **MIPC** into intuitionistic predicate logic in Example 5.3 is of Type 2 (if we ignore $\Diamond$).
Ono proves completeness of the logic $\mathbf{L}_0 + A_i$ for structures of type i and $\mathbf{L}_0 + A_3 + A_j$, for structures of type (3, j), $j = 1, 2$.
Ono continues to add $\Diamond$ to his systems. He considers the following axioms for $\Diamond$, which we call A_*:

(a) $A \Rightarrow \Diamond A$
(b) $\Diamond\Diamond A \Rightarrow \Diamond A$
(c) $\Box(A \Rightarrow B) \Rightarrow (\Diamond A \Rightarrow \Diamond B)$
(d) $\Diamond(A \vee B) \Rightarrow (\Diamond A \vee \Diamond B)$
(e) $\Diamond A \Rightarrow \Box\Diamond A$
(f) $\Diamond\Box A \Rightarrow \Box A$.

Ono shows that $\mathbf{L}_0 + A_2 + A_*$ is a conservative extension of $\mathbf{L}_0 + A_2$ and $\mathbf{L}_0 + A_2 + A_3 + A_*$ is a conservative extension of $\mathbf{L}_0 + A_2 + A_3$. Note that $\mathbf{L}_0 + A_2 + A_*$ is equivalent to Bull's **MIPC**.
Semantically, the truth condition for $\Diamond$ in type 2 structures is defined by

$$t \vDash \Diamond A \text{ iff for some } s \text{ such that } tRs \wedge sRt \text{ we have } s \vDash A.$$

3. **Discussion of Ono systems in our approach**
The Ono system $\mathbf{L}_0$, for modality $\Box$ only, is nothing but dovetailing intuitionistic logic with the **S4** modality $\Box$. The dovetailing condition on the structure (DV1)
(DV1) $\quad x \leq y \wedge yRz \rightarrow xRz$
reduces to the Ono condition because of reflexivity of R. Conversely, the Ono condition implies (DV1) because of the transitivity of R. Thus $\mathbf{L}_0$ is the dovetailed logic with $\Box$ of **S4**. The additional Ono axioms A_1–A_4 can be regarded from our point of view as interaction axioms to be studied. Whether they are interesting or not depends on whether they correspond to interesting and meaningful semantic conditions. His axiom A_4 corresponds to a pure condition on R (R is symmetric) and we can therefore immediately conclude that his system $\mathbf{L}_0 + A_4$ is the dovetailing of intuitionistic logic with the modal $\Box$ of $\mathbf{K} + \Diamond\Box A \rightarrow A$. It is interesting to note that in the intuitionistic context it can be axiomatised without $\Diamond$.
One can try and see whether some of the Ono systems $\mathbf{Y}$ can be related to some dovetailed systems $\mathbf{X}$ by presenting them as $\mathbf{Y} = \mathbf{X} + B$ with some simpler interaction axioms B. Thus our dovetailing can give familiar reference points to the intuitionistic modal logic landscape.
Ono adds $\Diamond$ to his systems but commits himself to a 'symmetric' truth condition for $\Diamond A$. In our framework there are clean ways of adding $\Diamond$. Here again we see the value of systematic methods for combining logics.
4. **Conclusion**
Some of Ono's systems are clear dovetailing of $\Rightarrow$ and $\Box$. It is obvious that our point of view helps place his other systems into a coherent map.

Decidability and finite the model property will also follow by Theorems 4.9 and 4.10.

Example 5.7 (The systems of W. B. Ewald). Ewald [1986] constructs intuitionistic tense logics. He adds to the connectives of propositional intuitionistic logic the temporal connectives F, P, G and H.

The structures have the form $(S, \leq, T^s, R^s, h^s), s \in S$, where $(S, \leq)$ is a partially ordered set and for each $s \in S, (T^s, R^s)$ is a temporal structure, with T^s the set of moments of time and R^s is a binary relation on T^s. We require

$$s_1 \leq s_2 \to T^{s_1} \subseteq T^{s_2}$$
$$s_1 \leq s_2 \to R^{s_1} \subseteq R^{s_2}.$$

h^s is an assignment in the temporal structure T^s, $(h^s(q) \subseteq T^s)$, and for atomic q we require

$$s_1 \leq s_2 \to h^{s_1}(q) \subseteq h^{s_2}(q).$$

Satisfaction is two dimensional:

- $(s,t) \vDash q$ iff $t \in h^s(q)$, for q atomic.
- $(s,t) \vDash A \wedge B$ iff $(s,t) \vDash A$ and $(s,t) \vDash B$.
- $(s,t) \vDash A \vee B$ iff $(s,t) \vDash A$ or $(s,t) \vDash B$.
- $(s,t) \vDash \neg A$ iff for all $s', s \leq s'$ implies *not* $(s',t) \vDash A$.
- $(s,t) \vDash A \Rightarrow B$ iff for all $s', s \leq s'$ and $(s',t) \vDash A$ imply $(s',t) \vDash B$.
- $(s,t) \vDash PA$ iff for some $t', t'R^st$ and $(s,t') \vDash A$.
- $(s,t) \vDash FA$ iff for some t', tR^st' and $(s,t') \vDash A$.
- $(s,t) \vDash HA$ iff for all $t', s', s \leq s'$ and $t' \in T^{s'}$ and $t'R^{s'}t$ imply $(s',t') \vDash A$.
- $(s,t) \vDash GA$ iff for all $s', t', s \leq s'$ and $t' \in T^{s'}$ and $tR^{s'}t'$ imply $(s',t') \vDash A$.
  One can show that for any A
- $(s,t) \vDash A$ and $s \leq s'$ imply $(s',t) \vDash A$.
- $A \vdash B$ iff for all structures and all (s,t), if $(s,t) \vdash A$ then $(s,t) \vDash B$.

Ewald offers a Hilbert axiom system for the logic. The axioms are the following:

1. all intuitionistic axioms
2. $G(A \Rightarrow B) \Rightarrow (GA \Rightarrow GB)$
3. $G(A \wedge B) \Leftrightarrow (GA \wedge GB)$
4. $F(A \vee B) \Leftrightarrow (FA \vee FB)$
5. $G(A \Rightarrow B) \Rightarrow (FA \Rightarrow FB)$
6. $GA \wedge FB \Rightarrow F(A \wedge B)$
7. $G\neg A \Rightarrow \neg FA$
8. $FHA \Rightarrow A$
9. $A \Rightarrow GPA$
10. $(FA \Rightarrow GB) \Rightarrow G(A \Rightarrow B)$
11. $F(A \Rightarrow B) \Rightarrow (GA \Rightarrow FB)$
12. The mirror image axioms of (2)–(11) above

13. Modus ponens for $\Rightarrow$ and necessitation for G and H.

Ewald notes that should we add the 'constant domain' condition on structures, i.e. for all $s_1, s_2, T^{s_1} = T^{s_2}$, we need the following additional rules

$$\frac{A \vdash B \vee C}{GA \vdash FB \vee GC}$$

and

$$\frac{A \vdash B \vee C}{HA \vdash PB \vee HC}.$$

Discussion

It is quite obvious that we have here an example of fibring temporal logic with intuitionistic logic.

Example 5.8 (The systems of J. M. Font). Font's (see his [1986]) starting point for intuitionistic modal logics is Gödel's proposal to consider $\Box$ as primitive **S4** modality and define $\Diamond$ as $\neg\Box\neg$. Font is not happy with the abundance of intuitionistic modal systems in the literature in which both $\Box$ and $\Diamond$ are independent primitive connectives. The paper is mainly algebraic in its approach, and is concerned with several systems and how many different modalities they have.

The basic system is **IM4** with the following axioms

1. $\Box A \Rightarrow A$
2. $\Box(A \Rightarrow B) \Rightarrow (\Box A \Rightarrow \Box B)$
3. $\Box A \Rightarrow \Box\Box A$
4. $\dfrac{\vdash A}{\vdash \Box A}$.

The author considers several extensions of **IM4** with axioms that classically yield **S5** if added to **S4**:

- The system **IM4W** with the axiom:
  $W: \quad \Diamond\neg\Diamond A \Rightarrow \neg\Diamond A$
  or using $\Box$
  $\neg\Box\neg\Box A \Rightarrow \neg\neg\Box\neg A.$
- The system **IM4M** with the axiom
  $M \quad \neg\Box A \Rightarrow \Box\neg\Box A$
  or equivalently the rule

  $$\frac{\Diamond p \Rightarrow q}{p \Rightarrow \Box q}.$$

  IM4M is a proper extension of **IM4W**.
- The system **IM4S** extends **IM4** with the extra axiom

  $$S: \quad \Box(\Box A \Rightarrow B) \Leftrightarrow (\Box A \Rightarrow \Box B).$$

  IM4S is a proper extension of **IM4M**.

- **IM5** is the extension of **IM4** with the axiom
  $\neg\Box\neg\Box A \to \Box A$
  IM5 is a proper extension of **IM4S**.

Font also shows that **IM5** is the extension of Bull's **MIPC** (see Example 5.3) with the axiom $\Diamond A \Leftrightarrow \neg\Box\neg A$.

Discussion
The basic system **IM4** of Font is the same as $\mathbf{L}_0$ of Ono: it is the dovetailing of intuitionistic logic with $\Box$ of **S4**.

The stronger systems we view as extensions. They seem arbitrary so why consider these extensions?

Since we have dovetailed semantics for the basic system, we can try and find conditions corresponding to the axioms. Recall that the structures for $\mathbf{L}_0$ (which is **IM4**) have the form $(S, \leq, R, h)$ with h hereditary, R reflexive and transitive satisfying (DV1) or the (Ono) condition

$$x \leq y \to xRy.$$

Consider the Font axiom:

M $\quad \neg\Box A \Rightarrow \Box\neg\Box A.$
My guess is that the corresponding condition for it is

(M) $\quad xRy \wedge y \leq u \to x \leq u.$
This should be compared with the Fischer-Servi condition:

(FS) $\quad xRy \wedge y \leq u \to xRu.$

The conditions for the other axioms may not be as simple.

It is obvious now that if we use the dovetailed logics as milestones and investigate natural conditions on the semantics as above, we get a 'coordinate system' of well understood logics which 'maps' the area. This was our original claim at the beginning of this section.

Example 5.9 (The systems of Nobu-Yuki Suzuki). Nobu-Yuki Suzuki [1988; 1990] considered the following problem: given an intuitionistic modal logic **X** and an intermediate monadic intuitionistic predicate logic **Y**, when do we have for any wff A

$$\mathbf{X} \vdash A \text{ iff } \mathbf{Y} \vdash [A]$$

where $[A]$ is the translation in Example 5.3.

The author gives several such pairs using algebraic analysis.

Example 5.10 (The systems of M. Božić and K. Došen). The authors set out to investigate, in [Došen, 1985; Božić and Došen, 1984], intuitionistic analogues of normal propositional modal logics. The approach is formal, an exercise in logical systems. It seems the authors were trying to organise the area a bit, facing an abundant variety of intuitionistic modal logics.

The basic system of the authors is **HK**$\Box$ which extends intuitionistic propositional logic with $\Box$ and the following axioms and rules

$\Box$1: $\Box A \wedge \Box B \Rightarrow \Box(A \wedge B)$
$\Box$2: $\Box(A \Rightarrow A)$
R$\Box$: $\dfrac{\vdash A \Rightarrow B}{\vdash \Box A \Rightarrow \Box B}$.

The above systems can be equivalently formulated by the axiom

$$\Box(A \Rightarrow B) \Rightarrow (\Box A \Rightarrow \Box B)$$

and the rule

$$\frac{\vdash A}{\vdash \Box A}.$$

The proposed structures for this system have the form (S, R_I, R_M, h), where S is non-empty, R_I is a reflexive and transitive relation on S, R_M is a binary relation on S and the following holds

(BK1) $R_I \circ R_M \subseteq R_M \circ R_I$
(*) tR_Is and $t \in h(q) \rightarrow s \in h(q)$.

Satisfaction is defined as follows

- $t \vDash q$ iff $t \in h(q), q$ atomic
  $t \vDash A \wedge B$ iff $t \vDash A$ and $t \vDash B$
  $t \vDash A \vee B$ iff $t \vDash A$ or $t \vDash B$
- $t \vDash A \Rightarrow B$ iff for all s, tR_Is and $s \vDash A$ imply $s \vDash B$
- $t \vDash \neg A$ iff for all s, tR_Is implies *not* $s \vDash A$
- $t \vDash \Box A$ iff for all s, tR_Ms implies $s \vDash A$.
- $A \vDash B$ iff for all structures and all t, if $t \vDash A$ then $t \vDash B$.

The authors prove soundness and completeness of **HK**$\Box$ for the proposed semantics.

The authors consider a bimodal logic with two modalities, one **S4** and one **K** modality and embed their system **HK**$\Box$ in the bimodal logic. The structures have the form (S, R_I, R_M, h) where R_I, R_M are two relations on S. If $\Box_I, \Box_M$ are the two modalities then the satisfaction conditions are

- $t \vDash \Box_I A$ iff $\forall s(tR_Is \rightarrow s \vDash A)$
- $t \vDash \Box_R A$ iff $\forall s(tR_Ms \rightarrow s \vDash A)$.

The authors point out the following correlation between interaction axioms and conditions on the semantics.

- $\Box_M\Box_I A \rightarrow \Box_I\Box_M A$
  corresponds to $R_I \circ R_M \subseteq R_M \circ R_I$
- $\Box_M A \rightarrow \Box_I\Box_M A$
  corresponds to $R_I \circ R_M \subseteq R_M$
- $\Box_M A \rightarrow \Box_M\Box_I A$
  corresponds to $R_M \circ R_I \subseteq R_M$.

Classical logic can be turned into bimodal logic by adding the following axioms for modalities.

$$
\begin{array}{ll}
\Box_I 1: & \Box_I A \wedge \Box_I B \to \Box_I (A \wedge B) \\
\Box_I 2: & \Box_I (A \to A) \\
\Box_I 3: & \Box_I A \to A \\
\Box_I 4: & \Box_I A \to \Box_I \Box_I A \\
\Box_I : & \dfrac{\vdash A \to B}{\vdash \Box_I A \to \Box_I B} \\
\Box_M 1: & \Box_M A \wedge \Box_M B \to \Box_M (A \wedge B) \\
\Box_M 2: & \Box_M (A \to A) \\
R\Box_M : & \dfrac{\vdash A \to B}{\vdash \Box_M A \to \Box_M B} \\
\Box S4K: & \Box_M \Box_I A \to \Box_I \Box_M A.
\end{array}
$$

The system with the above axioms and rules is named **S4K**.

If we add the axiom

$$\Box S4Kc: \quad \Box_M A \to \Box_I \Box_M A$$

we get the system the authors call **S4Kc**.

If we add to **S4K** the axiom

$$\Box S4Sc: \quad \Box_M A \to \Box_M \Box_I A$$

we get the system **S4Ksc**.

The authors claim completeness of the bimodal systems for the respective semantic conditions.

HK□ can be translated into **S4K** as follows, where t is the translation:

$$
\begin{array}{l}
t(q) = \Box_I q, q \text{ atomic} \\
t(A \wedge B) = t(A) \wedge t(B) \\
t(A \vee B) = t(A) \vee t(B) \\
t(A \Rightarrow B) = \Box_I (t(A) \to t(B)) \\
t(\neg A) - \Box_I \sim t(A) \\
t(\Box A) = \Box_M t(A).
\end{array}
$$

The authors show the translation is faithful into **S4K** or any of its extensions presented above.

This means the **HK□** semantics can be strengthened.

The authors consider also the intuitionistic possibility system **HK◊**. $\Diamond$ is added to intuitionistic logic with the following axioms:

$$
\begin{array}{ll}
\Diamond 1: & \Diamond (A \vee B) \Rightarrow \Diamond A \vee \Diamond B \\
\Diamond 2; & \neg \Diamond \neg (A \Rightarrow A) \\
R\Diamond : & \dfrac{\vdash A \Rightarrow B}{\vdash \Diamond A \Rightarrow \Diamond B}.
\end{array}
$$

The structures for **HK◊** have the form (S, R_I, R_M, h), where R_I is a reflexive and transitive binary relation on S and R_M is a binary relation on S satisfying:

$$(BK2): \quad R_I^{-1} \circ R_M \subseteq R_M \circ R_I^{-1}.$$

h satisfies (*) as before. The truth condition for $\Diamond$ is the following

- $t \vDash \Diamond A$ iff for some $s, tR_M s$ and $s \vDash A$.

Conditions (BK1) and (BK2) are needed for intuitionistic heredity

- $t \vDash A$ and $tR_I s \to s \vDash A$.

Soundness and completeness is proved.

Similar embedding theorems into classical bimodal logic can be given.

The authors consider now intuitionistic modal logic **KH**$\Box\Diamond$ with both modalities. One possibility for semantics is structures with three possible world relations of the form (S, R_I, R_M, R'_M, h) where (S, R_I, R_M, h) satisfies the **HK**$\Box$ conditions and (S, R_I, h) satisfies the **HK**$\Diamond$ conditions. Another possibility is to give truth conditions for $\Box$ and $\Diamond$ in terms of R_I, R_M alone. This may require connecting axioms for $\Box$ and $\Diamond$.

The author suggests extending **KH**$\Box$ with the following axioms

$$\Box\Diamond 1 : \Diamond A \vee \Box\neg A$$
$$\Box\Diamond 2 : \neg(\Diamond A \wedge \Box\neg A).$$

The system is complete for the following additional conditions on R_I and R_M:

- $R_I \circ R_M = R_M \circ R_I = R_M$
- $R_I^{-1} \circ R_M \subseteq R_M \circ R_I^{-1}$.

The authors proceed to consider normal extensions of **KH**$\Box$. The following axioms are studied:

$$\Box D: \quad \neg\Box\neg(A \Rightarrow A)$$
$$\Box T: \quad \Box A \Rightarrow A$$
$$\Box 4: \quad \Box A \Rightarrow \Box\Box A$$
$$\Box 5: \quad (\Box A \Rightarrow \Box B) \Rightarrow \Box(\Box A \Rightarrow \Box B)$$
$$\Box 5.1: \neg\Box A \Rightarrow \Box\neg\Box A.$$

These axioms are used to define the following systems (which are increasingly stronger):

$$\mathbf{HD}\Box = \quad \mathbf{HK}\Box + \Box D$$
$$\mathbf{HT}\Box = \quad \mathbf{HK}\Box + \Box T$$
$$\mathbf{HS4}\Box = \quad \mathbf{HK}\Box + \Box T + \Box 4$$
$$\mathbf{HS5.1}\Box = \mathbf{HK}\Box + \Box T + \Box 4 + \Box 5.1$$
$$\mathbf{HS5}\Box = \quad \mathbf{HK}\Box + \Box T + \Box 5$$

The logics are proved complete for the following additional conditions on **HK**$\Box$ structures (S, R_I, R_M, h) where $R_\Box = R_M \circ R_I$.

KD$\Box$	corresponds to $R_\Box$ serial
HKT$\Box$	corresponds to $R_\Box$ reflexive
HKS4$\Box$	corresponds to $R_\Box$ reflexive and transitive
HKS5.1$\Box$	corresponds to $R_\Box$ reflexive, transitive and satisfying $R_\Box \subseteq R_I \circ R_\Box^{-1}$

HKS5□ corresponds to $R_\Box$ reflexive and satisfying $xR_\Box y \wedge yR_\Box z \rightarrow \exists t(xR_I t \wedge tR_\Box z \wedge yR_\Box t)$.

Similarly the authors consider extensions of **HK◊**. The axioms considered are:

$$\begin{array}{ll} \Diamond D: & \Diamond(A \Rightarrow A) \\ \Diamond T: & A \Rightarrow \Diamond A \\ \Diamond 4: & \Diamond\Diamond A \Rightarrow A \\ \Diamond 5: & \Diamond(\Diamond A \Rightarrow \Diamond B) \Rightarrow (\Diamond A \Rightarrow \Diamond B) \\ \Diamond 5.1: & \Diamond\neg\Diamond A \Rightarrow \neg\Diamond A. \end{array}$$

The following extensions of **HK◊** are defined by the authors in order of increasing strength

$$\begin{array}{ll} \mathbf{HD}\Diamond = & \mathbf{HK}\Diamond + \Diamond D \\ \mathbf{HT}\Diamond = & \mathbf{HK}\Diamond + \Diamond T \\ \mathbf{HS4}\Diamond = & \mathbf{HK}\Diamond + \Diamond T + \Diamond 4 \\ \mathbf{HS5.1}\Diamond = & \mathbf{HK}\Diamond + \Diamond T + \Diamond 4 + \Diamond 5.1 \\ \mathbf{HS4}\Diamond = & \mathbf{HK}\Diamond + \Diamond T + \Diamond 5. \end{array}$$

The authors offer corresponding semantics for these systems. Let $R_\Diamond$ be $R_M \circ R_I^{-1}$. We have the following correspondence:

HD◊ corresponds to $R_\Diamond$ serial

HT◊ corresponds to $R_\Diamond$ reflexive

HS4◊ corresponds to $R_\Diamond$ reflexive and transitive

HS5.1◊ corresponds to $R_\Diamond$ reflexive and transitive and satisfying $R_\Diamond \subseteq R_\Diamond^{-1} \circ R_I^{-1}$

HS5◊ corresponds to $R_\Diamond$ reflexive and transitive and satisfying $xR_\Diamond y \rightarrow \exists t$ $(yR_I t \wedge tR_\Diamond x \wedge xR_\Diamond t)$.

Let us compare these systems and semantics with the dovetailed case. The dovetailed structures have the form $(S, \leq R, h)$. $\leq$ corresponds to R_I and R corresponds to R_M. The authors satisfaction notion is the same as the dovetailed case. All we need to check are the additional conditions imposed on the relation, i.e. compare (DV1), (DV2) with the authors' (BK1) and (BK2).

(BK1): $xR_I y \wedge yR_M z \rightarrow \exists u(xR_M u \wedge uR_I z)$

(BK2): $yR_I x \wedge yR_M z \rightarrow \exists u(xR_M u \wedge zR_I u)$.

To compare the conditions, define R' in the dovetailed semantics by

$$xR'y \text{ iff } \exists u(xRu \wedge u \leq y)$$

i.e. $R'_M = R \circ \leq$.

Then $\leq, R'_M$ satisfies (BK1). This shows that **HK□** is the same as the dovetailed system. The relation with (BK2) is not immediate, but looking at the axioms systems I believe we have the dovetailed logic. Because of heredity, truth value in the dovetailed structure does not change if we use R'_M instead of R.

The bimodal systems proposed are either direct dovetails of two modalities or are obtained from such using simple interaction axioms.

The axioms added to **HK**$\Box$ and **HK**$\Diamond$ yield only conditions on $R_\Box$ (the accessibility relation) and correspond to the dovetailing of these systems.

For further discussion see the end of Section 15.3.

5.2 Multimodal Logics and Systems of Knowledge and Belief

Multimodal logics are systems which arise from attempts of modelling several agents. Each agent has its own modality. $\Box_i A$ reads something like 'A holds after agent i takes action'. They also arise in many agent's belief systems. $B_i A$ reads 'agent i believes A'. Usually an agreed standard modal logic is chosen for the agent (e.g. **K** or **K45**) and the semantics is a dovetailed version of several copies of the system.

Systems of knowledge and belief will have two operators KA and BA, dovtailed together with a simple interaction axiom such as $KA \to BA$.

We give below one typical example from the literature which is slightly more sophisticated than usual.

Example 5.11 (The system of Fagin, Halpern and Vardi). The authors set out to solve the logical omniscience problem. Technically, one needs to block the derivation

$$\frac{KA; \vdash A \to B}{KB}$$

where K is a knowledge operator.

The authors solve the problem by turning $\neg$ into a non-standard negation.

The models have the form $(S, R_1, \ldots, R_m, *, h)$ where S is the set of possible states, R_i are binary relations for the agent i knowledge operator K_i and $*$ is a unary function for negation $\neg$, satisfying $t^{**} = t$ for all t. h is the assignment.

Satisfaction is defined as follows:

- $t \vDash q$ iff $t \in h(q)$, for atomic q.
- $t \vDash A \to B$ iff $t \vDash A$ implies $t \vDash B$
- $t \vDash A \wedge B$ iff $t \vDash A$ and $t \vDash B$
- $t \vDash A \vee B$ iff $t \vDash A$ or $t \vDash B$
- $t \vDash K_i A$ iff for all s such that $tR_i s$ we have $s \vDash A$
- $t \vDash \neg A$ iff *not* $t^* \vDash A$
- $A \vDash B$ iff in every model and every t, if $t \vDash A$ then $t \vDash B$.

The authors show that this system is essentially equivalent to the systems of Lakemeyer [Lakemeyer, 1987] and Levesque [Levesque, 1984]. They define a new implication $A \Rightarrow B$ by letting $A \Rightarrow B$ be $\neg A \vee B$.

The authors axiomatise their system by taking **K** axioms for K_i and the following for $\neg$.

- $\neg\neg A \leftrightarrow A$
- $\neg(A \to B) \leftrightarrow ((\neg B \to \neg A) \to \bot)$

- $\neg(A \wedge B) \leftrightarrow (\neg A \rightarrow \bot) \rightarrow \neg B$

together with modus ponens and necessitation.

Discussion
The above system is exactly the dovetailing of several copies of modal **K** with the $\neg$ fragment of relevance logic.

In fact, the $\neg$ fragment of relevance logic is obtainable from a modality for a possible world constellation of the form $\{t_1, t_2, t_1 R t_2, t_2 R t_1\}$ with $\neg A$ = def.$\Box \sim A$ (where $\sim A$ is classical negation).

The system is therefore decidable and has the finite model property by Theorems 3.13 and 3.27.

Example 5.12 (The system DL of Jones and Pörn). A. J. I. Jones and I. Pörn, set out to solve some difficulties in formalising the intuitions and paradoxes involving what ought to be and what actually is. In a series of papers [Jones and Pörn, 1985; Jones and Pörn, 1986; Jones, 1993] the authors formulated a bimodal system with two operators O and O' as follows.

The models have the form $(S, R_O, R_{O'}, h)$ where S is a set of possible worlds and R_O and $R_{O'}$ are two serial binary relations on S. h is the assignment. The intended meaning of R_O and $R_{O'}$ is the following:

- $tR_O s$ iff everything which is ideally the case in t is actually the case in s.
- $tR_{O'} s$ iff s is a sub-ideal version of t. Not everything that is ideally the case in t is actually the case in s.

The following semantic conditions must hold

(C1) $R_O \cap R_{O'} = \varnothing$

(C2) $tR_O t \vee tR_{O'} t$ holds for all t

Satisfaction is defined as expected:

- $t \vDash OA$ iff for all s, $tR_O s$ implies $s \vDash A$
- $t \vDash O'A$ iff for all s, $tR_{O'} s$ implies $s \vDash A$.

The authors use the above as a starting point to define and present what they believe as an adequate notion of 'ought'.

Discussion
Obviously the system DL is based on the dovetailing of two serial modalities. It is the extension of the dovetailed combination with the two semantic interaction conditions.

I suspect that these correspond to the following axioms:

$$
\begin{aligned}
&OA \wedge O'A \rightarrow A \\
&\sim A \wedge OA \wedge O'B \rightarrow B \\
&\sim A \wedge O'A \wedge OB \rightarrow B.
\end{aligned}
$$

6

INTRODUCING SELF-FIBRING

6.1 Introduction and Background

The purpose of this chapter is to introduce a coherent semantics for the following self-fibring problem.

Self-Fibring problem

Let $\mathbb{L}$ be a predicate language and let $\mathbb{L}_0, \mathbb{L}_1$ be sublanguages of $\mathbb{L}$. Let $\mid\!\sim_0, \mid\!\sim_1$ be two consequence relations for $\mathbb{L}_0$ and $\mathbb{L}_1$ respectively, and let $\mathcal{K}_i$ be their respective semantics.

Let $\mathbb{L}_0^1[\mathbb{L}_1]$ be the new language obtained by allowing substitutions of formulas and terms of $\mathbb{L}_1$ for the variables of $\mathbf{L}_0$. Let, for $m > 1$, $\mathbb{L}_0^m[\mathbb{L}_1]$ be $\mathbb{L}_0^1(\mathbb{L}_0^{m-1}[\mathbb{L}_1])$. The problem is to find good semantics and proof theory for $\mathbb{L}_0[\mathbb{L}_1] = \bigcup_m \mathbb{L}_0^m[\mathbb{L}_1]$, in terms of the semantics and proof theory of $\mid\!\sim_0$ and $\mid\!\sim_1$.

Note that this is a general problem, and its solution must provide a methodology applicable to arbitrary logics. $\mathbb{L}_i$ and $\mid\!\sim_i$ may be classical logic or any other logic, such as intuitionistic predicate logic.

We need to give solid and meaningful semantics to expressions of the form $A(t, \varphi$ '$(f(\alpha))$ where $A(x, y)$ is a formula of $\mathbb{L}_0$ and $\alpha, \varphi(x)$ are formulas of $\mathbb{L}_1$ and $f(x)$ a function symbol of $\mathbb{L}_1$. As one can see, we have substituted formulas within other formulas and other function symbols. We thus face two problems. The first problem is to give meaning to expressions of the form $A(t, \varphi)$, where φ is a formula. Current classical semantics can handle terms not formulas. The second problem is to find a consequence relation for the resulting fibred system. The meaning of φ is governed by $\mid\!\sim_1$ and the meaning of A is governed by $\mid\!\sim_0$. What do we get when we put them together? We must offer a reasonable and highly intuitive view.

Such expressions and languages are widely used in logic, philosophy, artificial intelligence and linguistics and have extensive and important applications. We mention a few typical examples:

- Logics of non-denoting singular terms; free logics.
- Highly interactive multimodal logics, where the modalities can be viewed as predicates applied to formulas.
- Conditional logics.
- In the logical analysis of language, we try and analyse statements like
  John said that Mary believed that he did not love her.
  Here we use the predicates, *say* (t, φ) and *believe* (t, φ).

- In logic programming there is a widespread use of metapredicates such as *Demo* (A, B)
  and
  Y if *Not*(X).

  Ordinary clauses for the transitive closure of a relation R are written as

  $$(cl(Z))(X,Y) \text{ if } Z(X,Y)$$
  $$(cl(Z))(X,Y) \text{ if } Z(X,V) \wedge ((cl(Z))(V,Y))$$

  or expressions like

  $$\forall x.\, \textit{believe}\ (\text{John, friend (John}, x)) \rightarrow \textit{exists}\ (y,\ \text{loves}\ (x,y))).$$

- Tarski truth predicate
  $$\mathbf{T}(\varphi) \leftrightarrow \varphi$$
  and the diagonalisation function symbol
  $$\mathbf{d}(\varphi) =\sim \varphi(\varphi).$$

- The Liar paradox.
- Ambivalent syntax.
- Generalised quantifiers and generalised first-order semantics.
- Language under specification.
- Channel theory.
- Context logics.

The basic tool we use is that of fibred semantics. We shall explain the intuitive idea of fibring by considering the case of classical logic and the need to give semantics to expressions of the form $P(\varphi)$, where φ is a classical formula and P a unary predicate, and to expressions of the form $P(c)$, where c is a non-denoting singular term.

Suppose we start with an ordinary model $\mathbf{m}$ of classical logic. This has a domain D. Let $P(x)$ be a unary predicate of the language. Then P is assigned a subset D_P of D. See Figure 6.1.

For any element $a \in D$ we can evaluate the 'expression' $P(a)$. The evaluation gives us a truth value $\top$ or $\bot$. If $a \in D_P$ the value is $\top$. If $a \notin D_P$ the value is $\bot$. Now consider the expression $P(\sim P(a))$ or the expression $P(c)$, where c is a non-denoting singular term.

We do not know how to evaluate it because the expression $y =\sim P(a)$, or $y = c$ inside the predicate is not assigned (in our model theory) an element of the domain. However, we do notice that all we need to have is an answer to the question: is the truth value of $P(y)$ $\top$, or is it $\bot$? Of course we can adopt a lazy view and say $P(\varphi)$ is always $\top$, but this approach does not take into account the meaning of φ and is not useful. To get a good widely applicable answer we use the 'fibring' methodology. We need a way of associating a value to the query $\mathbf{m} \vDash^{?} P(y)$ which is dependent on the

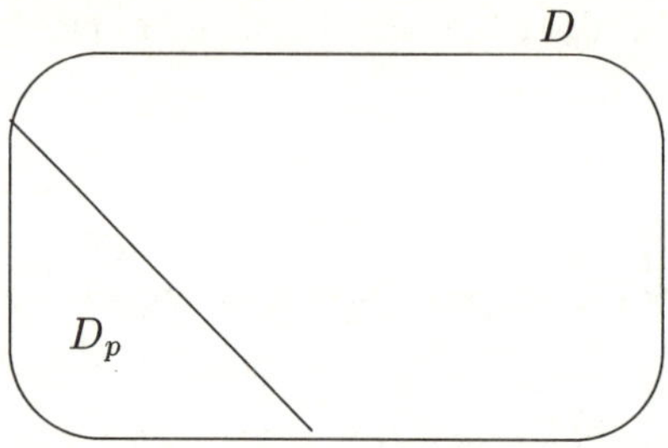

FIG. 6.1.

two 'parameters', P and y. The 'parameter' P can be identified with its extension D_P. Our idea is to associate with D_P another model $\mathbf{F}(D_P)$, of the appropriate kind, which can help us get a value. The function $\mathbf{F}$ giving us the model is called the *fibring function*.

For the case of a constant $y = c$, it is sufficient that the fibring function $\mathbf{F}$ assigns a new model $\mathbf{n} = \mathbf{F}(D_P)$, for the language with c based on the domain $D' \supseteq D$. $\mathbf{n}$ of course assigns a value for c in D'. Thus $\mathbf{n} \vDash P(c)$ can be evaluated so we can let $\mathbf{m} \vDash P(c)$ iff $\mathbf{F}(D_P) \vDash P(c)$.[16]

In the case of $y =\sim P(a)$, the model $\mathbf{n} = \mathbf{F}(D_P)$ is taken as another classical model, in which the expression y can be evaluated, because y is a formula. So we can let $\mathbf{m} \vDash P(\sim P(a))$ iff $\mathbf{F}(D_P) \vDash\sim P(a)$. This way we get our truth value.

Is our approach reasonable and intuitive? It is clear that formally the approach is the right one. The value $\mathbf{m} \vDash P(y)$ depends on three components, $\mathbf{m}$, P and y. It is also quite clear that a policy of associating an arbitrary value to $\mathbf{m} \vDash P(y)$, say a function $\mathbf{f}(\mathbf{m}, P, y)$, is too general and does not take account of the meaning and internal structure of the individual components involved. Our policy is to identify P with its extension D_P in $\mathbf{m}$, and to associate with P a model $\mathbf{n} = \mathbf{F}(D_P)$ (or $\mathbf{n} = \mathbf{F}(D_P, \mathbf{m})$ and to evaluate y in $\mathbf{n}$. This way we are certainly taking a structural account of the parameters involved.

Let us check some examples to see how intuitive the approach is.

Consider the predicate $P_a(x)$ reading *agent a sees x*. Here we have a relation of some sort between objects. The agent can see many things and then generalise. He can see, for example, lots of people happy on the beach at a holiday resort and then generalise to some statement A (*sunshine and water makes one happy?*). We can then give meaning to $P_a(A)$. This certainly depends on the extension of P_a and on the models the agent is imagining in which A is true as a rule.

Consider now the predicate *agent a believes A*. We present it as $B_a(A)$. Here the basic meaning of $B_a(x)$ is intended to be for formula substitution. However, if we allow

[16]We mention in passing that fibring methods should also be able to handle the case of ambiguous predicates. If the extension of P is ambiguous then we cannot give value to $P(c)$, even though we do have a value for c. Again we may need to go to another fibred model. We shall investigate underspecification of predicates elsewhere.

for names of sentences in the language then the extension D_{B_a} of B_a in the model $\mathbf{m}$ is a set of names. Clearly then in this case it is natural to have

$$\mathbf{F}(D_{B_a}) = \{A \mid \text{ name of } A \text{ is in } D_{B_a}\},$$

which again leads us to a fibred model.

A similar consideration applies to the predicate a *says* y. In this case it makes more sense to fibre a family of models $\mathbf{F}(say, \mathbf{m})$, one for each wff which was said.

Some predicates have a more intimate connection between the extensions and the formulas involved. Consider a *loves* y. a may love Mary not only for her looks but also (and in my opinion, mainly) for her qualities and properties. Thus if we adopt the view that an object (Mary) is the sum of its properties, then a *loves Mary* holds because a loves some properties and Mary realises them. So again, by looking at the set of all those elements that a loves we can get an idea as to what properties A he loves.

A more direct example for the predicate $P(B)$ is the subjunctive conditional $P > B$. One of the most widely studied semantics for the conditional is the interpretation:

- $t \vDash P > B$ iff in all possible worlds s that are P-near t, we have $s \vDash B$.

P-nearness is given by a function $\mathbf{f}(P, t)$. But then this is exactly the truth condition for our $P(B)$.

The reader should note the strong connection between the two cases, $y = c$ and y a formula. Indeed the connection is with free logic. Free logic systems allow for non-denoting terms such as Pegasus, thus there are constants y in the language where $P(y)$ gets a truth value but y does not get assigned an element of the domain. For this kind of logic the rule $\forall x P(x) \to P(y)$ does hold but $P(y) \to \exists x P(x)$ does not hold. Depending on the system of free logic we might even not have $\forall x \forall y$ and $\forall y \forall x$ being the same. See Bencivenga [1986].

Our fibred semantics can also be of service to free logic, because we have the domain D of existing elements and also the non-existing 'elements' y being the formulas of the language, or some other non-denoting constants.

The reader should note and observe that our method is not restricted to classical logic. We can equally study formulas of the $P(c), c$ non-denoting, or of the form $A(t, \varphi(s), f(B))$ in intuitionistic logic. The same approach applies as we shall see in a later section.

6.2 Free Logic

The first example we deal with involving two languages $\mathbb{L}_0$ and $\mathbb{L}_1$ is a version of free logic. Let $\mathbb{L}$ be a predicate language with the connectives $\sim, \wedge, \vee, \to, \bot$ and $\top$ and the quantifiers $(x), (Ex) =\sim (x) \sim, (\forall x)$ and $(\exists x) =\sim (\forall x) \sim$.

Our formal view of free logic is that of a predicate logic with some terms that do not denote. A typical helpful example is the sentence

The King of France is bald

which we can symbolise as $P(d)$. If there is no King of France, what do we do with such a sentence? We will look at two ways of generating free logics. One where the non-denoting part of the language is recognised syntactically, say as a separate set of constants $\mathbb{T}$. The other is that there is no syntactical distinction but at each model some constants get a denotation and some do not and it varies from model to model (e.g. some models have a King of France, and some don't). Thus our language needs two quantifiers, (x) which ranges over existing object and $(\forall x)$ which ranges over all terms.

Free logic is a large area of research with strong roots in philosophy and linguistics. See [Bencivenga, 1986]. Our interest in this topic in this section stems from the similarity of $P(d)$ and $P(\varphi)$. In both cases we are trying to apply P to an object (King of France, the wff φ) for which it does not 'know' what truth value to give.

Furthermore, when we axiomatise systems of self-fibred predicate logic, the underlying predicate axioms will be those of a version of free logic and so it is wise for us to present the needed versions of free logic first. We also see how free logic can be translated into self-fibred predicate logic.

Let us now proceed with technical definitions.

Let $\mathbb{T}$ be a set of constants and function symbols not in $\mathbb{L}$. Let $\mathbb{L}_{\mathbb{T}}$ be the extension of $\mathbb{L}$ with the additional function symbols of $\mathbb{T}$. Let *Free*($\mathbb{T}$) be the set of all terms which are in $\mathbb{L}_{\mathbb{T}}$ but not in $\mathbb{L}$. Thus the terms of *Free*($\mathbb{T}$) contain at least one function symbol from $\mathbb{T}$.

Definition 6.1 (Classical models). A classical model for $\mathbb{L}$ has the form $\mathbf{m} = (D, h, g)$, where D is a non-empty domain and h is an interpretation function, assigning to each n-place predicate P of $\mathbb{L}$ a set $h(P) \subseteq D^n$ and $g = (g_1, g_2)$ is a pair of functions assigning to each n-place function symbol c of $\mathbb{L}$ a function $g_1(c) : D^n \mapsto D$ and to each free variable x of $\mathbb{L}$ an element $g_2(x) \in D$. We use the notation $g(c)$ and $g(x)$, with the understanding that the appropriate component of g is applied. g can be extended to all terms of $\mathbb{L}$ by the inductive clause:

$$g(c(t_1, \ldots, t_n)) = g(c)(g(t_1), \ldots, g(t_n)).$$

We use the notation $g[x/d]$ for $d \in D$ to denote the pair of functions $(g_1, g_2[x/d])$ where $g_2[x/d](y) = g_2(y)$ if $y \neq x$ and $g_2[x/d](x) = d$.

Similarly, we use the notation $g[x_1/d_1, \ldots, x_n/d_n]$.

Satisfaction is defined in the traditional manner:

- $\mathbf{m} \vDash_g (P(t_1, \ldots, t_n)$ iff $(g(t_1), \ldots, g(n_n)) \in h(P)$
- $\mathbf{m} \vDash_g (x)A(x)$ iff for all $d \in D$, $\mathbf{m} \vDash_{g[x/d]} A(x)$.
- We are not yet ready to define satisfaction for $(\forall x)$.

We may, for convenience, suppress the assignment g, in which case we write $\mathbf{m} \vDash A[x/d]$ instead of $\mathbf{m} \vDash_{g[x/d]} A(x)$.

The model $\mathbf{m}$ can give values to the formulas of the language $\mathbb{L}$ but not to the language $\mathbb{L}_{\mathbb{T}}$, because the assignment g does not interpret the function symbols of $\mathbb{T}$. Thus our models $\mathbf{m}$ cannot give values to any atomic formulas $P(t_1, \ldots, t_n)$, where at least one term t_i is not from $\mathbb{L}$.

If we could arbitrarily assign truth values to such atoms, then inductively, we can give a truth value to arbitrary formulas of $\mathbb{L}_{\mathbb{T}}$.

K. Lambert and B. C. van Fraassen [1972] have introduced the notion of a *story* $\mathbf{S}$ to do the job of assigning truth values to atomic expressions of $\mathbb{L}_{\mathbb{T}}$, which are not in

$\mathbb{L}$. Let $P(t_1(\bar{y}), \ldots, t_n(\bar{y}))$ be an atomic formula with terms $t_1, \ldots, t_n$ containing the variables $\bar{y} = (y_1, \ldots, y_k)$. Assume that at least one of the terms contains a function symbol from $\mathbb{T}$ and hence the model $\mathbf{m}$ cannot evaluate it.

Consider the expression

$$g(P(t_1, \ldots, t_n)) = \text{(definition) } P(t_1, \ldots, t_n)[y_1/g(y_1), \ldots, y_k/g(y_k)].$$

This expression can be given a value $\{0, 1\}$ by the story $\mathbf{S}$.[17]

Using $\mathbf{m}$ and $\mathbf{S}$ we can evaluate an arbitrary formula of $\mathbb{L}_{\mathbb{T}}$ as follows:

- $(\mathbf{m}, \mathbf{S}) \vDash_g P(t_1, \ldots, t_n)$ if $\mathbf{m} \vDash_g P(t_1, \ldots, t_n)$ for atomic wff in $\mathbb{L}$.
- $(\mathbf{m}, \mathbf{S}) \vDash_g P(t_1, \ldots, t_n)$ if $\mathbf{S}(g(P(t_1, \ldots, t_n)) = 1$ for atomic wff not in $\mathbb{L}$.
- $(\mathbf{m}, \mathbf{S}) \vDash_g (x)A(x)$ if for all $d \in D$, $(\mathbf{m}, \mathbf{S}) \vDash_{g[x/d]} A(x)$.
- $(\mathbf{m}, \mathbf{S}) \vDash_g \forall x A(x)$ if for all terms t of $\mathbb{L}_{\mathbb{T}}$ we have $(\mathbf{m}, \mathbf{S}) \vDash_g A(t)$.
- Note however that in $(x)A(x)$, x ranges over D only and in $(\forall x)A(x)$, x ranges over all terms of $\mathbb{L}_{\mathbb{T}}$. Thus we may have any of the following happen, for a constant $t \in \mathbb{T}$:
- $(\mathbf{m}, \mathbf{S}) \vDash_g A(t)$ but $(\mathbf{m}, \mathbf{S}) \nvDash_g (Ex)A(x)$.
- $(\mathbf{m}, \mathbf{S}) \vDash_g \forall x A(x)$ but $(\mathbf{m}, \mathbf{S}) \nvDash_g (x)A(x)$, unless for every $d \in D$ there exists a term t of $\mathbb{L}$ such that $g(t) = d$.

The connection of fibred semantics with free logic can be seen on several fronts:

1. When we allow self-fibring in the syntax, we get expressions like $P(\alpha)$, where α is a formula. We can regard this as a free logic $\mathbb{L}_{\mathbb{T}}$, where $\mathbb{T}$ is the set of all wffs. Thus a model could be of the form $(\mathbf{m}, \mathbf{S_m})$, where $\mathbf{S_m}$ is a story for $\mathbf{m}$, being a function from wffs into $\{0, 1\}$. The 'story' is not as simple, however, since these terms are wffs and hence have logical relations among them. To coherently allow for this logical relation among our 'terms', we can, for example, let $\mathbf{S_m}(\alpha)$ be obtained by evaluating α in some model $\mathbf{M_m}$. Thus we have

$$\mathbf{S_m}(\alpha) = 1 \text{ iff } \mathbf{M_m} \vDash \alpha.$$

   Since we allow for repeated self-substitution (i.e. $\alpha = B(\beta)$, we need to repeat the process and get $\mathbf{S}' = \mathbf{S}_{\mathbf{M_m}}$ for $\mathbf{M_m}$. This already suggests the notion of fibred semantics.
2. Even for the pure case of free logic, where the set $\mathbb{T}$ is a set of terms, our fibred semantics approach can generalise the notion of a story $\mathbf{S}$ by deriving it from fibring models. We need a value for the satisfaction predicate $\vDash_g P(t_1, \ldots, t_n)$ in the case where $g(P(t_1, \ldots, t_n))$ is not in the language $\mathbb{L}$, i.e. the terms in this expression contain symbols from $\mathbb{T}$.

If for every predicate P we fibre a model $\mathbf{F}(P) = (D_P, h_P, g_P)$ of the language $\mathbb{L}_{\mathbb{T}}$, then we can let

[17] Note that van Fraassen allows for the domain D to be empty. $\mathbf{S}$ is a function giving $\{0, 1\}$ values to such expressions.

- $(\mathbf{m}, \mathbf{F}) \vDash_g P(t_1, \ldots, t_n)$ iff $\mathbf{F}(P) \vDash_g P(t_1, \ldots, t_n)$.

The story $\mathbf{S}$ can now be defined by

- $\mathbf{S}(g(P(t_1, \ldots, t_n)) = 1$ iff $\mathbf{F}(P) \vDash_g P(t_1, \ldots, t_n)$.

We shall see from the completeness theorem for $\mathbf{C}_0$ below that the language $\mathbb{L}_\mathbb{T}$ is weak enough to allow for $\mathbf{F}(P)$ to be assumed independent of P, namely $D_P = D \cup \mathbb{T}, g_P = g$ and $h_P(Q) = h_F(Q)$.

Definition 6.2 (Fibred $\mathbf{C}_0$ models).

1. A fibred $\mathbf{C}_0$ model for $\mathbb{L}$ and $\mathbb{T}$ has the form $\mathbf{n} = (D, h, g, \mathbf{F})$ where (D, h, g) is a classical model for $\mathbb{L}$ and $\mathbf{F}$ is a fibring function associating with each atomic predicate P of $\mathbb{L}$ a classical model $\mathbf{F}(P) = (D_P, h_P, g_P)$ of the language $\mathbb{L}_\mathbb{T}$. We require that $D \subseteq D_P$ and that g_P agrees with g and h_P agrees with h on the language $\mathbb{L}$.
2. Satisfaction is defined as follows:
    - $\mathbf{n} \vDash_g P(t_1, \ldots, t_n)$ if $(g(t_1), \ldots g(t_n)) \in h(P)$, for atoms of $\mathbb{L}$
    - $\mathbf{n} \vDash_g P(t_1, \ldots, t_n)$ iff $(g_P(t_1), \ldots, g_P(t_n)) \in h_P(P)$
    - $\mathbf{n} \vDash_g (x)A(x)$ iff for all $d \in D, \mathbf{n} \vDash_{g[x/d]} A(x)$
    - $\mathbf{n} \vDash_g \forall x A(x)$ iff for all terms t of $\mathbb{L}_\mathbb{T}$ we have $\mathbf{n} \vDash_g A(t)$, and for all $d \in D$ we have $\mathbf{n} \vDash_{g[x/d]} A(x)$.

The above considerations dealt with the case where the language was syntactically divided into denoting and non-denoting terms. A more intuitively pleasing way of looking at the situation is to assume a language $\mathbb{L}$, and models (D, h, g) where the assignment g gives values in D only to some of the terms of $\mathbb{L}$, and g is arbitrary otherwise. Thus the language is not divided syntactically a priori into denoting and non-denoting terms but the division changes from model to model. This can be formalised by the following definition.

Definition 6.3 (Free $\mathbf{C}_1$ models). Let $\mathbb{L}$ be a language with (x) and $(\forall x)$. A free $\mathbf{C}_1$ model for $\mathbb{L}$ has the form (D, D', h, g) where $D \cup D' \neq \varnothing, D \cap D' = \varnothing$ and $\mathbf{n} = (D^*, h, g)$ is a classical model of $\mathbb{L}$, where $D^* = D \cup D'$.

Recall that $g = (g_1, g_2)$. h and g_1 are assumed to be giving values in D^* but not necessarily g_2. Thus if for a term $t, g(t) \in D'$ then we say t is non-denoting.

Satisfaction is defined by the clauses:

- $\mathbf{n} \vDash (x)A(x)$ iff for all $d \in D, \mathbf{n} \vDash A(x/d)$
- $\mathbf{n} \vDash (\forall x)A(x)$ iff for all $d \in D^*, \mathbf{n} \vDash A(x/d)$.

The system $\mathbf{C}_0$ axiomatises the fibred semantics and the system $\mathbf{C}_1$ axiomatises the free semantics. These two systems will be used later to axiomatise the monadic self-fibred predicate semantics.

Example 6.4 (Monadic language). Consider a monadic language $\mathbb{M}$ without equality or function symbols (but possibly with individual constants) and let $\mathbb{T}$ be another set of constants. The fibred models in this case can have the form $\mathbf{n} = (D, h, g, \mathbf{F})$, where $D \neq \varnothing$ and for each unary predicate P, $\mathbf{F}(P)$ is a subset of $\mathbb{T}$.

Satisfaction is defined by

- $\mathbf{n} \models P(t)$ if $g(t) \in h(P)$, for t in $\mathbb{M}$.
- $\mathbf{n} \models P(t)$ if $t \in \mathbf{F}(P)$, for $t \in \mathbb{T}$.
- $\mathbf{n} \models (x)A(x)$ if for all $d \in D, \mathbf{n} \models A(d)$.
- $\mathbf{n} \models \forall x A(x)$ iff for all $x \in D$ and all $x \in \mathbb{T}$ we have $\mathbf{n} \models A(x)$.

We let $\mathbb{M}^0_{\mathbb{T}}$ be the language with the quantifier (x) only.

Definition 6.5 (The logic $\mathbf{C}_0$). Let $\mathbf{C}_0$ be the following axiomatic system for the language $\mathbb{L}_{\mathbb{T}}$ (with (x) and $(\forall x)$ quantifiers). The axioms are separated so the axioms with (x) only axiomatise the fragment of the language $\mathbb{L}^0_{\mathbb{T}}$, the part of $\mathbb{L}_{\mathbb{T}}$ with (x) only, and the axioms with $(\forall x)$ axiomatise the fragment $\mathbb{L}^1_{\mathbb{T}}$, with $\forall$ only.

- All substitution instances of truth functional tautologies.
- Traditional axioms for equality.
- If A and $A \to B$ are theorems then so is B.
- $\forall x A(x) \to (x)A(x)$.
- $(x)A(x) \to A(u)$, u is a variable or term of $\mathbb{L}$, properly substituted for x.
- $\forall x A(x) \to A(u)$, u is a variable or term of $\mathbb{L}_{\mathbb{T}}$, properly substituted for x.
- $\dfrac{\vdash A \to B(x)}{\vdash A \to (x)B(x)}$, x a variable of $\mathbb{L}$ not free in A.
- $\dfrac{\vdash A \to B(t)}{\vdash A \to \forall x B(x)}$, t is a constant of $\mathbb{T}$ not in A.

Let $\mathbf{C}^i_0$ be the fragment of $\mathbf{C}_0$ with the axioms for $\mathbb{L}^i_{\mathbb{T}}$.

Theorem 6.6 (Completeness of $\mathbf{C}_0$). *For any wff B of $\mathbb{L}^i_{\mathbb{T}}$: $\mathbf{C}^i_0 \vdash B$ iff for any fibred model $\mathbf{m}$, $\mathbf{m} \models B$.*

Proof

1. Soundness is easy to verify.
2. We show completeness.

Let B be a consistent wff. We shall construct a model $(D, h, g, \mathbf{F})$ for B. Our construction is separated so it works for each fragment separately. Let $A_0, A_2, \ldots,$ be an enumeration of all wffs of $\mathbb{L}_{\mathbb{T}}$. Let $\Delta_0 = \{B\}$. Assume we have defined Δ_n and it is finite and consistent. We define Δ_{n+1}. Consider A_n. We claim either $\Delta_n \cup \{A_n\}$ or $\Delta_n \cup \{\sim A_n\}$ is consistent

Let B_n be one among them which is consistent. We distinguish several cases.

1. $B_n = (Ex)C_n(x)$.
   Let u be the first variable from $\mathbb{L}$ which does not appear in Δ_n. We claim $\Delta_n \cup \{(Ex)C_n(x); C_n(u)\}$ is consistent. Otherwise

$$\begin{array}{l}\vdash \bigwedge \Delta_n \wedge (Ex)C_n(x) \to \sim C_n(u)\\ \vdash \bigwedge \Delta_n \wedge (Ex)C_n(x) \to (x) \sim C_n(x)\\ \vdash \bigwedge \Delta_n \to (x) \sim C_n(x),\end{array}$$

contrary to our assumption that $\Delta_n \cup \{B_n\}$ is consistent.
Let $\Delta_{n+1} = \Delta_n \cup \{B_n, C_n(u)\}$.

2. $B_n = (\exists x)C_n(x)$. Let u be the first constant from $\mathbb{T}$ that does not appear in Δ_n and proceed as in the previous case, using $(\forall x)$ instead of (x).
3. Let $\Delta_{n+1} = \Delta_n \cup \{B_n\}$, otherwise.

Let $\Delta = \bigcup_n \Delta_n$.

Define a fibred model **m** as follows.

- Let D = all terms of $\mathbb{L}$. D_Q = all terms of $\mathbb{L}_{\mathbb{T}}$.
- Let $h(P) = \{(d_1, \ldots, d_n) \mid d_i \in D$ and $P(d_1, \ldots, d_n) \in \Delta\}$.
- Let $h_Q(P) = \{(t_1, \ldots, t_n) \mid t_i$ are terms of $\mathbb{L}_{\mathbb{T}}$ and $P(t_1, \ldots, t_n) \in \Delta\}$.
- Let $g(x) = x$ for all terms x, similarly for g_Q.
- Let $\mathbf{F}(Q) = (D_Q, h_Q, g_Q)$.

Since $h_Q(P)$ depends only on P and $D_Q = D$ and$g_Q = g$, we can identify a story **S** by

- $\mathbf{S}(P(t_1, \ldots, t_n)) = 1$ iff $P(t_1, \ldots, t_n) \in \Delta$.

We claim that in $\mathbf{m}$, $\mathbf{m} \vDash A$ iff $A \in \Delta$.

The above holds for atomic formulas by definition. The classical connectives give no difficulties. We consider the case of $(x)A(x)$; the case of $(\forall x)$ is similar.

- Assume $(Ex)A(x) \in \Delta$. Then at some stage $(Ex)A(x)$ was put in Δ and so for some u, $A(u)$ was put in Δ and hence by the induction hypothesis $\mathbf{m} \vDash (Ex)A(x)$.
- Assume $(x) \sim A(x) \in \Delta$. Then for any u, a variable or term of $\mathbb{L}$, $\sim A(u) \in \Delta$ and hence $\mathbf{m} \vDash (x)A(x)$.

The reader can check that equality is a congruence relation such that the other conditions of the semantics are all satisfied.

This completes the proof of the theorem. □

Definition 6.7 (The logic $\mathbf{C}_1$). Let $\mathbf{C}_1$ be the following separated axiomatic system for the language $\mathbb{L}$ and for the language $\mathbb{L}^0$, being the language $\mathbb{L}$ with the quantifier (x) only, and $\mathbb{L}^1$, the fragment with $\forall$ only.

- All substitution instances of truth functional tautologies.
- Traditional axioms for equality.
- $\vdash A$ and $\vdash A \to B$ imply $\vdash B$.
- $\vdash (x)(A(x) \wedge B(x)) \to (x)A(x) \wedge (x)B(x)$.
- $\vdash (x)A(x) \wedge (x)(A(x) \to B(x)) \to (x)B(x)$.
- $\dfrac{\vdash A \to B(x)}{\vdash A \to (x)B(x)}$ where x is not free in A.
- $\vdash (x)(y)A \to (y)(x)A$.
- $(u)(B \wedge (x)A(x) \to A(u))$.
- $\vdash (\forall x)A(x) \to (x)A(x)$.

- $\vdash (\forall x)A(x) \to A(t)$.
- $\dfrac{\vdash A \to B(u)}{\vdash A \to (\forall u)B(u)}$, u not free in A.
- $\vdash A(t) \to (\exists x)A(x)$.

Theorem 6.8 (Completeness of $\mathbf{C}_1$). *For any wff B of $\mathbb{L}$ we have: $\mathbf{C}_1 \vdash B$ iff for any free model $\mathbf{m}, \mathbf{m} \vDash B$.*

Similarly for the respective (x) and $(\forall x)$ fragments.

Proof

1. Soundness
   This is easy to check.
2. Completeness
   Let B be a consistent wff. We seek a model $\mathbf{m} = (D, D', h, g)$ such that $\mathbf{m} \vDash B$. Let $A_0, A_1, \ldots$, be an enumeration of all wffs of $\mathbb{L}$. Let $\Delta_0 = \{B\}, D_0 = \varnothing$.

Assume Δ_n, D_n have been defined and that Δ_n satisfies the property (†) below

(†) If $(x)\varphi(x) \in \Delta_n$ and $y \in D_n$ then $\varphi(y) \in \Delta_n$.

We now define Δ_{n+1}, D_{n+1}, with the same properties.

Consider A_n. We claim either $\Delta_n \cup \{A_n\}$ or $\Delta_n \cup \{\sim A_n\}$ is consistent. Let B_n be the one among them which is consistent. We distinguish several cases.

1. $B_n = (Ex)C_n(x)$.
   Let u be the first variable from $\mathbb{L}$ not in Δ_n nor in D_n. We claim $\Delta_n \cup \{B_n, C_n(u)\} \cup \{\varphi(u) \mid (x)\varphi(x) \in \Delta_n\}$ is consistent. For otherwise $\vdash \bigwedge \Delta_n \to (\bigwedge_i \varphi_i(u) \to \sim C_n(u))$, hence $\vdash \bigwedge \Delta_n \to (x)(\bigwedge_i \varphi_i(x) \to \sim C_n(x))$ and hence $\vdash \bigwedge \Delta_n \to (\bigwedge_i (x)\varphi_i(x) \to (x) \sim C_n(x))$ and so $\Delta_n \vdash (x) \sim C_n(x)$, contradicting the assumption that $\Delta_n \cup \{B_n\}$ is consistent. Let Δ_{n+1} be $\Delta_n \cup \{B_n, C_n(u)\}, D_{n+1} = D_n \cup \{u\}$.
2. $B_n = (x)C_n(x)$.
   We claim $\Delta_n \cup \{B_n\} \cup \{C_n(y) \mid y \in D_n\}$ is consistent. Otherwise $\vdash \bigwedge \Delta_n \wedge (x)C_n(x) \to \sim \bigwedge_{y \in D_n} C_n(y)$.
   Assume $D_n = \{y_1, \ldots, y_k\}$. Hence

$$\vdash (y_1, \ldots, y_k)(\bigwedge \Delta_n \wedge (x)C_n(x) \to \sim \bigwedge_{i=1}^{k} C_n(y))$$

but, for $i = 1, \ldots, k$

$$\vdash (y_i)(\bigwedge \Delta_n \wedge (x)C_n(x) \to C_n(y_i))$$

hence

$$\vdash \bigwedge_i (y_i)(\bigwedge \Delta_n \wedge (x)C_n \to C_n(y_i))$$

hence

$$\vdash (y_1, \dots, y_k) \bigwedge_i (\bigwedge \Delta_n \wedge (x)C_n(x) \to C_n(y_i))$$

hence

$$\vdash (y_1, \dots, y_k)(\bigwedge \Delta_n \wedge (x)C_n(x) \to \bigwedge_{i=1}^{k} C_n(y_i))$$

which is a contradiction.

Thus let $\Delta_{n+1} = \Delta_n \cup \{B_n\} \cup \{C_n(y) \mid y \in D_n\}$ and let $D_{n+1} = D_n$.

3. $B_n = (\exists x)C_n(x)$
   Let u be a new variable as before. We claim $\Delta_n \cup \{C_n(u)\}$ is consistent. Otherwise $\bigwedge \Delta_n \vdash \sim C_n(u)$ and so $\Delta_n \vdash (\forall u) \sim C_n(u)$, a contradiction.
4. In any other case let

$$\Delta_{n+1} = \Delta_n \cup \{B_n\}, D_{n+1} = D_n.$$

We now define the model $\mathbf{m} = (D, D', h, g)$:

- Let $D = \bigcup_n D_n, \Delta = \bigcup_n \Delta_n$
- Let D' = all variables of the language not in D
- Let g be the identity
- Let $h(P) = \{(x_1, \dots, x_n) \mid P(x_1, \dots, x_n) \in \Delta\}$.

We show for any A

$$\mathbf{m} \vDash A \text{ iff } A \in \Delta.$$

This holds for atoms. The propositional connectives present no problem, since for any A, $A \in \Delta$ or $\sim A \in \Delta$. The case of $\forall$ presents no problems. We now examine the case of (x).

Assume $(x)A(x) \in \Delta$. We show that for all $u \in D, A(u) \in \Delta$. u was put into D at some stage n because of some $B_n = (Ex)C_n(x)$. $(x)A(x)$ was put in at some stage m. Note that we cannot have $m = n$.

If $m < n$, then $A(u)$ was put in as well at stage n. If $n < m$, then $A(u)$ was put in at stage m.

In any case $A(u) \in \Delta$ and hence $\mathbf{m} \vDash (x)A(x)$.

Assume $(x)A(x) \notin \Delta$. Then $(Ex) \sim A(x) \in \Delta$ and so for some $u \in D$, $\sim A(u) \in \Delta$ and so $\mathbf{m} \vDash \sim A(u)$ and so $\mathbf{m} \vDash \sim (x)A(x)$.

This completes the proof of the theorem. □

Theorem 6.9 (Decidability). *The monadic version of the systems* $\mathbf{C}_0$ *and* $\mathbf{C}_1$ *are decidable.*

Proof Their semantics can be faithfully translated into the semantics of classical monadic logic which is decidable.

Let (Ax) be the classical quantifier

1. For the system $\mathbf{C}_1$, let D, D' be a special unary predicate. Translate $[(x)\varphi]^*$ into $(Ax \in D)[\varphi]^*$, $[(\forall x)\varphi]^*$ into $(Ax \in D \cup D')[\varphi]^*$.
   Then $\mathbf{C}_1 \vdash \varphi$ iff $\vdash (D \neq \varnothing) \wedge (D \cap D' = \varnothing) \to [\varphi]^*$ in classical logic.

2. In the case of $\mathbf{C}_0$ we need to add to the antecedent of the translation $\bigwedge_{t_i} D'(t_i)$ where $\{t_i\}$ are the $\mathbb{T}$ constants appearing in φ. □

Remark 6.10 (Comparing $\mathbf{C}_0$ and $\mathbf{C}_1$). Consider a formula $A(x, y)$ with the free variables x and y. More specifically, let

$$A(x, y) = (z)P(z) \to P(x) \wedge P(y).$$

As a formula of $\mathbf{C}_0$, A is a theorem because the variables x, y get assignments in the domain D. The terms that do not necessarily get assigned elements of the domain are in $\mathbb{T}$.

As a formula of $\mathbf{C}_1$ it is not a theorem because in $\mathbf{C}_1$ there is no understanding that variables get values in D.

Consider now $\varphi_0(x, y)$, a formula of $\mathbf{C}_1$ with two free variables. Let c, d be two non-denoting constants of the language, i.e. we extend $\mathbf{C}_1$ by constants which we syntactically say are non-denoting.

Then the following rule is always valid.

$$(*)\ \frac{\vdash \varphi_0(x, y)}{\vdash \varphi_0(c, d)}.$$

The reason being that x, y are not committed to being denoting and so $\vdash \varphi_0$ means that even when they are arbitrary non-denoting we get a theorem.

Let $\mathbf{C}_1(\mathbb{T})$ be the extension of $\mathbf{C}_1$ with a set $\mathbb{T}$ of non-denoting terms. Semantically we are committed to assigning values to elements of $\mathbb{T}$ in D'. $\mathbf{C}_1(\mathbb{T})$ is a mixture of $\mathbf{C}_0$ and $\mathbf{C}_1$. It has the non-denoting terms but there is no commitment that its variables denote, just like in $\mathbf{C}_1$. It can be axiomatised by adding to $\mathbf{C}_1$ the rule (*) above.

It is easy to see that there is a simple translation between $\mathbf{C}_0$ and $\mathbf{C}_1$:

1. Let $\varphi(x_i)$ be a wff of $\mathbf{C}_1$ with free variables x_i. Let $c_i \in \mathbb{T}$ be individual constants, then let $\varphi(x_i/c_i)$ be the translation of $\varphi(x_i)$ into $\mathbf{C}_0$.
2. Conversely, let $\varphi(c_i)$ be a wff of $\mathbf{C}_0$ with constants $c_i \in \mathbb{T}$. Let $\varphi(c_i/x_i)$ be its translation into $\mathbf{C}_1$.

We have

$$\vdash_{\mathbf{C}_0} \varphi(c_i) \text{ iff } \vdash_{\mathbf{C}_1} \varphi(x_i).$$

Remark 6.11 (Stronger free systems). It is clear from the completeness proof for $\mathbf{C}_1$ that the following additional axioms correspond to the following semantic conditions.

1. $(Ex)(P(x) \to P(x))$ corresponds to the condition $D \neq \varnothing$. An equivalent axiom schema is $(x)A(x) \to (Ex)A(x)$.
2. $(x)A(x) \to A(y)$ corresponds to the condition $D = D \cup D'$, and yields classical logic.

7

SELF-FIBRING OF PREDICATE LOGICS

7.1 Self-fibred Semantics for Substitution into Monadic Logic without Function Symbols

This section will introduce the notion of a fibred model and provide semantics for expressions of the form $P(A)$, where P is a monadic predicate and A is a formula (which may contain equality and function symbols). An axiom system and a completeness theorem will also be given. Later sections will deal with more general languages.

7.1.1 *Discussion of our options*

This section deals with a fibred monadic language. By this we mean the language $\mathbb{L}_M[\mathbb{L}]$, where $\mathbb{L}$ is a predicate logic and $\mathbb{L}_M$ is its monadic fragment. In other words, we allow (repeated) substitution of formulas only in monadic predicates. Note that the language $\mathbb{L}$ itself need not be monadic and may contain n-place predicates, equality and function symbols. The definitions and theorems of the section are valid for this language. Of course the theorems on decidability assume that the language is totally monadic, i.e. $\mathbb{L}_M[\mathbb{L}_M]$.

The notion $\mathbb{L}_1[\mathbb{L}_2]$, although simple, does require some caution. Consider the formula $\varphi_0 = P(z)$. This is a monadic formula with a free variable z. We can substitute for z the formula $\alpha(x,y) = P(x) \wedge P(y)$ and get the formula $\varphi(x,y) = \varphi_0(z/\alpha) = P(P(x) \wedge P(y))$. φ has two free variables and we can therefore quantify and create $\psi = (x)(Ey)\varphi$.

It is important to observe that the formula $\alpha(x,y)$ was substituted for the variable z even though α had free variables other than z. We can impose the requirement that we substitute for variables z only wffs that contain at most z as a free variable. This requirement gives rise to the notions of *tight* substitution and *tight* formula.

The formula ψ is not a *tight formula*; it cannot be obtained by tight substitutions. These distinctions are important. For example, it is known [Gabbay, 1976] that monadic modal **K** with modalities $\Box$ and $\Diamond$ with one unary predicate P is not decidable. It is possible to interpret a classical reflexive and symmetric relation R in the system by letting $xRy = \Diamond(P(x) \wedge P(y))$.

The self-fibred monadic theory of one letter P would also be undecidable because we can let

$$xRy = \alpha(x,y) = P(P(x) \wedge P(y)).$$

If we restrict ourselves to tight formulas only, we can prove decidability for the tight monadic case.

If the following axioms are valid $(P(A \wedge B) \leftrightarrow P(A) \wedge P(B))$ and $\sim P(A) \leftrightarrow P(\sim A)$, then of course ψ can be obtained by tight substitution and the undecidability interpretation will not work.

We also have several options with regard to semantics. If we start with a model $\mathbf{m} = (D_\mathbf{m}, h_\mathbf{m}, g_\mathbf{m})$, then for each $X \subseteq D_\mathbf{m}$, we can fibre one or more other models. See Definition 7.2. Let $\mathbf{F}(X, \mathbf{m})$ be the set of these fibred models. In the general case, this set can also be empty. Let $\mathbf{n} = (D_\mathbf{n}, h_\mathbf{n}, g_\mathbf{n})$ be one of the fibred models. In the most general case there is no connection between the domains, interpretations or assignments of $\mathbf{m}$ and $\mathbf{n}$. However, many applications do enjoy some connections. Here is a list of some of the options.

- Constant domains $D_\mathbf{m} = D_\mathbf{n} = D$.
- Constant assignment $g_\mathbf{m} = g_\mathbf{n} = g$.
- Context free fibring. $\mathbf{F}$ depends only on subsets of the domain and not on the model (i.e. not on h and g).

There is another way of defining fibred models: these are the *generalised fibred structures* of Definition 7.9.

For notational convenience we can imagine a family of models indexed by labels $t \in S$. Our model $\mathbf{m}$ is associated with a label t taken from the set S. In this notation, we can present $\mathbf{m}$ as $\mathbf{m}_t = (D_t, h_t, g_t)$. The fibring function can be presented as a set function giving for each $X \subseteq D_t$ a set $\mathbf{F}(X, t)$ of labels from S. Thus $\mathbf{m}_s$ is fibred to $\mathbf{m}_t$ iff $s \in \mathbf{F}(X, t)$. This notation is visibly more compatible with modal logic notation.

The two kinds of semantics of Definitions 7.2 and 7.9 are equivalent; they give rise to the same logic. In fact in Definition 7.10 we refine the semantics into *tree models*, which make a better use of labels (and is still an equivalent semantics).

For the case where D is independent in t (constant domains with $D_t = D$) it can be convenient to identify $D = S$ when possible. Such models are called *reflective* and are very useful. In such a case a special locality predicate $W(x)$ can be used to identify the label through the condition

- $\mathbf{m}_t \vDash W(x)$ iff $g_t(x) = t$.

We also have two options with our quantifiers. We can consider $(x)A(x)$ ranging over the domain D_t and $(\forall x)A(x)$ ranging over the domain D_t and all formulas of the language.

7.1.2 *Basic semantics and completeness theorems*

Definition 7.1 (Fibred monadic syntax). Let $\mathbb{L}$ be a predicate language which also contains monadic predicates, the classical connectives and quantifiers. We allow for two kinds of universal quantifiers, $(\forall x)$ and (x). $(\forall x)$ is intended as quantifying over terms and formulas and (x) quantifies only over terms (i.e. is a sort of 'free logic' quantifier). We can define $(\exists x) =\sim (\forall x)$ and $(Ex) =\sim (x) \sim$.

We have no function symbols, but we do allow individual constants. We define the notion of a fibred wff (f-wff) of the language as follows.

1. A is a f-wff of level 0 if A is an ordinary free classical wff of $\mathbb{L}$ (built up with $\forall x$ and (x)).
2. A is a f-wff of level $\leq n+1$ if there exists a free classical wff $\varphi(x_1, \ldots, x_n, y_1, \ldots, y_m)$ of the monadic fragment of $\mathbb{L}$ with free variables $x_1, \ldots, x_n, y_1, \ldots, y_m$ and f-wffs $A_1, \ldots, A_n$ of level $\leq n$ such that $A = \varphi(x_1/A_1, \ldots, x_n/A_n, y_1, \ldots, y_m)$ where x_i/A_i denotes the simultaneous substitution of A_i for x_i in φ. The free variables of A are $y_1, \ldots, y_m$ together with all the free variables of $A_1, \ldots, A_n$. We assume the substitution is carefully made and does not result in incorrect binding of variables.
3. A is a *tight* formula of level $\leq n+1$ if it is obtained by substitution from tight formulas of level $\leq n$ as in (2) above but with the restriction that A_i has at most x_i as its free variable.
4. Level $\leq n$ is closed under the connectives and quantifiers.

The reader will know from context whether we assume $\mathbb{L}$ to contain only $(\forall x)$ or only (x) or both.

The fibred semantics for monadic logic is similar in spirit to the semantics for free logic given in the previous section. So before we present our definition of fibred monadic models, let us present in more detail the classical free logic models of the previous section.

1. A classical (free logic) model for the language $\mathbb{L}$ has the form $\mathbf{m} = (D, D', h, g)$ where $D \neq \varnothing$ and $D \cap D' = \varnothing$. D is the domain of existing elements ((x) ranges over D) and D' is the domain of non-existing elements ($(\forall x)$ ranges over $D^* = D \cup D'$). h is an interpretation function assigning to each m-place predicate P a subset $h(P) \subseteq D^{*m}$. g is a function assigning to each variable x a value $g(x) \in D$ and to each constant c value $g(c) \in D^*$ (thus $g(x)$ always exists in D but $g(c)$ may not exist and be in D') and to each n place function symbol f a function $g(f) : D^{*n} \mapsto D^*$. g can be extended to all terms in the obvious way. We assume that for $x_i \in D, g(f)(x_i) \in D$.
   In our self-fibred semantics D' is intended to be the set of all wffs.
   Note that the semantics described here corresponds to Free $\mathbf{C}_1$ models where $D \neq \varnothing$, as described in Definition 6.3. This semantics can be axiomatised for the language with (x) only by the logic $\mathbf{C}_1$ of Definition 6.7 together with the axiom $(Ex)(P(x) \to P(x))$ or equivalently $(x)A(x) \to (Ex)A(x)$.
2. Traditional classical models are obtained by letting $D' = \varnothing$. In such a case we present the model as (D, h, g).
   A fibred model is obtained by letting D' = set of all closed f-wffs of the language. In fact D' can be formally taken as all f-wffs—with free variables. The assignment g assigns elements of D to the variables which will fix the domain D' to be all wffs of the language based on the ' constants' $\{g(x_1), g(x_2), \ldots\}$. With this understanding we can define satisfaction for self-fibred models later on.
3. Satisfaction for classical models is defined using the mixed notation $\varphi(x)[x/\alpha]$ to mean the formula $\varphi(x)$ with the intended value α for x, where α is any value

element of D^* (or as we shall see in the case of fibred models where α is indeed a formula of the language).

With the above notation we let

- $\mathbf{m} \vDash P(\bar{x})[\bar{x}/\bar{d}]$, for $\bar{d} \in D^{*m}$ iff $\bar{d} \in h(P)$
- $\mathbf{m} \vDash P(\bar{t})$ iff $g(\bar{t}) \in h(P)$.

  The above can also be written as

$$\mathbf{m} \vDash P(\bar{t})[t/g(\bar{t})].$$

- $\mathbf{m} \vDash (x)\varphi(x)$ iff $\mathbf{m} \vDash \varphi(x)[x/d]$ for all $d \in D$
- $\mathbf{m} \vDash \forall x\varphi(x)$ iff $\mathbf{m} \vDash \varphi(x)[x/d]$, for all $d \in D^*$.

In the self-fibred model, D' is the set of all f-wffs and so $\forall x$ quantifies over real elements and formulas while (x) quantifies over real elements only.[18]

- $\mathbf{m} \vDash t_1 = t_2$ iff $g(t_1) = g(t_2)$.

Definition 7.2 (Fibred monadic models).

1. A fibred model has the form $\mathbf{n}_0 = (D, h_0, g, \mathbf{F})$, where (D, h_0, g) is a classical model and $\mathbf{F}$ is a function giving for each interpretation h and each subset $X \subseteq D$ a new interpretation $h' = \mathbf{F}(X, h)$.

   Note that the assignment g remains rigid. See however item 5 below.
2. A relation $\vDash$ is said to be a *satisfaction* relation between f-wffs A and fibred models $\mathbf{n} = (D, h, g, \mathbf{F})$ if the following holds:
   - $\mathbf{n} \vDash P(\bar{t})$ iff $g(\bar{t}) \in h(P)$, for $\bar{t}$ a tuple of the form $\bar{t} = (t_1, \ldots, t_m)$ where each t_i is a variable or constant and P atomic.
   - $\mathbf{n} \vDash t_1 = t_2$ iff $g(t_1) = g(t_2)$.
   - $\mathbf{n} \vDash P(\bar{x})[\bar{x}/\bar{d}], \bar{d} \in D^m$ iff $\bar{d} \in h(P)$.
   - $\mathbf{n} \vDash \forall x A(x)$ iff for all $\alpha \in D$ and all f-wffs $\alpha, \mathbf{n} \vDash A(x)[x/\alpha]$.
   - $\mathbf{n} \vDash (x)A(x)$ iff for all $\alpha \in D, \mathbf{n} \vDash A(x)[x/\alpha]$.

     Note that we obtain D' as the set of all f-wffs of the language.
   - $\mathbf{n} \vDash P(\varphi)$ iff $(D, h', g, \mathbf{F}) \vDash \varphi$ for $h' = \mathbf{F}(h(P), h)$.
   - $\mathbf{n} \vDash (A \wedge B)[x/\alpha]$ iff $\mathbf{n} \vDash A[x/\alpha]$ and $\mathbf{n} \vDash B[x/\alpha]$.
   - $\mathbf{n} \vDash (\sim A)[x/\alpha]$ iff $\mathbf{n} \nvDash A[x/\alpha]$.
3. The function $\mathbf{F}$ need not be defined for all subsets X of D. It is sufficient to have it defined for all subsets of the form $h(P)$, for atomic P of the language. This would make the model more effectively computable.
4. We can relinquish the requirement that $\mathbf{F}$ is a function and let $\mathbf{F}$ be a relation. Thus $\mathbf{F}(X, h)$ would be a set of assignments and the truth condition for $P(\varphi)$ would be:
   - $\mathbf{n} \vDash P(\varphi)$ iff $(D, h', g, \mathbf{F}) \vDash \varphi$ for all $h' \in \mathbf{F}(h(P), h)$.

   To make our notation more uniform we can write $\mathbf{F}(X, h) \vDash \varphi$ iff (definition) for all $h' \in \mathbf{F}(X, h), h' \vDash \varphi$.

[18] See, however, Remark 7.5 below.

Thus the clause for $h \vDash P(\varphi)$ can be the same, irrespective of whether $\mathbf{F}$ is a set function or not.

This version of $\mathbf{F}$ is of particular interest for the interpretation of $P(\varphi)$ as a modality on φ. We shall discuss this later. Note that we allow $\mathbf{F}(h(P), h)$ to be empty.

5. Note that we need not keep the assignment g to the variables and constants unchanged. We can allow $\mathbf{F}$ to change g as well, in which case $\mathbf{F}$ would be a function giving a set of new $(h', g') \in \mathbf{F}(X, h, g)$. The appropriate fibring satisfaction clause would then be
   - $\mathbf{n} \vDash P(\varphi)$ iff $(D, h', g', \mathbf{F}) \vDash \varphi$ for all $(h', g') \in \mathbf{F}(h(P), h, g)$.

   For many applications it is convenient to keep g *rigid*. See, however, the section on generalised assignment models.

Note that the above definition makes $\forall x P(x)$ range over all $x \in D$ and over all wffs φ as well.

Also note that for the case of $\forall$ the above definition of $\vDash$ is an implicit and not a recursive definition. Consider $\mathbf{m} \vDash \forall x P(x)$. We have to evaluate $\mathbf{m} \vDash P(\forall x P(x))$ and move to a new model $\mathbf{n} \vDash \forall x P(x)$. The variable x is instantiated also to the value $\forall x P(x)$. This causes an infinite recursive process. If we only have the quantifier (x) which ranges over $x \in D$, then $\vDash$ is recursively defined.

Further note that a more sophisticated satisfaction clause for $\mathbf{n} \vDash P(\varphi)$ can be the following

•• $\mathbf{n} \vDash P(\varphi)$ iff $(D, h', g', \mathbf{F}) \vDash \Psi_P(\varphi)$ for all $(h', g') \in \mathbf{F}(h(P), h, g)$ and $\Psi_P(Q)$ is a classical wff involving a predicate Q and $\Psi_P(\varphi)$ is the result of substituting φ for Q.[19]

The previous satisfaction clause is obtained by letting $\Psi_P(Q) = Q$.

The definition of the fibring function $\mathbf{F}$ makes it dependent on the interpretation h, the assignment g and subsets of D. The dependence on h and g makes a difference. If $\mathbf{F}$ were independent of h and g and dependent only on subsets the following f-wff would be valid.

- $(x)[P_1(x) \leftrightarrow Q(P_2(x))] \rightarrow [P_1(\varphi) \leftrightarrow Q(P_2(\varphi))]$.

The f-wff says that if P_1 in $\mathbf{m}$ has the same extension as P_2 in $\mathbf{m}_Q$ then $\mathbf{m} \vDash P_1(\varphi)$ iff $\mathbf{m}_Q \vDash P_2(\varphi)$, for arbitrary φ.

From now on, unless otherwise stated, we assume g is rigid in our fibred models.

Theorem 7.3 (Axiomatisation of the system with $\forall$ only). *Consider the logic with all classical axioms and rules together with the following additional axioms and rules:*

- $P(\forall x A(x)) \leftrightarrow \forall x P(A(x))$
- $P(A \wedge B) \leftrightarrow P(A) \wedge P(B)$
- $t_1 = t_2 \rightarrow P(t_1 = t_2)$

[19]The most general way of viewing $\mathbf{n} \vDash P(\varphi)$ is to think of it as invoking an information channel to another family of models $\mathcal{K} = \mathbf{F}(h(P), h, g)$. The formula φ is to be evaluated at the family or 'semantics' $\mathcal{K}$. We need two formulas $\Psi_{\mathbf{F}}$ and Ψ_P. Ψ_P says how to evaluate φ at each model $\mathbf{n} \in \mathcal{K}$. $\Psi_{\mathbf{F}}$ looks at the set $\mathcal{K}_\varphi = \{\mathbf{n} \in \mathcal{K} \mid \mathbf{n} \vDash \Psi_P(\varphi\}$ and states a condition on it. In our case $\Psi_{\mathbf{F}}(X)$ says $X = \mathcal{K}$.

- $t_1 \neq t_2 \rightarrow P(t_1 \neq t_2)$
- $\vdash A$ *implies* $\vdash P(A)$
- $\sim P(A) \leftrightarrow P(\sim A)$.

Then the logic is complete for satisfaction in rigid (i.e. g is rigid) fibred models where the universal quantifier $\forall$ ranges over all elements in the domain and all wffs as well.[20]

If we drop the last axiom ($\sim P(A) \leftrightarrow P(\sim A)$) we get completeness for the fibring of set function **F**.

Proof

1. Soundness
   This is easy to verify.
2. Completeness
   Let A be a consistent f-wff. Extend $\{A\}$ to a consistent, saturated and complete theory Δ_A.[21] So the following is assumed to hold.
   (*) If for all terms and wffs x, $\Delta_A \vdash \varphi(x)$ then $\Delta_A \vdash \forall y \varphi(y)$
   (**) For all α, $\alpha \in \Delta_A$ or $\sim \alpha \in \Delta_A$. □

Lemma 7.4. *Let Θ be any theory satisfying (*) above and assume that $\Theta \cup \{\alpha\}$ is consistent. Then $\Theta \cup \{\alpha\}$ satisfies (*) above.*

Proof Assume for all x $\Theta \cup \{\alpha\} \vdash \varphi(x)$. Then for all $x, \Theta \vdash \alpha \rightarrow \varphi(x)$ and so $\Theta \vdash \forall y(\alpha \rightarrow \varphi(y))$ for y not in α and so $\Theta \vdash \alpha \rightarrow \forall y \varphi(y)$ and so $\Theta \cup \{\alpha\} \vdash \forall y(\varphi(y)$. □

Construction

Let Θ be consistent and complete theory satisfying (*). We construct for each atomic predicate P a theory Θ_P as follows:

Let $\Theta_P = \{A \mid \Theta \vdash P(A)\}$.

Note that Θ_P is consistent, complete and satisfies (*). For assume $\Theta_P \vdash \varphi(x)$, for all x. Then for each x there exists $A_1, \ldots, A_n$ such that $\vdash A_1 \wedge \ldots \wedge A_n \rightarrow \varphi(x)$. Hence $\vdash P(A_1) \wedge \ldots \wedge P(A_n) \rightarrow P(\varphi(x))$. Hence $\Theta \vdash P(\varphi(x))$ for each x, hence $\Theta \vdash \forall x P(\varphi(x))$, hence $\Theta \vdash P(\forall x \varphi(x))$ and so $\forall x \varphi(x) \in \Theta_P$.

In the case where **F** is a set function, we do not have the axiom $\sim P(A) \leftrightarrow P(\sim A)$. Our construction is slightly different, but is quite familiar from modal logic. For each α

[20] Notice the similarity of the axioms with that of a modal operator. In this case the modal operator is a *next* operator. The axioms without $\sim P(A) \leftrightarrow P(\sim A)$ define a **K** modality and correspond to a fibring **F** which is a set function.

[21] This is done by enumerating all wffs of the language as $\alpha_0, \alpha_1, \alpha_2, \ldots$. Let $\Delta_0 = \{A\}$ or indeed any consistent theory such that there is an infinite number of variables not appearing in Δ_0 and assume that Δ_n is defined, is consistent and that there is still an infinite number of variables not appearing in Δ_n. Consider the wff α_n. Either $\Delta_n \cup \{\alpha_n\}$ is consistent or $\Delta_n \cup \{\sim \alpha_n\}$ is consistent. Let $\beta \in \{\alpha_n, \sim \alpha_n\}$ be the wff which is chosen and is consistent. If β has the form $\sim \forall x \beta_0(x)$, then let y be a new variable not free in Δ_n or in β and let $\Delta_{n+1} = \Delta_n \cup \{\beta, \sim \beta_0(y)\}$. If β is not of that form let $\Delta_{n+1} = \Delta_n \cup \{\beta\}$. Let $\Delta_A = \bigcup_n \Delta_n$.

Note that just like in modal logic, the notion of $\Delta \vdash A$ means there is a proof of A from Δ without the use of necessitation ($\vdash X$ imply $\vdash P(X)$) but with the help of all instances of the theorems of the logic. For such a consequence relation the deduction theorem holds.

such that $\Theta \vdash\sim P(\alpha)$, we construct a theory $\Theta_{\sim P(\alpha)}$ as follows. Let $\Theta_{0,\alpha} = \{A \mid \Theta \vdash P(A)\} \cup \{\sim \alpha\}$. We show $\Theta_{0,\alpha}$ is consistent and satisfies (*). It is consistent because otherwise $\vdash A_1 \wedge \ldots \wedge A_n \to \alpha$ and so $\vdash P(A_1) \wedge \ldots \wedge P(A_n) \to P(\alpha)$ and we get $\Theta \vdash P(\alpha)$.

We now show that $\Theta_{0,\alpha}$ satisfies (*). For assume $\Theta_{0,\alpha} \vdash \varphi(x)$. Then for each x there are $A_1, \ldots, A_n$ such that $\vdash A_1 \wedge \ldots \wedge A_n \wedge \sim \alpha \to \varphi(x)$. Hence $\vdash \bigwedge_i P(A_i) \vdash P(\sim \alpha \to \varphi(x))$. Hence $\Theta \vdash P(\sim \alpha \to \varphi(x))$ for each x.

Hence $\Theta \vdash \forall x P(\sim \alpha \to \varphi(x))$ for x not appearing in α. Hence $\Theta \vdash P(\sim \alpha \to \forall x \varphi(x))$ and so $\sim \alpha \to \forall x \varphi(x) \in \Theta_{0,\alpha}$ and so $\Theta_{0,\alpha} \vdash \forall x \varphi(x)$.

Thus we established that $\Theta_{0,\alpha}$ is consistent and satisfies $(*)$. We can continue with our construction. Now enumerate all wffs of the language as $\varphi_1, \varphi_2, \ldots$. Assume $\Theta_{n,\alpha}$ has been defined and is consistent and satisfies (*). Then either $\Theta_{n,\alpha} \cup \{\varphi_{n+1}\}$ or $\Theta_{n,\alpha} \cup \{\sim \varphi_{n+1}\}$ is consistent. Let ψ be φ_{n+1} if it is consistent and $\sim \varphi_{n+1}$ otherwise. Let $\Theta'_{n+1,\alpha} = \Theta_{n,\alpha} \cup \{\psi\}$. If ψ has the form $\sim \forall y \psi_0(y)$ then for some t, $\Theta'_{n+1,\alpha} \cup \{\sim \psi_0(t)\}$ is consistent (otherwise by (*) $\Theta'_{n+1,\alpha} \vdash\sim \psi$). Let $\Theta_{n+1,\alpha} = \Theta'_{n+1,\alpha} \cup \{\sim \psi_0(t)\}$. It is easy to check that it satisfies (*).

We claim $\Theta_{\sim P(\alpha)} = \bigcup_n \Theta_{n,\alpha}$ is consistent complete and satisfies (*). We check the latter. If $\Theta_{\sim P(\alpha)} \vdash\sim \forall y \varphi(y)$, then for some n, $\Theta_{n,\alpha}$ was enlarged with $\sim \forall y \varphi(y)$ but then for some t, $\Theta_{n,\alpha}$ was enlarged with $\sim \varphi(t)$.

This completes the construction of $\Theta_{\sim P(\alpha)}$.

We now present a model for Δ_A. Let D be the set of all terms of the language. Let Θ be any consistent complete and saturated theory satisfying (*). Let $h_\Theta(P)$ be defined as $\{\bar{t} \in D^m \mid \Theta \vdash P(\bar{t})\}$, let g be the identity with $g(x) = x$ for variable x and $g(f)(t_1, \ldots, t_n) = f(t_1, \ldots, t_n)$. The axioms for equality ensure that g is rigid and can be defined in the above manner.

For any h_Θ, any P let, for the case of $\mathbf{F}$ being a function:

- $\mathbf{F}(h_\Theta(P), h_\Theta) = h_{\Theta_P}$
  and for the case $\mathbf{F}$ is a set function:
- $\mathbf{F}(h_\Theta(P), h_\Theta) = \{h_{\Theta_{\sim P(\alpha)}} \mid \Theta \vdash\sim P(\alpha)\}$.
  We claim $(D, h_{\Delta_A}, g, \mathbf{F})$ is a model of Δ_A.
  We prove by induction that
- $(D, h_\Theta, g, \mathbf{F}) \vDash A$ iff $\Theta \vdash A$.

The above holds for atomic A. The cases of $\wedge$ and $\sim$ present no difficulties.

The case of $\forall$ also presents no difficulties since Θ satisfies (*).

We now check the case of $A = P(\alpha)$.

Assume $\Theta \vdash P(\alpha_0)$, then by construction $\alpha \in \Theta_P$ (or resp. $\alpha \in \Theta_{\sim P(\beta)}$, for all β) hence $(D, h_{\Theta_P}, \mathbf{F}) \vDash \alpha$ (resp. for all β, $(D, h_{\Theta_{P(\beta)}} \mathbf{F}) \vDash \alpha$). Hence $(D, h_\Theta, \mathbf{F}) \vDash P(\alpha)$.

If $\Theta \nvdash P(\alpha)$, then $\Theta \vdash\sim P(\alpha)$ and so $\sim \alpha \in \Theta_P$ (resp. $\sim \alpha \in \Theta_{\sim P(\alpha)}$) and so $(D, h_{\Theta_P}, \mathbf{F}) \nvDash \alpha$ (resp. $D, h_{\Theta_{\sim P(\alpha)}}, \mathbf{F}) \nvDash \alpha$) and so $(D, h_\Theta, \mathbf{F}) \nvDash P(\alpha)$.

This concludes the completeness proof.

Remark 7.5. The logic with $\forall$ only is satisfactory in one respect and not satisfactory in another. We do like the rule $\forall x \varphi(x) \to \varphi(t)$ to be valid for t any term including f-wffs. After all, this is the idea of allowing expressions like $P(A)$, A a formula. However, the drawback is that $\forall x P(x)$, P atomic is always false, since $\perp$ is a wff and $P(\perp)$ is false in the semantics. In case **F** is a set valued function then if the set is empty then $\forall x P(x)$ is always true. In either case the $\forall$ quantifier is trivial.

We may try and let $\forall x$ range over a restricted class Γ of wffs, say all positive ones, or maybe find another way out. At any rate, this requires further thought and we shall not address it in this section. We will, however, study the other extreme, the logic with (x) only.

Theorem 7.6 (Completeness theorem for fibred monadic logic with (x)).

1. *Consider the following axioms and rules (of classical free logic $\mathbf{C}_0$) augmented by the additional rule $\mathbb{R}$:*
   - *all substitution instances of truth functional tautologies*
   - $(x)A(x) \to A(u)$, *u a variable or (non-formula) term, properly substituted for x*
   - $\vdash A; \vdash A \to B$ *imply* $\vdash B$
   - $\dfrac{\vdash A \to B(x)}{\vdash A \to (x)B(x)}$, *$x$ is a variable not in A*

   $\mathbb{R}$ $\quad \dfrac{\vdash B(x_i/A_i)}{\vdash B(x_i/P(A_i))}$, *$B$ a Boolean expression in x_i.*

   Then the resulting system is complete for the rigid fibred monadic semantics with **F** *a point function.*[22]
2. *For the semantics with* **F** *a set function, the rule $\mathbb{R}$ has to be weakened. We adopt the following rules, $\mathbb{K}$ (based on modal* **K***):*
   - $\dfrac{\vdash \bigwedge A_i \to A}{\vdash \bigwedge P(A_i) \to P(A)}$
   - $\vdash (x)P(A(x)) \to P((x)A(x))$.

Proof

1. Soundness
   The soundness of $\mathbb{R}$ and $\mathbb{K}$ respectively is straightforward.
2. Completeness for $\mathbb{R}$
   We assume that $\nvdash \varphi$ and find a countermodel for φ. This we prove by structural induction on the level of φ. Assume by induction that for any consistent Boolean combination $B(z_i/A_i), i = 1, \ldots, k$, with A_i of level up to n, we can find a fibred model. Assume $\varphi = \varphi_0(x_i/A_i, y_j)$ where φ_0 is of level 1, and that φ_0 has

[22] To see how to use $\mathbb{R}$, consider $B(z_1, z_2) = z_1 \to\sim z_2$. Then $\vdash B(z_1/\sim A, z_2/A)$. Hence $\vdash B(z_1/P(\sim A), z_2/P(A))$ namely $\vdash P(\sim A) \to\sim P(A)$. Similarly we get

$$P(A \wedge B) \leftrightarrow P(A) \wedge P(B)$$
$$P(\sim A) \leftrightarrow\sim P(A).$$

Consider the substitution $z_1/(x)A(x), z_2/\sim A(y)$. We get $\vdash P((x)A(x)) \to P(A(y))$ hence $\vdash P((x)A(x)) \to (x)P(A(x))$. Similarly we get $\vdash (x)P(A(x)) \to P((x)A(x))$.

no free variables other than x_i, y_j, $i = 1, \ldots, k, j = 1, 2, \ldots, k'$. We can also assume that x_i do not appear in any of $A_1, \ldots, A_k$ nor do they appear anywhere else in φ_0 except as indicated. Since $\not\vdash \varphi$, we can assume that $\not\vdash \varphi_0$. We regard φ_0 as a $\mathbf{C}_0$ formula with $\mathbb{T} = \{x_i\}$. By Theorem 6.6, we can let $(\mathbf{m}, \mathbf{S})$, with $\mathbf{S}$ a story and $\mathbf{m} = (D, h, g)$ a $\mathbf{C}_0$ countermodel to φ_0. Consider the atoms $P_j(x_i)$ in $\varphi_0, j \in J$. Under $\mathbf{S}$ we get a value for $\mathbf{S}(P_j(x_i))$. When we substitute x_i/A_i, we want to find an $\mathbf{F}$ such that $P_j(A_i)$ gets the same value, i.e. A_i gets the same value in $\mathbf{F}(h(P_j), h)$. This will turn the model into a countermodel for φ. Let $X \subseteq D$ and $\{P_j\}$ for $j \in J_X \subseteq J$ be atoms such that $h(P_j) = X$ and consider $P_j(A_i), j \in J_X, \ldots, i = 1, \ldots$. We would like a model $\mathbf{m}_X$ such that for all i and for all j such that $h(P_j) = X$ we have $\mathbf{m}_X \vDash A_i$ iff $\mathbf{S}(P_j(x_i)) = 1$.
If this is always possible to find (at least for all subsets of the form $h(P_j)$, P_j predicates of the language) we let $\mathbf{F}$ be defined by

$$\mathbf{F}(X, h) = \mathbf{m}_X.$$

Otherwise for every countermodel $(\mathbf{m}, \mathbf{S})$ there exists an X, J_X and $\{P_j\}$ such that for $j \in J_X$, $h(P_j) = X$ and a Boolean combination $B(z_{ji})$ such that $B_{(\mathbf{m},\mathbf{S})}(z_{ji}/P_j(x_i))$ holds in $\mathbf{S}$ but $B_{(\mathbf{m},\mathbf{S})}(z_{ji}/A_i)$ cannot be realised by a fibred model. Hence $\vdash\sim B_{(\mathbf{m},\mathbf{S})}(z_{ji}/A_i)$ and $\vdash\sim B_{(\mathbf{m},\mathbf{S})}(z_{ji}/P(A_i))$. P is not one of the P_j but a new letter. In each model $(\mathbf{m}, \mathbf{S})$ a possibly different subset $J_X^{(\mathbf{m},\mathbf{S})}$ of the $\{P_j\}$ is involved. For each $(\mathbf{m}, \mathbf{S})$, X, we can write a formula

$$B^*_{(\mathbf{m},\mathbf{S})} = B_{(\mathbf{m},\mathbf{S})}(z_{i,j}/P_j(x_i)) \wedge \bigwedge_{j,j' \in J_X} (y)(P_j(y) \leftrightarrow P_{j'}(y))$$

whose negation is provable.
Since the number of possible Boolean combinations and sets J_X is finite we have for some family $\{B^*_r\}$ that

$$\vdash (\bigwedge_r \sim B^*_r(z_{ji}/P_j(x_i)) \rightarrow \varphi_0(P_j(x_i))$$

hence $\vdash \varphi$ a contradiction.

3. Completeness for $\mathbb{K}$
For the case of the semantics where $\mathbf{F}$ is a set function we need to modify slightly the proof in (2) above. We consider the inductive case and consider $\varphi = \varphi_0(x_i/A_i)$. Let P_j be all atomic unary in φ_0 and let $(\mathbf{m}, \mathbf{S})$ be a countermodel to φ_0. Let $X \subseteq D$ and let J_X be the set of all j such that $h(P_j) = X$.
We want to define a family of models $\{\mathbf{m}^1_X, \mathbf{m}^2_X, \ldots\}$ such that for all A_i such that for some $j \in J_X, \mathbf{S}(P_j(x_i)) = 1$ holds, we have for all k, $\mathbf{m}^k_X \vDash A_i$. We also need to have that for any A_{i_0} such that $h(P_j)(g(x_{i_0}))$ does not hold we have that for some $j \in J_X$ we have for some k, $\mathbf{m}^k_X \nvDash A_i$.

For this end consider

$$\Delta_i = \{A_i \mid \mathbf{S}(P_j(x_i)) = 1 \text{ holds for some } j \in J_X\} \\ \cup\{\sim A_{i_0} \mid \text{ for some } j_0 \in J_X, \mathbf{S}(P_{j_0}(x_{i_0})) = 0\}.$$

If Δ_{i_0} is consistent for all X and all i_0, we get our countermodel. Otherwise, for some X and i_0 we have $\vdash \bigwedge A_i \to A_{i_0}$ and hence $\vdash \bigwedge P(A_i) \to P(A_{i_0})$ where P is a new atomic letter. This means that we also have

$$\vdash B^*_{(\mathbf{m},\mathbf{S})} = \bigwedge_{j,j' \in J_X} (y)(P_j(y) \leftrightarrow P_{j'}(y) \to (\bigwedge P_j(A_i) \to P_{j_0}(A_{i_0})).$$

Since the number of combinations for $B^*_{(\mathbf{m},\mathbf{S})}$ is finite we get that for some finite number $\{B^*_r\}$

$$\vdash \bigwedge_r B^*_r \to \varphi_0(P_j(x_i))$$

which leads to a contradiction. □

We now turn to the completeness theorem for the case where (x) and $(\forall x)$ are both present. The proof goes along the lines of the case of $(\forall x)$ alone, except that we need a bit of a twist. Suppose $(\forall x)B(x)$ does not hold. We need a term t, being a variable u or a formula α such that $\sim B(t)$. When only $(\forall x)$ is present, we can always choose t to be a new variable u, because there is no way to say $(u)B(u)$, as the quantifier (u) is not available. However, when (u) is also present, we must choose a proper α, say an atomic $\alpha = Q(u)$. But then to prove consistency we need the rule

$$\mathbb{G} \quad \frac{\vdash A \to B(Q(u)) \wedge (x)B(x)}{\vdash A \to (\forall x)B(x)}$$

where $Q(u)$ is an atomic wff such that neither Q nor u appear in A.

So let us state our theorem:

Theorem 7.7 (Completeness theorem for the language with (x) and $(\forall x)$). *Consider the following axiom systems:*

- *all substitution instances of truth functional tautologies*
- *traditional axiom for equality*
- $\vdash A \to B; \vdash A$ *imply* $\vdash B$
- $(x)A(x) \to A(u)$, *u a variable or a non-formula term, properly substituted for x*
- $(\forall x)A(x) \to A(u)$, *u any term or formula, properly substituted for x*
- $(\forall x)A(x) \to (x)A(x)$
- $\dfrac{\vdash A \to B(x)}{\vdash A \to (x)B(x)}$, *$x$ a variable not free in A.*

Formula universal generalisation (Rule $\mathbb{G}$):

- $\mathbb{G}\ \dfrac{\vdash A \to B(\alpha) \wedge (x)B(x)}{\vdash A \to (\forall x)B(x)}$,
  α a formula built up from atoms and variables not appearing in A which is neither a theorem nor a contradiction.

The following P axioms:

- $t_1 = t_2 \to P(t_1 = t_2)$
- $t_1 \neq t_2 \to P(t_1 \neq t_2)$
- $P(\forall x A(x)) \leftrightarrow (\forall x)P(A(x))$
- $P((x)A(x)) \leftrightarrow (x)P(A(x))$
- $\dfrac{\vdash \bigwedge_i A_i \to A}{\vdash \bigwedge_i P(A_i) \to P(A)}$.

Then the system is complete for the rigid fibred monadic semantics, with **F** *a set function.*

Proof For technical reasons we assume the language has an infinite number of unary predicates

1. Soundness can be verified
2. For completeness we rely on the proof steps already described in the individual proof for the case of $(\forall x)$. The presence of (x) in the language will require some adjustments.

Let A be a consistent wff. We can extend $\{A\}$ to a consistent saturated and complete theory Δ_A, satisfying the following:

(*) 1. If for terms and wffs t (i.e. $t = x$ or $t = \alpha$) $\Delta_A \vdash \varphi(t)$ then $\Delta_A \vdash (\forall x)(\varphi(x)$.
   2. If for all terms t (i.e. t is not a wff) we have $\Delta_A \vdash \varphi(t)$ then $\Delta_A \vdash (x)\varphi(x)$.

(**) for all $\alpha, \alpha \in \Delta_A$ or $\sim \alpha \in \Delta_A$.

Δ_A can be constructed as follows: let $\alpha_0, \alpha_1, \alpha_2, \ldots$ be an enumeration of all wffs of the language. Let $\Delta_0 = \{A\}$. Assume Δ_n is defined and consider $\Delta_n \cup \{\alpha\}$ and $\Delta_n \cup \{\sim \alpha_n\}$. One of these is consistent. Say $\beta \in \{\alpha_n, \sim \alpha_n\}$ is the one such that $\Delta_n \cup \{\beta\}$ is consistent. If β has the form $\sim (x)\beta_0(x)$, let u be a new variable not appearing in Δ_n and let $\Delta_{n+1} = \Delta_n \cup \{\sim \beta_0(u), \beta\}$. Our axioms and rules allow us to show that Δ_{n+1} is also consistent. If $\beta =\sim (\forall x)\beta_o(x)$, we first try to add to Δ_n the formulas $\{\sim \beta_0(u), \beta\}$, as before. If it works then fine. Otherwise $\Delta_n \wedge \beta \vdash (x)\beta_0(x)$. In this case we choose a new atomic predicate $Q(u)$, with u a new variable and let $\Delta_{n+1} = \Delta_n \cup \{\sim \beta_0(Q(u)), \beta\}$.

Again our rules allow us to show that Δ_{n+1} is consistent. For otherwise we have

$$\Delta_n \wedge \beta \vdash (x)\beta_0(x) \wedge \beta_0(Q(u)).$$

Hence by our rules

$$\Delta_n \wedge \beta \vdash (\forall x)\beta_0(x).$$

a contradiction.

If β is not a universally quantified formula then we let $\Delta_{n+1} = \Delta_n \cup \{\beta\}$.

We let $\Delta_A = \bigcup_n \Delta_n$.

This theory satisfies (*) and (**).

From here on our proof continues as in the proof of Theorem 7.3. We can do that because we have shown for our case the key construction, namely, that property (*) holds for Δ_A for both quantifiers $(\forall x)$ and (x). □

Remark 7.8 (Decidability). We get immediately from the proof that the tight monadic fragment with $\mathbb{K}$ and the monadic system with $\mathbb{R}$ is *decidable*, since $\mathbf{C}_0$ is decidable. The decision process is as follows.

We first observe that for a fragment to be tight means that every formula of level $\leq n$ can be obtained from a closed formula of level 1 of the form $\varphi_0(x_i)$ (the x_i are bound in φ_0) via a tight substitution $x_i/A_i(x_i)$ of level $\leq n-1$ formulas $A_i(x_i)$ for x_i. This is possible only if we are dealing with tight wffs.

We shall prove decidability using induction based on the above property.

For the monadic logic with rule $\mathbb{R}$ we know that tight substitutions generate the whole fragment and we get decidability.

We give the proof for the case of $\mathbb{R}$. The case of tight fragment of $\mathbb{K}$ is similar.

Case level 0
Decision as in the $\mathbf{C}_0$ monadic theory.

Case level $n+1$
Assume we have a decision procedure for any Boolean combination of f-wffs of level $\leq n$. Consider $\psi = \varphi(x_i/A_i, y_j)$, where φ is a closed classical monadic and A_i are of level $\leq n$. We use the notation of the previous theorem:

1. If φ is a theorem then so is ψ.
2. Consider all possible Boolean combinations of the form $B(z_{ji}), j = 1, \ldots, k, i = 1, \ldots, n$ and find for which ones $\vdash B(z_{ji}/A_i)$ is a theorem. This is decidable by the induction hypothesis.

For those B_r for which $\vdash B_r(z_{ji}/A_i)$ are theorems consider whether there is a conjunction of them such that in $\mathbf{C}_0$

$$\vdash (x_1, \ldots, x_n)(\bigwedge_r \sim B_r^*(z_{ji}/P_j(x_i)) \rightarrow \varphi)$$

is a theorem.

If some such expression is a theorem then ψ is a theorem; if none is a theorem then ψ is not a theorem.

7.1.3 *Refinements of the semantics*

The notion of a fibred model as defined in item 4 of Definition 7.2 is not the most general one. In Definition 7.2, all fibred models have the same domain D and the fibring function $\mathbf{F}$ is defined on the interpretations and assignments. To address more complex applications, it is more convenient to introduce labels naming the fibred models and to allow the different models to have different domains. The following is a more general definition.

Definition 7.9 (General fibred structures). A general fibred structure has the form $\mathbf{n} = (S, D, a, \mathbf{F}, h, g)$ where

S is a non-empty set of labels naming classical models.

D is a function associating with each $t \in S$ a non-empty domain D_t.

$a \in S$ is the actual label.

$\mathbf{F}$ is a fibring function associating with each $t \in S$ and each subset $X \subseteq D_t$ a set of labels $\mathbf{F}(X, t) \subseteq S$. This set may be empty.

h is an interpretation associating with each m-place atomic predicate P and each t a subset $h(t, P) \subseteq D_t^m$.

g is an assignment associating with each t and each variable and constant x of the language an element $g_t(x) \in \bigcup_t D_t$.

Note that if $g_t(x) \notin D_t$ then x is non-denoting at label t. This is avoided if we adopt the more traditional approach and require the following:

- $g_t(x) \in D_a$, for all x.
- *Persistence*
  $s \in \mathbf{F}(X, t)$ implies $D_t \subseteq D_s$.
  Persistence is a traditional requirement in modal logics.

Satisfaction is defined in the traditional manner.

- $t \vDash P(\bar{x})$ iff $g_t(\bar{x}) \in h(t, P)$, for $\bar{x}$ denoting, i.e. $g_t(\bar{x}) \in D_t^m$.
- $t \vDash (x)A(x)$ iff for all $y \in D_t, t \vDash A(y)$.
- $t \vDash P(\varphi)$ iff for all $s \in \mathbf{F}(h(t, P), t)$ we have $s \vDash \varphi$.
  Note if $\mathbf{F}(h(t, P), t) = \varnothing$ then $P(\varphi)$ holds for all φ.

We say $\mathbf{n} \vDash A$ iff $a \vDash A$.

We can either assume persistence and hence have no problem with non-denoting terms or add a clause for getting a value in the non-denoting case.

We shall assume persistence and that g is rigid, i.e. does not depend on t.

These models can be further normalised and streamlined.

Definition 7.10 (Tree models). Let $\mathbf{n} = (S, D, a, \mathbf{F}, h, g)$ be a structure as in Definition 7.9.

1. By a *chain* we mean a sequence $t_1, X_1, t_2, X_2, \ldots$ such that for each i, $t_i \in S, X_i \subseteq D_{t_i}$ and $t_{i+1} \in \mathbf{F}(X_i, t_i)$.
   Let P be an atomic predicate. The chain is said to be a P-chain, if for each t_i, $X_i = h(t_i, P)$.
2. We write tRs (resp. tR_Ps) if there exists a chain (resp. a P-chain) $t = t_1, \ldots, t_n = s$.
3. $\mathbf{n}$ is said to be a tree model if the following two conditions hold:
   - (S, R, a) is a tree with root a, i.e. for any $s \in S$ there exists a *unique* chain $a = t_1, \ldots, t_n = s$.
   - If $s_i \in \mathbf{F}(X_i, t), i = 1, 2$, then $D_{s_1} \cap D_{s_2} = D_t$.

4. $\mathbf{n}$ is said to satisfy the *irreflexivity assumption* if there exists a unary locality predicate $W(x)$ such that for each t there exists a unique $x_t \in D_a$ such that for all s, $s \vDash W(x_t)$ iff $t = s$.
5. $\mathbf{n}$ is said to be *well-founded* if it is tree-like and every R chain $t_1, t_2, \ldots, t_n, \ldots$ is finite.
6. A point $t \in S$ is of rank $\leq n$ if every R increasing chain beginning with t is of length $\leq n$.
7. $\mathbf{n}$ is said to be of *constant domains* if for all $t \in S$, $D_t = D$, for some set D.

Theorem 7.11 (Completeness theorem for tree models of constant domains). *Let Δ_0 be a consistent theory in the logic with (x) and $(\forall x)$ and all the axioms of Theorem 7.7. Then Δ_0 has a constant domains tree model.*

Proof

1. Soundness can be readily verified.
2. Completeness
   Let Δ_0 be a consistent theory. We can assume there is an infinite number of variables of the language not in Δ_0. Construct Δ_A as in the proof of Theorem 7.7. We give a tree model for Δ_A. We use the construction in the proof of Theorem 7.3. There we associated with each theory Θ and a $\sim P(\alpha) \in \Theta$ a theory $\Theta_{\sim P(\alpha)} \supseteq \{\sim \alpha\} \cup \{\beta \mid \Theta \vdash P(\beta)\}$.
   Let S be the tree of all finite sequences of natural numbers including the empty sequence $\varnothing$, and $\beta_n =\sim P_{i_n}(\alpha_{j_n})$ be an enumeration of all wffs of this form in the language. Define a theory $\Theta_{(n_1,\ldots,n_k)}$ for each sequence $t_{(n_1,\ldots,n_k)}$ as follows:

$$\Theta_\varnothing = \Delta_A.$$

   Assume Θ_t is defined and consider the sequence (t, n). Let $\Theta_{(t,n)} = (\Theta_t)_{\beta n}$.
   For each t, let $\mathbf{m}_t$ be the following model.
   - D_t = set of all variables and constants of the language
   - $h(t, P)$ $\{\bar{x} \mid P(\bar{x}) \in \Theta_t\}$
   - g is the identity
   - $s \in \mathbf{F}(X, t)$ if for some P and α we have $h(t, P) = X$ and $\sim P(\alpha) = \beta_n$ and $s = (t, n)$.

   It is easy to show that in this tree model we have:
   - $t \vDash B$ iff $B \in \Theta_t$ □

7.2 Self-fibring and Equality

This section will consider the role of equality in self-fibring. Let $\mathbb{L}$ be a predicate logic with predicates, equality $=$ and constants only (no function symbols)[23] and let $\mathbb{L}_=$ be the fragment with equality and constants only. We consider the language $\mathbb{L}_=[\mathbb{L}]$, i.e. we

[23] There is no logical reason for not allowing function symbols in this section, beyond the fact that we have not yet dealt with them. Their turn comes in the next chapter.

allow for formulas of the form $A(x, \varphi)$, where $A(x, y)$ is a wff of the pure theory of equality and $A(x, \varphi)$ is obtained by substituting φ for y, φ another formula from $\mathbb{L}$.

Thus we can write expressions like $x = \varphi$ and also $y = (x = \varphi)$ and $(x = (\varphi = \psi)) = \varphi$. It is helpful to give some interpretation to the expression '$x = \varphi$'. The best way to think of it is as an indexed modality $\Box_x \varphi$. The significance of this reading is that we are getting a multimodal logic where we can quantify over the modalities. Indexed modalities have many applications in computer science (e.g. dynamic logic). Furthermore, we shall see in a later section that when self-fibring intuitionistic logic with equality then $x = \varphi$ can be read as 'the term x resides in the type φ'.

Let us define our syntax

Definition 7.12 (Fibred equality syntax). Let $\mathbb{L}$ be a language with equality $=$, the classical connectives and quantifiers. We allow for two kinds of universal quantifiers, $(\forall x)$ and (x). $(\forall x)$ is intended as quantifying over terms and formulas and (x) quantifiers only over terms (i.e. is a sort of 'free logic' quantifier). We can define $(\exists x) =\sim (\forall x) \sim$ and $(Ex) =\sim (x) \sim$.

We have no function symbols, but we do allow individual constants. We define the notion of a fibred wff (f-wff) of the language as follows

1. A is a f-wff of level 0 if A is an ordinary free classical wff of $\mathbb{L}$ (built up with $\forall x$ and (x)).
2. A is a f-wff of level $\leq n+1$ if there exists a free classical wff $\varphi(x_1, \ldots, x_n, y_1, \ldots, y_m)$ with equality only, with free variables $x_1, \ldots, x_n, y_1, \ldots, y_m$ and f-wffs $A_1, \ldots, A_n$ of level $\leq n$ such that $A = \varphi(x_1/A_1, \ldots, x_n/A_n, y_1, \ldots, y_m)$, where x_i/A_i denotes the simultaneous substitution of A_i for x_i in φ. The free variables of A are $y_1, \ldots, y_m$ together with all the free variables of $A_1, \ldots, A_n$. We assume the substitution is carefully made and does not result in incorrect binding of variables.

Let $\bar{u} = (u_1, \ldots, u_n)$ be a finite list of terms. We can define a predicate $P_{\bar{u}}$ as follows

$$P_{\bar{u}}(x) = (\text{definition}) \bigvee_{i=1}^{n} u_i = x.$$

In any model the extension of $P_{\bar{u}}$ is

$$h(P_{\bar{u}}) = \{g(u_1), \ldots, g(u_n)\}.$$

This gives us a clue of how to define our fibred equality models.

For simplicity we opt for constant domains.

Definition 7.13 (Fibred equality models).

1. A constant domain fibred equality model has the form $\mathbf{n} = (S, D, a, \mathbf{F}, h, g)$ where S is a non-empty set of labels, naming worlds, $D \neq \varnothing$ is a domain, $a \in S$ is the actual world and $\mathbf{F}$ is the fibring function, giving for each $X \subseteq D$ and t a set of labels $\mathbf{F}(X, t)$. g is an assignment giving each t and each constant and variable x an element $g_t(x) \in D$, and h gives for each $t \in S$ and m-place predicate symbol P, an extension $h(t, P) \subseteq D^m$.

2. Satisfaction is defined as follows
   - $t \vDash P(x_1, \ldots, x_m)$ iff $(g_t(x_1), \ldots, g_t(x_n)) \in h(t, P)$
   - $t \vDash x = y$ iff $g_t(x) = g_t(y)$
   - $t \vDash x = \varphi$ iff for all $s \in \mathbf{F}(\{g_t(x)\}, t), s \vDash \varphi$
   - $t \vDash (x)A(x)$ iff for all $d \in D, t \vDash A(d)$
   - $t \vDash \forall x A(x)$ iff for all $d \in D, t \vDash A(d)$ and for all wffs φ, $t \vDash A(\varphi)$
   - $t \vDash \varphi = \psi$ iff for all $s \in \mathbf{F}(\varnothing, t), s \vDash \varphi \leftrightarrow \psi$.
3. A model $\mathbf{n}$ is said to be self-reflective if $S = D$.
4. Let $W(x)$ be a special unary predicate for self-reflective models such that for all x and t
   - $t \vDash W(x)$ iff $t = x$.
5. A model is said to be a pure equality model if it is a rigid (i.e. $g_t(x)$ independent of t) constant domain model for the language with $=$ and W alone.

Example 7.14. Let us evaluate $x = \varphi$.

1. $t \vDash (x = \varphi)$ iff for all $s \in \mathbf{F}(\{g_t(x)\}, t), s \vDash \varphi$. The above expression can be viewed as a unary predicate on x, i.e. $P_\varphi(x) =$ (definition) $x = \varphi$. Thus in the spirit of the previous section we get:
   - $t \vDash P_\varphi(\psi)$ iff for all $s \in \mathbf{F}(Y, t), s \vDash \psi$, where $Y = \{g_t(x) \mid t \vDash x = \varphi\}$.

   But syntactically, $P_\varphi(\psi)$ is $\psi = \varphi$ (with φ substituted first).
   Similarly we can look at $P_\psi(\varphi)$ and get
   - $t \vDash P_\psi(\varphi)$ iff for all $s \in \mathbf{F}(\{g_t(x) \mid t \vDash x = \psi\}, t), s \vDash \varphi$.

   It is natural to define $t \vDash_1 \varphi = \psi$ as $t \vDash P_\varphi(\psi) \wedge P_\psi(\varphi)$.
   This in general will not be the same as $t \vDash \varphi = \psi$.
2. Consider now the case $t \vDash x = \varphi(x)$.
   We have:
   - $t \vDash x = \varphi(x)$ iff for all $s \in \mathbf{F}(\{g_t(x)\}, t)$ we have $s \vDash \varphi(x)$.

   When φ is atomic, we get
   - $t \vDash x = \varphi(x)$ iff for all $s \in \mathbf{F}(\{g_t(x)\}, t), g_s(x) \in h(t, \varphi)$.

   Let for example $\varphi(x)$ be $x = y$. We get
   - $t \vDash x = (x = y)$ iff for all $s \in \mathbf{F}(\{g_t(x)\}, t)$ we have $g_s(x) = g_s(y)$.

   Note that equality acts as quantification. Let $g_t R_x g_s$ be defined as

   $$g_t R_x g_s \text{ iff } s \subset \mathbf{F}(\{g_t(x)\}, t).$$

   Then we can view the above as:
   - $g_t \vDash x = \varphi(x)$ iff for all g_s such that $g_t R_x g_s, g_s \vDash \varphi(x)$.

Compare with Section 10.3. See also Section 20.2.

Theorem 7.15 (Completeness theorem for pure equality models). *Consider the following axiom system for the theory of pure equality models.*

1. *All axioms and rules in the system of Theorem 7.7, where the P axiom group is understood to be taken for the predicate of the form*

$$P_u(x) = \textit{definition } (u = x).$$

2. *W axioms*

(1)	$(Ex)W(x)$
(2)	$W(x) \wedge W(y) \to x = y$
(3m)	$(x)(y_1, \ldots, y_m)(W(x) \to y_1 = (y_2 = \ldots = y_m =\sim W(x))\ldots)$
(4m)	$(x)(y_1, \ldots, y_m)(y_1 \neq (y_2 \neq \ldots y_m \neq W(x))\ldots) \to\sim W(x))$.

Proof

1. Soundness can be verified.
2. To show completeness, let Δ_0 be a consistent theory containing every instance of the W-axioms. We can assume that there is an infinite number of variables not in Δ_0. We construct a tree model as in the proof of Theorem 7.7.

We get a tree model $\mathbf{n} = (S, D, a, \mathbf{F}, h, g)$. We now use the W axioms to identify S and D. Axiom (1) and (2) ensure that for each node t there exists exactly one x such that $t \vDash W(x)$. Axioms (3m) and (4m) ensure that it is unique. We can therefore denote it by t_x. Thus the mapping $t \mapsto t_x$ maps uniquely S into D.

There may be elements d of D which are not of the form $t_x = d$, i.e. that there is no world t such that $t \vDash W(d)$. If we assume S is infinite (which we can always do by duplicating worlds) we can rename D in such a way that the mapping is onto conditional logic. □

7.3 Fibring Predicate Logics with Binary Relations

So far we have dealt with self-fibring of monadic predicate logics without function symbols. We have not dealt with the case of atomic f-wffs of the form $R(A, B)$, with A, B f-wffs or $P(f(\alpha))$, with α a wff and $f(x)$ a unary function symbol.

Let us consider the binary relation case first. For notational simplicity we assume the assignment g is rigid. To deal with the binary case properly we need to assume that in classical logic (or in the base logic for which we want to do self-fibring) we have the notion of a model satisfying a *pair* of wffs, in symbols

$$\mathbf{m} \vDash (A, B).$$

There is no such clear-cut notion in classical model theory. There are, however, some options:

- $\mathbf{m} \vDash (A, B)$ iff $\mathbf{m}$ has the form $\mathbf{m}_1 \otimes \mathbf{m}_2$ and $\mathbf{m}_1 \vDash A$ and $\mathbf{m}_2 \vDash B$.
- $\mathbf{m} \vDash (A, B)$ iff at some other model $\mathbf{m}_A$, dependent on A, we have $\mathbf{m}_A \vDash B$.
- $\mathbf{m} \vDash (A, B)$ iff $\mathbf{m} \vDash \psi(A, B)$ where $\psi(X, Y)$ is a higher order formula in the 'formula' variables X and Y.

Following the above discussion we can define the fibred semantics for $R(A, B)$ as follows:

- $(D, h, \mathbf{F}) \vDash R(A, B)$ iff $(D, h', \mathbf{F}) \vDash (A, B)$ where $\mathbf{F}$ is a fibring function, associating a new model $h' = \mathbf{F}(h(R), h)$ with each model h and a binary relation R.

The above definition is not complete and does not cover all possibilities for the case of binary relations. We need also consider the cases of $R(t, A)$ and $R(A, t)$ for terms t. For a fixed t, $R(x, t)$ is a unary predicate. We know how to evaluate the substitution of a wff A into a unary predicate. Thus we can give meaning to $R(A, t)$, for arbitrary t. But the latter is now a unary predicate in t and so we know how to give meaning to $R(A, B)$ by substituting B in the unary predicate $R(A, x)$. This new point of view must correctly relate to the previous one. Are they compatible?

Let us proceed more formally

Consider $R(x, t)$. For fixed t, we get a unary predicate in x. Let us use the notation $\lambda x R_t(x)$ for this predicate. Given a wff A, we can semantically evaluate $R_t(A)$ in the unary fibred models of the previous Section 7.1.

We can now consider $R_t(A)$, for A fixed, as a predicate in t, i.e. consider $\{t \mid \mathbf{m} \vDash R_t(A)\}$ for this predicate. Given a wff A, we can semantically evaluate $R_t(A)$ in the unary fibred models

$$\mathbf{m} \vDash R_t(A) \text{ iff } \mathbf{F}(\lambda x R_t(x), h) \vDash A.$$

Hence, for a wff B, we can evaluate $R_B(A)$

$$\mathbf{m} \vDash_1 R_B(A) \text{ iff } \mathbf{F}(\{t \mid \mathbf{m} \vDash R_t(A)\}, h) \vDash B.$$

Summarising, we get

$$\mathbf{m} \vDash_1 R(A, B) \text{ iff } \mathbf{F}(\{t \mid \mathbf{F}(\lambda x R(x, t), h) \vDash A\}, h) \vDash B.$$

We obtained $\vDash_1$ by looking at $\lambda x R(x, t)$ first. If we were to start with $\lambda t R(x, t)$ as our initial parametrised unary predicate, we get

$$\mathbf{m} \vDash_2 R(A, B) \text{ iff } \mathbf{F}(\{x \mid \mathbf{F}(\lambda t R(x, t), h) \vDash B\}, h) \vDash A.$$

The two definitions are not the same. $R(A, B)$ is syntactically symmetrical and we do not know whether A was substituted first and then B or the other way round, or perhaps both were substituted simultaneously.

Thus we have three ways of evaluating $R(A, B)$, namely $\vDash_1, \vDash_2$ and $\vDash$. We choose to opt for the solution whereby binary predicates are made sensitive to order of substitution. Thus $R(A, B)$ is not a meaningful syntactical predicate substitution for us even in the case $A = B$. We must know whether A was substituted first or B first or perhaps both were substituted simultaneously.

This way of looking at things seems to be connected with the concept of *lazy evaluation*.

We shall now formulate the logic we need for our purpose, a logic sensitive to order of substitution. We shall call this logic *slot logic*, a logic where the atomic predicates are sensitive to order of substitution. We begin with an example involving a binary predicate R. Traditionally R is a set of pairs over the domain D. In slot logic, R is viewed differently. If we view the variables in $R(x,y)$ as indicating slots, then $R(x,y)$ can become $R_x(y)$, a unary predicate parametrised by x, or $R_y(x)$, a unary predicate parametrised by y or $R(x,x)$. If we write R as $R[x,y]$, indicating it has two slots, then the predicates we can get are

$$\begin{array}{lll} \lambda y R_a(y) & \text{denoted by} & \mathcal{S}^x_a R[x,y] \\ \lambda x R_b(x) & \text{denoted by} & \mathcal{S}^y_b R[x,y] \\ \lambda z R(z,z) & \text{denoted by} & \mathcal{S}^{x;y}_{z,z} R[x,y] \\ & \text{or by} & \mathcal{S}^{y;x}_{z,z} R[x,y]. \end{array}$$

If σ is the substitution $\sigma(x) = z$, $\sigma(y) = z$ then we can also view $\lambda z R(z,z)$ as $R_\sigma(z)$.

Consider now a formula of predicate logic built up using R. Take for example

$$\varphi(u,v) = \forall z(R(z,v) \to\sim R(u,z)).$$

φ has two slots. We indicate this by writing $\varphi[u,v]$. Then we have

$$\varphi[u,v] = \forall z(\mathcal{S}^{x;y}_{z,v} R(x,v) \to\sim \mathcal{S}^{y;x}_{z,u} R(x,y)).$$

Semantically, we need to give $R(x,y)$ three functions.

$$R_1 : D \mapsto 2^D$$

as a predicate in y parametrised by x, i.e. $\mathcal{S}^x_a R(x,y)$

$$R_2 : D \mapsto 2^D,$$

for $\mathcal{S}^y_b R(x,y)$ and

$$R_{1=2} \subseteq D,$$

for $R_\sigma(z)$.

Let us now evaluate $\varphi(u,v)$ in such a model:

$$\mathbf{m} \vDash \varphi(u,v) \text{ iff for all } d \in D, [R_1(d)(v) \to\sim R_2(d)(u)] \text{ holds in } \mathbf{m}.$$

Note that $\varphi(u,v)$ happens to be independent of the order of substitution of u and v. However, if we consider

$$\psi(u,v) = \varphi(u,v) \wedge R(u,v)$$

then it would be dependent.

We thus adopt the policy that all formulas must indicate order of substitution. The view we choose to take is as follows.

We begin with a slot matrix which we denote by $R[x,y]$. This matrix has two ordered slots (x,y). We can obtain more matrices by performing several operations.

1. We can use a substitution σ, where $\sigma(x) = u\ \sigma(y) = v$. σ transforms R to R_σ, which is still a slot matrix. If $u = v = z$, we get a matrix with one slot which we write as $R_\sigma[z]$.
2. We can perform parametric reduction, where we obtain

$$\mathcal{S}^x_u R, \quad \mathcal{S}^y_v R, \quad \mathcal{S}^{x,y}_{u,v} R \quad \text{and} \quad \mathcal{S}^{y,x}_{v,u} R.$$

The first two are parametrised slot matrices with one slot and the last two are predicates in the parametric variables u and v, where the order of substitution matters. We present the parameters as the sequence (u, v) (u first then v). The semantic interpretation of $\mathcal{S}^{x,y}_{u,v} R$ is the function R_1 and for $\mathcal{S}^{y,x}_{v,u} R$ is R_2. The semantic interpretation for R_σ is $R_{1=2}$. We can be more pedantic and insist that since $R_\sigma[z]$ is a matrix we need to write $\mathcal{S}^z_w R_\sigma$ to turn it into a predicate.

Thus, for example, the classically written ambiguous expression $R(d, d)$ for $d \in D$, can be understood in three ways:

1. Substitution $(u, v) = (d, d)$ in $\mathcal{S}^{x,y}_{u,v} R$, in which case it holds iff $d \in R_1(d)$.
2. Substitution $(v, u) = (d, d)$ in $\mathcal{S}^{y,x}_{v,u} R$, in which case it holds iff $d \in R_2(d)$.
3. Substitution $(u) = (d)$ in $\mathcal{S}^z_u R_\sigma[z]$, in which case it holds iff $d \in R_{1=2}$.

We need not assume that these three cases give the same answers. Of course we do get the same answers for standard classical logic semantics, in which the extension of R is a subset $||R|| \subseteq D^2$ and R_1, R_2 and $R_{1=2}$ are derived from $||R||$ in the obvious way.

Let us consider what can happen in practice. Let

$$A[x, y, z] = R[x, y] \wedge R[z, y] \wedge R[z, z].$$

We want to evaluate $A[\varphi_1, \varphi_2, \varphi_3]$. Substituting we get

$$R[\varphi_1, \varphi_2] \wedge R[\varphi_3, \varphi_2] \wedge R[\varphi_3, \varphi_3].$$

We know the above values depend on order of substitution. Let us agree that we substitute for x first, then for y, then for z. To represent this agreement we shall write $\mathcal{S}^{x,y,z}_{u_1,u_2,u_3} A$. Distributing $\mathcal{S}$ across the connectives we get

$$\mathcal{S}^{x,y}_{u_1,u_2} R \wedge \mathcal{S}^{y,z}_{u_2,u_3} R \wedge \mathcal{S}^z_{u_3} R_\sigma[z].$$

The last conjunct was ambiguous, and we chose one of the possibilities. We can agree that if we have a general $R[\ldots, z, \ldots, z, \ldots]$ we interpret it as an R_σ for the appropriate σ.[24]

[24] It makes a lot of sense to insist that even in the case of $R[z, z]$, we should know which slot was filled in first. There are applications which are sensitive to such order of substitution.

Consider a well-known problem of PROLOG involving the query

$$\leftarrow \underline{\text{var}}\ (x) \wedge \underline{\text{solve}}\ (p(x))$$

as compared with the query

$$\leftarrow \underline{\text{solve}}\ (p(x)) \wedge \underline{\text{var}}\ (x).$$

Now consider the formula $(z)A[x, y, z]$. This has x, y as free slots. Let us decide that we want to view it as $\mathcal{S}^{x,y}_{u_1,u_2}(z)A$. We have

$$(z)A = R[x, y] \wedge (z)R[z, y] \wedge (z)R[z, z].$$

The second conjunct, $(z)R[z, y]$, has one free variable. Does this mean that z is substituted first and then $y = u_2$ or do we substitute $y = u_2$ and then quantify over z?

In the first case we use R_1 in the semantical evaluation of the expression and in the second case we use R_2. Given $(z)R[z, \varphi]$, it makes a difference whether we view it as $(z)R_z[\varphi]$ or $(z)R_\varphi[z]$.

The difference can be understood syntactically as the difference between the following two reduction strategies.

$$\mathcal{S}^{x_1,\ldots,x_n}_{u_1,\ldots,u_n}(z)B = (y)\mathcal{S}^{x_1,\ldots,x_n,z}_{u_1,\ldots,u_n,y}B$$
$$\text{or}$$
$$= (y)\mathcal{S}^{z,x_1,\ldots,x_n}_{y,u_1,\ldots,u_n}B.$$

We choose the more intuitive strategy that z is substituted last, i.e. the first reduction option.

Definition 7.16 (Slot logic). Let $\mathbb{L}$ be a predicate language. Add to $\mathbb{L}$ the slot operator $\mathcal{S}$ and the slot variables $\{u_1, u_2, \ldots,\}$, to form the new slot language $\mathbb{L}_\mathcal{S}$, defined as follows.

1. Let $A(x_1, \ldots, x_n)$ (written $A(\bar{x})$) be a wff of $\mathbb{L}$ with the ordered sequence of slot variables $\bar{x} = (x_1, \ldots, x_n)$. Then $\mathcal{S}^{x_1,\ldots,x_n}_{u_1,\ldots,u_n}A$ is a basic slot formula with the ordered sequence of free variables $(u_1, \ldots, u_n)$. We also write it as $\mathcal{S}^{\bar{x}}_{\bar{u}}A$. The variable u_n is said to be ready for the next quantification.
   The order on $(x_1, \ldots, x_n)$ is just to identify the slots in A and there is no requirement to substitute values for x_1 first, x_2 second, etc. The order on $(u_1, \ldots, u_n)$ indicates the order of substitution of values. Note that the last variable in the ordering is the one to next accept quantification.
2. A slot formula is defined by the following clauses:
   - A basic slot formula is a slot formula.
   - If A and B are slot formulas whose free variable ordering are compatible (i.e. are relating the same on the shared variables), then so are $A \wedge B$, $A \vee B$, $\sim A$, $A \to B$, and the last variable in the combined ordering is the one ready for the next quantification.
   - $(u)A$ and $(Eu)A$ are slot formulas if A is such that u is the free variable in A, ready for the next quantification. The ordering of variables in the quantified formula is inherited and the last variable in this ordering (next to the last in the original) is ready for the next quantification.

Since a PROLOG interpreter evaluates the conjuncts sequentially, we get severe semantic problems with var, if we stick to traditional classical logic. If the program consists solely of the clause $p(a)$ (a a constant) then (using the leftmost literal computation rule) the first goal succeeds, while the second goal fails.

As a slot formula the query will have to specify which occurrence of x we want to instantiate first and therefore there is hope for an object level semantics for var and not necessarily only a metalevel semantics, at the cost of having to define unification in the metalevel. See [Hill and Lloyd, 1994].

3. Substitution
Let $B(u_1, \ldots, u_n) = \mathcal{S}_{\bar{u}}^{\bar{x}} A$ be a slot formula and $t_1, \ldots, t_n$ closed terms. Then we define $B(u_1/t_1, \ldots, u_n/t_n)$ to be $\mathcal{S}_{\bar{t}}^{\bar{x}} A$.

We are now ready for the definition of slot semantics. To explain the basic idea consider a three-place atomic relation $R[x_1, x_2, x_3]$. The sequence $\bar{x}$ of slot variables just indicates the three places in the slot predicate. The semantic value $||R||$ of R depends on the order of substitution into the slot. So for each permutation π of $(1, 2, 3)$ (or of (x_1, x_2, x_3)) we need a function $||R||_\pi^3 : D \mapsto (D \mapsto (D \mapsto \{0, 1\}))$ where D is the domain. So suppose we want to check whether $R(a_1, a_2, a_3)$ holds. This we cannot do because we do not know in what order the values a_i were substituted. In fact, $R(x_1, x_2, x_3)$ is not a predicate because it does not indicate in what order it accepts values. Consider $\mathcal{S}_{u,v,w}^{x_2,x_1,x_3} R$. This is now a predicate in (u, v, w) wanting the value for x_2 first, then for x_1, then for x_3.

Thus $\mathcal{S}_{u,v,w}^{x_2,x_1,x_3} R$ can be evaluated. Let π be the permutation $(1, 2, 3) \mapsto (2, 1, 3)$. Then the above holds for (u, v, w) iff $||R||_\pi^3(u)(v)(w) = 1$.

Let us now consider the matrix $R(y, x, y)$. This is obtained from $R(x_1, x_2, x_3)$ by the substitution σ, where $\sigma(x_1) = y$, $\sigma(x_2) = x$ and $\sigma(x_3) = y$. We view it as yielding a new matrix $R_\sigma(x, y)$.

The question is, if we want to know the semantical value $||R_\sigma||$ of R_σ, can we derive it from the semantical value for R, or do we need to provide new and unrelated semantics for R_σ? If we know the order of substitution into R, i.e. whether $\sigma(x_1) = y$ came before $\sigma(x_3) = y$, etc., then we know the permutation π involved, and we can derive $||R_\sigma||_\pi$. Otherwise we cannot derive a unique semantical value for R_σ.

There is always the traditional, classical logic, case where all functions $||R||_\pi^n$ are derived from one subset $||R|| \subseteq D^3$. We have in this case:

$$||R||_\pi(a_1)(a_2)(a_3) = 1 \text{ iff } \pi^{-1}(a_1, a_2, a_3) \in ||R||$$

where π^{-1} is the inverse permutation.

In this case there is no real dependence on the order of substitution, at least not for substitutions from D. Note that even in this case, for formula substitution, the order does matter, as we have already seen.

Definition 7.17 (Slot models). Let $\mathbb{L}$ be a predicate language and let D be a domain and let $R(x_1, \ldots, x_n)$ be an n-place predicate of $\mathbb{L}$.

We define the semantic extension of the predicate letter R as a family of functions as follows.

1. For each permutation π of $\{1, \ldots, n\}$ let $||R||_\pi^n$ be a function $||R||_\pi^n : D \mapsto (D \mapsto \ldots (D \to \{0, 1\}) \ldots)$ (n times, i.e. $||R||_\pi^n \subseteq D^n$).
2. R is said to have a normal semantics if there exists an $||R|| \subseteq D^n$ such that

$$||R||_\pi^n(a_1)(a_2) \ldots (a_n) = 1 \text{ iff } \pi^{-1}(a_1, \ldots, a_n) \in ||R||.$$

where $\pi^{-1}\bar{a} = (a_{\pi^{-1}}(i))$

3. A model $\mathbf{m}$ has the form $(D, ||R||^n_\pi, h)$, where D is a non-empty domain and $||\ ||$ is an assignment of semantical values to the atomic slot predicates. h assigns values from D to the individual variables $\{u_1, u_2, \ldots\}$ and constants.
4. Satisfaction is defined as follows:
   - $\mathbf{m} \vDash S^{\pi x_1,\ldots,\pi x_n}_{u_1,\ldots,u_n} R$ iff $||R||^n_\pi(h(u_1))\ldots(h(u_n)) = 1$
   - $\mathbf{m} \vDash A \wedge B$ iff $\mathbf{m} \vDash A$ and $\mathbf{m} \vDash B$
   - $\mathbf{m} \vDash \sim A$ iff $\mathbf{m} \nvDash A$
   - $\mathbf{m} \vDash (z)A(x)$ iff for all $d \in D, \mathbf{m} \vDash A(d)$.

   Note that $(z)A(z)$ is well formed only if z is a variable to take its value first in the formula $A(z)$.

There is another way of looking at slot predicates. Consider an n-place atomic predicate $R(x_1, \ldots, x_n)$. Let us turn it into an $(n+1)$-place predicate depending on a parameter τ. So we can write the new predicate as $R(\tau, x_1, \ldots, x_n)$. Such a procedure is common in logic. For example, τ may be a time variable, turning R into a time dependent predicate. Given a tuple $(a_1, \ldots, a_n)$ we cannot evaluate whether R holds until we are given the time τ. When we know τ, we can evaluate $R(\tau, a_1, \ldots, a_n)$. In the general case, the parameter can be anything, any member of some set of parameters. Let us try a special set of parameters. Let Π_n be the set of all the permutations of the sequence $(1, \ldots, n)$.

Thus for each $\pi \in \Pi_n$, we have a predicate $R(\pi, x_1, \ldots, x_n)$, which we can write as $R_\pi(x_1, \ldots, x_n)$.

To get a value we need to give a tuple $(a_1, \ldots, a_n)$ and a parameter (permutation) π. The a_i are elements of the domain; they can be easily given. To give π we need to give a one to one function, $\pi(i) = k_i, i = 1, \ldots, n$,

$$\pi : \{1, \ldots, n\} \mapsto \{1, \ldots, n\}.$$

Here is the trick that makes this logic into slot logic. If we give the $(a_1, \ldots, a_n)$ as a permuted sequence, i.e. $(a_{k_1}, \ldots, a_{k_n})$, then we can read the permutation π from the sequence.

Thus to say that $R(x_1, \ldots, x_n)$ is sensitive to the order of substitution is equivalent to saying that we have a parametrised family of predicates $R(\pi, x_1, \ldots, x_n)$, $\pi \in \Pi_n$ and that we agree to read π from the order in which the values $a_1, \ldots, a_n$ fill the slots $x_1, \ldots, x_n$.

Of course the meaning of a formula like

$$\forall x \forall y R(x, y)$$

is now not the same as

$$\forall y \forall x R(x, y).$$

This is so because of hidden information in the order in which the values are substituted in R.

$\forall x \forall y R(x,y)$ stands for the unordered quantified wff $(\forall xy)R(\pi_1,x,y)$ while $\forall y \forall x R(x,y)$ stands for the unordered quantified wff $(\forall xy)R(\pi_2,x,y)$ where $\pi_1(1,2) = (1,2)$ and $\pi_2(1,2)) = (2,1)$.

Theorem 7.18 (Completeness theorem for slot logic). *Let $\mathbb{L}_\mathcal{S}$ be the slot language for Definition 7.16. Consider the following axiom system for the language.*

1. *All substitution instances of truth functional tautologies.*
2. *The following reduction rules, where $\bar{x}$ presents all the free variables of both A and B:*
   - $\mathcal{S}^{\bar{x}}_{\bar{u}}(A \wedge B) = \mathcal{S}^{\bar{x}}_{\bar{u}} A \wedge \mathcal{S}^{\bar{x}}_{\bar{u}} B.$
   - $\mathcal{S}^{\bar{x}}_{\bar{u}} \sim A = \sim \mathcal{S}^{\bar{x}}_{\bar{u}} A.$
   - $\mathcal{S}^{\bar{x}}_{\bar{u}}(z)A(z,\bar{x}) = (y)\mathcal{S}^{\bar{x},z}_{\bar{u},y} A(z,\bar{x}).$
   - $\mathcal{S}^{\bar{y}}_{\bar{v}} \mathcal{S}^{\bar{x}}_{\bar{u}} A = \mathcal{S}^{\bar{y},\bar{x}}_{\bar{v},\bar{u}} A.$
   - $\mathcal{S}^{x}_{u} A = A$*, if x is not free in A.*
   - $\mathcal{S}^{\pi_1 \bar{x}}_{\bar{u}} \mathcal{S}^{\pi_2 \bar{y}}_{\bar{x}} A = \mathcal{S}^{\pi_1 \pi_2 \bar{y}}_{\bar{u}} A.$

   Obviously $\mathcal{S}$ can be pushed down to atomic formulas.
3. $\vdash (y)\mathcal{S}^{\bar{x},z}_{\bar{u},y} A \to \mathcal{S}^{\bar{x},z}_{\bar{u},t} A.$
4. $\dfrac{\vdash A \to \mathcal{S}^{\bar{x},z}_{\bar{u},y} B}{\vdash A \to (y)\mathcal{S}^{\bar{x},z}_{\bar{u},y} B}$ *,y not in A.*

Note that $\mathcal{S}^{\pi_1 \bar{x}}_{\bar{u}} A$ is not necessarily equivalent to $\mathcal{S}^{\pi_2 \bar{x}}_{\bar{u}} A$, unless $\pi_1 = \pi_2$.
Then the system is complete for the slot semantics of Definition 7.17.

Proof

1. Soundness can be verified.
2. Completeness
   We first use the reduction rules for $\mathcal{S}$ to push all $\mathcal{S}^{\bar{x}}_{\bar{u}}$ operators right next to the atoms. Thus given a slot wff A it is equivalent to another slot wff B where all slot operators $\mathcal{S}^{\bar{x}}_{u}$ are next to atomic predicates $R(\bar{y})$, where $\bar{y} = \pi\bar{x}$, for some permutation π.
   We can now consider B as a formula of ordinary predicate logic built up from the atomic predicates $R_\pi(\bar{u})$ (representing $\mathcal{S}^{\bar{x}}_{\bar{u}} R(\bar{y})$). The axioms (of predicate logic) available to us allow us to build a model for B. This model can be viewed as a slot model because it gives extensions $||R_\pi||$ to the atoms R_π. If, for some π, R_π is not represented in B, let $||R_\pi||$ be arbitrary.

This completes the proof. □

8

SELF-FIBRING WITH FUNCTION SYMBOLS

8.1 Fibring Predicate Logics with Function Symbols

We now turn to the case where function symbols are available for fibring into the language. Again we assume that the assignments g are rigid. We assume *unary* function symbols $e_i(x), i = 1, 2, 3, \ldots$, as well as unary predicates. Let $e(x)$ be a term. It has the form $e_1^{k_1}(e_2^{k_2}\ldots(e_n^{k_n}(x))\ldots)$. The atomic expressions have the form $P(e(x))$. We thus have to give semantics to expressions of the form $e(\varphi)$ and $P(e(\varphi))$.

There are two possibilities for $e(\varphi)$. It can be either another formula $\varphi_1 = e(\varphi)$ or a term. Both options make sense. The function e can either be like a unary operator and give a formula value or like a naming function and give a term. Of course there is the mixed possibility where $e(\varphi)$ can be a formula for some φ and a terms for others.

We shall address the pure case where $e(\varphi)$ yields a formula as value for a formula φ. The reader should note that later on we devote a section to ambivalent syntax, where $p(x)$ can be either a function symbol or a predicate. This makes the above distinctions less urgent.

We adopt the view that two formulas are the same if they behave the same in the system. Thus $\alpha = \beta$ iff for all $P, P(\alpha) = P(\beta)$. Thus if we denote by φ_1 the formula $e(\varphi)$, then we can identify the nature of φ_1 by looking at the behaviour of $P(\varphi_1) = P(e(\varphi))$ for all predicates P.

Our problem now is to understand the meaning of $P(e(\varphi))$.

The idea is to treat $P(e(\varphi))$ as $P_1(\varphi)$ where $P_1(x)$ is defined as $P(e(x))$. We therefore have for any P

$$P(\varphi_1) = P(e(\varphi)) = P_1(\varphi).$$

Semantically we need to have

$$\mathbf{m} \vDash P(\varphi_1) \text{ iff } \mathbf{m}_P \vDash \varphi_1$$

but we also have

$$\begin{aligned} \mathbf{m} \vDash P(\varphi_1) &\text{ iff } \mathbf{m} \vDash P_1(\varphi) \\ &\text{ iff } \mathbf{m}_{P_1} \vDash \varphi \end{aligned}$$

but $\hat{x}P_1(x) = \hat{x}P(e(x))$. Therefore

$$\mathbf{m}_P \vDash e(\varphi) \text{ iff } \mathbf{m}_{e^{-1}(P)} \vDash \varphi,$$

where e is now regarded as a function from predicates to predicates

$$e^{-1}(P)(x) = P(e(x)).$$

Equivalently, for $X \subseteq D$, let

$$e^{-1}(X) = \{y \mid e(y) \in X\}.$$

We therefore get the following recipe for evaluating $\mathbf{n} \vDash e(\varphi)$. Let P be arbitrary such that $\mathbf{n} = \mathbf{m}_P$. Then by the above considerations we have

$$\mathbf{n} = \mathbf{m}_P \vDash e(\varphi) \text{ iff } \mathbf{m}_{e^{-1}(P)} \vDash \varphi.$$

Thus we can now intuitively define:

- $\mathbf{n} \vDash e(\varphi)$ iff for all P and all $\mathbf{m}$ such that $\mathbf{n} = \mathbf{m}_P$, we have $\mathbf{m}_{e^{-1}(P)} \vDash \varphi$.

Thus we have given $e(\varphi)$ meaning by indicating how it behaves in the semantics.

We are now ready for the formal definitions.

Definition 8.1 (Fibred syntax with unary functions). Let $\mathbb{L}$ be the classical language of Definition 7.1, and assume that $\mathbb{L}_M$ now also contains besides the unary predicates also the additional unary function symbols $\{e_1, e_2, e_3, \ldots\}$. The notion of a well-formed formula of level 0 is modified accordingly.

The notion of f-wff of level $n+1$ is defined similarly with the added clause that if A is of level n and $e(x)$ is a term then $e(A)$ is a wff of level $n+1$.

Definition 8.2 (Semantics for fibred unary functions). We modify Definition 7.2. Since our language contains function symbols, the assignment g assigns to each function symbol e a function $g(e) : D \mapsto D$. With each set $X \subseteq D$ and each $g(e)$, we let $g(e)^{-1}(X) = \{y \mid g(e)(y) \in X\}$.

Notice that all terms are rigid designators!

Satisfaction is defined inductively as before, with the following additional inductive clause for the function symbol e. Recall that if $\mathbf{F}$ is a set function then $\mathbf{F}(X, t) \vDash A$ means for all $s \in \mathbf{F}(X, t), s \vDash A$.

- $(D, h, g, \mathbf{F}) \vDash e(\varphi)$ iff for all $Y \subseteq D$ and all h_1 such that $h \in \mathbf{F}(Y, h_1)$ we have that $(D, h', g, \mathbf{F}) \vDash \varphi$ for any $h' \in \mathbf{F}(g(e)^{-1}Y, h_1)$.

The meaning of the clause is as follows.

We want to evaluate $h \vdash e(\varphi)$. We essentially look at all h_1 and Q such that $h \in \mathbf{F}(h_1(Q), h_1)$ and evaluate $h_1 \vDash Q(e(\varphi))$.

Remark 8.3.

Note that in the previous definition, when evaluating $e(\varphi)$, we assumed e is a primitive function symbol. So if $\varphi = e_2(\varphi_2)$ then we apply the definition first to e and then to e_2 (option 1) and not to $e \circ e_2$ in one step (option 2).

Let us compare the two options. Figures 8.1 and 8.2 illustrate the situation:

Option 1
$h \vDash e(\varphi)$ iff for all h_1 and Q such that $h \in \mathbf{F}(h_1(Q), h_1)$, we have $h_1 \vDash Q_1(\varphi)$, where $Q_1(x) = Q(e(x))$.

Option 2
$h \vDash e(e_2(\varphi_2))$ iff for all h_2 and Q such that $h \in \mathbf{F}(h_2(Q), h_2)$ we have $h_2 \vDash Q_2(\varphi_2)$, where $Q_2(x) = Q(e(e_2(x)))$.

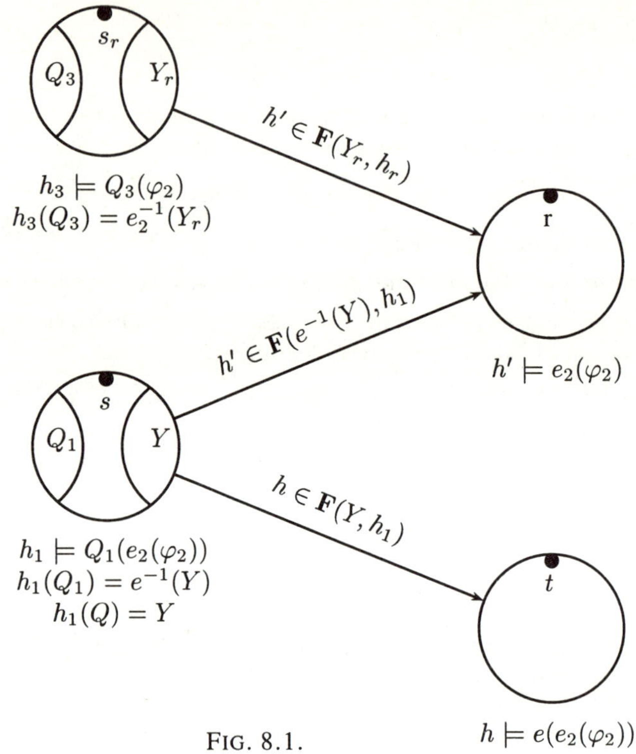

FIG. 8.1.

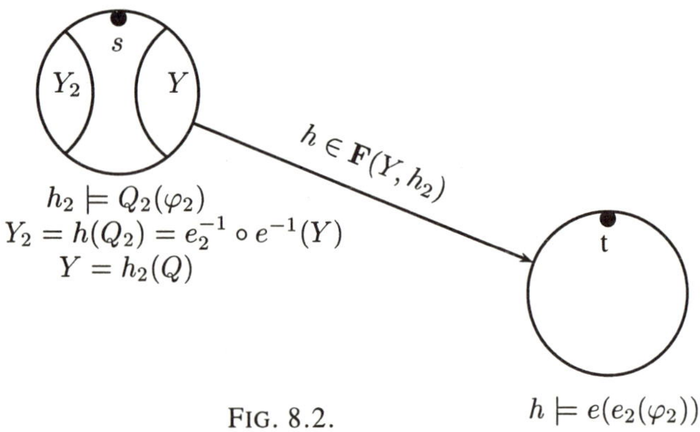

FIG. 8.2.

To check option 1 in detail we evaluate $h_1 \models Q_1(\varphi)$. This holds iff $\mathbf{F}(h_1(Q_1), h_1) \models e_2(\varphi_2)$ iff for all h_3, Q' such that $\mathbf{F}(h_3(Q'), h_3) \cap \mathbf{F}(h_1(Q_1), h_1) \neq \varnothing$ we have $h_3 \models Q_3(\varphi_2)$, where $Q_3(x) = Q'(e_2(x))$

In particular for $h_3 = h_1$ and $Q' = Q_1$ we get

$$h_1 \models Q_1(e_2(\varphi_2)) = Q(e(e_2(\varphi_2))).$$

thus if option 1 holds, so does option 2.

The converse does not hold in general, but it does hold in tree models (see Definition 7.10), which have a uniqueness property.

Let t be a point in a tree model. Then there exists a unique pair (Y, s), with $Y \subseteq D_s$ such that $t \in \mathbf{F}(Y, s)$.

We now evaluate $t \vDash e(e_2(\varphi_2))$ in two ways and verify that they are equivalent.

We start with option 2, see Figure 8.1:

$$t \vDash e(e_2(\varphi_2)) \text{ iff for all } s \text{ and } Q \text{ such that } t \in \mathbf{F}(h(s,Q), s)$$

we have $s \vDash Q_2(\varphi_2)$, where $Q_2(x) = Q(e(e_2(x)))$.

Let $Y = h(s, Q)$. Then the above s and Y are unique. Hence

$$h(s, Q_2) = e_2^{-1} \circ e^{-1}(Y) = Y_2.$$

Hence

$$t \vDash e(e_2(\varphi_2)) \text{ iff } \mathbf{F}(Y_2, s) \vDash \varphi_2$$

(recall that this means for all $r \in \mathbf{F}(Y_2, s), r \vDash \varphi$), where $Y_2 = e_2^{-1} \circ e^{-1}(Y)$ and s and Y are the unique predecessors of t (i.e. $t \in \mathbf{F}(Y, s)$).

To compare the evaluation according to option 1 we proceed as follows (Figure 8.2):

$t \vDash e(e_2(\varphi_2))$ iff for the unique Y and s such that $t \in \mathbf{F}(Y, s)$ we have $\mathbf{F}(e^{-1}(Y), s) \vDash e_2(\varphi_2)$.

We continue: iff for each $r \in \mathbf{F}(e^{-1}(Y), s)$, we have the unique s_r and Y_r such that $r \in \mathbf{F}(Y_r, s_r)$ we have $\mathbf{F}(e_2^{-1}(Y_r), s_r) \vDash \varphi_2$.

But because of uniqueness we have $s = s_r$ and $e^{-1}(Y) = Y_r$ and hence we get $t \vDash e(e_2(\varphi_2))$ iff for the unique s, Y such that $t \in \mathbf{F}(Y, s)$, we have $\mathbf{F}(e_2^{-1} \circ e^{-1}(Y), s) \vDash \varphi_2$.

Thus, for tree models, option 1 and option 2 lead to the same result.

Theorem 8.4 (Completeness theorem for fibred unary functions). *Let $\mathbb{L}_m[\mathbb{L}]$ be the self fibred language with the quantifiers (x) and $(\forall x)$, unary predicates and unary functions. Consider the following axiom system in the language.*

1. *All axioms and rules for the language of unary predicates with (x) and$(\forall x)$, as given in Theorem 7.7.*
2.
   - $l_1 \neq l_2 \to e(l_1 \neq l_2)$
   - $e((\forall x)A(x)) \leftrightarrow (\forall x)e(A(x))$
   - $e((x)A(x)) \leftrightarrow (x)e(A(x))$
   - $e(A \wedge B) \leftrightarrow e(A) \wedge e(B)$
   - $\vdash A \Leftrightarrow \vdash e(A) \Leftrightarrow \vdash Q(e(A))$, *where Q is not in A*

Then the above system is sound and complete for the tree semantics of Definition 7.10.

Proof

1. Soundness

   This can be verified. The rule $\vdash A \Rightarrow \vdash e(A)$ and $\vdash Q(A)$ holds in the tree semantics because for every tree model with actual world a, another tree model can be created with additional sequence of points which are related to a.

2. Completeness
   Assume A is consistent. Then $\sim Q \sim A$ is consistent. We can therefore assume that our consistent formula is such that every subformula of the form $e(A)$ is within a scope of a predicate Q. Let $Q_e(x)$ abbreviate $Q(e(x))$. A can be equivalently rewritten as a formula A^* which has no instances of fibring into any function symbols, but with the additional predicates of the form $Q_e(x)$. Let Δ_A be the following theory:

$$\Delta_A = \{A^*\} \cup \{\bar{P}((x)(Q_e(x) \leftrightarrow Q(e(x)))) \mid \text{ where } \bar{P} \text{ is a string of unary predicates and } e \text{ is an arbitrary string of function symbols}\}.$$

   Δ_A is consistent and is in the language $\mathbb{L}_M[\mathbb{L}]$.

We can use the completeness proof for unary predicates (Theorem 7.7) to construct a model for Δ_A provided we show that the predicates $Q_e(x)$ do behave like unary predicates and satisfy the axioms in Theorem 7.7. The axioms to check are

- $Q_e(\forall x B(x)) \leftrightarrow \forall x Q_e(B(x))$
- $Q_e((x)B(x)) \leftrightarrow (x)Q_e(B(x))$
- $Q_e(B \wedge C) \leftrightarrow Q_e(B) \wedge Q_e(C)$
- $\dfrac{\vdash \bigwedge A_i \to B}{\vdash \bigwedge Q_e(A_i) \to Q_e(B)}$.

The above hold because they hold for e.

We thus get a fibred tree model forΔ_A.

Define $t \vDash e(\varphi)$ using the clauses in Definition 8.2.

We need to show that we get a model for A. Since it is a model for Δ_A, it is also a model for A. □

8.2 Fibred Semantics for Ambivalent Syntax

So far our syntax was not ambivalent. We have predicate logic with unary predicates $p(x)$ and unary functions $e(x)$ and it is clear for each symbol $p(x)$ whether it is a predicate or a function. There are many application areas where the language is ambivalent, i.e. where $p(x)$ can be treated either as a function or as a predicate. See [Karlsbeek, 1995] for details and further references.[25] Our aim is to give a definition for $t \vDash P(\varphi)$

[25] The term ambivalent syntax is used by M. Karlsbeek. This kind of syntax was first seriously addressed in [Gabbay, 1992]. The language was called **HFP**, hereditarily finite predicates. The paper [Karlsbeek and Jiang, 1995] gives a good idea of what Karlsbeek has done. The syntax indeed allows for (what we called self-fibred) expressions of the form $p(p)$ but the semantics is different. It is nothing like our fibred semantics. The authors call it *closed term semantics*.

In this semantics the models are just pairs (D, T), where D, the domain, is the set of closed expressions of the language and the model is obtained via the subset T of the set of all closed atomic expressions of the language. Thus for atomic P, and a closed formula φ, $P(\varphi)$ can get an arbitrary truth value, which does not depend on the logical structure of φ, (depending only on whether $P(\varphi) \in T$). The authors are aware of this limitation. They write:

which is technically identical to that of $t \vDash e(\varphi)$ and thus enable ourselves to give a uniform definition for $t \vDash p(x)$, not caring any more whether p is a unary function or a unary predicate. This, we believe, will be a better semantics for ambivalent syntax.

Let us begin by examining the definitions of $t \vDash e(\varphi)$ and $t \vDash P(\varphi)$, with a view of making them formally the same.

1. $t \vDash e(\varphi)$ iff for all (Q, s) such that $t \in \mathbf{F}(Q, s)$ we have $s \vDash Q_e(\varphi)$, where $Q_e = (\lambda x)Q(e(x))$.
2. $t \vDash P(\varphi)$ iff for all $x \in \mathbf{F}(P, t), x \vDash \varphi$

Let us change the definition of $t \vDash P(\varphi)$ to the following

2*. $t \vDash_* P(\varphi)$ iff for all (Q, s) such that $t \in \mathbf{F}(Q, s)$ we have $s \vDash Q_P(\varphi)$, where $Q_P = \lambda x Q(P(x))$.

If we compare $\vDash_*$ and $\vDash$ we see that

$t \vDash_* P(\varphi)$ iff for all (Q, s) such that $t \in \mathbf{F}(Q, s)$ and for all $r \in \mathbf{F}(Q_P, s)$ we have $r \vDash \varphi$

Let $\mathbf{F}^*$ be defined as follows (see Figure 8.3):

- $r \in \mathbf{F}^*(P, t)$ iff for some (Q, s) we have $t \in \mathbf{F}(Q, s)$ and $r \in \mathbf{F}(Q_P, s)$.

Therefore all valid formulas of the self-fibred logic under definition (2) are also valid under definition (2*). We will show later in this section that we do not get any new theorems under the new definition (2*). For the moment, let us continue with our flow of ideas and give fibred semantics for ambivalent syntax.

Definition 8.5 (Ambivalent syntax with one-place symbols). Let $\mathbb{LA}$ be a language with a stock of individual constants and variables and a stock of unary symbols $\{p, q, \ldots, \}$.

1. An atomic expression has the form $p(x)$, where p is a unary symbol and x a variable (which we say is free in $p(x)$) or a constant.
2. An expression of level 0 is either an atomic expression or has the form $A \wedge B$ or $\sim A$ or $A \Rightarrow B$ or $A \vee B$, in which case the free variables of the combined expression are all the free variables of each, or has the form $(x)A(x)$ or $(\forall x)A(x)$ or $(\exists x)A(x)$ or $(Ex)A(x)$, in which case the free variables of the new expression are those of A less the variable x.
3. An expression of the form $A(x_1/C_1, \ldots, x_n/C_n, y_1, \ldots, y_k)$ is said to be of level $\leq n+1$ if it is obtained from an expression of level 0 of the form $A(x_1, \ldots, x_n, y_1, \ldots, y_k)$ by substitution of C_i for $x_i, i = 1, \ldots, n$, where C_i are expressions of level $\leq n$. The free variables of the new expressions are $y_1, \ldots, y_k$ and all the

The truth values of atoms are, in principle, not related to the structure of expressions which occur in them as subterms.

It should be noted that 'similar' expressions, like, for example, $\forall x.f(x)$ and $\forall y.f(y)$ constitute different, and unrelated, objects in the domain of models. That is, the truth values of the closed expressions $t(\forall x.f(x))$ and $t(\forall y.f(y))$ need not be the same.

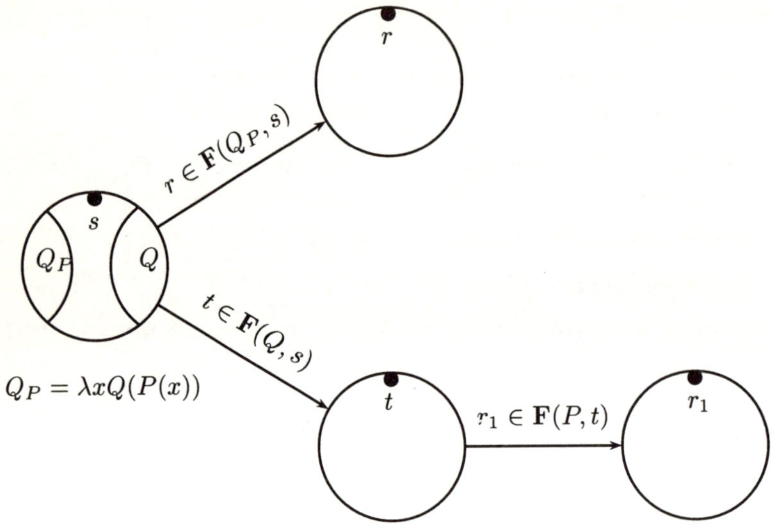

FIG. 8.3.

free variables of $C_i, i = 1, \ldots, n$. (It is assumed the substitution is done properly and does not bind any variables in C_i.)

Definition 8.6 ($\mathbb{LA}$-models).

1. An $\mathbb{LA}$-model has the form $\mathbf{m} = (S, D, \mathbf{1}, g, \mathbf{F})$ where S is a set of labels, D is a domain, $\mathbf{1} \in D$ and g assigns elements of D to each variable and constant of the language, and for each $t \in S$ and unary symbol p, $g(t,p)$ is a function from D into D.
2. $\mathbf{F}$ is a fibring function associating with each $t \in S$ and each function $f : D \mapsto D$ a set $\mathbf{F}(f,t) \subseteq S$.
3. Let f and e be two functions in D^D. Let $[f \circ e]$ denote the function $\lambda x(f(e(x)))$.
4. Satisfaction is defined by induction as follows
   - $t \vDash p(x)$ iff $g(t,p)(g(x)) = \mathbf{1}$, for atomic unary p and variable or constant x.
   - $t \vDash A \wedge B$ iff $t \vDash A$ and $t \vDash B$.
   - $t \vDash \sim A$ iff $t \nvDash A$
   - $t \vDash (x)A(x)$ iff for all $d \in D, t \vDash A(d)$.
   - $t \vDash \forall x A(x)$ iff for all x such that $x \in D$ or x is a syntactical expression we have $t \vDash A(x)$.
   - $t \vDash p(\varphi)$ iff for all $s \in S$ and all q such that $t \in \mathbf{F}(g(s,q), s)$ we have that for all $r \in \mathbf{F}([g(s,q) \circ g(s,p)], s)$ we have $r \vDash \varphi$.
5. We write $\mathbf{m} \vDash A$ iff for all $t \in S, t \vDash A$.

Note that in the previous definition we interpreted variables as ranging over elements of D, i.e. $g(x) \in D$, and thus in the expression $t \vDash p(x)$ we needed to interpret p as a predicate. Since the extension of p at t is a function $g(t,p) = p_t$, we turned it into a predicate by saying $t \vDash p(x)$ iff $p_t(g(x)) = \mathbf{1}$. This definition is somewhat arbitrary. We could have defined the extension of variables x also as functions $g(t,x) \in D^D$ and then we could have started our inductive definition with $t \vDash x$. Again we could have arbitrarily decided that $t \vDash x$ holds iff, say for example, $g(t,x)(\mathbf{1}) = \mathbf{1}$.

The details do not matter so much as the main novelty that our approach to ambivalent syntax resides in the clause evaluating $t \vDash p(\varphi)$ and in the very idea of the fibred semantics.

We shall conclude this section by proving the theorem we promised.

Theorem 8.7. *Let* $\mathbf{m} = (S, D, a, \mathbf{F}, h, g)$ *be a tree model with* D *infinite as defined in Definition 7.10. Then there exists a fibred model* $\mathbf{m}' = (S', D, a, \mathbf{F}', h', g)$, *with the same* D, a *and* g *satisfying the following:*

1. *For all formulas* A
$$\mathbf{m} \vDash A \textit{ iff } \mathbf{m}' \vDash_* A$$
where $\vDash_*$ *is the satisfaction relation in (2*) above.*
2. *If* $y \in S' - S$, *then for some* $X \subseteq D$, *we have* $\mathbf{F}(X,y) \subseteq S$.
3. $\mathbf{F}$ *and* h *are the respective restrictions of* $\mathbf{F}'$ *and* h' *to* S.

Proof We prove the theorem by induction on the level of the formulas A.

Case $n = 0$
In this case A is a wff of the classical language $\mathbb{L}$, and so we can let $\mathbf{m}' = \mathbf{m}$.

Case $n + 1 > 0$
Let $T = \{t \in S \mid \text{ for some } X \subseteq D, t \in \mathbf{F}(X,a)\}$. T is the set of all labels immediately accessible from a through any subset $X \subseteq D$. Let $\mathbf{m}_t = (S_t, D, t, \mathbf{F}_t, h_t, g)$, where $S = \{s \mid tRs\}$. Clearly $t_1 \neq t_2$ implies $S_{t_1} \cap S_{t_2} = \varnothing$. $\mathbf{F}_t$ and h_t are the respective restrictions of $\mathbf{F}$ and h to S_t.

By the induction hypothesis there exist models of the form

$$\mathbf{m}'_t = (S'_t, D, t, \mathbf{F}'_t, h'_t, g)$$

satisfying the conditions of the theorem for every formula of level $\leq n$. We can assume that $t_1 \neq t_2$ implies $S'_{t_1} \cap S'_{t_2} = \varnothing$. Let b be a new point, not in $S \cup \bigcup_{t \in T} S'_t$. Define $h'(b,Q)$ for any Q by letting $h'(b,Q) = X_0$, for some fixed $X_0 \subseteq D$. Since D is infinite and the number of atomic Q is countable, we can assume that for all Q we have $h(a,Q) \neq X_0$.

Let $\mathbf{F}'(X_0, b) = \{a\}$, and let $\mathbf{F}'(X,b) = \mathbf{F}(X,a)$, for $X \neq X_0$.

Let $\mathbf{m}'$ be the model with

$$S' = \{b,a\} \cup \bigcup_{t \in T} S'_t.$$

Let $\mathbf{m}' = (S', D, a, \mathbf{F}', h', g)$ where $\mathbf{F}'$ and h' are as defined above for b, are the same as $\mathbf{F}$ and h for a and are the same as $\mathbf{F}'_t$ and h'_t for S'_t.

We claim that $\mathbf{m}'$ is the required model.

Let us first check the set $\{x \mid b \vDash Q(P(x))\}$, for arbitrary Q and P.

$$b \vDash Q(P(x))\text{iff } \mathbf{F}'(h'(b,Q),b) \vDash P(x) \text{ iff } a \vDash P(x).$$

Thus the extension in b of $\lambda x Q_P(x)$ is the set $h(a,P)$. Therefore

$$\mathbf{F}'(h'(Q_P,b),b) = \mathbf{F}(h(a,P),a).$$

Hence for all α of level $\leq n$ and all P we have
$a \vDash_* P(\alpha)$ in $\mathbf{m}'$ iff for all t in $\mathbf{F}(h(a,P),a)$, $t \vDash_* \alpha$ in $\mathbf{m}'_t$ iff for all t in $\mathbf{F}(h(a,P),a)$, $t \vDash \alpha$ in $\mathbf{m}$ iff $a \vDash P(\alpha)$ in $\mathbf{m}$.

Since the above holds for all $P(\alpha)$, it holds for all wffs of level $\leq n+1$.

This completes the induction step and therefore our theorem is proved. □

Theorem 8.8. *The semantics with satisfaction according to (2*) above yields the same set of theorems as the ordinary semantics.*

Proof We already saw that it can only give possibly more theorems. We show that this does not happen.

If A has a general fibred countermodel with satisfaction evaluated according to (2), then it also has a tree countermodel. By the previous theorem it has a general fibred countermodel where satisfaction is evaluated according to (2*). □

Theorem 8.9 (Completeness theorem for ambivalent syntax). *Consider the language of ambivalent syntax with one place symbols of Definition 8.5 and its proposed semantics of Definition 8.6. The following axiom system is sound and complete for the semantics.*

- *All substitution instances of truth functional tautologies.*
- $(x)A(x) \to A(u)$, *u any term properly substituted for x.*
- $(\forall x)A(x) \to A(u)$, *u any term or any ambivalent syntactical expression properly substituted for u.*
- $(\forall x)A(x) \to (x)A(x)$.
- $\mathbb{G}\ \dfrac{\vdash A \to B(\alpha); \vdash A \to (x)B(x)}{\vdash A \to (\forall x)B(x)}$.
  α an ambivalent syntactical expression built up from atoms and variables not appearing in A. We assume neither $\vdash \alpha$ nor $\vdash\sim \alpha$ hold.
- $\dfrac{\vdash A \to B(x)}{\vdash A \to (x)A(x)}$, *$x$ not in A.*
- $\vdash p((\forall x)A(x)) \leftrightarrow (\forall x)p(A(x))$.
- $\vdash p((x)A(x)) \leftrightarrow (x)p(A(x))$.
- $\dfrac{\vdash \bigwedge_i A_i \to A}{\vdash \bigwedge_i p(A_i) \to p(A)}$.

- $\vdash A \leftrightarrow \vdash p(A)$.

Proof

1. Soundness can readily be verified.
2. Completeness
   We observe that the proposed axioms are the same as the axioms put forward in Theorem 7.7 for monadic logic with $(\forall x)$ and (x). We can formally define the canonical model as in the proof of Theorem 7.7. In this model the domain D is the set of all variables of the language. In fact we can assume completeness for tree models as in Definition 7.10.
   The truth definition for $t \vDash p(A)$ is as clause (2) on page 153, namely
3. $t \vDash p(\varphi)$ iff for all $x \in \mathbf{F}(p, t), x \vDash \varphi$.

We want clause (2^*) on the same page.

(2^*) $t \vdash_* p(\varphi)$ iff for all (q, s) such that $t \in \mathbf{F}(q, s), s \vDash (q \circ p)(\varphi)$.

This change we can achieve as proved in Theorem 8.7.

We have now almost obtained the right semantics except that in the model, we have assignments h such that $h(t, p) \subseteq D$, for p atomic, i.e. p is a predicate and not a function. This is not a problem. We can choose a distinguished element $\mathbf{1} \in D$ and define $g(t, p)(x) = \mathbf{1}$ if $h(t, p) = 1$ and $g(t, p) = d \neq \mathbf{1}$ if $h(t, p) = 0$.

The above gives us a model as required. □

Example 8.10. In Definition 8.6 the syntax contained unary symbols only. It is not difficult to generalise the semantics for n-place symbols. We can, however, allow for the arity of the symbols to be ambiguous. Assume p can have any arity, depending on context. Thus if we write the expression $p(p) \vee p$, then in the second disjunct p behaves like a propositional letter, taking value 0, 1. In the first disjunct, p behaves like a unary relation applied to p itself, as a formula.

To give semantics to such syntax, we need to define the extension $g(t, p)$ of p at t to be a function f from D^* into D, where D^* is the set of all finite sequences of elements of D including the empty sequence.

The fibring function $\mathbf{F}$ can still have the form $\mathbf{F}(f, t)$ associating a subset of S with each function $f : D^* \mapsto D$ and each t.

We can now evaluate $t \vDash p(p) \vee p$ as follows:

$$t \vDash p \text{ iff } g(t, p)(\varnothing) = \mathbf{1}.$$

$t \vDash p(p)$ is defined as in the case $t \vDash p(\varphi)$ in item 4 of Definition 8.6.

9

SELF-FIBRING OF INTUITIONISTIC LOGIC

9.1 Introduction

The methods of the previous sections are not specific to classical logic. We can self-fibre any logic. This section deals with self-fibring in intuitionistic logic. Consider the language of intuitionistic predicate logic with the connectives $\{\neg, \wedge, \vee, \Rightarrow, \perp\}$ and the quantifiers $\forall$ and $\exists$. We expand the language by allowing substitution of formulas for the free variables of other formulas. Thus we can form expressions of the form $\varphi(x, A(y, \varphi))$ out of $\varphi(x, z)$ and $A(y, z)$. Our purpose in this section is to develop fibred semantics and proof theory for such expressions.

Definition 9.1 (Self-fibred intuitionistic syntax). Let $\mathbb{H}$ be the language of full intuitionistic predicate logic with the connectives $\{\neg, \wedge, \vee, \Rightarrow\}$ and the two quantifiers $(\forall x), (\exists x), (x)$ and (Ex) and possibly equality and function symbols. Let $\mathbb{H}_f$ be a sub-language, say the monadic fragment. We define the fibred language $\mathbb{H}_f[\mathbb{H}]$ as follows:

1. A formula φ of intuitionistic logic language $\mathbb{H}$ is an f-formula of level 0.
2. If φ_i are f-formulas of level $\leq n$ and $A(x_i)$ is a formula of $\mathbb{H}_f$, then $A(x_i/\varphi_i)$ is a formula of level $\leq n+1$.

We recall the semantics for intuitionistic logic.

Definition 9.2 (Kripke semantics for intuitionistic predicate language with quantifiers (x) and (Ex).).

1. A Kripke model has the form $\mathbf{m} = (S, \leq, a, D, h, g)$ where $(S, \leq, a)$ is a partially ordered set of worlds, with first element a (which can be assumed to be a tree with root a), with $a \in S$, and D is a function giving for each $t \in S$ a non-empty domain D_t. h is the assignment function, giving for each n-place predicate P a subset $h(t, P) \subseteq D_t^n$. g gives to each variable, constant and n-place function symbol f in the language and each t a function $g(t, f) : D_t^n \mapsto D_t$. The following persistence conditions must hold:
   - $t \leq s$ implies $D_t \subseteq D_s$.
   - $t \leq s$ implies $h(t, P) \subseteq h(s, P)$, for all atomic P.
   - $t \leq s$ implies $g(s, f) \restriction D_t^n = g(t, f)$, for all function symbols f.
   - Variables x get values $g(x) \in D_a$.
   - g can be extended to all terms in the obvious way.
2. Satisfaction is defined by induction as follows.
   - $s \vDash P(t_1, \ldots, t_n)$, for t_i terms such that $g(s, t_i) \in D_s$, iff $(g(s, t_1), \ldots, g(s, t_n)) \in h(s, P)$

- $s \vDash t_1 = t_2$ iff $g(s, t_1) = g(s, t_2)$
- $s \vDash A \wedge B$ iff $s \vDash A$ and $s \vDash B$
- $s \vDash \neg A$ iff for all $s' \geq s, s' \nvDash A$
- $s \vDash A \vee B$ iff $s \vDash A$ or $s \vDash B$
- $s \vDash A \Rightarrow B$ iff for all $s' \geq s, s' \vDash A$ implies $s' \vDash B$
- $s \vDash (Ex)A(x)$ iff for some $a \in D_s$, $s \vDash A(a)$
- $s \vDash (x)A(x)$ iff for all $s' \geq s$ and all $a \in D_{s'}$ we have $s' \vDash A(a)$
- we say $\mathbf{m} \vDash A$ iff $a \vDash A$
- we define $A_1, \ldots, A_n \vDash B$ iff for all $\mathbf{m}$, if $\mathbf{m} \vDash A_i, i = 1, \ldots, n$ then $\mathbf{m} \vDash B$.

Theorem 9.3 (Completeness theorem for intuitionistic logic). *Intuitionistic predicate logic with equality* $=$ *and the additional axiom for terms* $t = s \vee \neg(t = s)$ *is sound and complete for the above semantics.*

Proof Well known, see [Gabbay, 1981c].

Also note that if we omit the axiom $t = s \vee \neg(t = s)$, the interpretation of $=$ will be that of a congruence relation $\equiv$ where elements may become congruent at some later stage in the ordering. □

We are now ready to define self-fibring for intuitionistic models. The basic idea should be the same as in the classical case, namely, with every subset X of the domain of every model t we associate several models s. This means we want a function $\mathbf{F}$ such that we have the basic relation $s \in \mathbf{F}(X, t)$. The definition is a bit more complex than in the classical case because we must accommodate persistence. This means that if $t_1 \leq t_2$ and for all x, $(P_1(x) \Rightarrow P_2(x))$ then $t_1 \vDash P_1(A)$ implies $t_2 \vDash P_2(A)$. The fibred models at t_1, t_2 (say $\{\mathbf{m}^i_{P,t_1}\}$, $\{\mathbf{m}^j_{P,t_2}\}$) must respect persistence, namely, if for all i $\mathbf{m}^i_{P_1,t_1} \vDash A$ then for all j, $\mathbf{m}^j_{P_2,t_2} \vDash A$.

Let us see what formal requirement corresponds to the above intuition. Note that for all x, $P_1(x) \Rightarrow P_2(x)$ means $P_1 \subseteq P_2$.

Let $\mathbf{m} = (S, \leq, a, D, h, g)$ be an intuitionistic model. What do we mean by a subset X of D, $(X \subseteq D)$? The answer is a function X associating with each $t \in S$ a set $X_t \subseteq D_t$, such that $t \leq s$ implies $X_t \subseteq X_s$. We similarly define the notion $X \subseteq Y$. Thus with each $t \in S$ we are going to associate a family of Kripke models $\mathbf{F}(X, t)$. Assume now that $t_1 \leq t_2$ and $X_1 \subseteq X_2$. We want a corresponding condition on $\mathbf{F}(X_1, t_1)$ and $\mathbf{F}(X_2, t_2)$ which will ensure persistence.

This can best be achieved by assuming that $\mathbf{F}(X_2, t_2)$ is part of $\mathbf{F}(X_1, t_1)$, i.e. if a model $\mathbf{m}$ is in the latter it is a submodel of a model $\mathbf{n}$ in the former. So if all the $\mathbf{n}$ models satisfy A then all the $\mathbf{m}$ models satisfy A.

In symbols: for every $\mathbf{m} \in \mathbf{F}(X_2, t_2)$ there exists an $\mathbf{n} \in \mathbf{F}(X_1, t_1)$ such that $\mathbf{m}$ is a branch of $\mathbf{n}$ (and hence for all wffs α, $\mathbf{n} \vDash \alpha$ implies $\mathbf{m} \vDash \alpha$).

The following fibring definition makes this intuition precise.

Definition 9.4 (Self-fibred intuitionistic models into the monadic fragment).

1. A self-fibred monadic intuitionistic model has the form $\mathbf{m} = (S, \leq, a, D, \mathbf{F}, h, g)$, where $(S, \leq, a)$ is a partially ordered set (which can be assumed to be a family of trees, i.e. for each $t \in S$, $(S^t, \leq^t, t)$ is a tree, where $S^t = \{x \mid t \leq x\}$ and $\leq^t = \leq \restriction S^t$), D is a function associating a non-empty domain D_t with each $t \in S$ and h and g are assignment functions such that the following holds:
   (a) For each t, $(S^t, \leq^t, t, D \restriction S^t, h \restriction S^t, g \restriction S^t)$ is an intuitionistic Kripke model.
   (b) For each $X \subseteq D$ and each t, $\mathbf{F}$ is a function associating a set of worlds $\mathbf{F}(X \restriction S^t, t)$ such that we have:
      - For all $t_1, t_2 \in S, X, Y \subseteq D$, $t_1 \leq t_2$ and $X \restriction S^{t_1} \subseteq Y \restriction S^{t_1}$ imply that for each $s_2 \in \mathbf{F}(Y \restriction S^{t_2}, t_2)$ there exists an $s_1 \in \mathbf{F}(X \restriction S^{t_1}, t_1)$ such that $s_1 \leq s_2$.
2. Satisfaction is defined as follows:
   - $s \vDash P(t_1, \ldots, t_n)$ iff $(g(t_1), \ldots, g(t_n)) \in h(s, P)$, for atomic P and terms $t_1, \ldots, t_n$.
   - The cases of $\vee, \wedge, \Rightarrow, \neg$ are defined as in Definition 9.2.
   - $s \vDash (Ex)A(x)$ iff for some $a \in D_s$, $s \vDash A(a)$.
   - $s \vDash (x)A(x)$ iff for all $s' \geq s$ and all $a \in D_{s'}$, $s' \vDash A(a)$.
   - $s \vDash (\forall x)A(x)$ iff for all $s' \geq s$ and all u such that either $u \in D_{s'}$ or u is a wff α we have $s' \vDash A(u)$.
   - $s \vDash (\exists x)A(x)$ iff for some u such that $u \in D_s$ or u is a wff α we have $s \vDash A(u)$.
   - $s \vDash P(\alpha)$, for P unary and α a formula, iff for all $s' \in \mathbf{F}(X, s)$ we have $s' \vDash \alpha$ where $X = h(s, P) \restriction S^s$.
   - We say $\mathbf{m} \vDash A$ iff $a \vDash A$.

The completeness theorem for the fibred semantics will be given in a later section. Before we consider full intuitionistic logic, we find it worthwhile to consider self-fibring for some fragments of the logic. This will help to illustrate our ideas. Let us assume further that we are dealing with the $\{\wedge, \Rightarrow, \forall, \bot\}$ fragment of intuitionistic logic. The advantage of this fragment is that it has a simple canonical syntactical Kripke model of the form $(S, D, \leq, \vDash)$. The set of worlds S in the Kripke structure are the (equivalence classes of provably equivalent) formulas themselves. The domain D is the set of all variables and constants of the language. The assignment to the variables is the identity. This means that the fragment has constant domains, which allows us to have simpler semantics for function symbols and equality.

Satisfaction for atomic A is defined by the clause:

- $\varphi \vDash A$ iff $\varphi \vdash A$.

The accessibility relation is defined by

- $\varphi \leq \psi$ iff $\psi \vdash \varphi$.

In this canonical Kripke model we can prove that for any wff B and any φ we have

(*) $\varphi \vDash B$ iff $\varphi \vdash B$.

See [Gabbay, 1981c] for details.

In case our fragment also contains the equality symbol $=$, its extension in the Kripke structure is interpreted as a congruence relation $\equiv$.

We have

- $\varphi \vDash x \equiv y$ iff $\varphi \vdash x = y$.

The traditional equality axioms ensure that if $\varphi \vDash x \equiv y$ then for all A, $\varphi \vDash A(x) \leftrightarrow A(y)$.

We do not have the property that $x \equiv y \vee x \not\equiv y$. This property corresponds in the fragment to the schema

$$((x = y) \Rightarrow A) \wedge (((x = y) \Rightarrow \bot) \Rightarrow A) \Rightarrow A$$

or equivalently using disjunction

$$x = y \vee x \neq y.$$

We now begin our examination of options for self-fibring for this fragment.

9.2 Intuitionistic Language with Unary Predicates

We begin with the case of unary predicates, no equality and no function symbols. In this context let us see how to extend the canonical Kripke semantics to the fibred language.

Consider a fibred formula of the form $P(\alpha)$, α a wff, and let us try and semantically evaluate it at a 'world' φ. This means we need to check $\varphi \vDash ?P(\alpha)$.

We do not know how to evaluate that, since α is a wff. Had α been a term of the language then of course we would check $\varphi \vdash P(\alpha)$. Let us try then and give meaning to the above in the case of α a wff. From previous sections we know the idea is to fibre a new model $\mathbf{F}(P, \varphi)$ at this node and have:

- $\varphi \vDash P(\alpha)$ iff $\mathbf{F}(P, \varphi) \vDash \alpha$.

Since we are dealing with the canonical model, where the nodes are formulas, another model $\mathbf{F}(P, \varphi)$ means just another formula. Let us denote such a formula by φ_P, i.e. we let

$$\mathbf{F}(P, \varphi) = \varphi_P.$$

Therefore we should have:

- $\varphi \vDash P(\alpha)$ iff $\varphi_P \vDash \alpha$ iff $\varphi_P \vdash \alpha$.

φ_P is arbitrary in the semantics, but in the canonical model it must be a certain choice. Let us try and identify φ_P.

We should have for all wffs X and Y:

- $\varphi \vdash P(X)$ iff $\varphi_P \vdash X$
- $\varphi \nvdash P(Y)$ iff $\varphi_P \nvdash Y$.

Suppose we let

- $\varphi_P = \{X \mid \varphi \vdash P(X)\}$.

(Here we assume $\vdash$ is extended somehow to the fibred language.)[26]

Then it is reasonable to assume (using experience from modal logic) that this mapping will do the job. Note that we get

- $\varphi \vdash \psi$ implies $\varphi_P \vdash \psi_P$.

In the general case, $\mathbf{F}(P, \varphi)$ can be a set of formulas $\{\varphi_{1,P}, \varphi_{2,P}, \ldots, \}$. However, in the case of our fragment intuitionistic logic, without $\vee$ and without $\exists$ we can assume without loss of generality that $\mathbf{F}(P, \varphi) = \{\varphi_P\}$. This is true because of the following considerations.

We must have for all X that

- $\varphi \vdash P(X)$ iff for all $i, \varphi_{i,P} \vdash X$.

Hence we can let, without loss of generality

- $\mathbf{F}(P, \varphi)\ \{\alpha \mid \alpha \vdash \varphi_P\}$

but in our intuitionistic fragment we have for any formula A, $\varphi_P \vdash A$ iff for all α ($\alpha \vdash \varphi_P$ implies $\alpha \vdash A$). Hence, without loss of generality we can assume $\mathbf{F}(P, \varphi) = \varphi_P$.

Example 9.5.

$$\neg\alpha \wedge P(\alpha) \vdash ?P(\neg\neg\alpha).$$

Let $\varphi = \neg\alpha \wedge P(\alpha)$. It looks like we should let $\varphi_P = \alpha$.

So we check

$$\alpha \vdash ?\neg\neg\alpha.$$

Definition 9.6 (The system IC).

1. Let $\mathbb{I}$ be the fragment of intuitionistic logic with $\{\wedge, \Rightarrow, \forall, \bot\}$ containing monadic predicates and unary function symbols. Let $\mathbb{I}_M$ be the fragment with unary predicates and let $\mathbb{I}_M[\mathbb{I}]$ be the self-fibred language defined as follows:
   - $\mathbb{I}^0_M[\mathbb{I}] = \mathbb{I}$
   - $\alpha \in \mathbb{I}^{n+1}_M[\mathbb{I}]$ iff for some $\varphi(x_1, \ldots, x_n, y_1, \ldots, y_n) \in \mathbb{I}_M$ and some $\alpha_1, \ldots, \alpha_n \in \bigcup_{m \leq n} \mathbb{I}^m_M[\mathbb{I}]$, we have $\alpha = \varphi(x_i/\alpha_i, y_j)$
   - $\mathbb{I}_M[\mathbb{I}] = \bigcup_n \mathbb{I}^n_M[\mathbb{I}]$.
2. Let **IC** be the following axiom system in any fibred language $\mathbb{I}_M[\mathbb{I}]$ with unary predicates (and possibly function symbols).

Axioms

- All substitution instances of intuitionistic propositional theorems
- $\forall x(A(x) \wedge B(x)) \Rightarrow \forall x A(x) \wedge \forall x B(x)$
- $\forall x A(x) \wedge \forall x(A(x) \Rightarrow B(x)) \Rightarrow \forall x B(x)$
- $\forall x \forall y A \Rightarrow \forall y \forall x A$
- $\forall x A(x) \Rightarrow A(t), t$ a variable or constant or a formula of $\mathbb{I}_M[\mathbb{I}]$

[26] A consequence relation for the fibred fragment with $\{\wedge, \Rightarrow, \forall, \bot\}$ can be viewed as intuitionistic if it satisfies the deduction theorem for $\Rightarrow$ and has the obvious rules for $\wedge$ and $\bot$ and possibly the $\mathbf{C}_1$ free rules for $\forall$. See the completeness theorem below.

- $\forall x P(\alpha(x)) \Rightarrow P(\forall x \alpha(x)), P$ atomic
- $P(\alpha \Rightarrow \beta) \Rightarrow (P(\alpha) \Rightarrow P(\beta))$.

Rules

- Modus ponens:
  $\vdash A$ and $\vdash A \Rightarrow B$ imply $\vdash B$.
- Universal generalisation

$$\frac{\vdash A \Rightarrow B(x)}{\vdash A \Rightarrow \forall x B(x)} \text{ where } x \text{ is not free in } A.$$

- Necessitation

$$\frac{\vdash A}{\vdash P(A)} \; P \text{ a unary predicate.}$$

- Inverse necessitation

$$\frac{Q(\alpha) \vdash Q(\beta)}{\alpha \vdash \beta}.$$

3. Let $A_1, \ldots, A_n \vdash_{\mathbf{IC}} B$ mean that there exists a sequence of wffs $B_1, \ldots, B_m = B$ such that each element in the sequence is either a theorem of **IC** or is an assumption A_i or is obtained from the previous two elements in the sequence by modus ponens.

Theorem 9.7. $\vdash_{\mathbf{IC}}$ *satisfies the deduction theorem.*

Proof The traditional proof works in our case, see [Gabbay, 1981c]. □

Definition 9.8 (Canonical IC model). We can now construct the self-fibred canonical model $\mathbf{M}$ as follows:

$$\mathbf{M} = (S, D, \leq, R_P, \vDash), P \text{ atomic},$$

where

- $S =$ the set of (equivalence classes of provably equivalent) wffs
- $D =$ set of all variables and terms of the language
- variables and constants are assigned themselves in the model
- $\varphi \leq \psi$ iff $\psi \vdash \varphi$
- $\varphi R_P \psi$ iff $\psi \vdash \varphi_P$, where $\varphi_P = \{\alpha \mid \varphi \vdash P(\alpha)\}$.
  In other words, $\varphi R_P \psi$ iff for all α, if $\varphi \vdash P(\alpha)$ then $\psi \vdash \alpha$.

Note that the following holds

- $\varphi \leq \psi$ and $\psi R_P \alpha$ imply $\varphi R_P \alpha$
- $\varphi R_P \psi$ and $\psi \leq \alpha$ imply $\varphi R_P \alpha$.

Let satisfaction be defined as follows:

- $\varphi \vDash P(x)$ iff $\varphi \vdash P(x)$, for P atomic
- $\varphi \vDash P(\alpha)$ iff for all $\psi, \varphi R_P \psi$ implies $\psi \vDash \alpha$.

Theorem 9.9 (Completeness theorem). *In the canonical model* **M** *the following holds:*

- $\varphi \vDash A$ *iff* $\varphi \vdash A$
- $\varphi \vDash P(A)$ *iff* $\varphi_P \vdash A$ *iff for all* ψ *such that* $\varphi R_P \psi$ *we have* $\psi \vdash A$.

Proof The proof is given by structural induction.

1. The atomic case and the cases of $\wedge, \Rightarrow$ and $\perp$ present no difficulties.
2. Assume $\varphi \vdash \forall x A(x)$. We need to show that for all terms or formula $t, \varphi \vdash A(t)$. This follows from the axiom $\vdash \forall x A(x) \Rightarrow A(t), t$ a term or formula.

Assume for all $t, \varphi \vdash A(t)$. In particular for a new variable y not in φ. But then by universal generalisation, $\varphi \vdash \forall x A(x)$.

3. Assume $\varphi_P \vdash A$. Then for some $\alpha_i \in \varphi_P, \vdash \bigwedge_i \alpha_i \Rightarrow A$. Hence $\vdash \bigwedge P(\alpha_i) \Rightarrow P(A)$. Hence $\varphi \vdash P(A)$.
   Assume $\varphi \vdash P(A)$. Then certainly $\varphi_P \vdash A$. □

9.3 Intuitionistic Language with Binary Predicates

Let us see how to extend our methods to binary predicates. Consider the intuitionistic predicate fragment with $\{\wedge, \Rightarrow, \forall, \perp\}$ and binary relations $\rho(x, y)$. Our problem is to give meaning to the semantic expression $\varphi \vDash \rho(A, B)$.

Following the classical case we can regard $\rho(x, y)$ as a predicate in x parametrised by y, i.e. $\rho_y(x)$.

We now know how to handle that, bearing in mind that in the canonical model $\varphi \vDash X$ means $\varphi \vdash X$. We have

$$\varphi \vdash \rho(A, y) \text{ iff } \varphi_{\lambda x \rho(x,y)} \vdash A$$

where $\varphi_{\lambda x \rho(x,y)} = \{\alpha \mid \varphi \vdash \rho(\alpha, y)\}$.

Now for a fixed $A, \rho(A, y)$ is a predicate in y and so we also have

$$\varphi \vdash \rho(A, B) \text{ iff } \varphi_{\lambda y \rho(A,y)} \vdash B$$

where $\varphi_{\lambda y \rho(A,y)} = \{\beta \mid \varphi \vdash \rho(A, \beta)\}$. But $\varphi \vdash \rho(A, \beta)$ iff $\{\alpha \mid \varphi \vdash \rho(\alpha, \beta)\} \vdash A$, hence for the case of substituting A first and then B, we have

$$\varphi \vdash_1 \rho(A, B) \text{ iff } \{\beta \mid \{\alpha \mid \varphi \vdash \rho(\alpha, \beta)\} \vdash A\} \vdash B.$$

The symmetrical image for the case of substituting B first and then A is:

$$\varphi \vdash_2 \rho(A, B) \text{ iff } \{\alpha \mid \{\beta \mid \varphi \vdash \rho(\alpha, \beta)\} \vdash B\} \vdash A.$$

Let us see what happens in the case of $P(A) = \rho(A, A)$, i.e. $A = B$.

We have three options:

$$\begin{aligned}&\varphi \vdash P(A) \text{iff } \varphi_P \vdash A \text{ iff } \{\alpha \mid \varphi \vdash \rho(\alpha, \alpha)\} \vdash A\\&\varphi \vdash_1 \rho(A, A) \text{ iff } \{\beta \mid \{\alpha \mid \varphi \vdash \rho(\alpha, \beta)\} \vdash A\} \vdash A\\&\varphi \vdash_2 \rho(A, A) \text{ iff } \{\alpha \mid \{\beta \mid \varphi \vdash \rho(\alpha, \beta)\} \vdash A\} \vdash A.\end{aligned}$$

It is not clear how $\vdash, \vdash_1$ and $\vdash_2$ relate to each other, and whether some interesting theorems can be proved for special cases.

There are two other points of view we can adopt for $\varphi \vdash \rho(A, B)$. One is to view $\rho(A, B)$ as some sort of a conditional with A helping to define the fibred world $\mathbf{F}(\rho, \varphi) = \varphi_\rho$. Thus we let $\varphi_\rho = \varphi_{\rho,A}$, and we let

$$\varphi \vdash \rho(A, B) \text{ iff } \varphi_{\rho,A} \vdash B.$$

An obvious choice is some formula or connective, $\Psi(\varphi, A)$ (e.g. we can have $\Psi(\varphi, A) = \varphi \wedge A$).

So we let

$$\varphi \vdash \rho(A, B) \text{ iff } \Psi(\varphi, A) \vdash B.$$

This option is similar to the $\vdash_1$ option.

The other option is to add a new connective $\sharp(A, B)$ to the language and let $\varphi \vdash \rho(A, B)$ iff $\varphi_\rho \vdash \sharp(A, B)$.

This option corresponds to an evaluation of a pair of wffs in a model, i.e. we identify $\alpha \vdash (A, B)$ as $\alpha \vdash \sharp(A, B)$.

The connective $\sharp$ must be a completely new connective with only minimal properties, dictated only by the fact that it should be 'intuitionistic'.

9.4 Intuitionistic Language with Unary Functions

Let us examine the case of unary functions. We want to give semantics to $\varphi \vDash e(\alpha)$, where e is a unary function symbol. In our case this means defining $\varphi \vdash e(\alpha)$. Let us check the definition we have in the classical case:

- $\varphi \vdash e(\alpha)$ iff for all ψ and all Q such that $\psi_Q \vdash \varphi$ we have $\psi \vdash Q(e(\alpha))$.

We also know that ψ_Q can be defined in our case as

$$\psi_Q = \{\beta \mid \psi \vdash Q(\beta)\}.$$

Let us assume, for the sake of getting a clear idea of what is happening, that ψ_Q is finite, i.e.

$$\psi_Q = \text{ the } \beta \text{ such that } \psi \vdash Q(\beta).$$

Since $\psi_Q \vdash \varphi$, we can safely assume that $\beta \vdash \varphi$. Thus the previous definition becomes

- $\varphi \vdash e(\alpha)$ iff for all ψ and Q such that $\psi \vdash Q(\varphi)$, we have $\psi \vdash Q(e(\alpha))$.

Now since the above has to hold for all ψ and Q we can 'safely' adopt the suggestion that $Q(\varphi) \vdash Q(e(\alpha))$.

Thus our final suggestion for a definition is:

•• $\varphi \vdash e(\alpha)$ iff for all Q, $Q(\varphi) \vdash Q(e(\alpha))$.

Let us now test this suggested definition for the case of e being a predicate P. Recall our definition for the classical case:

- $\varphi \vdash P(\alpha)$ iff for all ψ and Q such that $\varphi = \psi_Q$ we have $\psi \vdash Q(P(\alpha))$.

Similar considerations to the case of $\varphi \vdash e(\alpha)$ will lead us to the following proposed definition.

•• $\varphi \vdash P(\alpha)$ iff for all Q, $Q(\varphi) \vdash Q(P(\alpha))$.

Since Q is arbitrary we have

$$\forall Q(Q(X) \vdash Q(Y)) \text{ iff } X \vdash Y$$

which is correct!

Thus we can see that our approach is coherent.

Definition 9.10 (Canonical model for fibred unary functions).

1. Let $\mathbb{I}_{MF}$ be the language fragment with unary predicates and unary function symbols. Let $\mathbb{I}_{MF}[\mathbb{I}]$ be the fibred language defined by similar clauses to those in Definition 9.6 together with the following additional clause for function symbols:
   - If $e(x)$ is a unary function and α is a wff in $\mathbb{I}^m_{MF}[\mathbb{I}]$ then $e(\alpha)$ is a wff in $\mathbb{I}^{m+1}_{MF}[\mathbb{I}]$.
2. Let **N** be the canonical model with the semantic evaluation of $\varphi \vDash e(\alpha)$ defined as follows:
   $\varphi \vDash e(\alpha)$ iff for all ψ and Q such that $\psi_Q = \varphi$, we have $\psi \vDash Q_e(\alpha)$, where $Q_e = \lambda x Q(e(x))$.

Theorem 9.11 (IC-completeness theorem for fibred unary functions). *Let $\mathbb{I}_{MF}$ be the fibred language with unary predicates and functions and let* **N** *be the canonical* **IC** *model for this language constructed in Definition 9.10.*

Then the following holds in **N***:*

- $\varphi \vDash A$ *iff* $\varphi \vdash A$
- $\varphi \vDash P(A)$ *iff* $\varphi_P \vdash A$
- $\varphi \vDash e(\alpha)$ *iff for all atomic* Q, $Q(\varphi) \vdash Q(e(\alpha))$.

Proof The first two bullets are proved as in the case of Theorem 9.9. We check the case of $\varphi \vDash e(\alpha)$.

1. First we show that $\varphi \vDash e(\alpha)$ iff $\varphi \vdash e(\alpha)$.
   Assume $\varphi \vDash e(\alpha)$. Let Q be atomic and consider $\psi = Q(\varphi)$. Then we have $\psi_Q = \{\beta \mid Q(\varphi) \vdash Q(\beta)\}$. Clearly $\psi_Q = \varphi$. Hence $\psi \vdash Q(e(\alpha))$, i.e. $Q(\varphi) \vdash Q(e(\alpha))$.
   Assume $\varphi \vdash e(\alpha)$. Let Q and ψ be such that $\psi_Q = \{\beta \mid \psi \vdash Q(\beta)\} = \varphi$. Then certainly $\psi \vdash Q(\varphi)$. Hence $\psi \vdash Q(e(\alpha))$.
2. Now by the inverse necessitation rule we have $\varphi \vdash e(\alpha)$ iff for all Q, $Q(\varphi) \vdash Q(e(\alpha))$, as required. □

9.5 Intuitionistic Language with Equality

Let us now check the method of self-fibring for the case where we have equality in the language. In this case we can write expressions like $t = A$ or $\alpha = A$, for terms t and formulas A and α. Our first step is to give an intuitive interpretation for these expressions. We find that the 'formula as type' view of intuitionistic logic is the most compatible with our fibred semantics approach. We understand $t = A$ as saying: 'the term t resides in (or supports the proof of) the type A'.

This view is compatible with our view of reading $t = A$ as $P_t(A)$ where $P_t = \lambda y(t = y)$. We use the notation φ_t for φ_{P_t}. We have:

- $\varphi \vdash t = A$ iff $\varphi_t \vdash A$ iff $\{\beta \mid \varphi \vdash t = \beta\} \vdash A$.

This interpretation provides us with an opportunity to take another look at unary predicates. Can we understand $P(A)$ as relating in some way the terms of P with the type A? In other words, can we identify a relation between $\{t \mid P(t)\}$ and $\{t \mid t \in A\} = \{t \mid t = A\}$?

Let us check. We already know that

$$\varphi \vdash P(A) \text{ iff } \varphi_P \vdash A$$

where $\varphi_P = \{\beta \mid \varphi \vdash P(\beta)\}$.

We can let $\varphi \vdash t = A$ iff $\varphi_t \vdash A$.

Hence

$$\varphi \vdash t = A \text{ iff } \{\beta \mid \varphi \vdash t = \beta\} \vdash A.$$

This can be understood to say:

'The set of all types which contain t can prove A.'

Let $Q(x)$ be an arbitrary unary predicate and let

$$P_Q(x) = (\text{definition}) \bigvee_{t \in Q} t = x.$$

The last line is a bit of an abuse of notation, but nevertheless, let us continue.

Therefore

$$\begin{aligned} \varphi \vdash Q(A) &\text{ iff (we expect) } \varphi_{P_Q} \vdash A \\ &\text{ iff } \{\beta \mid \varphi \vdash P_Q(\beta)\} \vdash A \\ &\text{ iff } \{\beta \mid \text{ for some } t, \varphi \vdash Q(t) \wedge (t = \beta)\} \vdash A. \end{aligned}$$

We can read the last line as:

'A can be proved from the set of all those types which φ can prove that they share a term with Q'.

If we have available the axiom:

$$Q(\beta) \leftrightarrow \exists x(Q(x) \wedge x = \beta)$$

(our fragment does not contain $\exists$ but never mind), then we can continue:

$$\begin{aligned} &\text{iff } \{\beta \mid \varphi \vdash Q(\beta)\} \vdash A \\ &\text{iff } \varphi_Q \vdash A. \end{aligned}$$

Thus with the above axiom we get the old definition back.

Let us now check the expression $\varphi \vdash \alpha = A$.
By definition, $\alpha = A$ is viewed as $P_A(\alpha)$ where $P_A(x)$ is the predicate $x = A$. Thus

$$\begin{aligned}\varphi \vdash \alpha = A \text{ iff } & \varphi \vdash P_A(\alpha)\\ \text{iff } & \{\beta \mid \text{ for some } t, \varphi \vdash P_A(t) \text{ and } \varphi \vdash t = \beta\} \vdash \alpha\\ \text{iff } & \{\beta \mid \text{ for some } t, \varphi \vdash t = A \wedge t = \beta\} \vdash \alpha.\end{aligned}$$

The last line can be read as: 'α is derivable from the set of all types which, according to what φ can prove, share a term with A'.

We note that the meaning of 'β shares a term with A' is that $\beta \wedge A$ is consistent. Thus we continue:

$$\text{iff } \{\beta \mid \varphi \text{ proves that } A \wedge \beta \nvdash \bot\} \vdash \alpha.$$

The only way a wff X can *prove* that Y is consistent is to have in every Kripke model and every world t such that $t \vDash X$ also that $t \nvDash \neg Y$. Let us introduce the operator MY with the following truth table:

- $t \vDash MY$ iff for some $s \geq t, s \vDash Y$.

Note that in the logic **LC** of Dummett, obtained by adding the axiom schema $(\alpha \Rightarrow \beta) \vee (\beta \to \alpha)$, which is complete for linear sets of possible worlds, we have: $M(X \wedge Y) \Leftrightarrow MX \wedge MY$.

We can now express the notion 'X proves that Y is consistent' as $X \vdash MY$.

We can therefore continue:

$$\text{iff } \{\beta \mid \varphi \vdash M(A \wedge \beta)\} \vdash \alpha$$

which in the case of **LC** will be:

$$\text{iff } \varphi \vdash M(A \wedge \alpha).$$

The connection with M is surprising. The operator M is connected with intuitionistic non-monotonic logic and default theory. The default rule $X : \frac{Y}{Y'}$ can be represented as $X \wedge MY \Rightarrow Y'$. See my papers [Gabbay and Clarke, 1987; Gabbay, 1982].

We can therefore continue:

$$\text{iff } \Delta = (\varphi, \{\varphi : \frac{A \wedge \beta}{\beta}\}) \vdash \alpha,$$

i.e. the default theory Δ proves α.

If we let $\varphi = \top$ we get:

$$\vdash \alpha = A \text{ iff } \Delta_0 = \{\frac{A \wedge \beta}{\beta}\} \vdash \alpha.$$

Thus a general default theory of the form $(\{\varphi_1, \ldots, \varphi_n\}, \{\alpha_i : \frac{\beta_i}{\beta_i}\})$ can be represented in our self-fibred system as:

$$\bigwedge_{j=1}^{n} \varphi_i \wedge \bigwedge_i (\alpha_i \Rightarrow (\beta_i = \top)).$$

See Section 10.7 for a case study on M.

Of course the above considerations show that, in the case of **LC**, the operator M is definable in our language, namely:

- $M\alpha =$ (definition) $\alpha = \top$.

Let us check the truth conditions of both sides of the above. We want to show that for any world φ in the canonical model we have

$$\varphi \vDash M\alpha \text{ iff } \varphi \vDash \alpha = \top.$$

We have:

- $\varphi \vDash M\alpha$ iff for some $\varphi', \varphi' \vdash \varphi$ and $\varphi' \vdash \alpha$.

This holds iff $\varphi \wedge \alpha$ is consistent, or equivalently, if $\varphi \nvdash \neg\alpha$.

- $\varphi \vDash \alpha = \top$ iff $\{\beta \mid \varphi$ proves that β is consistent$\} \vdash \alpha$ iff $\{\beta \mid \varphi \wedge \beta$ is consistent$\} \vdash \alpha$.

The last line holds iff $\varphi \wedge \alpha$ is consistent (because of compactness).

We summarise that in the canonical model we have

- $\varphi \vDash \alpha = A$ iff $\varphi \wedge \alpha \wedge A$ is consistent.

We certainly have

- $\varphi \vDash \alpha = \alpha$
- $\varphi \vDash \alpha = A \Rightarrow \varphi \vDash A = \alpha$.

But we do not have transitivity. $\varphi \vDash \alpha = A$ and $\varphi \vDash A = B$ does not imply $\varphi \vDash \alpha = B$, as to be expected from the non-monotonic meaning of M.

We are now ready for formal definitions.

Definition 9.12 (Equality axioms and semantics).

1. The system $\mathbf{IC}_=$ comprises the system **IC** of Definition 9.6, together with the following additional equality axioms:
   - $\forall x(x = x)$
   - $\forall xy((x = y) \Rightarrow y = x))$
   - $\forall xyz((x = y) \Rightarrow ((y = z) \Rightarrow x = z))$
   - $\forall xy((x = y) \Rightarrow (A(x) \Rightarrow A(y)))$
   - $\forall x(\forall y(x = A(y)) \Rightarrow (x = \forall y A(y))$
   - $(x = (\alpha \Rightarrow \beta) \Rightarrow ((x = \alpha) \Rightarrow (x = \beta))$
   - $\dfrac{\vdash A}{\vdash x = A}$
   - $\dfrac{\vdash (x - A) \rightarrow (x - B)}{\vdash A \Rightarrow B}$, x not free in A or B.
2. Let $\mathbb{I}_{M,=}[\mathbb{I}]$ be the self-fibred language with equality defined in a similar way to Definition 9.6. The semantic condition for the predicate $t = A$, with t a term and A a formula is defined by looking at the predicate $\lambda y(x = y)$. It is therefore (in the canonical model) the following:
   - $\varphi \vdash (t = A)$ iff $\varphi_t \vdash A$ where $\varphi_t = \{\beta \mid \varphi \vdash (t = \beta)\}$.

3. The canonical model **M** for the language with equality as well is defined in a similar manner to Definition 9.8. We need to define φ_t, for terms t. This we can do by regarding φ_t as $\varphi_{\lambda y(t=y)}$.

Theorem 9.13 (Equality completeness theorem). *The clauses of Theorem 9.9 all hold together with the following additional clause:*

- $\varphi \vDash (t = A)$ *iff* $\varphi_t \vdash A$ *iff for all* ψ *such that* $\psi \vdash \varphi_t$ *we have* $\psi \vdash A$ *iff for all* ψ *such that (for all* $\beta, \varphi \vdash t = \beta$ *implies* $\psi \vdash \beta$*) we have* $\psi \vdash A$.

Proof We check the additional condition in the canonical model. This holds because of the special axioms turning $x = A$ into the predicate $\lambda y(t = y)[A]$. □

9.6 Fibring Kripke Intuitionistic Models

This section deals with self-fibring of full intuitionistic logic. We have already defined fibred models in Definition 9.4; it is now time to prove a sample completeness theorem. We select full intuitionistic logic fibred into its monadic fragment, with both quantifiers.

Definition 9.14 (Hilbert axioms for full intuitionistic self-fibred logic).

Consider the following axiom systems for fibred intuitionistic logic in $\mathbb{H}_M[\mathbb{H}]$, where $\mathbb{H}_M$ is the monadic fragment.

1. All substitution instances of intuitionistic propositional theorems.
2. *Quantifier axioms*
   For '(Qx)' being either '$(\forall x)$' or '(x)' and for '$(\bar{Q}x)$' being either '$(\exists x)$' or '(Ex)', respectively, we have
   - $(Qx)(A(x) \wedge B(x)) \Leftrightarrow (Qx)A(x) \wedge (Qx)B(x)$
   - $(Qx)(A \vee B(x)) \Leftrightarrow A \vee (Qx)B(x)$, x not free in A
   - $(Qx)(Qy)A \Leftrightarrow (Qy)(Qx)A$
   - $(Qx)A(x) \wedge (Qx)(A(x) \Rightarrow B(x)) \Rightarrow (Qx)B(x)$
   - $(\forall x)A(x) \Rightarrow (x)A(x)$
   - $(Ex)A(x) \Rightarrow (\exists x)A(x)$
   - $(\forall x)A(x) \Rightarrow A(t)$, for t a term or a wff α
   - $(x)A(x) \Rightarrow A(t)$, for t a term
   - $A(t) \Rightarrow (Ex)A(x)$, for t a term
   - $A(t) \Rightarrow \exists x A(x)$, for t a term or a wff α.
3. *P axioms and rules*
   - $(Qx)P(A(x)) \Leftrightarrow P((Qx)A(x))$, for P atomic unary
   - $\dfrac{\vdash \bigwedge A_i \rightarrow B}{\vdash \bigwedge P(A_i) \Rightarrow P(B)}$, P atomic unary.
4. *Equality axioms*
   - $(x)(x = x)$
   - $(x)(y)(x = y \Rightarrow y = x)$
   - $(x)(y)(z)(x = y \wedge y = z \Rightarrow x = z)$
   - $(x)(y)(x = y \vee \neg(x = y))$

- $(x)(y)(x = y \Rightarrow (A(x) \Rightarrow A(y))$
- $x = y \Rightarrow P(x = y)$, P atomic unary
- $\neg(x = y) \Rightarrow P(\neg(x = y))$, P atomic unary.

5. *Quantifier rules*
   - $\dfrac{\vdash A(x) \Rightarrow B}{\vdash (Ex)A(x) \Rightarrow B}$, x not in B
   - $\dfrac{\vdash (Ex)A(x) \Rightarrow B; \vdash A(\alpha) \Rightarrow B}{\vdash (\exists x)A(x) \Rightarrow B}$,
     α a contingent wff with symbols not in A nor in B.

Definition 9.15.

1. (a) A theory is a pair (Δ, Γ).
   (b) A theory is consistent if for no $A_i \in \Delta, B_i \in \Gamma$ do we have $\bigwedge A_i \vdash \bigvee B_j$.
   (c) Δ, Γ is saturated if the following holds:
   - $A \vee B \in \Delta$ implies $A \in \Delta$ or $B \in \Delta$
   - $(Ex)A(x) \in \Delta$ implies for some term t, $A(t) \in \Delta$,
   - $(\exists x)A(x) \in \Delta$ implies for some term t or some wff α, $A(t) \in \Delta$ or $A(\alpha) \in \Delta$.
   - if $(\Delta, \Gamma \cup \{(x)\beta(x)\}$ is consistent then for some term t, $(\Delta, \Gamma \cup \{\beta(t)\})$ is also consistent
   - if $(\Delta, \Gamma \cup \{(\forall x)\beta(x)\})$ is consistent, then for some x being a term t or x being a wff α we have that $(\Delta, \Gamma \cup \{\beta(x)\})$ is also consistent

   (d) we say $(\Delta, \Gamma) \vdash A$ iff $(\Delta, \Gamma \cup \{A\})$ is inconsistent.

Theorem 9.16 (Completeness theorem for fibred full intuitionistic logic). *Let (Δ_0, Γ_0) be a consistent self-fibred theory of the fibring language $\mathbb{H}_M[\mathbb{H}]$. Then (Δ_0, Γ_0) has a model.*

Proof

1. Soundness is readily verified.
2. To show completeness, assume (Δ_0, Γ_0) is a consistent theory. We can assume that there are infinitely many variables not in (Δ_0, Γ_0).

Lemma 9.17 (Saturation lemma). *Assume (Δ_0, Γ_0) is as above. Then it can be extended to a saturated and consistent theory (Δ, Γ), with $\Delta_0 \subseteq \Delta, \Gamma_0 \subseteq \Gamma$.*

Proof Enumerate all wffs β of one of the following forms: $A \vee B, (\exists x)\beta_1(x)$, $(Ex)\beta_1(x), (x)\beta_1(x)$ and$(\forall x)\beta_1(x)$.

We define by induction consistent theories $(\Delta_n, \Gamma_n), n \geq 1$, such that $\Delta_n - \Delta_0$, $\Gamma_n - \Gamma_0$ are finite, $\Delta_n \subseteq \Delta_{n+1}, \Gamma_n \subseteq \Gamma_{n+1}$, as follows.

Assume (Δ_n, Γ_n) has been defined for $n \geq 0$. Consider the $(n+1)$th formula β in the enumeration.

Case 1:
If $\beta = \beta_1 \vee \beta_2$ and $(\Delta_n \Gamma_n) \vdash \beta$, then one of β_i can be put in Δ_n to form a consistent Δ_{n+1}. Let $\Gamma_{n+1} = \Gamma_n$.

Case 2:
If $\beta = (Ex)\beta_1(x)$, and $(\Delta_n, \Gamma_n) \vdash \beta$, we can add to Δ_n the formula $\beta_1(u)$, where u is a completely new variable. $(\Delta_n \cup \{\beta, \beta_1(u)\}, \Gamma_n)$ is consistent, for otherwise, for some $\delta_i \in \Delta_n, \gamma_j \in \Gamma_n$, we have

$$\vdash \bigwedge \delta_i \wedge \beta \wedge \beta_1(u) \Rightarrow \bigvee \gamma_j.$$

Hence $\bigwedge \delta_i \wedge \beta \vdash \bigvee \gamma_j$, but since $(\Delta_0, \Gamma_0) \vdash \beta$ we get by cut that (Δ_n, Γ_n) is inconsistent. Let $(\Delta_{n+1}, \Gamma_{n+1})$ be this theory.

Case 3:
If $\beta = (\exists x)\beta_1(x)$ and it is provable from (Δ_n, Γ_n) and it is not possible to add consistently a $\beta_1(u)$, then let $\delta'_i \in \Delta_n, \gamma'_j \in \Gamma_n$ be such that $\vdash \bigwedge \delta'_i \wedge (Ex)\beta_1(x) \rightarrow \bigvee_j \gamma'_j$.

Let α be a formula built up from symbols not in (Δ_n, Γ_n) (for example we can take α to be $u_1 = u_2$ for u_1, u_2 not in (Δ_n, Γ_n)), which is neither a theorem nor a contradiction. Consider $(\Delta_n \cup \{\beta, \beta_1(\alpha)\}, \Gamma_n)$. If this is inconsistent then for some $\delta_i \in \Delta_n, \gamma_j \in \Gamma_n$ we have

$$\vdash \bigwedge \delta_i \wedge (\exists x)\beta_1(x) \wedge \beta_1(\alpha) \Rightarrow \bigvee_j \gamma_j.$$

Hence using the rule

$$\frac{A \wedge \beta_1(\alpha) \vdash B; A \wedge (Ex)\beta_1(x) \vdash B}{A \wedge (\exists x)\beta_1(x) \vdash B}, \alpha \text{ contingent, not in } A, B, \beta$$

we get a contradiction with $(\Delta_n, \Gamma_n) \vdash (\exists x)\beta_1(x)$. Let $(\Delta_{n+1}, \Gamma_{n+1})$ be this theory.

Case 4:
If $\beta = (x)\beta_1(x)$ and $(\Delta_n, \Gamma_n) \nvdash \beta$, then for some completely new u $(\Delta_n, \Gamma_n) \nvdash \beta_1(u)$. Let $\Delta_{n+1} = \Delta_n, \Gamma_{n+1} = \Gamma_n \cup \{\beta_1(u)\}$.

Case 5:
If $\beta = (\forall x)\beta_1(x)$ and$(\Delta_n, \Gamma_n) \nvdash \beta$ and $(\Delta_n, \Gamma_n) \vdash (x)\beta_1(x)$, then let α be a completely new contingent formula. We claim $(\Delta_n, \Gamma_n) \nvdash \beta_1(\alpha)$. For otherwise for some

$$\delta_i \in \Delta_n, \delta'_i \in \Delta_n, \gamma_j \in \delta_n, \gamma'_j \in \Gamma_n$$

we have

$$\bigwedge \delta_i \vdash \bigvee \gamma_j \vee (x)\beta_1(x)$$
$$\bigwedge \delta'_i \vdash \bigvee \gamma'_j \vee \beta_1(\alpha)$$

Hence by the relevant rule

$$\bigwedge \delta_i \wedge \bigwedge \delta'_i \vdash \bigvee \gamma_j \vee \bigvee \gamma'_j \vee (\forall x)\beta_1(x)$$

a contradiction. □

Lemma 9.18. *Let* (Δ, Γ) *be a consistent and saturated theory and assume* $(\Delta, \Gamma \cup \{P(\alpha)\})$ *is consistent. Then the following theory* (Δ_P, Γ_P) *is consistent, where*

$$\Delta^0_{P,\alpha} = \{X \mid (\Delta, \Gamma) \vdash P(x)\}$$
$$\Gamma^0_{P,\alpha} = \{\alpha\}.$$

Proof Otherwise $\vdash \bigwedge X_i \Rightarrow \alpha$.

Hence $\vdash \bigwedge P(X_i) \Rightarrow P(\alpha)$ a contradiction. □

Lemma 9.19. $(\Delta^0_{P,\alpha}, \Gamma^0_{P,\alpha})$ *can be extended to a saturated and consistent theory* $(\Delta_{P,\alpha}, \Gamma_{P,\alpha})$.

Proof Enumerate all wffs β of the forms mentioned in the proof of Lemma 9.17. Assume $(\Delta^n_{P,\alpha}, \Gamma^n_{P,\alpha})$ has been defined. We follow the same cases as in the proof of Lemma 9.17 for β the $n+1$ formula in the sequence. We assume $\Delta^n_{P,\alpha} - \Delta^0_{P,\alpha}$ and $\Gamma^n_{P,\alpha} - \Gamma^0_{P,\alpha}$ are finite and that $\Delta^i_{P,\alpha} \subseteq \Delta^{i+1}_{P,\alpha}, \Gamma^i_{P,\alpha} \subseteq \Gamma^{i+1}_{P,\alpha}, i = 0, 1, \ldots$. We now follow the case analysis of Lemma 9.17.

Cases 1–3:
are the same as before.

Case 4:
We claim we cannot have $(\Delta^n_{P,\alpha}, \Gamma^n_{P,\alpha} \nvdash (x)\beta_1(x)$ while for every term t we have $(\Delta^n_{P,\alpha}, \Gamma^n_{P,\alpha}) \vdash \beta_1(t)$. Otherwise, let $A = \bigwedge(\Delta^n_{P,\alpha} - \Delta^0_{P,\alpha}), B = \bigvee(\Gamma^n_{P,\alpha} - \Gamma^0_{P,\alpha})$ and we get that for each term t there exists $X_1, \ldots, X_{m(t)} \in \Delta^0_{P,\alpha}$ such that

$$\bigwedge X_i \Rightarrow (A \Rightarrow B \vee \alpha \vee \beta_1(t)).$$

Hence

$$\vdash \bigwedge P(X_i) \Rightarrow P(A \Rightarrow B \vee \alpha \vee \beta_i(t)).$$

Hence $\Delta \vdash P(A \Rightarrow B \vee \alpha \vee \beta_1(t))$ for all terms t.

Since (Δ, Γ) is saturated, we get $\Delta \vdash (x)P(A \Rightarrow B \vee \alpha \vee \beta_1(t))$ and so $\Delta \vdash P((x)(A \Rightarrow B \vee \alpha \vee \beta_1(x))$. Hence $(x)(A \Rightarrow B \vee \alpha \vee \beta_1(x)) \in \Delta^0_{P,\alpha}$ and hence $\Delta^0_{P,\alpha} \vdash (x)(B \vee \alpha \vee \beta_1(x))$ and $\Delta^0_{P,\alpha} \vdash B \vee \alpha \vee (x)\beta_1(x))$. Hence $(\Delta^n_{P,\alpha}, \Gamma^n_{P,\alpha}) \vdash (x)\beta_1(x)$, a contradiction.

Hence for some t, we have indeed that $(\Delta^n_{P,\alpha}, \Gamma^n_{P,\alpha} \cup \{\beta_1(t)\})$ is consistent. Let this theory be $(\Delta^{n+1}_{P,\alpha}, \Gamma^{n+1}_{P,\alpha})$.

Case 5:
This case is similar to case 4. We assume $(\Delta^n_{P,\alpha}, \Gamma^n_{P,\alpha}$ proves$(x)\beta_1(x)$ as well as $\beta_1(\varphi)$ for any contingent wff φ and show that it must prove $(\forall x)\beta_1(x)$.

Since it proves$(x)\beta_1(x)$, we get that for some $X_i \in \Delta^0_{P,\alpha}$ we have

$$\bigwedge X_i \wedge A \vdash B \vee \alpha \vee (x)\beta_1(x).$$

Hence $\Delta \vdash P(A \Rightarrow B \vee \alpha \vee (x)\beta_1(x))$. Hence $\Delta \vdash (x)P(A \Rightarrow B \vee \alpha \vee \beta_1(x))$. We can also get by reasoning as in case 4 that

$$\Delta \vdash P(A \Rightarrow B \vee \alpha \vee \beta_1(\varphi))$$

for all wffs φ.

Hence by saturation

$$\Delta \vdash (\forall x)P(A \Rightarrow B \vee \alpha \vee \beta_1(x))$$

and this, as in case 4, will yield that $(\Delta^n_{P,\alpha}, \Gamma^n_{P,\alpha}) \vdash (\forall x)\beta_1(x)$.

Thus if $(\forall x)\beta_1(x)$ is not provable, then for some φ, $(\Delta^n_{P,\alpha}, \Gamma^n_{P,\alpha} \cup \{\varphi\})$ is consistent and so let$(\Delta^{n+1}_{P,\alpha}, \Gamma^{n+1}_{P,\alpha})$ be this theory.

We now let $\Delta_{P,\alpha} = \bigcup_n \Delta^n_{P,\alpha}$ and $\Gamma_{P,\alpha} = \bigcup_n \Gamma^n_{P,\alpha}$ and this gives us the required consistent and saturated extension and the proof of the lemma is complete. □

We now continue the completeness proof.

Let (Δ_0, Γ_0) be a consistent theory. By the saturation lemma we can assume it is saturated. Let S be the set of all consistent saturated theories $t = (\Delta, \Gamma)$ of the language. Let $t_1 \leq t_2$ be defined as $\Delta_1 \subseteq \Delta_2$. Let U be the set of all terms of the language. For each t let $D_t = U$ (we have constant domains). Let g be the identity and let for each t and P, $h(t, P)$ be $\{(a_1, \dots, a_n) \mid t \vdash P(a_1, \dots, a_n)\}$.

We now need to define $\mathbf{F}$. Let $X \subseteq D$ and assume that X is the extension of a unary predicate P, i.e. $a \in X_t$ iff $t \vdash P(a)$. We define $\mathbf{F}(X, t)$ to be the indices $\{t_{P,\alpha} = (\Delta_{P,\alpha}, \Gamma_{P,\alpha}) \mid$ for α such that $t \nvdash P(\alpha)\}$. For $X \subseteq D$, which is not the extension of a unary predicate $\mathbf{F}$ can be arbitrary.

The definition of our model is now complete. We can now prove

Lemma 9.20. *In the model just defined, for any A we have:*

$$t \vDash A \textit{ iff } t \vdash A.$$

Proof By induction on A, using the properties of the theories constructed. □

The above lemmas prove the theorem. □

10

APPLICATIONS OF SELF-FIBRING

10.1 Connection with Multimodal Logic and Translation into Classical Logic

We have observed in Section 7.1 that $P(\alpha)$ can be viewed as a modal operator $\Box_P$ applied to the wff α. The aim of this section is to study this connection further. Note that officially we have $\Box_P$ for atomic P only. We can extend the language and allow for modalities $\Box_\varphi$ for arbitrary wffs φ.

To get an intuitive idea of what kind of modal operators are involved, consider the expression *SoonA*. In temporal setting this means 'A will be true soon'. The meaning of *Soon* may be time dependent itself. In ancient times *Soon* could mean within three months, while in modern times *Soon* can mean within a few days.

Thus we have $t \vdash \mathit{Soon}A$ iff for some s such that tR_ts we have $s \vDash A$, where xR_ty means that y comes soon after x according to the views held at time t.

Of course we can let $xR_{\mathit{Soon}}y$ iff xR_xy and we are back to ordinary modal logic, but then we lose the true picture of what *Soon* means.

We refer to such a connective as 'context dependent' and to the modal logics involved with such connectives as *context modal logics*.

Definition 10.1 (Context modal structures). A context multimodal structure has the form $(S, R_c, a, U, h, g, \mathcal{C})$ where S is the set of possible worlds, $a \in S$ is the actual world, $R_c, c \in \mathcal{C}$ is a binary relation on S, and U_t is the universe at t, and for each n-place atomic predicate P and each $t \in S$, $h(t, P) \subseteq U_t^n$. g is an assignment giving to each variable or constant x and each t a value $g_t(x) \in U_t$. $\mathcal{C}$ is the set of contexts.

Our modal language is assumed to contain modalities of the form $\Box_\varphi$, for functions $\varphi : S \mapsto \mathcal{C}$.

Satisfaction is defined in the traditional manner:

- $t \vDash P(u_1, \ldots, u_n)$ iff $(g_t(u_1), \ldots, g_t(u_n)) \in h(t, P)$, for atomic P and variables or constants u_i.
- Satisfaction for the classical connectives is defined as usual.
- $t \vDash \Box_\varphi A$ iff $\forall y(tR_{\varphi(t)}y \to y \vDash A)$.

We can now connect the multimodal model with our self-fibring of classical logic. Let D be a set and let $\mathcal{C} = 2^D$, i.e. the contexts are all subsets of D.

For each $s \in S$ let $\mathbf{m}_s$ be a classical model with domain D, sharing the same assignment g to the free variables. Define a fibring set function $\mathbf{F}$ by

$$s' \in \mathbf{F}(Q, s) \text{ iff } sR_Qs'.$$

Let $\varphi(x)$, x free in φ, be a wff of classical logic. We can regard $[x : \varphi]$ as a function from S to 2^D defined by

$$[x : \varphi](t) = \{d \in D \mid \mathbf{m}_t \vDash \varphi(d)\}.$$

Thus we can associate a modality $\Box_{[x:\varphi]}$ with this function.

Note that if φ has only one free variable x we can write $\Box_\varphi$ instead of $\Box_{[x:\varphi]}$.

We have now defined a proper context modal logic with constant domains and rigid g. We can translate the monadic self-fibring language with the quantifier (x) ranging over D only as follows:

Let $*$ be the translation

- $\alpha^* = \alpha$, for α of level 0
- $P(A)^* = \Box_{[x:P(x)]}A^*$
- $*$ commutes with the classical connectives and the quantifier (x).

Context multimodal logic can be translated into classical logic through its semantics. By combining the translations of the self-fibred predicate logic into context multimodal logic and that of the context multimodal logic into classical logic, we can obtain a direct translation of the self-fibred classical logic into classical logic itself. This means that expressions like $P(\alpha)$ can be translated into classical logic in a semantically meaningful way (as opposed to a formal syntactical translation such as '*Apply* ("P", "α")'.

Let us see how it is done.

Consider a modal logic with $\Box_\varphi$ and atomic predicates $Q(u_1, \ldots, u_n)$. We translate the semantics notion "$t \vDash A$" by turning it into a formula $A^\sharp(t)$ of first-order logic. So for atomic $Q(u_i)$ we let $t \vDash Q(u_1, \ldots, u_n)$ be $Q^\sharp(t, u_1, \ldots, u_n)$.

The semantic relation $xR_cy, c \in \mathcal{C}$, can be represented by a three-place relation $R(c, x, y)$, where we need a unary function φ representing the map from S into $\mathcal{C}$. Thus for any $t, \varphi(t)$ is a context. So $t \vDash \Box_\varphi A$ can be evaluated as $\forall y(R(\varphi(t), t, y) \to y \vDash A)$. The translation will therefore be $\forall y(R(\varphi(t), t, y) \to A^\sharp(y)$.

Note that if $\mathcal{C}$ is just a set without special structure, it can be taken as S.

The following is a formal definition, for the constant domain case with rigid g.

Definition 10.2 (Translation of context multimodal logic into classical logic).

Let $\{\Box_\varphi\}$ be a context multimodal logic. Consider a two-sorted classical logic with a t-variable sort and a u-variable sort. Assume we have t-sort unary function symbols $\{\varphi\}$, two sorted atomic predicates $Q^\sharp(t, u_i)$ and t-sort ternary relation $R(c, x, y)$. We translate from modal logic as follows:

- $Q(u_1, \ldots, u_n)^\sharp_t = Q^\sharp(t, u_1, \ldots, u_n)$.
- $(\Box_\varphi A)^\sharp_t = \forall y(R(\varphi(t), t, y) \to A^\sharp_y)$.
- $\sharp$ commutes with the classical connectives and quantifiers.

Note that if $\mathcal{C}$ has more structure (e.g. a Boolean algebra or $\mathcal{C} = 2^U$) then it cannot be identified with S and the translation must be into a three-sorted classical logic, where we have a $\mathcal{C}$ sort as well.

We now want to translate the monadic self-fibring language with the domain quantifer (x) only, back into classical logic.

Our problem is that the modalities $\Box_P$ and their relations R_P are dependent on the set extension of P at the world (model) at t. So really we need a ternary relation of the form $R_P(P_t, t, y)$, where $P_t = \{x \mid t \vDash P(x)\}$. However, we can make use of the fact that this set is already identifiable from the relation $P^\sharp(t, x)$ which means $t \vDash P(x)$.

We therefore introduce for each atom $P_i(x)$ the relation $R_i(t, y)$ with the additional axiom

$$(\sharp) \quad \forall x[P_i^\sharp(t, x) \leftrightarrow P_j^\sharp(t, x)] \to \forall y[R_i(t, y) \leftrightarrow R_j(t, y)].$$

We are ready for the translation.

Definition 10.3 (Translation of self-fibred logic into classical logic). Let **L** be the self-fibred monadic logic with the quantifier (x) ranging over the domain only. Let there be a version of classical two-sorted logic with binary $R_i(t, s)$, and binary $P_i^\sharp(t, x)$. Assume as axioms that R_i and P_i satisfy $(\sharp)$ above.

We now translate as follows

- $P(\bar{x})_t^\sharp = P^\sharp(t, \bar{x})$, for atomic $\bar{x}$ place P.
- $P_i(A)_t^\sharp = (y)(R_i(t, y) \to A_y^\sharp)$, for the ith monadic predicate.
- Let $\sharp$ commute with the connectives and quantifiers.

The above considerations regarded the atoms $P_i(x)$ as modalities. More complex formulas are classically reduced to the atoms. Consider the formula $A(x) = P(x) \wedge Q(x)$, with P and Q atomic unary predicates.

Let $\mathbf{m}$ be a model and consider the sets

$$\begin{aligned} D_A &= \{x \mid \mathbf{m} \vDash A(x)\} \\ D_P &= \{x \mid \mathbf{m} \vDash P(x)\} \\ D_Q &= \{x \mid \mathbf{m} \vDash Q(x)\}. \end{aligned}$$

To evaluate $\mathbf{m} \vDash A(\alpha)$, we proceed, under current definitions, to evaluate $\mathbf{m} \vDash P(\alpha)$ and $\mathbf{m} \vDash Q(\alpha)$ and get

$$\mathbf{m} \vDash A(\alpha) \text{ iff } \mathbf{F}(D_P, h) \vDash \alpha \text{ and } \mathbf{F}(D_Q, h) \vDash \alpha.$$

We do *not* regard A as 'atomic' and evaluate

$$\mathbf{F}(D_A, h) \vDash \alpha.$$

The two possibilities will coincide if for all α, D_1, D_2 we have

$$\mathbf{F}(D_1 \cap D_2, h) \vDash \alpha \text{ iff } \mathbf{F}(D_1, h) \vDash \alpha \text{ and } \mathbf{F}(D_2, h) \vDash \alpha.$$

This is a rather strong condition.

We can introduce a new box operator into the language and write expressions like $\Box^x_{A(x)}\alpha$, where x is a variable and A and α are wffs. The meaning of $\Box^x_{A(x)}\alpha$ is $A(\alpha)$ evaluated as if A were atomic. Thus

- $\mathbf{m} \vDash \Box^x_{A(x)}\alpha$ iff $\mathbf{F}(\hat{x}A(x), h) \vDash \alpha$ where $\hat{x}A(x) = \{x \mid \mathbf{m} \vDash A(x)\}$.

Thus

- $\Box^x_{P(x)}\alpha = P(\alpha)$, for P atomic.

We now give a formal definition.

Definition 10.4 (Box self-fibring). Let $\mathbb{L}$ be a predicate language. We define the box self-fibred language $\Box\mathbb{L}$ as follows.

1. Any wff of $\mathbb{L}$ is a b-wff of $\Box\mathbb{L}$.
2. If $\varphi(x)$ and α are b-wffs then so is $\varphi(x/\alpha)$.
3. If $\varphi(x)$ and α are b-wffs, so is $\Box^y_{\varphi(y)}\alpha$. The free variables of the new wff are all the free variables of φ and α except y which is bound.

Definition 10.5 (Semantics for monadic box self-fibred language). Let $\mathbf{m}$ be a fibred model as in Definition 7.2. We define satisfaction as in Definition 7.2 with the additional clause

- $\mathbf{m} \vDash \Box^y_{\varphi(y)}\alpha$ iff $\mathbf{F}(\hat{y}\varphi(y), h) \vDash \alpha$.

Example 10.6. To see the applicability of our self-fibred predicate language consider the problem of translating into classical logic an arbitrary applied logical system. Consider a system with a connective $O(A \mid B)$ meaning that given A, it is obligatory to have B. Such systems are widely used in modelling and formalisation of legal and normative reasoning. From our point of view it is important to note that it is very likely that a logic $\mathbf{L}_1$ be put forward for the operator $O(A \mid B)$ as a Hilbert system (based on an intuitive motivation) without a possible world semantics!

Given $\mathbf{L}_1$, further application-based considerations may motivate us to add a temporal component to the logic. (This is plausible because some obligations may take time to implement.) We are thus led to a combined system $\mathbf{L}_1 \otimes \mathbf{L}_2$, where $\mathbf{L}_1$ is a deontic logic, $\mathbf{L}_2$ a temporal logic and possibly some additional interaction axioms (between the deontic and temporal operators).

The crucial point is that the new logic $\mathbf{L}_1 \otimes \mathbf{L}_2$ may not have a clear semantics. Our question is how are we going to translate it into classical logic? Traditional translations use the semantics. For example the temporal logic part can be translated through the possible world semantics (flow of time) but we cannot translate the deontic part because we have no semantics for it.

The only translation we can offer for the whole system under the circumstances is a term translation where the formulas are translated as terms.

If we use our self-fibred logic, we can use a mixed translation. The temporal connectives (of $\mathbf{L}_2$) can be translated semantically while the deontic connective can be translated as a binary predicate.

To be more specific, consider the sample interaction axiom

$$\alpha = O(A \mid B) \to G(A \to FB)$$

where

$$\begin{aligned} G\varphi &= \text{always (in the future) } \varphi \\ F\varphi &= \text{sometimes (in the future) } \varphi. \end{aligned}$$

This axiom says that, if given A we ought to have B, then always in the future when A holds then the next day we have B.

We can have a mixed translation as

$$\alpha_t^* = R_0(A(t), B(t)) \to \forall s(t < s \wedge A(s) \to \exists s'(s < s' \wedge B(s')).$$

The formula $O(\top, \alpha)$ will have the mixed translation $R_0(\top, \alpha_t^*)$.

10.2 Fixed Point Self-reference

The self-fibred languages we have studied so far do not allow for fixed point operations. Although we can write, for a wff $\varphi(x)$ the formula $\varphi(\varphi(x))$, we are just mechanically substituting the 'pattern' $\varphi(x)$ inside itself. What we would like is to have a fixed point construction $(\mu x)\varphi(x)$, denoting (the unique? the smallest?) wff α of the language such that $\alpha \leftrightarrow \varphi(\alpha)$ holds. Such an α may not exist!

The purpose of this section is to explore the possibilities of adding a fixed point operator μ to the language. We assume rigid assignments.

To get some feel of what to expect, let us examine Tarski's truth predicate.

Example 10.7 (Tarski's truth predicate). Now that we have fibred semantics for both unary functions and unary predicates, let us see whether we can get a paradox using Tarski's truth predicate.

Recall that we need a theory Δ, a unary predicate $\mathbf{T}$, a naming function $\varphi \mapsto$ "φ" and a diagonalisation function symbol $\mathbf{d}(x)$ satisfying the following for all φ

- $\Delta \vdash \mathbf{T}(\text{“}\varphi\text{”}) \leftrightarrow \varphi$
- $\Delta \vdash \mathbf{d}(\text{“}\varphi(x_0)\text{”}) = \text{“} \sim \varphi(\text{“}\varphi(x_0)\text{”})\text{”}$

where x_0 is a fixed variable. Tarski's theorem says that if $\mathbf{T}$ and $\mathbf{d}$ are definable and the above holds, then we get a contradiction.

Let us recall how a contradiction is obtained.

Let $\alpha(x_0) = \mathbf{T}(\mathbf{d}(\text{“}\mathbf{T}(\mathbf{d}(x_0))\text{”}))$. Let $\varphi(x_0) = \mathbf{T}(\mathbf{d}(x_0))$.

Then since

$$\mathbf{d}(\text{“}\varphi(x_0)\text{”}) = \text{“} \sim (\varphi(\text{“}\varphi(x_0)\text{”}))\text{”}$$

we get that Δ proves that

$$\begin{aligned} \alpha(x_0) &= \mathbf{T}(\mathbf{d}(\text{“}\varphi(x_0)\text{”}) \\ &= \mathbf{T}(\text{“} \sim \varphi(\text{“}\varphi(x_0)\text{”})\text{”}) \\ &\leftrightarrow \sim \varphi(\text{“}\varphi(x_0)\text{”}) \\ &\leftrightarrow \sim \alpha(x_0). \end{aligned}$$

In our syntax we have "φ" $= \varphi$ for all φ. So we have a naming function available. We also have a natural diagonal function, at least in the metalevel. We can let $\mathbf{d}(\varphi(x_0)) = \sim \varphi(\varphi(x_0))$ for all φ.

Let us now ask whether we can define in our fibred language a truth predicate $\mathbf{T}$ and a representation of the function $\mathbf{d}$ satisfying the above properties and thus generate a paradox. Since we have a clear-cut semantics for our language we will try and identify the properties of the semantics required for the above conditions on $\mathbf{T}$ and on $\mathbf{d}$.

First note that to have $\varphi \leftrightarrow \mathbf{T}(\varphi)$ for all φ we must restrict our semantics to models $\mathbf{m} = (D, h, g, \mathbf{F})$ satisfying $\mathbf{F}(h(\mathbf{T}), h) = h$.

Definition 10.8 (Function representation). Let $\mathbf{e}$ be a syntactical meta-function associating with each formula φ another formula $\mathbf{e}(\varphi)$. Let $\mathbf{m}$ be a model of the language. We say that $\mathbf{e}$ is *realisable* in $\mathbf{m}$ through a function e of the language $\mathbb{L}$ iff for all φ

$$h \vDash e(\varphi) \text{ iff } h \vDash \mathbf{e}(\varphi).$$

Note that the extension of e is decided by the model, while $\mathbf{e}(\varphi)$ is a meta-syntactical function.

We say that e represents $\mathbf{e}$ iff e realises $\mathbf{e}$ in every model.

Theorem 10.9. *Let* $\mathbf{d}$ *be the diagonal meta-function* $\mathbf{d}(\varphi(x_0)) =\sim \varphi(\varphi(x_0))$. *Then there is no* d *realising* $\mathbf{d}$ *in any model.*

Proof Let us now check what conditions we need for an arbitrary $d(x_0)$, to satisfy for all φ, $d(\varphi(x_0)) \leftrightarrow\sim \varphi(\varphi(x_0))$.

We note that the meaning of $h \vDash d(\varphi(x_0))$ is fixed by the semantics. We cannot stipulate it. We need to calculate it and check under what conditions it happens to be the same as $h \vDash\sim \varphi(\varphi(x_0))$.

We shall show that no function d can exist. Take an arbitrary φ

1. $h \vDash d(\varphi(x_0))$ iff for all Q and h_1 such that $h = \mathbf{F}(h_1(Q), h_1)$ we have $h_1 \vDash Q(d(\varphi))$. Let $Q_1(x_0) = Q(d(x_0))$, then $h_1 \vDash Q_1(\varphi)$ iff $\mathbf{F}(h_1(Q_1), h_1) \vDash \varphi$.
2. $h \vDash\sim \varphi(\varphi(x_0))$ iff $\mathbf{F}(h(\varphi), h) \vDash\sim \varphi$.

Take $\varphi = \top$ (e.g. $P(x) \to P(x)$). Then (1) holds while (2) cannot hold.

Thus $\mathbf{d}$ cannot be realised in the general case. Note, however, that if we exclude $\varphi = \top$ or $\varphi = \bot$ then there are models where (1) is equivalent to (2). Let D contain only one element $\{a\}$, and the language contains only one atomic predicate $P(x)$. Then all wffs of the language are $P(a), \sim P(a), \top, \bot$. Let $d(x) = x$ and let $\mathbf{m}_1 = \{P(a)\}, \mathbf{m}_2 = \{\sim P(a)\}$ and $\mathbf{F}(\mathbf{m}_1) = \mathbf{m}_2$ and $\mathbf{F}(\mathbf{m}_2) = \mathbf{m}_1$. □

The next definition is intended to identify a language where for any formula $A(x, y)$, there always exists a solution to the equation $\alpha \leftrightarrow A(x, \alpha)$, i.e. there exists some α such that α is equivalent to $A(x, \alpha)$. This α is denoted by the symbol $\mu y A(x, y)$.

Thus $\mu y A(x, y)$ is *the* wff α such that $\alpha \leftrightarrow A(x, \alpha)$ holds.

To define the language inductively we need to have counters for two kinds of nested operations.

1. One is the usual nestings of self-reference. Thus in the formula $P_1(\mu y A(P_2(\mu x \beta(x), y)))$ we have two nestings of self-reference (P_1, P_2)
2. The other nesting is the application of the μ-fixed point operator. In the formula above we have two nestings of μ.

Together we need a counter (m, n) for nestings. n for P_i and m for μ.

Definition 10.10 (Self-fibred predicate logic with fixed points). Let $\mathbb{L}$ be a predicate language. We define the language $\mathbb{L}_\mu$ as follows

1. Any wff of $\mathbb{L}$ is a level $(0,0)$ μ-formula.
2. If $A(x_i, y_k)$ is a level $(0,0)$ μ-formula and C_j are level (m,n) μ-formulas then $A(x_i, y_j/C_j)$ is a level $(m, n+1)$ μ-wff.
3. If $A(x_i, y)$ is a level (m,n) μ-wff then $\mu y A(x_i, y)$ is a level $(m+1, n)$ μ-formula. The intended meaning of $\mu y A(y)$ is the weakest (if it exists) μ-formula α such that $\alpha \leftrightarrow A(\alpha)$ holds.

We need to show that $\mathbb{L}_\mu$ can have a fairly general semantics. For this purpose we consider the class of all well-founded tree models of Definition 7.10.

Such models have the form $\mathbf{n} = (S, D, a, \mathbf{F}, h, g)$ where (S, R, a) is a tree and every R_P chain is finite, for every predicate P. R_P is defined in Definition 7.10, item 2.

Let B be a formula of $\mathbb{L}_\mu$. We want to give meaning to $t \vDash B$. B is built up from a finite number of atomic unary predicates, say $P_1, \ldots, P_k$. The syntactic construction process that leads to B involves the operations of substitution of subformulas α into atomic predicates $P(x)$ to form $P(\alpha)$ and involves the fixed point operations $\mu y \varphi(y, x)$. The latter construct is the problem, as we need to define its value at points t. Let us consider each subformula of the form $\mu y \varphi(y, x)$ as a new atom $C_{\mu y \varphi(y)}(x)$, with x free.

Using these atoms, we get a new formula B^* of the self-fibred language without the operator μ, but with additional (possibly non-unary) predicates of the form $C_{\mu y \varphi(y)}$. If we give semantics to B^* in such a way that $C_{\mu y \varphi(y)} = \varphi(C_{\mu y \varphi(y)}(x), x)$, then we have succeeded in giving semantics for B.

This we now proceed to do. Our aim is to define the extension of $C_{\mu y \varphi(y)}$ in any well-founded tree model $\mathbf{n}$. This is done by induction on the complexity of $\mu y \varphi(y)$. We first assume that φ is of level $(0, n)$, i.e. there are no occurrences of $\mu y \alpha(y)$ inside φ. We show how to give a semantical extension to $\mu \varphi(y)$.

Then we show how to get semantics to $\mu y \varphi(y)$ where φ is of level $(m, n), m > 0$.

Recall that a tree model has the form $\mathbf{n} = (S, D, a, \mathbf{F}, h, g)$, where S is the set of labels, $a \in S$, D is the domain, h is an assignment giving for each $y \in S$ and for each atomic P of B a subset $h(y, P) \subseteq D$. g is an assignment to the constants and variables of $\mathbb{L}_\mu$ giving elements from D.

We now recall the notion of $t \vDash^h_g A(x)$, for a subformula A of B^*.

- $t \vDash P(x)$ iff $g(x) \in h(t, P)$.
- $t \vDash A \wedge B$ iff $t \vDash A$ and $t \vDash B$.
- $t \vDash \sim A$ iff $t \nvDash A$.
- $t \vDash (x)A(x)$ iff for all $a \in D$, $t \vDash A(a)$.
- $t \vDash P(\alpha)$ iff for all s such that $tR_P s$, we have $s \vDash \alpha$.
- $t \vDash \mu y \varphi(y, x)$ is yet to be defined.

Let $\varphi(y, x)$ be of level $(0, n)$. Assume φ is built up from the atomic predicates $(P_1, \ldots, P_k)$. Let t be a point of rank m. We define $t \vDash \mu y(y, x)$ by induction on m.

Case $m = 1$
In this case, we let $t \vDash \mu y\varphi(y)$ hold iff $t \vDash \varphi(P_i/\top)$ hold, where $\varphi(P_i/\top)$ is obtained from $\varphi(P_i)$ by substituting $\top$ for $P_i, i = 1, \ldots, k$.

Case $m > 1$
We let $t \vDash \mu y\varphi(y)$ hold iff $t \vDash \varphi(y/\mu y\varphi(y))$ holds. We need to observe that the latter can be evaluated. The variable y is in the range of $P_i(y)$. Thus the evaluation of $t \vDash \varphi(y/\alpha)$ for any α reduces to the evaluation of $t \vDash P_i(\alpha)$.

But

$$t \vDash P_i(\alpha) \text{ iff for } s \text{ such that } tR_P s, s \vDash \alpha.$$

Since s has lower rank than t, for $\alpha = \mu y\varphi(y)$ $s \vDash \mu y\varphi(y)$ can be evaluated by the induction hypothesis.

I am not able to give a completeness theorem for a self-fibred predicate logic with μ. I think such systems are not RE.

Example 10.11 (Common knowledge). Common knowledge is a formula φ which says 'I am what everybody knows'. Let $K(x, \alpha)$ mean x knows α. Then we are looking for $\mu y\forall xK(x, y)$. This is the solution to the equation

$$\alpha = \forall xK(x, \alpha).$$

10.3 Németi's Generalised Assignment Models

A surprising application of our fibred semantics for predicate logics is the connection with the Andreka–van Benthem–Németi *generalised first-order semantics* and *generalised assignment models*, see [Andréka *et al.*, 1995; Németi, 1995; Németi, 1996].

The Budapest and Amsterdam schools of logic have noticed that one can get decidable fragments of classical logic if one restricts the language a little bit in the spirit of modal logic.

If (D, h, g) is a classical model for the classical predicate logic, with g being the assignment to the variables, then the traditional Tarskian truth definition for the universal quantifier (x) is

- $g \vDash (x)A(x)$ iff for all g' such that $g =_x g'$ we have $g' \vDash A(x)$.

where $g' =_x g$ means that g' agrees with g on all variables $y \neq x$, and is possibly different on x. The Tarskian semantics allows $g'(x)$ to assume any value in the domain D.

The generalised first-order semantics is obtained by restricting the range of choice of g' in the clause above. For each variable x, let R_x be a binary relation on assignments. We modify the truth definition for (x) by letting

- $g \vDash (x)A(x)$ iff for all g' such that gR_xg' and $g =_x g'$ we have $g' \vDash A(x)$.

The above definition turns the quantifier (x) into a modality dependent on x.

Formally we define

Definition 10.12 (Generalised first-order models).

1. • A generalised model has the form (D, R_x, h, g), where (D, h, g) is a classical model and for each variable x, R_x is a binary relation on assignments.
   • Satisfaction is defined in the traditional manner, with the clause below for (x):
   $g \vDash (x)A(x)$ iff for all g' such that $gR_x g', g' \vDash A(x)$.
2. If each R_x is an equivalence relation then the model is called a generalised assignment model.
3. A generalised assignment model can also be presented in the form $(\mathcal{G}, D, h, g)$ where (D, h, g) is a classical model and $\mathcal{G}$ is a family of assignments, with $g \in \mathcal{G}$. Satisfaction is defined through the usual clauses, with the one for (x) being:
   $g \vDash (x)A(x)$ iff for all $g' \in \mathcal{G}$ such that $g =_x g'$ we have $g' \vDash A(x)$.

The following sample theorem is proved in Németi [1995].

Theorem 10.13. *The predicate logic defined by the class of generalised assignment models is decidable.*

We now show how to obtain the generalised assignment models as a special case of the fibred semantics.

Definition 10.14 (Fibred assignment models). Let $\mathbf{n} = (S, D, a, \mathbf{F}, h, g)$ be a fibred equality model as in Definition 7.13. $\mathbf{n}$ is said to be a fibred assignment model if the following hold.

- $S = D$.
- There exists a set $\mathcal{G}$ of assignments to the variables and constants of the language such that for all $t \in S$, $g_t \in \mathcal{G}$ and the following holds for all $X \subseteq D$

$$s \in \mathbf{F}(X, t) \text{ iff } g_s =_X g_t$$

  where $g =_X g'$ iff for any y such that $g(y) \notin X$ or $g'(y) \notin X$ we have $g'(y) = g(y)$.
- Each $g \in \mathcal{G}$ is one to one.
- h is independent of t for all predicates.

It is therefore possible to present $\mathbf{n}$ as $(\mathcal{G}, D, h, g)$.

Definition 10.15 (Translation into self-fibred classical logic with equality). We propose the following translation $*$:

- $A^* = A$, for A atomic
- $(A \wedge B)^* = A^* \wedge B^*$
- $(\sim A)^* = \sim A^*$
- $((x)A(x))^* = (x = A^*(x))$'.

Lemma 10.16. *Let $\mathbf{n} = (\mathcal{G}, D, h, g)$ be a fibred assignment model. The following holds*

1. *For $g \in \mathcal{G}$, let $g(x) = d$. Then for all g' we have $g' =_{\{d\}} g$ iff $g' =_x g$.*

2. *For any translated wff A^* we have*

$$t \vDash A^* \textit{ in } \mathbf{n} \textit{ iff } g_t \vDash^* A^* \textit{ in } \mathbf{n}$$

where $\vDash^*$ evaluates as a generalised assignment model.

Proof

1. holds because all $g \in \mathcal{G}$ are one-to-one.
2. is proved by structural induction. □

Remark 10.17. The reader should note that $(x)\varphi(x)$ is translated as $x = \varphi(x)$. If we let $\mathcal{G}$ be the set of all assignments, we get back classical logic, in a quantifier-free syntactical form.

Thus $(x)(y)A(x,y) \to (y)(x)A(x,y)$ is translated as

$$x = (y = A(x,y)) \to y = (x = A(x,y)).$$

I find this aspect of our semantics rather surprising. I am sure it has some cylindric algebraic meaning.

Note that since our considerations involve general self-fibring principles only, it is clear that the idea of generalised semantics can be immediately exported to any logic, e.g. intuitionistic logic.

The same translations can be applied and the results investigated.

10.4 Generalised Quantifiers

Another application of our fibring methods is a new point of view and a natural translation of generalised quantifiers into our self-fibred logic with equality. Generalised quantifiers are studied as possible world semantics operators in [Alechina, 1995; van Lambalgen, 1996; Alechina and van Lambalgen, 1996]. The basic idea is to consider models of the form $\mathbf{m} = (D, R, h, g)$, where (D, h, g) is a classical model and R is a relation between finite subsets of D and elements of D. The intepretation of a generalised quantifier $(Qx)A(x, d_1, \ldots, d_k)$ is as follows:

- $\mathbf{m} \vDash (Qx)A(x, d_1, \ldots, d_k)$ iff for all $d \in D$ such that $\{d_1, \ldots, d_k\}Rd$ we have $\mathbf{m} \vDash A(d, d_1, \ldots, d_k)$.

Different axioms on Q correspond to different conditions on R.

The task of this section is to study these quantifiers and their relations to self-fibred logic. Our approach is to view these quantifiers from the perspective of free logic. In Section 6.2 we introduced the free logics $\mathbf{C}_0$ and $\mathbf{C}_1$. We presented a completeness proof for the quantifier $(x)A(x)$ where the range of (x) was restricted to 'existing' elements. This was reflected in the form of our free models, (D, D', h, g), where the quantifier ranges over D while the domain is $D \cup D'$. We adopt the same view for generalised quantifiers. A quantifier of the form $(Qx)A(x, u_1, \ldots, u_k)$, where $x, u_1, \ldots, u_k$ are all the free variables of A, is viewed as the free quantifier $(x)A(x)$, where the range of x is restricted in different ways. Different restrictions yield different quantifiers. The most general system $\mathbf{Q}_0$ for the quantifier $(Qx)A(x, \bar{u})$ considers $(u_1, \ldots, u_n)$ as a sequence $\bar{u}$ of elements (perhaps the sequence determined by the order of instantiation) and restricts the range of $(x)A(x)$ through a relation R between sequences and elements. Thus $(x)A(x, \bar{u})$ holds iff for all d such that $\bar{u}Rd$ we have that $A(d, \bar{u})$ holds. The generalised quantifiers in the literature [Alechina, 1995; van Lambalgen, 1996; Alechina and van Lambalgen, 1996] use a relation R dependent on $\bar{u}$ as a set and not as a sequence.

We now begin our study of various aspects of quantifiers and fibring.

We can see from the completeness proof for $\mathbf{C}_1$ (Theorem 6.8) that it is possible to generalise the Henkin construction of the free semantics involved in the proof. When dealing with the case of $(Ex)C_n(x)$, we added to the theory Δ_n the formula $C_n(u)$, for u a new variable and in addition added $\{\varphi(u)\}$ for all φ such that $(x)\varphi(x)$ was in Δ_n. This can be generalised.

We can associate with every variable x a range of quantification V_x. Thus the rule becomes

(*) $(x)A(x) \to A(u)$, if $u \in V_x$.

We have to be careful how to read (*). There are two main options:

(*1) To read the dependence as syntactical. We can imagine a metalevel relation R between variables, where xRy means $y \in V_x$. We can read $(x)A(x)$ as for all $y(xRy \to A(y))$.
This is the approach of 'modal quantification' studied in [Alechina, 1995] (and see references there). Semantically when evaluating an expression $(Ex)A(d, x)$, where $d \in D$, we need to know for which variable y was d substituted. The following axiom does not hold

$$(t_1 = t_2) \wedge A(t_1) \to A(t_2).$$

See [Alechina, 1995, page 16], and [van Lambalgen, 1996] on generalised quantifiers.

(*2) The second option is to have a semantical dependence. In this option $B = (x)$ $A(x, \bar{u})$ is read as ranging over some domain (range) associated with the formula B. Thus $(x)A(x)$ means for all $d \in D_B, A(d)$.
There are various ways to have this dependence. We can make D_B depend on the instantiations of $\bar{u}$, the free variables of B. We can even be more sophisticated and make D_B dependent on the set $\{(d, \{\bar{y} \mid A(d, \bar{y}) \text{ holds}\}) \mid d \in D\}$. We shall discuss the more general options later.

Let us define a logic for the second option $(*2)$ where the dependence is on the set of instantiations for the free variables of the quantified formula.

The system $\mathbf{C}_2$

Definition 10.18 ($\mathbf{C}_2$ models).

1. A $\mathbf{C}_2$ model has the form (D, R, h, g), where (D, h, g) is a classical model and R is a relation between finite sets of elements of D and elements of D, i.e. R has the form, $R = \{(D_0, d) \mid D_0 \subseteq D, D_0 \text{ finite and possibly empty and } d \in D\}$.

2. Satisfaction for the language $\mathbb{L}$ is defined in the usual manner for the classical connective and for the atomic case. The case of the quantifiers is as follows.
   - $\mathbf{m} \vDash (Ex)A(x, u_1, \ldots, u_k)$ iff for some d such that $\{g(u_1), \ldots g(u_k)\}Rd$ we have $\mathbf{m} \vDash A(x/d, u_1, \ldots, u_k)$, where $u_1, \ldots, u_k$ are all the free variables of A.
   - $\mathbf{m} \vDash (x)A(x, u_1, \ldots, u_k)$ iff for all d such that $\{g(u_1), \ldots, g(u_k)\}Rd$ we have $\mathbf{m} \vDash A(x/d, u_1, \ldots, u_k)$, where $u_1, \ldots, u_k$ are all the free variables of A.
3. A $\mathbf{C}_2$ model is said to be reflexive (or a $\mathbf{C}_3$-model) if R satisfies D_0Rd for all $d \in D_0$.
4. A $\mathbf{C}_2$ model is said to be strictly reflexive (or a $\mathbf{C}_4$ model) iff R satisfies:
   - D_0Rd iff $d \in D_0$.

Definition 10.19 (The system $\mathbf{C}_2$).

1. The system $\mathbf{C}_2$ is defined by the following axioms and rules, where $\bar{u} = (u_1, \ldots, u_k)$.
   - All substitution instances of truth functional tautologies.
   - $\vdash A$ and $\vdash A \to B$ imply $\vdash B$.
   - $\dfrac{\vdash A \to B(x)}{\vdash A \to (x)B(x)}$, where x is not free in A.
   - $\vdash (x)\varphi(x, \bar{u}) \wedge (Ex)A(x, \bar{u}) \to (Ex)(\varphi(x, \bar{u}) \wedge A(x, \bar{u}))$.
   - $(x)(\varphi(x, \bar{u}) \to \psi(x, \bar{u})) \to ((x)\varphi(x, \bar{u}) \to (x)\psi(x, \bar{u}))$.
   - $(x)(\varphi(x, \bar{u}) \wedge \psi(x, \bar{u})) \leftrightarrow ((x)\varphi(x, \bar{u}) \wedge (x)\psi(x, \bar{u}))$.
2. The logic $\mathbf{C}_3$ is obtained from $\mathbf{C}_2$ by adding the axiom
   - $(u_1, \ldots, u_k)((x)\varphi(x, u_1, \ldots, u_k) \to \bigwedge_{i=1}^{k} \varphi(u_i, u_1, \ldots, u_k))$.
3. The logic $\mathbf{C}_4$ is obtained from $\mathbf{C}_2$ by adding
   - $(x)\varphi(x, u_1, \ldots, u_k) \leftrightarrow \bigwedge_{i=1}^{k} \varphi(u_i, u_1, \ldots, u_k)$.

Theorem 10.20 (Completeness for $\mathbf{C}_2, \mathbf{C}_3$ and $\mathbf{C}_4$). *The system $\mathbf{C}_i$ is complete for the class of $\mathbf{C}_i$ models, for $i = 2, 3, 4$.*

Proof

1. Soundness
   This can be verified.
2. Completeness
   Let B be a consistent $\mathbf{C}_i$ wff. We want to construct a $\mathbf{C}_i$ model $\mathbf{m}$ such that $\mathbf{m} \vDash B$. Let $A_1, A_1, A_2, \ldots$ be an enumeration of all wffs of $\mathbb{L}$. We define (Δ_n, R_n), Δ_n and R_n finite, as follows:

Case $n = 0$
We distinguish several subcases

(a) B has the form $(Ex)A(x, u_1, \ldots, u_k)$ and the logic is $\mathbf{C}_2$ or $\mathbf{C}_3$.

Let u be a new variable not appearing in B. We claim $\{B, A(u, u_1, \ldots, u_k)\}$ is consistent. For otherwise

$$\vdash B \to\sim A(u, u_1, \ldots, u_k)$$

and hence

$$\vdash B \to (x) \sim A(x, u_1, \ldots, u_k),$$

a contradiction to the consistency of B.

In this case, let $\Delta_0 = \{B, A(u, u_1, \ldots, u_k)\}$ and let $R_0 = \{(\{u_1, \ldots, u_k\}, u)\}$.

(b) As in the previous case and the logic is $\mathbf{C}_4$.

We claim for some i, $\{B, A(u_i, u_1, \ldots, u_k)\}$ is consistent. For otherwise

$$\vdash B \to\sim A(u_i, u_1, \ldots, u_n), i = 1, \ldots, k$$

hence

$$\vdash B \to\sim \bigvee_{i=1}^{k} A(u_i, u_1, \ldots, u_k),$$

but this contradicts the axiom

$$\vdash (u_1, \ldots, u_k)(Eu)A(u, u_1, \ldots, u_k) \leftrightarrow \bigvee_{i=1}^{k} A(u_i, u_1, \ldots, u_k)\}.$$

Let $\Delta_0 = \{B, A(u_i, u_1, \ldots, u_k)\}$ for the first $i \in \{1, \ldots, k\}$ for which the set is consistent. Let $R_0 = \{(\{u_1, \ldots, u_k\}, u_i) \mid i = 1, \ldots, k\}$.

(c) B has the form $(x)A(x, u_1, \ldots, u_k)$ and the logic is $\mathbf{C}_2$.

Let

$$\Delta_0 = \{B\}, R_0 = \varnothing.$$

(d) As in the previous case and the logic is $\mathbf{C}_3$ or $\mathbf{C}_4$.

Let

$$\Delta_0 = \{B\} \cup \{A(u, u_1, \ldots, u_k) \mid u \in \{u_1, \ldots, u_k\}\}.$$

We claim this set is consistent. Otherwise

$$\vdash B \to\sim \bigwedge_{i=1}^{k} A(u_i, u_1, \ldots, u_k))$$

hence

$$\vdash (u_1, \ldots, u_k)[(x)A(x, u_1, \ldots, u_k) \to\sim \bigwedge_{i=1}^{k} A(u_i, u_1, \ldots, u_k)].$$

But we have the axiom

$$\vdash (u_i)((x)A(x, u_1, \ldots, u_k) \to \bigwedge_{i=1}^{k} A(u_i, u_1, \ldots, u_k)),$$

a contradiction.

(e) B is neither universal nor existential and the logic is $\mathbf{C}_i, i > 1$. Let $\Delta_0 = \{B\}$ and $R_0 = \varnothing$.

Case $n+1, n \geq 0$

Assume Δ_n, R_n have been defined and Δ_n is consistent and finite. We now define Δ_{n+1}, R_{n+1}.

Consider A_n and $\sim A_n$. One of these two is consistent with Δ_n. Denote by B_n the one which is consistent. We distinguish several subcases, according to the form of B_n and the logic $\mathbf{C}_i, i > 1$, involved.

(a) Case $B_n = (Ex)A(x, u_1, \ldots, u_k)$ and the logic is $\mathbf{C}_2$ or $\mathbf{C}_3$.
We let u be a new variable not in Δ_n or R_n and let
$\Delta_{n+1} = \Delta_n \cup \{B_n, A(u, u_1, \ldots, u_k)\} \cup \Gamma_n$ where

$$\Gamma_n = \{\varphi(u, u_1, \ldots, u_k) \mid (x)\varphi(x, u_1, \ldots, u_n) \in \Delta_n\}.$$

We let $R_{n+1} = R_n \cup \{(\{u_1, \ldots, u_k\}, u)\}$.
We claim Δ_{n+1} is consistent. Otherwise

$$\vdash \bigwedge \Delta_n \wedge B_n \rightarrow [\bigwedge_{j=1}^{r} \varphi_j(u, u_1, \ldots, u_k) \rightarrow \sim A(u, u_1, \ldots, u_k))]$$

hence

$$\vdash \bigwedge \Delta_n \wedge B_n \rightarrow (x)[\bigwedge_{j=1}^{r} \varphi_j(x, u_1, \ldots, u_k) \rightarrow \sim A(x, u_1, \ldots, u_k))];$$

using the axiom

$$\vdash (x)[\alpha(x, u_1, \ldots, u_k) \rightarrow \beta(x, u_1, \ldots, u_k)] \rightarrow [(x)\alpha(x, \bar{u}) \rightarrow (x)\beta(x, \bar{u})]$$

we get a contradiction.

(b) B_n as in the previous case and the logic is $\mathbf{C}_4$.
We claim for some i, $\Delta_n \cup \{B_n, A(u_i, u_1, \ldots, u_k)\}$ is consistent. Otherwise

$$\vdash \bigwedge \Delta_n \wedge B_n \rightarrow \sim \bigvee_{i=1}^{j} A(u_i, u_1, \ldots, u_k)$$

and we get a contradiction as in the case $n = 0$ subcase (b).
Let

$$\begin{aligned} \Delta_{n+1} &= \Delta_n \cup \{B_n, A(u_i, u_1, \ldots, u_k)\}, \\ R_{n+1} &= R_n \cup \{(\{u_1, \ldots, u_k\}, u_i) \mid i = 1, \ldots, k\}. \end{aligned}$$

(c) B_n has the form $(x)A(x, u_1, \ldots, u_k)$ and the logic is $\mathbf{C}_2$ or $\mathbf{C}_3$.
Let $\Delta_{n+1} = \Delta_n \cup \{B_n\} \cup \{A(y, u_1, \ldots, u_k) \mid \{u_1, \ldots, u_k\} R_n y\}$.

Let $R_{n+1} = R_n$. We claim Δ_{n+1} is consistent. Otherwise for some $y_1, \ldots, y_r$ we have

$$\vdash \bigwedge \Delta_n \wedge B_n \rightarrow \sim \bigwedge_{j=1}^{r} A(y_j, u_1, \ldots, u_k).$$

Each y_j was put into Δ_n because of some $B_j = (Ex)A_j(x, u_1, \ldots, u_k)$ and the wff $A_j(y_j, u_1, \ldots, u_k)$ is also in Δ_n. Let us regroup our formulas and get

$$\vdash \bigwedge \Delta'_n \wedge [(x)A(x, u_1, \ldots, u_k) \wedge \bigwedge_{j=1}^{r}(Ex)A_j(x, u_1, \ldots, u_k)] \rightarrow \\ \sim \bigwedge_{j=1}^{r}(A(y_j, u_1, \ldots, u_k) \wedge A_j(y_j, u_1, \ldots, u_k))$$

where $\bigwedge \Delta'_n$ does not contain any free $y_j, j = 1, \ldots, r$.
We can now universally quantify over $y_1, \ldots, y_r$. We now use the axiom $\vdash (x)A(x, \bar{u}) \wedge (Ex)A_j(x, \bar{u}) \rightarrow (Ex)(A(x, \bar{u}) \wedge A_j(x, \bar{u}))$, to get that

$$\vdash \bigwedge \Delta'_n \wedge [(x)A(x, \bar{u}) \wedge \bigwedge_{j=1}^{r}(Ex)A_j(x, u)] \rightarrow \\ \bigwedge_{j=1}^{r}(Ex)(A(x, \bar{u}) \wedge A_j(x, \bar{u})) \wedge \\ (y_1, \ldots, y_r) \sim \bigwedge_{j=1}^{r}(A(y_r, \bar{u}) \wedge A_j(y_r, \bar{u})),$$

a contradiction.

(d) B_n has the form $(x)A(x, u_1, \ldots, u_k)$ and the logic is $\mathbf{C}_4$.
Let Δ_{n+1} be $\Delta_n \cup \{A(u_i, u_1, \ldots, u_k) \mid i = 1, \ldots, k\}$. $R_{n+1} = R_n$.
Δ_{n+1} is consistent by virtue of the $\mathbf{C}_4$ axiom.

(e) B_n is neither existential nor universal. In this case let $\Delta_{n+1} = \Delta_n \cup \{B_n\}$ and $R_{n+1} = R_n$.

We are now ready to define the model. Let

$$\Delta = \bigcup_n \Delta_n, R = \bigcup_n R_n.$$

Let D_n be the set of all variables and constants appearing in Δ_n and R_n. Let $\mathbf{m}_n = (D_n, R_n, h_n, g_n)$ be the finite model, where D_n and R_n are as above and where g_n is the identity and where h_n is defined by $h_n(P) = \{\bar{u} \mid P(\bar{u}) \in \Delta_n\}$, P atomic. Let $\mathbf{m}_\Delta = (D, R, h, g)$, where D is the set of variables and constants of the language. R is defined above, g is the identity and for atomic P let $h(P) = \{\bar{u} \mid P(\bar{u}) \in \Delta\}$.

We claim

(*) $\mathbf{m}_\Delta \vDash A$ iff $A \in \Delta$ iff there exists a number n such that $\mathbf{m}_k \vDash A$ for all $k \geq n$.

The proof is by induction on the formula. The atomic case holds by definition. We check the quantifier case.

Assume $(x)A(x, \bar{u}) \in \Delta$ and let $\bar{u}Rd$ hold. We want to show $A(d, \bar{u}) \in \Delta$. This holds because it was ensured by construction.

Assume $(Ex)A(x, \bar{u}) \in \Delta$. We are looking for a d such that $\bar{u}Rd$ holds and $A(d, \bar{u}) \in \Delta$. Again this was ensured by construction. □

Translating $\mathbf{C}_2$ into Self-fibred Logic

The language of self-fibred logic is rich enough to afford good translations of generalised quantifiers into it. Our task is to choose a natural one. Let us examine our options. Consider a tree generalised model of Definition 7.10. This has the form $\mathbf{m} = (S, D, a, \mathbf{F}, h, g)$.

In contrast, a generalised quantifiers model has the form $\mathbf{m}' = (D', R', h', g')$. R' is a relation between sets or sequences of elements of D' and elements of D'. $\mathbf{m}$ is a more complex model. To bring it nearer to $\mathbf{m}'$ we can assume $S = D$. We can also assume that h and g are not dependent on points in S, since $\mathbf{m}'$ has only one h' and one g'. Thus we can assume $h' = h, g' = g, D = S = D'$ and that they are all fixed. It is also obvious that $\mathbf{F}$ is destined to be R'. This seems likely since $\mathbf{F}$ depends on points of D and on subsets of D. So for the case of $\mathbf{C}_2$ where R' deals with subsets of D, our course of action is obvious. We make $\mathbf{F}$ independent of S, and we let

$$XR'x \text{ iff } x \in \mathbf{F}(X).$$

All we need now is a syntactical translation of $(x)A(x)$ into the fibred language, which, for the special fibred models with $S = D$ and $\mathbf{F}, g, h$ independent of S, yields the right semantics for the generalised quantifier. This is indeed what we do below for the case of $\mathbf{C}_2$.

Now that we understand our options, let us proceed to show how to translate generalised quantifiers into our fibred system. This will demonstrate the expressive power of predicate self-fibring with equality.

We first present a translation of the logic $\mathbf{C}_2$ of Definition 10.18 into $\mathbb{L}_=[\mathbb{L}]$. We saw from the semantics of $\mathbf{C}_2$ models that $(x)A(x, u_1, \ldots, u_k)$ is understood as: for all y such that $\bar{u}Ry$, $A(y, \bar{u})$ holds. This looks very much like $P_{\bar{u}}((x)A(x, \bar{u}))$.

The next series of definitions will lead to a translation.

Definition 10.21.

1. Let $\mathbb{L}$ be a predicate language with monadic predicates as well as equality and constants but no function symbols. Let $\mathbb{L}_{M,=}$ be the monadic fragment with equality and constants and the quantifiers $(x), (\forall x)$. Then $\mathbb{L}_{M,=}[\mathbb{L}]$ is the language obtained by repeated substitution of wffs of $\mathbb{L}$ into $\mathbb{L}_{M,=}$.
2. The fibred equality models of the form $\mathbf{n} = (S, D, a, \mathbf{F}, h, g)$ of Definition 7.2.2 are suitable semantics for the above language, with the added semantical clause $t \vDash P(\varphi)$ iff for all $s \in \mathbf{F}(h(t, P), t)$ we have $s \vDash \varphi$.
3. We assume we have a special *locality predicate*, denoted by $W(x)$, whose purpose is to identify each world internally, i.e. for each $t \in S$ there exists a *unique* d_t such that the following holds
   - $t \vDash W(x)$ iff $g_t(x) = d_t$.
4. A model is said to be an $\mathbf{E}$-model if the following holds.
   - $S = D$.
   - $\mathbf{F}$ is independent of t.

- h and g are independent of t for all atomic predicates except the locality predicate $W(x)$. In the case of $W(x)$ we have $d_t = t$.

5. Let $\bar{u} = \{u_1, \ldots, u_k\}$. We let $P_{\bar{u}}(x)$ be the predicate satisfying

$$P_{\bar{u}}(x) \leftrightarrow \bigvee_{i=1}^{k} x = u_i.$$

6. Let $\mathbf{E}$ be the logic defined by the class of all $\mathbf{E}$ models.
7. An $\mathbf{E}$ model can be presented as $\mathbf{n} = (S, \mathbf{F}, h, g)$.

We are ready to translate the syntax of the free logic $\mathbf{C}_2$ into that of $\mathbb{L}_{M,=}[\mathbb{L}]$.

Definition 10.22 (Translation of $\mathbf{C}_2$ into E). We translate any wff A of $\mathbf{C}_2$ into a wff $A^\sharp$ of $\mathbf{E}$ as follows:

1. $A^\sharp = A$, for A atomic
2. $(\sim A)^\sharp = \sim A^\sharp$
   $(A \wedge B)^\sharp = A^\sharp \wedge B^\sharp$
3. $((x)A(x, u_1, \ldots, u_k))^\sharp = P_{\bar{u}}((x)(W(x) \to A^\sharp(x, \bar{u})))$
   where $\bar{u} = (u_1, \ldots, u_k)$.

Let us see what the definition says. Let $\mathbf{m} = (D, R, h, g)$ be a $\mathbf{C}_2$ model. We can turn $\mathbf{m}$ into a uniform $\mathbf{E}$ model $\mathbf{n}$ by letting $S = D, \mathbf{F} = R$, $g_t = g$ and $h(t, P)$ (in $\mathbf{n}) = h(P)$. We let $\mathbf{n} = (S, \mathbf{F}, h, g)$. Then we have:

$$\mathbf{m} \vDash (x)A(x, \bar{u}) \text{ iff for all } d \text{ such that } \bar{u}Rd, \text{ we have } \mathbf{m} \vDash A(d, \bar{u}).$$

Let t be any point in $\mathbf{n}$. Then $t \vDash P_{\bar{u}}((x)(W(x) \to A^\sharp(x, \bar{u})))$ iff for all $d \in \mathbf{F}(\bar{u}), d \vDash (x)(W(x) \to A^\sharp(x, \bar{u}))$ iff for all $d \in \mathbf{F}(\bar{u}), d \vDash A^\sharp(d, \bar{u})$.

The translation is faithful because of the following lemma.

Lemma 10.23. *In the model* $\mathbf{n}$, *the value of* $s \vDash A^\sharp$ *is not dependent on* s.

Proof By induction on $A^\sharp$. □

The results above should be compared with van Lambalgen's treatment of quantifiers in [van Lambalgen, 1996]. Consider a language with the classical quantifier $\forall$ and the generalised quantifier Q. The following choice of axioms is offered for this quantifier.

Q0 Axioms and inference rules for classical logic.

Q1 $\sim (Qx)(x \neq x)$

Q2 $(Qy)(x \neq y)$

Q3 $(Qx)A(\ldots, x, \ldots) \to (Qy)A(\ldots, y, \ldots)$, provided y is free for x in A

Q4 $(Qx)A \wedge \forall x(A \to B) \to (Qx)B$

Q5 $(Qx)A \wedge (Qx)B \to (Qx)(A \wedge B)$

Q6 $(Qx)(Qy)A \to (Qy)(Qx)A$.

The semantics in [van Lambalgen, 1996] does involve a relation R like the one we have here. Van Lambalgen gives conditions on R which yield completeness for Q, as well as conditions for other quantifier systems.

Such conditions can of course be given for **F** so that the translation above remains faithful.

Example 10.24. Let $\mathbf{n} = (S, D, a, \mathbf{F}, h, g)$ be an equality model with g constant and $S = D$. Consider the translation of the quantifier $(Qx)A(x, \bar{u})$ as defined above. Let us examine what semantical conditions on **F** are required to validate various axioms on Q. For convenience we abuse notation and assume $g(\bar{u}) = \bar{u}$. Recall the definition of $Q(x)$:

- $t \vDash (Qx)A(x, \bar{u})$ iff $t \vDash P_{\bar{u}}((x)(W(x)\ to A(x, \bar{u}))$ iff for all $d \in \mathbf{F}(\{\bar{u}\}, t), d \vDash A(d, \bar{u})$.

Let $(\mathbb{D}x)A(x)$ be defined as a special diagonal operator, with

- $(\mathbb{D}x)A(x, \bar{u}) = \{d \in S \mid d \vDash A(d, \bar{u})\}$.

Hence we get

- $t \vDash (Qx)A(x, \bar{u})$ iff $\mathbf{F}(\{\bar{u}\}, t) \subseteq (\mathbb{D}x)A$.

Let us now examine the axioms **Q0–Q6**.

1. $t \vDash \sim (Qx)(x \neq x)$ iff $t \nvDash (Qx)(x \neq x)$
   iff $\exists d \in \mathbf{F}(\varnothing, t) d \vDash d = d$
   iff $\mathbf{F}(\varnothing, t) \neq \varnothing$.
2. $t \vDash (Qy)(x \neq y)$ iff for all $d \in \mathbf{F}(\{g(x)\}, t), d \vDash d \neq g(x)$.
   Since g is independent of points of S the condition is clearer:
   $$\text{iff } g(x) \notin \mathbf{F}(\{g(x)\}, t).$$
   Therefore, since g can be arbitrary, the corresponding condition should be:
   $$d \notin \mathbf{F}(\{d\}, t), \text{ for all } d, t \in S.$$
3. $t \vDash (Qx)A(x) \to (Qy)A(y)$.
   This clearly always holds.
4. $t \vDash (Qx)A(x, \bar{u}) \wedge (x)(A(x) \to B(x)) \to (Qx)B(x, \bar{u})$ iff for all $d \in \mathbf{F}(\{\bar{u}\}, t), d \vDash A(d, \bar{u})$ and for all s, $t \vDash A(s)$ implies $t \vDash B(s)$ both imply for all $d \in \mathbf{F}(\{\bar{u}\}, t), d \vDash B(d, \bar{u})$.
   This condition can be written as follows
   Let $h(t, \lambda x A(x)) = \{d \mid t \vDash A(d)\}$. Then
   $h(t, \lambda x A(x)) \subseteq h(t, \lambda x B(x))$ and $\mathbf{F}(\{\bar{u}\}, t) \subseteq (\mathbb{D}x)A$ imply $\mathbf{F}(\{\bar{u}\}, t) \subseteq (\mathbb{D}x)B$.
   Note that if h happens to be independent of t, the condition seems to be:
   $$(x)(A(x) \to B(x)) \to (\mathbb{D}x)A \subseteq (\mathbb{D}x)B$$
   i.e. $\mathbb{D}$ is monotonic.

5. $t \vDash (Qx)A \wedge (Qx)B \to (Qx)(A \wedge B)$
iff $\mathbf{F}(\{\bar{u}\}, t) \subseteq (\mathbb{D}x)A \cap (\mathbb{D}x)B \to \mathbf{F}(\{\bar{u}\}, t) \subseteq (\mathbb{D}x)(A \wedge B)$.
This will be satisfied if

$$(\mathbb{D}x)A \cap (\mathbb{D}x)B \subseteq (\mathbb{D}x)(A \wedge B).$$

6. $t \vDash (Qx)(Qy)A(x, y, \bar{u}) \to (Qy)(Qx)A(x, y, \bar{u})$
iff $\mathbf{F}(\{\bar{u}\}, t) \subseteq (\mathbb{D}x)(Qy)A(x, y, \bar{u} \to \mathbf{F}(\{\bar{u}\}, t) \subseteq (\mathbb{D}y)(Qx)A(x, y, \bar{u})$.
The above holds if

$$(\mathbb{D}x)(Qy)A \subseteq (\mathbb{D}y)(Qx)A.$$

By symmetry it is clear that we must have equality. Let us expand this condition:

$$\begin{aligned}(\mathbb{D}x)(Qy)A(x, y, \bar{u}) &= \{d \mid d \vDash (Qy)A(d, y, \bar{u})\}\\ &= \{d \mid \mathbf{F}(\{\bar{u}, d\}, d) \subseteq (\mathbb{D}y)A(d, y, \bar{u})\}.\end{aligned}$$

Similarly

$$(\mathbb{D}y)(Qx)A(x, y, \bar{u}) = \{d \mid \mathbf{F}(\{\bar{u}, d\}, d) \subseteq (\mathbb{D}x)A(x, d, \bar{u})\}.$$

For the required equality to hold we must have for all d,

- $(\mathbb{D}z)A(d, z, \bar{u}) = (\mathbb{D}z)A(z, d, \bar{u})$.

The above considerations show that in the context of fibred semantics we get very natural and meaningful conditions for the quantifier Q in terms of the fibring function $\mathbf{F}$ and the diagonal operator $\mathbb{D}$.

The R-semantics can be viewed as a translation into classical logic, while our own translation can be viewed as a translation into self-fibred classical logic. Our translation can give more than the translation into classical logic. The self-fibred classical logic is a much richer language (in an intuitive way, not necessarily in a purely formal way) and allows for more expressions.

For example, the quantifiers so far considered are 'dependent' on the set of variables free in the formula. Thus $(Qx)A(x, \bar{u})$ depends on the variables $\bar{u}$, through the relation $\bar{u}Rx$. We have:

$$\mathbf{m} \vDash (Qx)A(x, \bar{u}) \text{ iff for all } x, \bar{u}Rx \text{ implies } \mathbf{m} \vDash A(x, \bar{u}).$$

Another way of looking at it is that the dependence is on the set

$$Q_{\bar{u}} = \hat{x}A(x, \bar{u}) = \{x \mid \mathbf{m} \vDash A(x, \bar{u})\}$$

and that the quantifier is said to hold if this set satisfies

$$Q_{\bar{u}} = \hat{x}A(x, \bar{u}) \supseteq \{x \mid \bar{u}Rx\}.$$

The most general quantifier of the form $(Qx)A(x, u_1, \ldots, u_k)$ is defined using a parametrised family $\mathbb{Q}_{u_1,\ldots,u_k}$ of subsets of the domain D. We say $(Qx)A$ holds if $Q_{\bar{u}} = \{x \mid \mathbf{m} \vDash A(x, u_1, \ldots, u_k)\} \in \mathbb{Q}_{u_1,\ldots,u_k}$.

This quantifier can be translated into self-fibred predicate logic as follows. We can assume that $\pm(\exists x)A(x, u_1, \ldots, u_k)$ are consistent, otherwise we are quantifiying over a contradiction or a truth and these can be dealt with separately. Having that assumption, translate $(Qx)A$ as $\Box^y_{A(x,\bar{u})}(\exists x)A(x, \bar{u})$.

$\Box^x_{A(x,\bar{u})}$ identifies the set $Q_{\bar{u}} = \{x \mid A(x, \bar{u})\}$ and therefore goes to a fibred model where the value $\exists x A(x, \bar{u})$ is checked. Since this wff and its negation are consistent, we can choose the right model according to whether $(Qx)A$ is supposed to hold or not.

Suppose we want a more complex dependence, say, on the family of sets $\hat{u}A(x, u) = \{u \mid A(x, u)\}$? Can we easily represent that in classical logic? We need to be able to express properties of the predicate (in x) $\hat{u}A(x, u)$ and this must be expressed in the translation. It cannot be done in first-order logic. In self-fibred classical logic this can be easily done. We replace '$P_{\bar{u}}$' by '$P_{\hat{u}A(x,u)}$', or using the notation of Section 10.1 (Definition 10.4) we replace '$P_{\bar{u}}$' by '$\Box^u_{A(x,u)}$' and the self-fibred language is rich enough to express properties of this predicate. See also [Gabbay, 1976].

10.5 McCarthy's and Buvač's Context Systems

This section will show how various formal systems of context can be naturally presented as self-fibring of predicate logics. A fuller discussion of context logics and their relation to labelled deductive systems and the general fibring of logics can be found in [Gabbay and Nossum, 1996].

Contexts were first introduced into AI by John McCarthy in his Turing Award lecture [McCarthy, 1987]. The notion was developed further in a series of papers, see [Buvač, 1996; Buvač and Mason, 1995; McCarthy and Buvač, 1994] for the latest position.

Our discussions here are based on material from these papers.

The reader will notice the strong similarity of context formalism with that of *labelled deductive systems* (LDS). In fact, the machinery of LDS can provide the formal logical framework for theories of context. See [Gabbay, 1996b].[27]

10.5.1 *The system* **B** *of context*

We begin by defining a certain self-fibred predicate logic which will naturally specialise to the system of [Buvač, 1996].

Let $\mathbb{L}_1$ be a copy of predicate logic and let $\mathbf{ist}(x, y)$ be a binary predicate of $\mathbb{L}$. Let $\mathbb{L}_0$ be the language of the predicate logic with the parametrised unary predicate $\lambda y \mathbf{ist}(x, y)$.

Consider the self-fibred language $\mathbb{L}_0[\mathbb{L}_1]$. This means we can repeatedly substitute fibred formulas for the y coordinate of $\mathbf{ist}(x, y)$.

Consider a fibred model for this language. According to Definition 7.2, 7.9 and the discussion in Section 8.1, we can take as models for this language structures of the form $\mathbf{n} = (S, D, a, \mathbf{F}, h, g)$, satisfying the conditions of Definition 7.10.

[27] Gabbay LDS position paper was presented in Logic Colloquium 90 [Gabbay, 1993b]. A first draft of the LDS book was available in 1989 [Gabbay, 1994b]. Context theories seem to be ideal for LDS formalisation. In fact, the notation and examples in [McCarthy and Buvač, 1994] and [McCarthy, 1996] are very close to LDS examples and notation. The notion of fibred semantics was introduced in 1991 to give semantics for LDS and from it the notions of fibring logics and self-fibring have emerged, grew and developed into this book.

In such models, we recall

- S is the set of labels naming classical models
- D is the domain which we assume for the purpose of this section to be constant (the same) for all labels (classical models)
- $a \in S$ is the actual label
- $\mathbf{F}$ is the fibring function, associating with each $X \subseteq D$ and $t \in S$ a set of labels $\mathbf{F}(X,t) \subseteq S$
- h is the interpretation associating for each $t \in S$ and each m-place predicate P a subset $h(t,P) \subseteq D^m$
- g is a rigid assignment giving for each variable or constant x of the language an element $g(x) \in D$.

Satisfaction is defined as in Definition 7.9 with the clause for $\mathbf{ist}(x,A)$ being:

- $t \vDash \mathbf{ist}(x,A)$ iff for all $s \in \mathbf{F}(X_{t,x},t)$, we have $s \vDash A$, where$X_{t,x} = \{y \mid t \vDash \mathbf{ist}(x,y)\}$.

Let $\mathfrak{M}^t_x = \mathbf{F}(X_{t,x},t)$. We have

- $t \vDash \mathbf{ist}(x,A)$ iff $s \vDash A$, for all $s \in \mathfrak{M}^t_x$.

Remark 10.25 (The system $\mathbf{B}_0$). Let $\mathbf{B}_0$ be the system semantically defined above. $\mathbf{B}_0$ can be axiomatised as in Theorem 7.6.

This gives us a straightforward basic system of context. The system of Buvač, which we shall call $\mathbf{B}_u$, is obtained by further simplifying assumptions on the semantics. This we discuss below.

We now examine what kind of simplifying assumptions we can have on $\mathbf{F}$ and on $h(t,\mathbf{ist})$.

Option 1
We can assume that $\mathfrak{M}^t_x$ is independent of t. This can be achieved by assuming that $X_{t,x}$ does not depend on t and further that $\mathbf{F}$ does not depend on t. The first assumption means that **ist** is a rigid predicate, i.e. there exists a relation $R_{\mathbf{ist}} \subseteq D^2$ such that for all t, $h(t,\mathbf{ist}) = R_{\mathbf{ist}}$. The second assumption means that $\mathbf{F}(X,t)$ is rigid, i.e. does not depend on t. To appreciate the implications of this option, consider now the evaluation of $t \vDash \mathbf{ist}(x_1,\mathbf{ist}(x_2,A))$. Expanding the definition we get iff $(\forall s \in \mathfrak{M}^t_{x_1})(\forall r \in \mathfrak{M}^s_{x_2})(r \vDash A)$, which we can write compactly as $(\forall r \in \mathfrak{M}^t_{(x_1,x_2)})(r \vDash A)$, where $r \in \mathfrak{M}^t_{(x_1,x_2)}$ iff (definition) there exists an s such that $(s \in \mathfrak{M}^t_{x_1} \wedge r \in \mathfrak{M}^s_{x_2})$.

Since we assumed that $\mathfrak{M}^t_x$ does not depend on t at all, we get

- $t \vDash \mathbf{ist}(x_1,\mathbf{ist}(x_2,A))$ iff for all $r \in \mathfrak{M}_{(x_1,x_2)}$, $r \vDash A$.

Let us take a closer look at $\mathfrak{M}_{(x_1,x_2)}$ in case $\mathfrak{M}$ does not depend on h. $r \in \mathfrak{M}_{(x_1,x_2)}$ iff $\exists s(s \in \mathfrak{M}_{x_1} \wedge r \in \mathfrak{M}_{x_2})$. Assuming that $\mathfrak{M}_{x_1}$ is not empty we get $\mathfrak{M}_{(x_1,x_2)} = \mathfrak{M}_{x_2}$.

This immediately entails that $\mathbf{ist}(x_1,\mathbf{ist}(x_2,A)) = \mathbf{ist}(x_2,A)$, as indeed Buvač has in his paper.

Option 2
We note the expression $t \vDash \mathbf{ist}(x,A)$. x ranges over D and t ranges over S. We can

opt to make **ist** a two-sorted predicate and let x range over S. This assumption does not imply any loss of generality since there is no other 'serious' substitution in the x-coordinate of **ist** and the second coordinate of **ist** is not affected. Choosing this option brings us closer to the [Buvač, 1996] system.

We can simplify further by letting $S = D$, and we need not have a two-sorted system. This can also be done without loss of generality. The two choices give the same system in essence.

If we adopt the above two options, and further assume no function symbols, what we get is the self-fibring of predicate logic into a family of unary predicates $\mathbf{ist}_t(y)$ indexed by the labels $t \in S$, whose models have the form $(S, D, a, \mathbf{F}, h, g)$ where both $\mathbf{F}$ and $h(t, \mathbf{ist}_s)$ are *rigid*.

Using the results of Section 7.1, we immediately get an axiomatisability and dedicability result. These follow from Theorem 7.6 and Remark 7.8 and item 5 in Definition 7.2.

We summarise our results in the definition and theorem below.

Definition 10.26 (The context system $\mathbf{B}_u$). Let $\mathbb{L}_1$ be the monadic language with a family of unary predicates $\mathbf{ist}_t$, $t \in C$ and $\mathbb{L}_2$ be the classical predicate language. Let $\mathbb{L}_1[\mathbb{L}_2]$ be the self-fibred language obtained by allowing repeated substitution from $\mathbb{L}_2$ into $\mathbb{L}_2$, as defined in Definition 7.10.

Let $\mathbf{B}_u$ be the Hilbert system for this language with the following axioms and rules.

1. All axioms and rules of free $\mathbf{C}_1(\mathbb{T})$. These include:
    - all substitution instances of truth functional tautologies
    - $\vdash A$ and $\vdash A \to B$ imply $\vdash B$
    - $\vdash (x)(A(x) \wedge B(x)) \to (x)A(x) \wedge (x)B(x)$
    - $\vdash (x)A(x) \wedge (x)(A(x) \to B(x)) \to (x)B(x)$
    - $\dfrac{\vdash A \to B(x)}{\vdash A \to (x)B(x)}$ where x is not free in A
    - $\vdash (x)(y)A \to (y)(x)A$
    - $(u)(B \wedge (x)A(x) \to A(u))$.
2. Modal rules $\mathbb{K}$:
    - $\dfrac{\vdash \bigwedge A_i \to A}{\vdash \mathbf{ist}(t, \bigwedge A_i) \to \mathbf{ist}(t, A)}$
    - $(x)\mathbf{ist}(t, A(x)) \to \mathbf{ist}(t, (x)A(x))$.
3. Axiom for the rigidity of $\mathbf{F}$:
    - $(x)[\mathbf{ist}(t_1, x) \leftrightarrow \mathbf{ist}(t, \mathbf{ist}(t_2, x))] \to [\mathbf{ist}(t_1, A) \leftrightarrow \mathbf{ist}(t, \mathbf{ist}(t_2, A))]$.
4. Axiom for the rigidity of **ist**:
    - $\mathbf{ist}(t, x) \to \mathbf{ist}(s, \mathbf{ist}(t, x))$
    - $\sim \mathbf{ist}(t, x) \to \mathbf{ist}(s, \sim \mathbf{ist}(t, x))$

Remark 10.27. The axioms and rules of $\mathbf{B}_u$ are just the axioms and rules from Theorem 7.6, Remark 7.8 and from item 5 of Definition 7.2.

We are just applying our ready-made machinery to the particular case of the context language.

Theorem 10.28 (Completeness theorem for $\mathbf{B}_u$). $\mathbf{B}_u$ *is complete for the class of fibred models with* **F** *and* **ist** *rigid.*

Proof From the general Theorem 7.6. □

Theorem 10.29. *The tight fragment of monadic* $\mathbf{B}_u$ *is decidable.*

Proof From Remark 7.8. □

Note that by monadic $\mathbf{B}_u$, we mean we have only monadic predicates in the classical language and by *tight* fragment we mean that we substitute in $\mathbf{ist}(x, y)$ in the y position only wffs with at most y free. Otherwise we get undecidability!

10.5.2 *Comparison with the literature*

We compare with [Buvač, 1996]. Our models for $\mathbf{B}_u$ are practically identical with those of Buvač. He has equality but that can be added without any problem. He uses the notation $\mathfrak{A}$ for models while we use letters t.

His evaluation has the form $t \vDash s : A$, as compared with our $t \vDash A$. This s is redundant as can be seen from Buvač evaluation.

Buvač offers the following axioms for his system:

1. Classical predicate calculus axioms and rules
2. Propositional properties of contexts:

(K) $k : \mathbf{ist}(k', A \to B) \to (\mathbf{ist}(k', A) \to \mathbf{ist}(k', B))$

(Δ) $k : \mathbf{ist}(k_1, \mathbf{ist}(k_2, A)) \vee \mathbf{ist}(k_1, \sim \mathbf{ist}(k_2, A))$

(Flat) $k : \mathbf{ist}(k_1, \mathbf{ist}(k_2, A)) \leftrightarrow \mathbf{ist}(k_2, A)$

(Enter) $\dfrac{s : \mathbf{ist}(t, A)}{t : A}$

(Exit) $\dfrac{t : A}{s : \mathbf{ist}(t, A)}.$

These two rules correspond to saying

1. $t : A$ is the same as $\mathbf{ist}(t, A)$
2. $t : (s : A)$ equals $s : A$
3. Quantificational properties of contexts:

(BF) $k : (x)(\mathbf{ist}(k', A) \to \mathbf{ist}(k', (x)A)$

($\mathbf{ist} =$) $k : t_1 = t_2 \leftrightarrow \mathbf{ist}(k', t_1 = t_2)$.

Let us compare this axiom system with our own system of Definition 10.5.2. Group axioms (1) of our system yields all of Buvač classical axioms. The $\mathbb{K}$ group of rules yield his context axioms (K), (Δ) and (BF). The axiom Flat is obtained from our rigidity axioms. The rules (Enter) and (Exit) are obviously conservative additional axioms and we can add them if desired. The reader should bear in mind that we did not invent our axioms for $\mathbf{B}_u$ as a specific system. We have a general self-fibring mechanism that is axiomatised and it specialises to certain axioms for the case of $\mathbf{B}_u$. It will specialise to other axioms should the reader choose a different context logic.

Let us now list what we believe our self-fibring methodology can offer to context theories.

1. We provide a wide framework for general self-fibring. This is useful since in AI logics and applications there is a lot of use of substitutions of formulas inside formulas.
2. More specifically for the McCarthy–Buvač systems with the $\mathbf{ist}(x, y)$ predicate, we can use the methodology to allow the use of expressions of the form $\mathbf{ist}(A, B)$, where the context is given by wffs of possibly another logic and language. We can describe context directly in a logic (rather than just name them) and reason logically about context. We can even allow statements on the context which may involve considerations of what happens in them, i.e. what kinds of wffs B hold in them (i.e. $\mathbf{ist}(A, B)$ holds).
3. Given two known logics one for A wffs and one for B wffs, we can let the A-language serve as context for the B-language, by using the $\mathbf{ist}(A, B)$ predicate. Our methods can try and say under what conditions known properties of the component systems (e.g. decision procedure, proof theory) transfer to the combined context system with $\mathbf{ist}(A, B)$.

10.6 Natural Language Quantifiers

10.6.1 *Introduction*

The fibring idea has a simple application to natural language quantifiers. The possibilities in this direction are now under investigation.

Consider the statement *John is tall*. This can be modelled in the most simple minded manner by the formula *Tall(John)* of predicate logic, with *Tall* being the predicate and *John* the noun phrase. Similarly, the sentence *Every man is tall* can be translated as

$$(\forall x)(\mathit{Man}(x) \to \mathit{Tall}(x)).$$

Some linguists prefer to represent this sentence as Tall(every man), treating the expression *every man* as a noun phrase. This calls for a uniform semantical treatment of *Tall (J)* and *Tall*$((\forall x)$ *Man*$(x))$. The idea of fibred semantics makes this possible. We start with a quick example and then offer a more general definition.

Example 10.30 (Syllogistic fibring). We consider a special fibring function $\mathbf{F}$, defined as follows:

$$\mathbf{F}(h(P), h) = h_P$$

where $h_P(Q) = (D - h(Q)) \cup h(P)$.

According to this fibring, we can give a natural meaning to Aristotle's expressions

- *John is tall* translates to T (*John*)
- *All girls are tall* translates into $T((x)G(x))$.

We have

$$h \vDash T((x)G(x)) \text{ iff } h_T \vDash (x)G(x) \text{ iff } h_T(G) = D$$
$$\text{iff } h(T) \cup (D - h(G)) = D \text{ iff } h(G) \subseteq h(T).$$

There is more to this example than just the style of writing. Consider the traditional translation of temporal logic into classical logic. The assertion $t \vDash P$ is translated into $P(t)$, where t is a moment of time and P is a predicate. If P happens to hold at several moments of time, say $t_1, \ldots, t_n$ then we express that by $\bigwedge_{i=1}^{n} P(t_i)$. Consider the case when P holds at all moments of time t identified by a predicate Q. Following the same notation we would like to write $P((x)Q(x))$ and make it mean $(t)(Q(t) \to P(t))$. The above fibring yields the result.

To illustrate our ideas further consider the language of first-order predicate logic with the connectives $\neg, \wedge, \to$ and the quantifiers $\forall$, $\exists$. Let the models be a family of classical models of the form $\mathbf{m}_t = (D_t, h_t, g_t)$, $t \in S$, with domain D_t, interpretation function h_t for predicates, and assignments g_t to terms. Let us expand the language by adding generalised quantifier symbols $Q \in \{most, many, few, the, ...\}$ subject to the same syntactic formation rule as $\forall$ and $\exists$. The semantics of the new symbols is most naturally given by characteristic functions $h_t(Q) = f_t^Q$ on the domain D_t of individuals. That is, for any classical model $\mathbf{m}_t$ we have

$$\mathbf{m}_t \models Q\,x\,\phi(x) \text{ iff } f_t^Q(\{x \mid \mathbf{m}_t \models \phi(x)\}) = 1.$$

Different quantifiers have different functions. The axioms of a quantifier Q reflect the properties of the corresponding f^Q.

It is well known that this kind of language and semantics is not appropriate to deal with natural language quantifiers. Formal logic quantifiers are treated as operators that apply to a single formula, out of which they form another formula, binding a variable in the process. Natural language quantifiers are more complex than that. As shown in [Barwise and Cooper, 1981], it is not possible to introduce into predicate logic — just as we did above — a new quantifier symbol which, like $\exists$ and $\forall$, applies to single formulas and which would capture the meaning of a natural language determiner like, e.g., **most.** The problem is that the truth of a sentence in which **most** occurs depends upon the *comparison* between (the sets of objects satisfying) *two* formulas, not on the property of (the set of objects satisfying) some one single formula. Thus, 'Most Ps are Qs' will be true if the set of Ps that are Qs contains more than half as many objects as the set of Ps. The concept of a *generalised quantifier* is, therefore, that of a relation between two sets.

We can of course introduce into the formal logic quantifiers that take two set-denoting expressions as arguments. And in fact this is the approach commonly taken in linguistics. However there is a lot to be said for the Curried version of this approach, namely that of treating a generalised quantifier as function from subsets of D_t to characteristic functions f_t^Q on D_t.

This functional view is implemented in Montague's translation of generalised quantifiers into the language of type theory. Determiners are translated into expressions denoting functions in $(2^{D_t})^{D_t}$, yielding NP denotations as functions in 2^{D_t}. Thus *Most girls are nice* is evaluated as $f^{most}(||girl||)(||nice||)$ giving the dependency of f^{most} on the two arguments. Syntactically this dependency looks like $most\, girl(nice)$ and not like $nice(most\ girl)$, which would be more appropriate from an innocent point of view. This latter structure is only found in Montague's treatment of direct objects. *John kisses a girl* is evaluated as $(||kiss||)(j, f^{\exists}(||girl||)$. Meaning postulates will then guarantee the existence of a first order relation between individuals such that $(||kiss||)(j, f^{\exists}$ $(||girl||)$ holds iff John and some girl stand in this relation.

Our view on predicate argument structure is inspired by Montague's treatment of direct objects. It differs from his approach, however, in that it (i) gives the same treatment for all arguments of the verb, and (ii) does not use a type hierarchy to deal with problems of self-referentiality that may occur at this point. Although $nice(most\, x\, girl(x))$, or $(||like||)(j, f^{a}(||girl||)$, as a whole will be evaluated to true or false, the type of $most\, x\, girl(x)$ can never be a truth value in Montague grammar. But how can we give a semantics to such self referential logical forms without making use of type theory? As we are going to see, our fibred semantics can do the job.

Let us now define the fibred language formally. It is this language and its fibred semantics that will serve as tool for translating natural language.

Definition 10.31 (Fibred syntax for generalised quantifiers). Let **FQ** be a language with predicates, the classical connectives and a variety of quantifiers of the form $Q\, x$ including the classical quantifiers.

We define the notion of a *fibred generalised quantifier formula* (*FQ-wff*) of the language as follows

1. A is a FQ-wff of level 0 if A is a wff that results from an ordinary wff by allowing any quantifier of **FQ** to replace the classical quantifiers.
2. A is a FQ-wff of level $n + 1$ if there exists an ordinary wff $\varphi(x_1, \ldots, x_n, y_1, \ldots, y_m)$ of **FQ**, with free variables $x_1, \ldots, x_n, y_1, \ldots, y_m$ and FQ-wffs $A_1, \ldots, A_n$ of level $\leq n$ such that $A = \varphi(x_1/A_1, \ldots, x_n/A_n, y_1, \ldots, y_m)$, where x_i/A_i denotes the simultaneous substitution of A_i for x_i in φ. The free variables of A are $y_1, \ldots, y_m$ together with all the free variables of $A_1, \ldots, A_n$. We assume the substitution is carefully made and does not result in incorrect binding of variables.

The reader will know from context what quantifiers we assume **FQ** to contain.

We thus represent the sentence *Most girls are nice* by a formula of the form $N(most$ $x\ G(x))$, where N stands for *nice* and G for *girls*. Let us now try and semantically evaluate $N(most\, x\, G(x))$ at a (classical) model $\mathbf{m}_{t_0}$. This means we need to check whether $\mathbf{m}_{t_0} \models N(most\, x\, G(x))$ holds. We do not know how to evaluate this since $most\, x\, G(x)$ is a wff. Therefore we fibre a new model $\mathbf{m}_s = \mathbf{F}(t_0, N)$ depending on both, t_0 and N (or, more explicitely, $\mathbf{m}_{t_0}$ and $h_t(N)$), and have $\mathbf{m}_{t_0} \models N(most\, x\, G(x))$ iff $\mathbf{m}_s \models most\, x\, G(x)$. What does $\mathbf{m}_s$ look like. Assume that $D_{t_0} = D_s$, $g_{t_0} = g_s$ (g is the assignment to variables) and $h_{t_0} = h_s$ except for what it assigns to $most$. Let us define the function f_s^{most} by

$$f_s^{most}(X) = 1 \text{ iff } X \in \{X \mid 2 \times |X \cap N_{t_0}| > |X|\}.$$

Then we get $\mathbf{m}_s \models most\ x\ G(x)$ iff $f_s^{most}(N_{t_0}) = 1$ iff $2 \times |G \cap N_{t_0}| > |G|$. Note that the dependency of f_s^{most} on N is accounted for by making the fibring function $\mathbf{F}$ dependent on N, and not the interpretation function f_s^{most} of the quantifier itself. We say that $\mathbf{F}$ *normalises Q to its restrictor.*

Definition 10.32 (Fibring the monadic FQ).

1. Let $\mathbf{m}_t = (D_t, h_t, g_t)$ be a classical model, where D_t is the domain, h_t is an assignment function assigning to each n-ary predicate P a subset of D_t^n and to each quantifier Q a function $f_t^Q : \{0,1\}^{D_t} \to \{0,1\}$, and $g_t = (g_t^1, g_t^2)$ is a pair of functions assigning to each n-place function symbol f a function $g_t(x) = g_t^1(f) : D_t^n \mapsto D$ and to each free variable x an element $g_t(x) = g_t^2(x) \in D_t$.
2. A *fibred model for FQ-formulas* has the form $M = (t_0, \mathbf{F})$, where $t_0 = (D_{t_0}, h_{t_0}, g_{t_0})$ is a classical model and $\mathbf{F}$ is a function giving for each model $t = (D_t, h_t, g_t)$ and each subset Y_t of D_t a model $\mathbf{F}(t, Y_t)$ such that
   (a) $\mathbf{m}_t \models P(\tau)$ iff $g(\tau) \in h_t(P)$, where τ is an n-tuple of terms
   (b) $\mathbf{m}_t \models P(Q\ x\ \phi(x))$ iff $\mathbf{F}(\mathbf{m}_t, Y_t) \models Q\ x\ \phi(x)$ for $Y_t = \{x \mid t \models P(x)\}$
   (c) $\mathbf{m}_t \models Q\ x\ P(x)$ iff $h_t(Q)(\{x \mid \mathbf{m}_t \models P(x)\}) = 1$.

With $\mathbf{m}_t = \mathbf{F}(\mathbf{m}_{t'}, Y)$ and $Y_{t'} = \{x \mid \mathbf{m}_{t'} \models Y(x)\}$ we have the following definitions.

1. $f_t^{\exists}(X) = 1$ iff $X \in \{X \mid X \cap Y_{t'} \neq \{\}\}$
2. $f_t^{\forall}(X) = 1$ iff $X \in \{X \mid Y_{t'} \subseteq X\}$
3. $f_t^{most}(X) = 1$ iff $X \in \{X \mid 2 \times |X \cap Y_{t'}| > |X|\}$.

10.6.2 *Dynamic treatment of quantifiers*

Definition 10.32 does not allow us to distinguish between different types of quantifiers on the basis of their anaphoric potential. Consider the discourse *A girl came in. She smiled.* with translation *came-in*$(some\,x\, girl(x)) \wedge smile(x)$. To evaluate the first sentence in a model $\mathbf{m}_t$ we ask for the truth value of $some\,x\, girl(x)$ in the fibred model $\mathbf{F}(t, \textit{came-in})$. This does not allow us to bind the free occurrence of x in the second sentence. Thus the answer to the query $some\,x\, girl(x)$ must — in addition to a truth value — provide more information. It must give us a witness for a girl that came in. Similarly processing the first sentence of *Most girls are nice. They please us.* must provide a witness set for the nice girls – but not a witness for any one element of this set. These witnesses can be provided by a change in the value assignment g_t. Let $g_t =^{[y]} g_{t'}$ mean that $g_{t'}$ differs from g_t at most with respect to the values it assigns to y. For the existential quantifier $\exists$ the values will be taken to be the elements of the singleton sets contained in $f^{\exists}_{\mathbf{F}(t,Y_t)}$. For genuine quantifiers like $most$ we take them to be the maximal sets of Y_t's in $f^{most}_{\mathbf{F}(t,Y_t)}$. We assume the domains to be lattices in the sense of [Link, 1984]. *Most girls are nice. They please us.* will be represented as $nice(most\,x\, girl(x)) \wedge please(us, X)$, the intended meaning being that X refers to the set of nice girls, i.e. $X = \lambda x\,(girl(x) \wedge nice(x))$. Let $\mathbf{m}_{t''} = (\mathbf{F}(t), Y_t)$, and let WS_t be the function that provides the set of witnesses in t'' in the sense just described, i.e.

$WS_{t''}(\exists) = \{x \mid \{x\} \in f_t^{\exists}$, and
$WS_{t''}(most) = \{X \mid X \in f_t^{\exists}$ such that $Y_t \cap X$ is maximal$\}$.

Definition 10.33 (Dynamic interpretation of quantifiers).

1. $(t, \mathbf{F}) \models \phi$ iff $\exists t' \, \langle t, t' \rangle \in ||\phi||_{(t,\mathbf{F})}$
2. (a) $||P(\tau)||_{(t,\mathbf{F})} = \{\langle t, t \rangle \mid g(\tau) \in h_t(P)\}$, where τ is an n-tuple of terms
   (b) $||P(\exists x\, \phi(x))||_{(t,\mathbf{F})} = \{\langle t, t' \rangle \mid \langle \mathbf{F}(t, Y_t), t' \rangle \in ||\exists x\, \phi(x)||_{(t,\mathbf{F})}\}$
   for $Y_t = \{x \mid t \models P(x)\}$, and
   $t = t'$ except that $g_{t'} =^{[x]} g_t$ for $\{g_{t'}(x)\} \in WS_{t''}(\exists)$
   (c) for $Q \neq \exists$:
   $||P(Q\, x\, \phi(x))||_{(t,\mathbf{F})} = \{\langle t, t' \rangle \mid \langle \mathbf{F}(t, Y_t), t' \rangle \in ||Q\, x\, \phi(x)||_{(t,\mathbf{F})}\}$
   for $Y_t = \{x \mid t \models P(x)\}$, and
   $t = t'$ except that $g_{t'} =^{[X]} g_t$ for $\{g_{t'}(X)\} \in WS_{t''}(Q)$

The reader familiar with DRT may notice that the function WS corresponds to DRT's abstraction operation over duplex conditions. It has been noticed by Hans Kamp that dynamic logic does not yet provide any means to deal with Abstraction. (This will be presented in more detail in [Gabbay and Reyle, forthcoming].)

10.6.3 *Conclusion and further applications*

We have shown that the fibring methodology gives a straightforward way to dynamically interpret formulas of the form $\psi(most\, x\, \phi(x))$. For reasons of space we couldn't include the extension of the approach to verbs of arbitrary arity. This will be the subject of future research, where we also address the problem of scope ambiguity. The following topics are the subject of ongoing work as well.

Example 10.34 (Lexical ambiguities). Let $\phi(P(a))$ be an utterance with a lexically ambiguous predicate P and some constant a. In order to evaluate $\phi(P(a))$ in a model $\mathbf{m}$ w.r.t. an assignment g we need to know whether $g(a)$ is in the extension of P, i.e. $g(a) \in h(P)$. There are two cases to be distinguished. First, either $g(a) \notin h(P_i)$ for all of P's readings P_i, or else $g(a) \in h(P_i)$ for all P_i $(i = 1, ..., n)$. Second, $g(a) \in h(P_i)$ for some reading P_i. Only the second case depends on the intended reading of the utterance $\phi(P(a))$. In both cases, however, it is the truth value of $P(a)$ that determines the evaluation of $\phi(P(a))$. To get this truth value we go to other models $\mathbf{F}(\mathbf{m})$ in which $h_{\mathbf{F}(\mathbf{m})}(P)$ has the fixed value of its intended meaning and decide $P(a)$ there.

Example 10.35 (Definite descriptions and proper names). Suppose the proper name John is used ambiguously. Treat it as if it where a definite description, i.e. go to the model that gives you the intended John. Conversely, a definite description that gives one and the same individual in all fibred models turns out to refer directly, and thus rigidly. It functions as it where a proper name.

10.7 Default and Non-monotonic Reasoning

10.7.1 *Introduction*

Default logic [Reiter, 1980], as well as other forms of non-monotonic reasoning, make use of the idea of extending theories Δ with additional information (other wffs A) which are consistent with Δ, provided certain conditions are satisfied and certain procedures are followed. The basic rule involved has the form

- Under the conditions of this rule if A is consistent with Δ then B can be added to Δ.

There are various ways of formally incorporating such a rule in a system. One can include it in the metalevel as a default rule of the form

$$C : \frac{A}{B}$$

as Reiter proposed, or one can add an object level unary operator $MA (MA \in \Delta$ means that A is consistent with $\Delta)$ and represent the rule in the object level as $C \wedge MA \Rightarrow B$, as [McDermott and Doyle, 1980] has done.

The notion of fibring, especially in the case of intuitionistic logic, seems to have an unexpected application to the non-monotonic area and the use of such consistency rules. We have already seen in Section 9.5 that a default theory of the form $(\{\varphi_1, \ldots, \varphi_n\}, \{\alpha_i : \frac{\beta_i}{\beta_i}, i = 1, \ldots, n\})$ can be represented in the self-fibred intuitionistic system with equality by

$$\bigwedge_{j=1}^{n} \varphi_i \wedge \bigwedge_{i=1}^{n} (\alpha_i \Rightarrow (\beta_i = \top)).$$

The logic we use in the translation is intuitionistic, not classical, which has its advantages.

In fact, [Gabbay, 1982] has already proposed that intuitionistic logic may be a better basis for formalising non-monotonic reasoning. See also [Gabbay and Clarke, 1987]. Gabbay's proposal was in response to some difficulties in the McDermott and Doyle system.

In this section, our case study will pursue the connection between self-fibring and the object level operator M of McDermott and Doyle and Gabbay. We shall not study the subject in detail in this book, but leave the study of Reiter's default for a future paper.

10.7.2 *The operator M*

We begin by extending the language of intuitionistic propositional logic with the unary connective M. We give M meaning through the following semantical clause, in a (propositional) Kripke model $(S, \leq, a, h)$ (see Definition 9.2)

- $t \vDash MA$ iff for some $s, t \leq s$ and $s \vDash A$.

In this context it is convenient to think of the set of possible worlds as time.

Note that in the language with M, although atoms and ordinary M-free formulas are persistent, MA and formulas involving M may not be persistent. In Figure 10.1 MA is true at t and t_1 but not at t_2. Note also that MA is not the same as $\neg\neg A$ (although in linear time it is).

Intuitionistic entailment $A \vDash B$ was defined in Definition 9.2. Intuitionistic provability $\vdash$ can then be defined in terms of axiomatic, natural-deduction or tableaux systems that can be proved sound and complete with respect to the semantics. Our notion of non-monotonic provability $\mid\!\sim$ will be based on $\vdash$, or equivalently on $\vDash$. But first we check the behaviour of the operator M.

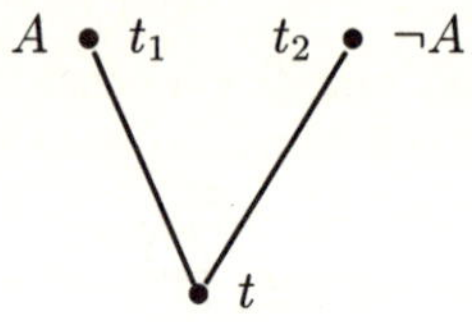

FIG. 10.1.

MA and $\neg A$ are semantically exclusive and exhaustive possibilities, so we have $\models MA \vee \neg A$. Also, $\neg MA$ is semantically equivalent to $\neg A$ and $A \models MA$ is valid. Furthermore, $MA \Rightarrow B$ is equivalent to $\neg A \vee B$; either $\neg A$ is true now or, if not, then MA is true now and hence B. Note that $C \vee \neg C$ is not a theorem, but is equivalent to $MC \Rightarrow C$. Also, $MC \Rightarrow \neg C$ is equivalent to $\neg C$.

We have seen that $\models MB \vee \neg B$, but that is not quite the same as $A \models MB$, which is equivalent to $A \not\models \neg B$. Consider the following canonical model. Let the time points T be the subset of all wffs generated by $\neg, \wedge$ and $\Rightarrow$ only, and let the ordering relation $A \geq B$ mean $\models A \Rightarrow B$. For atomic p let $h(A,p) = 1$ iff $A \models p$. For any formula B it can then be shown inductively that $h(A, B) = 1$ iff $A \models B$. Now $A \not\models \neg B$ iff $A \wedge B$ is consistent. But $A \wedge B \geq A$ since $A \wedge B \models A$, so if $A \not\models \neg B$ then there is a state, namely $A \wedge B$, in the future of A in which B is established, and conversely if there is such a state B in the future of A then $A \not\models \neg B$. So $A \models MB$ iff $A \not\models \neg B$.[28]

10.7.3 *The non-monotonic component*

We have not yet said what constitutes non-monotonic deduction in our system. We define non-monotonic deduction iteratively in terms of chained single-step default rules.

The single-step non-monotonic rules are of the form $A_1, A_2, \ldots, A_n \rightsquigarrow B$ if, for some $P_1, \ldots, P_k$ such that $A_1, A_2, \ldots, A_n, MP_1, \ldots, MP_k$ are consistent we have $A_1, A_2, \ldots, A_n, MP_1, \ldots, MP_k \models B$. We can also say as a special case that $A_1, A_2, \ldots, A_n \rightsquigarrow B$ if $A_1, A_2, \ldots, A_n \models B$ with no default.

As defined here, the intuitionistic deduction of B is semantically based. The corresponding proof might involve several basic steps such as modus ponens, whereas earlier

[28]By the way, McDermott and Doyle's [1980] system, also discussed in [Gabbay, 1982], for the operator M, is based on classical logic and thus has some difficulties. Among several problematic cases for their logic, McDermott and Doyle cite the following:

1. $\{(MC \Rightarrow D), \neg D\}$ is inconsistent;
2. MC does not follow from $M(C \wedge D)$;
3. $\{MC, \neg C\}$ is not inconsistent;
4. $\{MC \Rightarrow \neg C\}$ is inconsistent, but $\{MC \Rightarrow \neg C, \neg C\}$ proves $\neg C$.

They find $\{(MC \Rightarrow D), \neg D\}$ inconsistent because $\neg C$ cannot be shown to follow, forcing the assumption of MC. Intuitionistically $MC \Rightarrow D$ is equivalent to $\neg C \vee D$, so $\{(MC \Rightarrow D), \neg D\}$ is equivalent to $\neg C$, justifying the intuition of McDermott and Doyle, who remedied the inconsistency of their classically based system by arbitrarily adding $\neg C$.

Also, MC does follow intuitionistically from $M(C \wedge D)$, and $\{MC, \neg C\}$ is semantically inconsistent, while $MC \Rightarrow \neg C$ is equivalent to $\neg C$. $\{MC \Rightarrow C\}$ proves $C \vee \neg C$, which says that either C is known now or $\neg C$ is known, but does not say which.

we implied that $\rightsquigarrow$ stood for only a single basic step. What is of interest, however, is whether $\rightsquigarrow$ satisfies the usual non-monotonic consistency and monotonicity properties (see [Gabbay, 1985a]).

We now define non-monotonic $\mid\!\sim$ as follows.

$A_1, A_2, \ldots, A_n \mid\!\sim X$ if for some $B_1, \ldots, B_m$

$$\begin{gathered} A_1, A_2, \ldots, A_n \rightsquigarrow B_1 \\ A_1, A_2, \ldots, A_n, B_1 \rightsquigarrow B_2 \\ \vdots \\ A_1, A_2, \ldots, A_n, B_1, B_2, \ldots, B_m \rightsquigarrow X \end{gathered}$$

The following are examples.

1. $\{MP \Rightarrow P\} \rightsquigarrow P$ because $\{(MP \Rightarrow P), MP\} \models P$
2. $\{(MP \Rightarrow P), (P \wedge MP \Rightarrow R)\} \mid\!\sim R$

because

$$\{(MP \Rightarrow P), (P \wedge MP \Rightarrow R), MP\} \models P$$

and

$$\{(MP \Rightarrow P), (P \wedge MQ \Rightarrow R), P, MQ\} \models R.$$

10.7.4 *Restricted monotonicity*

This is the rule:

$$A \mid\!\sim B \text{ and } A \mid\!\sim X \text{ implies } A, B \mid\!\sim X.$$

In [Gabbay, 1985a] it is shown that if then consistency and restricted monotonicity properties hold for single-step rules then they can be shown to be hereditary along chains of such rules. Thus it suffices to consider the single-step rules, and for simplicity we take these to be of the following form:

$A \rightsquigarrow B$ if $A, MP \models B$ for some P such that A, MP are consistent.

As a preliminary, the following properties can be easily established:

1. A, MP are consistent iff $A \not\models \neg P$;
2. if $A, MP \models B$ then $A \models B \vee \neg P$;
3. if A, MP are consistent and $A, MP \vdash B$ then A, B are consistent.

Now strict monotonicity $\frac{A \rightsquigarrow X}{A, B \rightsquigarrow X}$ does not hold for $\rightsquigarrow$. A counterexample is $(MP \Rightarrow P) \rightsquigarrow P$ but $\{(MP \Rightarrow P), \neg P\} \models \neg P$. Suppose, however, that we take a B that is 'more consistent' with A. This is the idea behind restricted monotonicity. Does

$$\frac{A \rightsquigarrow B \quad A \rightsquigarrow X}{A, B \rightsquigarrow X}$$

always hold? Again, the answer in general is no. A counterexample is $MP \Rightarrow P \rightsquigarrow P$ and $MP \Rightarrow P \rightsquigarrow \neg P$, because $\{(MP \Rightarrow P), M(\neg P)\} \models \neg P$, but $\{(MP \Rightarrow P), P, M(\neg P)\}$ is inconsistent. (It can be shown, however, that if P and Q are not the same then $(MP \Rightarrow Q) \rightsquigarrow Q$ but $(MP \Rightarrow Q) \not\rightsquigarrow \neg Q$).)

Now suppose that $\{A, MP\} \models B$ and $\{A, MQ\} \models X$, where A, B and A, MQ are separately consistent, but A, B, MQ are inconsistent. Then $A, MQ \models \neg B$. So we have $A \models B \vee \neg P$ and $A \models \neg B \vee \neg Q$. This is possible in branching time, so restricted monotonicity need not hold, but in linear time i.e. the system **LC** of Dummett (see Section 9.5), the schema $(\alpha \Rightarrow \beta \vee \gamma) \Rightarrow (\alpha \Rightarrow \beta) \vee (\alpha \Rightarrow \gamma)$ is provable, so $\models (A \Rightarrow B) \vee (A \Rightarrow \neg P)$ and $\models (A \Rightarrow \neg B) \vee (A \Rightarrow \neg Q)$. But $A \not\models \neg P$, so there is a state, namely $A \wedge P$, in the future of A in which P is established. Therefore, $h(t, A \Rightarrow \neg P) = 0$ for all t and hence $h(t, A \Rightarrow B) = 1$ for all t. By a similar argument on Q, we also have $h(t, A \Rightarrow \neg B) = 1$ for all t. So in linear time the assumption that $\{A, B, MQ\}$ is inconsistent leads to a contradiction and $\dfrac{A \rightsquigarrow B \quad A \rightsquigarrow X}{A, B \rightsquigarrow X}$ holds.[29]

10.7.5 *Connection with fibring*

The relation $A_1, \ldots, A_n \mid\!\sim B$ was defined inductively in Subsection 10.7.3. We want to express it in the self-fibring fragment of Section 9.5. The most convenient system is Dummett's **LC** of linear time. We saw in Section 9.5 that $M\alpha$ can be expressed, using self-fibred intuitionistic equality, as $\alpha = \top$. Thus we have: $(\alpha = \top) \wedge (\beta = \top) \Leftrightarrow \alpha \wedge \beta = \top$. Therefore in the system $\mathbf{IC}_=$ of Definition 9.12 we can represent the notion $A_1, \ldots, A_n \mid\!\sim B$ using $\mathbf{IC}_=$ quantifiers as follows:

1. A single step $A_1, \ldots, A_n \rightsquigarrow B$ can be defined as
   for some $\alpha_1, \ldots, \alpha_k$ such that $\beta = \bigwedge_i A_i \wedge \bigwedge_j (\alpha_j = \top)$ is consistent, we have $\beta \models B$.
   This can be written as
   $$\vdash \exists x_1, \ldots, x_k (\bigwedge_i A_i \wedge \bigwedge_j (x_j = \top) \Rightarrow B).$$
   In **LC** we can assume $k = 1$.
   Note that if all of x_j are terms then $x_j = \top$ is $\top$ and we get $\bigwedge A_i \models B$, which is still OK.
2. The inductive case is
   $A_1, \ldots, A_n \mid\!\sim B$ iff for some $B_1, \ldots, B_m$ we have for each $1 \leq r < m \vdash \exists x_1^r, \ldots, \exists_{k_r}^r (\bigwedge_i A_i \wedge \bigwedge_{j \leq r} B_r \wedge \bigwedge_{k=1}^{k_r} x_k^r = \top \Rightarrow B_{r+1})$.
   For the case of **LC** we can assume $k_r = 1$. We get in this case a simplified version $A_1, \ldots, A_n \mid\!\sim B$ iff $\exists x (\bigwedge A_i \wedge (x = \top) \Rightarrow B)$.

10.8 The Subjunctive Conditional

Self-fibring can also be applied to the analysis of the subjunctive conditional.

Traditional semantics for the (propositional) conditional $A > B$ offers models of the form $\mathbf{m} = (S, a, \mathbf{f}, h)$, where S is a set of worlds, $a \in S$ is the actual world, $\mathbf{f}$ is a nearness function associating with each $t \in S$ and each subset $X \subseteq S$ a set of points $\mathbf{f}(t, X)$. h is an assignment giving truth values to the atoms. Thus for each $t \in S$ and atomic q, $h(t, q) \in \{0, 1\}$. Satisfaction is defined by

[29] Note also that in linear time MA can be taken as $\neg\neg A$.

- $t \vDash q$ if $h(t, q) = 1$
- satisfaction for $\sim, \wedge, \vee, \rightarrow$ is defined as in classical logic
- $t \vDash A > B$ iff for all $y \in \mathbf{f}(t, \hat{A})$ we have $y \vDash B$, where $\hat{A} = \{s \mid s \vDash A\}$.
- $\mathbf{m} \vDash A$ iff $a \vDash A$

We perceive $\mathbf{f}(t, \hat{A})$ as the set of points A-near to t and we need $A > B$ to mean that if we were to change the current world t slightly to accommodate A (i.e. pass to a $y \in \mathbf{f}(t, \hat{A})$) then B would hold. Thus $\mathbf{f}(t, \hat{A}) \subseteq \hat{A}$ holds.

The function $\mathbf{f}$ is required by different authors and their conditional systems to satisfy various restrictions, for example

- $t \in \mathbf{f}(t, X)$.

We need not be concerned with these refinements. The basic idea of using an $\mathbf{f}$ is what we need in order to present an application of self-fibring.

Let us try to define a traditional translation of the conditional language into classical logic, along the lines of the translation of modal logic. This would mean that any wff A of the conditional language is translated into a monadic formula of classical logic with one free variable of the form $[A]^*_t$ (meaning semantically $t \vDash A$) as follows:

First we associate a classical monadic predicate $P_i(x)$ with each propositional atom p_i. We let

- $[p_i]^*_t = P_i(t)$
- $[A \wedge B]^*_t = [A]^*_t \wedge [B]^*_t$
- $[\sim A]^*_t = \sim [A]^*_t$.

Here comes our problem; we want to translate $A > B$:

- $[A > B]^*_t = \forall y([y \in \mathbf{f}(t, \hat{A})]^* \rightarrow [B]^*_y)$.

We are writing $[y \in \mathbf{f}(t, \hat{A})]^*$ to mean that we need to translate this expression. However this is not a first-order expression. For comparison, recall the translation of $\Box A$ in modal logic. Here we can use the accessibility relation R and translate

- $[\Box A]^*_t = \forall y(tRy \rightarrow [A]^*_y)$.

So how are we going to translate $y \in \mathbf{f}(t, \hat{A})$? The answer is that we need to translate it into self-fibred monadic logic.

Recall the truth condition for $t \vDash P(\varphi)$ in a general fibred model $\mathbf{n} = (S, D, a, \mathbf{F}, h, g)$ as defined in Definition 7.9:

- $t \vDash P(\varphi)$ iff for all $s \in \mathbf{F}(h(t, P), t)$ we have $s \vDash \varphi$.

The model $\mathbf{n}$ is monadic, where D is the domain, S is the set of worlds and monadic predicates $P(x)$ get extensions $h(t, P) \subseteq D$, for each $t \in S$.

Call such a model self-reflective if $S = D$ i.e. we identify the set of possible worlds with the domain D. Under this assumption, the model $\mathbf{n}$ becomes $\mathbf{n} = (S, a, \mathbf{F}, h, g)$. If we assume no function symbols (only the constant a), the model can be written as $(S, a, \mathbf{F}, h)$ (with g ignored). $\mathbf{F}$ becomes a function from $2^S \times S$ into 2^S, i.e. $\mathbf{F}(X, t) \subseteq S$. We can now rename $\mathbf{F}(X, t)$ as $\mathbf{f}(t, X)$ and compare the truth conditions of $P > \varphi$

as a conditional with that of $P(\varphi)$ as a self-fibred formula. We find them identical. Thus we can now continue our translation of conditional logic into (self-fibred) monadic logic

- $[A > B]^*_t = [A]^*_{\forall s(W(s) \to [B]^*_s)}$

where $W(s)$ is the locality predicate of Definitions 10.21 and 7.13. $W(x)$ satisfies

- $t \vDash W(x)$ iff $x = t$.

Note that self-fibred logic with $W(x)$ as well as modal logic with $W(x)$ are axiomatisable. There are additional first-order axioms on W which can give it the correct meaning. See Theorem 7.15.[30]

10.9 Discussion and Conclusion

What is the message of this self-fibring? Put very simply, it is that predicate logics can very easily and intuitively be self-fibred yielding a very useful and expressive language. The semantics is very simple and natural and the expressive power that is gained is surprisingly rich. I believe that self-fibred classical logic can really play the role of a general universal logic. In practice in application areas such as artificial intelligence it is already used implicitly. The various sections in this chapter have shown how a variety of systems from diverse applications are naturally embeddable in it.

The emphasis is on the word 'naturally'. Let us trace our steps:

1. We started by considering the question of giving semantics to $P(\varphi)$, where $P(x)$ and φ are wffs.
2. We naturally formulated the fibred semantics.
3. The natural formulation of the fibred semantics looked like multimodal and context logics and so we obtained a natural connection there.
   We further notice that the well-known semantic condition for the conditional $A > B$ looks very much the same as the fibred semantics for $A(B)$.
4. The next step was to consider $R(x, \varphi)$ as a parametrised unary predicate and the natural fibred semantics for that immediately followed. We got slot logic as a bonus with a strong indication of applications.
5. Then of course we looked at $x = \varphi$ and immediately got the generalised quantifiers and generalised assignment models, as well as connections with default logics.
6. The semantics for function symbols and ambivalent syntax are also the most natual to define.

So the progression of ideas (1)–(6) is almost forced upon us, and the translations of the various systems into the self-fibred predicate logic are not technically contrived but completely natural.

This naturalness has additional methodological rewards:

[30]The translation of $A > B$ as $[A > B]^*_t$ works only for atomic A. This is so because we fibre according to the extensions of atoms not one place wffs. We really need the context modal logic of Section 10.1 and translate $[A > B]^*_t$ as $\Box_{[t:[A]^*_t]}\forall s(W(s) \to [B]^*_s)$.

All the above considerations apply to any logic and our section on the intuitionistic fragment demonstrates this fact. Thus we can apply self-fibring to, for example, relevance logic. If we develop steps (1)–(6) to the case of relevance logic, we immediately get systems like relevant context modal logics, relevant generalised quantifers etc. The relevant logic community have been seeking a satisfactory definition of these concepts for two generations! We believe we can offer one solid way of introducing these concepts through self-fibring.[31]

Furthermore, by rich expressive power we do not only mean the ability to express a variety of logics within one natural formalism. We also have a bit of second-order capability through the fibring function **F**. This is reminiscent of the ability of modal formulas to express second-order properties of structures (correspondence theory).

There is a wide scope for further research:

- solving decision problems for some of our systems
- developing proof theory
- understanding better the emerging new logics (e.g. slot logic)
- studying new generalised quantifiers for classical and other logics
- develop algebraic logic for our systems
- developing modal logics with formula dependent modalities
- investigate fibring and predicate underspecification and ambiguity
- study correspondence theory
- develop the self-fibred machinery for other logics such as substructural logics
- study the connections with channel theory
- and more.

This is only the beginning!

[31] This is to be done in a subsequent paper. It will take time and a lot of pages but the basic ideas, the beginning of the road ahead, are now available.

11

CONDITIONAL IMPLICATIONS AND NON-MONOTONIC CONSEQUENCE

11.1 Introduction

The study of non-monotonic reasoning through properties of its consequence relation was first proposed by Gabbay [1985a]. Let $\mathbb{L}$ be a language and let $\mid\!\sim$ be a relation between sets of well-formed formulas (wffs) Δ and single wffs A, of the form $\Delta \mid\!\sim A$. Gabbay proposed three minimal conditions on a relation $\mid\!\sim$ to be considered as a non-monotonic logic. These are as follows.

1. Reflexivity:
$$\Delta, A \mid\!\sim A$$
2. Cut:
$$\frac{\Delta, A \mid\!\sim B; \Delta \mid\!\sim A}{\Delta \mid\!\sim B}$$
3. Restricted monotonicity:
$$\frac{\Delta \mid\!\sim A; \Delta \mid\!\sim B}{\Delta, A \mid\!\sim B}.$$

Note that if classical conjunction is available in the language then $\Delta, A \mid\!\sim B$ is the same as $\bigwedge \Delta \wedge A \mid\!\sim B$.

$\mathbb{L}$ can be any language, e.g. the language of classical propositional logic, modal logic, temporal logic, and so on. The minimal consequence relation on $\mathbb{L}$ satisfying the above conditions is denoted by $\mathbf{C}[\mathbb{L}]$ ($\mathbf{C}$ for *cumulative*). For $\mathbb{L}$ being the language of classical logic, $\mathbf{C}[\mathbb{L}]$ was called $\mathbf{C}$ by Kraus *et al.* [1990].

Gabbay put forward his case for his system $\mathbf{C}$ using proof-theoretic considerations. Makinson [1989] proposed a semantics for Gabbay's logic and showed completeness. Makinson also studied the properties of the consequence relations associated with many well-known non-monotonic systems. Kraus *et al.* [1990] used the preferential semantics of Shoham [1987; 1988] to give semantics for a family of axiomatic extensions of Gabbay's $\mathbf{C}$ logic for the language of classical logic. Thus a new area of research activity has emerged, namely the syntactic and semantic study of non-monotonic logics through their consequence relations.

This area is closely connected (at least formally) with the well-established syntactic and semantic investigations into conditional logics. There are formal similarities between the notion of non-monotonic consequence $A \mid\!\sim B$ and that of conditional implication $A > B$.

Conditional logic augments the classical connectives with the binary connective $>$ and gives some axioms for it. Non-monotonic logic, in comparison, operates in the metalevel by endowing the language of classical logic with a new consequence relation $A \mid\!\sim B$, postulating some properties governing it.

Formally, given a system $\mathbb{C}$ of conditional logic and two wffs of classical logic, we can define a consequence relation $\mid\!\sim_{\mathbb{C}}$ by (*):

(*) $A \mid\!\sim_{\mathbb{C}} B$ iff $\mathbb{C} \vdash A > B$.

In fact, the consequence relation system **P** (of Kraus *et al.* [1990]) obtained by adding to Gabbay's **C** the rule

$$\frac{A \mid\!\sim C; B \mid\!\sim C}{A \vee B \mid\!\sim C}$$

turns out to be obtainable in the above manner through (*) from the conditional logic systems of Burgess [1981] and Veltman [1986]. System **P** is also connected in a similar way to the probabilistic approach to indicative conditionals of Adams [1975].

The similarities between conditional logic and non-monotonic consequence relations seem to be purely formal, through (*). The intended semantics for the conditional is through a ternary relation $R(x, y, z)$, with the semantic truth table for $>$ being as follows:

$t \vDash A > B$ iff for all x, y we have that $R(t, x, y)$ and $x \vDash A$ imply $y \vDash B$.

The intended meaning for $A > B$ is that if A were the case in a situation very similar to the existing one then B would have been the case.

On the other hand, the intended interpretation of $A \mid\!\sim B$ is that A non-mono- tonically implies B under normal circumstances. Semantically, B should hold in all minimal preferable models in which A holds.

Thus the two intended interpretations for $\mid\!\sim$ and $>$ do not exactly match.

Nevertheless, some attempts at connecting the two concepts were made by Ginsberg [1986], Delgrande [1988], and Fariñas del Cerro *et al.* [1994]. See [Katsuno and Sato, 1995] for a unified point of view of consequence relation, belief revision, and conditional logic.

This chapter has three objectives.

1. **Creating non-monotonic systems**
   We describe a semantic process whereby for any given language $\mathbb{L}$ and a class of models $\mathcal{K}$ for this language, we can define a non-monotonic consequence relation $\mid\!\sim_{\mathcal{K}}$ on $\mathbb{L}$. Of course, given any logical system **k** on $\mathbb{L}$, complete for a class of structures $\mathcal{K}$ (i.e. $\vdash_{\mathbf{k}}$ is equal to the semantically defined consequence relation $\vDash_{\mathcal{K}}$), we can then define a non monotonic consequence $\mid\!\sim_{\mathbf{k}}$ (or $\mid\!\sim_{\mathcal{K}}$).
2. **Reflecting a consequence relation into the object level**
   Given a consequence relation $\mid\!\sim$ on a language $\mathbb{L}$, we would like to add a new binary connective $>$ to the language forming $\mathbb{L}^* = \mathbb{L} \cup \{>\}$ and define the minimal consequence relation $\mid\!\sim^*$ on $\mathbb{L}^*$ such that the following holds:
   (a) $\mid\!\sim^*$ is a conservative extension of $\mid\!\sim$ on $\mathbb{L}$
   (b) for all A, B of $\mathbb{L}^*$, $A \mid\!\sim^* B$ iff $\mid\!\sim^* A > B$.

The above two objectives and the methods used to achieve them will give us a clearer idea on the connections between non-monotonic consequence and conditional logics. A third objective, of equal importance, will emerge as a side effect from the previous ones.

3. **General semantics for conditionals**
   The semantics for $\mid\!\sim^*$ of the previous item is such that it provides a new general two-dimensional semantics for conditional logics.

11.2 The Non-monotonic Companion to a Logical System

Let $\mathbb{L}$ be a language, say the language of classical logic with $\{\wedge, \vee, \rightarrow, \perp\}$ or the language of modal logic with $\{\wedge, \vee, \rightarrow, \perp, \Box, \Diamond\}$ or any other language. Let $\vdash$ be a monotonic consequence relation on $\mathbb{L}$. Let $\mathcal{K}$ be a class of structures for $\mathbb{L}$ and assume that $\vdash$ is complete for $\mathcal{K}$. In other words we have

$$A \vdash B \text{ iff for all structures } \mathbf{m} \in \mathcal{K} \text{ if } \mathbf{m} \vDash A \text{ then } \mathbf{m} \vDash B.$$

Examples of such systems and classes are abundant. Here are some familiar ones.

1. $\vdash$ is the classical logic consequence and $\mathcal{K}$ is the class of all classical models.
2. $\vdash$ is the intuitionistic logic consequence and $\mathcal{K}$ is the class of all intuitionistic Kripke models.
3. $\vdash$ is the modal logic consequence for the modal logic **S4** and $\mathcal{K}$ is the class of all Kripke models with accessibility relation which is reflexive and transitive.
4. $\vdash$ is a modal intuitionistic consequence relation (there are many such systems, say we choose one of them) and $\mathcal{K}$ is a suitable class of Kripke models with two accessibility relations for which $\vdash$ is complete.

Thus our starting point is a language $\mathbb{L}$ and a logic $\vdash$ for $\mathbb{L}$ which has a suitable semantics $\mathcal{K}$. We now define a *companion* non-monotonic consequence relation for $\vdash$ as follows. The reader should compare with Section 2.2.

Definition 11.1.

1. By a *language* $\mathbb{L}$ we mean a propositional language with atomic propositions $\{q_1, q_2, \ldots,\}$ and a family of connectives $\{\sharp_1, \sharp_2, \ldots,\}$, each $\sharp_i$ is a r_i-place connective. The notion of a well-formed formula is defined in the usual manner. We assume that classical $\wedge$ and $\vee$ are present.
2. By a *semantics* $\mathcal{K}$ we mean a class of structures $\mathbf{m} \in \mathcal{K}$ and a satisfaction relation between structures $\mathbf{m}$ and wffs A of the language. We write $\mathbf{m} \vDash A$ to mean that $\mathbf{m}$ satisfies A. The classical $\wedge, \vee$ have their traditional meaning.
3. By a compact monotonic consequence relation on $\mathbb{L}$ we mean a relation between sets of wffs and single formulas of the form $\Delta \vdash B$ such that the conditions of reflexivity, monotonicity, cut, and compactness hold, where

   Compactness: $\Delta \vdash A$ iff for some *finite* $\Delta_0 \subseteq \Delta, \Delta_0 \vdash A$.
4. A monotonic consequence relation $\vdash$ is said to be *complete* for a class of cumulative structures $\mathcal{K}$ iff the following holds:

$A \vdash B$ iff $\forall \mathbf{m} \in \mathcal{K}(\mathbf{m} \vDash A$ implies $\mathbf{m} \vDash B)$.

Definition 11.2.

1. Let $\mathbb{L}$ be a language and let $\mathcal{K}$ be a class of structures for the language. We define a new *companion* class of structures $\mathbb{K}$, symbiotic on $\mathcal{K}$ and $\mathbb{L}$. The structures of $\mathbb{K}$ have the form $(S, \prec, \mathbf{l})$ where S is a set of states, $\prec$ is a binary preference relation on S satisfying the smoothness condition below, and $\mathbf{l}$ is a function associating with each $s \in S$ a non-empty set of structures from $\mathcal{K}$.
   Define for $s \in S$ and wff A of $\mathbb{L}$ the following:
   - $s \vDash A$ iff for all $\mathbf{m} \in \mathbf{l}(s), \mathbf{m} \vDash A$
   - $\hat{A} = \{s \mid s \vDash A\}$.

   Then the following holds.

   Smoothness: $\forall x \in \hat{A}\, \exists y \in \hat{A}[y \prec x$ or $y = x$, and there does not exist a $u \in \hat{A}$ such that $u \neq y \wedge u \prec y$; in other words, y is minimal in the $\prec$ ordering of $\hat{A}$].
2. Define $A \mid\!\sim_{\mathbb{K}} B$ iff for all structures $\mathbf{n} = (S, \prec, \mathbf{l}) \in \mathbb{K}$, we have $A \mid\!\sim_{\mathbf{n}} B$ where

   $A \mid\!\sim_{\mathbf{n}} B$ iff for all $x \in \hat{A}$ such that x is minimal in $\hat{A}$, we have $x \vDash B$.
3. Define $\mid\!\sim_{\mathbf{n}} B$ iff we have $x \vDash B$ for all minimal points x in S. Similarly define $\mid\!\sim_{\mathbb{K}} B$.
   We need to assume smoothness for the set S itself.
4. A stronger condition to assume on the structures of $\mathbb{K}$ is that each set $\hat{A}$ as well as the set S itself has a minimum. In other words, there exists an $x \in \hat{A}$ such that for all $y \neq x$ we have $x \prec y$. Such structures are called *strongly cumulative*. It is sometimes convenient to present them as $\mathbf{n} = (S, \prec, \mathbf{l}, s_0)$, where $s_0 \in S$ is the minimum in S.

Example 11.3. Let $\mathbb{L}$ be the language of classical propositional logic and $\mathcal{K}$ be the class of all classical models. Then $\mathbb{K}$ is the class of all Kraus–Lehman–Magidor cumulative models for the logic $\mathbf{C}$ of Gabbay, as presented in [Kraus *et al.*, 1990]. In fact, the completeness theorem in [Kraus *et al.*, 1990] shows that we can have completeness for the class of *strong cumulative models*, namely those which satisfy the following condition:

For all A, $\hat{A}$ has a minimum.

We shall use this fact in comparing our semantics with that of the subjunctive conditional.

Theorem 11.4. *Let $\mathbb{L}$ be a language and $\mathbf{k}$ a logical system for $\mathbb{L}$, complete for a class of models $\mathcal{K}$ for $\mathbb{L}$.*

Let $\mathbb{K}$ be the companion class of strongly cumulative models and let $\mid\!\sim_{\mathbb{K}}$ be the consequence relation corresponding to $\mathbb{K}$. Then $\mid\!\sim_{\mathbb{K}}$ is the smallest consequence relation on $\mathbb{L}$ satisfying the following conditions and is denoted by $\mathbf{C}[\mathbf{k}]$.

1. *Reflexivity*

$$A, B \mid\!\sim B$$

2. *Left logical equivalence*

$$\frac{A \vdash_{\mathbf{k}} B; B \vdash_{\mathbf{k}} A; A \mid\!\sim C}{B \mid\!\sim C}$$

3. *Right weakening*

$$\frac{\bigwedge A_i \vdash_{\mathbf{k}} B; C \mid\!\sim A_i}{C \mid\!\sim B}$$

4. *Cut*

$$\frac{A, B \mid\!\sim C; A \mid\!\sim B}{A \mid\!\sim C}$$

5. *Restricted monotonicity*

$$\frac{A \mid\!\sim B; A \mid\!\sim C}{A, B \mid\!\sim C}$$

Proof

1. Soundness can easily be proved.
2. We show that if $A_0 \not\mid\!\sim B_0$ then there exists a model $(S, \prec, \mathbf{l})$ which verifies A_0 but not B_0.

Let $\Delta_A = \{X \mid A \mid\!\sim X\}$. We claim $\Delta_{A_0} \nvdash_{\mathbf{k}} B_0$.

Otherwise, by compactness for some $A_i \in \Delta_{A_0}$ we have $\bigwedge A_i \vdash_{\mathbf{k}} B_0$. From left logical equivalence and right weakening we find that $\wedge$ behaves classically relative to $\mid\!\sim$ and using that and right weakening we obtain $A_0 \mid\!\sim B_0$.

Let $A \equiv B$ iff $A \mid\!\sim B$ and $B \mid\!\sim A$ and let $S = \mathrm{WFF}_{/\equiv}$. Let $\prec$ be defined as $A_{/\equiv} \prec B_{/\equiv}$ iff $B \mid\!\sim A$. Let $\mathbf{l}(A_{/\equiv})$ be the class of all $\mathcal{K}$ models of Δ_A.

Consider $\mathbf{n} = (S, \prec, \mathbf{l})$. It is easy to show that this is a strong cumulative model in $\mathbb{K}$. The minimal point of $\hat{A}$ is $A_{/\equiv}$. S itself may not have a minimum point, unless we have a symbol $\top$ for truth in the language, with $\top \mid\!\sim A$ holding. Otherwise we can augment S with the empty set $\varnothing$ and let $\Delta_\varnothing = \{X \mid \varnothing \mid\!\sim X\}$ and $\mathbf{l}(\varnothing) = \{\mathbf{m} \in \mathcal{K} \mid \mathbf{m} \vDash \Delta_\varnothing\}$.

In this model, $A_0 \not\mid\!\sim_{\mathbf{n}} B_0$ since $A_{0/\equiv} \vDash A_0$ but not B_0, since there exists in $\mathbf{l}(A_{0/\equiv})$ a model of Δ_{A_0} in which B_0 does not hold. □

Theorem 11.5. *Let $\mid\!\sim$ be a non-monotonic consequence relation on a language $\mathbb{L}$. Assume that $\mathbb{L}$ contains conjunction $\wedge$ and that $\wedge$ behaves classically under $\mid\!\sim$, namely the following hold:*

- *$C \mid\!\sim A \wedge B$ iff $C \mid\!\sim A$ and $C \mid\!\sim B$*
- *$A, B \mid\!\sim C$ iff $A \wedge B \mid\!\sim C$.*

Then there exists a compact monotonic consequence relation $\vdash$ with semantics $\mathcal{K}$ such that $\mid\!\sim$ is $\mid\!\sim_{\mathbb{K}}$.

Proof Let $\vdash$ be the smallest consequence relation containing $\mid\!\sim$ such that $\vdash$ is compact and monotonic and $\vdash$ satisfies left logical equivalence and right weakening relative to $\mid\!\sim$. Such a $\vdash$ exists because the class of all such consequence relations is closed under intersections and is non-empty.

By a standard completeness theorem for monotonic consequence relations there exists a class of models $\mathcal{K}$ such that

$$\vdash = \models_{\mathcal{K}} .$$

Let $\mid\!\sim_{\mathbb{K}}$ be the consequence relation, defined as in Definition 11.2, from $\vdash$ and $\mathcal{K}$. We want to show that $\mid\!\sim = \mid\!\sim_{\mathbb{K}}$.

1. We first show that $\mid\!\sim_{\mathbb{K}} \subseteq \mid\!\sim$. This we achieve by proving that $\mid\!\sim$ and $\vdash$ satisfy left logical equivalence and right weakening. Since $\mid\!\sim_{\mathbb{K}}$ is the minimal consequence relation satisfying the above we get $\mid\!\sim_{\mathbb{K}} \subseteq \mid\!\sim$.
   To show that $\mid\!\sim$ indeed satisfies these properties, observe that $\vdash$ was defined as the smallest compact monotonic consequence relation satisfying these conditions with $\mid\!\sim$.
2. We now show that $\mid\!\sim \subseteq \mid\!\sim_{\mathbb{K}}$.
   Assume $A \not\mid\!\sim_{\mathbb{K}} B$. Then there is a model $(S, \prec, \mathbf{l})$ and a state $s \in S$ and an $\mathbf{m} \in \mathbf{l}(s)$ such that $\mathbf{m} \in \mathcal{K}, \mathbf{m} \models A$ hold but $\mathbf{m} \not\models B$. This means that $A \nvdash B$. Since $\mid\!\sim \subseteq \ \vdash$, we obtain that $A \not\mid\!\sim B$.

This completes the proof of the theorem. □

11.3 Reflecting a Consequence Relation into the Object Level

Let $\mid\!\sim$ be a consequence relation for a language $\mathbb{L}$. Assume without loss of generality (by Theorem 11.5) that it is complete for a class of models of the form $\mathbb{K}$, a companion semantics to a monotonic logic $\mathbf{k}$, and its semantics $\mathcal{K}$. We would like to add to the language $\mathbb{L}$ an additional connective $>$ and define a consequence relation $\mid\!\sim^*$ on $\mathbb{L}^* = \mathbb{L} + \{>\}$ such that the following holds.

1. For any Δ, A of $\mathbb{L}$, $\Delta \mid\!\sim A$ iff $\Delta \mid\!\sim^* A$.
2. For any A, B of $\mathbb{L}^*$, $A \mid\!\sim^* B$ iff $\mid\!\sim^* A > B$.

These two conditions mean that $>$ formalises $\mid\!\sim$ and $\mid\!\sim^*$ in the object language.

Note that condition (2) above states essentially the deduction theorem for $>$ only for formulas or it can also be viewed as stating the Ramsey test for $>$ relative to the empty theory. A more general condition would be as follows;

3. For any Δ, A, B of $\mathbb{L}^*$ we have $\Delta, A \mid\!\sim^* B$ iff $\Delta \mid\!\sim^* A > B$.

This is a stronger condition and is not needed for reflecting $\mid\!\sim$ into the object language.[32]

[32] Some concepts need to be clarified here. Let $\mathbb{L}$ be a language with several connectives including $>$ and a binary connective $\otimes$. We can understand $\otimes$ as an object level representation of an input function *Input*(Δ, A) allowing us to input wffs A into theories Δ. Thus every finite theory Δ can be seen as the result of successive inputs into the empty theory $\emptyset$. If we write a generalised deduction theorem for $>$ as:

Our strategy in this section is to define $\mid\!\sim^*$ semantically, through the notion of fibred semantics and then give a completeness proof. We shall first carry out the construction for the relatively easier case of the language $\mathbb{L}$ of classical logic and the Kraus–Lehmann–Magidor semantics for system **C**.

We begin by explaining the basic idea of fibring. Suppose that we are given a non-monotonic consequence relation $\mid\!\sim$ for **C** and semantics for it. We would like to add to the language a connective $>$ which means the same as $\mid\!\sim$. This means that we can now write statements like $A\mid\!\sim(B > C)$. We need a semantics for this new language, and we would like to construct it from the known semantics of $\mid\!\sim$. Let $(S, \prec, \mathbf{l})$ be a model for $\mid\!\sim$. $A\mid\!\sim(B > C)$ has the form $A\mid\!\sim D$, and so we can formally evaluate it in the model. $A\mid\!\sim D$ holds in the model iff in all minimal states s in $\hat{A}$ we have that D holds. D holds at a state s if in all models $\mathbf{m} \in \mathbf{l}(s), \mathbf{m} \vDash D$. The problem is that $\mathbf{m}$ is a model for the language without $>$ and it does not know how to give a value for a wff of the form $B > C$. We note, however, that $B > C$ is really supposed to mean $B\mid\!\sim C$. So if each model $\mathbf{m}$ in $\mathbf{l}(s)$ were itself of the form $(S^{\mathbf{m}}, \prec^{\mathbf{m}}, \mathbf{l}^{\mathbf{m}})$ then $B\mid\!\sim C$ can be evalated. Note that the models $\mathbf{m}$ must be able to give values for both $\mathbf{m} \vDash A$ for A a classical formula and $\mathbf{m} \vDash B > C$. Thus $\mathbf{m}$ must be both 'classical' and 'cumulative'. Obviously we need the right definitions for constructing such models. To achieve that we need to look at the **C** models slightly differently to enable us to *fibre* them and obtain the new semantics.

We begin by looking at a cumulative **C** model more closely.

Let $\mathbf{n} = (S, \prec, \mathbf{l})$ be such a structure. Here S is a set of states, $\prec$ is the preferential relation, and $\mathbf{l}$ is a function associating with each $s \in S$ a non-empty set $\mathbf{l}(s)$ of classical models. Our first inessential change is to write $\mathbf{n}$ as

$$\mathbf{n} = (W^{\mathbf{n}}, S^{\mathbf{n}}, \prec^{\mathbf{n}}, \mathbf{l}^{\mathbf{n}}, \mathbf{m}^{\mathbf{n}})$$

where $W^{\mathbf{n}}$ is a set of labels, and $\mathbf{l}^{\mathbf{n}}(s)$, for $s \in S^{\mathbf{n}}$, is a non-empty subset of $W^{\mathbf{n}}$. $\mathbf{m}^{\mathbf{n}}$ is a function associating with each $u \in W^{\mathbf{n}}$ a classical model $\mathbf{m}^{\mathbf{n}}(u)$. The above is essentially the same semantics as before, except that now we are using the intermediary labels $W^{\mathbf{n}}$. This device allows us to attach other structures to any $u \in W^{\mathbf{n}}$. So if we

$$\Delta\mid\!\sim A > B \text{ iff } \Delta \otimes A\mid\!\sim B$$

then, depending on what $\otimes$ does, we obtain different 'deduction theorems' or 'Ramsey tests' for $>$. Take for example concatenation logic [Gabbay, 1996b] where theories are sequences of wffs and $>$ is substructural implication $\rightarrow$; then

$$(A_1, \ldots, A_n) \vdash A \rightarrow B \text{ iff } (A_1, \ldots, A_n, A) \vdash B$$

is a deduction theorem for an associative $\otimes$ where $(A_1, \ldots, A_n)$ is generated as $A_1 \otimes \ldots \otimes A_n$.

If $\Delta \otimes A$ represents a consistent update (i.e. $\Delta \otimes A$ is the result of input and then consistency maintenance), then we obtain the Ramsey test. In the case of $\mathbf{C}[\mathbf{k}], \otimes$ is $\wedge$.

The connective $\otimes$ can be used to define a new concept of restricted monotonicity:

$$\frac{\Delta\mid\!\sim A; \Delta\mid\!\sim B}{\Delta \otimes A\mid\!\sim B}.$$

More details of this belong to the area of structured consequence relations and the area of labelled deductive systems [Gabbay, 1996b].

need a truth value for a classical A at the label u, we can evaluate using $\mathbf{m^n}(u)$, and if we want a value for $B > C$, we can use another suitable structure associated with u.

The Kraus–Lehmann–Magidor semantics contains essentially two semantic evaluation functions: first, the evaluation $\mathbf{m^n}(u) \vDash A$, which is classical and second the evaluation $s \vDash A$, which is not classical. We have

$$s \vDash A \text{ iff for all } u \in \mathbf{l^n}(s), \mathbf{m^n}(u) \vDash A.$$

We will distinguish between the two by writing the name of the model as an index for one and a function notation for the other.

Thus we write

$$s \vDash_{\mathbf{n}} A \text{ iff for all } u \in \mathbf{l^n}(s), \mathbf{m^n}(u)(A) = 1.$$

The difference can be seen in the case of disjunction. The condition

$$s \vDash_{\mathbf{n}} A \vee B \text{ iff } s \vDash_{\mathbf{n}} A \text{ or } s \vDash_{\mathbf{n}} B$$

does not necessarily hold. $\mathbf{l^n}(s)$ may contain two models, $\mathbf{m^n}(u_1) \vDash A \wedge \sim B$ and $\mathbf{m^n}(u_2) \vDash B \wedge \sim A$.

In fact, the semantic condition $\mathbf{l^n}(s) = \{s\}$, containing exactly one point (which can be identified with s itself), is shown in [Kraus *et al.*, 1990] to give rise to a stronger system called **P**, satisfying the rule

$$\frac{A \mid\!\sim C; B \mid\!\sim C}{A \vee B \mid\!\sim C}.$$

The semantics for it is called *preferential semantics*.

P is the system connected with the conditionals of Adams [1975], Pearl and Geffner [1988] and Burgess [1981].

Now let us see what happens when we allow $>$ into the language. Let A, B be two wffs of $\mathbb{L}$. Assume that they have the form

$$\begin{aligned} A &= A[x_1, \ldots, x_n, y_1, \ldots, y_m] \\ B &= B[x_1, \ldots, x_n, y_1, \ldots, y_m] \end{aligned}$$

where the display of atoms means that the formula is built out of xs and ys.

Let P_i, Q_i be arbitrary formulas of $\mathbb{L}$ and consider

$$\begin{aligned} A^* &= A[y_i/P_i > Q_i] \\ B^* &= B[y_i/P_i > Q_i] \end{aligned}$$

obtained, as the above notation suggests, by substituting $P_i > Q_i$ for y_i.

A^*, B^* have now the general form of a wff in $\mathbb{L}^*$.

Suppose that we want to check whether $A^* \mid\!\sim_{\mathbf{n}} B^*$. **n** is not a model for the language $\mathbb{L}$ but let us evaluate the above anyway. We need to check whether in all minimal states $s \vDash A^*$, we have also $s \vDash B^*$.

We have a problem here. To decide whether s is minimal in $\{t \mid t \vDash A^*\}$, we have to obtain a value for the semantic function $t \vDash A^*$. The recipe for this is to check the value of $\mathbf{m^n}(u)(A^*)$ for each $u \in \mathbf{l^n}(t)$. $\mathbf{m^n}(u)$ is an assignment to the atoms x_i, y_j, but it has no value for $P_i > Q_i$ which appear in A^*. If we have values for the above wffs, for each u, then we can obtain a value for $s \vDash A^*$.

Our intention in the fibred semantics is that $P_i > Q_i$ be the same as $P|\!\!\sim Q$. Thus if we attach another Kraus–Lehmann–Magidor model $\mathbf{n'} = \mathbf{F^n}(u)$ to u, we can obtain a value for $P_i > Q_i, i = 1, \ldots, m$, as $P_i|\!\!\sim_{\mathbf{n'}} Q_i$.

$\mathbf{F^n}$ is a fibring function. For each $\mathbf{n}$, each $s \in S^{\mathbf{n}}$, and each $u \in \mathbf{l^n}(s)$, it associates a model $\mathbf{n'} = \mathbf{F^n}(u)$.

Consider now the function $g^{\mathbf{n}}(u)$ defined by

$$g^{\mathbf{n}}(u)(y_i) = 1 \text{ iff } P_i|\!\!\sim_{\mathbf{n'}} Q_i \text{ holds.}$$

Thus the assignment $h^{\mathbf{n}}(u)$, giving $\mathbf{m^n}(u), i = 1, \ldots, n$, to x_i and the value $g^{\mathbf{n}}(u)$ $(y_i), i = 1, \ldots, m$, to y_i, can give a value to $t \vDash A^*$. We can now obtain a value for $A^*|\!\!\sim_{\mathbf{n}} B^*$.

Therefore we are led to the following intuitive definition.

Definition 11.6 (Graded fibred semantics).

1. A graded fibred structure of level 0 has the form $\mathbf{n} = (W^{\mathbf{n}}, S^{\mathbf{n}}, \prec^{\mathbf{n}}, \mathbf{l^n}, \mathbf{m^n})$ as discussed above.
2. A graded fibred structure of level $n + 1$ has the form $\mathbf{n} = (W^{\mathbf{n}}, S^{\mathbf{n}}, \prec^{\mathbf{n}}, \mathbf{l^n}, \mathbf{m^n}, \mathbf{F^n})$, where $\mathbf{F^n}$ is a function which, for each $s \in S^{\mathbf{n}}$ and $u \in \mathbf{l^n}(s)$, gives a graded fibred structure $\mathbf{F^n}(u)$ of level n.
3. We can now define satisfaction of a formula of $\mathbb{L}^*$ in a graded fibred structure. We assume that the level of the structure is no less than the maximal number of nested occurrences of '$>$' in the formula. We define three semantic valuations, $s \vDash A, A \vDash_{\mathbf{n}} B$, and $u \vDash A$, for a label u and state s:
   - $A \vDash_{\mathbf{n}} B$ means that the structure $\mathbf{n}$ validates $A|\!\!\sim B$.
   - $s \vDash A$ means that A holds in s of the structure.
   - $u \vDash A$ means A holds at the label u.

   We also have that:
   - $A \vDash_{\mathbf{n}} B$ iff $s \vDash B$ for all s that are minimal in $\hat{A}$.
   - $s \vDash A$ iff for all $u \in \mathbf{l}(s) u \vDash A$.

   The above clauses are the same as those of cumulative semantics. The only twist is that now we have $>$ in the language and we need to look at the structure of A to determine how many nested occurrences of $>$ it has. To evaluate $u \vDash A$ we need to ask if A is classical atomic or if it is A of the form $P > Q$; the inductive clauses differ for each case. To evaluate $u \vDash P > Q$, we associate with each u another model $\mathbf{F^n}(u)$. We define $u \vDash P > Q$ to be $P \vDash_{\mathbf{F^n}(u)} Q$, which inductively we assume that we know how to evaluate.

   For a given A^*, we know that A^* is built up from atoms q and 'atoms' of the form '$P_i > Q_i$'. Since we do not know what connectives the language $\mathbb{L}$ has, the structural induction used in defining $u \vDash A^*$ is more complex. We assume that A^* is obtained from $A(q, y_i)$, which has no occurrences of $>$ by substitution, i.e. $A^* = A(q, y_i/P_i > Q_i)$ and the inductive definition of $u \vDash A^*$ is done in terms of $u \vDash A$ and $u \vDash P_i > Q_i$.

We now give the relevant inductive definitions.

(a) For a formula A without $>$ and label u, and state s

$$u \vDash A \text{ iff } \mathbf{m^n}(u)(A) = 1$$
$$s \vDash A \text{ iff for all } u \in \mathbf{l^n}(s), u \vDash A.$$

(b) For formulas A and B without $>$,
$A \vDash_{\mathbf{n}} B$ iff for all $s \in S^{\mathbf{n}}$ minimal in $\hat{A}^{\mathbf{n}} = \{t \in S \mid t \vDash A\}$ we have $s \vDash B$.

(c) $u \vDash P > Q$ iff $P \vDash_{\mathbf{F^n}(u)} Q$.

(d) $s \vDash P > Q$ iff for all $u \in \mathbf{l^n}(s)$, $u \vDash P > Q$.

(e) Let $A[x_1, \ldots, x_n, y_1, \ldots, y_m], B[x_1, \ldots, x_n, y_1, \ldots, y_m]$, be two wffs of $\mathbb{L}$. Let $P_i, Q_i, i = 1, \ldots, m$, be two wffs of $\mathbb{L}^*$ and let $A^* = A[y_i/P_i > Q]$ and $B^* = B[y_i/P_i > Q_i]$. Then

- $u \vDash A^*$ iff $h^{\mathbf{n}}(u)(A) = 1$, where $h^{\mathbf{n}}(u)(x_i) = \mathbf{m^n}(u)(x_i), i = 1, \ldots, n$, and $h^{\mathbf{n}}(u)(y_i) = 1$ iff $P_i \vDash_{\mathbf{F^n}(u)} Q_i, i = 1, \ldots, m$.
- $s \vDash A^*$ iff for all $u \in \mathbf{l^n}(s), u \vDash A^*$.

(f) $A^* \vDash_{\mathbf{n}} B^*$ iff for all minimal $s \in \hat{A}^{*\mathbf{n}} = \{t \in S^{\mathbf{n}} \mid t \vDash A^*\}$ we have $s \vDash B^*$.

(g) $\vDash_{\mathbf{n}} A$ iff in all minimal points $x \in S^{\mathbf{n}}, s \vDash A$.

4. Note that since **C** is complete for the strong cumulative semantics (with the property that each $\hat{A}$ has a minimum), we can similarly define the notion of *strong graded fibred semantics*.

Remark 11.7 (Refined graded fibred semantics). In the next section we are going to compare the fibred semantics with the traditional semantics for conditional logic. For this purpose we need to refine the graded fibred semantics to enable us to transform it into the semantics of conditional logic. This is the task of the present remark.

Our starting point is the level 0 models of the form $\mathbf{n} = (W, S, \prec, \mathbf{l}, \mathbf{m})$. We can assume without loss of generality that all level 0 models have countable disjoint sets of labels and of states, i.e. if $\mathbf{n}_i^0 = (W_i^0, S_i^0, \prec_i^0, \mathbf{l}_i^0, \mathbf{m}_i^0), i = 1, 2, 3, \ldots$, is an enumeration of all models, then the sets $W_i^0, S_i^0, W_j^0, S_j^0, i \neq j$, are all pairwise disjoint. We assume that each model is isomorphically copied and repeated infinitely often. We can assume that these sets are sets of 4-tuples of numbers. We can also identify the model name $\mathbf{n}_i^0$ with a new 4-tuple natural number which we also denote by $\mathbf{n}_i^0$ and $W_0' = \{\mathbf{n}_i^0\}$, and let

$$W^0 = W_0' \cup \bigcup_i W_i^0 \cup \bigcup_j S_j^0.$$

Here are the details of how it is done.

Let $\mathbb{N}$ be the set of natural numbers $\{0, 1, 2, \ldots, \}$. To do the coding properly, we use as labels 4-tuples $w = (0, i, j, k) \in \mathbb{N}^4$. The first component 0 is the level. Let the names of all level 0 models be $\mathbf{n}_i^0 = (o, i, o, o)$. Let the elements in S_i^0 be all from the set $\{(0, i, j, 0) \mid j \geq 1\}$ and let the elements of $\mathbf{l}_i^0((0, i, j, 0))$ be all from the set $\{(0, i, j, k) \mid k \geq 1\}$.

Let $W_i^0 = \bigcup_{j \geq 1} \mathbf{l}_i^0((0, i, j, 0))$. W_i^0 is the set of all level 0 labels. Note that $W^0 \subseteq \{0\} \times \mathbb{N}^3$.

Thus we have coded all models of level 0, using 4-tuples $w = (0, i, j, k)$.

Let us now code models of level 1. These have the form $\mathbf{n} = (W, S, \prec, \mathbf{l}, \mathbf{m}, \mathbf{F})$, where $(W, S, \prec, \mathbf{l}, \mathbf{m})$ is a model of level 0, say $\mathbf{n}_i$, and $\mathbf{F}$ is a fibring function, associating with each label $u \in W$ another model $\mathbf{F}(u)$ of level 0. We want to code all models of level 1 in such a way that the labels used for level 1 are different from the labels used for level 0. For this purpose, let us change the labels $(0, i, j, k)$ of level 0 models into $(1, i, j, k)$ of level 1. The total set of labels is now denoted by W^1.

In general, let

$$W^n = \{(n, i, j, k) \mid (0, i, j, k) \in W^0\}$$

and similarly define $S_i^n, W_i^n, W_n', \mathbf{n}_i^n$. The functions $\mathbf{l}_i^0$, $\mathbf{m}_i^0$, and $\prec_i^0$ can also be isomorphically renamed $\mathbf{l}_i^n, \mathbf{m}_i^n$ and $\prec_i^n$ respectively.

Thus any original level 0 model can now be isomorphically presented using labels from W^n with the respective functions as required. A model of level 0 can be turned into a model of level 1 in many different ways by adding different fibring functions $\mathbf{F}$. It is for this reason that we have assumed that in the level 0 enumeration each model appears infinitely often. This allows us to form all needed variation of models of level 1 with different fibring function $\mathbf{F}$. Thus we can assume that the sequence of models with names $\mathbf{n}_i^1$ of the form $(W_i^1, S_i^1, \prec_i^1, \mathbf{l}_i^1, \mathbf{m}_i^1, \mathbf{F}_i^1)$ enumerates all level 1 models and has the following properties.

- All models are isomorphically repeated infinitely often.
- $\mathbf{F}_i^1$ is a function giving each label $u \in W_i^1$ a name of a level 0 model $\mathbf{F}_i^1(u) \in W_0'$.
- Different labels u receive different models $\mathbf{F}_i^1(u)$. We can ensure this because each model of level 0 is isomorphically repeated infinitely often.

Inductively we can assume the above properties for models of level $n+1$, with $\mathbf{F}_i^{n+1}(u)$, $u \in W_i^n$ being a value in W_n'.

We now let $W = \bigcup_n W^n$, and we note that for any $w = (n, i, j, k) \in W$ there is a uniquely identifiable model where it resides. It is a level n model and its name is $\mathbf{n}_i^n$. If $j = 0$ then also $k = 0$ and the model name is $w = \mathbf{n}_i^n$; otherwise w is a state in S_i^n when $j > 0$ and $k = 0$, or it is a label in the state $s = (n, i, j, 0)$. For each such w, let $S^w = S_i^n, \mathbf{l}^w = \mathbf{l}_i^n, \prec^w = \prec_i^n, \mathbf{m}^w = \mathbf{m}_i^n$, and $\mathbf{F}^w = \mathbf{F}_i^n$.

In view of the above, we can now present all our structures in uniform form, namely

$$w = (W, S^w, \prec^w, \mathbf{l}^w, \mathbf{m}^w, \mathbf{F}^w)$$

where $S^w \subseteq W$, $\prec^w$ is an ordering on S^w, and $\mathbf{l}^w$ is a function giving each $s \in S^w$ a subset $\mathbf{l}^w(s) \subseteq W$. For each $u \in \mathbf{l}^w(s), \mathbf{m}^w(u)$ is a classical model and the function $\mathbf{F}^w(u)$ gives us another model index w'. Thus $\mathbf{F}^w(u) = w'$, and the fibred structure $\mathbf{F}^w(u)$ has the form

$$w' = (W, S^{w'}, \prec^{w'}, \mathbf{l}^{w'}, \mathbf{m}^{w'}, \mathbf{F}^{w'}).$$

The above careful construction allows us further simplification and refinement of the semantics.

First let us assume that we are dealing with strong graded models such that each $\hat{A}$ has a minimum. If we look at the coding, we can identify the name of the model $\mathbf{n}_i^n = (n, i, 0, 0)$ with the minimum point of S_i^n. So we can assume that $\mathbf{n}_i^n \in S_i^n$ and is the minimum.

We further note that the fibred models $\mathbf{F}^w(u)$ are of lower level than the level of u itself and so the value of $u \vDash A$ is independent of the values or even the existence of other points of the same level.

Further $\mathbf{m}_i^n(y)$ depends on y only since y already determines n and i uniquely.

The term 'graded' is no longer needed because the function $\mathbf{F}$ eliminates the need for the notion of level. In fact, the function $\mathbf{F}$ itself can be eliminated, with u being identified with $w' = \mathbf{F}^w(u)$, i.e. we can require $\mathbf{F}^w(u) = u$ for all u. This we can do because, for different us, $\mathbf{F}$ assigns different $\mathbf{F}(u)$s.

The semantics can be simplified further for the following reasons. Let us assume that $A \not\mid\!\sim B$. Then there exists a countermodel

$$w = (W, S^w, \prec^w, \mathbf{l}^w, \mathbf{m}^w, \mathbf{F}^w)$$

such that, for the minimum $w_0 \in \hat{A}^w$, $w_0 \vDash A$ but $w_0 \nvDash B$. We can let S^{w_0} be $\hat{A}$ and consider the model $w_0 = (W, S^{w_0}, \prec^{w_0}, \mathbf{l}^{w_0}, \mathbf{m}^{w_0}, \mathbf{F}^{w_0})$ where $\prec^{w_0}, \mathbf{l}^{w_0}, \mathbf{m}^{w_0}, \mathbf{F}^{w_0}$ are the respective restrictions of $\prec^w, \mathbf{l}^w, \mathbf{m}^w, \mathbf{F}^w$ to S^{w_0}. Note that indeed the model name w_0 is minimum in S^{w_0}. The values $w_0 \vDash A$, $w_0 \nvDash B$ are not changed in the restricted model because the fibring at each label u goes to lower levels and never goes back to the model w.

Thus our models can be refined to the following.

A refined fibred structure has the form

$$(W, S^w, \prec^w, \mathbf{l}^w, \mathbf{m}, w_0)$$

where, for each w, $w \in S^w$ is minimal in S^w and w_0 is minimal in S^{w_0}. In this refined semantics completeness reads

$A \mid\!\sim B$ iff in all refined models, if $w_0 \vDash A$ then $w_0 \vDash B$.

We give our definition of the refined semantics for an arbitrary language $\mathbb{L}$ and arbitrary semantics $\mathcal{K}$.

Definition 11.8 (Fibred semantics).

1. A fibred structure has the form $\mathbf{n} = (W, S^w, \prec^w, \mathbf{l}^w, \mathbf{m}^w, w_0)$, where W is a set of labels, $w_0 \in W$, and for each $w \in W, S^w \subseteq W, S^w \neq \varnothing$ is the set of states of the model labelled by w, $\prec^w$ is a preference relation on $S^w, \mathbf{l}^w$ is a function, giving for each $x \in S^w$ a subset $\mathbf{l}^w(x) \subseteq W$ and $\mathbf{m}^w$ is a function giving for each $y \in \mathbf{l}^w(x)$ a structure $\mathbf{m}^w(y)$ in $\mathcal{K}$.
   Satisfaction for wffs A is defined by induction on the number n of occurrences of $>$, for pairs (w, y) such that $y \in S^w$.

(a) *Case* $n = 0$
$(w, x) \models A$ iff $\mathbf{m}^w(y) \models A$, for all $y \in \mathbf{l}^w(x)$.
(b) Let $\hat{A}^w = \{x \in S^w \mid (w, x) \models A\}$. Define $A \mid\!\sim_w B$ iff for all $\prec^w$ minimal points $x \in \hat{A}^w$, we have $(w, x) \models B$.
(c) *Case* $n > 0$
Assume $(w, x) \models A$ is defined for A_1, A_2 of less than n nested $>$ in it. We define $(w, x) \models A_1 > A_2$.
First, let $\hat{A}^w$ be $\hat{A}^w = \{y \in S^w \mid (w, y) \models A\}$. $\hat{A}^w$ is ordered by $\prec^w$ and may have minimal points.
Let $(w, x) \models A_1 > A_2$ iff for all $y \in \mathbf{l}^w(x)$, we have $A_1 \mid\!\sim_y A_2$, where $A_1 \mid\!\sim_y A_2$ means that for all minimal points $u \in \hat{A}_1^y$ we have $(y, u) \models A_2$.

2. We write $w \models A$ to mean for all minimal $y \in S^w$, $(w, y) \models A$, i.e. $\top \mid\!\sim_w A$.
3. We say that A holds in the model if $w_0 \models A$.
4. We assume for each A and w that $\hat{A}^w$ is smooth, or for the strong semantics, that it has a minimum. We can assume for the strong semantics that w is the minimum of S^w.

In the case of $\mathbb{L}$ being the language of classical logic and $\mathcal{K}$ the class of all classical models, we have as our starting point the usual Kraus–Lehmann–Magidor semantics. The models $\mathbf{m}^w(y), w \in W, x \in S^w, y \in \mathbf{l}^w(x)$ can be a classical assignment (more conveniently written as $h^w(y)$) of truth-values $\{0,1\}$ to the atoms.

Theorem 11.9 (Completeness of fibred semantics for C + $\{>\}$). *Let* $\mid\!\sim$ *be a consequence relation for the language* $\mathbb{L}$ *of classical propositional logic extending the system* **C**. *Let* $>$ *be a new connective and let* $\mid\!\sim^*$ *be the smallest extension of* $\mid\!\sim$ *in the language* $\mathbb{L}^* = \mathbb{L} + \{>\}$ *satisfying the following.*

1. *If* A^*, B^* *are substitution instances of* A, B *of* $\mathbb{L}$ *then*

$$\frac{A \mid\!\sim B}{A^* \mid\!\sim^* B^*}$$

2. *For any* A, B

$$A \mid\!\sim^* B \text{ iff } \mid\!\sim^* A > B$$

3.

$$\frac{A, C \to D \mid\!\sim^* B; C \mid\!\sim^* D}{A \mid\!\sim^* B}.$$

Then $\mid\!\sim^*$ *is complete for the fibred semantics based on the Kraus–Lehmann–Magidor semantics for* $\mid\!\sim$.

Proof We use graded fibred semantics for convenience.

Soundness

1. Assume $A^* \not\mid\!\sim^* B^*$, where

$$A^* = A[y_i/C_i > D_i] \qquad B^* = B[y_i/C_i > D_i]$$

and where $A[y_i], B[y_i]$ are wffs of $\mathbb{L}$, C_i, D_i are arbitrary wffs of $\mathbb{L}^*$ and the notation suggests that A^*, B^* are obtained from A, B by substituting $C_i > D_i$ for y_i. A and B may have other atoms (say x_i). We want to show that $A \not\mid\!\sim B$.
Let $\mathbf{n} = (W, S^{\mathbf{n}}, \prec^{\mathbf{n}}, \mathbf{l}^{\mathbf{n}}, \mathbf{m}^{\mathbf{n}}, \mathbf{F}^{\mathbf{n}})$ be a countermodel; then we have that $A^* \not\models_{\mathbf{n}} B^*$.
This means that for some minimal point $x_0 \in \hat{A}^{*\mathbf{n}}$ we have $x_0 \not\models B^*$.
Let $x \in S^{\mathbf{n}}$ be arbitrary and let $u \in \mathbf{l}^{\mathbf{n}}(x)$. Define an assignment $g^{\mathbf{n},u}(y_i)$ on the atoms y_i of A and B by letting

$$g^{\mathbf{n},u}(y_i) = 1 \text{ iff } C_i \models_{\mathbf{F}^{\mathbf{n}}(u)} D_i.$$

Extend $g^{\mathbf{n},u}$ to the other atoms y of A, B by letting $g^{\mathbf{n},u}(y) = \mathbf{m}^{\mathbf{n}}(u)(y)$.
We call $g^{\mathbf{n},u}$ the *surface assignment* of $\mathbf{n}$. We now define a countermodel to $A \mid\!\sim B$.
Let $\mathbf{n}' = (S, \prec, \mathbf{l})$ be defined as follows:

$$S = S^{\mathbf{n}}$$
$$\prec = \prec^{\mathbf{n}}$$
$$\mathbf{l}(y) = \{g^{\mathbf{n},u} \mid u \in \mathbf{l}^{\mathbf{n}}(y)\}.$$

We have for all y, that

$$y \models A^* \text{ iff } y \models A \text{ in } \mathbf{n}$$
$$y \models B^* \text{ iff } y \models B \text{ in } \mathbf{n}.$$

2. We want to show that $A \mid\!\sim^* B$ iff $\mid\!\sim^* A > B$.
This means that the following two statements are equivalent:
   - $A^* \models_{\mathbf{n}} B^*$ holds in every model $\mathbf{n}$ of the appropriate level.
   - $\top \models_{\mathbf{n}} A^* > B^*$ holds in every model of the appropriate level.
3. Assume $A^* \not\models_{\mathbf{n}} B^*$; then for some $x_0 \in S^{\mathbf{n}}, x_0 \models_{\mathbf{n}} A^*, x_0 \not\models_{\mathbf{n}} B^*$ and x_0 is minimal in $\hat{A}^{*\mathbf{n}}$. Thus for some $u_0 \in \mathbf{l}^{\mathbf{n}}(x_0)$ and some atoms $y_i, i = 1, \ldots, r$, we have

$$A^* = A[y_i/C_i > D_i]$$
$$B^* = B[y_i/C_i > D_i]$$

and A, B contain no $>$. Let

$$g^{\mathbf{n},u_0}(y_i) = 1 \text{ iff } C_i \models_{\mathbf{F}^{\mathbf{n}}(u_0)} D_i$$
$$g^{\mathbf{n},u_0}(y_i) = \mathbf{m}^{\mathbf{n}}(u_0)(y) \text{ for } y \neq y_i.$$

Let $\mathbf{n}' = (S', \prec', \mathbf{l}', \mathbf{F}')$ be the structure with $S' = \{u\}.\ u', \prec' u', \mathbf{l}'(u') = u_0$, and $\mathbf{F}(u_0) = \mathbf{F}^{\mathbf{n}}(u_0)$. (We need not worry about W'.) Then $u' \not\models_{\mathbf{n}'} A > B$.

4. Assume that $\mathbf{n}$ is such that for some minimal $x_0 \in S^{\mathbf{n}}, x_0 \not\models_{\mathbf{n}} A^* > B^*$. Then for some $u \in \mathbf{l}^{\mathbf{n}}(x_0), A^* \not\models_{\mathbf{F}^{\mathbf{n}}(u)} B^*$. Hence $A^* \not\mid\!\sim^* B^*$.
5. We show that $A, C \to D \mid\!\sim^* B$ and $C \mid\!\sim^* D$ imply $A \mid\!\sim^* B$.
   Assume that $A \not\mid\!\sim^* B$ and let $\mathbf{n}$ be a countermodel. Thus $A \not\models_{\mathbf{n}} B$. Hence there exists some state $x_0 \in S$ such that x_0 is minimal in $\hat{A}^{\mathbf{n}}$, we have $x_0 \models_{\mathbf{n}} A$ but $x_0 \not\models_{\mathbf{n}} B$, we claim $x_0 \models C \to D$. If not then for some $u \in \mathbf{l}^{\mathbf{n}}(x_0)$, we have $u \not\models_{\mathbf{n}} C \to D$.
   We can present C and D as $C = C_0[y_i/P_i > Q_i]$ and $D = D_0[y_i/P_i > Q_i]$ and if we take the surface assignment $g^{\mathbf{n},u}(y_i) = 1$ iff $P_i \models g_u Q_i$, $g^{\mathbf{n},u}(y) = \mathbf{m}^{\mathbf{n}}(u)(y)$ for $y \neq y_i$, then we obtain $g^{\mathbf{n},u}(C_0) = 1$ and $g^{\mathbf{n},u}(D_0) = 0$.
   Consider the structure $\mathbf{n}'$ with $S' = \{u_0\}, u_0 \prec u_0, \mathbf{l}^{\mathbf{n}'}(u_0) = \{u\}\mathbf{n}^{\mathbf{n}'}(u) = \mathbf{F}^{\mathbf{n}}(u)$. In this structure $C \not\models_{\mathbf{n}'} D$.
   Hence we obtain $x_0 \models_{\mathbf{n}} A, x_0 \models_{\mathbf{n}} C \to D$ but $x_0 \not\models_{\mathbf{n}} B$.
   Thus rule (3) is valid.

Completeness
We show that if $A^* \not\mid\!\sim^* B^*$ then there exists a graded countermodel. The proof is by structural induction on the maximal number of nested $>$ in A^* and B^*.

Assume completeness for any wffs $A^* \not\mid\!\sim^* B^*$ for A^*, B^* with less than n nested occurrences. Let A^*, B^* be obtained from an A, B of $\mathbb{L}$ by substitution for some of the atoms of conditionals of the form $P_i > Q_i$, where P_i, Q_i do not contain $>$. Thus

$$\begin{aligned} A &= A[x_1, \ldots, x_n, y_1, \ldots, y_m] \\ B &= B[x_1, \ldots, x_n, y_1, \ldots, y_m] \\ A^* &= A[y_i/P_i > Q_i] \\ B^* &= B[y_i/P_i > Q_i]. \end{aligned}$$

Assume $A^* \not\mid\!\sim^* B^*$. We seek a fibred countermodel. It follows that $A \not\mid\!\sim B$. Hence there exists a countermodel $\mathbf{n}_1 = (S, \prec, \mathbf{l}, \mathbf{m})$ and a minimal state $s \in \hat{A}$ such that $s \not\models B$. That is, some label $u \in \mathbf{l}(s)$ such that $\mathbf{m}(u) \models A, \mathbf{m}(h) \not\models B$.

$\mathbf{m}(u)$ is an assignment to the atoms x_i and y_j of A and B. If we can realise the values $\mathbf{m}(u)(y_i)$ by a fibred model $\mathbf{n} = (W^{\mathbf{n}}, S^{\mathbf{n}}, \prec^{\mathbf{n}}, \mathbf{l}^{\mathbf{n}}, \mathbf{m}^{\mathbf{n}}, \mathbf{F}^{\mathbf{n}})$ such that

$$P_i \mid\!\sim_{\mathbf{n}} Q_i \text{ iff } \mathbf{m}(y_i) = 1,$$

then by fibring this $\mathbf{n}$ in the model $(S, \prec, \mathbf{l})$ at u and fibring arbitrary other models elsewhere, we can obtain a fibred countermodel for $A^* \not\mid\!\sim^* B^*$.

The problem arises when this assignment cannot be realised. Perhaps then we can use another countermodel. Let us go through all countermodels $\mathbf{n}_1, \mathbf{n}_2, \ldots$ for $A \not\mid\!\sim B$. It may be the case that for each i and each minimal point $s_i \in \hat{A}^{\mathbf{n}_i}$ and each $u_i \in \mathbf{l}_i(s_i)$ such that $\mathbf{m}_i(u_i) \models A$ and $\mathbf{m}_i(u_i) \not\models B$, the assignment cannot be realised by a graded fibred model.

Thus there exist a finite number of assignments $h_1, \ldots, h_k$ which cannot be realised by a graded fibred countermodel but whose disjunction needs to be realised in any countermodel.

In other words, let X_i for each $i = 1, \ldots, k$, be a set of indices $X_i \subseteq \{1, \ldots, r\}$. Then the assignment requires $h_i(y_j) = 1$ iff $j \in X_i$. The fact that h_i cannot be realised means that there is no graded model $\mathbf{n}$ such that $P_j \mid\!\sim_{\mathbf{n}} Q_j$ iff $j \in X_i$, for $j = 1, \ldots, r$, i.e.

$$\bigwedge_{j \in X_i} P_j > Q_j \mid\!\sim^* \bigvee_{j \notin X_i, j \leq r} P_j > Q_j.$$

Let

$$C_i = \bigwedge_{j \in X_i} y_j \text{ and } D_i = \bigvee_{j \notin X_i, j \leq r} y_j.$$

Let $C_i \to D_i, i = 1, \ldots, k$, be the sequents corresponding to these assignments. Therefore we have that

$$A, \bigwedge_i (C_i \to D_i) \mid\!\sim B.$$

We also have $C_i^* \mid\!\sim^* D_i^*$ where $C_i^* = C_i[y_i/P_i > Q_i]$ and $D_i^* = D_i[y_i/P_i > Q_i]$.

Therefore by the properties of $\mid\!\sim^*$ we must also have $A^* \mid\!\sim^* B^*$, which is a contradiction. This proves completeness. □

11.4 General Semantics for the Subjunctive Conditional

The fibred semantics of the previous section suggests a general semantics for the subjunctive conditional. Formally, the language of the conditional is the language of classical logic $\mathbb{L}$, augmented with the binary connective $A > B$. Formally we do have semantics for such a language, namely the fibred semantics of the previous section. The fibred structures have the form

$$(W, S^w, \prec^w, \mathbf{l}^w, \mathbf{m}^w, \mathbf{F}^w, w_0).$$

For the special case of preferential semantics, we can let $\mathbf{l}^w(x) = \{x\}$. In this case, semantic evaluation is two dimensional, as follows, for $x \in S^w$:

- $(w, x) \vDash q, q$ atomic iff $\mathbf{m}^w(x)(q) = 1$
- $(w, x) \vDash A \wedge B$ iff $(w, x) \vDash A$ and $(w, x) \vDash B$
- $(w, x) \vDash A \vee B$ iff $(w, x) \vDash A$ or $(w.x) \vDash B$
- $(w, x) \vDash \neg A$ iff $(w, x) \nvDash A$
- $(w, x) \vDash A > B$ iff for all $y \in S^{x'}$ such that y is minimal in $\{u \mid (x', u) \vDash A\}$ we have $(x', y) \vDash B$, where $x' = \mathbf{F}^w(x)$
- A is valid in the semantics iff for all structures and all indices (w, x) we have $(w, x) \vDash A$.

The above semantics can be reduced to the traditional semantics of the subjunctive conditional. To see this more clearly, let us present the fibred structures above in the traditional notation of conditional logic [Nute, 1984]. Our starting point is the refined fibred structures of the form $(W, S^w, \prec^w, \mathbf{l}^w, \mathbf{m}, w_0)$. We can assume that the following holds in these structures:

- w is the $\prec^w$ minimal point in S^w
- $\mathbf{F}^w(x) = x$ for all x
- for preferential models (the system **P**) we have already assumed that $\mathbf{l}^w(x) = \{x\}$.

Define the following notation:

- wRy iff (by definition) $y \in S^w$
- $\mathbf{f}(w, A) =$ (by definition) the set of $\prec^w$ minimal points in $\hat{A}^w$.

A model can now be presented as $(W, R, \mathbf{f}, \mathbf{l}^w, \mathbf{m})$, where R is reflexive and **f** is a partial function satisfying the following:

- $\mathbf{f}(w, X)$ is defined for $X \subseteq \{y \mid wRy\}$
- $\mathbf{f}(w, X) \subseteq X$
- $X \subseteq Y$ implies $\mathbf{f}(w, Y) \cap X \subseteq \mathbf{f}(w, X)$.

The ordering $\prec^w$ and indeed R can be retrieved from **f** by letting[33]

- wRy iff $\mathbf{f}(w, \{y\})$ is defined
- for s, s' such that wRs and wRs'

$$s \prec^w s' \text{ iff (def) } \mathbf{f}(w, \{s, s'\}) = \{s\}.$$

For preferential models **l** is not needed.

We thus summarise our semantics for the preferential case as a class of models of the form $(W, R, \mathbf{f}, \mathbf{m})$. Evaluation at a point becomes the following (compare with Definition 11.8):

- $x \vDash q$ iff $\mathbf{m}(x)(q) = 1$
- $x \vDash A > B$ iff for all $y \in \mathbf{f}(x, A)$ we have $y \vDash B$
- the inductive clauses for the classical connectives are the usual ones.

We would like to impress upon the reader the conceptual value of this reduction. The traditional semantics for the conditional, involving the function **f**, was developed through philosophical considerations about similarity of worlds, minimal change, and class selection functions. The non-monotonic preferential semantics has a different basis. It is known in the literature that the two are technically connected, see for example [Katsuno and Sato, 1995]. However, in this chapter, we have presented a conceptual connection between the two semantics based on a meaningful and independently motivated fibring construction. Thus by fibring the cumulative semantics with itself for the purpose of reflecting a consequence relation into the object-level, we were able to derive technically the same traditional semantics for the conditional. The full conceptual conclusions to be drawn from this reduction still remain to be investigated. Full discussion of the ideas of fibred semantics and combination of logics can be found in [Gabbay, 1995b].

The reader should compare also with Section 10.8 where $A > B$ is interpreted as the self-fibred formula $A(B)$.

[33] The presentation of cumulative $\prec$ using **f** is an observation of K. Schlechta.

12

HOW TO MAKE YOUR LOGIC FUZZY

12.1 Introduction and Motivating Examples

The aim of this chapter is to provide a methodology for turning a known crisp (i.e. $\{0,1\}$ valued) logic into a fuzzy system. We require of the methodology that it be meaningful in general terms, using processes which are independent of the notion of fuzziness, and that it yield a considerable number of known fuzzy systems.

To appreciate the need for such a methodology, consider for example the $\{\wedge, \vee, \Box, \Diamond\}$ fragment of modal propositional logic **K**, and let us examine our options for turning it into a fuzzy system. This logic is complete for the crisp Kripke semantics. Kripke models have the form $\mathbf{m} = (S, R, a, h)$, where $S \neq \varnothing$ is a set of possible worlds, $R \subseteq S \times S$ is a crisp binary relation (of the form $R : S \times S \mapsto \{0,1\}$), $a \in S$ is the actual world, and h is a binary function assigning to each $t \in S$ and each atomic q a crisp value $h(t,q) \in \{0,1\}$.

h can be extended to all wffs in the usual way with the inductive evaluation of $h(t, \Box A)$ being

$h(t, \Box A) = 1$ iff for all y such that tRy we have $h(y, A) = 1$. Or equivalently,

$h(t, \Box A) = \text{Inf}\,\{h(y, A) \mid tRy\}$. Similarly

$h(t, \Diamond A) = 1$ iff for some y, tRy and $h(y, A) = 1$. We say $\mathbf{m} \vDash A$ iff $h(a, A) = 1$.

Let us try and turn this logic fuzzy!

Working intuitively, one may turn modal logic into a fuzzy modal logic in several ways [Fitting, 1991; Thiele, 1993]:

1. changing the function $h(t,q)$ into a fuzzy function $h^*(t,q) \in [0,1]$ (obtaining real number values)
2. changing the crisp relation R into a fuzzy one $R^* : S^2 \mapsto [0,1]$
3. making $a \in S$ fuzzy
4. any combination of the above.

Let us proceed and consider a particular option. We take the real closed interval $[0,1]$ as our set of fuzzy values and choose 0 as truth $\top$ and 1 as falsity $\bot$.[34] The Łukasiewicz tables for the connectives are the following:

- $x \to y = \max(0, y - x)$

[34] This is more convenient for the mathematical considerations of fibring. We can view fuzzy logic as a possible world substructural logic with $[0,1]$ as its set of possible worlds.

- $x \wedge y = \max(x, y)$
- $x \vee y = \min(x, y)$
- $\neg x = 1 - x$.

We can turn ordinary modal logic into a fuzzy modal logic by introducing functions h^* and R^* and defining the following evaluation in the fuzzy model $\mathbf{m}^* = (S, R^*, a, h^*)$

(*1) $h^*(t, \Diamond A) = \text{Inf}_y\{\max(R^*(t, y), h^*(y, A))\}$

(*2) $h^*(t, \Box A) = \text{Sup}_y\{\max(0, h^*(y, A) - R^*(t, y))\}$.

These definitions are proposed and discussed by H. Thiele [1993, Definition 6] and seem reasonable, since for the classical two-valued case they give the right result (remember we take 0 as truth!).

The definition is in the right form, since classically satisfaction is written as

$$t \vDash \Diamond A \text{ iff } \exists y(tRy \wedge y \vDash A)$$
$$t \vDash \Box A \text{ iff } \forall y(tRy \to y \vDash A)$$

and we are turning it fuzzy by reading the right hand sides as formulas in Łukasiewicz's logic.

Although the above seems reasonable, it is still an arbitrary choice. We might take a different option. We might adopt the view that since $\exists y(tRy \wedge y \vDash A)$ classically says that both conjuncts should hold, we can read it in the many-valued case as

$$\text{Inf}_{\{y|R^*(t,y)=0\}}\{h^*(y, A)\}.$$

In other words, we 'average' (using Inf) over those worlds where we are 'sure' they are accessible. H. Thiele [1993] discusses other options, considered by other authors. It is safe to say that the choices are theoretically arbitrary and the preference of one choice to another can be determined only by application needs and not by any methodological considerations.

Can we offer some methodological considerations for making a logic fuzzy or do we just go from logic to logic and make fuzzy whatever semantical component we find?

Obviously the latter alternative is not desirable. What if we use a different semantics for **K** and make fuzzy the functions involved in that semantics? Do we get yet another batch of fuzzy modal logics?

Clearly we could use some general principles.

The methodology we use is that of combining two logics together through the fibring of their semantics. There are general ways of doing fibring that are independent of the logics themselves, provided they satisfy some simple assumptions, as we saw in earlier chapters.

Let $\mathbb{L}_1$ and $\mathbb{L}_2$ be two logic languages, say for example that $\mathbb{L}_1$ is of modal logic **K** and $\mathbb{L}_2$ is of intuitionistic logic. Our methodology allows for several methods of combinations:

1. one layer $\mathbb{L}_1[\mathbb{L}_2]$ allowing for substituting intuitionistic formulas for the atoms of **K**;

2. one layer $\mathbb{L}_2[\mathbb{L}_1]$ allowing the substitution of modal **K** formulas for the atoms of intuitionistic logic;
3. more fibred layers say $\mathbb{L}_1[\mathbb{L}_2[\mathbb{L}_1]]$ and so on;
4. full fibring $\mathbb{L}_{1,2} = \mathbb{L}_1 \otimes \mathbb{L}_2$.

Methods for obtaining semantics for such fibred logics from the semantics of the components were described earlier in this book. See also [Gabbay, 1993a] and [Finger and Gabbay, 1992] and [Gabbay, 1995b; Finger and Gabbay, 1996; Gabbay, 1996c]. The results are independent of the semantics employed, are logic independent, and yield a huge number of known combined logics, logics existing in the literature in the past thirty years.

In our particular example, let $Ł_\infty$ be Łukasiewicz infinite valued logic (with values in $[0, 1]$) and let us apply our fibring machinery to **K** and $Ł_\infty$. We get the following:

1. the fibred semantics for $\mathbf{K}[Ł_\infty]$ is a fuzzy semantics with h fuzzy, R crisp;
2. the fibred semantics for $Ł_\infty[\mathbf{K}]$ is a semantics with R fuzzy and h crisp;
3. the semantics for $Ł_\infty[\mathbf{K}[Ł_\infty]]$ is a semantics where both R and h are fuzzy;
4. additional reasonable considerations arising from the fibring methodology will allow and lead us towards variations in the fuzzy semantics obtained.

So in short, when you ask me how to make your logic $\mathbf{L}_1$ fuzzy, I would answer—take a pure fuzzy logic $\mathbf{L}_2$ (e.g. $Ł_\infty$ or any other) and fibre it to $\mathbf{L}_1$ in different ways, and follow the natural fibring options.

Our plan is as follows. The rest of this section will give a preparatory example of a parametrised family of modal logics which will explain and prepare the reader for the fibring construction.

The next section explains intuitively, via examples, how *fuzzling* (making fuzzy by fibring) is done and how it relates to the general theory of combining logics. Later sections will develop the formal machinery.

Consider a parametrised family of bi-modal models of the form $\mathbf{m}_t = (S, \rho_t, R_t, a, h_t)$ where t ranges over $[0, 1]$ with the binary relations ρ_t and R_t where only ρ_t, R_t and h_t are dependent on t. These satisfy the following

- $t \leq s$ implies for all atomic q and $x \in S$, $h_t(x, q) = 1$ implies $h_s(x, q) = 1$
- $t \leq s$ implies $\rho_t \subseteq \rho_s$
- $t \leq s$ implies $R_t \supseteq R_s$.

The above are three persistence principles. The first one says if q is true at a point it continues to be true and the second says that if x and y are ρ connected at t they continue to be ρ connected. The third one says if x and y are R not connected, then they continue to be not connected.

We can define

- $\rho^\sharp(x, y) = \text{Inf}_t\{x\rho_t y \text{ holds}\}$
- $R^\sharp(x, y) = \text{Sup}_t\{xR_t y \text{ holds}\}$
- $h^\sharp(x, q) = \text{Inf}_t\{h_t(x, q) \text{ holds}\}$.

We can assume that both Inf and Sup are actually attained!
Thus we have

- $x\rho_t y$ holds iff $\rho^\sharp(x,y) \leq t$
- $h_t(x,q)$ holds iff $h^\sharp(x,q) \leq t$
- $xR_t y$ holds iff $t \leq R^\sharp(x,y)$.

We imagine we have a language with two modalities, $\Box$ and $\Diamond$. $\Box$ uses R_t and $\Diamond$ uses ρ_t. Thus $\Box$ and $\Diamond$ are not interdefinable unless R and ρ are properly related.

Let us denote by $\vDash_t$ satisfaction in the model $\mathbf{m}_t$, and let us consider the question of $x \vDash_t \Diamond q$.

Clearly

($\sharp$1) $x \vDash_t \Diamond q$ iff $\exists y(x\rho_t y \wedge y \vDash_t q)$ iff $\exists y(\rho^\sharp(x,y) \leq t \wedge h^\sharp(x,y) \leq t)$
($\sharp$2) $x \vDash_t \Box q$ iff $\forall y(t \leq R^\sharp(x,y) \rightarrow h^\sharp(x,y) \leq t)$.

So far we have done nothing 'fuzzy' except that the family of parametrised models $\mathbf{m}_t, t \in [0,1]$ can be *presented* as a single *fuzzy-looking* model.

$$\mathbf{m}^\sharp = (S, \rho^\sharp, R^\sharp, a, h^\sharp).$$

This model looks like the fuzzy modal model $\mathbf{m}^* = (S, R^*, a, h^*)$ considered earlier, but the evaluation is done differently. However, we can motivate the earlier model $\mathbf{m}^*$ by using the family $\mathbf{m}_t$, through the following considerations:

(1) *Case of* $\Diamond$
Consider the parameter $t = 0$. We want to check where $x \vDash_0 \Diamond q$. Choose a y. It may be that $x\rho_0 y$ does not hold. In fact, $x\rho_t y$ starts to hold only for $t \geq \rho^\sharp(x,y)$. Similarly, $h_t(y,q)$ starts to hold only after $h^\sharp(y,q)$. We can say that $y \vDash_0 q$ is numerically $h^\sharp(y,q)$ 'far' from holding. Similarly, $x\rho_0 y$ is numerically $\rho^\sharp(x,y)$-far from holding. Together then $(x\rho_0 y \wedge y \vDash_0 q)$ is far from holding to an amount which depends on both. It is up to us to choose how to 'combine' these two amounts. We can add them together, or we can take the maximum (recall 0 is truth). Either option is well known from fuzzy logic. We choose in this case to take the maximum. Thus we say that the value of $x\rho_0 y \wedge y \vDash_0 q$ is

$$\max(\rho^\sharp(x,y), h^\sharp(y,q))$$

far from holding.

We can therefore reasonably say that

$$x \vDash_0 \Diamond q,$$

which means,

$$\exists y(x\rho_0 y \wedge y \vDash_0 q)$$

is

$$\text{Inf}_y\{\max(\rho^\sharp(x,y), h^\sharp(y,q))\}$$

far from holding.

(2) *Case of* $\Box$
For this case, consider $x \vDash_t \Box q$. We have

$x \vDash_t \Box q$ iff for all y (xR_ty implies $y \vDash_t q$).

Choose an arbitrary y. If $t > R^\sharp(x,y)$, then xR_ty does not hold and y makes no contribution towards making $x \vDash_t \Box q$ false. If $t \leq R^\sharp(x,y)$ then xR_ty holds and the distance from the truth of $y \vDash_t q$ is $\max(0, h^\sharp(x,q) - t)$.

Taking a generous point of view this overall distance is minimised when $t = R^\sharp(x,y)$. Thus the contribution of y towards falsifying $x \vDash_t \Box q$ is $\max(0, h^\sharp(x,q) - R^\sharp(x,y))$ and is now not dependent on t. We can take Sup over y. Therefore we can say that

$$x \vDash \Box q \text{ is } \mathrm{Sup}_y\{\max(0, h^\sharp(x,q) - R^\sharp(x,y))\}$$

far from holding.

In (1) and (2) above we made arbitrary, though reasonable choices. This is, however, enough to motivate the fuzzy $(*)$ evaluation, which is itself a reasonable though still arbitrary choice. Our considerations above do show that the landscape of possible fuzzy choices can be methodologically motivated by what we can call the $[0,1]$ *parametrised family of modal models.*

The next section will show that such a parametrised family of models arises naturally in the fibring process. We thus motivate our choice of making fuzzy by fibring considerations: the fibring of modality with some chosen many-valued logics.

12.2 Fibring with Łukasiewicz Logic

This section reintroduces the general process of fibring and in particular discusses fibring of modal logic with many-valued logic. We begin by motivating the process of fibring.

In many application areas there arises the formal need of combining two languages together. The most well known in applied logic is the use of temporal logic to describe and verify the temporal behaviour of systems. Here $\mathbb{L}_2$ is the language for describing a static system and $\mathbb{L}_1$ is a temporal language. By substituting sentences of $\mathbb{L}_2$ as 'atoms' within $\mathbb{L}_1$ we can express temporal properties of the system. This is called 'temporalising' of $\mathbb{L}_2$, see Chapter 13. Another well known family of examples is multimodal logics, such as logics of knowledge and belief, logics of action and dynamic logics. The need to combine systems also arises in languages that are self-referential and can talk about themselves. If $A(x,y)$ is a formula with the free variables x, y, we may form the self-referential formula $B(x,y) = A(x, A(x,y))$. We need to provide semantics for such expressions and fibred semantics can provide for our needs.

The next example shows how to combine two modal logics.

Example 12.1 (Fibring two modalities). Let $\mathbb{L}_1, \mathbb{L}_2$ be two modal propositional languages built on the same atoms, with modalities $\Box_1$ and $\Box_2$ respectively. Assume $\mathbb{L}_i$ is complete for the class of models $\mathcal{K}_i$. The models are of the form $\mathbf{m} = (S, R, a, h)$, where S is the set of possible worlds, $R \subseteq S^2$ is the accessibility relation, $a \in S$ is the actual world and h is the assignment function, associating with each $t \in S$ and atomic q a crisp value $h(t, q) \in \{0, 1\}$. We can assume the model satisfies the following:

$$S = \{x \mid aR^n x, \text{ for some } n\}$$

where xR^0y iff $x = y$ and, $xR^{n+1}y$ iff $\exists z(xRz \wedge zR^n y)$.

Satisfaction $t \models A, t \in S, A$ a wff, is defined in the traditional manner:

- $t \models q$ iff $h(t, q) = 1$, for q atomic;
- $t \models A \wedge B$ iff $t \models A$ and $t \models B$;
- $t \models \sim A$ iff $t \not\models A$;
- $t \models \Box A$ iff for all s such that tRs we have $s \models A$;
- $t \models \Diamond A$ iff for some s, tRs and $s \models A$;
- $\mathbf{m} \models A$ iff $a \models A$;
- $\mathcal{K} \models A$ iff $\mathbf{m} \models A$ for all $\mathbf{m} \in \mathcal{K}$.

The *level (1, 2)* fibred language $\mathbb{L}_{(1,2)} = \mathbb{L}_1[\mathbb{L}_2]$ allows for wffs of the form $B = B_1(x_1/C_1, \ldots, x_n/C_n, y_1, \ldots, y_m)$ where $B_1(x_1, \ldots, x_n, y_1, \ldots, y_m)$ is a wff of $\mathbb{L}_1$. $C_1, \ldots, C_n$ are wffs of $\mathbb{L}_2$ and B is obtained by the simultaneous substitution in B_1 of C_i for x_i respectively.

Let $\alpha = (e_1, \ldots, e_n)$ be a sequence of alternating numbers from $\{1, 2\}$. We define the fibred language $\mathbb{L}_\alpha$ by induction:

$$\mathbb{L}_{(e)} = \mathbb{L}_e, e = 1, 2$$
$$\mathbb{L}_{(e_1, \ldots, e_n)} = \mathbb{L}_{e_1}[\mathbb{L}_{(e_2, \ldots, e_n)}].$$

Let $\mathbb{L}_\infty = \bigcup_\alpha \mathbb{L}_\alpha$ denote the full fibred language.

We now define fibred semantics for $\mathbb{L}_\alpha$. A level (1, 2) fibred model has the form $\mathbf{n} = (S, R, a, h, \mathbf{F})$ where $\mathbf{m} = (S, R, a, h)$ is a Kripke model of $\mathcal{K}_1$ and $\mathbf{F}$ is a function on S giving for each $t \in S$ a model $\mathbf{F}(t) = (S_t, R_t, a_t, h_t)$ in $\mathcal{K}_2$.[35]

We can assume that S and all the S_t are pairwise disjoint.

Satisfaction is defined in the usual manner, with the crucial fibring clause for $\Box_i$ (and similarly for $\Diamond_i$) being the following:

$$w \models \Box_i A \text{ iff}$$

1. $w \in S_t$ for some t, and $i = 2$ and for all $s \in S_t(wR_ts \to s \models A)$.
2. $w \in S$ and $i = 1$ and for all $s(wRs \to s \models A)$.
3. $w \in S$ and $i = 2$ and $a_w \models \Box_i A$.
4. Undefined,, when $w \in S_t$ and $i = 1$.

In other words, when, at a point $w \in S$, we want to evaluate $\Box_2 A$, we go to $\mathbf{F}(w)$ and continue the evaluation of $\Box_2 A$ at the model $\mathbf{m}_t(S_t.R_t, a_t, h_t)$. If $\mathbf{F}(t)$ is a family of models then we have $t \models \Box_2 A$ iff for all $\mathbf{m}_t \in \mathbf{F}(t), \mathbf{m}_t \models \Box_2 A$.

[35] There is no reason not to allow $\mathbf{F}(t)$ to be a family of models or to be empty.

A level $\alpha = (e_1, \ldots, e_n)$ fibred model is defined by induction. It has the form $(S, R, a, h, \mathbf{F})$ where (S, R, a, h) is a model in $\mathcal{K}_{e_1}$ and for each $t \in S$, $\mathbf{F}(t)$ is a model in $\mathcal{K}_{(e_2,\ldots,e_n)}$, i.e. a model of level $(e_2, \ldots, e_n)$.

Let $\mathcal{K}_\infty = \bigcup_\alpha \mathcal{K}_\alpha$.

The models of $\mathcal{K}_\alpha$ can provide semantics for $\mathbb{L}_\alpha$.

Figure 12.1 shows what the models in $\mathbb{L}_{(1,2,1)}$ look like.

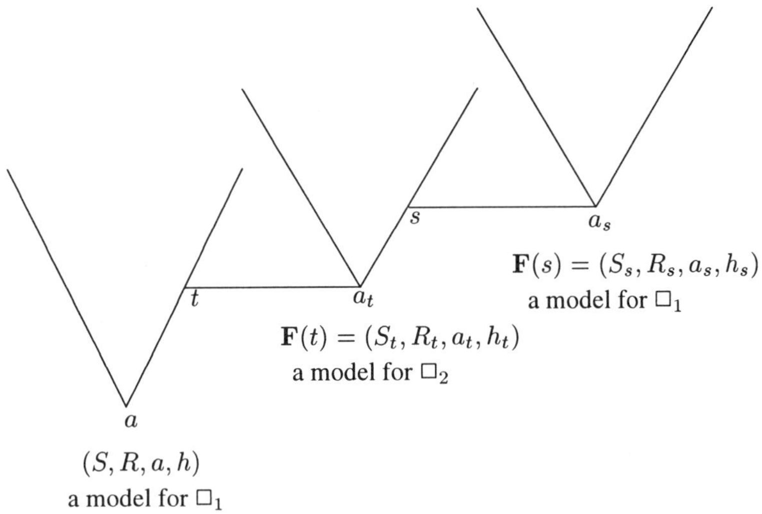

FIG. 12.1.

The next example shows how such models can be presented more nicely.

Example 12.2 (Simplified fibred models). Consider a model in $\mathcal{K}_{(1,2)}$. There is another way of looking at this fibred model. Since S and $S_t, t \in S$ are all pairwise disjoint, let $\mathbf{n}^* = (S^*, W_a, R^*, a, h^*, \mathbf{F}^*)$ be the model with:

$$
\begin{aligned}
&S^* = S \cup \textstyle\bigcup_{t \in S} S_t \\
&R^* = R \cup \textstyle\bigcup_t R_t \\
&W_a = \{a\} \cup \{a_t \mid t \in S\} \\
&h^* = h \cup \textstyle\bigcup_t h_t \\
&\mathbf{F}^* : S^* \mapsto S^* \text{ be the function with} \\
&\mathbf{F}^*(x) = \begin{cases} a_x, & x \in S \\ x & \text{otherwise.} \end{cases}
\end{aligned}
$$

Consider a language $\mathbb{L}^*$ with the modalities $\Box$, $\Diamond$ and a jump operator $\mathbb{J}$. Satisfaction is defined by

$$
\begin{aligned}
&x \vDash \Box A \text{ iff } \forall y (xRy \to y \vDash A) \\
&x \vdash \Diamond A \text{ iff } \exists y (xRy \wedge y \vDash A) \\
&x \vDash \mathbb{J} A \text{ iff } \mathbf{F}^*(x) \vDash A.
\end{aligned}
$$

Let A be a wff of the mixed language. Translate $\Box_1$ as $\Box$ and $\Box_2$ as $\mathbb{J}\Box$. Let A^* be the translation. Then

$$a \vDash A \text{ iff } a \vDash^* A^*$$

where $\vDash$ is satisfaction in $(S, R, a, h, \mathbf{F})$ and $\vDash^*$ s satisfaction in $(S^*, W_a, R^*, a, h^*, \mathbf{F}^*)$.

The sets S and S_t can be retrieved by

$$S = \{x \in S^* \mid aR^{*n}x, \text{ for some } n\}$$
$$S_t = \{x \in S^* \mid a_t R^{*n}x, \text{ for some } n\}$$

and the models $\mathbf{F}(t)$ and $\mathbf{m}$ can be retrieved by restricting R^* and h^* to S and S_t.

In fact, let $(S^*, R^*, a, h^*, \mathbf{F}^*)$ be a Kripke model with $\mathbf{F}^*$ a function from S^* to S^*. We call this model a *simplified fibred model* (*SFM model*) iff the following holds:

Let $W_a = \{a\} \cup \{y \mid \exists x \in S^*(y = \mathbf{F}^*(x)\}$.

For $y \in W_a$, let $S_y = \{t \mid yR^{*n}t, \text{ for some } n\}$.

Then

1. $y_1 \neq y_2 \to S_{y_1} \cap S_{y_2} = \varnothing$
2. $\mathbf{F}^*(x) \neq x \wedge \mathbf{F}^*(y) \neq y \to \mathbf{F}^*(x) \neq \mathbf{F}^*(y)$
3. $S^* = \bigcup_{y \in W_a} S_y$
4. An SFM model is said to be of *depth* $n+1$ if for all x, $\mathbf{F}^*(\mathbf{F}^{*n}(x)) = F^{*n}(x)$, where $\mathbf{F}^{*0}(x) = x$.

The SFM models of depth n are models for $\mathbb{L}_{(x_1,\ldots,x_n)}$. General SFM models provide semantics for $\mathbb{L}_\infty$. See [Gabbay, 1993a] for more details.

The next example will motivate the shift in view of looking at fibred models as fuzzy models.

Example 12.3 (Motivating fuzzy values). We now give a concrete example of an SFM model of level 1. Figure 12.2 shows a $\Box_1$ Kripke model.

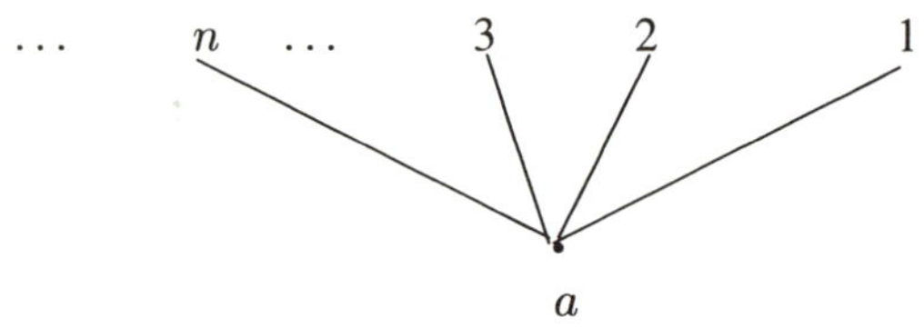

FIG. 12.2.

Here $S = \{a\} \cup \{1, 2, 3, \ldots\}$ with aRn holding, for $n = 1, 2, \ldots$.

Assume $h(a, q) = 0$ and $h(n, q) = 1$ for $n = 1, 2, \ldots$. Try to evaluate $\Diamond_1 \Box_2 q$.

$a \vDash \Diamond_1 \Box_2 q$ iff for some n, $n \vDash \Box_2 q$. Since $\Box_2$ is in the $\mathbb{L}_2$ language, we cannot continue to evaluate. We need an $\mathbb{L}_2$ model to get a value at n. The fibring function $\mathbf{F}(n)$ gives an $\mathbb{L}_2$ model (S_n, R_n, a_n, h_n). Let $S_m = \{a_m\} \cup \{(m, n) \mid n = 1, 2, 3, \ldots\}$. Let R_m be defined by

$$xR_my \text{ iff } \begin{cases} x = a_m \\ \text{or} \\ x = (m, n_1) \text{ and } y = (m, n_2) \text{ and } n_2 \leq n_1 \end{cases}$$

and let

$$h_m(a_m, q) = 0$$

and

$$h_m((m, n), q) = 1 \text{ iff } m \leq n.$$

To complete the picture, let $\mathbf{F}(a) = \mathbf{F}(1)$. Thus $\Box_2 q$ is false at a_n in all the models $\mathbf{F}(n)$, but we have

$$(m, n) \vDash \Box_2 q \text{ iff } m \geq n.$$

This particular fibred model has a special feature which is important. All the models $\mathbf{F}(n)$ have isomorphic *frames*; they are isomorphic to $(T = \{0, 1, \frac{1}{2}, \frac{1}{3}, \ldots\}, \leq, 0)$, through π_m, where $\pi_m(a_m) = 0$, and $\pi_m(m, n) = \frac{1}{n}$, and they differ only in the assignment $h_m(q)$. The image of the truth set $h_m(\Box_2 q) = \{y \mid y \vDash \Box_2 q\}$ is projected on $\{0, 1, \ldots, \frac{1}{n} \mid n = 1, 2, 3, \ldots\}$ and this image gets larger and larger as m increases. In the limit we have

$$\bigcup_m \pi_m h_m(\Box_2 q) = \{\frac{1}{n} \mid n = 1, 2, \ldots\}.$$

Since we are interested in $a \vDash (\Diamond_1 \Box_2 q)$, where the table for $\Diamond_1$ is existential, we need to find a point m such that $h_m(\Box_2 q) = S_m$, or equivalently $\pi_m h_m(\Box_2 q) = \{\frac{1}{n} \mid n = 1, 2, \ldots\}$. No such m exists, but as m gets bigger and bigger, $\pi_m h_m(\Box_2 q)$ approaches the set $\{\frac{1}{n} \mid n = 1, 2, \ldots\}$. Therefore in this sense, we can say that $\Diamond_1 \Box_2 q$ *almost* holds; its evaluation (through using π_m) *approaches the* 'fuzzy' (or 'modal-$\mathbb{L}_2$') truth set $\{\frac{1}{n} \mid n = 1, 2, \ldots\}$.

This is quite a conceptual jump. The model (S, R, a, h) is a model of $\mathbb{L}_1$ (in which $\Diamond_1 \Box_2 q$ is false at the point a). This model has no business getting set values from the set T via the mappings π_m of the models of $\mathbb{L}_2$. However, since all the fibred models $\mathbf{F}(t), t \in S$ are based on isomorphic frames, we can extend the evaluation from the fibred models back into the $\mathbb{L}_1$ language.

It is important to note that the way we extended the evaluation from the fibred model to $\Diamond_1$ of $\mathbb{L}_1$ was arbitrary. We chose a way of doing it which was reasonable, but nevertheless it was a choice. We could have said, let us take as value for $a \vDash \Diamond_1 \Box_2 q$, not the union of $\pi_m h_m(\Box_2 q)$ but the maximum or some other reasonable definition.

Having adopted a good definition, we now consider the expanded model $(S, R, a, h, \mathbf{F}, T, \pi)$.

We can define an $\mathbb{L}_2$-fuzzy value $\mu_t(A)$, for $t \in S$ and any A as follows:

- $\mu_t(A) = \pi_t h_t(A) = \{\pi_t(s) \mid s \in S_t \text{ and } s \vDash A\}$ for A in $\mathbb{L}_2$ or A atomic
- $\mu_t(A \wedge B) = \mu_t(A) \cap \mu_t(B)$
- $\mu_t(\sim A) = T - \mu_t(A)$

- $\mu_t(\Diamond_1 A) = \bigcup_{\{s|tRs\}} \mu_s(A)$
- $\mu_t(\Box_1 A) = \bigcap_{\{s|tRs\}} \mu_s(A)$.

What we have done can be best understood in algebraic terms. Let $\mathbb{B}$ be the boolean algebra of the set $T = \{0, 1, \frac{1}{2}, \frac{1}{3}, \ldots\}$ with the interior operation $Q^{\Box_2}$, for $Q \subseteq T$ being

$$Q^{\Box_2} = \{x \in T \mid \text{ for all } y \geqq x, y \in Q\}.$$

Assign to each atom q and $t \in S$ the 'fuzzy' algebraic subset $\mu_t(q) \subseteq T$.
In our particular model we assign

$$\mu_n(q) = \{1, \tfrac{1}{2}, \ldots, \tfrac{1}{n}\}$$
$$\mu_a(q) = \mu_1(q).$$

We extend the assignment by

- $\mu_t(\Box_2 A) = (\mu_t(A))^{\Box_2}$
- $\mu_t(\Box_1 A) = \bigcap_{\{s|tRs\}} \mu_s(A)$.

The next example brings the idea forward even more clearly.

Example 12.4 (Many-valued modal logic). This is an example of fibring semantical models (modal logic) with algebraic models (Łukasiewicz many-valued logic). We consider the modal language $\mathbb{L}_1$ with $\Box$ and the many-valued language $\mathbb{L}_2$, with $\{\wedge, \vee, \rightarrow, \neg\}$ and with truth values in the real interval $[0, 1]$. We study $\mathbb{L}_1[\mathbb{L}_2]$. The algebraic models of $\mathbb{L}_2$ are linearly ordered Abelian groups which are embeddable in $[0, 1]$. So it is sufficient to consider assignments μ of values and truth table for values in $[0, 1]$. The following are the algebraic functions:

- the domain is $[0, 1]$
- $\leq$ is numerical $\leq$
- $\top = \{0\}$ (0 is truth)
- $\bot$ is 1 (1 is falsity)
- $f_\wedge(x, y) = \max(x, y)$
- $f_\vee(x, y) = \min(x, y)$
- $f_\neg(x) = 1 - x$
- $f_\rightarrow(x, y) = \max(0, y - x)$.

We now turn to fibring.

Let $\mathbf{m} = (S, R, a, h)$ be a Kripke model for $\Box$ and $\Diamond$. The fibring function $\mathbf{F}$ associates with each $t \in S$ an algebraic model $\mathbf{a}_t = (A_t, \leq, f_\wedge, f_\vee, f_\rightarrow, f_\neg, \{0\}, \mu_t)$. Since $A_t = [0, 1]$, fibring algebras $\mathbf{a}_t$ to t is nothing more than associating with each t an arbitrary many-valued assignment μ_t to the atoms of the modal language.

Let us now evaluate $\Box(q \rightarrow p), q, p$ atomic, at the fibred model $(\mathbf{m}, \mathbf{F})$.

- $a \vDash \Diamond(q \rightarrow p)$ iff for some $t \in S$ we have aRt and $t \vDash q \rightarrow p$
- $a \vDash \Box(q \rightarrow p)$ iff for all $t \in S$ such that $aRt, t \vDash q \rightarrow p$
- Since the main connective of $q \rightarrow p$ is many-valued, we have $t \vDash q \rightarrow p$ iff $\mathbf{a}_t \vDash q \rightarrow p$ iff $\mu_t(q \rightarrow p) = 0$ iff $\max(0, \mu_t(p) - \mu_t(q)) = 0$ iff $\mu_t(p) \leq \mu_t(q)$.

Notice that the value of $a_t \vDash q \to q$ is crisp; it is true if $\mu_t(p) \leq \mu_t(q)$ and is false otherwise. We could make a choice and make it fuzzy by giving it the value $\mu_t(p) - \mu_t(q)$ when false.

Similarly, to evaluate $t \vDash \Box q$, q atomic, we can consider $a_t \vDash q$ and use μ_t. $a_t \vDash q$ is true iff $\mu_t(q) = 0$. Again we get a crisp answer unless we make a choice to make it fuzzy.

This fuzzy choice is discussed in the next remark.

Remark 12.5 (Fuzzy modality). We would like to highlight a point which will be of importance later. Consider the fibring in the previous example. We start with $\mathbf{m} = (S, R, a, h)$. Then with each $t \in S$, we fibre an algebra $\mathbf{a}_t$. Since all the algebras have the same domain, the fibring reduces to μ_t, the assignment. Let us pause at this stage and consider the entity (S, R, a, h, μ) and let us try to evaluate $t \vDash \Box q$. Since $\Box q$ contains no many-valued connectives, we get $t \vDash \Box q$ holds iff $\forall s\ (tRs$ implies $s \vDash q)$ iff $\forall s(tRs$ implies $h(t, q) = 1)$. Consider the wff $Iq = \text{def}\ (q \to q) \to q$. Really Iq is q but it is formally a many-valued wff. So we have to evaluate it at the algebra $\mathbf{a}_t$. We have $\mathbf{a}_t \vDash I(q)$ iff $\mu_t(q) = 0$. $t \vDash \Box I(q)$ iff for all $s(tRs$ implies $\mu_t(q) = 0)$.

To summarise, consider $t \vDash \Box q$; we have two ways of looking at it.

1. Regard 'q' as an atom of the modal language, in which case

$$t \vDash \Box q \text{ iff for all } s, tRs \text{ implies } h(s, q) = 1.$$

2. Regard 'q' as an atom of the many-valued language, in which case

$$t \vDash \Box q \text{ iff for all } s, tRs \text{ implies } \mu_s(q) = 0.$$

The two evaluations need not give the same result.

We now have the opportunity to make $t \vDash \Box q$ fuzzy (i.e. 'fuzzle' the satisfaction $\vDash$, or in other words, 'fuzzle' the modal logic) by extending μ_t to $\Box q$ and $\Diamond q$:

($\sharp$) • $\mu_t(\Box q) = \text{Sup}_{\{s|tRs\}} \mu_s(q)$.

• $\mu_t(\Diamond q) = \text{Inf}_{\{s|tRs\}} \mu_s(q)$.

The reader should note that this definition is a chosen one and we could have chosen some other 'averaging' function.

Using ($\sharp$) we can now fuzzle any wff of the modal logic and extend μ_t to all wffs, by taking the many-valued table for $\wedge, \vee, \neg$ and $\to$. We have thus by understable intuitive definition, through ($\sharp$), turned (S, R, a, μ) into a sort of modal many-valued logic by changing the crisp $\{0,1\}$ assignment h into a fuzzy μ. Note that what we are getting is not fibring, it is something new.

Example 12.6 (Persistence). This example will fibre modal logic to the intermediate logic Dummett's **LC**. It will serve to prepare the ground for fibring in the presence of persistence. Let $\Rightarrow$ be intuitionistic implication. **LC** is the extension of intuitionistic logic with the axiom schema

$$(p \Rightarrow q) \vee (q \Rightarrow p)$$

or if disjunction is not available, we can write an implicational axiom schema

$$(p \Rightarrow q) \Rightarrow (((q \Rightarrow p) \Rightarrow r) \Rightarrow r).$$

Let $\mathbb{L}_1$ be the language with $\{\Rightarrow, \wedge, \vee, \bot\}$ and let $\mathbb{L}_2$ be modal logic with $\Box$. Consider the intuitionistic **LC** model with $U = [0, 1]$ (unit real number interval) of the form $(U, \leq, 0, h)$. Since we are dealing with an intuitionistic model, we must have persistence, i.e. for all atomic q and any $t, s \in U$.

(*) $t \leq s$ and $h(t, q) = 1$ imply $h(s, q) = 1$.

(**) We also require, for technical reasons, that for all q, $h(1, q) = 1$.

Satisfaction is defined as follows:

- $t \vDash A \wedge B$ iff $t \vDash A$ and $t \vDash B$
- $t \vDash A \vee B$ iff $t \vDash A$ or $t \vDash B$
- $t \vDash A \Rightarrow B$ iff $\forall s(t \leq s \wedge s \vDash A$ imply $s \vDash B)$
- $t \vDash \bot$ iff $t = 1$.

The reader familiar with **t**-conorms (see [Gottwald, 1993]) can view the above as follows:

For each atomic q let

$$\mu(q) = \text{ Inf } \{t \mid h(t, q) = 1\}.$$

We have (because of persistence) that μ can be extended to all wffs as follows:

- $\mu(A \wedge B) = \max(\mu(A), \mu(B))$
- $\mu(A \vee B) = \min(\mu(A), \mu(B))$
- $\mu(A \to B) = \text{ Inf } \{t \mid \max\,(t, \mu(A)) \geq \mu(B)\}$.

For each $t \in U$, let $\mathbf{F}(t) = (S_t, R_t, a_t, h_t)$ be a modal model of $\Box$. Note that $\Diamond$ is not intuitionistically definable from $\Box$ and so we have to explicitly include $\Diamond$ if we want, in which case a relation ρ_t needs to be added for it. Here we assume we have $\Box$ only.

By general fibring principles, we must have persistence for modal formulas as well, for example, for $\Box^k A$.

[*] $\mathbf{F}(t) \vDash \Box^k A$ and $t \leq s$ imply $\mathbf{F}(s) \vDash \Box^k A$.

This means that

$$t \leq s \to [\forall y[a_t R_t^k y \to y \vDash A] \to \forall y[a_s R_s^k y \to y \vDash A]].$$

It is possible to show that we can assume without loss of generality (i.e. without changing the semantic consequence relation) that:

(†) $t \leq s \wedge x R_s y \to x R_t y$.[36]

In fact if we let $S = \bigcup_t S_t$ we can assume that the fibred models are

$$\mathbf{F}(t) = (s, R_t, a_t, h_t).$$

We are going to assume the following additional properties:

[36] This condition is for $\Box$. For $\Diamond$ we need $t \leq s \wedge x\rho_t y \to x\rho_s y$.

$a_t = a$ for some fixed a
and
$t \leq s$ and $h_t(x, q) = 1$ imply $h_s(x, q) = 1$ for all $x \in S$ and atomic q.

We believe one can show that such assumptions can be made without loss of generality.

So the models differ only in their accessibility relation R_t which satisfies (†) above, and the assignment h_t.

Define functions $h^\sharp(x, q) \in U, q$ atomic, $x \in S$ and $R^\sharp : S^2 \mapsto U$ by letting

$$\begin{aligned} h^\sharp(x, q) &= \text{Inf}\ \{t \mid h_t(x, q) = 1\} \\ R^\sharp(x, y) &= \text{Sup}\ \{t \mid xR_t y\}. \end{aligned}$$

(Let us assume the Inf and Sup are attained.)

Consider the system $(U, \leq, 0, \mu, S, R^\sharp, a, h^\sharp)$. We can view this system in two ways:

1. An **LC** model $(U, \leq, 0, \mu)$ with a fibring of modal models (S, R_t, a, h_t), where $xR_t y$ holds iff $R^\sharp(x, y) \geq t$, and $h_t(x, q) = 1$ iff $t \geq h^\sharp(x, q)$.
2. A fuzzy model $(S, R^\sharp, a, h^\sharp)$ where the accessibility relation $R^\sharp$ and the assignment $h^\sharp$ are fuzzy and where the fuzzy truth set is $(U, \leq, 0, \mu)$ and evaluation is done using the **t**-conorm *max*, as indicated above.[37]

Let us explore further the fuzzy model $(S, R^\sharp, a, h^\sharp)$. Consider, for $x \in S$, the statement $x \vDash_t A$, i.e. $x \vDash A$ in the model $\mathbf{F}(t)$. Because of persisitence, we can define (we shall understand the Infimum of the empty set as 1):

$$\mu^\sharp(x, A) = \text{Inf}\ \{t \mid x \vDash_t A\}.$$

Consider $\mu^\sharp(x, \Box A)$

$$\begin{aligned} \mu^\sharp(x, \Box A) &= \text{Inf}\ \{t \mid x \vDash_t \Box A\} \\ &= \text{Inf}\ \{t \mid \forall y(xR_t y \text{ implies } y \vDash_t A)\} \end{aligned}$$

but $xR_t y$ holds iff $t \leq R^\sharp(x, y)$ and $y \vDash_t A$ holds iff $\mu^\sharp(y, A) \leq t$.

Hence

$$\mu^\sharp(x, \Box A) = \text{Inf}\ \{t \mid \forall y(t \leq R^\sharp(x, y) \text{ implies } \mu^\sharp(y, A) \leq t)\}.^{38}$$

[37] If we choose a different **t**-conorm, say

$$\begin{aligned} \mu(A \wedge B) &= \text{min}\ (1, \mu(A) + \mu(B)) \\ \mu(A \to B) &= \text{max}\ (0, \mu(B) - \mu(A)) \\ \mu(A \vee B) &= \text{max}\ (0, \mu(A) + \mu(B) - 1) \end{aligned}$$

we get evaluation which makes the accessibility relation Łukasiewicz fuzzy.

[38] For the case of $\Diamond$ and ρ, let $\rho^\sharp(x, y) = \text{Inf}_t\{\rho_t(x, y) \text{ holds})\}$. We get

$$\begin{aligned} \mu^\sharp(x, \Diamond A) &= \text{Inf}\{t \mid x \vDash_t \Diamond A\} \\ &= \text{Inf}\{t \mid \exists y(x\rho_t y \text{ and } y \vDash_t A\} \\ &= \text{Inf}\{t \mid \exists y(\rho^\sharp(x, y) \leq t \text{ and } \mu^\sharp(y, A) \leq t)\}. \end{aligned}$$

Note that $\mu^\sharp(x, \Diamond A)$ actually gives the lowest point t above which $x \vDash \Diamond A$ holds. So if the Inf is taken over the empty set the value $\mu^\sharp(x, \Diamond A) = 1$, and $x \vDash_t \Diamond A$ is always false.

Our semantics may look fuzzy but it can still be viewed as possible world semantics. We are not adopting the view that since $x \vDash_0 \Diamond A$ means $\exists y(xR_0y \wedge y \vDash_0 A)$, we do a traditional fuzzy evaluation and let

- $\nu_0(x, \Diamond A) = \text{Inf}_y\{\max(R^\sharp(x,y), \mu_y(A)\}$

The model $(S, R^\sharp, a, h^\sharp)$ should be compared with the parametrised model $(S, \rho^\sharp, R^\sharp, a, h^\sharp)$ of Section 12.1. We saw there how we can choose to turn a parametrised model into a fuzzy model.

The previous two examples show that modal and many-valued logic can be put together in two different ways. If we start with a modal model (S, R, a, h) then we can fuzzle (make fuzzy) h by changing it into a many-valued assignment μ and extend to the entire modal language. If we start with a many-valued model μ then we can fuzzle μ by changing it into a function into elements of a modal algebra. This turned out to be equivalent to looking at modal models where the possible world relation is fuzzy but the assignment is crisp. i.e. models of the form $(S, R^\sharp, a, \mu)$ where $R^\sharp(x,y) \in [0,1]$, while μ is a $\{0,1\}$ assignment. μ can be extended to all wffs, in which case it becomes a $[0,1]$ valued function.

The obvious combination of the two approaches is to make both $R^\sharp$ and $\mu^\sharp$ fuzzy. This leads us to the following definition.

Definition 12.7. An algebraic fuzzled many-valued modal model has the form $(S, R^\sharp, a, \mu^\sharp)$, where $R^\sharp; S^2 \mapsto [0,1]$ is a fuzzy possible world relation and for each $s \in S$ and atomic q, $\mu^\sharp(q) \in [0,1]$.

$\mu^\sharp_s$ can be extended to arbitrary formulas as follows:

$$\mu^\sharp_s(A * B) = f_*(\mu^\sharp_s(A), \mu^\sharp_s(B))$$

where $* \in \{\wedge, \vee, \rightarrow, \neg\}$ and f_* is the many-valued truth table for $*$:

$$\mu^\sharp_x(\Box A) = \mathit{Inf}_t\{\text{for all } y, R^\sharp(x,y) \geq t \text{ implies } \mu^\sharp_y(A) \leq t\}.^{39}$$

Summary 12.8. We summarise the ideas of this section.

- Making fuzzy is identical with fibring in a special way.
- Any logic $\mathbf{L}_1$ can be 'made fuzzy' by fibring it with $\mathbf{L}_2$ as $\mathbf{L}_1[\mathbf{L}_2]$.
- If $\mathbf{L}_1$ is the Łukasiewicz infinite valued logic and $\mathbf{L}_2$ is modal logic then $\mathbf{L}_1[\mathbf{L}_2]$ can be understood as modal logic with fuzzy accessibility but crisp assignment to atoms while $\mathbf{L}_2[\mathbf{L}_1]$ is modal logic with fuzzy assignment to atoms (but crisp accessibility).

Query: What about $\mathbf{L}_1[\mathbf{L}_2[\mathbf{L}_1]]$ and $\mathbf{L}_2[\mathbf{L}_1[\mathbf{L}_2]]$? Do we get fuzzy accessibility and fuzzy assignment?

The next section will deal with the case study of fuzzy modal logic in detail.

[39] $\mu^\sharp_x(\Diamond A) = \text{Inf}\{t \mid \text{for some } y,\ \max(\rho^\sharp(x,y), \mu^\sharp_y(A)) \leq t\}$.

12.3 Fuzzy Modal Logic

Let U be the unit interval $[0, 1]$. We consider 0 as $\top$ and 1 as $\bot$. Let $\mathbf{s}$ be a triangular conorm on $[0, 1]$ and let φ be its ϕ-operator. This means that $\mathbf{s}$ and φ are binary functions on U and the following holds:

(s0) $\mathbf{s}$ is associative.

(s1) $\mathbf{s}$ is commutative.

(s2) $\mathbf{s}$ is non-decreasing in both arguments.

(s3) $\mathbf{s}$ has 0 as the neutral element, i.e. $x\mathbf{s}0 = x$, for all $x \in U$.

(φ 1) $v \geq w$ implies $u\varphi v \geq u\varphi w$.

(φ 2) $(u\varphi v)\mathbf{s}u \geq v$.

(φ3) $v \geq u\varphi(v\mathbf{s}u)$.

Under suitable conditions the following holds

$$u\varphi v = \text{ Inf}\{w \mid w\mathbf{s}u \geq v\}.$$

See [Gottwald, 1993] for details.

We can define a many-valued logic with connectives $\otimes$ and $\twoheadrightarrow$ based on $\mathbf{s}$.

Definition 12.9 (Fuzzy logic). Consider a language with binary $\{\otimes, \twoheadrightarrow\}$ and consider $U = [0, 1]$ and cotriangular norm $\mathbf{s}$ and its associated ϕ-operator φ.

1. A model has the form $\mathbf{m} = (U, \leq, 0, \mu)$, where μ is an assignment to the atoms satisfying
$$\mu(t, q) = 1 \wedge t \leq s \rightarrow \mu(s, q) = 1.$$
2. Satisfaction is defined as follows.
   - $t \vDash q$ iff $\mu(t, q) = 1, q$ atomic.
   - $t \vDash A \otimes B$ iff for some x, y such that $x \vDash A$ and $y \vDash B$ we have $x\mathbf{s}y \leq t$.
   - $t \vDash A \twoheadrightarrow B$ iff for all y such that $y \vDash A$, we have $t\mathbf{s}y \vDash B$.
3. We say $\mathbf{m} \vDash A$ iff $0 \vDash A$.
4. We say $A_1, \ldots, A_n \vDash_{\mathbf{m}} B$ iff for all t, if $t \vDash A_1 \otimes \cdots \otimes A_n$ then $t \vDash B$.
5. To stress that $\mathbf{m}$ uses the conorm $\mathbf{s}$, we may write $\mathbf{m_s}$.
6. Let $\mathcal{K}$ be a class of models (based on possibly different conorms $\mathbf{s}$). We write $A_1, \ldots, A_n \vDash_{\mathcal{K}} B$ to mean that for all $\mathbf{m} \in \mathcal{K}, A_1, \ldots, A_n \vDash_{\mathbf{m}} B$.

Lemma 12.10. *Let $\mathbf{m}$ be a model. Let $\mu^\sharp(q) = \mathit{Inf}\{t \mid \mu(t, q) = 1\}$, for q atomic. Extend $\mu^\sharp$ to arbitrary wffs as follows:*

- $\mu^\sharp(A \otimes B) = \mu^\sharp(A)\mathbf{s}\mu^\sharp(B)$
- $\mu^\sharp(A \twoheadrightarrow B) = \mu^\sharp(A)\varphi\mu^\sharp(B)$.

Then for any t and A we have

(*) $t \vDash A$ iff $\mu^\sharp(A) \leq t$.

Proof For atomic q this holds by definition.

Case of $\otimes$
Assume $t \vDash A \otimes B$. Then for some $x, y, x \vDash A, y \vDash B$ and $x\mathrm{s}y \leq t$. By the induction hypothesis $\mu^\sharp(A) \leq x$ and $\mu^\sharp(B) \leq y$, and since s is non-decreasing, we get

$$\mu^\sharp(A \otimes B) = \mu^\sharp(A)\mathrm{s}\mu^\sharp(B) \leq x\mathrm{s}y \leq t.$$

Conversely, assume $\mu^\sharp(A \otimes B) \leq t$. Let $x = \mu^\sharp(A), y = \mu^\sharp(B)$. Clearly by the induction hypothesis $x \vDash A$ and $y \vDash B$ and $x\mathrm{s}y \leq t$ and so $t \vDash A \otimes B$.

Case of $\twoheadrightarrow$
Assume $t \vDash A \twoheadrightarrow B$. Hence for all $y, y \vDash A$ implies $t\mathrm{s}y \vDash B$. This means that for all y:

$$\mu^\sharp(A) \leq y \to \mu^\sharp(B) \leq t\mathrm{s}y$$

in particular, for $y = \mu^\sharp(A)$ we get

$$\mu^\sharp(B) \leq t\mathrm{s}\mu^\sharp(A).$$

We also have that

$$\mu^\sharp(A)\varphi\mu^\sharp(B) = \mathrm{Inf}\{w \mid \mu^\sharp(B) \leq w\mathrm{s}\mu^\sharp(A)\}.$$

For the choice $w = t$, we get

$$\mu^\sharp(A)\varphi\mu^\sharp(B) \leq t.$$

Assume now that $\mu^\sharp(A)\varphi\mu^\sharp(B) \leq t$. We want to show that $t \vDash A \twoheadrightarrow B$, i.e. we want to show that for all y

$$\mu^\sharp(A) \leq y \to \mu^\sharp(B) \leq t\mathrm{s}y.$$

We have by (φ2) that:

$$\mu^\sharp(B) \leq (\mu^\sharp(A)\varphi\mu^\sharp(B))\mathrm{s}\mu^\sharp(A) \leq t\mathrm{s}\mu^\sharp(A).$$

But also

$$\mu^\sharp(A) \leq y \to t\mathrm{s}\mu^\sharp(A) \leq t\mathrm{s}y.$$

Hence

$$\mu^\sharp(A) \leq y \to \mu^\sharp(B) \leq t\mathrm{s}y$$

as required. □

The previous lemma turns many-valued logic into a substructural, possible world semantics, logic. It was essentially observed by D. Scott for Łukasiewicz infinite valued logic. See [Bolc and Borowik, 1992]. Note that we did not use the commutativity of s in the proof.

We are now ready to fibre.

Definition 12.11 (General fibred models). Let $\mathbb{L}_1$ be the language of modal logic with $\Box$ and let $\mathbb{L}_2$ be the fuzzy language with $\otimes$ and $\twoheadrightarrow$. Let $\mathcal{K}_1, \mathcal{K}_2$ be respective classes of models for the languages. It is convenient for us in the many-valued case to allow for many disjoint copies of the interval $[0, 1]$. Let

$$U_m = \{m\} \times [0, 1],$$

and let

$$\begin{array}{l}(m, x) \leq (m, y) \text{ iff } x \leq y \\ (m, x)\mathbf{s}(m, y) = (m, x\mathbf{s}y) \\ (m, x)\varphi(m, y) = (m, x\varphi y).\end{array}$$

For convenience we write O_m instead of $(m, 0)$.

The above gives us an infinite number of disjoint copies of our fuzzy models in $\mathcal{K}_2$.

We are going to build a fuzzled $\mathbb{L}_i$ structure of level n, for arbitrary n. We are going to use distinct copies of U_k during the process, so that all models involved are disjoint.

1. (a) The fuzzled structures of $\mathcal{K}_1^1$ are all structures of $\mathcal{K}_1$.
   (b) The fuzzled structures of $\mathcal{K}_1^2$ are all structures of $\mathcal{K}_2$. Note that each structure $(U, \leq, 0, \mu) \in \mathcal{K}_2$ has many isomorphic copies in $\mathcal{K}_2$.

   Given a structure in $\mathcal{K}_i^1$ and a possible world t in it we say that t is of level 1 in the structure.
2. Assume that the fuzzled structures of level n have been defined. Let us use the notation $\mathcal{K}_1^n$ and $\mathcal{K}_2^n$ for the families of these structures. Assume that the notion of a possible world t appearing in such a structure is also defined and that each such point has a clearly defined unique level $k, 1 \leq k \leq n$. We now define $\mathcal{K}_i^{n+1}, i = 1, 2$.

   The structures of $\mathcal{K}_1^{n+1}$ have the form

   $$\mathbf{m} = (S, R, a, h, \mathbf{F}_1^n).$$

   For the case $n = 1, \mathbf{F}_1^0$ is interpreted as identity.

   The structure

   $$\mathbf{n} = (S, R, a, h) \in \mathcal{K}_1$$

   is called the *surface model* of $\mathbf{m}$.

   $\mathbf{F}_1^n$ is a function giving for each $t \in S$ a structure $\mathbf{F}_1^n(t) \in \mathcal{K}_2^n$.

   The points in S are called the *surface points* of $\mathbf{m}$ and are of level $n + 1$. We assume that S and all points in $\mathbf{F}_1^n(t)$, for all $t \in S$, are all pairwise disjoint.

   The structures in $\mathcal{K}_2^{n+1}$ have the form

   $$\mathbf{m} = (U_k, \leq, (k, 0), \mu_k, \mathbf{F}_2^n),$$

   where

   $$\mathbf{n} = (U_k \leq, (k, 0), \mu_k)$$

   is a copy model of $\mathcal{K}_2$ (called the *surface model* of $\mathbf{m}$) and for all $u \in U_k, \mathbf{F}_2^n(u)$ is a model of $\mathcal{K}_1^n$ of the form

   $$\mathbf{F}_2^n(u) = (S, R_u, a, h_u, \mathbf{F}_1^{n-1}).$$

   We assume that $u \leq u'$ implies $R_u \supseteq R_{u'}$, and that for all $x \in S$ and atomic $q, h_u(x, q) = 1$ implies $h_{u'}(x, q) = 1$.

We can also assume that k is chosen in such a way that U_k and $\mathbf{F}_2^n(u), u \in U_k$ are all pairwise disjoint.

The points of U_k are said to be of level $n+1$ and are called the *surface points* of $\mathbf{m}$.

Note that S, a and $\mathbf{F}_1^{n-1}$ do not depend on u.
We now define the wffs of the language $\mathbb{L}_i^n, n = 1, 2, \ldots$.

- The wffs of $\mathbb{L}_i^1$ are the wffs of $\mathbb{L}_i$.
- The wffs of $\mathbb{L}_i^{n+1}$ have the form $B = B_0(x_1/C_1, \ldots, x_n/C_n, y_1, \ldots, y_m)$ where $B_0(x_1, \ldots, x_n, y_1, \ldots, y_m)$ is a formula of $\mathbb{L}_i$, and $C_1, \ldots, C_n$ are formulas of $\mathbb{L}_j^m, j \neq i, m \leq n$, and B is the result of the simultaneous substitution of C_k for x_k in B_0.

Definition 12.12 (Satisfaction). We now define a satisfaction $t \vDash_n A$ in a model $\mathbf{m}$ of $\mathcal{K}_i^n$ of a formula of $\mathbb{L}_i^{n'}, n' \leq n$.

1. For models of $\mathcal{K}_i$ and wffs of $\mathbb{L}_i$, satisfaction $\vDash_n, n \geq 1$, is as in modal or many-valued logic.
2. Assume satisfaction $t \vDash_n C$ has been defined in models of $\mathcal{K}_i^n$ for wffs of $\mathbb{L}_i^{n'}$ and surface point t of level $n, n' \leq n$.

Consider a model $\mathbf{m} \in \mathcal{K}_i^{n+1}$, and a formula $B = B_0(x_k/C_k, y_r)$, $k = 1, \ldots, m, r = 1, \ldots, m'$, of $\mathbb{L}_i^{n'+1}$, $n' + 1 \leq n + 1$. If the surface model $\mathbf{n}$ of $\mathbf{m}$ is a modal model, then $B_0(x_k, y_r)$ is a modal formula and $C_k, k = 1, \ldots, m$, are $\mathbb{L}_2^{n_k}$ formulas with $n_k \leq n'$. In case the surface model of $\mathbf{m}$ is an $\mathbb{L}_2$ model, then B_0 is an $\mathbb{L}_2$ formula and $C_k, k = 1, \ldots, m$, are $\mathbb{L}_1^{n_k}$ formulas for $n_k \leq n'$.

Let t be a surface point of $\mathbf{m}$. For any subformula C of B_0, $t \vDash_{n+1} C$ can be evaluated. We need only to say how to evaluate $t \vDash_{n+1} C_k$, since C_k is not in the language of the surface model. In this case we stipulate that $t \vDash_{n+1} C_k$ iff $\mathbf{F}^n(t) \vDash_n C_k$.

Example 12.13. Consider a model in $\mathcal{K}_2^2$. This has the surface structure of the form, say,

$$\mathbf{m}_2 = (U_2, \leq, (2, 0), \mu_2).$$

$\mathbf{m}_2$ is a fuzzy model, with satisfaction defined by

$$u \vDash q \text{ iff } \mu_2(q) \leq u.$$

The points of U_2 are of level 2.

With each $u \in U_2$, we fibre a level 1 modal model of the form

$$\mathbf{n}_u = (S_1, R_u, a_1, h_u).$$

We have seen in Example 12.6 that the model $\mathbf{m}_2$ can be presented in the form

$$\mathbf{m}_2^\sharp = (S_1, R_1^\sharp, a_1, h_1^\sharp)$$

where satisfaction for $x \in S_1$ is defined as follows

$$x \vDash_u q \text{ iff } h_u(x,q) = 1, q \text{ atomic}$$
$$x \vDash_u \Box A \text{ iff } \forall y(xR_u y \to y \vDash_u A)$$
$$x \vDash_u \Diamond A \text{ iff } \exists y(xR_u y \wedge y \vDash_u A).$$

Note that since $u_1 \leq u_2 \wedge h_{u_1}(x,q) = 1 \to h_{u_2}(x,q) = 1$, we can let

$$h_1^\sharp(x,q) = \text{ Inf }\{u \mid h_u(x,q) = 1\}.$$

We assume $h_1^\sharp(x,A)$ is defined for A and that

$$x \vDash_u A \text{ iff } h_1^\sharp(x,A) \leq u.$$

We extend $h_1^\sharp$ to all wffs by

- $h_1^\sharp(x,\Box A) = \text{ Inf }\{u \mid x \vDash_u \Box A\}$
  $= \text{ Inf }\{u \mid \text{ for all } y, xR_u y \to y \vDash_u A\}$
  $= \text{ Inf }\{u \mid \text{ for all } y, u \leq R_1^\sharp(x,y) \to h_1^\sharp(y,A) \leq u\}$
- $h_1^\sharp(x,\Diamond A) = \text{ Inf}\{u \mid \exists y(u \leq R_1^\sharp(x,y) \wedge h_1^\sharp(y,A) \leq u)\}$
- $h_1^\sharp(x,A \wedge B) = h_1^\sharp(x,A)\mathbf{s}h_1^\sharp(x,B)$
- $h_1^\sharp(x,A \twoheadrightarrow B) = h_1^\sharp(x,A)\varphi h_1^\sharp(x,B)$.

Consider now a model $\mathbf{n} \in \mathcal{K}_1^3$. This has a surface model (S_2, R_3, a_3, h_3) with a fibring function $\mathbf{F}_1^2$, giving for each $t \in S_3$ a $\mathcal{K}_2^2$ model $\mathbf{m}_t$. By the discussion above, $\mathbf{m}_t$ can be presented as a model

$$\mathbf{m}_t^\sharp = (S_t, R_t^\sharp, a_t, h_t^\sharp).$$

Thus we can now view our $\mathcal{K}_1^3$ models as a fibred model of the form $(S_3, R_3, a_3, h_3, \mathbf{F}^\sharp)$, where $\mathbf{F}^\sharp(t) = (S_t, R_t^\sharp, a_t, h_t^\sharp)$, for each $t \in S_3$.

Consider now a model in $\mathcal{K}_2^4$. This has the level 4 surface fuzzy model $(U_4, \leq, (4,0), \mu_4, \mathbf{F}_2^3)$, where for each $u \in U_4, \mathbf{F}_2^3(u)$ is a $\mathcal{K}_1^3$ model.

The latter has the level 3 form $(S_3, R_u, a_3, h_u, \mathbf{F}^\sharp)$, where $\mathbf{F}^\sharp$ is as we have seen above.

Again by the method of Example 12.6 and considerations as above the level 4 and level 3 surface models can be presented as

$$(S_3, R^\sharp, a_3, h_3^\sharp),$$

and the general model in $\mathcal{K}_2^4$ has the form

$$(S_3, R_3^\sharp, a_3, h_3^\sharp, \mathbf{F}^\sharp).$$

The above considerations show that the basic modal fuzzy model can be taken to be of the form $(S, R^\sharp, a, h^\sharp)$, with $R^\sharp, h^\sharp$ both fuzzy. These models can be fibred together to form the family $\mathcal{K}_\sharp^n$, see Definition 12.15 below. Note that $\mathcal{K}_\sharp^n$ is identical with $\mathcal{K}_2^{2n}$ viewed as n level fibring of models of the form $(S, R^\sharp, a, h^\sharp)$. Thus $\bigcup_n \mathcal{K}_\sharp^n$ is the family of all models in $\mathcal{K}_2^n, n = 2, 4, 6, \ldots$.

A model in $\mathcal{K}_2^{2n+1}$ has a model of $\mathcal{K}_1$ as its level 1 model. We note that any model of $\mathcal{K}_1$ can be regarded as a model of $\mathcal{K}_1^2$, where the fuzzy model of level 1 is a crisp $\{0, 1\}$ model.

Thus any model of $\mathcal{K}_2^{2n+1}$ can be regarded also as a model of $\mathcal{K}_2^{2n+2}$. Therefore we conclude that

$$\bigcup_n \mathcal{K}_2^n = \bigcup_n \mathcal{K}_\sharp^n.$$

Furthermore, a model **m** in $\mathcal{K}_1$ can be regarded as a model **n** in $\mathcal{K}_2^2$ where the surface model of **n** is the $\{0, 1\}$ crisp logic. Thus all models in $\mathcal{K}_1^n$ can be regarded as models in $\mathcal{K}_2^{n+1}$ and so we conclude also that

$$\bigcup_n \mathcal{K}_1^n = \bigcup_n \mathcal{K}_\sharp^n.$$

Definition 12.14. Let $\mathcal{K}_\sharp^1$ be the class of all models of the form $(S, R^\sharp, a, h^\sharp)$ where $R^\sharp$ and $h^\sharp$ are fuzzy functions into some fuzzy model $(U, \leq, 0, \mathbf{s}, \varphi)$.

Let $\mathcal{K}_\sharp^{n+1}$ models have the form $(S, R^\sharp, a, h^\sharp, \mathbf{F}^\sharp)$ where $(S, R^\sharp, a, h^\sharp)$ is a $\mathcal{K}_\sharp^1$ surface model and $\mathbf{F}^\sharp(t)$, for $t \in S$ is in $\mathcal{K}_\sharp^n$.

Definition 12.15 (Simplified fuzzy modal models). Let $(U, \leq, 0, \mathbf{s}, \varphi)$ be a fixed fuzzy structure with conorm **s** and ϕ-operator φ. Let $\mathcal{K}$ be a family of modal Kripke structures of the form (S, R, a, h). A simplified uniform fuzzy modal model (**MZ**-model) has the form $(W, R^\sharp, a, h^\sharp, \mathbf{F}^\sharp)$ where W is a set of worlds. $\mathbf{F}^\sharp$ is a fibring function $\mathbf{F}^\sharp : W \mapsto W$, $R^\sharp : W \times W \mapsto U$ is a fuzzy relation and $h^\sharp$ is a fuzzy assignment, giving for each $t \in W$ and each atomic q the value $h^\sharp(t, q) \in U$.

Consider a language with $\Box, \mathbb{J}, \twoheadrightarrow$ and $\otimes$.

Evaluation in the structure is as follows

- $\mu_t(q) = h^\sharp(t, q)$, for q atomic
- $\mu_t(A \otimes B) = \mu_t(A)\mathbf{s}\mu_t(B)$
- $\mu_t(A \twoheadrightarrow B) = \mu_t(A)\varphi\mu_t(B)$
- $\mu_t(\Box A) = \text{Inf}\{u \mid \text{ for all } y, u \leq R^\sharp(t, y) \text{ implies } \mu_y(A) \leq u\}$
- $\mu_t(\mathbb{J}A) = \mu_{\mathbf{F}^\sharp(t)}(A)$
- Let A hold in the model iff $\mu_a(A) = 0$
- Let $A \mid\!\sim B$ hold in the model iff $\mu_a(A) \geq \mu_a(B)$.

The reader should note that the presence of $\mathbb{J}$ (i.e. the fibring function **F**) allows us a multitude of ways of making a formula fuzzy. Consider an atom q, and a $t \in W$. We can evaluate q at t in many ways:

$$\begin{array}{l} t \vDash_1 q \text{ iff } t \vDash q \\ t \vDash_2 q \text{ iff } t \vDash \mathbb{J}q \text{ iff } \mathbf{F}(t) \vDash q \\ \vdots \\ t \vDash \mathbb{J}_q^n \text{ iff } \mathbf{F}^n(t) \vDash q. \end{array}$$

Let A be a modal formula. We define the set of wffs *Fuzz* (A) in the modal language with $\mathbb{J}$ as follows:

$$\begin{aligned}&\mathit{Fuzz}\,(q) = \{q, \mathbb{J}q, \mathbb{J}^2 q, \ldots\}\\&\mathit{Fuzz}\,(A \wedge B) = \{X \wedge Y \mid X \in \mathit{Fuzz}\,(A), Y \in \mathit{Fuzz}\,(B)\}\\&\mathit{Fuzz}\,(\sim A) = \{\sim X \mid X \in \mathit{Fuzz}\,(A)\}\\&\mathit{Fuzz}\,(\Box A) = \{\mathbb{J}^n \Box X \mid X \in \mathit{Fuzz}\,(A), n = 0, 1, 2, \ldots\}.\end{aligned}$$

For any wff A and any $X \in$ *Fuzz* (A), we can see X as a way of making A fuzzy.

Let $\Box_n = \mathbb{J}^n \Box$. The set $\{\Box_n\}$ is a family of modalities. For any wff A containing $\Box$, we have the option of reading each box as $\Box_n$ for some n. All of these options give rise to the various wffs in *Fuzz*(A). Let μ^* be a function giving for each natural number n a fuzzy value $\mu^*(n) \in [0, 1]$. We can use μ^* to 'choose' which $\Box_n$ the connective $\Box$ is to be.

We can let

- $\mu^*_t(\Box A) = \text{Sup}_n\{\mu^*(n) \cdot \mu_t(\Box A)\}$.

The above consideration amounts to making the syntax itself (namely '$\Box$') fuzzy.

12.4 Making a Fuzzy Logic Even More Fuzzy

This section deals with the question of how to make a fuzzy logic even more fuzzy—by fibring it with yet another fuzzy logic.

We begin with an illustrative example.

Example 12.16. Let $\mathbf{L}_1$ be a Łukasiewicz many-valued logic and let $\mathbb{L}_1$ be its language. The values are from $U = [0, 1]$ with $0 = \top, 1 = \bot$ and the conorm and ϕ-operators are

- $x \circ y = \min(1, x + y)$
- $x \twoheadrightarrow y = \max(0, y - x)$
- $x \wedge y = \max(x, y)$.

A model is an assignment μ_1 giving each atomic q a value $\mu_1(q)$. μ_1 can be extended to all wffs using the tables above. As we saw in Lemma 12.10, this can be turned into a possible world model $(U, \leq, 0, h, t)$ by defining a crisp assignment h and satisfaction as follows:

- $h_t(q) = \top$ iff $\mu_1(q) \leq t$, for q atomic.
- $t \vDash A \wedge B$ iff $t \vDash A$ and $t \vDash B$.
- $t \vDash A \twoheadrightarrow B$ iff for all $y \vDash A$ we have $t o y \vDash B$.

By Lemma 12.10 we have for all wffs A

$$t \vDash A \text{ iff } \mu_1(A) \leq t.$$

Let us now take another logic $\mathbf{L}_2$, in the language $\mathbb{L}_2$ based on $[0, 1]$, with co-norm s and ϕ-operator φ, and fibre this logic to $\mathbf{L}_1$, to form $\mathbf{L}_{(1,2)} = \mathbf{L}_1[\mathbf{L}_2]$.

We want to study what we get. The models of $\mathbf{L}_{(1,2)}$ have the form $(U, \leq, 0, h_t, \mu^2_t)$ where $(U, \leq, 0, h_t)$ is as before and $\mu^2_t(q) \in [0, 1]$, for atomic q. Following general fibring principles, we require that the fibred models respect $\leq$, i.e.

- $t \leq s \to \mu_t^2(q) \geq \mu_s^2(q)$.

μ_t^2 is defined for any wff α of $\mathbb{L}_2$. It is not defined on wffs of $\mathbb{L}_1$. Our objective is to extend μ_t^2 to the language $\mathbb{L}_1$ in some natural way.

We observe that in the model $(U, \leq, 0, h_t)$ the assignment h_t is crisp. We fibre μ_t^2 to the model in order to make it fuzzy. We get the model $(U, \leq, 0, \mu_t^2)$.

Thus μ_t^2 replaces h_t, i.e. the crisp $t \vDash q$ is replaced by $\mu_t(q)$. The question is how do we evaluate $\mu_t^2(A \twoheadrightarrow B)$, for the $\twoheadrightarrow$ of $\mathbb{L}_1$?

We know that $t \vDash A \twoheadrightarrow B$ is evaluated as

$$\forall y(y \vDash A \text{ implies } t \circ y \vDash B).$$

If we assume by induction that $x \vDash A$ is now replaced by $\mu_x^2(A)$, then we want to evaluate $\forall y[\mu_y^2(A) \text{ implies } \mu_{toy}^2(B)]$.

At this juncture it is up to us to decide how to extend μ_t^2. We can choose the most general view and take a completely new conorm $*$ with its ϕ-operator φ_* and let

$$\mu_t^{*2}(A \twoheadrightarrow B) = \text{Sup}_y^*\{\mu_y^{*2}(A)\varphi_*\mu_{toy}^{*2}(B)\}$$

where Sup* is defined on *sets* of elements from U by:

$$\text{Sup}^*\{x_1, \ldots, x_n\} = x_1 * \ldots * x_n$$

(recall that $*$ is commutative and associative) and

$$\text{Sup}^* X = \text{Sup}\{\text{Sup}^* Y \mid Y \subseteq X, Y \text{ is finite}\}.$$

Thus 'implies' is evaluated by 'φ_*' and '$\forall y$' by 'Sup$_y^*$'.[40]

It is more sensible to take the conorm o of the logic $\mathbf{L}_1$ itself and its ϕ operator $\twoheadrightarrow$ and extend μ_t^2 using them. We get

$$\begin{aligned}\mu_t^2(A \twoheadrightarrow B) &= \forall y(\mu_y(A) \twoheadrightarrow \mu_{toy}(B)) \\ &= \text{Sup}_y^\circ \max(0, \mu_{toy}(B) - \mu_y(A))\end{aligned}$$

Note that since

$$t \leq s \to \mu_t^2(q) \geq \mu_s^2(q)$$

we also get by structural induction

$$t \leq s \to \mu_t^2(A) \geq \mu_s^2(A)$$

for all wffs of $\mathbb{L}_1$.

Now let us assume that all the μ_t are equal to some fixed μ_2.

[40]To be precise, the logic $\mathbf{L}_1$ may already have a way of evaluating the universal quantifier '$\forall y$'. We should in such a case use that same definition. Our proposal here corresponds to the Sup$_y^*$ evaluation for '$\forall y$'. Another possibility is just to take Sup$_y$, regarding '$\forall$' as a big conjunction '$\bigwedge$'.

Then we get

$$\mu_t^2(A \twoheadrightarrow B) = \mu_2(A) \twoheadrightarrow \mu_2(B) = \mu_2(A \twoheadrightarrow B)$$

i.e. we get the fuzzy model $(U, \leq, o, \mu_2)$. If we choose $\mu_2 = \mu_1$, we get the original model.

Thus we can take our basic fibred fuzzy concept of model to be $(U, \leq, \rightarrow, \mu_t)$, where μ_t is a family of assignments, with the inductive evaluation being

- $\mu_t(A \twoheadrightarrow B) = \text{Sup}_y^o\{\mu_y(A) \twoheadrightarrow \mu_{toy}(B))$.

For the case of $\mu_t = \mu$ for all t, this gives us the old unfibred definition of a model.

The reader should note that Sup_y^* operates on sets and not on sequences. Thus

$$\text{Sup}_y^*\{f(y_1), f(y_2)\}$$

will be $f(y_1)$ if $f(y_1) = f(y_2)$ and not $f(y_1) * f(y_2)$. It might be simpler to use our ordinary Sup instead of Sup^*.

The notion of a self-fibred model was obtained from trying to make a fuzzy logic even fuzzier. We took an ordinary $(U, \leq, o, \mu_1)$. We looked at it as a possible world model $(U, \leq, o, h_t)$, where h_t is crisp and we have

$$h_t(q) = \top \text{ iff } \mu(q) \leq t.$$

We then replaced the crisp h_t by a fuzzy μ_t^2. This gave us a new notion of a self-fibred model $(U, \leq, o, \mu_t)$. μ_t itself is a standard model (indexed by t). If we iterate the process, we would have to replace the models μ_t^2 by the new concept of self-fibred models. This means replacing μ_t^2 by a family $\mu_{t,s}^2, t, s \in U$.

Thus the iterated concept gives us the notion of level n self-fibred fuzzy model as $(U, \leq, o, \mu_{t_1,\ldots,t_n}), (t_1, \ldots, t_n) \in U^n$.

The value of $\mu_{t_1,\ldots,t_n}(A \twoheadrightarrow B)$ can be computed in n different ways, depending on which coordinate we use:

$$\mu_{t_1,\ldots,t_n}^i(A \twoheadrightarrow B) = \text{Sup}_y^\circ\{\mu_{t_1,\ldots,y,\ldots,t_n}^i(A)\varphi_\circ\mu_{t_1,\ldots,t_i oy,\ldots,t_n}(B)\}$$

i.e. we use coordinate i.

We are thus led to the following definition.

Definition 12.17 (Self-fibred fuzzy logic). A fibred fuzzy logic has the form $(U, \leq, o, \mu_t, \mathbf{s}, \varphi_\mathbf{s})$ where $(U, \leq, o)$ is a set of values (e.g. [0, 1]), $\mathbf{s}$ and $\varphi_\mathbf{s}$ are a conorm and its ϕ-operator and μ_t, for $t \in U$, is a fuzzy assignment giving each atom q a value $\mu_t(q) \in U$.

We require

$$t \leq s \rightarrow \mu_t(q) \geq \mu_s(q).$$

We extend μ_t to all wffs by

$$\mu_t(A \twoheadrightarrow B) = \text{Sup}_y^\mathbf{s}\{\mu_y(A)\varphi_1\mu_{t\mathbf{s}y}(B)\}.$$

A self-fibred model of the level n has the form $(U, \leq, o, \mu_{t_1,\ldots,t_n})$, where $\mu_{t_1,\ldots,t_n}$ is monotonic down in $t_1, \ldots, t_n$ and where evaluation of $A \twoheadrightarrow B$ can be done in n different ways, with

$$\mu^i_{t_1,\ldots,t_n}(A \twoheadrightarrow B) = \mathrm{Sup}^{\mathbf{s}}_y(\mu^i_{t_1,\ldots,y,\ldots,t_n}(A)\varphi_{\mathbf{s}}\mu^i_{t_1,\ldots,t_i\mathbf{s}y,\ldots,t_n}(B)).$$

We can add *mode* connectives $\mathbb{J}_r$ $r = 1, \ldots, n$, and let

$$\mu^i_{t_1,\ldots,t_n}(\mathbb{J}_r A) = \mu^r_{t_1,\ldots,t_n}.$$

We start with an implicational logic in a language $\mathbf{L}_1$ with the connectives $\otimes$ and $\twoheadrightarrow$ and with possible world semantics of the form $(S, \circ, \leq, a, h_t)$ where

- $h_t(q) = \top$ and $t \leq s \to h_s(q) = \top$
- $t \vDash A \otimes B$ iff for some $x, y, x \vDash A$ and $y \vDash B$ and $x \circ y \leq t$
- $t \vDash A \twoheadrightarrow B$ iff for all $y, y \vDash A$ implies $t \circ y \vDash B$.

We make no assumptions of the binary operation '$\circ$' except possibly associativity and monotonicity in the first coordinate:

- $t \leq s \to t \circ y \leq s \circ y$.

Lemma 12.18.

$$t \vDash A \wedge t \leq s \to s \vDash A.$$

Proof By induction on A. The atomic case and the case of $A \circ B$ are clear (since $\leq$ is transitive).

Assume $\forall y(y \vDash A \to t \circ y \vDash B)$. Assume $t \leq s$. We want to show $\forall y(y \vDash A \to s \circ y \vDash B)$. This follows from the monotonicity of $x \circ y$ in the variable x.

Let $\mathbf{L}_2$ be another fuzzy logic, say a many-valued logic with values in $[0, 1]$, with conorm $\mathbf{s}$ and ϕ-operator φ.

We consider $\mathbf{L}_{(1,2)} = \mathbf{L}_1[\mathbf{L}_2]$. This is a fibred system and so its models have the form $(S, \leq, \circ, a, h_t, \mathbf{F})$, where for each $t \in S, \mathbf{F}(t)$ is an $\mathbf{L}_2$ model, say $h^\sharp_t(q)$ giving values to the atoms q in $[0, 1]$. We can write the model as $(S, \leq, \circ, a, h_t, h^\sharp_t)$.

According to the principles of fibring, $h^\sharp_t$ must respect the $\leq$ order, namely (remember $0 = \top, 1 = \bot$) the following holds.

- $t \leq s \to h^\sharp_t(q) \geq h^\sharp_s(q)$.

We would like to extend $h^\sharp_t$ to the language of $\mathbf{L}_1$. The truth definition for $\mathbf{L}_1$ is $t \vDash A \twoheadrightarrow B$ iff for all y $(y \vDash A \to t \circ y \vDash B)$.

If A, B are atomic then $y \vDash A$ means $h^\sharp_y(A) = 0$. So the condition becomes

- $t \vDash_1 A \twoheadrightarrow B$ iff for all y, $(h^\sharp_y(A) = 0 \to h^\sharp_{t \circ y}(B) = 0)$.

$\vDash_1$ is not satisfactory. It does not take full account of the range of values of $h^\sharp_t$. It is up to us to choose a better definition. We can try

- $t \vDash_w A \twoheadrightarrow B$ iff for all y, $(h^\sharp_y(A) \geq h^\sharp_{t \circ y}(B))$

or, following the previous example, we can let

- $h^\sharp(t, A \twoheadrightarrow B) = \mathrm{Sup}_y(\max(0, h^\sharp(toy, B) - h^\sharp(y, A))$.

If we let $\rightarrow_2$ denote Łukasiewicz implication and let $\wedge$ denote Łukasiewicz conjunction then

- $h^\sharp(t, A \twoheadrightarrow B) = \bigwedge_y(h^\sharp(y, A) \rightarrow_2 h^\sharp(toy, B))$. □

12.5 Comparison with the Literature

Let us now compare our method with existing fuzzy systems from the literature.

Example 12.19 (System of Gabbay, 1968). It is easy to turn any logic into a many-valued system. This example recalls how Gabbay, in a letter to A. N. Prior, defined an intuitionistic many-valued logic. The Kripke structures for any logic (which allows for Kripke semantics) have the form $(S, R_1, R_2, \ldots, R_n, h)$ where R_i are various relations and h is an assignment. If we consider h as a function $h(t, q) \in \{0, 1\}$ for $t \in S$ and q atomic, it is easy to turn the semantics into many-valued semantics. We make h many valued, e.g. have $h(t, q) \in [0, 1]$, the unit interval. 0 is $\top$, 1 is $\bot$. When this is done for intuitionistic logic we get the following structures, $(S, \leq, h)$, where $(S, \leq)$ is a partially ordered set and h is a function such that

- $h(t, q) \in [0, 1]$
- $t \leq s \rightarrow h(t, q) \geq h(s, q)$.

h can be extended to all wffs as follows:

- $h(t, A \vee B) = \min\{h(t, A, h(t, B)\}$
- $h(t, A \wedge B) = \max\{h(t, A), h(t, B)\}$
- $h(t, A \Rightarrow B) = \{\max(0, h(s, B) - h(s, A))\}$
- $h(t, \neg A) = \mathrm{Sup}_{s \geq t}\{1 - h(s, A)\}$
- $A \vdash B$ iff for all structures and all $t, h(t, A) \geq h(t, B)$.

No axiom system was offered.

Example 12.20 (The systems of Fitting). [41]

Melvin Fitting [1991] observed that although there were many proposals for many-valued modal logics in the literature, they all

> 'retain the general notion of possible world semantics, while allowing formulas to have values in many-valued space. What seems not to have been considered is allowing the accessibility relation between possible worlds itself to be many valued.'

Fitting presents such a system. He takes a space of truth values which is a finite lattice $\mathcal{T}$, with ordering $\leq$. The language fitting uses contains propositional constants naming elements of $\mathcal{T}$. The connectives are $\wedge, \vee$ and $\rightarrow$. A valuation $\nu(A)$ gives lattice values to atoms and the constants (constants get what they name) and ν is extended naturally to all wffs. $\nu(A \wedge B) = \nu(A) \wedge \nu(B), \nu(A \vee B) = \nu(A) \vee \nu(B)$.

For the case of $\nu(A \rightarrow B)$, we have several options. We must make sure that

- $\nu(A \rightarrow B) = 1$ iff $\nu(A) \leq \nu(B)$

[41] Note that Fitting takes the value 1 as $\top$ and 0 as $\bot$. To compare with our systems, exchange '$\leq$' and '$\geq$' and 'max' and 'min'.

but the exact definition of $\nu(A \to B)$ is still optional. Since $\mathcal{T}$ is finite, it having the property of being a pseudo-complemented lattice is equivalent to its being distributive, and so we can define

$$\begin{aligned}\nu(A \to B) = & \max\{x \mid \nu(A) \wedge x \leq \nu(B)\} \\ & = \nu(A) \twoheadrightarrow \nu(B).\end{aligned}$$

Fitting gives a Gentzen system for the above and uses the above to define two modal many-valued logics.

Add $\Box$ to the language and let (s, R, ν) be a Kripke structure with ν a function $\nu(t, q) \in \mathcal{T}$.

ν can be extended to all wffs pointwise as before with

- $\nu(t, \Box A) = \bigwedge\{\nu(s, A) \mid tRs\}$
- $\nu(t, \Diamond A) = \bigvee\{\nu(s, A) \mid tRs\}$.

Fitting axiomatises this system.

Note that we have names in the language for all elements of $\mathcal{T}$.

In his second version Fitting allows R to be many-valued, thus for each $t, s \in S$, $R(t, s)$ is a value in $\mathcal{T}$.

ν is extended to $\Box$ as follows:

$$\nu(t, \Box A) = \bigwedge_s \{R(t, s) \twoheadrightarrow \nu(s, A)\}.$$

Fitting gives a complete axiom system for this semantics.

In a continuation paper, [Fitting, 1994], Fitting investigates the case where the many-valued lattice $\mathcal{T}$ is a Heyting algebra.

Remark 12.21 (Discussion of Fitting and Gabbay). The previous examples are clear-cut cases of 'fuzzling'. The first option is to fibre some many-valued logic to modal logic and make the assignment fuzzy. Fitting makes his modal logic fuzzy by using the lattice $\mathcal{T}$ as values. The accessibility relation remains 'crisp' two-valued.

On the other hand, if we take many-valued logic and fibre modal logic to it (giving values in modal logic), we get the system where R the accessibility relation becomes fuzzy. This is what Fitting does and this is what we illustrated in Section 2.

If we do both steps, we get a logic where both the accessibility relation and the assignment to worlds are fuzzy.

The important point is that these procedures are examples of general fibring and can be done uniformly for any pair of logics.

Example 12.22 (The system of Finger–Gabbay). Finger and Gabbay [1992] were motivated from the application area of time varying systems. Given a system S described in a logic $\mathbf{L}_1$, its state may change in time. Thus we need a way to express changes in state of the system and one of the ways of doing so is to add to $\mathbf{L}_1$ a temporal component.

This is done by allowing atomic formulas of $\mathbf{L}_1$ also to be formulas of a logic $\mathbf{L}_2$, where $\mathbf{L}_2$ is a temporal logic. Thus the wffs of the new logic $\mathbf{L}_1[\mathbf{L}_2]$ are obtained by substituting in wffs of $\mathbf{L}_1$ formulas of $\mathbf{L}_2$ for the atoms.

This process is the same as giving 'new values' to the atoms of $\mathbf{L}_1$. The ideas of this paper were adopted by the logic and natural language community. There are ways of adding 'features' to 'types' to generate what are known as feature logics. These can be viewed as making (fuzzling) a type logic (say the Lambek calculus) into a 'many valued' logic where the 'values' are the features.

Example 12.23 (The system of H. Thiele). H. Thiele proposes in his paper [Thiele, 1993] to change the assignment h into a fuzzy one as well as the accessibility relation R. He considers all three options, making h fuzzy and R crisp, R fuzzy and h crisp and both R and h fuzzy.

We already discussed in Section 12.1 how to get the Thiele definition from a parametrised family of modal models. We saw in Section 12.2 that such a family can be obtained from fibring with Łukasiewicz logic.

12.6 Case Study: Fuzzy Automata

Let us apply our fibring ideas and see what notion of a fuzzy automaton we can get. The reader should bear in mind that the definition of a fuzzy automaton should arise automatically from the idea of fibring. We are not free to define it at will by making fuzzy any component we choose in the traditional definition of an automaton.

Definition 12.24 (Traditional non-deterministic automaton).

1. A non-deterministic automaton has the form $\mathfrak{A} = (S, \Sigma, M, s_0, Q)$ where S is a (finite) set of states, Σ a finite set of atomic alphabet letters, $s_0 \in S$ is the initial state and $Q \subseteq S$ is a set of the accepting states.
   M is a function giving for each $\sigma \in \Sigma$ and $s \in S$ a (possibly empty) set $M(\sigma, s) \subseteq S$, being the set of all states accessible to s when encountering the input σ.
2. A string of words $(\sigma_1, \ldots, \sigma_n)$ is said to be accepted by the automaton, if there exists a sequence of states $(s_1, \ldots, s_n)$ such that
   - $s_1 \in M(\sigma_1, s_0)$
   - $s_i \in M(\sigma_i, s_{i-1}), 1 \leq i \leq n$
   - $s_n \in Q$.

We now try to find a definition of a fuzzy automaton arising directly from the idea of fibring. The next example will establish a connection between automata and modal logic.

Example 12.25. Let $\mathfrak{A}$ be an automaton. Let $tR_\sigma s$, for $\sigma \in \Sigma$ and $t, s \in S$ be defined as $s \in M(\sigma, t)$. R_σ can be viewed as a modal accessibility relation for the modality $\Diamond_\sigma$. Let q be a propositional constant and consider the modal logic with $\{\Diamond_\sigma \mid \sigma \in \Sigma\}$ and a single atomic proposition q. Consider the special model $\mathbf{m} = (S, R_\sigma, s_0, h)$ where $h(s, q) = 1$ iff $s \in Q$.

We have the following

(*) $(\sigma_1, \ldots, \sigma_n)$ is accepted by the automaton $\mathfrak{A}$ iff $\mathbf{m}_\mathfrak{A} \vDash \Diamond_{\sigma_1} \ldots \Diamond_{\sigma_n} q$.

Having established the connection between the automaton and modal logic, we now know how to make a fuzzy automaton. We make h fuzzy and each R_σ fuzzy. This corresponds to the two ways of fibring, as discussed in previous sections. Making h fuzzy means making Q a fuzzy function: $Q^\sharp : S \mapsto [0, 1]$. Making $R_\sigma(x, y)$ fuzzy means making M into a fuzzy function $M^\sharp(\sigma, s, t) : \Sigma \times S \times S \mapsto [0, 1]$.

Thus we can define a fuzzy evaluation as discussed in Section 12.1:

- $\mu_t(q) = Q^\sharp(t)$
- $\mu_x(\Diamond_\sigma A) = \text{Inf}\{y \mid \max \mu_y(A), M^\sharp(\sigma, x, y))\}$.

Thus a fuzzy automaton can accept a string $(\sigma_1, \ldots, \sigma_n)$ with a fuzzy value $\mu_{s_0}(\Diamond_{\sigma_1}, \ldots, \Diamond_{\sigma_n} q)$.

The above definition corresponds to what was proposed in the literature, see [Santos, 1968].

Definition 12.26 (Fuzzy automaton).

1. A fuzzy automaton has the form $\mathfrak{A}^\sharp = (S, \Sigma, M^\sharp, s_0, q^\sharp)$ where S is the set of states, Σ the alphabet, $s_0 \in S$ is the initial state and $Q^\sharp, M^\sharp$ are functions as follows: $Q^\sharp : S \mapsto [0, 1]$ and $M^\sharp : \Sigma \times S \times S \mapsto [0, 1]$.
2. We define the notion of $\mathfrak{A}(\sigma, s) = x$ for a state $s, \sigma = (\sigma_1, \ldots, \sigma_n)$ and $x \in [0, 1]$ (reading the automaton $\mathfrak{A}$ accepts σ when at s, with value x) by induction on σ:
   - $\mathfrak{A}(\varnothing, s) = Q^\sharp(s)$, for the empty word $\varnothing$
   - $\mathfrak{A}(\sigma * (\sigma_1), s) = \text{Inf}\{t \mid \max(\mathfrak{A}(\sigma, t), M^\sharp(\sigma_1, s, t))\}$.
   - Let $\mathfrak{A}(\sigma)$ be $\mathfrak{A}(\sigma, s_0)$.

13

COMBINING TEMPORAL LOGIC SYSTEMS

13.1 Introduction

Temporal logic is in itself a fibred system. It is a bimodal system with two **K4** modalities $\Box_1$ and $\Box_2$ and with the interaction axiom xR_1y iff yR_2x. This gives rise to one-dimensional temporal logic. It is also very useful in applications involving time to consider two-dimensional systems, i.e. systems with evaluation in two indices. We can read $t, s \vDash A$ to mean 'from the point of view of context t, A holds in s'.

In this chapter we are interested in describing systems in which two distinct temporal 'points of view' coexist. Descriptions of temporal systems under a single point of view, i.e. one-dimensional temporal systems, abound in the literature. These one-dimensional temporal logics differ from each other in several ways. They differ in the form the logic is presented, whether a proof theoretical, model theoretical or algebraic presentation. They differ in the ontology of time adopted, whether time is to be represented as a set of points, intervals or events. They can also differ in the properties assigned to flows of time, whether linear or branching time, discrete or dense, continuous or allowing for gaps. In this chapter we consider both proof- and model-theoretical presentations of temporal logics on a point-based ontology. Most of the results presented assume that the flow of time is linear.

The systems considered in this chapter arise from the study of applications of two-dimensional temporal logics [Finger, 1992]. Finger was aware of Venema's negative results concerning the unaxiomatisability of two-dimensional temporal logics over the upper semi-plane of $\mathbb{N}\times\mathbb{N}$, $\mathbb{Z}\times\mathbb{Z}$ and $\mathbb{R}\times\mathbb{R}$ [Venema, 1990] (see also Proposition 13.17 below), and he used a weaker language which was sufficient for the applications. It then became clear that this weaker two-dimensional language could be generalised and a family of languages resulting from adding a (second) temporal dimension *externally* to a temporal logic system was thus obtained. This process was formalised in [Finger and Gabbay, 1992], where it was called *temporalisation*, and several results were obtained concerning the transfer of logical properties from the component logical system to the combined one. As a result, a family of temporalised logic systems was obtained, the properties of which can be derived from the properties of the component logic systems via the transfer results.

The purpose of this chapter is to study the general ways in which two temporal logic systems can be combined, thus generating several families of combined two-dimensional temporal systems. Different combination methods may be presented by the distinct interactions between related parts of the two logic systems involved, leading to two-dimensional systems based on different languages with different semantical structure, expressive power and other properties (that may be transferred or not from the component systems).

We begin with several examples in which two distinct temporal dimensions (or temporal 'points of view') can coexist, motivating several methods for combining two temporal logics. We will also relate these methods to earlier chapters of this book, and to [Gabbay, 1995b; Gabbay, 1996b].

First case: external time

One temporal point of view can be *external* to the other. The external point of view is seen as describing the temporal evolution of a system $\mathcal{S}$, when system $\mathcal{S}$ is itself a temporal description. Suppose $\mathcal{S}$ is described using a temporal logic $\mathbf{T}$ and suppose that the external point of view is given in a possibly distinct logic $\bar{\mathbf{T}}$. For example, consider an agent A, whose temporal beliefs are expressed in logic $\bar{\mathbf{T}}$, that we want to allow to reason about the temporal beliefs of an agent B, which are expressed in a possibly distinct logic $\mathbf{T}$. This is illustrated in Figure 13.1.

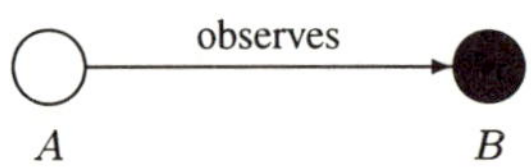

Fig. 13.1 One agent externally observing the other

Agent A's beliefs are external to agents B's beliefs, so that $\bar{\mathbf{T}}$ is externally describing the evolution of $\mathbf{T}$. The external temporal point of view $\bar{\mathbf{T}}$ is then applied to the internal system $\mathbf{T}$, in a process called *temporalisation* or *adding a temporal dimension to a logic system*, defined in [Finger and Gabbay, 1992]. The resulting combined logic system $\bar{\mathbf{T}}(\mathbf{T})$ is illustrated in Figure 13.2.

The temporalisation associates every time point in $\bar{\mathbf{T}}$ with a temporal description in $\mathbf{T}$, where those $\mathbf{T}$-descriptions need not be all identical. Given the logical properties of $\mathbf{T}$ and $\bar{\mathbf{T}}$, what can be said about the logical properties of $\bar{\mathbf{T}}[\mathbf{T}]$?

In terms of a generic combination of logics, the temporalisation method can be matched with a process called *'fuzzling'* or *layering*, which is characterised by the fact that the formulae of system $\mathbf{T}$ can be substituted for the atoms of system $\bar{\mathbf{T}}$. In (Kripke-) semantical terms, this means that every possible world of $\bar{\mathbf{T}}$ is associated to a whole model of $\mathbf{T}$; as we saw in earlier chapters.

Second case: independent agents

Suppose now that agent A has the ability of referring to agent B's temporal beliefs and vice versa. The agents are therefore observing each other, as illustrated in figure 13.3.

The agents' beliefs are then capable of interacting with each other through several levels of cross-reference, as in the sentence 'A believes that B believes that A believes that ...'. A new combination method for $\mathbf{T}$ and $\bar{\mathbf{T}}$ is needed in order to represent such a sentence as a formula; this is called the *independent combination*, $\bar{\mathbf{T}} \oplus \mathbf{T}$. Since a formula of $\bar{\mathbf{T}} \oplus \mathbf{T}$ has a finite nature, it can be unravelled in a finite number of alternating temporalisation as illustrated in Figure 13.4.

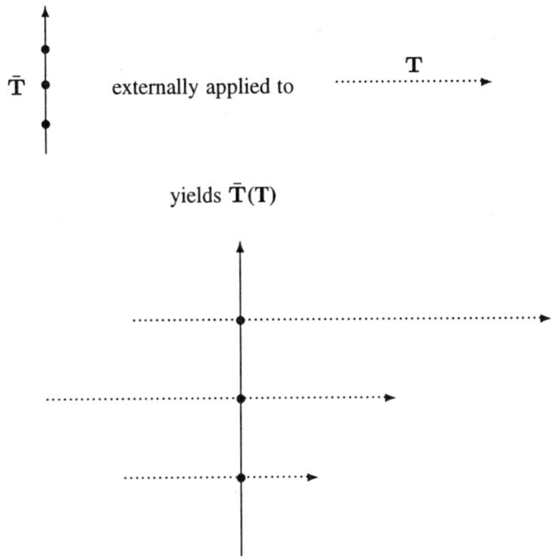

Fig. 13.2 The combined flow of time resulting from temporalisation

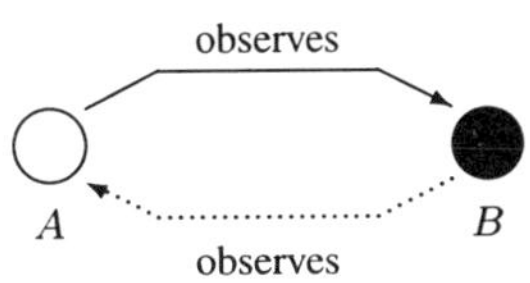

Fig. 13.3 Independent interaction of agents

Figure 13.4 suggests a way of analysing the properties of the independent combination method using the temporalisation method as an intermediary step. It will turn out that the independent combination method is the (infinite) union of all finite alternated temporalisations. An illustration of a possible independently combined flow of time is presented in Figure 13.5.

In terms of a generic combination of two logics, this process can be matched to the *dovetailing* process of Chapters 3 and 4, whereby atoms of $\mathbf{T}$ can be substituted by formulae of $\bar{\mathbf{T}}$ and vice versa. The semantical counterpart is obtained, by providing each possible world with two distinct accessibility relations, $<$ and $\overline{<}$, so that from every possible world it is possible to reach another possible world either via $<$ or via $\overline{<}$.

Third case: two-dimensional plane

Yet another distinct situation can be found where we have the coexistence of two distinct temporal 'points of view'. This time a single agent with temporal reasoning capabilites is considered, and we want to be able to describe the evolution of his own beliefs. This is perhaps better illustrated by considering the agent as a temporal database where each

Fig. 13.4 Unravelling the independent combination

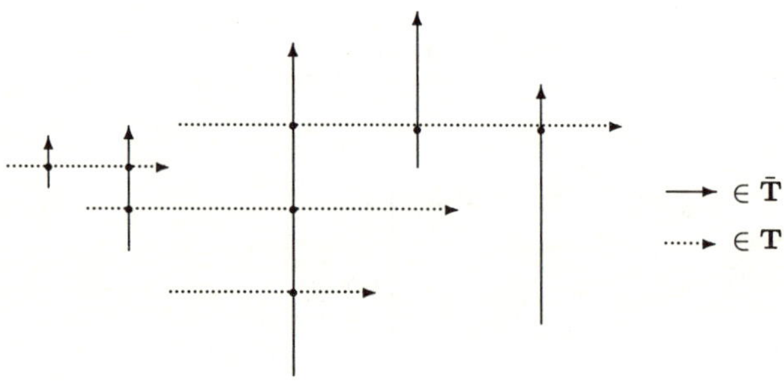

Fig. 13.5 Independently combined flow of time

piece of information is associated to a validity time (or interval). For example, consider the traditional database relation $employee(Name, Salary, Manager)$. Suppose the following is in the database at March 94

Name	*Salary*	*Dept*	*Start*	*End*
Peter	1000	R&D	Apr 93	Mar 94

where the attributes start and end represent the end-points of the validity interval associated with the information. We assume that Peter's salary has not changed since Apr 93. Suppose in Apr 94 Peter receives a retroactive promotion dating back to the beginning of the year, increasing his salary to 2000. The whole database evolution is illustrated at Figure 13.6, where only the value of Peter's salary is indicated at each point.

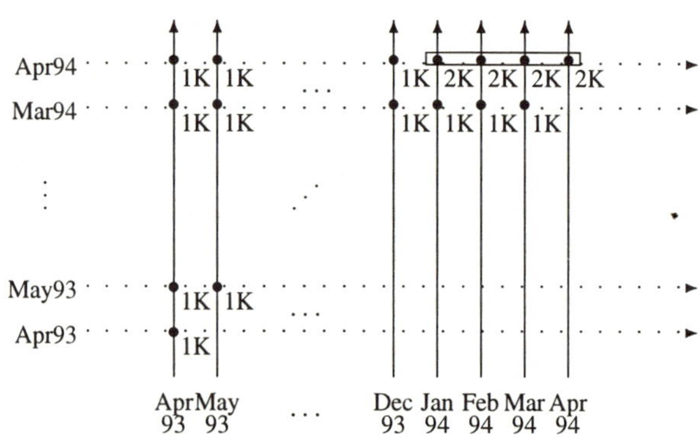

Fig. 13.6 Two-dimensional temporal database evolution

If $\mathbf{T}$ represents valid-time and $\bar{\mathbf{T}}$ represents transaction-time, we have guaranteed a two-dimensional plane $\bar{\mathbf{T}} \times \mathbf{T}$ in order to represent the database evolution.

Another application of the two-dimensional plane (or its NW-semi-plane) is in the representation of intervals on a line, as presented in [Venema, 1990].

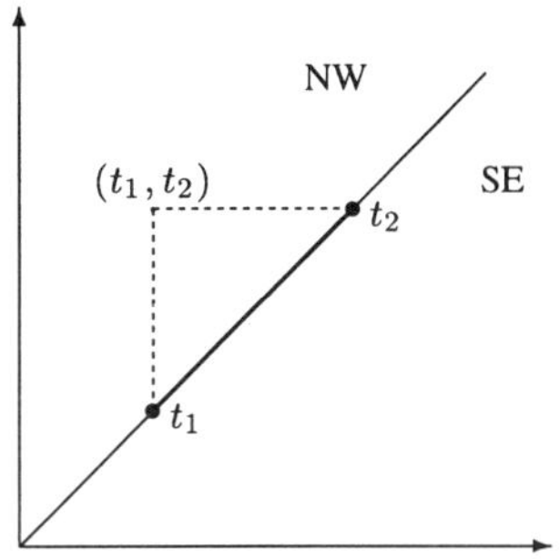

Fig. 13.7 Two-dimensional representation of intervals

In Figure 13.7 we can see a line considered as the diagonal of a two-dimensional plane and that an interval $[t_1, t_2]$ on that line is represented by the point (t_1, t_2) on the NW-semi-plane.

The combination between two temporal systems that generates a combined flow of time that is isomorphic to a two-dimensional plane is called the *join* of two logical systems. We have adopted the term *join of logics* here (instead of the previously used *interlacing of logics* in [Finger, 1994]) to be in accordance with the concept as defined in the larger context of generic combination of logics [Gabbay, 1995b]. Although the language generated in this process is the same as that of the independent combination (for the case of two temporal logics), the semantic interaction between $\mathbf{T}$ and $\bar{\mathbf{T}}$ is a lot stronger; this is due to the fact that the temporal operators of the two logics are commutative in the join. As will be seen in Section 13.7, it is necessary to restrict this interaction to obtain transfer of logical properties. The restriction will be applied to the type of operators allowed in one of the logics involved in the restricted join.

Aims of this chapter

In this chapter we study those three situations of coexistence of 'two temporal points of view' as the result of a combination of two linear, one-dimensional temporal logics. We explore several methods for systematically combining two temporal logics, $\mathbf{T}$ and $\bar{\mathbf{T}}$, thus generating for each method a new family of *two- dimensional temporal logics*.

A great number of (one-dimensional) temporal logics exist in the literature to deal with the great variety of properties one may wish to express about flows of time. When building two-dimensional temporal logics, the combination of two classes of flows of time generates an even greater number of possible systems to be studied. Furthermore, as we will see, there are several distinct classes of temporal logics, that may be considered two-dimensional, each generated by a distinct combination method. It is, therefore, desirable to study if it is possible to *transfer the properties* of long known and studied (one-dimensional) temporal logic systems to the two-dimensional case.

So the main goal of this chapter is to study, for each combination method, the transfer of logical properties from component one-dimensional temporal systems to a combined two-dimensional one.

We concentrate on the transfer of three basic properties of logic systems, namely soundness, completeness and decidability. This by no means implies that those are the only properties whose transfer deserve to be studied, but, as has already been noted in [Finger and Gabbay, 1992] for the temporal case, and in [Kracht and Wolter, 1991; Fine and Schurz, 1991] for the monomodal case, the transfer of completeness serves as a basis for the transfer of several other properties of logical systems.

As for the methods for combining two temporal logics, we consider the following:

(a) the temporalisation method, i.e. the external application of a temporal logic to another temporal system, also known as adding a temporal dimension to a logic system;

(b) the independent combination of two temporal systems;

(c) the full join of two temporal systems, where flows of time are considered over a two-dimensional plane;

(d) the restricted join of two temporal systems, a combination method that restricts the previous one but generates nice transfer results.

We proceed as follows. Section 13.2 presents the basic notions of one- and two-dimensional temporal logics. Section 13.3 discusses combinations of logics in general terms, so that in the rest of the chapter we can present special cases of combination methods. Section 13.4 briefly examines the transfer results obtained for the temporalisation method in [Finger and Gabbay, 1992]. Section 13.5 studies the method of independent combination. Section 13.6 deals with the full join method and Section 13.7 with its restricted version. Section 13.8 analyses the properties of a two-dimensional diagonal on the model generated by the full and restricted join methods. In Section 13.9 we discuss the results of this work.

13.2 Preliminaries

For the purposes of this chapter, a logic system is composed of three elements:

(a) a language, normally given by a set of formation rules generating well formed formulae over a signature and a set of logical connectives.

(b) An inference system, i.e. a relation, $\vdash$, between sets of formulae, normally represented by upper case Greek letters $\Delta, \Gamma, \Sigma, \Psi, \Phi$ and a single formula, normally represented by upper case letters $A, B, C, ...$; the fact that A is inferred from a set Δ is indicated by $\Delta \vdash A$. When Δ is a singleton, $\Delta = \{B\}$, the notation is abused and we write $B \vdash A$.

(c) The semantics of formulae over a class $\mathcal{K}$ of model structures. The fact that a formula A is true of or holds at a model $\mathbf{m} \in \mathcal{K}$ is indicated by $\mathbf{m} \models A$.

In providing a method for combining two logics into a third one, it will be necessary to provide three submethods that combine the languages, inference systems and semantics of the component logic systems. The component systems considered in this chapter will be one-dimensional linear **US**-temporal logics. Their language is built from a countable signature of propositional letters $\mathcal{P} = \{p_1, p_2, \ldots\}$, the Boolean connectives $\wedge$ (conjunction) and $\neg$ (negation), the two-place temporal operation U (until) and S (since), possibly renamed, and the following formation rules:

- every propositional letter is a formula
- if A and B are formulae, so are $\neg A$ and $A \wedge B$
- if A and B are formulae, so are $U(A, B)$ (reads 'until A is time in the future, B will be time') and $S(A, B)$ (reads 'since A was time in the past, B has been time')
- nothing else is a formula.

The *mirror image* of a formula is another temporal formula obtained by swapping all occurrences of U and S, e.g. the mirror image of $U(A, S(B, C))$ is $S(A, U(B, C))$.

The other Boolean connectives $\vee$ (disjunction), $\rightarrow$ (material implication), $\leftrightarrow$ (material bi-implication) and the constants $\bot$ (false) and $\top$ (true) can be derived in the standard way. Similarly, the one-place temporal operators F ('sometime in the future'), P ('sometime in the past'), G ('always in the future') and H ('always in the past') can be defined in terms of U and S.

To provide the semantics of temporal formulae we have to consider a (one-dimensional) *flow of time*, $\mathcal{F} = (T, <)$, where T is a set of time points and $<$ is an order over T. A temporal valuation $h : T \rightarrow 2^{\mathcal{P}}$ associates every time point with a set of propositional letters, i.e. $h(t)$ is the set of propositions that are true at time t. (Equivalently, and perhaps more usually, a valuation could be defined as a function $h : \mathcal{P} \rightarrow 2^T$, associating every propositional letter to a set of time points in which it holds true; see [Burgess, 1984; Gabbay *et al.*, 1994].) A model structure $\mathbf{m} = (T, <, h)$ consists of a flow of time $(T, <)$ and a temporal assignment h and, for the purposes of combination of logics, we consider a 'current world' $t \in T$ as part of the model. $\mathbf{m}, t \models A$ reads 'A is true at t over model $\mathbf{m}$'. Classes of models are normally defined by restrictions over the order relation $<$ of the flow of time.

The semantics of temporal formulae is given by:

$\mathbf{m}, t \models p$	iff $p \in \mathcal{P}$ such that $p \in h(t)$.
$\mathbf{m}, t \models \neg A$	iff it is not the case that $\mathbf{m}, t \models A$.
$\mathbf{m}, t \models A \wedge B$	iff $\mathbf{m}, t \models A$ and $\mathbf{m}, t \models B$.
$\mathbf{m}, t \models S(A, B)$	iff there exists an $s \in T$ with $s < t$ and $\mathbf{m}, s \models A$ and for every $u \in T$, if $s < u < t$ then $\mathbf{m}, u \models B$.
$\mathbf{m}, t \models U(A, B)$	iff there exists an $s \in T$ with $t < s$ and $\mathbf{m}, s \models A$ and for every $u \in T$, if $t < u < s$ then $\mathbf{m}, u \models B$.

The following restriction will be applied throughout this presentation. Flows of time will always be considered to have the properties:

(a) irreflexivity: $\forall t \neg (t < t)$

(b) transitivity: $\forall t, s, u(t < s \wedge s < u \rightarrow t < u)$

(c) totality: $\forall t, s(t = s \vee t < s \vee s < t)$.

The class of all flows respecting the restrictions above is the class $\mathcal{K}_{lin}$ of linear flows of time. We also represent the class of all models based on linear flows as $\mathcal{K}_{lin}$. Further restrictions can be applied to the nature of flows of time so that several other linear subclasses can be formed, e.g. the classes of dense ($\mathcal{K}_{dense}$), discrete ($\mathcal{K}_{dis}$), $\mathbb{Z}$-like, $\mathbb{Q}$-like and $\mathbb{R}$-like flows of time. The linearity property allows for the definition of the 'at all times' temporal connective $\Box$

$$\Box A = A \wedge GA \wedge HA.$$

In case of discrete flows of time, the operator 'next time', $\bigcirc$, and 'previous time', $\bullet$ are also defined:

$$\bigcirc A = U(A, \bot)$$
$$\bullet A = S(A, \bot).$$

The inference systems will be considered to be finite axiomatisations, i.e. a pair $(\Sigma, \mathcal{I})$ where Σ is a finite set of formulae called *axioms* and $\mathcal{I}$ is a set of inference rules. Consider the Burgess–Xu axiomatisation for $\mathcal{K}_{lin}$ [Burgess, 1982; Xu, 1988] consisting of the following axioms:

A0 all classical tautologies

A1a $G(p \rightarrow q) \rightarrow (U(p, r) \rightarrow U(q, r))$

A2a $G(p \rightarrow q) \rightarrow (U(r, p) \rightarrow U(r, q))$

A3a $(p \wedge U(q, r)) \rightarrow U(q \wedge S(p, r), r)$

A4a $U(p, q) \rightarrow U(p, q \wedge U(p, q))$

A5a $U(q \wedge U(p, q), q) \rightarrow U(p, q)$

A6a $(U(p, q) \wedge U(r, s)) \rightarrow$
$$(U(p \wedge r, q \wedge s) \vee U(p \wedge s, q \wedge s) \vee U(q \wedge r, q \wedge s))$$

plus their mirror images (**b** axioms). The inference rules are:

Subst Uniform substitution, i.e. let $A(q)$ be an axiom containing the propositional letter q and let B be any formula. Then from $\vdash A(q)$ infer $\vdash A(q/B)$ by substituting all appearances of q in A by B.

MP Modus ponens: from $\vdash A$ and $\vdash A \rightarrow B$ infer $\vdash B$.

TG Temporal generalisation: from $\vdash A$ infer $\vdash HA$ and $\vdash GA$.

A formula A is deducible from the set of formulae Δ, $\Delta \vdash A$, if there exist a finite sequence of formulae $B_1, \ldots, B_n = A$ such that every B_i is either

(a) a formula in Δ; or

(b) an axiom; or

(c) obtained from previous formulae in the sequence through the use of an inference rule.

We write $\vdash A$ for $\varnothing \vdash A$, i.e. only items (b) and (c) above are used in the deduction of A, in which case A is said to be a *theorem*. A set of formulae Δ is inconsistent if $\Delta \vdash \bot$, otherwise it is *inconsistent*. A formula A is consistent if $\{A\}$ is consistent.

On the semantical side, a set of formulae Δ is *satisfiable* over a class of models $\mathcal{K}$ if there exist a model $\mathbf{m} \in \mathcal{K}$ (and a $t \in T$) such that, for every $B_i \in \Delta$, $\mathbf{m}, t \models B_i$. A formula A is *valid* over $\mathcal{K}$, $\mathcal{K} \models A$ if, for every model $\mathbf{m} = (T, <, h) \in \mathcal{K}$ (and $t \in T$), $\mathbf{m}, t \models A$. The expression $\Delta \models A$ represents the notion that every model satisfying Δ also satisfies A.

An inference system is *sound* with respect to a class of models $\mathcal{K}$ iff every theorem is a valid formula, i.e. $\vdash A$ implies $\mathcal{K} \models A$. An inference system is *(weakly)* complete over $\mathcal{K}$, if every theorem $\vdash A$ is valid, $\mathcal{K} \models A$, or equivalently, if every consistent formula is satisfied over $\mathcal{K}$. Let $\mathbf{L} = \langle \mathbb{L}, \vdash, \models \rangle$ be a logic system with language $\mathbb{L}$, inference system $\vdash$ and semantics $\models$. We say that $\mathbf{L}$ is *decidable* if there exists an algorithm (decision procedure) that determines, for every $A \in \mathcal{L}$, whether A is a theorem or not. The *validity problem* for $\mathbf{L}$ is to determine whether some $A \in \mathbb{L}$ is a valid formula or not.

We have the following results.

Theorem 13.1. *Burgess–Xu [Burgess, 1982; Xu, 1988] The Burgess–Xu axiomatisation presented above is sound and complete over the class $\mathcal{K}_{lin}$.*

Theorem 13.2. *Rabin [1969] The logic $\mathbf{US} = \langle \mathbb{L}_{US}, \vdash_{\mathbf{US}}, \models_{\mathbf{US}} \rangle$ is decidable over $\mathcal{K}_{lin}$.*

13.3 Combining Logics

As we have mentioned earlier, the combination of two one-dimensional temporal logics will generate a two-dimensional temporal logic. Throughout this presentation, we refer to one of the temporal dimensions as the *horizontal dimension* and the other one as the *vertical dimension*; the symbols related to the vertical dimension are normally obtained by putting a bar on top of the corresponding horizontal ones, e.g. $\mathbf{T}$ and $\bar{\mathbf{T}}$, $F$and $\overline{F}$, $<$ and $\overline{<}$.

There are two distinct criteria for defining a modal/temporal logic system as two-dimensional:

(i) If the alphabet of the language contains two non-empty, disjoint sets of corresponding modal or temporal operators, Φ and $\overline{\Phi}$, each set associated to a distinct flow of time, $(T, <)$ and $(\overline{T}, \overline{<})$, then the system is two-dimensional.

(ii) If the truth value of a formula is evaluated with respect to two time points, then the system is two-dimensional. In this case, we even have the distinction between strong and weak interpretation of formulae that, as a consequence, generates different notions of valid formulae (a formula is valid if it holds in all models for all pairs of time points). Under the *strong interpretation*, the truth value of atoms depends on both dimensions, giving rise to the notion of *strongly valid formulae* when the evaluation of formulae is inductively extended to all connectives. In the *weak interpretation*, the truth value of atoms depends only on one dimension, e.g. the horizontal dimension, giving rise to the notion of *weakly valid formulae*.

Usually for this notion of two-dimensionality, both time points refer to the same flow of time, so we may also have the notion of (weak/strong) *diagonally valid* formulae by restricting validity to the case where both dimensions refer to the same point, i.e. A is diagonally valid iff $\mathbf{m}, t, t \models A$ for all $\mathbf{m}$ and t; see [Gabbay *et al.*, 1994] for more details.

Criterion (i) will be called the *syntactic criterion* for two-dimensionality, although it is not completely syntactic, i.e. it depends on the semantic notion of flows of time; criterion (ii) will be called the *semantic criterion* for two-dimensionality.

Note that both cases can yield, as an extreme case, one-dimensional temporal logic. In (i), by making $T = \overline{T}$ and $\overline{<} = (<)^{-1} = (>)$, i.e. by taking two flows with the same set of time points such that one order is the inverse of the other; the future operators $\Phi = \{F, G, U\}$ are associated with $(T, <)$ and the past operators $\overline{\Phi} = \{P, H, S\}$ are associated with $(T, >)$. In (ii), by fixing one dimension to a single time point so that it becomes redundant.

These two distinct approaches to the two-dimensionality of a system are independent. In fact, we will see in Section 13.5 a system that contains two distinct sets of operators over two classes of flows of time, but its formulae are evaluated at a single point. On the other hand, there are several temporal logics in the literature satisfying (ii) but not (i), containing a single set of temporal operators in which formulae are evaluated according to two or more time points in the same flow, e.g. [Åqvist, 1979; Kamp, 1971; Gabbay *et al.*, 1994].

A logic system that respects both the syntactic and the semantic criteria for two-dimensionality is called *broadly two-dimensional*, and this will be the kind of system we will be aiming to achieve through combination methods; we consider in this work only strong evaluation and validity; the weak interpretation generates systems with the expressivity of only monadic first-order language [Gabbay *et al.*, 1994], but for broadly two-dimensional systems we are interested in the expressivity of dyadic first-order language, although it is known that no set of temporal operators can be expressively complete over dyadic first-order language [Venema, 1990]. (A modal/temporal language is *expressively complete* over a class of first-order formulae if, for any first-order formula A in that class, there exists a modal/temporal formula B such that A is first-order equivalent to B^*, where B^* is the standard first-order translation of B; see [Gabbay *et al.*, 1994].) Venema's [1990] two-dimensional temporal logic, Segerberg's [1973] two-dimensional modal logic and the temporalisation of a temporal logic are all broadly two-dimensional; so are the combined logics in Sections 13.6 and 13.7.

In the study of one-dimensional temporal logics (1DTLs) several classes of flows of time are taken into account. When we move to 2DTLs, the number of such classes increases considerably, and every pair of one-dimensional classes can be seen as generating a different two-dimensional class. The study of 2DTLs would benefit much if the properties known to hold for 1DTLs could be systematically transferred to 2DTLs, avoiding the repetition of much of the work that has been published in the literature. This is a strong motivation to consider methods of combination of 1DTLs into 2DTLs and studying the transfer of logical properties through each method. Also in favour of such an approach is the fact that the results concerning 2DTLs are then presented in a general, compact and elegant form.

In providing a method to combine two 1DTLs $\bar{\mathbf{T}}$ and $\mathbf{T}$ we have to pay attention to the following points:

(a) A method for combining logics $\bar{\mathbf{T}}$ and $\mathbf{T}$ is composed of three submethods, namely a method for combining the languages of $\bar{\mathbf{T}}$ and $\mathbf{T}$, a method for combining their inference systems and a method for combining their semantics.

(b) We study the combined logic system with respect to the way certain logical properties of $\bar{\mathbf{T}}$ and $\mathbf{T}$ are transferred to the two-dimensional combination. We focus here on the properties of soundness, completeness and decidability of the combined system given those of the component ones.

(c) The combined language should be able to express some properties of the interaction between the two dimensions; otherwise the combination is just a partial one, and the two systems are not fully combined. For example, it is desirable to express formulae like $F\bar{F}A \leftrightarrow \bar{F}FA$ and $P\bar{F}A \leftrightarrow \bar{F}PA$ that are not in the temporalised language of $\bar{\mathbf{T}}[\mathbf{T}]$.

(d) If we want to strengthen the interaction between the two systems, some properties of the interaction between the two-dimensions are expected to be theorems of the combining system, e.g. the commutativity of horizontal and vertical future operators such as $F\bar{F}A \leftrightarrow \bar{F}FA$ and $P\bar{F}A \leftrightarrow \bar{F}PA$. Those are called the *interaction axioms* in Chapters 3 and 4.

(e) We want the combination method to be as independent as possible from the underlying flows of time.

All methods of combination must comply with item (a). The method for combining the languages of $\bar{\mathbf{T}}$ and $\mathbf{T}$ includes the choice of which sublanguage of $\bar{\mathbf{T}}$ and $\mathbf{T}$ is going to be part of the combined two-dimensional language, as well as the way in which this combination is done; in this presentation we will work, in the most general case, with the standard languages of S and U, $\bar{S}$ and $\bar{U}$, but we also consider some sublanguages, e.g. the sublanguage generated by a set of derived operators, such as the vertical 'previous' ($\bar{\bullet}$) and 'next' ($\bar{\bigcirc}$) in Section 13.7. In combining the inference systems of $\bar{\mathbf{T}}$ and $\mathbf{T}$, we will assume that they are both an extension of classical logic and that they are presented in the form of a regular, normal axiomatic system $(\Sigma, \mathcal{I})$, where Σ is a set of axioms and $\mathcal{I}$ is a set of inference rules. One important requirement is that the combined system be a conservative extension of the two components. The conservativeness property states that if A is a formula in the language of $\mathbf{L}$ and $\mathbf{L}^*$ is a logic system extending $\mathbf{L}$ (i.e. the language of $\mathbf{L}$ is a sublanguage of the language of $\mathbf{L}^*$) then A will be a theorem of $\mathbf{L}^*$ only if it is already a theorem of $\mathbf{L}$; conservativeness guarantees that no new information about the original system $\mathbf{L}$ is present in the extended one $\mathbf{L}^*$.

The combined semantics has to deal with the structure of the combined model, the evaluation of two-dimensional formulae over those structures and also with the combinations of classes of flows of time.

Items (b), (c), (d) and (e) may conflict with each other. In fact, the rest of this chapter shows that this is the case, as we try to compromise between expressivity, independence of the underlying flow of time and the transfer of logical properties.

13.4 Temporalising a Logic

The first of the combination methods, known as 'adding a temporal dimension to a logic system' or simply 'temporalising a logic system', has been extensively discussed in [Finger and Gabbay, 1992].

Temporalisation is a methodology whereby an arbitrary logic system **L** can be enriched with temporal features to create a new system **T[L]**. The new system is constructed by combining **L** with a pure propositional temporal logic **T** (such as linear temporal logic with 'Since' and 'Until') in a special way.

Although we are only interested here in temporalising an already temporal system, so as to generate a 2DTL, the original method is more general and is applicable to a generic logic **L**; **L** is constrained to be an extension of classical logic, i.e. all propositional tautologies must be valid in it, but such a constraint does not affect us, for we are assuming that both temporal systems **T** and **L** are extensions of $\mathbf{US}/\mathcal{K}_{lin}$ The language of a temporalised system is based on the US language and is a subset of the language of **L**, $\mathbb{L}_\mathbf{L}$. The set $\mathbb{L}$ is partitioned in two sets, $\mathbb{CL}_\mathbf{L}$ and $\mathbb{ML}_\mathbf{L}$. A formula $A \in \mathbb{L}_\mathbf{L}$ belongs to the set of *Boolean combinations*, $\mathrm{BC}_\mathbf{L}$, iff it is built up from other formulae by the use of one of the Boolean connectives $\neg$ or $\wedge$ or any other connective defined only in terms of those; it belongs to the set of *monolithic formula* $\mathbb{ML}_\mathbf{L}$ otherwise.

The result of temporalising over $\mathcal{K}$ the logic system **L** is the logic system **T[L]**/$\mathcal{K}$. The alphabet of the temporalised language uses the alphabet of **L** plus the two-place operators S and U, if they are not part of the alphabet of **L**; otherwise, we use $\overline{S}$ and $\overline{U}$ or any other proper renaming.

Definition 13.3 (Temporalised formulae). The set $\mathbb{L}_{\mathbf{T}[\mathbf{L}]}$ of formulae of the logic system **L** is the smallest set such that:

1. if $A \in \mathbb{ML}_\mathbf{L}$, then $A \in \mathbb{L}_{\mathbf{T}[\mathbf{L}]}$;
2. if $A, B \in \mathbb{L}_{\mathbf{T}[\mathbf{L}]}$ then $\neg A \in \mathbb{L}_{\mathbf{T}[\mathbf{L}]}$ and $(A \wedge B) \in \mathbb{L}_{\mathbf{T}[\mathbf{L}]}$;
3. if $A, B \in \mathbb{L}_{\mathbf{T}[\mathbf{L}]}$ then $S(A, B) \in \mathbb{L}_{\mathbf{T}[\mathbf{L}]}$ and $U(A, B) \in \mathbb{L}_{\mathbf{T}[\mathbf{L}]}$.

Note that, for instance, if $\Box$ is an operator of the alphabet of **L** and A and B are two formulae in $\mathbb{L}$, the formula $\Box U(A, B)$ is *not* in $\mathbb{L}_{\mathbf{T}[\mathbf{L}]}$. The language of **T[L]** is independent of the underlying flow of time, but not its semantics and inference system, so we must fix a class $\mathcal{K}$ of flows of time over which the temporalisation is defined; if $\mathbf{m}_\mathbf{L}$ is a model in the class of models of **L**, $\mathcal{K}_\mathbf{L}$, for every formula $A \in \mathbb{L}_\mathbf{L}$ we must have either $\mathbf{m}_\mathbf{L} \models A$ or $\mathbf{m}_\mathbf{L} \models \neg A$. In the case that **L** is a temporal logic we must consider a 'current time' o as part of its model to achieve that condition.

Definition 13.4 (Semantics of the temporalised logic). Let $(T, <) \in \mathcal{K}$ be a flow of time and let $g : T \to \mathcal{K}_\mathbf{L}$ be a function mapping every time point in T to a model in the class of models of **L**. A model of **T[L]** is a triple $\mathbf{m}_{\mathbf{T}[\mathbf{L}]} = (T, <, g)$ and the fact that A is true in $\mathbf{m}_{\mathbf{T}[\mathbf{L}]}$ at time t is written as $\mathbf{m}_{\mathbf{T}[\mathbf{L}]}, t \models A$ and defined as:

$\mathbf{m_{T[L]}}, t \models A, A \in \mathbb{ML}_\mathbf{L}$	iff $g(t) = \mathbf{m_L}$ and $\mathbf{m_L} \models A$.
$\mathbf{m_{T[L]}}, t \models \neg A$	iff it is not the case that $\mathbf{m_{T[L]}}, t \models A$.
$\mathbf{m_{T[L]}}, t \models (A \wedge B)$	iff $\mathbf{m_{T[L]}}, t \models A$ and $\mathbf{m_{T[L]}}, t \models B$.
$\mathbf{m_{T[L]}}, t \models S(A, B)$	iff there exists $s \in T$ such that $s < t$ and $\mathbf{m_{T[L]}}, s \models A$ and for every $u \in T$, if $s < u < t$ then $\mathbf{m_{T[L]}}, u \models B$.
$\mathbf{m_{T[L]}}, t \models U(A, B)$	iff there exists $s \in T$ such that $t < s$ and $\mathbf{m_{T[L]}}, s \models A$ and for every $u \in T$, if $t < u < s$ then $\mathbf{m_{T[L]}}, u \models B$.

Figure 13.2 illustrates a temporalised model. The inference system of $\mathbf{T[L]}/\mathcal{K}$ is given by the following.

Definition 13.5 (Axiomatisation for T[L]). An axiomatisation for the temporalised logic **T[L]** is composed of:

- the axioms of **T**/$\mathcal{K}$;
- the inference rules of **T**/$\mathcal{K}$;
- for every formula A in $\mathbb{L}_\mathbf{L}$, if $\vdash_\mathbf{L} A$ then $\vdash_{\mathbf{T[L]}} A$, i.e. all theorems of **L** are theorems of **T[L]**. This inference rule is called **Persist**.

Example 13.6 (Temporalising propositional logic). Consider classical propositional logic $\mathbf{PL} = \langle \mathbb{L}_\mathbf{PL}, \vdash_\mathbf{P,L}, \models_\mathbf{P,L} \rangle$. Its temporalisation generates the logic system $\mathbf{T[PL]} = \langle \mathbb{L}_{\mathbf{T[PL]}}, \vdash_{\mathbf{T[PL]}}, \models_{\mathbf{T[PL]}} \rangle$.

It is not difficult to see that $\mathbb{L}_{\mathbf{T[PL]}} = \mathbb{L}_\mathbf{US}$ and $\vdash_{\mathbf{T[PL]}} = \vdash_\mathbf{US}$, i.e. the temporalised version of **PL** over any $\mathcal{K}$ is actually the temporal logic $\mathbf{T} = \mathbf{US}/\mathcal{K}$. With respect to $\mathbf{m_{T[L]}}$, the function g actually assigns, for every time point, a **PL** model.

Example 13.7 (Temporalising US-temporal logic). If we temporalise over $\mathcal{K}$ the one-dimensional logic system **US**/$\mathcal{K}$ we obtain the two-dimensional logic system $\mathbf{T[US]} = \langle \mathbb{L}_{\mathbf{T[US]}}, \vdash_{\mathbf{T[US]}}, \models_{\mathbf{T[US]}} \rangle = \mathbf{T^2[PL]}/\mathcal{K}$. In this case we have to rename the two-place operators S and U of the temporalised alphabet to, say, $\overline{S}$ and $\overline{U}$.

In order to obtain a model for **T[US]**, we must fix a 'current time', o, in $\mathbf{m_{US}} = (T_1, <_1, g_1)$, so that we can construct the model $\mathbf{m_{T[US]}} = (T_2, <_2, g_2)$ as previously described. Note that, in this case, the flows of time $(T_1, <_1)$ and $(T_2, <_2)$ need not be the same. $(T_2, <_2)$ is the flow of time of the upper-level temporal system whereas $(T_1, <_1)$ is the flow of time of the underlying logic which, in this case, happens to be a temporal logic.

The logic system we obtain by temporalising US-temporal logic is the two-dimensional temporal logic described in [Finger, 1992].

Example 13.8 (N-dimensional temporal logic). If we repeat the process started in the last two examples, we can construct an n-dimensional temporal logic $\mathbf{T^n[PL]}/\mathcal{K}$ (its alphabet including S_n and U_n) by temporalising an $(n-1)$-dimensional temporal logic.

Every time we add a temporal dimension, we are able to describe changes in the underlying system. Temporalising the system **L** once, we are creating a way of describing the history of **L**; temporalising for the second time, we are describing how the history of **L** is viewed in different moments of time. We can go on indefinitely, although it is not clear what is the purpose of doing so.

From now on we restrict the logic systems to **L** = **US**/$\mathcal{K}$ and **T** = $\bar{\mathbf{U}}\bar{\mathbf{S}}/\overline{\mathcal{K}}$, where $\mathcal{K}, \overline{\mathcal{K}} \subseteq \mathcal{K}_{lin}$. We write $\bar{\mathbf{U}}\bar{\mathbf{S}}$**[US]** instead of **T[L]** and the generated class of models is referred to as $\overline{\mathcal{K}}[\mathcal{K}]$. For this system, we enumerate a series of results that are proved in [Finger and Gabbay, 1992]. Those results will be useful for the discussion of the independent combination method.

Theorem 13.9 (Transference via temporalisation). *Let* $\bar{\mathbf{U}}\bar{\mathbf{S}}/\overline{\mathcal{K}}$ *and* $\mathbf{US}/\mathcal{K}$ *be logic systems such that* $\overline{\mathcal{K}}, \mathcal{K} \subseteq \mathcal{K}_{lin}$.

1. *If* $\bar{\mathbf{U}}\bar{\mathbf{S}}$ *is sound with respect to* $\overline{\mathcal{K}}$ *and* **US** *is sound with respect to* $\mathcal{K}$*, then* $\bar{\mathbf{U}}\bar{\mathbf{S}}$**[US]** *is sound w.r.t.* $\overline{\mathcal{K}}[\mathcal{K}]$.
2. *If* $\bar{\mathbf{U}}\bar{\mathbf{S}}$ *is complete w.r.t.* $\overline{\mathcal{K}}$ *and* **US** *is complete w.r.t.* $\mathcal{K}$ *then* $\bar{\mathbf{U}}\bar{\mathbf{S}}$**[US]** *is complete w.r.t.* $\overline{\mathcal{K}}[\mathcal{K}]$.
3. *If* $\bar{\mathbf{U}}\bar{\mathbf{S}}$ *is complete w.r.t.* $\mathcal{K}$*, then* $\bar{\mathbf{U}}\bar{\mathbf{S}}$**[US]** *is a conservative extension of both* $\bar{\mathbf{U}}\bar{\mathbf{S}}$ *and* **US**.
4. *If* $\bar{\mathbf{U}}\bar{\mathbf{S}}$ *is complete and is decidable over* $\overline{\mathcal{K}}$ *and* **US** *is complete and decidable over* $\mathcal{K}$ *then* $\bar{\mathbf{U}}\bar{\mathbf{S}}$**[US]** *is decidable over* $\overline{\mathcal{K}}[\mathcal{K}]$.

13.5 Independent Combination

We have seen in the previous section how to add a temporal dimension to a logic system. In particular, if a temporal logic is itself temporalised we obtain a two-dimensional temporal logic. Such a logic system is, however, very weakly expressive; if **US** is the internal (horizontal) temporal logic in the temporalisation process (F is derived in **US**), and $\bar{\mathbf{U}}\bar{\mathbf{S}}$ is the external (vertical) one ($\overline{F}$ is defined in $\bar{\mathbf{U}}\bar{\mathbf{S}}$), we cannot express the idea that vertical and horizontal future operators commute,

$$F\overline{F}A \leftrightarrow \overline{F}FA.$$

In fact, the subformula $F\overline{F}A$ is not even in the temporalised language of $\bar{\mathbf{U}}\bar{\mathbf{S}}$**[US]**, nor is the whole formula. In other words, the interplay between the two-dimensions is not expressible in the language of the temporalised $\bar{\mathbf{U}}\bar{\mathbf{S}}$**[US]**.

The idea is then to define a new method of combination of logic systems that puts together all the expressivity of the two component logic systems in an independent way; for that we assume that the language of a system is given by a set of formation rules.

Definition 13.10. Let $Op(\mathbf{L})$ be the set of non-Boolean operators of a generic logic **L**. Let $\bar{\mathbf{T}}$ and **T** be logic systems such that $Op(\mathbf{T}) \cap Op(\bar{\mathbf{T}}) = \varnothing$. The *fully combined language* of logic systems $\bar{\mathbf{T}}$ and **T** over the set of atomic propositions $\mathcal{P}$ is obtained by the union of the respective set of connectives and the union of the formation rules of the languages of both logic systems.

Let the operators U and S be in the language of **US** and $\overline{U}$ and $\overline{S}$ be in that of $\bar{\mathbf{U}}\bar{\mathbf{S}}$. Note that the renaming of the temporal operator is done prior to the combination, so that the combined systems contains the set of Boolean operators $\{\neg, \wedge\}$ coming from both components, plus the set of temporal operators $\{U, S, \overline{U}, \overline{S}\}$. Their fully combined language over a set of atomic propositions $\mathcal{P}$ is given by

- every atomic proposition is in it;
- if A, B are in it, so are $\neg A$ and $A \wedge B$;
- if A, B are in it, so are $U(A, B)$ and $S(A, B)$.
- if A, B are in it, so are $\overline{U}(A, B)$ and $\overline{S}(A, B)$.

In general, we do not want any non-Boolean operator to be shared between the two languages, for this may cause problems when combining their axiomatisations. For example (this example is due to Ian Hodkinson), if a generic operator $\Box$ belongs to both temporal logic system such that $\mathbf{T}$ contains axiom $q \leftrightarrow \Box q$ and system $\bar{\mathbf{T}}$ contains axiom $\neg q \leftrightarrow \Box q$, the union of their axiomatisations will result in an inconsistent systems even though each system might have been itself-consistent. To avoid such behaviour the restriction $Op(\mathbf{T}) \cap Op(\bar{\mathbf{T}}) = \varnothing$ was imposed on the fully combined language of $\bar{\mathbf{T}}$ and $\mathbf{T}$. Not only are the two languages taken to be independent of each other, but the set of axioms of the two systems are supposed to be disjoint; so we call the following combination method the *independent combination* of two temporal logics.

This new method of combination is called *independent* because it takes the independent union of the axiomatisation of its two component systems, and it is based on their fully combined language.

Definition 13.11. Let **US** and $\bar{\mathbf{U}}\bar{\mathbf{S}}$ be two US-temporal logic systems defined over the same set $\mathcal{P}$ of propositional atoms such that their languages are independent. The *independent combination* $\mathbf{US} \oplus \bar{\mathbf{U}}\bar{\mathbf{S}}$ is given by the following:

- The fully combined language of **US** and $\bar{\mathbf{U}}\bar{\mathbf{S}}$.
- If $(\Sigma, \mathcal{I})$ is an axiomatisation for **US** and $(\overline{\Sigma}, \overline{\mathcal{I}})$ is an axiomatisation for $\bar{\mathbf{U}}\bar{\mathbf{S}}$, then $(\Sigma \cup \overline{\Sigma}, \mathcal{I} \cup \overline{\mathcal{I}})$ is an axiomatisation for $\mathbf{US} \oplus \bar{\mathbf{U}}\bar{\mathbf{S}}$. Note that, apart from the classical tautologies, the set of axioms Σ and $\overline{\Sigma}$ are supposed to be disjoint, but not the inference rules.
- The class of independently combined flows of time is $\mathcal{K} \oplus \overline{\mathcal{K}}$ composed of biordered flows of the form $(\tilde{T}, <, \overline{<})$ where the connected components of $(\tilde{T}, <)$ are in $\mathcal{K}$ and the connected components of $(\tilde{T}, \overline{<})$ are in $\overline{\mathcal{K}}$, and $\tilde{T}$ is the (not necessarily disjoint) union of the sets of time points T and $\overline{T}$ that constitute each connected component; such a biordered flow of time has been discussed in Chapters 3 and 4 and [Kracht and Wolter, 1991] for the case of the independent combination of two mono-modal systems.

A model structure for $\mathbf{US} \oplus \bar{\mathbf{U}}\bar{\mathbf{S}}$ over $\mathcal{K} \oplus \overline{\mathcal{K}}$ is a 4-tuple $(\tilde{T}, <, \overline{<}, g)$, where $(\tilde{T}, <, \overline{<}) \in \mathcal{K} \oplus \overline{\mathcal{K}}$ and g is an assignment function $g : \tilde{T} \rightarrow 2^{\mathcal{P}}$. An independently combined model is illustrated in Figure 13.5.

The semantics of a formula A in a model $\mathbf{m} = (\tilde{T}, <, \overline{<}, g)$ is defined as the union of the rules defining the semantics of **US**/$\mathcal{K}$ and $\bar{\mathbf{U}}\bar{\mathbf{S}}/\overline{\mathcal{K}}$. The expression $\mathbf{m}, t \models A$ reads that the formula A is true in the (combined) model $\mathbf{m}$ at the point $t \in \tilde{T}$. The semantics of formulae is given by induction in the standard way:

$\mathbf{m}, t \models p$ iff $p \in g(t)$ and $p \in \mathcal{P}$.
$\mathbf{m}, t \models \neg A$ iff it is not the case that $\mathbf{m}, t \models A$.
$\mathbf{m}, t \models A \wedge B$ iff $\mathbf{m}, t \models A$ and $\mathbf{m}, t \models B$.
$\mathbf{m}, t \models S(A, B)$ iff there exists an $s \in \tilde{T}$ with $s < t$ and $\mathbf{m}, s \models A$ and for every $u \in \tilde{T}$, if $s < u < t$ then $\mathbf{m}, u \models B$.
$\mathbf{m}, t \models U(A, B)$ iff there exists an $s \in \tilde{T}$ with $t < s$ and $\mathbf{m}, s \models A$ and for every $u \in \tilde{T}$, if $t < u < s$ then $\mathbf{m}, u \models B$.
$\mathbf{m}, t \models \bar{S}(A, B)$ iff there exists an $s \in \tilde{T}$ with $s \overline{<} t$ and $\mathbf{m}, s \models A$ and for every $u \in \tilde{T}$, if $s \overline{<} u \overline{<} t$ then $\mathbf{m}, u \models B$.
$\mathbf{m}, t \models \bar{U}(A, B)$ iff there exists an $s \in \tilde{T}$ with $t \overline{<} s$ and $\mathbf{m}, s \models A$ and for every $u \in \tilde{T}$, if $t \overline{<} u \overline{<} s$ then $\mathbf{m}, u \models B$.

Note that, despite the combination of two flows of time, formulae are evaluated according to a single point. The independent combination generates a system that is two-dimensional according to the first criterion but fails the second one, so it is not broadly two-dimensional.

The following result is due to [Thomason, 1980] and is more general than the independent combination of two **US**-logics.

Proposition 13.12. *With respect to the validity of formulae, the independent combination of two modal logics is a conservative extension of the original ones.*

Note that we have previously defined conservative extension in proof theoretical terms; completeness for the independently combined case will lead to the conservativeness with respect to derivable theorems.

As usual, we will assume that $\mathcal{K}, \overline{\mathcal{K}} \subseteq \mathcal{K}_{lin}$, so $<$ and $\overline{<}$ are transitive, irreflexive and total orders; similarly, we assume that the axiomatisations are extensions of **US**/$\mathcal{K}_{lin}$.

The temporalisation process will be used as an inductive step to prove the transfer of soundness, completeness and decidability for $\mathbf{US} \oplus \bar{\mathbf{U}}\bar{\mathbf{S}}$ over $\mathcal{K} \oplus \overline{\mathcal{K}}$. Let us first consider the *degree of alternation* of a $(\mathbf{US} \oplus \bar{\mathbf{U}}\bar{\mathbf{S}})$-formula A for **US**, $dg(A)$, and $\bar{\mathbf{U}}\bar{\mathbf{S}}$, $\overline{dg}(A)$.

$dg(p) = 0$	$\overline{dg}(p) = 0$
$dg(\neg A) = dg(A)$	$\overline{dg}(\neg A) = \overline{dg}(A)$
$dg(A \wedge B) = max\{dg(A), dg(B)\}$	$\overline{dg}(A \wedge B) = max\{\overline{dg}(A), \overline{dg}(B)\}$
$dg(S(A, B)) = max\{dg(A), dg(B)\}$	$\overline{dg}(\bar{S}(A, B)) = max\{\overline{dg}(A), \overline{dg}(B)\}$
$dg(U(A, B)) = max\{dg(A), dg(B)\}$	$\overline{dg}(\bar{U}(A, B)) = max\{\overline{dg}(A), \overline{dg}(B)\}$
$dg(\bar{S}(A, B)) = 1 + max\{\overline{dg}(A), \overline{dg}(B)\}$	$\overline{dg}(S(A, B)) = 1 + max\{dg(A), dg(B)\}$
$dg(\bar{U}(A, B)) = 1 + max\{\overline{dg}(A), \overline{dg}(B)\}$	$\overline{dg}(U(A, B)) = 1 + max\{dg(A), dg(B)\}$

Any formula A of $\mathbf{US} \oplus \bar{\mathbf{U}}\bar{\mathbf{S}}$ can be seen as a formula of some finite number of alternating temporalisations of the form **US[$\bar{\mathbf{U}}\bar{\mathbf{S}}$[US[...]]]**; more precisely, A can be seen as a formula of $\mathbf{US[L}_n\mathbf{]}$, where $dg(A) = n$, $\mathbf{US[L_0] = US}$, $\bar{\mathbf{U}}\bar{\mathbf{S}}[\mathbf{L}_0] = \bar{\mathbf{U}}\bar{\mathbf{S}}$, and $\mathbf{L}_{n-2i} = \bar{\mathbf{U}}\bar{\mathbf{S}}[\mathbf{L}_{n-2i-1}]$, $\mathbf{L}_{n-2i-1} = \mathbf{US}[\mathbf{L}_{n-2i-2}]$, for $i = 0, 1, \ldots, \lceil \frac{n}{2} \rceil - 1$. This fact is illustrated in Figure 13.4. The following lemma actually allows us to see the independent combination as the (infinite) union of a finite number of alternating temporalisations of **US** and $\bar{\mathbf{U}}\bar{\mathbf{S}}$; it will also be used in the proof of transfer of completeness and decidability (given completeness) for $\mathbf{US} \oplus \bar{\mathbf{U}}\bar{\mathbf{S}}$.

Lemma 13.13. *Let* **US** *and* $\bar{\mathbf{U}}\bar{\mathbf{S}}$ *be two complete logic systems. Then, A is a theorem of* $\mathbf{US} \oplus \bar{\mathbf{U}}\bar{\mathbf{S}}$ *iff it is a theorem of* $\mathbf{US[L}_n\mathbf{]}$*, where* $dg(A) = n$.

Proof If A is a theorem of $\mathbf{US[L}_n\mathbf{]}$, all the inferences in its deduction can be repeated in $\mathbf{US} \oplus \bar{\mathbf{U}}\bar{\mathbf{S}}$, so it is a theorem of $\mathbf{US} \oplus \bar{\mathbf{U}}\bar{\mathbf{S}}$.

Suppose A is a theorem of $\mathbf{US} \oplus \bar{\mathbf{U}}\bar{\mathbf{S}}$; let $B_1, \ldots, B_m = A$ be a deduction of A in $\mathbf{US} \oplus \bar{\mathbf{U}}\bar{\mathbf{S}}$ and let $n' = max\{dg(B_i)\}$, $n' \geq n$. We claim that each B_i is a theorem of $\mathbf{US[L}_{n'}\mathbf{]}$. In fact, by induction on m, if B_i is obtained in the deduction by substituting into an axiom, the same substitution can be done in $\mathbf{US[L}_{n'}\mathbf{]}$; if B_i is obtained by temporal generalisation from B_j, $j < i$, then by the induction hypothesis, B_j is a theorem of $\mathbf{US[L}_{n'}\mathbf{]}$ and so is B_i; if B_i is obtained by modus ponens from B_j and B_k, $j, k < i$, then by the induction hypothesis, B_j and B_k are theorems of $\mathbf{US[L}_{n'}\mathbf{]}$ and so is B_i.

So A is a theorem of $\mathbf{US[L}_{n'}\mathbf{]}$ and, since **US** and $\bar{\mathbf{U}}\bar{\mathbf{S}}$ are two complete logic systems, by Theorem 13.9, each of the alternating temporalisations in $\mathbf{US[L}_{n'}\mathbf{]}$ is a conservative extension of the underlying logic; it follows that A is a theorem of $\mathbf{US[L}_n\mathbf{]}$, as desired. □

The transfer of soundness, completeness and decidability follows directly from this result.

Theorem 13.14 (Independent combination). *Let* **US** *and* $\bar{\mathbf{U}}\bar{\mathbf{S}}$ *be two sound and complete logic systems over the classes* $\mathcal{K}$ *and* $\overline{\mathcal{K}}$*, respectively. Then their independent combination* $\mathbf{US} \oplus \bar{\mathbf{U}}\bar{\mathbf{S}}$ *is sound and complete over the class* $\mathcal{K} \oplus \overline{\mathcal{K}}$*. If* **US** *and* $\bar{\mathbf{U}}\bar{\mathbf{S}}$ *are complete and decidable, so is* $\mathbf{US} \oplus \bar{\mathbf{U}}\bar{\mathbf{S}}$.

Proof Soundness follows immediately from the validity of axioms and inference rules. For completeness, suppose that A is a consistent formula in $\mathbf{US} \oplus \bar{\mathbf{U}}\bar{\mathbf{S}}$; by Lemma 13.13, A is consistent in $\mathbf{US[L}_n\mathbf{]}$, so we construct a temporalised model for it, and we obtain a model $(\tilde{T}_1, <_1, g_1, o_1)$ over $\mathcal{K}[\overline{\mathcal{K}}[\mathcal{K}[\ldots]]]$, where o_1 is the 'current time' necessary for the successive temporalisations. We show now how it can be transformed into a model over $\mathcal{K} \oplus \overline{\mathcal{K}}$.

Without loss of generality, suppose that **US** is the outermost logic system in the multi-layered temporalised system $\mathbf{US}[\bar{\mathbf{U}}\bar{\mathbf{S}}[\mathbf{US}[\ \ldots\ ]]]$, and let n be the number of alternations. The construction is recursive, starting with the outermost logic. Let $i \leq n$ denote the step of the construction; if i is odd, it is a **US**-temporalisation, otherwise it is a $\bar{\mathbf{U}}\bar{\mathbf{S}}$-temporalisation. At every step i we construct the sets $\tilde{T}_{i+1}$, $<_{i+1}$ and $\overline{<}_{i+1}$ and the function g_{i+1}.

We start the construction of the model at step $i = 0$ with the temporalised model $(\tilde{T}_1, <_1, g_1, o_1)$ such that $(\tilde{T}_1, <_1) \in \mathcal{K}$, and we take $\overline{<}_1 = \varnothing$. At step $i < n$, consider the current set of time points $\tilde{T}_i$; according to the construction, each $t \in \tilde{T}_i$ is associated to:

- a temporalised model $g_i(t) = (\tilde{T}^t_{i+1}, <^t_{i+1}, g^t_{i+1}, o^t_{i+1}) \in \mathcal{K}$ and take $\overline{<}^t_{i+1} = \varnothing$, if i is even; or
- a temporalised model $g_i(t) = (\tilde{T}^t_{i+1}, \overline{<}^t_{i+1}, g_{i+1}, o^t_{i+1}) \in \overline{\mathcal{K}}$ and take $<^t_{i+1} = \varnothing$, if i is odd.

The point t is made identical to $o^t_{i+1} \in \tilde{T}^t_{i+1}$, so as to add the new model to the current structure; note that this preserves the satisfiability of all formulae at t. Let $\tilde{T}_{i+1}$ be the (possibly infinite) union of all $\tilde{T}^t_{i+1}$ for $t \in \tilde{T}_i$; similarly, $<_{i+1}$ and $\overline{<}_{i+1}$ are generated. And finally, for every $t \in \tilde{T}_{i+1}$, the function g_{i+1} is constructed as the union of all g^t_{i+1} for $t \in \tilde{T}_i$.

Repeating this construction n times, we obtain a combined model over $\mathcal{K} \oplus \overline{\mathcal{K}}$, $\mathbf{m} = (\tilde{T}_n, <_n, \overline{<}_n, g_n)$, such that for all $t \in \tilde{T}_n$, $g_n(t) \subseteq \mathcal{P}$. Since satisfiability of formulae is preserved at each step, it follows that $\mathbf{m}$ is a model for A, and completeness is proved.

For decidability, suppose we want to decide whether a formula $A \in$ **US** $\oplus$ **ŪS̄** is a theorem. By Lemma 13.13, this is equivalent to deciding whether $A \in$ **US[L$_n$]** is a theorem, where $n = dg(A)$. Since **US**/$\mathcal{K}$ and **ŪS̄**/$\overline{\mathcal{K}}$ are both complete and decidable, by successive applications of Theorem 13.9(b) and (d), it follows that the following logics are decidable: **US[ŪS̄]**, **ŪS̄[US[ŪS̄]]** = **ŪS̄[L$_2$]**, ..., **ŪS̄[L$_{n-1}$]**= **L$_n$**; so a last application of Theorem 13.9(b) and (d) yields that **US[L$_n$]** is decidable. □

13.6 Full Join

With respect to the generation of two-dimensional systems, the method of independent combination has two main drawbacks. First, it generates logic systems whose formulae are evaluated at one single time point, not generating a broadly two-dimensional logic. Second, since the method independently combines the two component logic systems, no interaction between the dimensions is provided by it. As a consequence, although a formula like $F\overline{F}A \leftrightarrow \overline{F}FA$ is expressible in its language, it will not be valid, as can easily be verified, for it expresses an interplay between the dimensions. We therefore introduce the notion of a *two-dimensional plane model*.

Definition 13.15. Let $\mathcal{K}$ and $\overline{\mathcal{K}}$ be two classes of flow of time. A *two-dimensional plane model* over the *fully combined class* $\mathcal{K} \times \overline{\mathcal{K}}$ is a 5-tuple $\mathbf{m} = (T, <, \overline{T}, \overline{<}, g)$, where $(T, <) \in \mathcal{K}$, $(\overline{T}, \overline{<}) \in \overline{\mathcal{K}}$ and $g : T \times \overline{T} \to 2^{\mathcal{P}}$ is a two-dimensional assignment. The semantics of the horizontal and vertical operators are independent of each other.

$\mathbf{m}, t, x \models S(A, B)$ iff there exists $s < t$ such that $\mathbf{m}, s, x \models A$ and for all u, $s < u < t$, $\mathbf{m}, u, x \models B$.

$\mathbf{m}, t, x \models \overline{S}(A, B)$ iff there exists $y \overline{<} x$ such that $\mathbf{m}, t, y \models A$ and for all z, $y \overline{<} z \overline{<} x$, $\mathbf{m}, t, z \models B$.

Similarly for U and $\overline{U}$, the semantics of atoms and Boolean connectives remaining the standard one. A formula A is (strongly) valid over $\mathcal{K} \times \overline{\mathcal{K}}$ if for all models $\mathbf{m} = (T, <, \overline{T}, \overline{<}, g)$, for all $t \in T$ and $x \in \overline{T}$ we have $\mathbf{m}, t, x \models A$.

With respect to the expressivity of fully combined two-dimensional languages, Venema [1990] has shown that no finite set of two-dimensional temporal operators is expressively complete over the class of linear flows with respect to dyadic first-order logic — despite the fact that **US**-temporal logic is expressively complete with respect to monadic first-order logic over $\mathbb{N}$ and over $\mathbb{R}$, and that, with additional operators (the Stavi operators), we can get expressive completeness over $\mathbb{Q}$ and $\mathcal{K}_{lin}$ [Gabbay, 1981a]. So expressive completeness is not transferred by the full join.

It is easy to verify that the following formulae expressing the commutativity of future and past operators between the two dimensions are valid formulae in two-dimensional plane models.

I1 $F\overline{F}A \leftrightarrow \overline{F}FA$
I2 $F\overline{P}A \leftrightarrow \overline{P}FA$
I3 $P\overline{F}A \leftrightarrow \overline{F}PA$
I4 $P\overline{P}A \leftrightarrow \overline{P}PA$.

Therefore, if we want to satisfy both the syntactic and the semantic criteria for two-dimensionality, we may define the method of *full join* containing the fully combined language of **US** and $\bar{\mathbf{U}}\bar{\mathbf{S}}$ and their fully combined class of models. The question is whether there is a method for combining their axiomatisations so as to generate a *fully joined axiomatisation* that transfers the properties of soundness, completeness and decidability. The answer, however, is no, not in general. In some cases we can obtain the transfer of completeness, in some other cases the transfer fails. To illustrate that, we consider completeness results over classes of the form $\mathcal{K}\times\mathcal{K}$.

We start by defining some useful abbreviations. Let p be a propositional atom; define:

$$hor(p) = \Box(p\wedge\overline{H}\neg p\wedge\overline{G}\neg p)$$
$$ver(p) = \overline{\Box}(p\wedge H\neg p\wedge G\neg p).$$

It is clear that $hor(p)$ makes p true along the horizontal line and false elsewhere; similarly for $ver(p)$ with respect to the vertical.

The axiomatisation of $\mathbf{US}\times\bar{\mathbf{U}}\bar{\mathbf{S}}$ over $\mathcal{K}_{lin}\times\mathcal{K}_{lin}$ extends that of $\mathbf{US}\oplus\bar{\mathbf{U}}\bar{\mathbf{S}}$ over $\mathcal{K}_{lin}\oplus\mathcal{K}_{lin}$ by including the join axioms **I1–I4** and the following inference rules:

IR1 if $\vdash hor(p)\rightarrow A$ and p does not occur in A, then $\vdash A$
IR2 if $\vdash ver(p)\rightarrow A$ and p does not occur in A, then $\vdash A$.

IR1 and **IR2** are two-dimensional extensions of the irreflexivity inference rule (IRR) defined in [Gabbay, 1981b] for the one-dimensional case: if $\vdash p\wedge H\neg p\rightarrow A$ and p does not occur in A, then $\vdash A$.

Theorem 13.16 (2D-completeness). *There is a sound and complete axiomatisation over the class of full two-dimensional temporal models over* $\mathcal{K}_{lin}\times\mathcal{K}_{lin}$.

The proof consists of a Henkin-style construction of a two-dimensional grids, where each point is a maximally consistent set. The basic step of the construction is the elimination of 'defects' from the grid, i.e. adding new points to the grid for a semantic condition that fails for the grid. The final model is obtained as the (infinite) union of all steps, and the grid thus constructed is shown to be a $\mathcal{K}_{lin}\times\mathcal{K}_{lin}$ model for an original consistent formula. The full details of the proof can be found in [Finger, 1994], but due to space limitations (the full proof takes up to 10 pages) we omit it from here. If $\mathcal{K}_{dis}$ is the class of all linear and discrete flows, [Finger, 1994] also shows completeness results for the classes $\mathcal{K}_{dis} \times \mathcal{K}_{dis}$, $\mathbb{Q} \times \mathbb{Q}$, $\mathcal{K}_{lin} \times \mathcal{K}_{dis}$, $\mathcal{K}_{lin} \times \mathbb{Q}$ and $\mathbb{Q} \times \mathcal{K}_{dis}$.

The negative result is the following.

Proposition 13.17 (2D-unaxiomatisability). *There are no finite axiomatisations for the (strongly) valid two-dimensional formulae over the classes* $\mathbb{Z}\times\mathbb{Z}$, $\mathbb{N}\times\mathbb{N}$ *and* $\mathbb{R}\times\mathbb{R}$.

This proposition follows directly from Venema's proof that the valid formulae over the upper half two-dimensional semi-plane are not enumerable for $\mathbb{Z} \times \mathbb{Z}$, $\mathbb{N} \times \mathbb{N}$ and $\mathbb{R} \times \mathbb{R}$, which in its turn was based on [Halpern and Shoham, 1986]. Since there are sound, complete and decidable **US**-temporal logics over $\mathbb{Z}$, $\mathbb{N}$ and $\mathbb{R}$ [Reynolds, 1992; Büchi, 1962; Rabin, 1969; Burgess and Gurevich, 1985], the general conclusion on full join is the following.

Theorem 13.18 (Full join). *Completeness and decidability do not transfer in general through full join.*

It has to be noted that two-dimensional temporal logics seem to behave like modal logics in the following sense. We can see the result of the independent combination of **US** and $\bar{\mathbf{U}}\bar{\mathbf{S}}$ as generating a 'minimal' combination of the logics, i.e. one without any interference between the dimensions. The addition of extra axioms, inference rules or an extra condition on its models has to be studied on its own, just as adding a new axiom to a modal logic or imposing a new property on its accessibility relation has to be analysed on its own.

The full join method illustrates the conflict between the generality of a method and its ability to achieve the transfer of logical properties. We next restrict the join method so as to recover the transfer of logical properties.

13.7 Restricted Join

The fact that the transfer of logical properties fails for the join of two **US**-temporal logics does not mean that the join of any two temporal logic systems fails to achieve this transfer. We restrict the vertical logic system to a temporal logic $\bar{\mathbf{N}}\bar{\mathbf{P}}$ with operators $\bar{\bigcirc}$ for **N**ext time and $\bar{\bullet}$ for **P**revious time; the formation rules for the formulae of $\bar{\mathbf{N}}\bar{\mathbf{P}}$ are the standard ones. This is a restriction of the $\bar{\mathbf{U}}\bar{\mathbf{S}}$-language for $\bar{\bigcirc}$ and $\bar{\bullet}$ can be defined in terms of $\bar{U}$ and $\bar{S}$, namely by

$$\bar{\bigcirc} A =_{def} \bar{U}(A, \bot)$$
$$\bar{\bullet} A =_{def} \bar{S}(A, \bot).$$

Not only is the expressivity of the language reduced in this way, but also the underlying flow of time is now restricted to a discrete one; in fact, we concentrate our attention on integer-like flows of time.

Let $h : \mathbb{Z} \to \mathcal{P}$ be a temporal assignment over the integers so that the semantics of $\bar{\mathbf{N}}\bar{\mathbf{P}}$ over the integers is the usual one for atoms and Boolean operators and

$$(\mathbb{Z}, <, h), t \models \bar{\bigcirc} A \quad \text{iff} \quad (\mathbb{Z}, <, h), t+1 \models A$$
$$(\mathbb{Z}, <, h), t \models \bar{\bullet} A \quad \text{iff} \quad (\mathbb{Z}, <, h), t-1 \models A.$$

An axiomatisation for **NP**/$\mathbb{Z}$ is given by the classical tautologies plus

NP1 $\bar{\bigcirc}\,\bar{\bullet}p\rightarrow p$
NP2 $\bar{\bigcirc}\neg p\leftrightarrow\neg\bar{\bigcirc}p$
NP3 $\bar{\bigcirc}(p\wedge q)\rightarrow\bar{\bigcirc}p\wedge\bar{\bigcirc}q$
NP4 The mirror image of **NP1–3** obtained by swapping $\bar{\bigcirc}$ and $\bar{\bullet}$.

The rules of inference are the usual substitution, modus ponens and temporal generalisation (from A infer $\bar{\bigcirc}A$ and $\bar{\bullet}A$).

The converse of each axiom can be straightforwardly derived, so the formulae on both sides of the $\rightarrow$-connective are actually equivalent. It follows that every $\bar{\mathbf{N}}\bar{\mathbf{P}}$-formula can be transformed into an equivalent one by 'pushing in' the temporal operators, e.g. by following the arrows in the axioms, and by 'cancelling' the occurrences of $\bar{\bigcirc}$ and $\bar{\bullet}$ in a string of temporal operators, e.g. $\bar{\bigcirc}\,\bar{\bullet}\,\bar{\bullet}\,\bar{\bigcirc}\,\bar{\bullet}p$ is equivalent to $\bar{\bullet}p$; the resulting $\bar{\mathbf{N}}\bar{\mathbf{P}}$-*normal form* formula is a Boolean combination of formulae of the form $\bar{\bigcirc}^k p$ and $\bar{\bullet}^l q$, where p and q are atoms, $k, l \in \mathbb{N}$ and $\bar{\bigcirc}^k$ is a sequence of $\bar{\bigcirc}$-symbols of size k, similarly for $\bar{\bullet}^l$; it is useful sometimes to consider k negative or 0, so we define $\bar{\bigcirc}^{-k}A = \bar{\bullet}^k A$ and $\bar{\bigcirc}^0 A = A$. As an example, the formula $\bar{\bigcirc}\,\bar{\bigcirc}(\bar{\bullet}\,\bar{\bullet}\,\bar{\bullet}(p\wedge q)\vee p)$ has normal form $(\bar{\bullet}p\wedge\bar{\bullet}q)\vee\bar{\bigcirc}\,\bar{\bigcirc}p$. The existence of such a normal form gives us very simple proofs for completeness and decidability of $\bar{\mathbf{N}}\bar{\mathbf{P}}$/$\mathbb{Z}$ that we outline next.

For completeness, let Σ be a possibly infinite consistent set of $\bar{\mathbf{N}}\bar{\mathbf{P}}$-formulae and assume all formulae in the set are in the normal form. Σ can be seen as a consistent set of propositional formulae where each maximal subformula of the form $\bar{\bigcirc}^k p$ is understood as a new propositional atom, so let h_0 be a propositional valuation assigning every extended atom into $\{true, false\}$. For $n \in \mathbb{Z}$, let $h(n) = \{p \in \mathcal{P} \mid h_0(\bar{\bigcirc}^n p) = true\}$. Clearly $(\mathbb{Z}, <, h)$ is a model for the original set.

For decidability, let A be a formula of $\bar{\mathbf{N}}\bar{\mathbf{P}}$ and let A^* be its normal form; clearly there exists an algorithm to transform A into A^*. By considering subformulae of the form $\bar{\bigcirc}^k p$ as new atoms, k possibly negative, we apply any decision procedure for propositional logic to A^*. A is an $\bar{\mathbf{N}}\bar{\mathbf{P}}$-valid formula iff A^* is a propositional tautology.

Definition 13.19. The *restricted join* of temporal logic systems **US**/$\mathcal{K}$ and $\bar{\mathbf{N}}\bar{\mathbf{P}}$/$\mathbb{Z}$ is the two-dimensional temporal logic system $\mathbf{US} \times \bar{\mathbf{N}}\bar{\mathbf{P}}$ given by:

- the fully combined language of **US** and $\bar{\mathbf{N}}\bar{\mathbf{P}}$;
- the two-dimensional plane model over $\mathcal{K} \times \mathbb{Z}$, equipped with the broadly two-dimensional semantics;
- the union of the axioms of **US**/$\mathcal{K}$ and $\bar{\mathbf{N}}\bar{\mathbf{P}}$/$\mathbb{Z}$ plus the join axioms

$$\bar{\bigcirc}U(p,q)\rightarrow U(\bar{\bigcirc}p,\bar{\bigcirc}q)$$
$$\bar{\bigcirc}S(p,q)\rightarrow S(\bar{\bigcirc}p,\bar{\bigcirc}q)$$

  plus their duals obtained by swapping $\bar{\bigcirc}$ with $\bar{\bullet}$; the inference rules are just the union of the inference rules of both component systems.

What has therefore been restricted in the interlacing process is the expressivity of the language over the vertical dimension, which also restricted the underlying flow of time to a discrete one. The following gives us a normal form for $\mathbf{US} \times \bar{\mathbf{N}}\bar{\mathbf{P}}$.

Lemma 13.20. *Let A be a formula of $\mathbf{US} \times \bar{\mathbf{N}}\bar{\mathbf{P}}$. There exists a normal form formula A^* equivalent to A, such that all the occurrences of $\bar{\bigcirc}$ and $\bar{\bullet}$ in it are in the form $\bar{\bigcirc}^k p$ and $\bar{\bullet}^l q$, where p and q are atoms.*

Proof First we show that the converse of the join axioms are theorems too. For that, note that U and S respect the *congruence property*, i.e. if $A \leftrightarrow C$ and $B \leftrightarrow D$ then $U(A,B) \leftrightarrow U(C,D)$ and $S(A,B) \leftrightarrow S(C,D)$. Also note that

$$\textbf{equiv} \quad (p \leftrightarrow \bar{\bigcirc}\bar{\bullet} p) \wedge (p \leftrightarrow \bar{\bullet}\bar{\bigcirc} p).$$

The transitivity of the $\rightarrow$-operator connects the steps in the proof of the formula $U(\bar{\bigcirc} p, \bar{\bigcirc} q) \rightarrow \bar{\bigcirc} U(p,q)$ below:

$$\begin{aligned} U(\bar{\bigcirc} p, \bar{\bigcirc} q) &\rightarrow \bar{\bigcirc}\bar{\bullet} U(\bar{\bigcirc} p, \bar{\bigcirc} q) && \text{by } \textbf{equiv} \\ &\rightarrow \bar{\bigcirc} U(\bar{\bullet}\bar{\bigcirc} p, \bar{\bullet}\bar{\bigcirc} q) && \text{by join axiom} \\ &\rightarrow \bar{\bigcirc} U(p,q) && \text{by } \textbf{equiv} \text{ and congruence.} \end{aligned}$$

It follows that $U(\bar{\bigcirc} p, \bar{\bigcirc} q) \leftrightarrow \bar{\bigcirc} U(p,q)$. It is completely analogous to show the converse of other join axioms, so we omit the details.

Given A in the language of $\mathbf{US} \times \bar{\mathbf{N}}\bar{\mathbf{P}}$, the equivalence between both sides of the join axioms allows for 'pushing in' the vertical operators $\bar{\bigcirc}$ and $\bar{\bullet}$, so a simple induction on the number of nested temporal operators in A shows an algorithmic way to generate an equivalent formula A^* in the desired normal form. □

Theorem 13.21 (Completeness via restricted join). *Let* **US** *be a logic system complete over the class $\mathcal{K} \subseteq \mathcal{K}_{lin}$. Then the two-dimensional system $\mathbf{US} \times \bar{\mathbf{N}}\bar{\mathbf{P}}$ is complete over $\mathcal{K} \times \mathbb{Z}$.*

Proof Consider a $\mathbf{US} \times \bar{\mathbf{N}}\bar{\mathbf{P}}$-consistent formula A and assume it is in the normal form. So we can see A as a **US**-formula over the extended set of atoms $\bar{\bigcirc}^k$, k possibly negative or 0. From the completeness of $\mathbf{US}/\mathcal{K}$ there exist a one-dimensional model $(T, <, h_{US})$ for A at a point $o \in T$, where $(T, <) \in \mathcal{K}$. Define the two-dimensional assignment

$$h(k,t) = \{p \in \mathcal{P} \mid \bar{\bigcirc}^k p \in h_{US}(t)\}.$$

Clearly, $(T, <, \mathbb{Z}, <_{\mathbb{Z}}, h)$ is a two-dimensional plane $\mathbf{US}\times\bar{\mathbf{N}}\bar{\mathbf{P}}$-model for A at $(o, 0)$. □

Corollary 13.22. *If $\mathbf{US}/\mathcal{K}$ is strongly complete, so is $\mathbf{US} \times \bar{\mathbf{N}}\bar{\mathbf{P}}/\mathcal{K} \times \mathbb{Z}$.*

Theorem 13.23 (Decidability via restricted join). *If the logic system* **US** *is decidable over $\mathcal{K}$, so is $\mathbf{US}\times\bar{\mathbf{N}}\bar{\mathbf{P}}$ over $\mathcal{K}\times\mathbb{Z}$.*

Proof The argument of the proof is the same as that of the decidability of **NP**; all we have to do is note that there exists an algorithmic way to convert a combined two-dimensional formula into its normal form, so it can be seen as a **US**-formula and we can apply the **US**-decision procedure to it. □

So by restricting the expressivity and the underlying class of flows of time, we can obtain the transfer of the basic logical properties via restricted join. It should not be difficult to extend these results to $\mathbb{N}$ instead of $\mathbb{Z}$, although we do not explore this possibility here.

It is also worth noting that the restricted join method answers a conjecture posed by Venema
[1990] on the existence of some expressively limited two-dimensional temporal logic over $\mathbb{Z}\times\mathbb{Z}$ that was 'well behaved' in the sense of having the completeness and decidability properties.

13.8 The Two-dimensional Diagonal

We now study some properties of the diagonal in two-dimensional plane models. The diagonal is a privileged line in the two-dimensional model intended to represent the sequence of time points we call 'now', i.e. the time points on which an historical observer is expected to traverse. The observer is, therefore, on the diagonal when he or she poses a query (i.e. evaluates the truth value of a formula) on a two-dimensional model. The diagonal is illustrated in Figure 13.8.

So let δ be a special atom and consider the formulae:

D1 $\Diamond\delta\wedge\overline{\Diamond}\delta$
D2 $\delta\rightarrow(G\neg\delta\wedge H\neg\delta\wedge\overline{G}\neg\delta\wedge\overline{H}\neg\delta)$
D3 $\delta\rightarrow(\overline{H}G\neg\delta\wedge\overline{G}H\neg\delta)$.

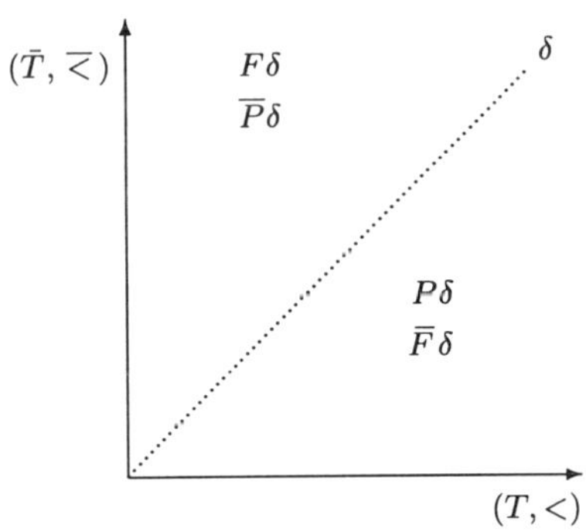

Fig. 13.8 The two-dimensional diagonal

Let $Diag = \Box\overline{\Box}(\mathbf{D1}\wedge\mathbf{D2}\wedge\mathbf{D3})$. The intuition behind $Diag$ is the following. **D1** implies that the two-dimensional diagonal can always be reached in both vertical and horizontal directions; **D2** implies that there are no two diagonal points on the same horizontal line and on the same vertical line and **D3** implies that the diagonal goes in the direction SW–NE. We say that $Diag$ characterises a two-dimensional diagonal in the following sense.

Proposition 13.24. *Let* $\mathbf{m} = (T, <, \overline{T}, \overline{<}, g)$ *be a full two-dimensional model over* $\mathcal{K}\times\overline{\mathcal{K}}$, $\mathcal{K}, \overline{\mathcal{K}} \subseteq \mathcal{K}_{lin}$, *and let* δ *be a propositional letter. Then the following are equivalent.*

(a) $\mathbf{m}, t, x \models Diag$, *for some* $t \in T$ *and* $x \in \overline{T}$.

(b) $\mathbf{m}, t, x \models Diag$, *for all* $t \in T$ *and* $x \in \overline{T}$.

(c) There exists an isomorphism $i : T \to \overline{T}$ *such that* $\mathbf{m}, t, x \models \delta$ *iff* $x = i(t)$.

Proof It is straightforward to show that (a) $\Longleftrightarrow$ (b) and (c) $\Longrightarrow$ (a); we show only (b) $\Longrightarrow$ (c). So assume that $\mathbf{m}, t, x \models Diag$, for all $t \in T$ and $x \in \overline{T}$. Define

$$i = \{(t, x) \in T\times\overline{T} \mid \mathbf{m}, t, x \models \delta\}.$$

All we have to show is that i is an isomorphism.

- i, i^{-1} are functions such that $dom(i) = T$ and $dom(i^{-1}) = \overline{T}$. Suppose that both (t, x_1) and (t, x_2) are in i; then $\mathbf{m}, t, x_1 \models \delta$ and $\mathbf{m}, t, x_2 \models \delta$. By linearity of $\overline{T}$, $x_1 = x_2$, $x_1 \overline{<} x_2$ or $x_2 \overline{<} x_1$, but **D2** eliminates the latter two; **D1** gives us that $dom(i) = T$. Similarly, the linearity of T and **D2** gives us that i^{-1} is a function and **D1** gives us that $dom(i^{-1}) = \overline{T}$.
- $i(t) = x$ iff $i^{-1}(x) = t$ follows directly from the definition. So i is a bijection.
- i preserves ordering. Suppose $t_1 < t_2$; by the linearity of $\overline{T}$ we have three possibilities:
  - $i(t_1) = i(t_2)$ contradicts i is a bijection;
  - $i(t_2) \overline{<} i(t_1)$ contradicts **D3**;
  - $i(t_1) \overline{<} i(t_2)$ is the only possible option.

Therefore i is an isomorphism, which proves the result. □

This result shows that adding **D1–D3** to the axiomatisation over the two-dimensional plane $\mathcal{K}_{lin}\times\mathcal{K}_{lin}$ of Section 13.6 gives us completeness over the class of models of the form $(T, <, T, <, g)$, where $(T, <) \in \mathcal{K}_{lin}$. It follows from [Halpern and Shoham, 1986], however, that such a logic system is undecidable.

The diagonal is interpreted as the sequence of time points we call 'now'. The diagonal divides the two-dimensional plane in two semi-planes. The semi-plane that is to the (horizontal) left of the diagonal is 'the past', and the formula $F\delta$ holds over all points of this semi-plane. Similarly, the semi-plane that is to the (horizontal) right of the diagonal is 'the future', and the formula $P\delta$ holds over all points of this semi-plane. Figure 13.8 puts this fact in evidence. If we assume that $Diag$ holds over $\mathbf{m}$ such that i is the isomorphism defined in Lemma 13.24, $t < s$ iff $i(t) \overline{<} i(s)$, then

$$\begin{aligned} \mathbf{m}, t, x \models F\delta &\text{ iff exists } s > t \text{ such that } \mathbf{m}, s, x \models \delta \text{ and } i(s) = x \\ &\text{ iff exists } y = i(t) \overline{<} x \text{ such that } \mathbf{m}, t, y \models \delta \\ &\text{ iff } \mathbf{m}, t, x \models \overline{P}\delta. \end{aligned}$$

Similarly, it can be shown that:

$$\mathbf{m}, t, x \models P\delta \text{ iff } \mathbf{m}, t, x \models \overline{F}\delta.$$

It follows that the following formula is valid for $\mathbf{US}\times\overline{\mathbf{US}}$ over $\mathcal{K}_{lin}\times\mathcal{K}_{lin}$:

$$Diag \rightarrow ((F\delta \leftrightarrow \overline{P}\delta) \wedge (P\delta \leftrightarrow \overline{F}\delta)).$$

As a consequence, $\overline{P}\delta$ holds over all points of the 'past' semi-plane and $\overline{F}\delta$ holds over all points of the 'future' semi-plane, as is indicated in Figure 13.8.

The formula $Diag$ is in the language of $\mathbf{US} \times \bar{\mathbf{U}}\bar{\mathbf{S}}$ but not in the language of $\mathbf{US} \times \bar{\mathbf{N}}\bar{\mathbf{P}}$, for $Diag$ contains the vertical temporal operators $\overline{G}$, $\overline{H}$, $\overline{\Box}$ and $\overline{\Diamond}$. To characterise a two-dimensional diagonal in $\mathbf{US} \times \bar{\mathbf{N}}\bar{\mathbf{P}}$ we do the following. We say that a formula A *holds over* or *is valid over* a two-dimensional model $\mathbf{m}$ if for every $t \in T$ and every $x \in \overline{T}$, it is the case that $\mathbf{m}, t, x \models A$. Consider the formulae

d1 $\Diamond\delta$
d2 $\delta \rightarrow (G\neg\delta \wedge H\neg\delta)$
d3 $\delta \leftrightarrow \bar{\bigcirc}\bigcirc\delta$

where δ is a proposition. Those formulae are all in the language of $\mathbf{US} \times \bar{\mathbf{N}}\bar{\mathbf{P}}$, for $Diag$ (so also in the language of $\mathbf{US} \times \bar{\mathbf{U}}\bar{\mathbf{S}}$) and they can characterise the two-dimensional diagonal due to the following property.

Proposition 13.25. *Let* $\mathbf{m}$ *be a two-dimensional plane model over* $\mathbb{Z} \times \mathbb{Z}$. *Then the formula* **D1**$\wedge$**D2**$\wedge$**D3** *holds over* $\mathbf{m}$ *iff* **d1**$\wedge$**d2**$\wedge$**d3** *holds over* $\mathbf{m}$.

Proof From Proposition 13.24 we know that **D1**$\wedge$**D2**$\wedge$**D3** holds over $\mathbf{m}$ iff the relation i defined as below

$$i = \{(t, x) \in \mathbb{Z} \times \mathbb{Z} \mid \mathbf{m}, t, x \models \delta\}.$$

is an isomorphism in $\mathbb{Z}$. So all we have to do is to prove that i as defined above is an isomorphism iff **d1**$\wedge$**d2**$\wedge$**d3** holds over $\mathbf{m}$. The *only if* part is a straightforward verification that for all x and t in $\mathbb{Z}$, $\mathbf{m}, t, x \models$ **d1**$\wedge$**d2**$\wedge$**d3**.

Assume **d1**$\wedge$**d2**$\wedge$**d3** holds over $\mathbf{m}$. Then:

(a) **d1** gives us that for every x there exists a t such that $\mathbf{m}, t, x \models \delta$;

(b) **d2** gives us that for every x, t, t', $t \neq t'$, $\mathbf{m}, t, x \models \delta$ implies $\mathbf{m}, t', x \not\models \delta$;

(c) **d3** give us that for every x, t, $\mathbf{m}, t, x \models \delta$ iff $\mathbf{m}, t+1, x+1 \models \delta$ iff for every $n \in \mathbb{Z}$, $\mathbf{m}, t+n, x+n \models \delta$.

The first two items give us that $i^{-1} : \mathbb{Z} \rightarrow \mathbb{Z}$ is a function. To show that i is also a function, suppose that $(t, x_1), (t, x_2) \in i$. By linearity of $\mathbb{Z}$, it follows that either $x_1 < x_2$ or $x_2 < x_1$ or $x_1 = x_2$. Let $x_1 - x_2 = m$; then, by the third item above, $(t + m, x_2 + m = x_1) \in i$, so $t = (t + m)$ and $m = 0$. It follows that $x_1 = x_2$, so $i : \mathbb{Z} \rightarrow \mathbb{Z}$ is a function. Directly by the definition of i, it follows that i is a bijection.

By the third item above, if $i(t_1) = x_1$ and $i(t_2) = x_2$, then $t_1 - t_2 = x_1 - x_2$. It follows that i is order preserving and hence an isomorphism, which finishes the proof. □

It would be desirable to generalise the idea of a diagonal as the sequence of 'now' moments to any pair of flows of time that are not necessarily isomorphic. For that, we would have to create an order between the points of the two flows, i.e. we would have to merge the flows.

So let $(T, <)$ and $(\overline{T}, \overline{<})$ be two flows of time such that T and $\overline{T}$ are disjoint. Then there always exists a flow $(T', <')$ and a mapping $f : T \cup \overline{T} \rightarrow T'$ such that f is one-to-one and order preserving. The *f-merge of* $(T, <)$ *and* $(\overline{T}, \overline{<})$ is the flow of time consisting of the image of f ordered by the restriction of $<'$ to the image of f. An example of an f-merge is shown in Figure 13.9, where $f(y)$ is made equal, via merge, to $f(\bar{x})$ and on the merged flow the order is preserved, i.e. originally $x < y$ and $\bar{x} \overline{<} \bar{y}$ and on the f-merged flow $f(x) <' f(y) = f(\bar{x}) <' f(\bar{y})$.

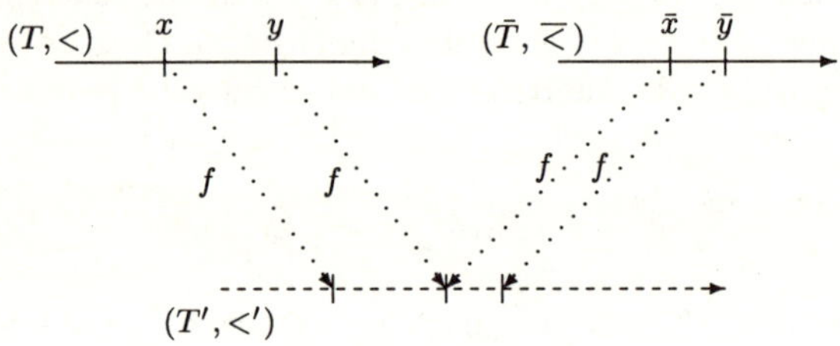

Fig. 13.9 The f-merge

We can then construct a two-dimensional model with two copies of the f-merged flow, in which we can define a diagonal over $(T', <') \times (T', <')$ as shown in Figure 13.10. Another particularly interesting situation arrises when the f-merged flow $(T', <')$ is identical to one of the component flows, e.g. $(\overline{T}, \overline{<})$, so that f is an *embedding* of $(T, <)$ into $(\overline{T}, \overline{<})$. In this case, the flow $(T, <)$ could be viewed as a more 'abstract' representation of $(\overline{T}, \overline{<})$wherein several details, i.e. pieces of information, points in time, are ignored.

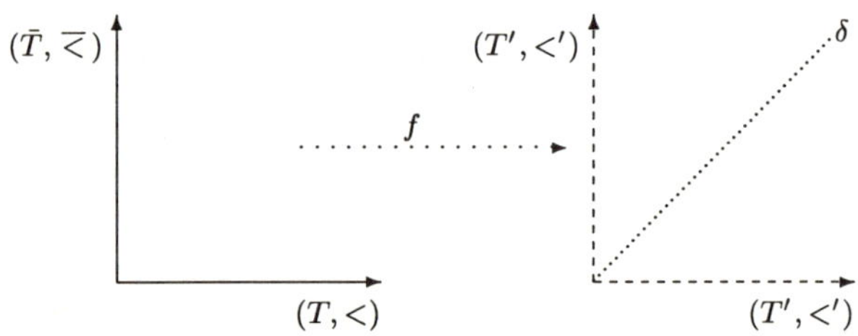

Fig. 13.10 The diagonal of two distinct flows

The f-merge construction serves as motivation for another method of combining two one-dimensional temporal logics, this time generating another one-dimensional logic. This could be achieved over the class of all f-merges of its two-component flows of time or subclasses of it. We could then study the transfer of logical properties in the same way as we have done in this and the previous section, but those matters remain beyond the scope of this chapter.

13.9 Conclusion

This chapter dealt with the combination of two logic systems in order to obtain a new logic system. The issues were:

- Several methods of combination of two logic systems were presented. Each combination involved at least one temporal logic system. Each method had a particular discipline for combining the language, the semantics and the inference system of two logic systems. Each combination generated a single logic system.
- The study of transfer of logical properties from the component systems into their combined form has been the major point in the analysis of combination methods. The basic logical properties whose transfer was analysed were soundness, completeness and decidability; for some combination methods, the transfer of other properties was also investigated such as conservativeness and the compactness property (in the form of strong completeness).
- The investigation of four basic methods has been accomplished. The temporalisation method and the independent combination method were shown to transfer all basic properties, although they do not generate an expressive enough system to be called fully two-dimensional. The full join method does generate a fully two-dimensional temporal system, but in many cases it failed to transfer even the completeness property. As a compromise, it was shown that a restricted join method, although generating two-dimensional temporal logic systems that were not as expressive and generic as the fully interlaced one, accomplishes the transfer of all basic logical properties.

Another contribution of our analysis was to answer a question raised by Venema [1990] on the existence of a fragment of the two-dimensional plane temporal logic that, in his own words, was 'better behaved' than the two-dimensional plane system with respect to completeness and decidability properties. We have shown that the two-dimensional temporal logic systems obtained by restricted join are an example of such fragments.

Another question raised by Venema in that work remains open, namely, whether there exists a complete axiomatisation over the two-dimensional model using only canonical inference rules, i.e. without using the special inference rules **IR1** and **IR2**. This problem seems to be a very hard one. Nevertheless we succeeded in extending Venema's completeness result, that originally holds for only two-dimensional flows built from two identical one-dimensional flows, to any two-dimensional flow built from any flow in the classes $\mathcal{K}_{lin}$, $\mathcal{K}_{dis}$, $\mathcal{K}_{dense}$ and $\mathbb{Q}$.

In comparison with earlier chapters, it seems that the present chapter concentrated on linear time. The reader should note that the focus on linear flows of time was due to database applications of two-dimensional temporal logics as in [Finger and Gabbay, 1992; Finger, 1994], but we believe that this restriction may be lifted without affecting the transfer results of the temporalisation and independent combination methods. These have to be further investigated and the transfer of any other logical property has to be analysed on its own.

The generalisation of combination methods other than the independent combination method to modal logics is another area for further work. As noted in [Finger and Gabbay, 1992], the temporalisation process is directly extensible to monomodal logics. It may even be the case that, for monomodal logics, the full join method achieves transfer of completeness over several classes of fully two-dimensional Kripke frames using only canonical inference rules, as suggested by the results in [Segerberg, 1973].

The complexity class of the decision problem for the combined logic is another interesting subject for study. For the independent combination of monomodal logics, such a study was done by Spaan [1993] and the conclusion was that the satisfiability problem of an independently combined logic is either reducible to that of one of the component logics, or it is PSPACE-hard or it is in NP. We believe a similar result can be obtained for the temporalisation and the independent combination of temporal logics, although the details have not yet been worked out. The complexity of the full and restricted join methods still have to be studied.

All the systems dealt with in this chapter are extensions of classical logic. Is it possible that the temporalisation process preserves its transfer properties even in the case when the underlying system is not an extension of classical logic? What if the external temporal logic is non-classical itself? The same question applies to other combination methods. Do they achieve transfer of logical properties when one or both of the combined temporal or modal logics is not classical? Chapter 20 (on fibring *labelled deductive systems*) will address that question in a very generic framework involving *labelled deductive systems* (LDS).

14

FIBRING IMPLICATION LOGICS

14.1 Introduction

The proof-theoretical side of our investigations into fibred semantics can be formulated as follows: how can we combine different proof systems into more general ones, possibly preserving the desirable properties of the originals?

We expect semantically based methods, e.g. semantic tableau methods, to play a crucial role in developing suitable proof methods based on fibred semantics. A systematic investigation of this problem will be carried out in Chapter 18. In this chapter and in the next one we shall focus on case studies which are interesting for their own sake: (i) combining ('substructural') implication logics, and (ii) adding modal operators to such logics. The main message of these two chapters is that the methodology of labelled deductive systems (LDS) ([Gabbay, 1996b]) provides a very natural framework for developing suitable proof-methods based on fibred semantics. So, we shall not concentrate too much on the process of generating the fibred semantics, but mainly on the structure that result from this process and on the way in which they can be 'imported' into the proof-theory in the spirit of the LDS methodology. We assume that the reader can fill the gaps in our semantical analysis by applying the principles which have been presented at length in the previous chapters.

The last two decades have seen a proliferation of different notions of implication proposed for a variety of different purposes, both theoretical and practical. These notions appear to satisfy the needs of different application areas or to capture different interpretations of the logical constant that we call 'implication' (or 'conditional') in each given context. In the light of the latest developments in the study of logical systems, it seems more appropriate to say that we are not dealing with conflicting interpretations of the same logical operator, but rather with a *class* of different operators bearing a family resemblance to each other, each of which may fit a different context or application. So, a crucial problem, both in pure and applied logic, is that of developing a method for combining different implication operators into one fibred system where the user can switch from one context to the other. In this chapter we apply the principles of fibred semantics and of the LDS methodology to this problem. As a case study we take the implication systems belonging to the family of *substructural* logics. The term 'substructural' is now of widespread use to indicate a family of logics weaker than classical logic (including intuitionistic, relevance [Anderson and Belnap Jr, 1975; Dunn, 1986] and linear logic [Girard, 1987; Avron, 1988]), which result from dropping some or all of the traditional structural rules of the classical sequent calculus (see [Došen, 1993] for an introduction), and are proving useful in several applications of logic to computer science and AI (see, for instance, [Garson, 1989] for an application of relevance logic to modular reasoning systems, and [Alexiev, 1994] for an overview of computational applications of linear logic).

14.2 Background on Substructural Implication

Let us briefly review the treatment of substructural implication systems (taken in isolation). The following exposition is loosely based on [D'Agostino and Gabbay, 1994] to which we refer the reader for further details.

14.2.1 *Consequence relations*

By a *consequence relation* over a given logical language $\mathbb{L}$ we intend a relation $\vdash$ between *sequences* of $\mathbb{L}$-formulae and $\mathbb{L}$-formulae satisfying the following two conditions:

(Identity) $$A \vdash A$$

(Surgical cut) $$\Gamma \vdash A \text{ and } \Delta, A, \Lambda \vdash B \Longrightarrow \Delta, \Gamma, \Lambda \vdash B.$$

Let $\mathbb{L}$ be a standard propositional language containing two binary operators $\rightarrow_1$ and $\rightarrow_2$. By an *implication logic* we mean a consequence relation over $\mathbb{L}$ characterised by the following universal conditions on the operators $\rightarrow_1$ and $\rightarrow_2$:

$C_{\rightarrow_1}$ $$\Gamma, A \vdash B \Longleftrightarrow \Gamma \vdash A \rightarrow_1 B$$

$C_{\rightarrow_2}$ $$A, \Gamma \vdash B \Longleftrightarrow \Gamma \vdash A \rightarrow_2 B.$$

Let us now consider the following *structural rules*:

$$\frac{\Gamma, A, B, \Delta \vdash C}{\Gamma, B, A, \Delta \vdash C}\text{Exchange} \qquad \frac{\Gamma, A, A, \Delta \vdash C}{\Gamma, A, \Delta \vdash C}\text{Contraction}$$

$$\frac{\Gamma, A, \Delta \vdash C}{\Gamma, A, A, \Delta \vdash C}\text{Expansion} \qquad \frac{\Gamma, \Delta \vdash C}{\Gamma, A, \Delta \vdash C}\text{Weakening.}$$

We are interested in the family of implication logics closed under different collections of these structural rules. We shall call the logical systems belonging to this family *substructural implication logics*. Given identity and surgical cut, it is easy to check that Conditions $C_{\rightarrow_1}$ and $C_{\rightarrow_2}$ above are equivalent to (a suitable variant of) the operational rules of Gentzen's sequent calculus **LJ**, namely:

$$\frac{\Gamma, A \vdash B}{\Gamma \vdash A \rightarrow B} \qquad \frac{\Gamma \vdash A \qquad \Delta, B, \Lambda \vdash C}{\Delta, A \rightarrow B, \Gamma, \Lambda \vdash C}.$$

Moreover, the structural rules given above are those of **LJ** except that the possibility of dropping the weakening rule leaves room for a new structural rule—called expansion—which is the inverse of the contraction rule.

Notice that in all the systems allowing *exchange* the two implications $\rightarrow_1$ and $\rightarrow_2$ collapse. In this context we shall use the symbol $\rightarrow$ without subscripts. Notice also that every system closed under *weakening* is, of course, closed also under *expansion*.

14.2.2 *Uniform semantics for substructural implications*

As is well known, intuitionistic implication can be characterised in terms of Kripke models. The other substructural implication operators can also be given Kripke-style semantics of some sort. So, a crucial preliminary step towards fibring all these semantics together consists in reducing them to the same format. In the next definition we shall define a class of models which allows us to provide a uniform characterisation of all the substructural implication logics.

Definition 14.1. A *s.o.m. model* (s.o.m. stands for 'semilattice ordered monoid') is a 5-tuple $\mathbf{m} = \langle W, \circ, 1, \leq, V \rangle$, where

- $\langle W, \circ, 1 \rangle$ is a monoid with identity 1
- $\leq$ is a (lower) semilattice ordering on W, i.e. if a and b are in W, $a \sqcap b$ also belongs to W, where $a \sqcap b$ is the greatest lower bound of a and b.
- $\circ$ is order-preserving on both arguments, i.e.

$$a_1 \leq b_1 \text{ and } a_2 \leq b_2 \Longrightarrow a_1 \circ a_2 \leq b_1 \circ b_2$$

- V is a monotonic valuation, i.e. a function of type $P \mapsto 2^W$, where P is the set of propositional letters, satisfying the following *persistence* condition:

$$(\forall a, b \in \mathbf{m})(\forall p \in P)(a \leq b \text{ and } a \in V(p) \Longrightarrow b \in V(p))$$

and the following (downward) *continuity* condition:

$$(\forall a, b \in \mathbf{m})(\forall p \in P)(a \in V(p) \text{ and } b \in V(p) \Longrightarrow a \sqcap b \in V(p)).$$

The 'forcing' relation $a \Vdash_{\mathbf{m}} A$ (read 'a forces A in $\mathbf{m}$', 'A is true at a in $\mathbf{m}$' or '$\mathbf{m}$ verifies A at a'), where $\mathbf{m}$ is a s.o.m. model, is defined as follows:

1. $(\forall p \in P)(\forall a \in \mathbf{m})(a \Vdash_{\mathbf{m}} p \Longleftrightarrow a \in V(p))$
2. $(\forall a \in \mathbf{m})(a \Vdash_{\mathbf{m}} A \rightarrow B \Longleftrightarrow (\forall b \in \mathbf{m})(b \Vdash_{\mathbf{m}} A \Longrightarrow (a \circ b \Vdash_{\mathbf{m}} B)))$.

We say that a formula A is *true* in a s.o.m. model $\mathbf{m}$, and write $\mathbf{m} \models A$, if and only if $1 \Vdash_{\mathbf{m}} A$.

It is easy to show, from the above definition, that both the persistence and the continuity condition must hold also for arbitrary formulas, that is for all $a, b \in \mathbf{m}$ and all formulas A:

$$a \leq b \text{ and } a \Vdash_{\mathbf{m}} A \Longrightarrow b \Vdash_{\mathbf{m}} A$$

and

$$a \Vdash_{\mathbf{m}} A \text{ and } b \Vdash_{\mathbf{m}} A \Longrightarrow a \sqcap b \Vdash_{\mathbf{m}} A.$$

We can now construe all the substructural implication systems as characterised by similar structures, that is different classes of s.o.m. models, each class being identified by a different subset of the following additional constraints:

C1 $(\forall a, b \in \mathbf{m})(a \circ b \leq b \circ a)$
C2 $(\forall a, b \in \mathbf{m})(a \leq a \circ b)$
C3 $(\forall a \in \mathbf{m})(a \circ a \leq a)$
C4 $(\forall a \in \mathbf{m})(a \leq a \circ a)$.

For instance, linear implication is characterised by the class of all s.o.m. models $\mathbf{m}$ such that $\leq$ and $\circ$ satisfy Condition C1 above, while for intuitionistic implication we take the class of all s.o.m. models satisfying C1, C2 and C3. Relevance implication corresponds to the constraints C1 and C3, mingle implication to C1, C3 and C4, etc.

There is a close correspondence between Conditions C1–C4 above and the structural rules of p. 284. Condition C1 corresponds to the exchange rule, Condition C2 to the weakening rule, condition C3 to the contraction rule, and condition C4 to the expansion rule. To stress this correspondence we shall call the conditions C1–C4 *structural constraints*. Just as the sequent calculus for some substructural implication system is obtained from the sequent calculus for intuitionistic implication by disallowing some of the structural rules, the class of models characterising it is obtained from the class of all s.o.m. models by waiving the corresponding structural constraints.

Variants of the semantics outlined in this section can be found in a large number of papers. We mention, in particular [Došen, 1988; Došen, 1989] and [D'Agostino and Gabbay, 1994]. The Kripke-style semantics of implication described above is nothing but a generalisation of Urquhart's semantics for relevant implication introduced in [Urquhart, 1972]. For related semantic investigations into substructural logics see also [Avron, 1988; Wansing, 1993; Ono, 1993; Sambin, 1993; Allwein and Dunn, 1993; Abrusci, 1991; MacCaull, 1996].

14.2.3 *Soundness and completeness*

Let us use the following abbreviations for the structural rules: E for the Exchange rule, C for Contraction, X for eXpansion and K for weaKening. Let us denote by E°, C°, X° and K° the corresponding structural constraints on s.o.m. models. By saying that a model is *of type* τ, where τ is a subset of $\{E^\circ, C^\circ, X^\circ, K^\circ\}$, we mean that it satisfies the constraints in τ.

Now, let $\mathbf{S}$ be an arbitrary subset of $\{E, C, X, K\}$. We say that an implication formula A is an $\mathbf{S}$-*theorem*, and write $\vdash_{\mathbf{S}} A$, if the sequent $\vdash A$ can be derived by using only the operational rules of **LJ** and the structural rules in $\mathbf{S}$. Let us denote by $\tau[\mathbf{S}]$ the set of structural constraints corresponding to the structural rules in $\mathbf{S}$ (the correspondence is summarised in Table 14.1).

Then, we can prove the following proposition.

Proposition 14.2. *A formula A is an* $\mathbf{S}$*-theorem if and only if A is true in all models of type* $\tau[\mathbf{S}]$.

Proof Let us say that a sequent $A_1, \ldots, A_n \vdash B$ is $\tau[\mathbf{S}]$-valid if the formula $A_1 \to (A_2 \to \cdots \to (A_{n-1} \to A_n) \cdots)$ is true in all models of type $\tau[\mathbf{S}]$. Then it can be easily checked that (i) all the sequents of the form $A \vdash A$ are $\tau[\mathbf{S}]$-valid, and (ii) the operational rules of **LJ** and the structural rules in $\mathbf{S}$ preserve $\tau[\mathbf{S}]$-validity. It follows that if A is an $\mathbf{S}$-theorem, it must be $\tau[\mathbf{S}]$-valid. This proves the only-if direction.

Table 14.1 *Correspondence between structural constraints and structural rules.*

$x \circ y \leq y \circ x$	$\dfrac{\Gamma, A, B, \Delta \vdash C}{\Gamma, B, A, \Delta \vdash C}$
$x \circ x \leq x$	$\dfrac{\Gamma, A, A, \Delta \vdash C}{\Gamma, A, \Delta \vdash C}$
$x \leq x \circ y$	$\dfrac{\Gamma, \Delta \vdash C}{\Gamma, A, \Delta \vdash C}$
$x \leq x \circ x$	$\dfrac{\Gamma, A, \Delta \vdash C}{\Gamma, A, A, \Delta \vdash C}$

For the if-direction, let us say that a set of formulas **U** is **S**-decreasing if it satisfies the following condition: $A \in \mathbf{U}$ and $A \to B$ is an **S**-theorem imply that $B \in \mathbf{U}$. (We call these sets 'decreasing' because $A \vdash B$ can be interpreted as 'the information carried by B is less than or equal to the information carried by A'.) Now, consider the 'canonical model' $\mathbf{m} = \langle W, \circ, 1, \leq, V \rangle$ where

- W is the set of all **S**-decreasing sets of formulas;
- $\circ$ is a binary operation defined as follows:

$$a \circ b =_{\text{def}} \{C \mid \exists A \in a, \exists B \in b, \vdash_{\mathbf{S}} A \to (B \to C)\};$$

- 1 is the set $\{A \mid \vdash_{\mathbf{S}} A\}$;
- the relation $\leq$ is set-inclusion;
- V is defined as follows:

$$a \in V(p) \Longleftrightarrow p \in a$$

  for all atomic formulas p and all a in W.

The reader can verify that the canonical model $\mathbf{m}$ is indeed a s.o.m. model of type $\tau[\mathbf{S}]$. For instance, the definition of $\circ$ immediately implies its associativity. To see that 1 is an identity element, reason as follows. First, $1 \circ a$ is equal, by definition, to $\{C \mid \exists A \in 1, \exists B \in a, \vdash_{\mathbf{S}} A \to (B \to C)\}$. Now, by the operational rules of **LJ**, $A \to (B \to C)$ is an **S**-theorem if and only if the sequent $A, B \vdash C$ is **S**-provable. So $1 \circ a$ is equal to the set of all C's such that $A, B \vdash C$ is **S**-provable, for some $A \in 1$ and some $B \in a$. Since 1 is the set of all **S**-theorems and the cut rule is admissible in **LJ**, $1 \circ a$ is equal to the set of all C's such that $B \vdash C$ is **S**-provable, i.e. $B \to C$ is an **S**-theorem, for some $B \in a$. But, since a is decreasing, such a set is equal to a itself. Hence, $1 \circ a = a$. A similar argument shows that $a \circ 1 = a$. All the other properties of models are easily seen to be satisfied by our canonical model. Moreover, it can be shown, by induction on the structure of A, that:

$$a \Vdash_{\mathbf{m}} A \Longleftrightarrow A \in a$$

holds true for every formula A. For the atomic case, this follows by the definition of V above and the standard definition of $\Vdash_{\mathbf{m}}$. Suppose that $A = B \to C$. Then, by definition of $\Vdash_{\mathbf{m}}$:

$$a \Vdash_{\mathbf{m}} B \to C \Longleftrightarrow (\forall b)(b \Vdash_{\mathbf{m}} B \Longrightarrow a \circ b \Vdash_{\mathbf{m}} C).$$

Let us denote by $[B]$ the set $\{D \mid \vdash_{\mathbf{S}} B \to D\}$. Clearly $[B] \in W$ and $B \in [B]$. So, by the inductive hypothesis, $[B] \Vdash_{\mathbf{m}} B$ and, therefore, $a \circ [B] \Vdash_{\mathbf{m}} C$. Again by the inductive hypothesis, $C \in a \circ [B]$ which implies, by definition of $\circ$, that $A \to (E \to C)$ is an **S**-theorem, i.e. the sequent $A, E \vdash C$ is **S**-provable, for some $A \in a$ and some $E \in [B]$. Since $B \vdash E$ is **S**-provable for all E in $[B]$, it follows by cut that $A, B \vdash C$ is **S**-provable, and therefore $A \to (B \to C)$ is an **S**-theorem, for some $A \in a$, which means that $B \to C \in a$. Thus we have shown that $a \Vdash_{\mathbf{m}} B \to C$ implies that $B \to C \in a$. The converse is easily shown by a similar argument.

Now, suppose that A is not an **S**-theorem. Then $1 \not\Vdash_{\mathbf{m}} A$ in the canonical model $\mathbf{m}$. So the canonical model is a countermodel to A of type $\tau[\mathbf{S}]$. This concludes our proof. □

The above argument shows that every substructural sequent system for implication is sound and complete with respect to the corresponding class of models. So, our task of formulating a uniform semantics for substructural implication systems is completed.

All the possible types of models, which can be defined by imposing different combinations of our four structural constraints $E^\circ, C^\circ, X^\circ$, and K°—and correspond to all the possible substructural implication systems contained in **LJ**—are summarised in Table 14.2, where each row corresponds to a type of model, while the first column indicates the logic characterised by each type.

Table 14.2 *Correspondence between implication systems and sets of structural constraints.*

	$x \circ y \leq y \circ x$	$x \circ x \leq x$	$x \leq x \circ x$	$x \leq x \circ y$
Lambek's implications				
Linear implication	•			
Relevant implication	•	•		
Mingle implication	•	•	•	
Intuitionistic implication	•	•	•	•
BCK implication	•		•	•

14.3 LKE

In this section, following the approach in [D'Agostino and Gabbay, 1994], we briefly show how the LDS methodology can be applied to construct a tableau-like refutation system for substructural implication logics. The latter consists of a generalisation of the classical refutation system **KE** ([D'Agostino and Mondadori, 1994]), where the labels can be interpreted as 'pieces of information' that may or may not verify a formula.

14.3.1 *The classical* **KE** *system*

The system **KE**, like the tableau method and resolution, is a refutation system for classical logic. Unlike resolution, however, **KE** is not restricted to clausal form and, unlike the tableau method, it includes a cut rule which *cannot*, in general, be eliminated. This classical cut rule is called PB (from Principle of Bivalence) and has the following forms, depending on whether we deal with signed or unsigned formulas:

$$\frac{}{\mathsf{T}\,A \mid \mathsf{F}\,A} \qquad\qquad \frac{}{A \mid \neg A}$$

The formula A introduced by an application of this rule is called a *PB-formula* or *cut formula*.[42] Once such a cut rule has been allowed, the branching elimination rules typical of the tableau method become unnecessarily strong and can be replaced by weaker *non-branching* rules (with two premises).

The rules of the **KE** system (for unsigned formulas) are illustrated in Table 14.3. The two-premiss elimination rules correspond to familiar principles of inference: *modus ponens, modus tollens, disjunctive syllogism* and its dual. The one-premiss elimination rules are the same as the tableau rules. A **KE**-refutation of a set of formulas Γ is, as usual, a closed tree of formulas constructed according to the rules of **KE** starting from formulas in Γ.

A crucial property of **KE** is the *analytic cut property*: the applications of the cut rule can be restricted to subformulas of the formulas occurring above in the branch without loss of completeness This property allows for systematic and efficient refutation procedures. Indeed, results in [D'Agostino and Mondadori, 1994] imply that any refutation procedure which can be formulated in terms of the tableau rules can be efficiently (linearly) simulated by means of the **KE** rules, but there are efficient and systematic **KE**-procedures which cannot be polynomially simulated by means of the tableau rules. One of these procedures is *the canonical procedure*, consisting in giving priority to the linear elimination rules over the cut rule, so that the cut rule is applied only when no elimination rule is further applicable, and the choice of the cut formulas is restricted to pairs $A,\neg A$ such that A is an immediate subformula of a formula of type β (in the Smullyan notation) occurring above in the branch.

14.3.2 **LKE** *for substructural implication*

An inferential characterisation of substructural implication logics can be obtained by turning the 'semantics' described in Section 14.2.2 into the rules of a labelled deductive system (in the sense of [Gabbay, 1996b]). This will take the form of a labelled generalisation of the classical system **KE** introduced in the previous scion. We shall call this

[42] To see that PB plays the same role as the cut rule in the classical sequent calculus, think of it as a rule that allows one to construct a closed **KE**-tree for Γ, given a closed **KE**-tree for Γ, A and a closed **KE**-tree for $\Gamma, \neg A$; for a discussion of this point see [D'Agostino and Mondadori, 1994].

Table 14.3 **KE**-*rules for unsigned formulas.*

Disjunction rules

$$\frac{\begin{array}{l}A \vee B\\ \neg A\end{array}}{B}\,\mathrm{E}\vee 1 \qquad \frac{\begin{array}{l}A \vee B\\ \neg B\end{array}}{A}\,\mathrm{E}\vee 2 \qquad \frac{\neg(A \vee B)}{\begin{array}{l}\neg A\\ \neg B\end{array}}\,\mathrm{E}\neg\vee$$

Conjunction rules

$$\frac{\begin{array}{l}\neg(A \wedge B)\\ A\end{array}}{\neg B}\,\mathrm{E}\neg\wedge 1 \qquad \frac{\begin{array}{l}\neg(A \wedge B)\\ B\end{array}}{\neg A}\,\mathrm{E}\neg\wedge 2 \qquad \frac{A \wedge B}{\begin{array}{l}A\\ B\end{array}}\,\mathrm{E}\wedge$$

Implication rules

$$\frac{\begin{array}{l}A \to B\\ A\end{array}}{B}\,\mathrm{E}{\to}1 \qquad \frac{\begin{array}{l}A \to B\\ \neg B\end{array}}{\neg A}\,\mathrm{E}{\to}2 \qquad \frac{\neg(A \to B)}{\begin{array}{l}A\\ \neg B\end{array}}\,\mathrm{E}\neg\to$$

Negation rule

$$\frac{\neg\neg A}{A}\,\mathrm{E}\neg\neg$$

Principle of bivalence

$$\frac{}{A \mid \neg A}\,\mathrm{PB}$$

generalisation **LKE**. The rules of **LKE** are tree-expansion rules which are immediately justified by—and are indeed equivalent to—our previous definitions (see [D'Agostino and Gabbay, 1994] for a systematic introduction to the **LKE** system and also [Broda *et al.*, 1998] for a detailed discussion of the route which leads from sequent-based presentation of substructural consequence relations to **LKE** via algebraic and Kripke-style semantics). The rules for the implication fragment are listed in Table 14.4 (see [D'Agostino and Gabbay, 1994] for a more detailed discussion). Notice that the use of *signed* formulas is crucial in this type of system.

Notice that the only branching rule can be seen as a labelled generalisation of the classical principle of bivalence. Observe also that a simple special case of the closure rule given in the table is the following:

Table 14.4 *The* **LKE** *rules for substructural implication.*

$$\frac{\begin{array}{l}\mathsf{T}\,A\rightarrow B:x\\ \mathsf{T}\,A:y\end{array}}{\mathsf{T}\,B:x\circ y}\qquad\qquad \frac{\begin{array}{l}\mathsf{T}\,A\rightarrow B:x\\ \mathsf{F}\,B:x\circ y\end{array}}{\mathsf{F}\,A:y}$$

$$\frac{\mathsf{F}\,A\rightarrow B:x}{\begin{array}{l}\mathsf{T}\,A:a\\ \mathsf{F}\,B:x\circ a\end{array}}\ \text{where } a \text{ is a } \textit{new} \text{ atomic label}$$

$$\frac{}{\mathsf{T}\,A:x\ \big|\ \mathsf{F}\,A:x}\qquad\qquad \frac{\begin{array}{l}\mathsf{T}\,A:x_1\\ \quad\vdots\\ \mathsf{T}\,A:x_n\\ \mathsf{F}\,A:y\end{array}}{\times}\ \text{if } \sqcap(x_1,\ldots,x_n)\leq y$$

$$\frac{\begin{array}{l}\mathsf{T}\,A:x\\ \mathsf{F}\,A:y\end{array}}{\times}\ \text{provided } x\leq y$$

Indeed this special case is sufficient for completeness except for the implication systems which do not allow weakening but do allow expansion (as, for instance, the system of mingle implication; see [D'Agostino and Gabbay, 1994] for more details).

In these rules the declarative units are not just signed formulas as in the classical **KE** system (or in the system of analytic tableaux) but *labelled signed formulas*, or *LS-formulas* for short.[43] The points of our s.o.m. models are turned into 'labels', while signs play the usual role, so that $\mathsf{T}\ A : x$ is interpreted as 'A is true at point x' and $\mathsf{F}\ A : x$ is interpreted as 'A is *not true* at point x'. Different types of s.o.m. models—defined in terms of the structural constraints which are imposed—correspond to different *labelling algebras*, i.e. sets of rules that can be used in manipulating the labelling terms to verify whether or not the condition for the application of the closure rule is satisfied.

For each implication system, the *labelling language* consists of a denumerable set of atomic labels, also called *parametric labels* denoted by 'a, b, c', etc. (possibly with subscripts), a distinguished atomic label '1', called *the unit label*, the identity symbol '$=$', and two binary operation symbols '$\circ$' and '$\sqcap$'. Complex labels are built up from the atomic ones by means of the binary operation symbols. The *basic labelling algebra*, which applies to all s.o.m. models, consists of the following ingredients:

1. the axioms expressing the fact that $\circ$ is a monoid operation with 1 as identity, i.e.

$$x\circ(y\circ z) = (x\circ y)\circ z \tag{14.1}$$

$$1\circ x = x\circ 1 = x; \tag{14.2}$$

2. the axioms expressing the fact that $\sqcap$ is a semilattice meet:

$$x\sqcap(y\sqcap z) = (x\sqcap y)\sqcap z \tag{14.3}$$

[43] A similar approach is used by Fitting in his 'prefixed' tableaux for classical modal logics [Fitting, 1983].

$$x \sqcap y = y \sqcap x \tag{14.4}$$
$$x \sqcap x = x; \tag{14.5}$$

3. the axioms expressing the distributivity of $\sqcap$ over $\circ$ (this is equivalent to the property that $\circ$ is order-preserving):

$$(x \sqcap y) \circ z = (x \circ z) \sqcap (y \circ z) \tag{14.6}$$
$$z \circ (x \sqcap y) = (z \circ x) \sqcap (z \circ y). \tag{14.7}$$

Clearly all the variables x, y, z are intended as being universally quantified. The partial ordering $\leq$ can be defined, as usual, by putting $x \leq y$ if and only if $x \sqcap y = x$. Observe that the above axioms implies that $\leq$ is order-preserving, i.e.

$$x \leq y \Longrightarrow x \circ z \leq y \circ z \qquad x \leq y \Longrightarrow z \circ x \leq z \circ y. \tag{14.8}$$

This basic labelling algebra augmented with a set τ of structural constraints, provides a specific labelling algebra sufficient to characterise the notion of validity in the class $\mathcal{M}_{\langle\tau\rangle}$, i.e. the class of all models satisfying the constraints in τ. We shall denote by **LKE**$\langle\tau\rangle$, the **LKE** system equipped with the labelling algebra obtained by adding the constraints in τ to the basic labelling algebra.

A proof of the validity of a formula A for the class $\mathcal{M}_{\langle\tau\rangle}$ consists in a refutation of the assumption that A is false at the identity element 1 of some model in $\mathcal{M}_{\langle\tau\rangle}$. Such a refutation is represented by a closed **LKE**$\langle\tau\rangle$-tree starting with the LS-formula $\mathsf{F}\ A : 1$, where the constraints in τ may be used together with the basic labelling algebra in order to close a branch. This means that the closure condition $\sqcap(x_1, \ldots, x_n) \leq y$ may be shown to hold on the basis of the labelling algebra. Whenever such a closed tree can be constructed, we say that A is a **LKE**$\langle\tau\rangle$-theorem.

We shall also denote by $\vdash_{\mathbf{LKE}\langle\tau\rangle}$ the (finitary) deducibility relation of **LKE**$\langle\tau\rangle$ defined as follows (where [ ] represents the empty list of formulas):

1. $[\,] \vdash_{\mathbf{LKE}\langle\tau\rangle} A$ *iff* A is an **LKE**$\langle\tau\rangle$-theorem;
2. $A_1, \ldots, A_n \vdash_{\mathbf{LKE}\langle\tau\rangle} B$ *iff* $[\,] \vdash_{\mathbf{LKE}\langle\tau\rangle} A_1 \rightarrow (A_2 \rightarrow \cdots \rightarrow (A_n \rightarrow B) \cdots)$.

We shall also write simply$\vdash_{\mathbf{LKE}\langle\tau\rangle}$ instead of $[\,] \vdash_{\mathbf{LKE}\langle\tau\rangle} A$. This definition implies that $A_1, \ldots, A_n \vdash_{\mathbf{LKE}\langle\tau\rangle} B$ if and only if there is a closed **LKE**$\langle\tau\rangle$-tree starting with the sequence of LS-formulas

$$\mathsf{T}\ A_1 : a_1, \ldots, \mathsf{T}\ A_n : a_n, \mathsf{F}\ B : a_1 \circ \cdots \circ a_n$$

where $a_1, \ldots, a_n$ are all *distinct atomic* labels.

In this approach the whole family of substructural implication logics is, therefore, characterised by *the same* tree-expansion rules, and different members of the family are identified by the different labelling algebras that can be employed to check branch-closure. This is what we call the *separation-by-closure* property of the **LKE** system. Another remarkable property of **LKE** is the *atomic closure property:* if there is a closed tree starting from a given set of initial LS-formulas, then there is also one starting from

Table 14.5 *The contraction axiom is valid for the class of contractive models (i.e. those satisfying the condition $a \circ a \leq a$, for all a). For, $b \circ b \leq b$ and, since $\circ$ is order-preserving, $a \circ b \circ b \leq a \circ b$. Therefore, this one-branch tree is closed. The axiom is not valid in Linear Logic which is characterised by a class of models which are not contractive.*

$$
\begin{array}{l}
\mathsf{F}\ (A \to (A \to B)) \to (A \to B) : 1 \\
\mathsf{T}\ A \to (A \to B) : a \\
\mathsf{F}\ A \to B : 1 \circ a (= a) \\
\mathsf{T}\ A : b \\
\mathsf{F}\ B : a \circ b \\
\mathsf{T}\ A \to B : a \circ b \\
\mathsf{T}\ B : a \circ b \circ b
\end{array}
$$

the same set of LS-formulas such that the closure rule is applied only with *atomic* LS-formulas, i.e. LS-formulas of the form $sp : x$ (where $s = T$ or F) with p atomic.

An example of an **LKE**-proof is given in Table 14.5.

This example does not make any use of the branching rule of 'generalised bivalence'. Indeed, this rule introduces a good deal of 'non-determinism' into the system, in that it allows for the use of (a) arbitrary formulas and (b) arbitrary labels in each rule application. However, this 'non-determinism' can be tamed to some extent. As for (a), it can be shown that the applications of PB can be restricted to *analytic* ones, i.e. involving only subformulas of formulas previously occurring in the branch. Let us say that an **LKE**-tree enjoys the *subformula property* if for every LS-formula $sA : x$ occurring in it, where s is one of the two signs T or F, the formula A is a subformula of some formula B such that $sB : y$, for some s and some y, occurs above in the same branch. Let us now say that two closed **LKE**-trees are *equivalent* if they are constructed starting from the same set of initial LS-formulae. Then it can be shown that:

Proposition 14.3. *There is a procedure to turn every closed* **LKE**-*tree into an equivalent one which enjoys the subformula property.*

The proof of this proposition (that we omit here) involves showing that all the applications of the labelled PB rule which violate the subformula property can be eliminated without loss. Indeed, it can also be shown that the applications of the PB rule can be restricted even further to *canonical* applications as in the canonical procedure for the classical system outlined in Section 14.3.1.

As for (b), it must be remarked that in order to obtain a fully mechanical refutation procedure we need also to 'tame' the non-determinism related to the use of labels in the generalised PB rule, i.e. we have to develop a procedure which terminates either with a closed tree or with a tree containing enough information to construct a counter-model. To solve this problem, the best strategy is to apply the PB rule with a *variable* as label. This means that branches may be closed under appropriate substitutions of the variables with terms of the labelling algebra. An example of this use of variable labels is given in Table 14.6. So, the closure of a branch ultimately depends on the solution of an inequality in the given algebra of the labels and the closure of the whole tree on the simultaneous solution of a system of inequalities. This is a well-defined algebraic problem which can be addressed via unification-like techniques.

Table 14.6 *This tree is closed for all commutative models under the substitution* $x = 1$.

$$F((A \to A) \to B) \to ((B \to C) \to C : 1$$

$$\mathsf{T}\ (A \to A) \to B : a$$

$$\mathsf{F}\ (B \to C) \to C : a$$

$$\mathsf{T}\ B \to C : b$$

$$\mathsf{F}\ C : a \circ b$$

$\mathsf{T}\ A \to A : x$	$\mathsf{F}\ A \to A : x$
$\mathsf{T}\ B : a \circ x$	$\mathsf{T}\ A : c$
$\mathsf{T}\ C : b \circ a \circ x$	$\mathsf{F}\ A : x \circ c$

The solution of this problem crucially depends on what structural constraints are allowed in the algebra of the labels and on the development of suitable algorithms for solving the systems of inequalities associated with **LKE**-trees which arise when variable labels are used in PB-applications. It seems plausible that efficient decision algorithms based on our method will have to be logic-specific, exploiting the computational properties of each given labelling algebra.

14.4 Fibring Substructural Implication Logics

Let us consider two implication systems S_1 and S_2, e.g. the implication fragments of linear logic and intuitionistic logic respectively. How can we combine these two logics into one 'fibred' system which is a conservative extension of both? Is the result of this fibring uniquely determined by the component logics or do we have options? To answer these question we need some preliminary investigation.

14.4.1 *Preliminary investigations*

Let $\mathbb{L}_1, \mathbb{L}_2, \ldots, \mathbb{L}_n$ be propositional languages each of which is built up in the usual way from an infinite stock of propositional variables (denoted as before by $p, q, r, \ldots$) and the binary operator $\rightarrow_i$. We can consider 'mixed' languages formed in the obvious way, for instance the language $\mathbb{L}_{13}$ defined as the smallest language containing both $\mathbb{L}_1$ and $\mathbb{L}_3$, and closed under the following conditions:

- if A and B are in $\mathbb{L}_{13}$, then $A \rightarrow_1 B$ is also in $\mathbb{L}_{13}$;
- if A and B are in $\mathbb{L}_{13}$, then $A \rightarrow_3 B$ is also in $\mathbb{L}_{13}$.

In the sequel we shall denote by $\mathbb{L}_s$, where s is any sequence of numerals, the language which arises from fibring the implication languages $\mathbb{L}_i$, with $i \in s$, as explained above.

In order to evaluate formulas belonging to a fibred language in a fibred structure we need to classify them according to their main operator. We can think of a fibred structure as the combination of several s.o.m. models of a given type, and an element of a s.o.m. model of type, say, τ_i 'understands' only two types of formulas: atomic formulas (propositional letters) or formulas whose main operator is $\rightarrow_i$. So, given a fibred language $\mathbb{L}_s$, we shall denote by $\mathbb{L}_s^i$ the sublanguage of $\mathbb{L}_s$ consisting only of all the atomic formulas and all the formulas with $\rightarrow_i$ as main operator.

In general, the fibring methodology involves the consideration of two types of functions:

1. 'point-to-model' functions, $\mathbf{F}^{\tau}_{\mathbf{m}}$, mapping each element of a model $\mathbf{m}$ to a model of type τ;
2. 'point-to-point' functions, $\psi^{\mathbf{m}'}_{\mathbf{m}}$, mapping each element of a model $\mathbf{m}$ to a point of another model $\mathbf{m}'$ (usually of different type).

More specifically, given a model $\mathbf{m}$, and a point $a \in \mathbf{m}$, in order to evaluate a formula $A \rightarrow_i B$ which is not 'understood' by $\mathbf{m}$, we call a function $F^{\tau_i}_{\mathbf{m}}$ which maps a to a model $\mathbf{m}'$ where $A \rightarrow_i B$ *is* understood. Next, since valuation of formulas takes place at a specific point of a model, we must pick up a point a' belonging to $\mathbf{m}'$ which is associated with a and satisfies certain persistence properties, e.g. a' must verify all the atomic formulas which are verified by a. For this purpose we consider a function $\psi^{\mathbf{m}'}_{\mathbf{m}}$ mapping every point of $\mathbf{m}$ to a point of $\mathbf{m}'$ where we can proceed with the evaluation of $A \rightarrow_i B$. This may, in turn, force us to invoke other point-to-model and point-to-point functions, until the bottom stage of atomic formulas (which are 'understood' by any point in any model) is reached.

The basic idea of fibring is captured by the following definition:

Definition 14.4. Let $\mathbb{L}_{1\ldots n}$ be a language containing n operators $\rightarrow_1, \ldots, \rightarrow_n$, where each $\rightarrow_i$ is characterised by the class $\mathcal{M}_{\langle \tau_i \rangle}$, i.e. the class of all s.o.m. models satisfying the structural constraints in τ_i. A *fibred structure* $\mathcal{S}$ for $\mathbb{L}_{1\ldots n}$ is a triple $\langle \mathcal{B}, \mathcal{F}, \Psi \rangle$, such that

- $\mathcal{B}$ is subset of $\mathcal{M}_{\langle\tau_1\rangle} \cup \ldots \cup \mathcal{M}_{\langle\tau_n\rangle}$, $\mathcal{F}$ is a set of point-to-model functions and Ψ is a set of point-to-point functions;
- for every $\mathbf{m} \in \mathcal{B}$ and every j such that $\mathbf{m} \notin \mathcal{M}_{\langle\tau_j\rangle}$, the set $\mathcal{F}$ contains a function $\mathbf{F}_{\mathbf{m}}^{\tau_j}$ of type $\mathbf{m} \mapsto \mathcal{M}_{\langle\tau_j\rangle}$;
- $\mathcal{B}$ is closed under all the functions in $\mathcal{F}$, i.e. for every $\mathbf{m} \in \mathcal{B}$, every $a \in \mathbf{m}$, and every $\mathbf{F}_{\mathbf{m}}^{\tau_j} \in \mathcal{F}$, the model $\mathbf{F}_{\mathbf{m}}^{\tau_j}(a)$ is also in $\mathcal{M}$;
- for every $\mathbf{m}, \mathbf{m}'$ such that $\mathbf{m}' = \mathbf{F}_{\mathbf{m}}^{\tau_i}(a)$ for some $a \in \mathbf{m}$ and some $\mathbf{F}_{\mathbf{m}}^{\tau_i}$ in $\mathcal{F}$, the set Ψ contains a function $\psi_{\mathbf{m}}^{\mathbf{m}'}$ of type $\mathbf{m} \mapsto \mathbf{m}'$ satisfying the following condition for all atomic formulas p and all $a \in \mathbf{m}$:

$$a \in V_{\mathbf{m}}(p) \Longleftrightarrow \psi_{\mathbf{m}}^{\mathbf{m}'}(a) \in V_{\mathbf{m}'}(p).$$

Now, we must define the notion of truth at a given point a of a fibred structure $\mathcal{S}$:

Definition 14.5. Let a be an element of a model $\mathbf{m}$ belonging to a fibred structure $\mathcal{S}$. The relation $a \Vdash_{\mathbf{m}}^{\mathcal{S}} A$ (read 'a verifies A relative to $\mathbf{m}$ and $\mathcal{S}$') is defined as follows:

1. for all atomic p

$$a \Vdash_{\mathbf{m}}^{\mathcal{S}} p \Longleftrightarrow x \in V(p);$$

2. for all formulas A, B

$$a \Vdash_{\mathbf{m}}^{\mathcal{S}} A \to_i B \Longleftrightarrow \begin{cases} (\forall b)(b \Vdash_{\mathbf{m}}^{\mathcal{S}} A \Longrightarrow a \circ_i b \Vdash_{\mathbf{m}}^{\mathcal{S}} B) & \text{if } \mathbf{m} \in \mathcal{S}_i \\ \psi_{\mathbf{m}}^{\mathbf{m}'}(a) \Vdash_{\mathbf{m}'}^{\mathcal{S}} A \to_i B & \text{otherwise} \end{cases}$$

where $\mathbf{m}' = \mathbf{F}_{\mathbf{m}}^{\tau_i}(a)$.

As usual, a formula A is true in a model $\mathbf{m}$ of a fibred structure $\mathcal{S}$ if it is true at the identity point of $\mathbf{m}$. Moreover, we say that formula A belonging to $\mathbb{L}_s^i$ is *true in a fibred structure* $\mathcal{S}$ if A is true in every $\mathbf{m} \in \mathcal{S}$ of type τ_i. Finally, A is *valid* if it is true in all fibred structures for the appropriate language.

Let us now examine the options we have in specifying our fibring functions. For the sake of clarity let us consider the example of a fibred semantics for a language consisting of two implication operators: $\to_1$, interpreted as linear implication, and $\to_2$ interpreted as intuitionistic implication.

Example 14.6. Suppose we are given a formula

$$p \to_1 (q \to_2 p).$$

Is this formula valid? i.e. is it true in all fibred structures for the given language?

Let $\mathcal{S} = \langle \mathcal{B}, \mathcal{F}, \Psi \rangle$ be an arbitrary fibred structure for the language under consideration and let $\mathbf{m}$ be a model of type τ_1 in $\mathcal{B}$. If the above formula is valid, it must be true at the identity of $\mathbf{m}$.

Now, the formula $p \to_1 (q \to_2 p)$ is true at the identity of $\mathbf{m}$ if and only if the following condition is satisfied:

$$(\forall a)(a \in V_{\mathbf{m}}(p) \Longrightarrow a \Vdash_{\mathbf{m}} q \to_2 p).$$

Since $\mathbf{m}$ does not 'understand' the language of intuitionistic implication, the above condition translates to the following one:

$$(\forall a)(a \in V_{\mathbf{m}}(p) \Longrightarrow \psi_{\mathbf{m}}^{\mathbf{m}'}(a) \Vdash_{\mathbf{m}'} q \to_2 p)$$

where $\mathbf{m}' = \mathbf{F}_{\mathbf{m}}^{\tau_2}(a)$.

According to the definition of $\psi_{\mathbf{m}}^{\mathbf{m}'}$, the points a and $\psi_{\mathbf{m}}^{\mathbf{m}'}(a)$ verify exactly the same atomic formulas. So, we have that $\psi_{\mathbf{m}}^{\mathbf{m}'}(a) \in V_{\mathbf{m}'}(p)$. Now, since $\mathbf{m}'$ is model for intuitionistic implication, it follows from the properties of such models that $\psi_{\mathbf{m}}^{\mathbf{m}'}(a) \Vdash_{\mathbf{m}'} q \to p$, whatever q may be. Therefore the above condition is always satisfied. This means that the formula $p \to_1 (q \to_2 p)$ is true in all models of type τ_1 of every fibred structure, i.e. it is a valid formula.

The above examples show that even without specifying any options for the fibring functions, we can characterise some formulas as valid. There are, however, some natural conditions that we may want to impose on the 'fibred' logic that force corresponding constraints on the fibring functions.

For instance, we may want the 'persistence' condition to keep holding in the fibred structures in its full generality. This means that for every fibred structure S, every $\mathbf{m} \in S$ and every point $a \in \mathbf{m}$ the following must hold true:

$$x \leq y \;\&\; x \Vdash_m^S A \to_i B \Longrightarrow y \Vdash_m^S A \to_i B$$

no matter the type of $\mathbf{m}$.

Now, when $\mathbf{m}$ is not of type τ_i, $\mathbf{F}_{\mathbf{m}}^{\tau_i}(x)$ and $\mathbf{F}_{\mathbf{m}}^{\tau_i}(y)$ may have different values, i.e. may pick up different models of type τ_i. And since the value of a conditional formula at a point crucially depends on the model in its globality, we may well obtain different values for $A \to_i B$ at the points $\psi_{\mathbf{m}}^{\mathbf{m}'}(x)$ and $\psi_{\mathbf{m}}^{\mathbf{m}''}(y)$.

A direct way of enforcing the persistence property consists in requiring that the function $\mathbf{F}_{\mathbf{m}}^{\tau_i}$ maps x and y to the same model whenever $x \leq y$. This, together with the given properties of the point-to-point function ψ, is sufficient to achieve the desired result.

Another requirement we probably want to consider is that logical validity be preserved under substitution of atomic formulas. This implies, to pursue the example above, that any substitution of a complex formula for p in the formula $p \to_1 (q \to_2 p)$ must yield a valid formula, for instance the formula

$$(p \to_1 r) \to_1 (q \to_2 (p \to_1 r))$$

must be valid.

Now, to be valid, this formula must be true at the identity element $1_{\circ_1}$ of every model $\mathbf{m}$ of type τ_1 in every fibred structure. This means that for every point $a \in \mathbf{m}$, we must have that if $a \Vdash_{\mathbf{m}} p \rightarrow_1 r$, then $a \Vdash_{\mathbf{m}} q \rightarrow_2 (p \rightarrow_1 r)$. Since a does not 'understand' $q \rightarrow_2 (p \rightarrow_1 r)$, we must invoke the fibring function $\mathbf{F}^{\tau_2}_{\mathbf{m}}$ to jump to a model $\mathbf{F}^{\tau_2}_{\mathbf{m}}(a) = \mathbf{m}'$ and then the point-to-point function $\psi^{\mathbf{m}'}_{\mathbf{m}}$ to pick up a point $\psi^{\mathbf{m}'}_{\mathbf{m}}(a)$. For such a point, call it b, we must have $b \Vdash_{\mathbf{m}'} q \rightarrow_2 (p \rightarrow_1 r)$. Now, since q is arbitrary, this is true, in general, only if $b \Vdash_{\mathbf{m}'} p \rightarrow_1 r$. Since b does not 'understand' the formula $p \rightarrow_1 r$, we must invoke the fibring function $\mathbf{F}^{\tau_1}_{\mathbf{m}'}$ to jump to a model $\mathbf{m}''$ of type τ_1 and then the function $\psi^{\mathbf{m}''}_{\mathbf{m}'}$ to pick up a specific point c in $\mathbf{m}''$. For this c we must have that $c \Vdash_{\mathbf{m}''} p \rightarrow_1 r$. There is no guarantee that this is the case unless we impose some special conditions on the fibring functions. A very direct way of ensuring the desired result is that $c = a$, in other words that the fibring functions bring us back to the original model and to the original point a.

Instead of trying to cope with the proliferation of models by imposing convoluted conditions on the fibring functions, it seems then better to simplify our fibred structures and require that all the different models are connected in a particularly strong way. Indeed, our discussion suggests that the best option for fibring substructural implication logics is to 'dovetail' models of different type yielding a *single* simplified model.

14.4.2 *Simplified fibred models*

We can generate fibred models, starting from pairwise disjoint component models, by applying algebraic fibring techniques similar to the ones which we discussed in Section 20.3. For the purpose of this chapter we shall concentrate on the resulting structures which can be seen as generalising the notion of s.o.m. model of Chapter 14 in a straightforward way.

Definition 14.7. A *fibred s.o.m. model* is a structure

$$\mathbf{m} = \langle W, \circ_1, \ldots, \circ_n, 1_{\circ_1}, \ldots, 1_{\circ_n}, \leq, V\rangle$$

where

- W is closed under all the compositions $\circ_i$, with $i = 1, \ldots, n$;
- each $\circ_i$ is order-preserving in both arguments (see Definition 14.1 above);
- $1_{\circ_i} \in W$ for each $i = 1, \ldots, n$;
- $\leq$ is a semilattice ordering of W;
- V is a valuation of atomic formulas satisfying the persistence and the (downward) continuity conditions of Definition 14.1;
- every substructure $\mathbf{m}_i = (W, \circ_i, 1_{\circ_i}, \leq, V)$ is a s.o.m. model of type τ_i.

The *type* of a fibred model $\mathbf{m}$ with n composition operators is an ordered n-tuple $\langle \tau_1, \ldots, \tau_n\rangle$ where each τ_i is the type of the i-th composition operator.

So, all the different types of models are replaced by a single class of fibred models.

We are now able to evaluate the formulas of a fibred language $\mathbb{L}_{1\ldots n}$ in a fibred model $\mathbf{m}$ of type $\langle \tau_1, \ldots, \tau_n\rangle$ in the expected way. Every formula belonging to $\mathbb{L}^i_s$ is evaluated by means of the usual clause:

$$a \Vdash_{\mathbf{m}} A \rightarrow_i B \Longleftrightarrow (\forall b \in \mathbf{m})(b \Vdash_{\mathbf{m}} A \Longrightarrow a \circ_i b \Vdash_{\mathbf{m}} B).$$

A formula A in $\mathbb{L}^i_s$ is *true* in a fibred model $\mathbf{m}$ if it is true at the appropriate identity element, i.e. if $1_{\circ_i} \Vdash_{\mathbf{m}} A$. As usual, a formula is *valid* if it is true in all fibred s.o.m. models.

Let us denote by $\mathcal{M}_{\langle\tau_1,\ldots,\tau_n\rangle}$ the class of all fibred models of type $\langle\tau_1,\ldots,\tau_n\rangle$, and by $\mathcal{L}_{\langle\tau_1,\ldots,\tau_n\rangle}$ the corresponding 'logic', i.e. the set of formulas which are valid in all models of this class.

Proposition 14.8. *Every logic $\mathbf{L}_{\langle\tau_1,\ldots,\tau_n\rangle}$ is a conservative extension of $\mathbf{L}_{\langle\tau_i\rangle}$, for each $i = 1,\ldots,n$.*

Proof It is not difficult to verify that for every model $\mathbf{m} \in \mathcal{M}_{\langle\tau_1,\ldots,\tau_n\rangle}$, every $a \in \mathbf{m}$ and every $B \in \mathbb{L}_i$, $a \Vdash_{\mathbf{m}} B$ if and only if $a \Vdash_{\mathbf{m}^i} B$, where $\mathbf{m}_i$ is the submodel of type τ_i of $\mathbf{m}$ (this is easily shown by induction on the complexity of B). Now, if A were false in a fibred model $\mathbf{m}$, we should have $1_{\circ_i} \nVdash_{\mathbf{m}} A$. Therefore $1_{\circ_i} \nVdash_{\mathbf{m}_i} A$. But, if A is true in all models of type τ_i, this is impossible. Hence, A must be true in all fibred models. □

14.4.3 *Structured consequence relations*

Let us consider a mixed language $\mathbb{L}_{1\ldots n}$ with n operators $\rightarrow_1,\ldots,\rightarrow_n$ denoting different substructural implications. From the previous section we know how to generate a class of fibred (or better dovetailed) models for this language and how to characterise a 'fibred logic' (intended as a set of valid formulas) which is a conservative extension of the component logics. We can now ask ourselves: what is a *consequence relation* in this new environment?

The consequence relation associated with each single implication system was expressed in terms of *sequents*, i.e. expressions of the form $\Gamma \vdash A$, where Γ is a *sequence* of formulas. In different consequence relations, however, the sequences of formulas on the left of the turnstile identify different types of *structures* whose properties are captured by the set of allowed structural rules. If we mix different notions of implication in the same language, all these different structures must learn to cohabit and co-operate without interfering with each other. For this purpose we need to update our notion of consequence relation.

As a preliminary step we augment our fibred language $\mathbb{L}_{1\ldots n}$ with n unary constants $\top_1,\ldots,\top_n$. These additional constants play the following role: each $\top_i$ implies every valid formula in $\mathbb{L}^i_{1\ldots n}$. Moreover, we replace the standard commas in the antecedent of a sequent with a variety of composition operators $\bullet_1,\ldots,\bullet_n$ corresponding to the variety of conditionals. The type *structure-of-formulas*, that we shall denote by $\mathcal{F}$, is defined as the smallest set closed under the following conditions:

1. all the constants $\top_1,\ldots,\top_n$ belong to $\mathcal{F}$;
2. every formula A of $\mathbb{L}_{1\ldots n}$ belongs to $\mathcal{F}$;
3. if X and Y are in $\mathcal{F}$, each $X \bullet_i Y$, for $i = 1,\ldots,n$, is also in $\mathcal{F}$.

For each constant $\top_i$ we have that $X \bullet_i \top_i = \top_i \bullet_i X = X$ for every structure X. We shall use the notation $X[Y]$ to indicate a structure that contains some occurrence of the structure Y. With the notation $X[Y/Z]$ we shall ambiguously denote any structure that results from replacing any occurrence of Y with Z in $X[Y]$.

Definition 14.9. A *structured consequence relation* is a relation $\vdash$ between structures of formulas and formulas satisfying:

(Identity) $$A \vdash A$$

(Surgical cut) $$X \vdash A \text{ and } Y[A] \vdash B \Longrightarrow Y[A/X] \vdash B.$$

We can now define the notion of a 'multi-implication' system.

Definition 14.10. A *multi-implication system* is a structured consequence relation satisfying the following additional condition, for every $\rightarrow_i$:

$C_{\rightarrow_i}$ $$X \vdash A \rightarrow_i B \Longleftrightarrow X \bullet_i A \vdash B$$

and such that each $\bullet_i$ satisfies a given subset of the following structural rules:

Exchange $$\frac{X[A \bullet_i B] \vdash C}{X[A \bullet_i B/B \bullet_i A] \vdash C}$$

Contraction $$\frac{X[A \bullet_i A] \vdash C}{X[A \bullet_i A/A] \vdash C}$$

Expansion $$\frac{X[A] \vdash C}{X[A/A \bullet_i A] \vdash C}$$

Weakening $$\frac{X[Y] \vdash C}{X[Y/Y \bullet_i A] \vdash C}.$$

In this new setting the variety of consequence relations is replaced by a single structured consequence relation with a variety of substructural implication operators.

14.4.4 *The fibred canonical model*

We now describe a concrete example of a fibred s.o.m. model defined in terms of the deducibility relation $\vdash$ associated with any given multi-implication system. We call this concrete model *the fibred canonical model.*

The points of the fibred canonical model are, as in the case of the canonical model for standard implication systems, decreasing sets of formulas, i.e. if A belongs to any such set and $A \vdash B$, then B also belongs to the set, where $\vdash$ is the deducibility relation of an arbitrary multi-implication system.

The *fibred canonical model* is a structure $(W, \circ_1, \ldots, \circ_n, 1_{\circ_1}, \ldots, 1_{\circ_n}, \leq, V)$ such that:

- W is the set of all decreasing sets of formulas with respect to the given consequence relation $\vdash$;

- each $\circ_i$ is a binary operation defined as follows:

$$a \circ_i b =_{\text{def}} \{C \mid \exists A \in a, \exists B \in b, \top_i \vdash A \rightarrow_i (B \rightarrow_i C)\}.$$

- $1_{\circ_i}$ is defined as the set $\{A \mid \top_i \vdash A\}$;
- the relation $\leq$ is set-inclusion;
- the *canonical valuation* V over M is defined as usual:

$$a \in V(p) \Longleftrightarrow p \in a.$$

It is not difficult to verify that the fibred canonical model satisfies all the clauses of the definition of a s.o.m. model.

By considering this canonical model, it is routine to prove the following:

Theorem 14.11. *Let S be a multi-implication system and let $\mathbb{L}_{1\ldots n}$ be the underlying fibred language. Let also τ_i be the set of structural constraints corresponding to the set of allowed structural rules for $\rightarrow_i$ (according to the correspondence outlined in Table 14.1) and $\mathcal{M}_{\langle\tau_1,\ldots,\tau_n\rangle}$ be the class of all models of type $\langle\tau_1,\ldots,\tau_n\rangle$. Then for every formula $A \in \mathbb{L}^i_{1\ldots n}$,*

$$\top_i \vdash_S A \Longleftrightarrow A \textit{ is valid for } \mathcal{M}_{\langle\tau_1,\ldots,\tau_n\rangle}.$$

14.5 LKE for Multi-implication Systems

It is easy to generalise the **LKE** system to cover multi-implication systems. First we extend the labelling language to include a variety of binary operation symbols, each with its identity element.

Definition 14.12. A *labelling language* $\mathbf{L}^n$ *of type* n consists of the following ingredients:

- a denumerable set of symbols called *parametric labels*;
- n special symbols called *unit labels*;
- n binary operation symbol called *composition operators*;
- a binary operation symbols, usually denoted by '$\sqcap$', called the *meet operator*;
- the identity symbol '$=$'.

We shall call *atomic labels* the set of symbols containing the parametric labels and the unit labels. A *term* of such a labelling language will be any expression built up from the atomic labels by means of the composition operators.

We shall use, in the metalanguage, the lower case roman letters 'a', 'b', 'c', ..., possibly with subscripts, to refer to the parametric labels of the given labelling language, the symbols '$1_{\circ_1}$',..., '$1_{\circ_n}$', to refer to the unit labels, and the symbols '$\circ_1$', ..., '$\circ_n$' to refer to the composition operators.

Each implication rule of Section 14.3.2 is replaced by a family of similar rules with $\rightarrow_i$ and $\circ_i$ instead of $\rightarrow$ and $\circ$. An **LKE** *tree for* S, where S is a given set of LS-formulas, is a tree of LS-formulas constructed in accordance with the implication rules starting from the LS-formulas in S.

The *labelling algebra* $\mathcal{A}_{\langle\tau_1,\ldots,\tau_n\rangle}$, corresponding to the class $\mathcal{M}_{\langle\tau_1,\ldots,\tau_n\rangle}$ of all the s.o.m. models of type $\langle\tau_1,\ldots,\tau_n\rangle$, is described by specifying, for each operation symbol $\circ_i$, a set of axioms and rules of the same form as (14.1)–(14.8) with $\circ_i$ in the place of $\circ$ and $1_{\circ_i}$ in the place of 1, plus the set of axioms corresponding to the structural constraints in τ_i.

We denote by **LKE**$\langle\tau_1,\ldots,\tau_n\rangle$ the **LKE** system equipped with the labelling algebra $\mathcal{A}_{\langle\tau_1,\ldots,\tau_n\rangle}$. A formula A belonging to the fibred language $\mathbb{L}_{1\ldots n}$ and such that its main operator is $\rightarrow_i$ is provable in **LKE**$\langle\tau_1,\ldots,\tau_n\rangle$, if there is a closed **LKE**$\langle\tau_1,\ldots,\tau_n\rangle$-tree for $\mathsf{F}A : 1_{\circ_i}$.

As remarked in Section 14.4.3, the main difference between an implication system and a multi-implication system lies in the notion of a formula A's being provable *from* something. In the latter, we no longer say that a formula A is provable from a given *sequence* of formulas but from a given *structure* of formulas. (For simplicity we shall here identify the structural operators which occur in the definition of a structure of formulas with the composition operators in the labelling algebra.) A formula A is *provable from* a structure of formulas X if there is a closed **LKE**$\langle\tau_1,\ldots,\tau_n\rangle$ tree for the set of LS-formulas $\{\mathsf{T}A_1 : a_1,\ldots,\mathsf{T}A_n : a_n\} \cup \{\mathsf{F}A : \alpha\}$, where α is a term obtained from the structure X by replacing each formula A_i with its label a_i.

Let $\vdash_{\mathbf{LKE}\langle\tau_1,\ldots,\tau_n\rangle}$ be the LKE$\langle\tau_1,\ldots,\tau_n\rangle$-provability relation just described. Moreover, for every formula A in $\mathbb{L}_s^i$, i.e. such that either A is atomic or the main operator of A is $\rightarrow_i$, we write $\vdash_{\mathbf{LKE}\langle\tau_1,\ldots,\tau_n\rangle} A$ instead of $\mathsf{T}_i \vdash_{\mathbf{LKE}\langle\tau_1,\ldots,\tau_n\rangle} A$. Then, it is not difficult to show that:

Theorem 14.13. $\vdash_{\mathbf{LKE}\langle\tau_1,\ldots,\tau_n\rangle} A$ *iff if A is valid for the class* $\mathcal{M}_{\langle\tau_1,\ldots,\tau_n\rangle}$.

It follows from Theorem 14.11 that **LKE** is adequate for all multi-implication systems.

14.5.1 *Examples*

We now illustrate some examples of refutations in our fibred **LKE** system. In all the examples $\rightarrow_0$ will denote the 'basic' notion of implication, typical of the system with no structural rules; $\rightarrow_1$ will denote the implication of linear logic (satisfying only the commutativity rule); $\rightarrow_2$ will denote relevant implication (satisfying commutativity and contraction), $\rightarrow_3$ will denote 'mingle' implication (satisfying commutativity, contraction and expansion), while $\rightarrow_4$ will denote intuitionistic implication (satisfying all the structural rules).

Example 14.14. In Table 14.7 we show that the formula

$$A \rightarrow_2 (A \rightarrow_2 B) \rightarrow_1 (((A \rightarrow_2 B) \rightarrow_1 C) \rightarrow_1 C)$$

is valid in all multi-implication structures. Let us denote by $\circ$ the composition operator associated with $\rightarrow_1$, by $\star$ the composition associated with $\rightarrow_2$ and by $1_\circ$ the unit element of $\circ$. So $\circ$ is commutative while $\star$ is both commutative and contractive.

The tree is closed for $x = a$, since: (i) $b \circ a \leq a \circ b$ (because $\circ$ is commutative), and (ii) $a \star b \star b \leq a \star b$ (because $\star$ is contractive). Notice that the formula becomes invalid if we replace $\rightarrow_1$ with $\rightarrow_0$.

Table 14.7 *Example 14.14.*

$$\mathsf{F}\ (A \to_2 (A \to_2 B)) \to_1 (((A \to_2 B) \to_1 C) \to_1 C) : 1_\circ$$

$$\mathsf{T}\ A \to_2 (A \to_2 B) : a$$

$$\mathsf{F}\ ((A \to_2 B) \to_1 C) \to_1 C : a$$

$$\mathsf{T}\ (A \to_2 B) \to_1 C : b$$

$$\mathsf{F}\ C : a \circ b$$

Left branch:

$$\mathsf{T}\ A \to_2 B : x$$

$$\mathsf{T}\ C : b \circ x$$

$$\times$$

Right branch:

$$\mathsf{F}\ A \to_2 B : x$$

$$\mathsf{T}\ A : b$$

$$\mathsf{F}\ B : x \star b$$

$$\mathsf{T}\ A \to_2 B : x \star b$$

$$\mathsf{T}\ B : x \star b \star b$$

$$\times$$

Example 14.15. In Table 14.8 we show that the formula

$$A \to_1 (((B \to_4 A) \to_1 C) \to_1 C)$$

is valid in all multi-implication structures. Let us denote by $\circ$ the composition operation associated with $\to_1$, and by $*$ the composition associated with the intuitionistic implication $\to_4$. This means that $\circ$ is commutative, while $*$ is commutative, contractive and monotonic.

The tree is closed for $x = a$, since: (i) $b \circ a \leq a \circ b$ (because $\circ$ is commutative), and (ii) $a \leq a * c$ (because $*$ is monotonic).

Example 14.16. Let's see (in Table 14.9) a slightly more complex example (as far as the labels are concerned). We want to show that the formula

$$(A \to_2 (A \to_2 B)) \to_0 (C \to_4 (A \to_2 B))$$

is valid in all multi-implication structures. Let us denote by $\circ$ the composition operation associated with $\to_0$, by $\star$ the composition operator associated with $\to_2$, and by $*$ the one associated with $\to_4$. So $\star$ is commutative and contractive, while $*$ is also monotonic.

Table 14.8 *Example 14.15.*

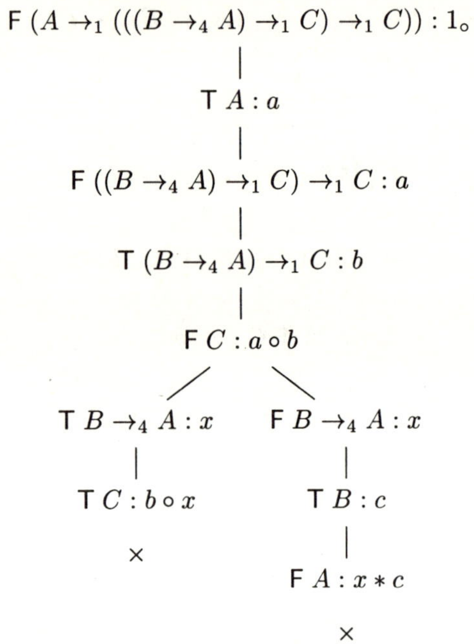

This one-branch tree is closed since: (i) $c \star c \leq c$ (because $\star$ is contractive), and (ii) $a \leq a * b$ (because $*$ is monotonic). Hence: $a \star c \star c \leq (a * b) \star c$.

Example 14.17. In our last example (in Table 14.10) we shall consider an argument which infers a formula from a structure. The inference is the following (we use square brackets for grouping the items of the structure, and round brackets for the formulas):

$$[A \star A \rightarrow_2 B \star A \rightarrow_2 C \star B \rightarrow_2 (C \rightarrow_2 D)] \circ D \rightarrow_1 E \circ (G \rightarrow_3 E) \rightarrow_1 H \vdash H.$$

In this example, $\circ$ is the composition operation associated with $\rightarrow_1$, $\star$ the one associated with $\rightarrow_2$ and $*$ the one associated with $\rightarrow_4$.

The tree is closed for $x = a_5 \circ \alpha$, where $\alpha = a_4 \star a_2 \star a_1 \star a_3 \star a_1$. For, $\alpha \leq a_1 \star a_2 \star a_3 \star a_4$ (since $\star$ is commutative and contractive), and this is sufficient to close the left-hand branch, given also that $\circ$ is commutative. Moreover, since $*$ is monotonic, $a_5 \circ \alpha \leq a_5 \circ \alpha * a_7$, and therefore the right-hand branch is also closed.

Table 14.9 *Example 14.16.*

$$\mathsf{F}\ (A \to_2 (A \to_2 B)) \to_0 (C \to_4 (A \to_2 B)) : 1_\circ$$

$$|$$

$$\mathsf{T}\ A \to_2 (A \to_2 B) : a$$

$$|$$

$$\mathsf{F}\ C \to_4 (A \to_2 B) : a$$

$$|$$

$$\mathsf{T}\ C : b$$

$$|$$

$$\mathsf{F}\ A \to_2 B : a * b$$

$$|$$

$$\mathsf{T}\ A : c$$

$$|$$

$$\mathsf{F}\ B : (a * b) \star c$$

$$|$$

$$\mathsf{T}\ A \to_2 B : a \star c$$

$$|$$

$$\mathsf{T}\ B : a \star c \star c$$

$$\times$$

Table 14.10 *Example 14.17*.

$$\mathsf{T}\, A : a_1$$

$$\mathsf{T}\, A \to_2 B : a_2$$

$$\mathsf{T}\, A \to_2 C : a_3$$

$$\mathsf{T}\, B \to_2 (C \to_2 D) : a_4$$

$$\mathsf{T}\, D \to_1 E : a_5$$

$$\mathsf{T}\, (G \to_4 E) \to_1 H : a_6$$

$$\mathsf{F}\, H : (a_1 \star a_2 \star a_3 \star a_4) \circ a_5 \circ a_6$$

$$\mathsf{T}\, B : a_2 \star a_1$$

$$\mathsf{T}\, C : a_3 \star a_1$$

$$\mathsf{T}\, C \to_2 D : a_4 \star a_2 \star a_1$$

$$\mathsf{T}\, D : a_4 \star a_2 \star a_1 \star a_3 \star a_1 (= \alpha)$$

$$\mathsf{T}\, E : a_5 \circ \alpha$$

$\mathsf{T}\, G \to_4 E : x$	$\mathsf{F}\, G \to_4 E : x$
$\mathsf{T}\, H : a_6 \circ x$	$\mathsf{T}\, G : a_7$
$\times$	$\mathsf{F}\, E : x * a_7$

15

GRAFTING MODALITIES ONTO SUBSTRUCTURAL IMPLICATION SYSTEMS

15.1 Introduction

When logical systems are used to formalise some application area, they may require the addition of modality to the language for a variety of reasons: to cater for changes of the system in time, or perhaps for the dependency of the system on the context, or even to bring metalevel notions into the object level. Be the reasons and motivations what they may, we must develop the logical capability of incorporating modalities into these systems, so adding a new dimension which can explicitly and naturally account for 'accessibility' relations involved in the processes that are being modelled. This chapter deals with an important case-study for our methodology: the hybrid systems which result from grafting the modal operators $\Box$ and $\Diamond$ onto the variety of 'substructural' implication logics discussed in the previous chapter. For the sake of simplicity we shall consider systems with one implication operator, but the reader should be aware that our approach applies to the fibred systems studied in the previous chapter without essential modifications.

Formally, we have a system with the conditional operator $\rightarrow$ and we want to add unary operators $\Box$ and $\Diamond$ which behave like modalities. We would like to do this in a way which is compatible with the spirit of the system, i.e. to define modalities from within the system in some natural way. Substructural logics can be characterised by structures of 'objects' which can be understood as 'pieces of information' or, sometimes, as 'resources'. A natural way of defining modalities within such semantics consists therefore in adding an accessibility relation between pieces of information. This is the approach we shall adopt in this chapter. By analogy with classical modal logics the intuitive idea is that the verification of a proposition of the form '$\Box A$' or '$\Diamond A$' by means of a given piece of information (or resource) a, depends on what is verified by other pieces of information (or resources) 'accessibile' from a. Unlike classical modal logics, however, the verification of non-modal formulae, such as a conditional formula, by a given piece of information a also depends on what is or is not verified elsewhere in the structure (as with the intuitionistic conditional). We shall not pursue any particular intuitive interpretation, because it may change from one logic to the other, and shall leave it to the reader to envisage his or her favourite application contexts.

In Sections 15.2, 15.3 and 15.4, we investigate the semantics of the logical systems obtained by introducing the modalities $\Box$ and $\Diamond$ into the family of substructural implication logics. This can be seen as extending the semantic analysis of modal intuitionistic logics of Chapter 4, (see also [Fischer-Servi, 1977; Božić and Došen, 1984; Plotkin and Stirling, 1986; Amati and Pirri, 1993; Simpson, 1993]. In Section 15.5, we present an alternative proof-theoretical characterisation of the same logics in terms of labelled refutation trees, extending the LKE method presented in Chapter 14.

15.2 Information Frames

In the previous chapter we defined a s.o.m. model as a 5-tuple $\mathbf{m} = \langle W, \circ, 1, \leq, V\rangle$. We can distinguish between two components of such a model: the underlying *frame*, namely the 4-tuple $\langle W, \circ, 1, \leq\rangle$ and the *valuation* V. In this way the same frame can support a variety of models depending on the valuation V. By a *s.o.m. frame* we shall therefore mean the underlying frame of a s.o.m. model.

The semantics based on s.o.m. models, outlined in Chapter 14, is adequate to deal with the substructural implication operators. In order to add a modal dimension to this semantics it is convenient to enrich it by moving from s.o.m. frames to what we shall call 'information frames' to stress the fact that they appear to have enough structure to model a large variety of information processes.

Definition 15.1. An *information frame* is a structure $\mathcal{F} = (W, \circ, 1, \leq)$ such that:

1. W is a non-empty set of elements called *pieces of information* or *information tokens*;
2. $\leq$ is a partial ordering which makes W into a complete lattice;
3. $\circ$ is a binary associative operation on W which is
   (a) distributive over $\bigsqcup$: for every non-empty family $\{c_i\} \subseteq W$,
   $$\bigsqcup\{c_i \circ a\} = \bigsqcup\{c_i\} \circ a \text{ and } \bigsqcup\{a \circ c_i\} = a \circ \bigsqcup\{c_i\}$$
   (b) distributive over $\bigsqcap$: for every non-empty family $\{c_i\} \subseteq W$,
   $$\bigsqcap\{c_i \circ a\} = \bigsqcap\{c_i\} \circ a \text{ and } \bigsqcap\{a \circ c_i\} = a \circ \bigsqcap\{c_i\};$$
4. $1 \in W$ and for every $a \in W$, $a \circ 1 = 1 \circ a = a$.

In other words, an information frame is a s.o.m. frame such that (i) the partial ordering $\leq$ is a complete lattice, and (ii) the monoid operation $\circ$ is fully distributive over $\bigsqcup$ and $\bigsqcap$. As we already noticed in Chapter 14, the properties of $\circ$ imply that this operation is order-preserving, i.e.

$$a_1 \leq a_2 \Longrightarrow a_1 \circ b \leq a_2 \circ b \text{ and } b \circ a_1 \leq b \circ a_2.$$

In analogy with what we did for s.o.m. models, we can define classes of information frames which satisfy additional structural constraints on the ordering $\leq$:

Definition 15.2. We say that an information frame is:

commutative	if	$a \circ b \leq b \circ a$
contractive	if	$a \circ a \leq a$
expansive	if	$a \leq a \circ a$
monotonic	if	$a \leq a \circ b$.

In this new context, a *model* will be defined as follows:

Definition 15.3. A *model* $\mathbf{m}$ is a pair $\langle \mathcal{F}, V \rangle$ where $\mathcal{F}$ is an information frame and V is a valuation of atomic formulae satisfying the persistence condition of Definition 14.1 and the following continuity conditions:

1. *(downward continuity)* For every atomic formula p and every non-empty set S of information tokens:

$$(\forall a \in S) a \in V(p) \Longrightarrow \bigsqcap S \in V(p);$$

2. *(upward continuity)* For every atomic formula p and every non-empty set S of information tokens:

$$(\forall a \in S) a \notin V(p) \Longrightarrow \bigsqcup S \notin V(p).$$

As far as the implication operator is concerned, the forcing relation is defined as in Chapter 14, namely:

1. $(\forall p \in P)(\forall a \in \mathbf{m})(a \Vdash_{\mathbf{m}} p \Longleftrightarrow a \in V(p))$
2. $(\forall a \in \mathbf{m})(a \Vdash_{\mathbf{m}} A \to B \Longleftrightarrow (\forall b \in \mathbf{m})(b \Vdash_{\mathbf{m}} A \Longrightarrow (a \circ b \Vdash_{\mathbf{m}} B)))$.

The reader can check that the continuity properties, as well as the persistence condition, hold also for arbitrary formulae, i.e.

$$a \leq b \text{ and } a \Vdash_{\mathbf{m}} A \Longrightarrow b \Vdash_{\mathbf{m}} A \tag{15.1}$$

and, for every non-empty set S of information tokens:

$$(\forall a \in S) a \Vdash_{\mathbf{m}} A \Longrightarrow \bigsqcap S \Vdash_{\mathbf{m}} A \tag{15.2}$$

$$(\forall a \in S) a \nVdash_{\mathbf{m}} A \Longrightarrow \bigsqcup S \nVdash_{\mathbf{m}} A. \tag{15.3}$$

It can also be shown that the move from s.o.m. models to information frames does not increase the stock of valid implication formulae However, in the sequel, it will prove convenient for an elegant and uniform treatment of the modal operators.

15.3 The Modal Operators

We now want to see how to introduce modality into the logic and obtain a combined system with substructural implication and the modalities $\Box$, $\Diamond$. We already have experience with the introduction of modalities. We have done it for intuitionistic logic, for Łukasiewicz infinitely valued logic as well as, implicitly, for strict implication (when fibring two modalities). In all of these cases we took two disjoint models of each logic and fibred or dovetailed them together to get one big model for the combined language. In the dovetailed case, a close look at the resulting structures shows that they can be presented as augmenting the implication semantics with an additional relation R which interacts with it in a way to accommodate persistence. This suggests that perhaps we can simplify the fibring process for the case of introducing modalities by simply starting with the semantics of implication and 'grafting' onto it a new 'accessibility' relation R satisfying certain conditions and define the semantics for the combined language in this manner. Although this approach may not work in general we believe it is viable for a wide variety of systems and constitutes a general shortcut for 'dovetailing'.

Let us introduce in our information frames a binary relation R between their points, called the *accessibility relation*. We assume that this relation is 'closed' under arbitrary $\bigsqcup$ and $\bigsqcap$, namely:

$$(\forall b \in S)aRb \Longrightarrow aR\bigsqcup S \text{ and } (\forall b \in S)aRb \Longrightarrow aR\bigsqcap S. \tag{15.4}$$

We then introduce into our language the two unary operators $\Box$ and $\Diamond$ intended as the usual modalities. The forcing relation extends to these modal operators in the expected way. For $\Box$ we have the following clause:

$$a \Vdash_{\mathbf{m}} \Box A \text{ iff } b \Vdash_{\mathbf{m}} A \text{ for all } b, \text{ such that } aRb \tag{15.5}$$

and for $\Diamond$:

$$a \Vdash_{\mathbf{m}} \Diamond A \text{ iff } b \Vdash_{\mathbf{m}} A \text{ for some } b, \text{ such that } aRb. \tag{15.6}$$

In order to preserve the persistence property expressed in (15.1) in the new setting, R must also satisfy the following conditions (see [Božić and Došen, 1984] and [Plotkin and Stirling, 1986]):

$$a \leq b \text{ and } aRc \Longrightarrow (\exists c')(bRc' \text{ and } c \leq c') \tag{15.7}$$

and

$$a \leq b \text{ and } bRc \Longrightarrow (\exists c')(aRc' \text{ and } c' \leq c). \tag{15.8}$$

Definition 15.4. A *modal information frame* is a pair $(\mathcal{F}, R)$ where $\mathcal{F}$ is an information frame and R is a binary accessibility relation defined on the domain of $\mathcal{F}$ and satisfying conditions (15.4), (15.7) and (15.8).

If R is defined as above, any atomic valuation V over an information frame can be extended to a forcing relation over a modal information frame by means of the forcing clauses for $\Box$ and $\Diamond$ preserving the persistence property.

Definition 15.5. A *modal implication model* is a triple $(\mathcal{F}, R, V)$, where $(\mathcal{F}, R)$ is a modal information frame and V is a valuation. The *forcing* relation over a modal implication model $\mathbf{m}$ is a relation between elements of $\mathbf{m}$ and formulae, satisfying the following conditions:

1. $(\forall p \in P)(\forall a \in \mathbf{m})(a \Vdash_{\mathbf{m}} p \Longleftrightarrow a \in V(p))$
2. $(\forall a \in \mathbf{m})(a \Vdash_{\mathbf{m}} A \to B \Longleftrightarrow (\forall b \in \mathbf{m})(b \Vdash_{\mathbf{m}} A \Longrightarrow (a \circ b \Vdash_{\mathbf{m}} B)))$
3. $a \Vdash_{\mathbf{m}} \Box A$ iff $b \Vdash_{\mathbf{m}} A$ for all b, such that aRb
4. $a \Vdash_{\mathbf{m}} \Diamond A$ iff $b \Vdash_{\mathbf{m}} A$ for some b, such that aRb.

We can restrict our attention, without loss of generality, to modal implication models with the additional property that any two points verifying exactly the same formulae are identical.

Definition 15.6. We say that a formula A is *verified* in a *modal implication model* M if it is verified at the identity point 1 of M. We also say that A is *verified* in a *frame* $\mathcal{F}$ if is verified in all modal implication models based on $\mathcal{F}$.

Observe that these definitions are essentially different from the classical definitions. In classical modal logics a formula is verified in a modal Kripke model if it is verified in *all* the worlds belonging to that model. Our definition is equivalent to the classical one only in the case of *monotonic* information frames. For, suppose a formula A is verified at the identity point 1 of a structure based on a monotonic frame, i.e. $1 \Vdash_{\mathbf{m}} A$. Then, by the monotonicity of the frame $1 \leq 1 \circ a (= a)$ for every point a in the frame. So, by persistence, $a \Vdash_{\mathbf{m}} A$ for all the points in the frame.

Our final aim is to incorporate the semantics of the modal operators into the proof-theoretical framework of Section 14.3.2. For this purpose we shall simplify this semantics by replacing the binary relation R with a pair of unary operators corresponding to the modalities. (These operators are reminiscent of the ones used in the algebraic semantics of classical modal logics, for which see [Bull and Segerberg, 1984].) Such a reformulation will allow us, in Section 15.5.1, to formulate an algebra of the labels which is a simple extension of the one characterising substructural implication systems obtained by adding to the structural constraints a (possibly empty) set of *modal constraints*, i.e. inequations of the form $\alpha \leq \beta$, where α and β are terms of the extended algebra including the new unary operators.

Let us denote by $S(a)$ *the sphere* of a, i.e. the set of all pieces of information accessible from a. We shall also use $\bigsqcup$ and $\bigsqcap$ to denote the usual lattice join and meet. We now define the two unary operators ! and ? as follows:

$$!a =_{\text{def}} \bigsqcap S(a) \tag{15.9}$$

and

$$?a =_{\text{def}} \bigsqcup S(a). \tag{15.10}$$

Now observe that (15.4) implies that the following property of R is satisfied in every frame:

$$(\forall a)(\exists b)aRb. \tag{15.11}$$

In other words, for all a, $S(a) \neq \varnothing$. In fact, for every point a, both $!a$ and $?a$ are accessible from a, that is

$$(\forall a)aR!a \text{ and } aR?a. \tag{15.12}$$

In the theory of classical modal logics a property like (15.11) is known as *seriality*. So, in our approach *all* frames are serial. Using the ! and ? operators, this assumption is expressed by the following inequation which holds for all modal information frames:

$$!a \leq ?a. \tag{15.13}$$

So, in our approach, the points $!a$ and $?a$ have a special status: $!a$ is a point in $S(a)$ that verifies all and only the formulae which are true at *all* points in $S(a)$, while $?a$ is a point in $S(a)$ that verifies all and only the formulae which are true at *at least one* point in $S(a)$.

It is not difficult to show that (15.7) and (15.8) above are respectively equivalent to the following conditions on ? and ! expressing the fact that the operators ? and ! are both *order-preserving*:

$$a \leq b \Longrightarrow ?a \leq ?b \tag{15.14}$$

$$a \leq b \Longrightarrow !a \leq !b. \tag{15.15}$$

Proposition 15.7. *In every modal information frame, (15.8) holds if and only if (15.15) holds, and (15.7) holds if and only if (15.14) holds.*

Proof We show only that (15.8) is equivalent to (15.15) and leave it to the reader to show the other equivalence. First, assume (15.8), and suppose $a \leq b$. By (15.12), we know that $bR!b$. By (15.8), $(\exists c')c' \leq !b$ and aRc'. Then, by definition of !, $!a \leq c'$ and, therefore, $!a \leq !b$. Now, assume (15.15) and suppose (i) $a \leq b$, (ii) bRc. Then $!b \leq c$ and, by (15.15), $!a \leq c$. Since $aR!a$, there exists c' such that aRc' and $c' \leq c$. □

Given our definitions of the operators ! and ?, the forcing clauses for □ and ◇ can be reformulated more concisely as below:

$$a \Vdash_{\mathbf{m}} \Box A \text{ iff } !a \Vdash_{\mathbf{m}} A \tag{15.16}$$

and

$$a \Vdash_{\mathbf{m}} \Diamond A \text{ iff } ?a \Vdash_{\mathbf{m}} A. \tag{15.17}$$

Let us now show the equivalence of clauses (15.16) and (15.17) with (15.5) and (15.6) respectively. First, we assume that the clause in (15.5) holds and show that (15.16) holds too.

For the only-if direction, let us assume $a \Vdash_{\mathbf{m}} \Box A$; then by (15.5), $b \Vdash_{\mathbf{m}} A$ for all $b \in S(a)$. Now, by the definition of valuation, A remains true at the greatest lower bound of $S(a)$, that is at $!a$. For the if-direction let us assume $a \Vdash_{\mathbf{m}} \Box A$. It follows that there exists a point $b \in S(a)$ such that A is false at b. Since, $!a \leq b$, A must be false at $!a$.

Now we show that (15.16) implies (15.5). For the only-if direction, suppose $a \Vdash_{\mathbf{m}} \Box A$. Then, by (15.16), $!a \Vdash_{\mathbf{m}} A$. So, A is true at every point b such that $!a \leq b$ and, therefore, at every point in $S(a)$. For the if-direction, assume that A is true at every point in $S(a)$. Suppose, that $\Box A$ is false at a. Then, by (15.16), $!a \Vdash_{\mathbf{m}} A$. So the assumption that A is true at every point $S(a)$ is contradicted. Hence, $\Box A$ must be true at a. A similar argument shows the equivalence of (15.6) and (15.17).

We can now exploit the new operators to express complex statements about the accessibility relation R concisely, in the form of simple inequalities of the form $\alpha \leq \beta$, where α and β are expressions built up from atomic terms by means of the operators $\circ$, ! and ?.

Let us consider some of the most familiar properties of the accessibility relation R.

Seriality A frame is *serial* if for every point a, $S(a) \neq \varnothing$, that is, for every point a there exists a b such that aRb. Observe that a frame is serial if and only if it verifies the formula:

$$\Box A \to \Diamond A.$$

As mentioned above, in our approach all frames are serial, as a consequence of (15.4), and seriality corresponds to the assumption that $!a \leq ?a$ for all a.
Reflexivity A frame is said to be *reflexive* if aRa for all a. Observe that a frame is reflexive if and only if it verifies the following formula:

$$\Box A \to A.$$

In our notation reflexivity can be expressed by the following condition:

(Reflexivity) $(\forall a)(!a \leq a)$.

The equivalence between the two formulations is immediately seen as follows. First, assume that $!a \leq a$. Then every formula true at $!a$ is true also at a. Suppose aRa does *not* hold for some a. This means that there exists a point c such that $\neg cRc$. Now, consider a model Bm and a formula A such that $a \Vdash_{\mathbf{m}} A$ for all $a \in S(c)$, i.e. $c \Vdash_{\mathbf{m}} \Box A$, but $c \nVdash_{\mathbf{m}} A$ (which is possible, since $c \notin S(c)$). For this Bm, we have that $!c \Vdash_{\mathbf{m}} A$ and, since $!c \leq c$, $c \Vdash_{\mathbf{m}} A$. This is a contradiction. Hence, if $!a \leq a$ for all a, then aRa for all a. Conversely, if aRa for all a, every formula that is true at all points accessible from a is true also at a, i.e. for all a, $!a \leq a$.
Transitivity A frame is said to be *transitive* if the following condition holds:

$$(\forall a)(\forall b)(\forall c)(aRb \text{ and } bRc \Longrightarrow aRc).$$

A frame is transitive if and only if it verifies the following formula:

$$\Box A \to \Box\Box A.$$

In our notation, transitivity is expressed by the following condition:

(Transitivity) $(\forall a)(!a \leq !!a)$.

Assume that $!a \leq !!a$ for all a. Now, suppose that the frame is not transitive. Then, there exists a point c, a model Bm and a formula A such that $c \Vdash_{\mathbf{m}} A$ but $c \nVdash_{\mathbf{m}} \Box\Box A$. This implies by (15.16) that $!c \Vdash_{\mathbf{m}} A$ and $!!c \nVdash_{\mathbf{m}} A$. But, since $!c \leq !!c$, this is impossible. Hence, if $!a \leq !!a$ for all a, the frame is transitive.

Conversely, suppose the frame is transitive. We show that $!a \leq !!a$. Let A be any formula which is true at $!a$. Then A is true at all points in $S(a)$. Moreover, $!a$ is accessible from a. Since the frame is transitive, if a point b is accessible from $!a$, it is also accessible from a. So, if A is true at all the points accessible from a it is also true at all the points accessible from $!a$, that is $!a \leq !!a$.

Symmetry A frame is said to be *symmetric* if the following condition holds:

$$(\forall a)(\forall b)(aRb \Longrightarrow bRa).$$

A frame is symmetric if and only if it verifies the following formula:

$$A \to \Box\Diamond A.$$

In our notation, symmetry is expressed by the following condition:

(Symmetry) $(\forall a)(a \leq ?!a)$.

For this property as well as for the next two, we leave it to the reader to verify the equivalence between the corresponding formulations.
Euclideanism A frame is said to be *Euclidean* if the following condition holds true:

$$(\forall a)(\forall b)(\forall c)(aRb \text{ and } aRc \Longrightarrow bRc).$$

A frame is Euclidean if and only if it verifies the following formula:

$$\Diamond A \to \Box\Diamond A.$$

The condition corresponding to this property is the one stated below:

(Euclideanism) $(\forall a)(?a \leq ?!a)$.

Directedness A frame is *directed* if the following condition holds true:

$$(\forall a)(\forall b)(\forall c)(aRb \text{ and } aRc \Longrightarrow (\exists d)(bRd \text{ and } cRd))$$

which corresponds to:

(Directedness) $(\forall a)(!?a \leq ?!a)$.

Observe that a frame is directed if and only if it verifies the following formula:

$$\Diamond\Box A \to \Box\Diamond A.$$

Our discussion so far amounts to a reformulation of the well-known correspondence theory in our new setting. This correspondence is summarised in Table 15.1.

We shall now state a crucial lemma. Let σ_1 and σ_2 be (possibly empty) strings of ! and ?. The *dual* of σ_i, denoted by σ_i' is the string obtained by interchanging ! and ? in σ_i. Moreover, let $\mathbf{M}^* \subseteq \mathbf{M}$, where $\mathbf{M}$ is the set of modal conditions listed above. We have the following *duality principle*:

Theorem 15.8 (Duality lemma). *Assume that $\sigma_1 a \leq \sigma_2 a$ holds for all a in all $\mathbf{M}^*$-frames, i.e. there is a chain of inequalities $\phi_0 \leq \cdots \leq \phi_n$ such that (i) $\phi_0 = \sigma_1 a$, (ii) $\phi_n = \sigma_2 a$, and (iii) each inequality $\phi_i \leq \phi_{i+1}$, with $i = 0, \ldots, n-1$, is one of of the primitive inequalities in $\mathbf{M}^*$. Then $\sigma_2' a \leq \sigma_1' a$ also holds for all a in all $\mathbf{M}^*$-frames.*

Table 15.1 *Correspondence between modal formulae, conditions on R and constraints on the modal information frames.*

Modal formulae	Conditions on R	Modal constraints
$\Box A \to A$	$(\forall a)aRa$	$!a \leq a$
$\Box\Box A \to \Box A$	$(\forall a)(\forall b)(\forall c)(aRb \text{ and } bRc \Longrightarrow aRc)$	$!a \leq !!a$
$\Box\Diamond A \to A$	$(\forall a)(\forall b)(aRb \Longrightarrow bRa)$	$a \leq ?!a$
$\Box\Diamond A \to \Diamond A$	$(\forall a)(\forall b)(\forall c)(aRb \text{ and } aRc \Longrightarrow bRc)$	$?a \leq ?!a$
$\Box\Diamond A \to \Diamond\Box A$	$(\forall a)(\forall b)(\forall c)(aRb \text{ and } aRc \Longrightarrow (\exists d)(bRd \text{ and } cRd))$	$?!a \leq !?a$

Proof The proof is an easy induction on the length of the chain $\phi_0 \leq \cdots \leq \phi_n$.

Base: $n = 1$. Then the inequality $\sigma_1 a \leq \sigma_2 a$ is an element of $\mathbf{M}^*$. The reader can easily verify that the lemma holds for all primitive inequalities in $\mathbf{M}$.

Step: $n = k + 1$. Let ρ be the prefix of ϕ_k, i.e. $\phi_k = \rho a$. By the ctive hypothesis, $\rho' a \leq \sigma_1' a$. Moreover, since $\rho a \leq \sigma_2 a$, again by the ctive hypothesis we have that $\sigma_2' a \leq \rho' a$. Hence, $\sigma_2' a \leq \sigma_1' a$. □

Let now $\mathbb{L}'$ be a propositional language containing the two binary operators $\to_1$ and $\to_2$ plus the two modal operators $\Box$ and $\Diamond$. (Recall that the two conditionals collapse in all the systems allowing the structural rule of *Exchange*, in which case we use the symbol $\to$ without subscripts.) By a *modal implication logic*— or **MIL** for short— we mean a (substructural) implication logic (see Section 14.2.1) over $\mathbb{L}'$ satisfying the following conditions on the operators $\Box$ and $\Diamond$:

$$\frac{A \vdash B}{\Box A \vdash \Box B} \tag{15.18}$$

and

$$\frac{A \vdash B}{\Diamond A \vdash \Diamond B}. \tag{15.19}$$

In addition to (15.18) and (15.19) above, different **MIL**'s may also satisfy different subsets of the following *modal axioms*:

1. $\Box A \to \Diamond A$
2. $\Box A \to A$
3. $\Box A \to \Box\Box A$
4. $A \to \Box\Diamond A$
5. $\Diamond A \to \Box\Diamond A$
6. $\Diamond\Box A \to \Box\Diamond A$.

In this chapter we restrict our attention to *serial* **MIL**'s, i.e. those satisfying the modal axiom $\Box A \to \Diamond A$.

Previous work on the specific topic of this section has been concentrating on intuitionistic modal logics, see Chapter 5. The implication fragments of these logics belong to the wide family of substructural modal systems presented here (i.e. those satisfying *all* the structural rules of Table 14.1). We consider in particular the works of Božić and Došen [1984], see Example 5.1.10, and Plotkin and Stirling [1986] (see also [Simpson, 1993] for an overview). Our experience in Chapters 3, 4 and 12 shows that the main problem in defining modal connectives on Kripke-style intuitionistic structures is to guarantee the intuitionistic hereditary property (here called 'persistence') for modal formulae. This can be achieved either by defining appropriate satisfaction clauses for the modal operators which embed some 'sort of' monotonicity feature (e.g. [Plotkin and Stirling, 1986]) or by imposing conditions between the accessibility intuitionistic relations (e.g. [Božić and Došen, 1984]). Božić and Došen propose [Božić and Došen, 1984] three different intuitionistic modal logics, one with only the necessity operators (i.e. $HK_\Box$), one with only the possibility operator (i.e. $HK_\Diamond$) and the third with both modalities (i.e. $HK_{\Diamond,\Box}$). In the $HK_\Box$ system, a frame is a possible world structure $(W, \leq, R)$ where the partial ordering $\leq$ and the binary relation R satisfy the following condition:

$$\text{if } a \leq b \text{ and } bRc \text{ then } \exists c' \text{ such that } aRc' \text{ and } c \leq c'. \qquad (15.20)$$

A class of models is obtained by adding on a frame $HK_\Box$ a valuation function V which satisfies the hereditary property for the propositional variables. Standard intuitionistic and modal satisfaction clauses are then considered. In the $HK_\Diamond$, the following condition is added to the same type of frame:

$$\text{if } a \leq b \text{ and } aRc \text{ then } \exists c' \text{ such that } bRc' \text{ and } c \leq c'. \qquad (15.21)$$

These are Conditions (15.8) and (15.7) in this chapter.

Plotkin and Stirling [1986] combine the two approaches described above in that they consider the condition 15.21 of Božić and Došen, which allows them to preserve the traditional (classical) satisfaction clause for $\Diamond$, but they define an alternative satisfaction clause for $\Box$

$$a \models \Box A \Leftrightarrow_{\text{def}} \forall b\; a \leq b\; \forall c \text{ if } bRc \text{ then } c \models A. \qquad (15.22)$$

In this way, Plotkin and Stirling generate intuitionistic modal logics in which modal operators are interpreted similarly to the intuitionistic quantifiers $\forall$ and $\exists$, and the two modal operators are not interdefinable. An extension of the work presented in this chapter to full substructural modal logics (including negation) will allow better comparisons with [Plotkin and Stirling, 1986] and others.

15.4 The canonical model

In this section we describe a concrete example of a modal implication model defined in terms of the deducibility relation $\vdash$ associated with any given (serial) **MIL**. We call this concrete model *the canonical model*.

In the canonical model the points of the underlying information frame are decreasing sets of formulae, i.e. if A belongs to any such set and $A \vdash B$, then B also belongs to the set, where $\vdash$ is the deducibility relation of an arbitrary **MIL** (we call these sets 'decreasing' because $A \vdash B$ can be interpreted as 'the information carried by A is less than or equal to the information carried by B').

Let $\vdash$ be the deducibility relation of an arbitrary **MIL**. The *canonical frame* generated by $\vdash$ is a structure $\langle M, \circ, 1, \subseteq, R\rangle$ such that

- M is the set of all decreasing sets of formulae with respect to the given $\vdash$;

- $\circ$ is a binary operation defined as follows:

$$a \circ b =_{\text{def}} \{C \mid (\exists A \in a)(\exists B \in b)A, B \vdash C\};$$

- 1 is the set $\{A \mid [\,] \vdash A\}$, where $[\,]$ denotes the empty list;
- the relation $\subseteq$ is set-inclusion.

Clearly, M is a complete lattice under $\subseteq$ with standard union and intersection as join and meet. Moreover, it is not difficult to verify that $\circ$ is a monoid operation with 1 as identity. Furthermore, it is easy to show that $\circ$ distributes over arbitrary joins and meets of non-empty subsets of M. Therefore, the structure $\langle M, \circ, 1, \subseteq\rangle$ is an information frame.

Let us now define two functions on the points of the canonical model:

$$h(a) =_{\text{def}} \{A | \Box A \in a\}$$

and

$$g(a) =_{\text{def}} \{A | \Diamond A \in a\}.$$

Notice that, for every a, $h(a)$ and $g(a)$ are themselves decreasing sets of formulae, because of the monotonicity properties of $\Box$ and $\Diamond$ (15.18) and (15.19). Therefore they also belong to the canonical model. Notice also that our 'seriality' assumption implies that for all a:

$$h(a) \subseteq g(b).$$

Accessibility between points of this canonical model is then defined as follows:

$$aRb \text{ iff } h(a) \subseteq b \subseteq g(a).$$

It follows from our definitions that

1. R satisfies the continuity property (15.4);
2. $a \subseteq b$ implies $h(a) \subseteq h(b)$;
3. $a \subseteq b$ implies $g(a) \subseteq g(b)$;
4. $a \subseteq b$ and aRc implies $\exists c', bRc'$ and $c \subseteq c'$; to see this take $c' = g(b)$; we have that bRc', because $h(b) \subseteq g(b) \subseteq g(b)$; moreover, since aRc, we have that $c \subseteq g(a)$, and since $g(a) \subseteq g(b)$, we can conclude $c \subseteq g(b)$. Therefore $c \subseteq c'$;
5. $a \subseteq b$ and bRc implies $\exists c', aRc'$ and $c' \subseteq c$; to see this take $c' = h(a)$; notice that $h(a) \subseteq h(a) \subseteq g(a)$, so aRc'; moreover, since $h(a) \subseteq h(b)$ and $h(b) \subseteq c$ (because bRc), we have that $c' = h(a) \subseteq h(b) \subseteq c$;
6. $\bigcap S(a) = h(a)$;
7. $\bigcup S(a) = g(a)$;
8. $\Box A \in a$ iff $A \in h(a)$;
9. $\Diamond A \in a$ iff $A \in g(a)$.

Properties 4 and 5 above show that R satisfies the conditions (15.7) and (15.8) and therefore the canonical frame is a modal information frame.

A *valuation* V over M can be defined simply as follows:

$$a \in V(p) \Longleftrightarrow p \in a.$$

Such a function V satisfies all the conditions on the definition of a valuation (Definition 15.3). As for the persistence property, observe that $a \in V(p)$ implies, by definition of V, that $p \in a$, which together with the condition $a \subseteq b$ gives that $p \in b$ and, therefore, $b \in V(p)$. As for the continuity properties, it is not difficult to check that V satisfies also conditions 1 and 2 (for this purpose notice that since the relation $\subseteq$ is set-inclusion, the $\bigsqcup$ and $\bigsqcap$ are the set operations $\bigcup$ and $\bigcap$ respectively).

The canonical frame together with the valuation V defined as above is a modal implication model. We shall call this model *the canonical model* (generated by the given **MIL** $\vdash$). It is not difficult to show that for every formula A and every element a of the canonical model:

$$a \Vdash A \text{ iff } A \in a. \tag{15.23}$$

The proof is left to the reader.

Let $\mathbf{C} = \mathbf{S} \cup \mathbf{M}$, where $\mathbf{S}$ is an arbitrary subset of the structural rules in Table 14.1 and $\mathbf{M}$ is an arbitrary subset of the modal axioms in Table 15.1. We denote by $\vdash_{\mathbf{C}}$ the smallest **MIL** satisfying all the rules and axioms in $\mathbf{C}$. Moreover, let $\mathbf{C}'$ be the set of constraints corresponding to the rules and axioms in $\mathbf{C}$ as outlined in Tables 14.1 and 15.1. By a $\mathbf{C}'$*-frame* we mean a frame satisfying all the constraints in $\mathbf{C}'$. Observe that the canonical frame generated by $\vdash_{\mathbf{C}}$ is a $\mathbf{C}'$-frame (the proof is left to the reader). Then, by considering the canonical model, it is easy to show the following.

Proposition 15.9. *If A is verified in all $\mathbf{C}'$-frames, then $\vdash_{\mathbf{C}} A$.*

Proof Suppose $\nvdash_{\mathbf{C}} A$. Then the canonical model generated by $\vdash_{\mathbf{C}}$ describes a modal implication model (based on a $\mathbf{C}'$ frame) which does not verify A. □

15.5 Adding Modalities to the LKE system

In this section we show how the LKE method of Section 14.3.2 can be extended to deal with the modal operators in a very natural way.

15.5.1 *Modal* LKE *rules*

We enrich the labelling algebra of Section 14.3.2 with the two unary operators ! and ? satisfying conditions 15.13, 15.15 15.14, and the duality principle (see the duality lemma above). Notice that this simple labelling algebra is sufficient for completeness, that is, as already stressed, we do not need all the complexity brought in by the potentially infinitary lattice operations which appear in the definition of information frames. The latter enter the picture only as intermediary for an explanation of the unary operators ! and ?. The completeness theorem stated below guarantees that they do not play any additional role with respect to characterising the set of valid formulas.

The valuation clauses (15.16) and (15.17) immediately imply simple **LKE**-style rules for the modal operators.

Elimination rules for $\Box$

$$\frac{\mathsf{F}\Box A : a}{\mathsf{F}A : !a}\ \mathsf{EF}\Box$$

Elimination rules for $\Diamond$

$$\frac{\mathsf{T}\Diamond A : a}{\mathsf{T}A : ?a}\ \mathsf{ET}\Diamond$$

The rules have a clear intuitive interpretation: if $\Box A$ is true at a point a, then it is true at all points accessible from it and, therefore, it is true at the special point $!a$ which verifies all and only the formulae which are true at all points in $S(a)$. Conversely, if $\Box A$ is false at a, then it must be false at $!a$.

Similarly, if $\Diamond A$ is true at a point a, then it is true at some point accessible from it and, therefore, at the special point $?a$ which verifies all and only the formulae which are true at at least one point in $S(a)$. Conversely, if $\Diamond A$ is false at a, then it must be false at $?a$.

By adding these rules to the implication rules we obtain an **LKE**-system for modal substructural implication. Different systems will be characterised by adding to the labelling algebra different sets of structural and modal constraints to be used in checking for branch-closure. So, this extended method preserves the 'separation-by-closure' property of the original method. Moreover, it also preserves the atomic closure property (see Section 14.3.2 above). Notice that, by virtue of the duality principle, whenever the constraint $\sigma_1 a \leq \sigma_2 b$ is derivable in the algebra of the labels, so is its dual $\sigma_2' b \leq \sigma_1' a$.

In the next subsection we shall see some examples of proofs in this modal extension of the **LKE** system.

15.5.2 *Some examples*

Example 15.10. The axiom T, $\Box A \to A$ is valid in all reflexive frames:

$$\mathsf{F}\Box A \to A : 1$$

$$|$$

$$\mathsf{T}\Box A : a$$

$$|$$

$$\mathsf{F}A : 1 \circ a (= a)$$

$$|$$

$$\mathsf{T}A : !a.$$

This one-branch tree is closed in all reflexive frames since $!a \leq a$. Observe that, by the duality principle, the constraint $a \leq ?a$ also holds in all reflexive frames and, by means of this, the dual of T, $A \to \Diamond A$ can be proved.

Example 15.11. The axiom D, $\Box A \rightarrow \Diamond A$ is valid in all serial frames:

$$
\begin{array}{c}
\mathsf{F}\Box A \rightarrow \Diamond A : 1 \\
| \\
\mathsf{T}\Box A : a \\
| \\
\mathsf{F}\Diamond A : a \\
| \\
\mathsf{T}A : !a \\
| \\
\mathsf{F}A : ?a.
\end{array}
$$

This tree is closed in all serial frames because $!a \leq ?a$.

Example 15.12. The axiom 5, $\Diamond A \rightarrow \Box\Diamond A$ is valid in all Euclidean frames:

$$
\begin{array}{c}
\mathsf{F}\Diamond A \rightarrow \Box\Diamond A : 1 \\
| \\
\mathsf{T}\Diamond A : a \\
| \\
\mathsf{F}\Box\Diamond A : 1 \circ a (= a) \\
| \\
\mathsf{T}A : ?a \\
| \\
\mathsf{F}\Diamond A : !a \\
| \\
\mathsf{F}A : ?!a.
\end{array}
$$

A similar tree, making use of the dual constraint $?a \leq !a$, shows the validity of the dual axiom $\Diamond\Box A \rightarrow \Box A$.

Example 15.13. The formula

$$(A \rightarrow (A \rightarrow B)) \rightarrow (\Box A \rightarrow B)$$

is valid in all contractive and reflexive frames:

$$\begin{array}{c}
\mathsf{F}(A \to (A \to B)) \to (\Box A \to B) : 1 \\
| \\
\mathsf{T}A \to (A \to B) : a \\
| \\
\mathsf{F}\Box A \to B : a \\
| \\
\mathsf{T}\Box A : b \\
| \\
\mathsf{F}B : a \circ b \\
| \\
\mathsf{T}A : !b \\
| \\
\mathsf{T}A \to B : a \circ !b \\
| \\
\mathsf{T}B : a \circ !b \circ !b.
\end{array}$$

In all contractive frames $!b \circ !b \leq !b$, so $a \circ !b \circ !b \leq a \circ !b$. Moreover, in all reflexive frames $!b \leq b$. So, in all contractive and reflexive frames $a \circ !b \circ !b \leq a \circ b$.

Example 15.14. The formula

$$\Box A \to (B \to \Box\Box A)$$

is valid in all frames which are both monotonic and transitive:

$$\begin{array}{c}
\mathsf{F}\Box A \to (B \to \Box\Box A) : 1 \\
| \\
\mathsf{T}\Box A : a \\
| \\
\mathsf{F}B \to \Box\Box A : a \\
| \\
\mathsf{T}B : b \\
| \\
\mathsf{F}\Box\Box A : a \circ b \\
| \\
\mathsf{T}A : !a \\
| \\
\mathsf{F}\Box A : !(a \circ b) \\
| \\
\mathsf{F}A : !!(a \circ b).
\end{array}$$

This one-branch tree is closed for all monotonic-transitive frames. For, $a \leq a \circ b$ by monotonicity; therefore, $!a \leq !(a \circ b)$ by (15.15); moreover, by transitivity, $!(a \circ b) \leq !!(a \circ b)$; so, $!a \leq !!(a \circ b)$ and the branch is closed.

Example 15.15. The formula

$$A \to (A \to \Box\Diamond A)$$

is valid in all frames which are expansive and symmetric:

$$\begin{array}{c} \mathsf{F}A \to (A \to \Box\Diamond A) : 1 \\ | \\ \mathsf{T}A : a \\ | \\ \mathsf{F}A \to \Box\Diamond A : a \\ | \\ \mathsf{F}\Box\Diamond A : a \circ a \\ | \\ \mathsf{F}\Diamond A :!(a \circ a) \\ | \\ \mathsf{F}A :?!(a \circ a). \end{array}$$

This one-branch tree is closed in all expansive-symmetric frames. For, $a \leq a \circ a$ by expansion; $(a \circ a) \leq ?!(a \circ a)$ by symmetry. These two inequalities imply that $a \leq ?!(a \circ a)$.

Example 15.16. The formula $\Diamond(A \to A)$ is valid in all monotonic frames:

$$\begin{array}{c} \mathsf{F}\Diamond(A \to A) : 1 \\ | \\ \mathsf{F}A \to A :?1 \\ | \\ \mathsf{T}A : a \\ | \\ \mathsf{F}A :?1 \circ a. \end{array}$$

In all monotonic frames, $a \leq ?1 \circ a$ and the branch is closed.

Example 15.17. The proof rule

$$\frac{\vdash A \to B}{\vdash \Box A \to \Box B}$$

is valid in all modal information frames.

This is shown by the following closed **LKE**-tree:

$$\mathsf{F}\Box A \to \Box B : 1$$
$$\mathsf{T}\Box A : a$$
$$\mathsf{F}\Box B : a$$
$$\mathsf{T}A :!a$$
$$\mathsf{F}B :!a$$
$$\mathsf{T}A \to B : 1 \qquad \mathsf{F}A \to B : 1$$
$$\mathsf{T}B :!a \qquad \mathcal{T}$$

where $\mathcal{T}$ is a closed **LKE**-tree for $\mathsf{F}A \to B : 1$ which exists by hypothesis.

Example 15.18. The proof rule

$$\frac{\vdash A}{\vdash \Box A}$$

is valid in all monotonic frames.

This is shown by the following closed **LKE**-tree:

$$\mathsf{F}\Box A : 1$$
$$\mathsf{F}A :!1$$
$$\mathsf{T}A : 1 \qquad \mathsf{F}A : 1$$
$$\mathcal{T}$$

where $\mathcal{T}$ is a closed **LKE**-tree for $FA : 1$ which exists by hypothesis. Here the left-hand brach is closed because $1 \leq 1\circ!1$ by the monotonicity of $\circ$.

Example 15.19 (Regular modal implication systems). We define the class of *regular* **MIL**'s, as the class of all **MIL**'s closed under the additional axiom

$$\Box(A \to B) \to (\Box A \to \Box B).$$

(Recall that all **MIL**'s considered in this chapter satisfy the other condition that—in the context of classical modal logic—is used to characterised regular modal logics, namely Condition 15.18.) The class of *regular frames*, corresponding to regular **MIL**'s, is characterised by the following constraint:

$$!a\circ!b \leq !(a \circ b).$$

Table 15.2 *This tree is closed for all regular and directed frames, under the substitution* $x = a$.

$$\mathsf{T}\Box(\Diamond A \to \Diamond B) : a$$

$$\mathsf{T}(\Box\Diamond A \to \Diamond\Box B) \to C : b$$

$$\mathsf{F}C : a \circ b$$

$$\mathsf{T}\Diamond A \to \Diamond B :!a$$

$\mathsf{T}\Box\Diamond A \to \Diamond\Box B : x$	$\mathsf{F}\Box\Diamond A \to \Diamond\Box B : x$
$\mathsf{T}C : x \circ b$	$\mathsf{T}\Box\Diamond A : c$
	$\mathsf{F}\Diamond\Box B : x \circ c$
	$\mathsf{T}\Diamond A :!c$
	$\mathsf{F}\Box B :?(x \circ c)$
	$\mathsf{F}B :!?(x \circ c)$
	$\mathsf{T}\Diamond B :!a\circ!c$
	$\mathsf{T}B :?(!a\circ!c)$

The reader can easily check that the above axiom becomes provable if this constraint is added to the algebra of the labels. The tree in Table 15.2 shows that the following formula

$$\vdash \Box(\Diamond A \to \Diamond B) \to (\Box\Diamond A \to \Diamond\Box B) \to C) \to C$$

is verified in all frames which are both *regular* and *directed* (we use x, y, z as label-variables and skip the first few nodes in the construction of the tree). Since this example involves an application of the branching rule of generalised bivalence, it can also serve the purpose of illustrating the use of *variables in the labels* hinted at the end of Section 14.3.2.

Since the left-hand branch of this tree contains the pair of LS-formulae $\mathsf{T}C : x \circ b$ and $\mathsf{F}C : a \circ b$, it is closed if $x \circ b \leq a \circ b$. On the other hand, the right-hand branch is closed if $?(!a \circ !c) \leq !?(x \circ c)$, because of the pair of LS-formulae $\mathsf{T}B : ?(!a \circ !c)$ and $\mathsf{F}B : !?(x \circ c)$. Hence, this tree is closed for a given class of frames if the system consisting of the two inequalities:

$$x \circ b \leq a \circ b \qquad\qquad ?(!a \circ !c) \leq !?(x \circ c)$$

has a solution in the corresponding algebra of the labels. In this simple case it is easy to work out the solution $x = a$ for the algebra of the labels corresponding to directed regular frames. To verify this is a solution of the inequality generated by the right-hand branch, it is sufficient to go through the following steps:

$!a \circ !c \leq !(a \circ c)$	by regularity
$?(!a \circ !c) \leq ?!(a \circ c)$	by the monotonicity of ?
$?!(a \circ c) \leq !?(a \circ c)$	by directedness and duality
$?(!a \circ !c) \leq !?(a \circ c)$	by transitivity of $\leq$.

A general technique for solving systems of inequality generated by **LKE**-trees (with variables) within a given algebra of the labels is crucial for developing decision algorithms based on our method. This problem will be discussed in a subsequent work. Here we only observe that our approach separates each proof in two components: (i) a logical component, which depends only on the 'universal' meaning of the logical operators as defined by the inference rules which are a straightforward generalisation of the classical ones, and (ii) an algebraic component, which consists in solving a system of inequalities within a given labelling algebra that mirrors the specific properties of the information system in which the inferential process takes place. While algorithms for the first component can be devised as straightforward extensions of known algorithms for classical logic (simply by shifting from formulae to labelled formulae), algorithms for the second component can be devised by generalising existing algebraic methods for solving similar problems in related areas (e.g., unification, term-rewriting, constraint programming, etc.).

15.5.3 *Completeness of modal* **LKE**

The following table summaries the constraints on the algebra of the labels we have been considering so far.

Structural constraints		Modal constraints	
commutativity	$a \circ b \leq b \circ a$	seriality	$!a \leq ?a$
contraction	$a \circ a \leq a$	reflexivity	$!a \leq a$
expansion	$a \leq a \circ a$	transitivity	$!a \leq !!a$
monotonicity	$a \leq a \circ b$	symmetry	$a \leq ?!a$
		Euclideanism	$?a \leq ?!a$
		directedness	$!?a \leq ?!a$

Let **C** be, as above, a set of structural rules and/or modal axioms, and let **C**′ be the corresponding constraints (according to the correspondence outlined in Tables 14.1 and

15.1). We call **LKE**($\mathbf{C}'$) the basic **LKE** system obtained by augmenting the basic labelling algebra (see page 285) with the additional constraints in $\mathbf{C}'$, and we denote by $\vdash_{\mathbf{LKE}(\mathbf{C}')}$ the deducibility relation associated with this system. Then we can prove the following:

Proposition 15.20. *If* $\vdash_{\mathbf{C}} A$, *then* $\vdash_{\mathbf{LKE}(\mathbf{C}')} A$.

Proof *(Sketch)* The relation $\vdash_{\mathbf{LKE}(\mathbf{C}')}$, intended as a (finitary) consequence relation, is a **MIL**—that is it satisfies the basic conditions for a consequence relation (identity and surgical cut), the conditions $\mathrm{C}{\rightarrow}_i$ of Section 14.2.1 and conditions (15.18) and (15.19)—and is closed under all the structural rules and modal axioms in **C**. □

Propositions 15.20 and 15.9 imply that **LKE** is complete with respect to modal implication models. To close the circle we only need to prove that **LKE** is *sound* with respect to this semantics.

Proposition 15.21. *If* $\vdash_{\mathbf{LKE}(\mathbf{C}')} A$, *then* A *is verified in all* $\mathbf{C}'$*-frames*

The proof is routine and is omitted. It immediately follows from Propositions 15.20, 15.9 and 15.21 that

Corollary 15.22. $\vdash_{\mathbf{C}} A$ *if and only if* $\vdash_{\mathbf{LKE}(\mathbf{C}')} A$ *if and only if* A *is verified in all* $\mathbf{C}'$*-frames.*

16

PRODUCTS OF MODAL LOGICS

16.1 Introduction and Background

Let $F_1 = (W_1, R_1), F_2 = (W_2, R_2)$ be two Kripke frames describing two independent systems of possible worlds (or 'states'). Suppose we need to construct a complex frame F combining them both. A natural way of doing that is the following: the states of F should be all possible pairs (i.e. elements of $W_1 \times W_2$), and possible accessibility of the states can go in two directions. Thus we come to the notion of the *product*

$$F_1 \times F_2 = (W_1 \times W_2, R_1^\vee, R_2^\vee),$$

where

$$R_1^\vee = \{((x,y),(z,y)) | xR_1z\}, R_2^\vee = \{((x,y),(x,z)) | yR_2z\}.$$

This frame corresponds to a bimodal language with two modal operators $\Box_1$ and $\Box_2$, and Kripke semantics gives us the following interpretation:

$$(x,y) \models \Box_1 A \Leftrightarrow \forall z (xR_1z \Rightarrow (z,y) \models A);$$

$$(x,y) \models \Box_2 A \Leftrightarrow \forall z (yR_2z \Rightarrow (x,z) \models A).$$

So $\Box_1$ and $\Box_2$ are interpreted respectively as 'in all horizontally accessible worlds' and 'in all vertically accessible worlds'.[44]

What kind of modal logics correspond to those frames? This question is the topic of this chapter. But before moving to the main body of the chapter, let us look at products in the general context of the book. We first recall the combining logic problem introduced in Section 1.2.

The Combining Logic Problem

Let $\mathbb{L}_1$ and $\mathbb{L}_2$ be two modal languages with modalities $\Box_1$ and $\Box_2$ respectively. Let $\vdash_1$ and $\vdash_2$ be consequence relations (or 'logics') for the respective languages; $\vdash_i$ can be presented as traditional Hilbert systems H_i, but not necessarily. Consider the language $\mathbb{L}$ having both modalities $\Box_1$ and $\Box_2$ and ask the following questions about a possible consequence relation $\vdash$ on $\mathbb{L}$.

1. (a) Identify all the consequence relations $\vdash$ on $\mathbb{L}$, which are conservative extensions of $\vdash_1$ and $\vdash_2$.

[44] No special names for $\Box_1, \Box_2$ were given so far. In the case when the original accessibility relations are linear, these modalities are sometimes called 'compass operators' (corresponding to the East and North respectively) [Marx and Venema, 1997].

   (b) Identify the smallest such consequence relation $\vdash_{\text{min}}$.
   (c) Identify all maximal such extensions.
2. If $\mathcal{K}_1$ and $\mathcal{K}_2$ are two classes of Kripke models for which $\vdash_1$ and $\vdash_2$ are (respectively) complete, identify natural semantic constructions $\mathbb{C}$ that would yield a new semantics $\mathbb{C}(\mathcal{K}_1, \mathcal{K}_2)$ for $\mathcal{L}$ and axiomatise the corresponding $\vdash_{\mathbb{C}}$.
3. Identify a construction $\mathbb{C}_{\text{min}}$ for $\vdash_{\text{min}}$.
4. (a) ('Axiomatisation problem') If $\vdash_1$ and $\vdash_2$ are presented as proof systems Π_1 and Π_2 (e.g. Π_i = Hilbert system H_i), find a combinatorial definition for combining them.
   (b) Similarly to (a) for directly combining $\vdash_1$ and $\vdash_2$ as consequence relations to yield a new $\vdash$ on $\mathcal{L}$.
5. ('Transfer problem') If $\vdash_1$ and $\vdash_2$ satisfy a property $\mathcal{P}$ (such as a finite model property, decidability, interpolation, etc.), under what conditions does $\vdash$ satisfy a related $\mathcal{P}'$?
6. Characterise known polymodal logics in terms of $\mathbb{C}(\mathcal{K}_1, \mathcal{K}_2)$ where $\mathcal{K}_i$ are semantics of their components and $\mathbb{C}$ is a suitable operation.
7. Under what assumption on Hilbert formulations H_i of $\vdash_i$ which are complete for $\mathcal{K}_i$ (respectively) can we find a $\mathbb{C}$ such that $H_1 \cup H_2$ as a Hilbert system characterises $\mathbb{C}(\mathcal{K}_1, \mathcal{K}_2)$?

We use the terminology 'fibring' or 'combining' logics for the general process of constructing a logic $\vdash$ out of $\vdash_1$ and $\vdash_2$.

16.1.1 *Joining (product) of Kripke semantics*

The product semantics of $\mathcal{K}_1$ and $\mathcal{K}_2$ is obtained by looking at models of the form

$$(S_n^1 \times S_m^2, (R_n^1)^\vee, (R_m^2)^\vee, (a_n^1, a_m^2), h_{n,m}),$$

where

$$(S_n^1, R_n^1, a_n^1, h_n^1) \in \mathcal{K}_1, (S_m^2, R_m^2, a_m^2, h_m^2) \in \mathcal{K}_2,$$

$$(x, y)(R_n^1)^\vee(x', y') \text{ iff } xR_n^1x' \wedge y = y',$$

$$(x, y)(R_n^2)^\vee(x', y') \text{ iff } yR_n^2y' \wedge x = x',$$

and $h_{n,m}$ is any assignment satisfying the following conditions

(i) for each $y \in S_m^2$, let $h_{n,m}^y(x, q) = h_{n,m}((x, y), q)$, then $(S_n^1, R_n^1, a_n^1, h_{n,m}^y) \in \mathcal{K}_1$;
(ii) a similar condition for $x \in S_n^1, h_{n,m}^x$.

We call these assignments *two-dimensional*.

Note that if the logic $\vdash_i \quad (i = 1, 2)$ is complete for a class of models

$$(S_n^i, R_n^i, a_n^i, h_n^i), \quad n = 1, 2, \ldots$$

then the above definition yields that the semantics of the product is also defined by the class of models[45]

$$(S_n^1 \times S_m^2, (R_n^1)^\vee, (R_m^2)^\vee, h_{n,m}, (a_n^1, a_m^2)).$$

Note that the product can be of two different or of the same logic $\vdash$ (we refer to this case as *self-product*). In the latter case we can also look at a strict product, which we call squaring of the logic. For *squaring*, we look at models of the form $(S \times S, (a, a), R_1, R_2, h)$ only. Squaring is discussed in Section 15.4 of [Gabbay and Shehtman, 1998].

On the other hand, we can look at the special case of dovetailed models (S, a, R_1, R_2, h), where R_1 and R_2 commute, and R_1^{-1} and R_2 'half-commute':

$$R_1^{-1} \circ R_2 \subseteq R_2 \circ R_1^{-1}.$$

Such models are called *commutative*. It turns out that commutative models can be represented as p-morphic images of products (Theorem 16.59 below). So in this sense products are a special case of dovetailing. Compare with Sections 1.2.4 and 3.3.

16.1.2 *Fusion of two logics*

This operation is done on the proof theoretic level when two proof theoretic systems are combined. The simplest example is as follows. Given two modal logics $\mathbf{L}_1, \mathbf{L}_2$ formulated as Hilbert systems with disjoint sets of modal operators, combine their languages and take the union of the axioms and rules.

The resulting system is called the *fusion* of $\mathbf{L}_1$ and$\mathbf{L}_2$ and denoted by $\mathbf{L}_1 * \mathbf{L}_2$ (or by $\mathbf{L}_1 \otimes \mathbf{L}_2$ in another notation). In the case when $\mathbf{L}_1, \mathbf{L}_2$ are normal 1-modal logics, fusion yields exactly the consequence relation $\vdash_{\min}$ mentioned in the combining logic problem on page 327. On the other hand, on the semantical level fusion of two normal 1-modal logics corresponds to dovetailing. In more precise words, if $\mathbf{L}_1$ and $\mathbf{L}_2$ are complete w.r.t. the classes of frames $\mathcal{C}_2$ and $\mathcal{C}_2$ respectively, the fusion $\mathbf{L}_1 * \mathbf{L}_2$ is complete w.r.t. the class of all frames obtained by iterated dovetailing constructions from $\mathcal{C}_1$ and $\mathcal{C}_2$. This result was proved in [Fine and Schurz, 1997] and independently in [Hughes and Cresswell, 1996].

16.1.3 *Background results*

1. General fibring and dovetailing was studied earlier in this book.
2. Fusion for normal extensions of $\mathbf{K}$ ws studied in [Fine and Schurz, 1997] and [Hughes and Cresswell, 1996]. These papers show that fusion preserves Kripke completeness, the finite model property, decidability and many other properties.

[45]Note that if $\mathcal{K}_1$ and $\mathcal{K}_2$ are each based on a class of frames, then so is the product.

Connections between fusion and dovetailing are considered also in [Gabbay, 1995b]. [Shehtman, 1978] studies the complexity of fusions and shows that in general the complexity goes up.

3. Enriching by universal modality. This is a way of combining a modal logic with **S5**. The operation is defined semantically: given a logic **L** determined by a class of Kripke frames $\mathcal{C}$, consider the logic $\mathbf{L}^U$ of the class of bimodal frames $(W, R, W \times W)$; cf. [Goranko, 1996; Goranko and Passy, 1992].
4. Enriching by 'difference modality'. This operation is also defined semantically: given a logic **L** determined by a class of Kripke frames $\mathcal{C}$, consider the logic $\mathbf{L}^D$ of the class of bimodal frames $(W, R, \neq_W)$, where $\neq_W$ is the inequality relation on W. Actually, this is the way of combining a given logic **L** with difference logic **DL**; cf. [de Rijke, 1993–94].
5. There are many other options of that sort; we can mention also 'temporalisation' [Wolter, 1994], and 'enriching by nominals' [Gargov and Goranko, 1993; Goranko, 1996; Goranko, 1994; Blackburn, 1993].
6. Products of frames and of modal logics were introduced and studied in [Shehtman, 1978], but products of two frames with equivalence accessibility relations appeared earlier, in [Segerberg, 1973]. Further results on products were stated in the abstract [Shehtman, 1987]. The paper [Gabbay and Shehtman, 1998] is an essential development of [Shehtman, 1978] and [Shehtman, 1987]. Various types of multi-dimensional modal logics and of some products were considered over the last 10 years by different authors (J. Halpern, M. Shoham, Y. Venema, A. Zanardo, E. Spaan, M. Marx and others) in connection with relation algebras, cylindric algebras, temporal and dynamic logics of different kinds; we refer the reader to [Marx and Venema, 1997] for corresponding references and an outline of these results.

Our chapter is based on [Gabbay and Shehtman, 1998]. Sections 16.2–16.4 contain some basic definitions and facts about Kripke models, fusions and products. Section 16.6 recalls (with a slight modification) the main construction and the theorem from [Shehtman, 1978] on axiomatising the class of all products of Kripke frames. Section 16.7 proves a more general result about products of PTC (pseudo-transitive closed) logics, and Section 16.8 demonstrates counterexamples to this result in the general case. In Section 16.5 we show that in many cases modal logics of products are recursively axiomatisable and have the countable frame property. Section 16.9 considers some more examples of products. Sections 16.10-16.13 prove decidability of products and some other polymodal logics via the finite model property (f.m.p.). In Sections 16.10, 16.12 we modify the method of normal forms introduced by Kit Fine [Fine, 1975]. In Section 16.11 we use a (probably, new) 'finite depth method', which gives a shorter proof, but a worse deciding algorithm. In Section 16.13 we use a modified filtration method from [Gabbay, 1972] and [Shehtman, 1993]. As a corollary in Section 16.14 we obtain decidability for fragments of modal and intermediate predicate logics with a single individual variable. Section 16.15 describes transition products which will be addressed in a sequel paper. A short outline of related results and questions is given in the final Section 16.16.

16.2 Basic Definitions and Notations

Consider a propositional language $\mathbb{L}_n$ with a countable set $PL = \{p_1, p_2, \ldots\}$ of propositional letters, classical connectives (let $\rightarrow, \bot$ be basic and $\wedge, \vee, \sim, \top, \equiv$ be derived) and additional one-place modal connectives $\Box_1, \ldots, \Box_n, n \geq 1$. As usual, $\Diamond_i$ denotes $\neg\Box_i\neg$. Formulas of this language are called *n-modal*. An *n-modal (Kripke) frame* is a pair $F = (W, R_1, \ldots, R_n)$, such that $W \neq \varnothing$, $R_i \subseteq W \times W$. Elements of W are called *worlds (or points)*. A *Kripke model* over the frame F is a pair $\mathbf{m} = (F, h)$, in which $h : PL \longrightarrow 2^W$ is a *valuation*. h is extended to all formulas in a standard way:

$$h(\bot) = \varnothing, h(A \rightarrow B) = (W - h(A)) \cup h(B), h(\Box_i A) = \{x | R_i(x) \subseteq h(A)\}.$$

We also write: $\mathbf{m}, w \models A$ instead of $w \in h(A)$, and say that A is *true at the world w of the model* $\mathbf{m}$.

A formula A is called *true in a model* $\mathbf{m}$ (notation: $\mathbf{m} \models A$) if it is true at every world of $\mathbf{m}$. We say that A is *satisfied in* $\mathbf{m}$ if it is true at some world of $\mathbf{m}$. A is called *valid in a frame* F (notation: $F \models A$) if it is true in every model over F. A set of formulas Γ is *valid* in F (notation: $F \models \Gamma$) if every $A \in \Gamma$ is valid. In the latter case we say also that F is a Γ*-frame.*

A formula A is *satisfied at the world w of the frame F* (or briefly, A is *satisfied at* F, w) if there exists a model $\mathbf{m}$ over F such that $\mathbf{m}, w \models A$.

We use the following notation referring to binary relations in a set W. $R \circ S$ denotes the composition of R and S; R^{-1} is the inverse relation to R; I_W is the equality relation in W. $R(x)$ denotes $\{y | xRy\}$. $R \upharpoonright V$ denotes the restriction of R to a subset V, i.e. $R \upharpoonright V = R \cap (V \times V)$.

R^n is the composition $\underbrace{R \circ \ldots \circ R}$ if $n > 0$, and I_W if $n = 0$. R^* is the reflexive transitive closure of R, i.e. $R^* = \bigcup\limits_{k \geq 0} R^k$.

A (normal) *n-modal logic* is a set of $\mathbb{L}_n$-formulas closed under the rules of substitution, modus ponens, necessitation $(A/\Box_i A), 1 \leq i \leq n$, and containing all classical tautologies and all formulas $\Box_i(p \rightarrow q) \rightarrow (\Box_i p \rightarrow \Box_i q)$ $(p, q \in PL, 1 \leq i \leq n)$. $\mathbf{K}_n$ denotes the minimal n-modal logic. The smallest logic containing a given logic $\mathbf{L}$ and a set of formulas Γ is denoted by $\mathbf{L} + \Gamma$; for a formula A, $\mathbf{L} + A$ is an abbreviation for $\mathbf{L} + \{A\}$. We say that the logic $\mathbf{K}_n + \Gamma$ *is axiomatised by* the set Γ.

If $n = 1$ we write $\Box$ instead of $\Box_1$, and $\mathbf{K}$ instead of $\mathbf{K}_1$. We use also standard notation for the following well known modal logics:

$$\mathbf{D} = \mathbf{K} + \Diamond\top, \mathbf{K4} = \mathbf{K} + \Box p \rightarrow \Box\Box p, \mathbf{T} = \mathbf{K} + \Box p \rightarrow p, \mathbf{D4} = \mathbf{K4} + \Diamond\top,$$

$$\mathbf{S4} = \mathbf{K4} + \Box p \rightarrow p, \mathbf{B} = \mathbf{K} + \Diamond\Box p \rightarrow p, \mathbf{S5} = \mathbf{K4} + \Diamond\Box p \rightarrow p$$

An *m-formula* is a formula having no occurrences of letters p_i with $i > m$. If $\mathbf{L}$ is a modal logic, $\mathbf{L}\lceil m$ denotes its restriction to m-formulas. The sets $\mathbf{L}\lceil m$ are called *restricted*, or *weak logics.*

If $\mathbf{L}$ is an n-modal logic and A is an n-modal formula, we say also that A is a *theorem* of $\mathbf{L}$ if $A \in \mathbf{L}$, and that A is $\mathbf{L}$*-consistent* if $\sim A \notin \mathbf{L}$. A set S of (n-modal) formulas is $\mathbf{L}$*-consistent* if each finite conjunction of elements is $\mathbf{L}$-consistent.

All formulas valid in a certain n-modal frame F constitute an n-modal logic $\mathbf{L}$(F) (*the modal logic of* F). The modal logic $\mathbf{L}(\mathcal{C})$ of a class of n-frames $\mathcal{C}$ is defined as the intersection $\bigcap\{\mathbf{L}(F)|F \in \mathcal{C}\}$.[46]

Dually, every formula A gives rise to a class of n-modal frames $\mathcal{V}(A) = \{F|F \models A\}$. The same notation is used for a set of formulas $\Gamma : \mathcal{V}(\Gamma) = \{F|F \models \Gamma\}$. A class of this kind is called a *modal variety* (or *modal axiomatic*).

An n-modal logic Λ is called *Kripke-complete* if $\Lambda = \mathbf{L}(\mathcal{C})$ for some class of n-modal frames $\mathcal{C}$. Obviously, this is equivalent to $\Lambda = \mathbf{L}(\mathcal{V}(\Lambda))$; also one can prove that Λ is Kripke-complete iff $\Lambda = \mathbf{L}(F)$ for some n-modal frame F.

Recall also some important constructions.

Definition 16.1. Let $\mathbf{m} = (F, h)$ be a Kripke model over a frame $F = (W, R_1, \ldots, R_n)$, $x \in W$. A subset $V \subseteq W$ is called *R-stable* if $\forall i \forall x (x \in V \Rightarrow R_i(x) \subseteq V)$. The *subframe of F generated by* x is defined as the frame $F^x = (W^x, R_1^x, \ldots, R_n^x)$, such that W^x is the intersection of all R-stable subsets of W, containing x, and (for every i) $R_i^x = R_i \restriction W^x$. It is clear that $W^x = (R_1 \cup \ldots \cup R_n)^*(x)$. The submodel of $\mathbf{m}$ generated by x is defined as $\mathbf{m}^x = (F^x, h^x)$, where (for any $p \in PL$) $h^x(p) = h(p) \cap W^x$. We say also that a frame F (or a Kripke model $\mathbf{m}$) is *generated by* x (*or has a root* x) if $F = F^x$ (respectively, if $\mathbf{m} = \mathbf{m}^x$).

The following is well known:

Lemma 16.2.
(1) For any $y \in W^x$, *for any* $\mathbb{L}_n$*-formula* A,
$\mathbf{m}^x, y \models A \Leftrightarrow \mathbf{m}, y \models A$.
(2) $\mathbf{L}(F) = \bigcap\{\mathbf{L}(F^x)|x \in W\}$.

Definition 16.3. A surjective map $f : V \longrightarrow W$ is called a *p-morphism* of a frame $F = (W, R_1, \ldots, R_n)$ onto a frame $G = (V, S_1, \ldots, S_n)$ if it satisfies the following two conditions:

1. $\forall x \forall y (x R_i y \Rightarrow f(x) S_i f(y))$ *(monotonicity)*;
2. $\forall x \forall y (f(x) S_i z \Rightarrow \exists y (x R_i y \wedge f(y) = z))$.
   f is called a *p-morphism* of a Kripke model $\mathbf{m} = (F, h)$ onto a model $\mathbf{n} = (G, h')$ if it satisfies also
3. $\forall p \in PL\ \forall x (\mathbf{m}, x \models p \Leftrightarrow \mathbf{n}, f(x) \models p)$.

A map f satisfying (1) and (2) which is not necessarily surjective is called a *morphism* of corresponding frames. One can easily check that a morphism of generated frames sending a root to a root is always surjective.

An arbitrary map satisfying (1) is called a *homomorphism.*

We use the notation $f : F \twoheadrightarrow G$ (respectively $f : \mathbf{m} \twoheadrightarrow \mathbf{n}$), to show that f is a p-morphism of frames (Kripke models).

The following is well known:

[46] We adopt also that $\mathbf{L}(\varnothing)$ is a contradictory logic i.e. the set of all n-modal formulas; thus $\mathbf{L}(\varnothing)$ depends on n.

Lemma 16.4. *(1) If $f : \mathbf{m} \twoheadrightarrow \mathbf{n}$ (for Kripke models $\mathbf{m}, \mathbf{n}$), then for any x, for any $\mathbb{L}_n$-formula $A, \mathbf{m}, x \models A \Leftrightarrow \mathbf{n}, f(x) \models A$.*

(2) If $f : F \twoheadrightarrow G$ (for frames F, G), then $\mathbf{L}(F) \subseteq \mathbf{L}(G)$.

Definition 16.5. The *canonical frame* for an n-modal logic $\mathbf{L}$ is $F_{\mathbf{L}} = (W_{\mathbf{L}}, R_{1,\mathbf{L}}, \dots, R_{n,\mathbf{L}})$, where $W_{\mathbf{L}}$ is the set of all maximal $\mathbf{L}$-consistent sets of n-modal formulas, $xR_{i,\mathbf{L}}y$ iff for any $A, \Box_i A \in x$ implies $A \in x$.

The *canonical model* for $\mathbf{L}$ is $\mathbf{m}_{\mathbf{L}} = (F_{\mathbf{L}}, h_{\mathbf{L}})$, where $h_{\mathbf{L}}(p_i) = \{x | p_i \in x\}$.

Quite analogously, the *canonical frame* for $\mathbf{L}\lceil m$ is $F_{\mathbf{L}\lceil m} = (W_{\mathbf{L}\lceil m}, R_{1,\mathbf{L}\lceil m}, \dots, R_{n,\mathbf{L}\lceil m})$, where $W_{\mathbf{L}\lceil m}$ is the set of all maximal $\mathbf{L}$-consistent sets of n-modal m-formulas,

$$xR_{i,\mathbf{L}\lceil m}y \text{ iff for any } A, \Box_i A \in x \text{ implies } A \in x.$$

The *canonical model* for $\mathbf{L}\lceil m$ is $\mathbf{m}_{\mathbf{L}\lceil m} = (F_{\mathbf{L}\lceil m}, h_{\mathbf{L}\lceil m})$, where

$$h_{\mathbf{L}\lceil m}(p_i) = \{x | p_i \in x\} \text{ if } i \le m. \tag{16.1}$$

The *canonical models* $\mathbf{m}_{\mathbf{L}\lceil m}$ and frames $F_{\mathbf{L}\lceil m}$ are called *restricted*, or *weak*.

The following is well known:

Theorem 16.6. *For any m-formula A,*

1. $\mathbf{m}_{\mathbf{L}\lceil m}, x \models A$ *iff* $\mathbf{m}_{\mathbf{L}}, x \models A$ *iff* $A \in x$;
2. $\mathbf{m}_{\mathbf{L}\lceil m} \models A$ *iff* $\mathbf{m}_{\mathbf{L}} \models A$ *iff* $A \in \mathbf{L}$.

Proposition 16.7. *Let Λ, Λ_1 be two n-modal logics, perhaps restricted by the same number m. Then $\Lambda \subseteq \Lambda_1$ implies $\mathbf{L}(F_\Lambda) \subseteq \mathbf{L}(F_{\Lambda_1})$.*

Proof Obviously, every Λ_1-consistent set is Λ-consistent, and thus $W_{\Lambda_1} \subseteq W_\Lambda$. Furthermore, if $x \in W_{\Lambda_1}, y \in W_\Lambda$ and $xR_{i,\Lambda}y$ then $y \in W_{\Lambda_1}$ too. For, assume that y is Λ_1-inconsistent, then (since y is closed under finite conjunctions) for some $A \in y$ we have $\sim A \in \Lambda_1$, and thus $\Box_i \sim A \in \Lambda_1$, which implies $\sim \Diamond_i A \in \Lambda_1$. On the other hand $xR_{i,\Lambda}y$ together with $A \in y$ implies $\Diamond_i A \in x$, and we get that x is Λ_1-inconsistent.

Therefore, $F^x_{\Lambda_1}$ is a generated subframe in F_Λ, and thus $\mathbf{L}(F^x_{\Lambda_1}) \supseteq \mathbf{L}(F_\Lambda)$. Now Lemma 16.2 (2) implies $\mathbf{L}(F_\Lambda) \subseteq \mathbf{L}(F_{\Lambda_1})$. □

Lemma 16.8. *(cf. [Fine, 1974b]). Let $\mathbf{L}$ be an n-modal logic. Consider the following equivalence relation in the restricted canonical model $\mathbf{m}_{\mathbf{L}\lceil m}$:*

$$x \approx y \Leftrightarrow \forall i(1 \le i \le m \to (\mathbf{m}_{\mathbf{L}\lceil m}, x \models p_i \Leftrightarrow \mathbf{m}_{\mathbf{L}\lceil m}, y \models p_i))\wedge$$
$$\forall j(1 \le j \le n \Rightarrow R_{j,\mathbf{L}\lceil m}(x) = R_{j,\mathbf{L}\lceil m}(y)).$$

Then actually $x \approx y \Leftrightarrow x = y$.

Proof By induction on the length of an m-formula A, one can obtain that $x \approx y$ implies $\mathbf{m}, x \models A \Leftrightarrow \mathbf{m}, y \models A$. In other words, $x \approx y$ provided that the same m-formulas are true in x and y. Hence $x = y$, due to Theorem 16.6 (1). □

Definition 16.9. (cf. [van Benthem, 1983]). A modal logic is called *canonical* if it is valid in each of its canonical frames (with extended set of propositional letters).

Corollary 16.10. *Every canonical logic is Kripke-complete.*

Proof Immediate from Theorem 16.6. □

16.3 Fusions and Products

Definition 16.11. Let $F_1 = (W, R_1, \ldots, R_n), F_2 = (W, S_1, \ldots, S_m)$ be frames with the same set of possible worlds. Their *fusion* is the frame

$$F_1 * F_2 = (W, R_1, \ldots, R_n, S_1, \ldots, S_m).$$

If $\mathcal{C}_1, \mathcal{C}_2$ are classes of n- and m-modal frames respectively, their *fusion* is the class of $(n+m)$-modal frames:

$$\mathcal{C}_1 * \mathcal{C}_2 = \{F_1 * F_2 | F_1 \in \mathcal{C}_1, F_2 \in \mathcal{C}_2\}.$$[47]

Definition 16.12. For an m-modal formula A, and $n \geq 0$, let A^{+n} be the $(n+m)$-modal formula, obtained from A by replacing each operator $\Box_j$ by $\Box_{j+n}$; for a set Γ of m-modal formulas let $\Gamma^{+n} = \{A^{+n} | A \in \Gamma\}$.

Definition 16.13. Let $\mathbf{L}_1$ be an n-modal logic, $\mathbf{L}_2$ be an m-modal logic. Their *fusion* is the $(n+m)$-modal logic $\mathbf{L}_1 * \mathbf{L}_2 = \mathbf{K}_{n+m} + \mathbf{L}_1 + \mathbf{L}_2^{+n}$.

It is easily proved that $\mathbf{K}_n * \mathbf{K}_m = \mathbf{K}_{n+m}$, and also that

$$(\mathbf{K}_n + \Gamma_1) * (\mathbf{K}_m + \Gamma_2) = \mathbf{K}_{n+m} + \Gamma_1 + \Gamma_2^{+n}.$$

Lemma 16.14.

1. *For any m-modal formula A, $F_1 \models A \Leftrightarrow F_1 * F_2 \models A^{+n}$.*
2. *For any n-modal logic $\mathbf{L}_1$, m-modal logic $\mathbf{L}_2$*

$$\mathcal{V}(\mathbf{L}_1 * \mathbf{L}_2) = \mathcal{V}(\mathbf{L}_1) * \mathcal{V}(\mathbf{L}_2).$$

In [Fine and Schurz, 1997], [Kracht and Wolter, 1991] the following is proved:

Theorem 16.15. *Fusion of logics preserves Kripke-completeness.*

So in this case by Lemma 16.14, we obtain that the fusion of frames corresponds to the fusion of logics:

Corollary 16.16. $\mathbf{L}_1 * \mathbf{L}_2 = \mathbf{L}(\mathcal{V}(\mathbf{L}_1) * \mathcal{V}(\mathbf{L}_2))$, *whenever* $\mathbf{L}_1, \mathbf{L}_2$ *are Kripke-complete.*

[47] Of course it may happen that $\mathcal{C}_1 * \mathcal{C}_2 = \varnothing$.

Definition 16.17. The *product* of an n-modal frame $F_1 = (W_1, R_1, \dots, R_n)$ and an m-modal frame $F_2 = (W_2, S_1, \dots, S_m)$ is the $(n+m)$-modal frame

$$F_1 \times F_2 = (W_1 \times W_2, R_1^{\vee}, \dots, R_n^{\vee}, S_1^{\vee}, \dots, S_m^{\vee}),$$

in which

$$R_i^{\vee} = \{((x,z),(y,z)) | xR_i y\}, S_j^{\vee} = \{((z,x),(z,y)) | xS_j y\}.$$

Definition 16.18. Let $\mathcal{C}_1, \mathcal{C}_2$ be classes of n- and m-modal frames respectively. Their *product* is the class of $(n+m)$-modal frames

$$\mathcal{C}_1 \times \mathcal{C}_2 = \{F_1 \times F_2 | F_1 \in \mathcal{C}_1, F_2 \in \mathcal{C}_2\}.$$

Now we can recall the problem considered in [Shehtman, 1978]:

Given two classes $\mathcal{C}_1, \mathcal{C}_2$ of n- and m-modal frames respectively and their logics $\mathbf{L}(\mathcal{C}_1), \mathbf{L}(\mathcal{C}_2)$, find the logic $\mathbf{L}(\mathcal{C}_1 \times \mathcal{C}_2)$.

A special case of this problem is when $\mathcal{C}_1, \mathcal{C}_2$ are modal varieties. It is better to reformulate it using the following notion.

Definition 16.19. Let $\mathbf{L}_1$ be an n-modal logic, $\mathbf{L}_2$ be an m-modal logic. Their *semantical product* is the $(n+m)$-modal logic

$$\mathbf{L}_1 \times \mathbf{L}_2 = \mathbf{L}(\mathcal{V}(\mathbf{L}_1) \times \mathcal{V}(\mathbf{L}_2)).$$

So the particular problem is:

Given two logics $\mathbf{L}_1, \mathbf{L}_2$, find the logic $\mathbf{L}_1 \times \mathbf{L}_2$.

As was proved in [Shehtman, 1978], in many cases this construction can be done as follows.

Definition 16.20. For an n-modal $\mathbf{L}_1$, m-modal $\mathbf{L}_2$ let

$$[\mathbf{L}_1, \mathbf{L}_2] = \mathbf{L}_1 * \mathbf{L}_2 + \{C_{ij}^1 | 1 \le i \le n, 1 \le j \le m\} + \{C_{ij}^2 | 1 \le i \le n, 1 \le j \le m\},$$

where

$$C_{ij}^1 = (\Box_i \Box_{n+j} p \equiv \Box_{n+j} \Box_i p), C_{ij}^2 = (\Diamond_i \Box_{n+j} p \to \Box_{n+j} \Diamond_i p).$$

The logics $\mathbf{L}_1, \mathbf{L}_2$ are called *product-matching* if $\mathbf{L}_1 \times \mathbf{L}_2 = [\mathbf{L}_1, \mathbf{L}_2]$.

In [Shehtman, 1978] a family of product-matching pairs was described; here this result will be somewhat improved (see Theorem 16.76 below). Let us begin with some other definitions and simple facts.

Lemma 16.21. *Let F be an $(n+m)$-modal frame. Then*

$$F \models C_{ij}^1 \Leftrightarrow R_i \circ R_{n+j} = R_{n+j} \circ R_i,$$

$$F \models C_{ij}^2 \Leftrightarrow R_i^{-1} \circ R_{n+j} \subseteq R_{n+j} \circ R_i^{-1}.$$

This lemma is a consequence of Sahlqvist's theorem (cf. e.g. [van Benthem, 1983]); a straightforward proof is also rather easy.

Definition 16.22. An $(n+m)$-modal frame is called (n,m)-*commutative* if

$$F \models C^1_{ij} \wedge C^2_{ij} \text{ for any } i \in \{1,\dots,m\}, j \in \{1,\dots,n\}.$$

Definition 16.23. Let $\mathcal{C}_1, \mathcal{C}_2$ be classes of n- and m-modal frames respectively. Their *commutator* is the class $[\mathcal{C}_1, \mathcal{C}_2] = \{F | F \in \mathcal{C}_1 * \mathcal{C}_2, F \text{ is } (n,m)\text{-commutative}\}$.

Definition 16.24. For a class $\mathcal{C}$ of n-modal frames, let $GS(\mathcal{C})$ be the class of all generated subframes of frames from $\mathcal{C}$, and let $\mathcal{C}^\sim = \{F | GS(\{F\}) \subseteq GS(\mathcal{C})\}$, i.e. a frame is in $\mathcal{C}^\sim$ iff each of its generated subframes can be generated by some frame from $\mathcal{C}$.

The following is an obvious consequence of Lemma 16.2:

Lemma 16.25. $\mathbf{L}(\mathcal{C}) = \mathbf{L}(GS(\mathcal{C})) = \mathbf{L}(\mathcal{C}^\sim)$.

Proposition 16.26. *Let $\mathcal{C}_1, \mathcal{C}_2$ be classes of n- and m-modal frames respectively. Then*

1. $\mathcal{C}_1 \times \mathcal{C}_2 \subseteq [\mathcal{C}_1^\sim, \mathcal{C}_2^\sim]$;
2. $[\mathbf{L}(\mathcal{C}_1), \mathbf{L}(\mathcal{C}_2)] \subseteq \mathbf{L}([\mathcal{C}_1, \mathcal{C}_2])$;
3. $[\mathbf{L}(\mathcal{C}_1), \mathbf{L}(\mathcal{C}_2)] \subseteq \mathbf{L}([\mathcal{C}_1^\sim, \mathcal{C}_2^\sim]) \subseteq \mathbf{L}(\mathcal{C}_1 \times \mathcal{C}_2)$.

Proof

1. Let $F_1 = (W_1, R_1, \dots, R_n) \in \mathcal{C}_1, F_2 = (W_2, S_1, \dots, S_m) \in \mathcal{C}_2$. Then $F_1 \times F_2$ is (n,m)-commutative. In fact, assume that $(x,y) R_i^\vee (x_1, y_1) S_j^\vee (x_2, y_2)$, i.e. that $x R_i x_1 = x_2$ and $y = y_1 S_j y_2$. Then there exists (u,v) such that $(x,y) S_j^\vee (u,v)$ $R_i^\vee (x_2, y_2)$; namely, we can take $u = x, v = y_2$.
This proves $R_i^\vee \circ S_j^\vee \subseteq S_j^\vee \circ R_i^\vee$; the converse is proved similarly. In the same way one can check that $(R_i^\vee)^{-1} \circ S_j^\vee \subseteq S_j^\vee \circ (R_i^\vee)^{-1}$.
Finally, note that $F_1 \times F_2 = G_1 * G_2$, where $G_1 = (W_1 \times W_2, R_1^\vee, \dots, R_n^\vee)$, $G_2 = (W_1 \times W_2, S_1^\vee, \dots, S_m^\vee)$, and $G_1^{(a,b)}$ is isomorphic to F_1^a by the map $(x,b) \mapsto x$; $G_2^{(a,b)}$ is isomorphic to F_2^b by the map $(a,y) \mapsto y$.
Therefore $F_1 \times F_2 \in [\mathcal{C}_1^\sim, \mathcal{C}_2^\sim]$.
2. First note that $\mathbf{L}(\mathcal{C}_1) * \mathbf{L}(\mathcal{C}_2) \subseteq \mathbf{L}(\mathcal{C}_1 * \mathcal{C}_2) \subseteq \mathbf{L}([\mathcal{C}_1, \mathcal{C}_2])$.
Hence we obtain that $[\mathbf{L}(\mathcal{C}_1), \mathbf{L}(\mathcal{C}_2)] \subseteq \mathbf{L}([\mathcal{C}_1, \mathcal{C}_2])$.
3. From (2) by Lemma 16.25, we get $[\mathbf{L}(\mathcal{C}_1), \mathbf{L}(\mathcal{C}_2)] \subseteq \mathbf{L}([\mathcal{C}_1^\sim, \mathcal{C}_2^\sim])$. Then notice that $[\mathcal{C}_1^\sim, \mathcal{C}_2^\sim] \supseteq \mathcal{C}_1 \times \mathcal{C}_2$ by (1), and thus $\mathbf{L}([\mathcal{C}_1^\sim, \mathcal{C}_2^\sim]) \subseteq \mathbf{L}(\mathcal{C}_1 \times \mathcal{C}_2)$. □

Corollary 16.27. $[\mathbf{L}_1, \mathbf{L}_2] \subseteq \mathbf{L}_1 \times \mathbf{L}_2$.

Proof $[\mathbf{L}_1, \mathbf{L}_2] \subseteq [\mathbf{L}(\mathcal{V}(\mathbf{L}_1)), \mathbf{L}(\mathcal{V}(\mathbf{L}_2))] \subseteq \mathbf{L}(\mathcal{V}(\mathbf{L}_1) \times \mathcal{V}(\mathbf{L}_2)) = \mathbf{L}_1 \times \mathbf{L}_2$. □

Lemma 16.28. $\mathcal{V}([\mathbf{L}_1, \mathbf{L}_2]) = [\mathcal{V}(\mathbf{L}_1), \mathcal{V}(\mathbf{L}_2)]$.

Proof Trivial from Lemma 16.14. □

Proposition 16.29. *If the logics $\mathbf{L}_1, \mathbf{L}_2$ are canonical then $[\mathbf{L}_1, \mathbf{L}_2]$ is canonical, and* $[\mathbf{L}_1, \mathbf{L}_2] = \mathbf{L}([\mathcal{V}(\mathbf{L}_1), \mathcal{V}(\mathbf{L}_2)])$.

Proof First observe that $\mathbf{L}_1 * \mathbf{L}_2$ is canonical, and thus by Proposition 16.7, it is valid in a canonical frame for $[\mathbf{L}_1, \mathbf{L}_2] \supseteq \mathbf{L}_1 * \mathbf{L}_2$. The axioms C^1_{ij}, C^2_{ij} are also valid, due to Sahlqvist's theorem. The last statement follows from Lemma 16.28. □

Corollary 16.30. $[\mathbf{K}_n, \mathbf{K}_m] = \mathbf{L}$(*all* (n,m)*-commutative frames).*

Let us prove some other useful facts.

Lemma 16.31. *Let F be an n-frame, G be an m-frame, $a \in F, b \in G$. Then $(F \times G)^{(a,b)} = F^a \times G^b$.*

Proof If $F = (W, R_1, \ldots, R_n), G = (V, S_1, \ldots, S_n)$, then for any x, y $(a(R_1 \cup \ldots \cup R_n)^* x$ and $b(S_1 \cup \ldots S_m)^* y) \Leftrightarrow (a,b)(R_1^\vee \cup \ldots \cup R_n^\vee \cup S_1^\vee \cup \ldots \cup S_m^\vee)^*(x,y)$.

Hence it follows that $(W \times V)^{(a,b)} = W^a \times V^b$. The relations in $(F \times G)^{(a,b)}$ and $F^a \times G^b$ are the same, because they come as restrictions from $F \times G$. □

Lemma 16.32. *Let $\mathcal{C}_1, \mathcal{C}_2$ be classes of n- and m-modal frames respectively. Then $GS(\mathcal{C}_1 \times \mathcal{C}_2) = GS(\mathcal{C}_1) \times GS(\mathcal{C}_2)$.*

Proof Immediate from Lemma 16.31. □

The simple proof of the following Lemma is left to the reader.

Lemma 16.33. *Let F_1, F_2 be n-frames, G_1, G_2 be n-frames, $f : F_1 \twoheadrightarrow F_2$, $g : G_1 \twoheadrightarrow G_2$, and let $h = f \times g$ (i.e. $h(x,y) = (f(x), g(y))$). Then $h : F_1 \times G_1 \twoheadrightarrow F_2 \times G_2$.*

16.4 Thickening and Unravelling

Now let us define some technical constructions.

Definition 16.34. Let $F = (W, R_1, \ldots, R_n)$ be a frame generated by u_0, and let V be an arbitrary non-empty set. Consider the frame $F \cdot V = (W \times V, R_1 \cdot V, \ldots, R_n \cdot V)$, where $(a,x)(R_i \cdot V)(b,t)$ iff aR_ib. Fix some $v_0 \in V$, and take the generated subframe $F_V = (F \cdot V)^{(u_0,v_0)} = (W_V, R_{1_V}, \ldots, R_{n_V})$. The frame F_V is called the *thickening* of F by V.

Obviously, up to an isomorphism, F_V does not depend on v_0. One can easily check that $W_V = W \times V$ if there exists a 'loop' in $F : u_0 R_{i_1} u_1 \ldots u_{m-1} R_{i_m} u_0$ (for some $m \geq 1$), and that $W_V = ((W - \{u_0\}) \times V) \cup \{(u_0, v_0)\}$ otherwise.

We skip the straightforward proof of the following lemma.

Lemma 16.35. *The first projection $(x,t) \mapsto x$ is a p-morphism $F_V \twoheadrightarrow F$.*

Definition 16.36. (cf. [Gabbay, 1976], [Sahlqvist, 1975], [van Benthem, 1983]). Let $F = (W, R_1, \ldots, R_n)$ be a frame generated by u_0. The *unravelling* of F is the frame $F^\sharp = (W^\sharp, R_1^\sharp, \ldots, R_n^\sharp)$, such that

$$W^\sharp = \{(u_0, j_0, u_1, \ldots, j_{n-1}, u_n) | n \geq 0 \wedge \forall i < n\, u_i R_{j_i} u_{i+1}\},$$

$$(u_0, j_0, u_1, \ldots, u_n) R_j^\sharp x \Leftrightarrow \exists u (x = (u_0, j_0, u_1, \ldots, u_n, j, u) \wedge u_n R_j u).$$

For the particular case $n = 1$ we have that $j_i = 1$ for any i, so we can reduce this construction to:

$$W^\sharp = \{(u_0, u_1, \ldots, u_n) | n \geq 0 \wedge \forall i < n\; u_i R_1 u_{i+1}\},$$

$$(u_0, u_1, \ldots, u_n) R_1^\sharp x \Leftrightarrow \exists u (x = (u_0, u_1, \ldots, u_n, u) \wedge u_n R_1 u).$$

Lemma 16.37. *The map $\pi : F^\sharp \to F$ such that $\pi(u_0, u_1, \ldots, u_n) = u_n$, is a p-morphism of $F^\sharp$ onto F.*

The proof is straightforward.

Definition 16.38. A frame $F = (W, R_1, \ldots, R_n)$ is called a *tree* (or an n-tree) *with a root* u_0, if the following holds

1. F is generated by u_0;
2. $\forall x \neq u_0\ \exists! i\ \exists! y\ y R_i x$;
3. $\forall j\ \forall x \sim x R u_0$.

From the definitions it is clear that unravelling is a tree.

Definition 16.39. A 1-tree $(W, \sqsubset)$ is called *standard* if its worlds are (some) finite sequences of natural numbers, its root is the void sequence λ, and $\alpha \sqsubset \beta$ iff β is a successor of α, i.e. iff β is obtained by adding a single element at the end of α.

The following is quite clear:

Lemma 16.40. *Every countable*[48] *n-tree is isomorphic to some standard n-tree.*

16.5 Recursive Axiomatisability and Countable Frame Property

In many situations it is possible to prove that a product of two logics is RE (i.e. recursively axiomatisable), without giving its explicit axiomatisation. Also in many cases products can be characterised by countable frames.

Definition 16.41. Consider the classical first-order language $\mathbb{L}1_n$ with equality and binary predicate letters $R_1, \ldots, R_n$. Then $\mathbb{L}1_n$-models are nothing but n-modal frames. $\mathbb{L}1_n$-formulas are denoted by small Greek letters. The classical truth of an $\mathbb{L}1_n$-formula φ in a frame F is denoted by $F \models \varphi$ as usual.

Definition 16.42. Recall that a class of models of a set of first-order sentences is called Δ-*elementary*; if this set is recursive, we call the class *R-elementary*. An n-modal logic Λ is called Δ-*(respectively, R-) elementary* if $\mathbf{V}(\Lambda)$ is Δ- (R-) elementary in the language $\mathbb{L}1_n$.

Definition 16.43. A modal logic Λ has a *countable frame property (c.f.p.)* if $\Lambda = \mathbf{L}(\mathcal{C})$ for some class $\mathcal{C}$ of countable frames.

Recall also how modal formulas can be translated as first-order formulas. This translation is well known; it was introduced first by G. Mints (1968).

Consider the language $\mathbb{L}1_{n1}$ obtained by adding countably many monadic predicate letters $(P_1, P_2, \ldots)$ to $\mathbb{L}1_n$. Every $\mathbb{L}_n$-formula A can be translated as an $\mathbb{L}1_{n1}$-formula $A^*(t)$ with at most one parameter t, so that

$p_i^* = P_i(t)$,

$\bot^*\ (t) = \bot$,

$(A \to B)^*(t) = A^*(t) \to B^*(t)$,

[48] We call a set *countable* if its cardinality is $\leq \aleph_0$.

$(\Box_k A)^*(t) = \forall x(R_k(t,x) \to A^*(x))$,
where $A^*(x)$ is obtained from $A^*(t)$ by renaming all bound occurrences of x by some new variable, and all free occurrences of t by x.

Lemma 16.44.

1. *Let* $\mathbf{m} = (F, h)$ *be a Kripke model over a frame* $F = (W, \varrho_1, \ldots, \varrho_n)$. *Let* $\mathbf{m}^* = (W, \varrho_1, \ldots, \varrho_n, h(p_1), h(p_2), \ldots)$ *be the corresponding* $\mathbb{L}1_{n1}$*-model. Then for any* $a \in W$, *for any* n*-modal formula* A,

$$\mathbf{m}, a \models A \Leftrightarrow \mathbf{m}^* \models A^*(a).$$

2. *For any* n*-modal frame* F, *for any* n*-modal formula* A

$$F \models A \Leftrightarrow \forall h (F, h)^* \models \forall t A^*(t).$$

Proof

1. Easy by induction. For example, in the case $A = \Box_k B$ we have:
   $\mathbf{m}, b \models A \Leftrightarrow$ for any $b \in \varrho_k(a)$ $\mathbf{m}, b \models B$
   $\Leftrightarrow$ for any $b \in \varrho_k(a)$ $\mathbf{m}^* \models B^*(b)$ (by the induction hypothesis)
   $\Leftrightarrow \mathbf{m}^* \models \forall x(R_k(a,x) \to B^*(x)) \Leftrightarrow \mathbf{m}^* \models (\Box_k B)^*(a)$.
2. $F \models A \Leftrightarrow \forall h \forall a (F, h), a \models A \Leftrightarrow \forall h \forall a (F, h)^* \models A^*(a)$ (by (1))
   $\Leftrightarrow \forall h (F, h)^* \models \forall t A^*(t)$. □

Proposition 16.45.

1. *If* $\mathcal{C}$ *is an* R*-elementary class of* $\mathbb{L}1_n$*-models then the logic* $\mathbf{L}(\mathcal{C})$ *is recursively axiomatisable.*
2. *If* $\mathcal{C}$ *is a* Δ*-elementary class of* $\mathbb{L}1_n$*-models then* $\mathbf{L}(\mathcal{C})$ *is has the c.f.p.*

Proof

1. Assume that $\mathcal{C}$ is a class of models of a recursive set Σ of $\mathbb{L}1_n$-sentences. Then we have:

   $A \in \mathbf{L}(\mathcal{C}) \Leftrightarrow \forall F \in \mathcal{C}, F \models A$ (due to Kripke-completeness)
   $\Leftrightarrow \forall F \in \mathcal{C}\ \forall h (F, h)^* \models \forall t A^*(t)$ (by Lemma 16.44)
   $\Leftrightarrow \Sigma \models \forall t A^*(t)$ (in the classical sense)
   $\Leftrightarrow \Sigma \vdash \forall t A^*(t)$

   (in classical predicate logic, by Gödel's Completeness Theorem).
2. The proof is almost the same. Assume that $\mathcal{C}$ is a class of models of a set Σ of $\mathbb{L}1_n$-sentences, and let $\mathcal{C}_0$ be the class of all countable models in $\mathcal{C}$. Then we have:

   $A \in \mathbf{L}(\mathcal{C}) \Leftrightarrow \Sigma \models \forall t A^*(t)$
   $\Leftrightarrow \forall F (F \models \Sigma$ and F is countable $\Rightarrow F \models \forall t A^*(t))$
   (by the Löwenheim–Skolem theorem)
   $\Leftrightarrow \forall F \in \mathcal{C}_0 (F, h)^* \models \forall t A^*(t) \Leftrightarrow \forall F \in \mathcal{C}_0 F \models A$ (by Lemma 16.44)
   $\Leftrightarrow A \in \mathbf{L}(\mathcal{C}_0)$, and thus $\mathbf{L}(\mathcal{C}) = \mathbf{L}(\mathcal{C}_0)$. □

Proposition 16.45 can also be deduced from Theorem 16.54 below.

Theorem 16.46. *Let $\mathcal{C}_1$, $\mathcal{C}_2$ be classes of n- and m-modal frames respectively.*

1. *If $\mathcal{C}_1$, $\mathcal{C}_2$ are R-elementary then the logic $\mathbf{L}(\mathcal{C}_1 \times \mathcal{C}_2)$ is recursively axiomatisable.*
2. *If $\mathcal{C}_1$, $\mathcal{C}_2$ are Δ-elementary then the logic $\mathbf{L}(\mathcal{C}_1 \times \mathcal{C}_2)$ has the c.f.p.*

Proof

1. Consider the language $\mathbb{L}1_{n+m}$ and its extension $\mathbb{L}1^{\times}_{n+m}$ by monadic functional symbols τ_1, τ_2. By assumption, $\mathcal{C}_1$ (respectively, $\mathcal{C}_2$) is a class of models of a recursive set of $\mathbb{L}1_n$-sentences (resp., $\mathbb{L}1_m$-sentences); those sets are denoted by S_1, S_2.
   Let α be the conjunction of the following $\mathbb{L}1^{\times}_{n+m}$-sentences:

$$\alpha_1 = \forall x \forall y(\tau_1(x) = \tau_1(y) \wedge \tau_2(x) = \tau_2(y) \rightarrow x = y);$$
$$\alpha_2 = \forall x \forall y \exists z(\tau_1(z) = \tau_1(x) \wedge \tau_2(z) = \tau_2(y));$$
$$\alpha_{3,i} = \forall x \forall y(x R_i y \equiv \tau_1(x) R_i \tau_1(y) \wedge \tau_2(x) = \tau_2(y))(i \leq n);$$
$$\alpha_{3,n+j} = \forall x \forall y(x R_{n+j} y \equiv \tau_2(x) R_{n+j} \tau_2(y) \wedge \tau_1(x) = \tau_1(y))(j \leq m).$$

Lemma 16.47. *Let $\mu = (W, t_1, t_2) \models \alpha$, $W_1 = \{t_1(x) | x \in W\}$, $W_2 = \{t_2(x) | x \in W\}$.*
Then the map $h : W \rightarrow W_1 \times W_2$ defined by $h(x) = (t_1(x), t_2(x))$ is a bijection.

Proof The injectivity of h is expressed by α_1, and the surjectivity by α_2. □

For an $\mathbb{L}1_n$-formula β let β' denote the result of replacing all occurrences of each variable x in every atomic subformula by $\tau_1(x)$. Similarly, for an $\mathbb{L}1_m$-formula γ let γ'' denote the result of replacing all occurrences of each variable x in every atomic subformula by $\tau_2(x)$, and every R_j by R_{n+j}. Let $S_1' = \{\beta' | \beta \in S_1\}, S_2'' = \{\beta'' | \beta \in S_2\}$.

Lemma 16.48. *Let $\mu = (W, R_1, \ldots, R_{n+m}, t_1, t_2)$ be an $\mathbb{L}1^{\times}_{n+m}$-model such that $\mu \models S_1' \cup S_2'' \cup \{\alpha\}$, and let $F = (W, R_1, \ldots, R_{n+m})$. Let W_1, W_2, h be the same as in 16.47. Consider the frames $F_1 = (W_1, R_1 \restriction W_1, \ldots, R_n \restriction W_1)$, $F_2 = (W_2, R_{n+1} \restriction W_2, \ldots, R_{n+m} \restriction W_2)$. Then $F_1 \models S_1$, $F_2 \models S_2$, and h is an isomorphism from F onto $F_1 \times F_2$.*

Proof By 16.47, h is a bijection. The isomorphicity of h is expressed by $\alpha_{3,i}, \alpha_{3,n+j}$. For the first two statements notice that

$$F_1 \models \beta(t_1(a_1), \ldots, t_1(a_k)) \Leftrightarrow F \models \beta'(a_1, \ldots, a_k)$$
$$\text{(for any } \mathbb{L}1_n\text{-formula } \beta \text{, for any } a_1, \ldots, a_k \in W); \tag{16.2}$$

$$F_2 \models \gamma(t_2(a_1), \ldots, t_2(a_k)) \Leftrightarrow F \models \gamma''(a_1, \ldots, a_k)$$
$$\text{(for any } \mathbb{L}1_m\text{-formula } \gamma \text{, for any } a_1, \ldots, a_k \in W). \tag{16.3}$$

Both equivalences are proved easily by induction. So $\mu \models S_1'$ implies $F \models S_1'$, and hence $F_1 \models S_1$; similarly we have $F_2 \models S_2$. □

Now consider the first-order theory $S = S_1' \cup S_2'' \cup \{\alpha\}$. Let $A^*(x)$ be the standard translation of an $\mathbb{L}1_{n+m}$-formula A into an $\mathbb{L}1_{(n+m)1}$-formula (cf. Lemma 16.44). We can observe:

Lemma 16.49. $A \in \mathbf{L}(\mathcal{C}_1 \times \mathcal{C}_2) \Leftrightarrow S \vdash \forall x A^*(x)$.

Proof $A \in \mathbf{L}(\mathcal{C}_1 \times \mathcal{C}_2) \Leftrightarrow$
$\forall F_1 \in \mathcal{C}_1\, \forall F_2 \in \mathcal{C}_2 F_1 \times F_2 \models A$ (by definition 16.19) $\Leftrightarrow$
$\forall F_1 \in \mathcal{C}_1\, \forall F_2 \in \mathcal{C}_2\, \forall h\, \forall a \in F_1 \times F_2 (F_1 \times F_2, h), a \models A \Leftrightarrow$
$\forall F_1 \in \mathcal{C}_1 \forall F_2 \in \mathcal{C}_2\, \forall h (F_1 \times F_2, h)^* \models \forall x A^*(x)$ (by Lemma 16.44) $\Leftrightarrow$
$S \models \forall x A^*(x)$.

Let us prove the latter equivalence.

($\Rightarrow$) Assume that $(F_1 \times F_2, h)^* \models \forall x A^*(x)$ holds for any $F_1 \in \mathcal{C}_1$, $F_2 \in \mathcal{C}_2$ and for any h. Assume also that $(F, t_1, t_2) \models S$. Then by Lemma 16.48, F is isomorphic to $F_1 \times F_2$, for some F_1, F_2 such that $F_1 \models S_1$, $F_2 \models S_2$. Then $F_1 \in \mathcal{C}_1$, $F_2 \in \mathcal{C}_2$ by the choice of S_1, S_2, and hence $(F_1 \times F_2, \Pi_1, \Pi_2, \ldots) \models \forall x A^*(x)$ for any interpretation $\Pi_1, \Pi_2, \ldots$ of $P_1, P_2, \ldots$ (due to our assumption). Then by isomorphism we get that for any interpretation $\sigma_1, \sigma_2, \ldots$ of $P_1, P_2, \ldots$ in F, $(F, \sigma_1, \sigma_2, \ldots) \models \forall x A^*(x)$. Thus $S \models \forall x A^*(x)$.

($\Leftarrow$) Assume $S \models \forall x A^*(x)$, and consider any $F_1 \in \mathcal{C}_1$, $F_2 \in \mathcal{C}_2$. Let $F = F_1 \times F_2$, and choose some $a_1 \in F_1$, $a_2 \in F_2$. If $t_2 : F \to \{a_1\} \times F_2$, $t_1 : F \to F_1 \times \{a_2\}$ are the standard projections, then obviously $(F, t_1, t_2) \models \alpha$. Also we have that $F_i \models S_i$ by the choice of S_i. Since $\{a_1\} \times F_2$ (as a restriction of F) is isomorphic to F_2, and $F_1 \times \{a_2\}$ is isomorphic to F_1, the equivalences (16.2) and (16.3) of the proof of Lemma 16.48 hold in this case too, and hence $(F, t_1, t_2) \models S_1' \cup S_2''$. Thus $(F, t_1, t_2) \models S$, and so by our assumption, $(F, \Pi_1, \Pi_2, \ldots) \models \forall x A^*(x)$ for any interpretation $\Pi_1, \Pi_2, \ldots$ of $P_1, P_2, \ldots$. Therefore $(F, h)^* \models \forall x A^*(x)$ for any valuation h.

To finish the proof of Lemma 16.49, note that $S \models \forall x A^*(x) \Leftrightarrow S \vdash \forall x A^*(x)$ by Gödel's completeness theorem. □

The statement (1) of Theorem 16.46 now follows immediately from Lemma 16.49.
(2) In this case Lemma 16.49 also holds, so we have

$$A \in \mathbf{L}(\mathcal{C}_1 \times \mathcal{C}_2) \Leftrightarrow S \models \forall x A^*(x)$$

$\Leftrightarrow S \models_0 \forall x A^*(x)$ by the Löwenheim–Skolem theorem (where $\models_0$ denotes the classical consequence relation w.r.t. countable models)
$\Leftrightarrow \forall F_1 \in (\mathcal{C}_1)_0\, \forall F_2 \in (\mathcal{C}_2)_0\, \forall h (F_1 \times F_2, h)^* \models \forall x A^*(x)$ (by the argument in the above proof of (1), in which F, F_1, F_2 should be supposed countable)
$\Leftrightarrow \forall F_1 \in (\mathcal{C}_1)_0\, \forall F_2 \in (\mathcal{C}_2)_0, F_1 \times F_2 \models A$ (by Lemma 16.44).

Therefore we have proved that

$$\mathbf{L}(\mathcal{C}_1 \times \mathcal{C}_2) = \mathbf{L}((\mathcal{C}_1)_0 \times (\mathcal{C}_2)_0), \tag{16.4}$$

thus showing that $\mathbf{L}(\mathcal{C}_1 \times \mathcal{C}_2)$ has the c.f.p. □

Corollary 16.50.

1. *A product of two R-elementary and Kripke-complete logics is recursively axiomatisable.*
2. *A product of two* Δ*-elementary and Kripke-complete logics has the c.f.p.*

By Sahlqvist's Theorem (cf. [van Benthem, 1983]), every modal logic with axioms of a special type is Kripke-complete and Δ-elementary. These logics will be called *Sahlqvist.* So we have

Corollary 16.51.

1. *A product of two RE Sahlqvist logics is RE.*
2. *A product of two Sahlqvist logics has the c.f.p.*

Theorem 16.46 can be extended also to products of several logics; we leave the details to the reader.

A similar idea is used to prove a more general result about logics on the rational plane. To state it we need some definitions concerning the neighbourhood semantics.

Consider the propositional language $\mathbb{L}$ with classical connectives $(\to, \bot)$ and the additional connectives $\Delta_k, k \in I, |I| \leq \omega$, Δ_k being r_k-place. A *neighbourhood (or Montague) frame* for $\mathbb{L}$ (or briefly, $\mathbb{L}$-frame) is a pair $F = (W, (\delta_k)_{k \in I})$, where $W \neq \varnothing$ (the set of *worlds*), $\delta_k : (2^W)^{r_k} \longrightarrow 2^W$ is an r_k-place operation on subsets (if $r_k = 0$ we assume that $\delta_k \subseteq W$). A pair $\mathbf{m} = (F, h)$, in which $h : PL \longrightarrow 2^W$ is a *valuation*, is called a *neighbourhood model* over F. The prolongation of h to all formulas is standard:

$$h(\bot) = \varnothing, h(A \to B) = (W - h(A)) \cup h(B),$$
$$h(\Delta_k(A_1 \ldots A_{r_k})) = \delta_k(h(A_1), \ldots, h(A_{r_k}))(\text{ or } h(\Delta_k) \subseteq W \text{ if } r_k = 0).$$

Notations: $\mathbf{m}, w \models A$, $\mathbf{m} \models A$, $F \models A$, $F \models \Gamma$ and their readings are defined similarly to the Kripke case.

The set $\mathbf{L}(F) = \{A | F \models A\}$ is called the ($\mathbb{L}$-)modal logic of F. In general, an $\mathbb{L}$-modal logic can be defined as the set of $\mathbb{L}$-modal formulas closed under substitution, modus ponens, and replacement:

$$A_1 \equiv B_1, \ldots, A_{r_k} \equiv B_{r_k} / \Delta_k(A_1 \ldots A_{r_k}) \equiv \Delta_k(B_1 \ldots B_{r_k}).$$

Now consider a particular type of neighbourhood frame. Let $\mathbb{N}1$ be a (countable) first-order language, μ be an $\mathbb{N}1$-model, with the individual domain M. For an $\mathbb{N}1$-formula $\Psi(x_1, \ldots, x_m)$ with m parameters, let

$$|\Psi|_{\mu,m} = \{(a_1, \ldots, a_m) \in W^m | \mu \models \Psi(a_1, \ldots, a_m)\}.$$

Furthermore m is fixed, so we omit it and write $|\Psi|_\mu$, and even $|\Psi|$ if both μ, m are clear from the context. As usual a set $W \subseteq M^m$ is called *first-order definable over* μ if $W = |\Psi|$ for some $\mathbb{N}1$-formula $\Psi(x_1, \ldots, x_m)$.

Definition 16.52. Let $\mathbb{N}1^m$ be the extension of $\mathbb{N}1$ by m-place predicate letters P_i, $i \geq 0$. An *m-dimensional first order definable $\mathbb{L}$-frame over μ* is an $\mathbb{L}$-frame $(W, (\delta_k)_{k\in I})$ such that

- $W \subseteq M^m$ is first-order definable over μ (by some formula Ψ)
- for every $k \in I$ there exists an $\mathbb{N}1^m$-formula $\alpha_k(P_1, \dots, P_{r_k}, x_1, \dots, x_m)$ with free variables among $x_1, \dots, x_m$, such that every P_i occurring in α_k is among $P_1, \dots, P_{r_k}$, and for any $U_1, \dots, U_{r_k} \subseteq W$ $\delta_k(U_1, \dots, U_{r_k}) = \{\bar{a} \in W^m | (\mu, U_1, \dots, U_{r_k}) \models \alpha_k(P_1, \dots, P_{r_k}, \bar{a})\}$.

Here $(\mu, U_1, \dots, U_{r_k})$ means the expansion of μ by interpreting $P_1, \dots, P_{r_k}$ as $U_1, \dots, U_{r_k}$ respectively.

We will denote this frame $(W, (\delta_k)_{k\in I})$ by $F(\mu, \psi, \bar{\alpha})$, where $\bar{\alpha}$ is an abbreviation for $(\alpha_k)_{k\in I}$.

Definition 16.53. Let $\mathcal{C}$ be a class of $\mathbb{N}1$-models. A class Ω of $\mathbb{L}$-frames is *m-dimensional first-order definable (f.o.d.) over $\mathcal{C}$* if $\Omega = \{F(\mu, \psi, \bar{\alpha}) | \mu \in \mathcal{C}\}$ for some $\psi, \bar{\alpha}$.

We will also use the notation $F(\mathcal{C}, \psi, \bar{\alpha})$ for such Ω.

Theorem 16.54.

1. *If the class $\mathcal{C}$ of $\mathbb{N}1$-models is R-elementary, then the logic of any m-dimensional f.o.d. class over $\mathcal{C}$ is recursively axiomatisable.*
2. *If the class $\mathcal{C}$ of $\mathbb{N}1$-models is Δ-elementary, then the logic of any m-dimensional f.o.d. class over $\mathcal{C}$ has the c.f.p.*

Proof Some more notation first. Given an $\mathbb{N}1^m$-formula $\alpha(P_1, \dots, P_r, \bar{x})$, with a list of parameters $\bar{x} = (x_1, \dots, x_m)$ and predicate letters $P_1, \dots, P_r$, and some other $\mathbb{N}1^m$-formulas $\beta_1, \dots, \beta_r$, let $\alpha(\beta_1, \dots, \beta_r, \bar{x})$ be the substitution instance of α obtained by replacing each occurrence of any $P_i(\bar{y})$ by $\beta_i(\bar{y})$. The following is easily checked by induction on α:

For any $\mathbb{N}1$-model $\mu, \bar{a} \in M^m$:

$$(\mu, |\beta_1|, \dots, |\beta_r|) \models \alpha(P_1, \dots, P_r, \bar{a}) \Leftrightarrow \mu \models \alpha(\beta_1, \dots, \beta_r, \bar{a}). \tag{16.5}$$

For a formula $A \in \mathbb{L}$ and a list $\bar{x} = (x_1, \dots, x_m)$, let $A^*(\bar{x})$ be the $\mathbb{N}1^m$-formula defined by induction:

$$\text{`}p_i^*(\bar{x}) = P_i(\bar{x}), \bot^* = \bot, (A \to B)^* = A \to B^*,$$
$$(\Delta_k(A_1 \dots A_{r_k}))^*(\bar{x}) = \alpha_k(A_1^*, \dots, A_{r_k}^*, \bar{x}),$$

where the latter formula is obtained from α_k by substituting each occurrence of $P_i(\bar{y})$ by $A_i^*(\bar{y})$, $i = 1, \dots, r_k$.

By induction we get then (for any $\mathbb{N}1$-model μ):

$$(F(\mu, \psi, \bar{\alpha}), h), \bar{a} \models A(p_1, ..., p_s) \Leftrightarrow (\mu, h(p_1), ..., h(p_s)) \models A^*(\bar{a}), \tag{16.6}$$

i.e. $h(A) = |A^*|_{\mu, h(p_1), \dots, h(p_s)}$.

In fact, if $A = p_i$ we have
$(F(\mu,\psi,\bar{\alpha}),h),\bar{a} \models p_i \Leftrightarrow \bar{a} \in h(p_i) \Leftrightarrow (\mu, h(p_1),...,h(p_s)) \models P_i(\bar{a})$.
If A is $\bot$ or an implication, the argument is trivial.
If $A = \Delta_k(A_1...A_{r_k})$ we have:

$(F(\mu,\psi,\bar{\alpha}),h),\bar{a} \models A \Leftrightarrow \bar{a} \in \delta_k(h(A_1),...,h(A_{r_k}))$ (where δ_k is the operation in $F(\mu,\psi,\bar{\alpha})$ corresponding to Δ_k)

$\Leftrightarrow (\mu, h(A_1),...,h(A_{r_k})) \models \alpha_k(P_1,...,P_{r_k},\bar{a})$ (by definition of δ_k)

$\Leftrightarrow (\mu, |A_1^*|,...,|A_{r_k}^*|) \models \alpha_k(P_1,...,P_{r_k},\bar{a})$ (by the induction hypothesis; for the sake of simplicity, we have omitted the subscripts $\mu, h(p_1),...,h(p_{r_k})$)

$\Leftrightarrow (\mu, h(p_1),...,h(p_{r_k})) \models \alpha_k(A_1^*,...,A_{r_k}^*,\bar{a})$ (by (16.5) above)
and note that $A^*(\bar{x}) = \alpha_k(A_1^*,...,A_{r_k}^*,\bar{x})$.

Let Σ be the $\mathbb{N}1$-theory of $\mathcal{C}$, Σ^+ be its expansion to the language $\mathbb{N}1^m$. We claim that

$A \in \mathbf{L}(F(\mathcal{C},\psi,\bar{\alpha})) \Leftrightarrow \Sigma^+ \vdash \forall\bar{x}(\psi(\bar{x}) \to A^*(\bar{x}))$, which yields the recursive enumerabilty of $\mathbf{L}(F(\mathcal{C},\psi,\bar{\alpha}))$.

In fact from (16.6) we have:

$$\begin{aligned}
&A = A(p_1,...,p_s) \in \mathbf{L}(F(\mathcal{C},\psi,\bar{\alpha}))\\
&\Leftrightarrow \forall\mu \in \mathcal{C}\ \forall h \forall \bar{a} \in |\psi|_\mu\ (\mu, h(p_1),...,h(p_s)) \models A^*(\bar{a})\\
&\Leftrightarrow \forall\mu \in \mathcal{C}\ \forall U_1,...,U_s \subseteq M\ (\mu, U_1,...,U_s) \models \forall\bar{x}(\psi(\bar{x}) \to A^*(\bar{x}))\\
&\Leftrightarrow \Sigma^+ \models \forall\bar{x}(\psi(\bar{x}) \to A^*(\bar{x})) \Leftrightarrow \Sigma^+ \vdash \forall\bar{x}(\psi(\bar{x}) \to A^*(\bar{x})).
\end{aligned}$$

The same argument yields that

$$A \in \mathbf{L}(F(\mathcal{C}_0,\psi,\bar{\alpha})) \Leftrightarrow \Sigma^+ \models_0 \forall\bar{x}(\psi(\bar{x}) \to A^*(\bar{x})),$$

where $\mathcal{C}_0$ denotes the class of all countable frames in $\mathcal{C}$, $\models_0$ is the logical consequence relation w.r.t. countable frames. Hence the claim (2) follows. □

Corollary 16.55. *Let k be an infinite cardinal, Σ be a k-categorical first-order theory without finite models. Let also $\mu \models \Sigma, |\mu| = k$. Then the logic $\mathbf{L}(F(\mu,\psi,\bar{\alpha}))$ has the c.f.p. If Σ is RE then $\mathbf{L}(F(\mu,\psi,\bar{\alpha}))$ is RE too.*

Proof By Vaught's criterion, Σ is complete, and thus $\Sigma = Th(\mu)$. Let $\mathcal{C}$ be the class of all models of Σ. To apply Theorem 16.54, it is sufficient to show that $\mathbf{L}(F(\mathcal{C},\psi,\bar{\alpha})) = \mathbf{L}(F(\mu,\psi,\bar{\alpha}))$ i.e. that for any $A = A(p_1,...,p_s)$:

$$\begin{aligned}
&\forall U_1,...,U_s \subseteq M(\mu, U_1,...,U_s) \models \forall\bar{x}(\psi(\bar{x}) \to A^*(\bar{x}))\\
&\Leftrightarrow \forall\mu' \in \mathcal{C}\ \forall V_1,...,V_s \subseteq M'(\mu', V_1,...,V_s) \models \forall\bar{x}(\psi(\bar{x}) \to A^*(\bar{x})).
\end{aligned}$$

In this equivalence the direction $\Leftarrow$ is clear. To prove the converse, suppose $(\mu', V_1, \ldots, V_s) \not\models \forall\bar{x}(\psi(\bar{x}) \to A^*(\bar{x}))$. Consider the theory $\Sigma' = Th(\mu', V_1,...,V_s)$. Since $\Sigma' \supseteq \Sigma, \Sigma'$ has no finite models, so it must have a model μ' of cardinality k, by the Löwenheim–Skolem–Tarski theorem. The $\mathbb{N}1$-restriction of μ' is isomorphic to μ, due to the k-categoricity of Σ. Hence $(\mu, U_1,...,U_s) \not\models \forall x A^*(x)$ for some $U_1,...,U_s \subseteq M$. □

This corollary can be applied in various situations. Here are two examples.

Example 16.56. Consider the modal logic of the rational plane with the following modalities ('modal relational logic'):

$(x, y) \models \Delta \Leftrightarrow x = y$;
$(x, y) \models \Lambda \Leftrightarrow x = y$;
$(x, y) \models A \circ B \Leftrightarrow \exists z((x, z) \models A \wedge (z, y) \models B)$;
$(x, y) \models \Box_1 A \Leftrightarrow \exists z > x(z, y) \models A$;
$(x, y) \models \Box_2 A \Leftrightarrow \exists z > y(x, z) \models A$.

This logic is RE, because the theory $Th(\mathbf{Q}, <)$ is ω-categorical.

Example 16.57. [cf. [Marx and Venema, 1997, Ch. 4].] We can restrict the previous example to the north-western halfplane $\{(x, y) \in \mathbf{Q}^2 | x \leq y\}$. The points of this halfplane correspond to intervals in $\mathbf{Q}$, and what we get is an interval temporal logic with a sort of earlier—later relation and the 'chop' modality $\circ$. The corresponding logic is also RE.

Example 16.58 ('Group modalities'). Consider the following modalities on the real line:

$x \models A + B \Leftrightarrow \exists y, z(y \models A \wedge z \models B \wedge z = x + y)$;
$x \models -A \Leftrightarrow -x \models A$;
$x \models O \Leftrightarrow x = 0$.

The corresponding logic is RE, because the theory of torsion-free divisible Abelian groups is ω_1-categorical. Also we obtain that it has the c.f.p.

The logics in the first two examples are known to be undecidable. The decidability of the third logic is an open problem.

16.6 Products of Minimal Modal Logics

Now let us recall the proof from [Shehtman, 1978] showing that every pair of minimal modal logics is product-matching. To clarify the main idea, we begin with the bimodal case.

Theorem 16.59. *Let $F = F^{u_0} = (W, R_1, R_2)$ be a (1,1)-commutative countable generated frame. Then F is a p-morphic image of a product of two 1-trees.*

Proof Consider corresponding 1-frames: $F_1 = (W, R_1)$ and $F_2 = (W, R_2)$, and for each of them take subsequently the generated subframe, thickening, and unravelling.

Thus we obtain two trees:

$$\Phi_1 = ((F_1^{u_0})_\omega)^\sharp, \Phi_2 = ((F_2^{u_0})_\omega)^\sharp.$$

Now let us construct a p-morphism $f : \Phi_1 \times \Phi_2 \twoheadrightarrow F$.

By Lemma 16.40, every world in Φ_i is a sequence $\alpha = (u_0, u_1, \dots, u_n)$. Then let $l(\alpha) = n$.

For a pair $(\alpha, \beta) \in \Phi_1 \times \Phi_2$, $f(\alpha, \beta)$ is defined by induction on $l(\alpha, \beta) = l(\alpha) + l(\beta)$. To simplify notation, we write u_0 instead of the sequence (u_0).

1. Let $f(u_0, u_0) = u_0$.

2. Let $f(\alpha, u_0) = \gamma_1(\pi_1(\alpha))$, where the p-morphisms

$$\pi_1 : ((F_1^{u_0})_\omega)^\sharp \twoheadrightarrow (F_1^{u_0})_\omega \text{ and } \gamma_1 : (F_1^{u_0})_\omega \twoheadrightarrow F_1^{u_0}$$

are standard (see Lemmas 16.35 and 16.37).

3. Similarly, let $f(u_0, \beta) = \gamma_2(\pi_2(\beta))$, where the p-morphisms

$$\pi_2 : ((F_2^{u_0})_\omega)^\sharp \twoheadrightarrow (F_2^{u_0})_\omega \text{ and } \gamma_2 : (F_2^{u_0})_\omega \twoheadrightarrow F_2^{u_0}$$

are standard.

Definition 16.60. Let $F = (W, \varrho_1, \dots, \varrho_n), G = (V, \sigma_1, \dots, \sigma_n)$ be two frames. We say that a morphism $f : F \to G$ is ω-fold if the following holds for any i, t :

$$(2.\omega)\ \forall z (f(t)\sigma_i z \Rightarrow |\varrho_i(t) \cap f^{-1}(z)| = \omega),$$

$|...|$ meaning the cardinality.

From the definitions it follows that γ_1, γ_2 are ω-fold. Returning to our construction, let $\Phi_1 = (W_1, S_1), \Phi_2 = (W_2, S_2)$. Let D_n be the restriction of $\Phi_1 \times \Phi_2$ to the set $\{t \in W_1 \times W_2 | l(t) \leq n\}$. By induction we construct the restriction $f \restriction D_n$ and prove that it is 'almost' an ω-fold p-morphism, namely that it is monotonic and satisfies $(2.\omega)$ for $i = 1, 2, l(t) \leq n - 1$.

If $l(\alpha, \beta) = n + 1, \alpha, \beta \neq u_0$, take the predecessors α_0, β_0 of α, β respectively. Suppose we have defined $f(\alpha_0, \beta_0) = u, f(\alpha, \beta_0) = x, f(\alpha_0, \beta) = y$. Since f satisfies $(2.\omega)$ at proper points, there exist bijections

$$e_{1x} : S_1^\vee(\alpha_0, \beta_0) \cap f^{-1}(x) \to \omega, e_{2y} : S_2^\vee(\alpha_0, \beta_0) \cap f^{-1}(y) \to \omega.$$

Also by the induction hypothesis we have that uR_1x, uR_2y, and thus, by commutativity, the set $T_{xy} = R_2(x) \cap R_1(y)$ must be non-empty and countable (since F is countable). Then let $g_{xy} : \omega \to T_{xy}$ be any surjective map.

Take any 'reproducing function' $\chi : \omega \times \omega \to \omega$, such that

1. $\forall a \forall b |\{c | \chi(a, c) = b\}| = \omega$
2. $\forall a \forall b |\{c | \chi(c, a) = b\}| = \omega$

Such a function exists. For example, we can put $\chi(a, b) = \theta(a + b)$, where $\theta : \omega \to \omega$ is any surjective map, such that $\theta^{-1}(t)$ is infinite for any t.

Now we can prolong the map f:

$$f(\alpha, \beta) = g_{xy}(\chi(e_{1x}(\alpha, \beta_0), e_{2y}(\alpha_0, \beta))).$$

Then immediately we have:

$$f(\alpha, \beta_0) = xR_2 f(\alpha, \beta), f(\alpha_0, \beta) = yR_1 f(\alpha, \beta),$$

and thus monotonicity holds for $f \restriction D_{n+1}$, so it remains to check the condition $(2.\omega)$ for $t = (\alpha, \beta_0), i = 2$, and for $t = (\alpha_0, \beta), i = 1$.

Both cases are similar, so we consider the first. Then $(2.\omega)$ means that

$$(2.\omega.1)\ \forall z(xR_2z \Rightarrow |S_2^{\vee}(\alpha,\beta_0)\cap f^{-1}(z)| = \omega).$$

If xR_2z then for some v, uR_2vR_1z, implying that $z \in T_{xv}$. Then $z = g_{xv}(b)$ for some b; and also $|\{c|\chi(a,c) = b\}| = \omega$ (where $a = e_{1x}(\alpha,\beta_0)$).

Now by the definition of e_{2v},

$\forall c\exists\beta_1(e_{2v}(\alpha_0,\beta_1) = c \wedge \beta_0S_2\beta_1)$.

Then we get immediately

$$|\{\beta_1|\beta_0S_2\beta_1 \wedge \chi(a, e_{2v}(\alpha,\beta_1)) = b\}| = \omega,$$

and hence

$$|\{(\alpha,\beta_1)|\beta_0S_2\beta_1 \wedge f(\alpha,\beta_1) = z\}| = \omega$$

showing that $(2.\omega.1)$ holds. Therefore f is a p-morphism. □

Finally, let us notice that in a particular case when one of the original relations (say R_1) is empty, we get that Φ_1 is a singleton tree $(\{\lambda\},0)$, and thus $\Phi_1\times\Phi_2$ is isomorphic to $(W_2,\varnothing,S_2)$ (i.e. to Φ_2 with the additional empty relation). In this case the items (1) and (3) are not used in the construction of f.

Theorem 16.59 can be rather easily extended to polymodal logics:

Theorem 16.61. *Let $F = F^{u_0} = (W, R_1,\ldots,R_{n+m})$ be an (n,m)-commutative-countable generated frame. Then F is a p-morphic image of a product of an n-tree and an m-tree.*

Proof Repeats the previous one, with slight changes. Let $F = F^{u_0} = (W, R_1,\ldots, R_{n+m})$ be the original frame. Consider the frames:

$$F_1 = (W, R_1,\ldots,R_n) \text{ and } F_2 = (W, R_{n+1},\ldots,R_{n+m}),$$

and take the trees:

$$\Phi_1 = ((F_1^{u_0})_\omega)^\sharp\ ,\ \Phi_2 = ((F_2^{u_0})_\omega)^\sharp.$$

Then let us define a p-morphism $f:\Phi_1\times\Phi_2 \twoheadrightarrow F$.

Again for a pair $(\alpha,\beta)\in\Phi_1\times\Phi_2$, $f(\alpha,\beta)$ is defined by induction on $l(\alpha,\beta) = l(\alpha)+l(\beta)$ (now we put $l(u_0,j_0,u_1,\ldots,j_{n-1},u_n) = n$).

1. Let $f(u_0,u_0) = u_0$.
2. Let $f(\alpha,u_0) = \gamma_1(\pi_1(\alpha))$, where the p-morphisms

$$\pi_1 : ((F_1^{u_0})_\omega)^\sharp \twoheadrightarrow (F_1^{u_0})_\omega \text{ and } \gamma_1 : (F_1^{u_0})_\omega \twoheadrightarrow F_1^{u_0}$$

are standard (see Lemmas 16.35 and 16.37).

3. Let $f(u_0,\beta) = \gamma_2(\pi_2(\beta))$, where the p-morphisms

$$\pi_2 : ((F_2^{u_0})_\omega)^* \twoheadrightarrow (F_2^{u_0})_\omega \text{ and } \gamma_2 : (F_2^{u_0})_\omega \twoheadrightarrow F_2^{u_0}$$

are standard.

Let $\Phi_1 = (W_1, S_{11}, \ldots, S_{1n}), \Phi_2 = (W_2, S_{21}, \ldots, S_{2m})$. Let D_k be the restriction of $\Phi_1 \times \Phi_2$ to the set $\{t \in W_1 \times W_2 | l(t) \leq k\}$. By induction we define the restriction $f \upharpoonright D_k$ and prove that it is monotonic and satisfies $(2.\omega)$ for $i = 1, 2, l(x) \leq k - 1$.

If $l(\alpha, \beta) = k + 1, \alpha, \beta \neq u_0$, take the predecessors α_0, β_0 of α, β respectively. Suppose $\alpha_0 R_i \alpha, \beta_0 R_j \beta, f(\alpha_0, \beta_0) = u, f(\alpha, \beta_0) = x, f(\alpha_0, \beta) = y$. Then there exist bijections

$$e_{ix} : S_i^{\vee}(\alpha_0, \beta_0) \cap f^{-1}(x) \to \omega, e_{jy} : S_j^{\vee}(\alpha_0, \beta_0) \cap f^{-1}(y) \to \omega,$$

and a surjection $g_{xy} : \omega \to R_j(x) \cap R_i(y)$.

We define $f(\alpha, \beta) = g_{xy}(\chi(e_{ix}(\alpha, \beta_0), e_{jy}(\alpha_0, \beta)))$, and the further proof is carried out in the same manner as that of Theorem 16.59. □

Theorem 16.62. $\mathbf{K}_n \times \mathbf{K}_m = \mathbf{L}(\mathcal{T}_n \times \mathcal{T}_m) = [\mathbf{K}_n, \mathbf{K}_m]$, *where $\mathcal{T}_l$ denotes the class of all l-trees.*

Proof Trivially, $\mathbf{K}_n \times \mathbf{K}_m \subseteq \mathbf{L}(\mathcal{T}_n \times \mathcal{T}_m)$. $[\mathbf{K}_n, \mathbf{K}_m] \subseteq \mathbf{K}_n \times \mathbf{K}_m$ by Corollary 16.27. It remains to show only that $\mathbf{L}(\mathcal{T}_n \times \mathcal{T}_m) \subseteq [\mathbf{K}_n, \mathbf{K}_m]$. So suppose $A \notin [\mathbf{K}_n, \mathbf{K}_m]$. Then by Corollary 16.30, A is non-valid in some (n, m)-commutative frame F. Since the class of all (n, m)-commutative frames is Δ-elementary, we can assume that F is countable (by Proposition 16.45). By Lemma 16.2, A is non-valid also in some generated subframe F^u, and F^u inherits (n, m)-commutativity from F. By Theorem 16.6.3, F^u is a p-morphic image of a frame from $\mathcal{T}_n \times \mathcal{T}_m$, and thus $A \notin \mathbf{L}(\mathcal{T}_n \times \mathcal{T}_m)$, by Lemma 16.2.4. □

Proposition 16.63. *For any $[\mathbf{K}, \mathbf{K}]$-consistent bimodal formula A there exist standard 1-trees T_1, T_2 such that A is satisfied at $T_1 \times T_2, (\lambda, \lambda)$.*

Proof In the previous proof take $\neg A$ instead of A. Since there is a p-morphism $f : \Phi_1 \times \Phi_2 \twoheadrightarrow F$ such that $f(\lambda, \lambda) = u_0$, and A is satisfied at F, u_0, we get that A is satisfied at $\Phi_1 \times \Phi_2, (\lambda, \lambda)$. □

16.7 Products of PTC-logics

Definition 16.64. An $\mathbb{L}_n$-formula is *closed* if it does not contain propositional letters. An n-modal logic is *closed* if it is axiomatised by a set of closed $\mathbb{L}_n$-formulas.

Lemma 16.65. *If F, G are n-modal frames and $f : F \twoheadrightarrow G$ then $F \models A \Leftrightarrow G \models A$ for any closed n-formula A.*

Proof This follows trivially from Lemma 16.4, because for closed A the truth value of A at any world does not depend on the assignment. □

Proposition 16.66. *Every closed logic is canonical.*

Proof By Theorem 16.6, every axiom of the logic is true in the canonical model. For a closed formula, this is equivalent to validity in the canonical frame. □

Theorem 16.67. *Every pair of closed logics is product-matching.*

Proof Following the line of the proof of Theorem 16.62 above, suppose $A \notin [\mathbf{L}_1, \mathbf{L}_2]$. Then by Proposition 16.66, Proposition 16.29 and Lemma 16.2, A is non-valid in some generated frame $F \in [\mathcal{V}(\mathbf{L}_1), \mathcal{V}(\mathbf{L}_2)]$. By Theorem 16.61 there exists $f : \Phi_1 \times \Phi_2 \twoheadrightarrow F$, and by Lemma 16.4, $\Phi_1 \times \Phi_2 \not\models A$. By Lemma 16.7.2, closed axioms of $[\mathbf{L}_1, \mathbf{L}_2]$ hold in $\Phi_1 \times \Phi_2$. Thus $\Phi_1 \times \Phi_2 \in \mathcal{V}(\mathbf{L}_1) \times \mathcal{V}(\mathbf{L}_2)$. □

Definition 16.68. A *pseudo-transitive modal formula* is that of the form $\nabla_1 \Box_k p \to \Delta$, where $p \in PL, \Delta = \Box_{i_1} \dots \Box_{i_r}, \nabla_1 = \Diamond_{j_1} \dots \Diamond_{j_t}$ are sequences of modal operators (maybe void). A *PTC-logic* is a modal logic axiomatised by a set of formulas, which are either pseudo-transitive or closed.

Definition 16.69. An $\mathbb{L}1_n$-formula φ *corresponds* to an n-modal formula A if for any n-modal frame $F, F \models A$ (modally) $\Leftrightarrow F \models \varphi$ (classically).

From Sahlqvist's theorem we obtain

Proposition 16.70.

1. *Consider a pseudo-transitive n-formula* $A = \nabla_1 \Box_k p \to \Delta, \Delta = \Box_{i_1} \dots \Box_{i_r}, \nabla_1 = \Diamond_{j_1} \dots \Diamond_{j_t}$. *Then A corresponds to the* $\mathbb{L}1_n$*-formula*

$$\forall x, v(\exists u(uR_{\nabla_1}x \wedge uR_{\nabla_2}v) \to xR_kv),$$

where

$$R_{\nabla_1} = R_{j_1} \circ \dots \circ R_{j_t}, R_{\nabla_2} = R_{i_j} \circ \dots \circ R_{i_r}.$$

(We adopt the convention that the composition of the void sequence is the equality relation I_W*.)*

2. *Every PTC-logic is canonical.*

Similarly to the well known transitive closure, there exists a 'closure' w.r.t. pseudo-transitivity. Let us describe this construction in a slightly more general context. We begin with some definitions.

Definition 16.71. Let $F_1 = (W_1, \varrho_1, \dots, \varrho_n), F_2 = (W_2, \sigma_1, \dots, \sigma_n)$ be two n-modal frames. We say that F_1 is a *weak subframe* of F_2 if $W_1 \subseteq W_2$ and (for any i, a, b) $a\varrho_i b$ implies $a\sigma_i b$ (or in other words, if the inclusion map is a homomorphism of F_1 into F_2). F_1 is called a *full weak subframe* of F_2 (notation: $F_1 \subseteq F_2$) if also $W_1 = W_2$.

Furthermore arbitrary lists of individual variables are denoted by $\bar{x}, \bar{y}$ etc. and arbitrary tuples of individuals by $\bar{a}, \bar{b}$ etc. As usual, we write $\varphi(x, y, \bar{z})$ to show that all parameters of the formula φ are among $x, y, \bar{z}$.

Definition 16.72. An $\mathbb{L}1_n$-formula of the form

$$\psi = \forall x \forall y \forall \bar{z}(\varphi(x, y, \bar{z}) \to R_k(x, y))$$

is a called a *universal strict Horn formula,* if $\varphi(x, y, \bar{z})$ is quantifier-free positive (i.e. if it is built from atomic formulas by using $\wedge, \vee$).

Proposition 16.73. *Let $F = (W, \varrho_1, \dots, \varrho_n)$ be an n-modal frame, and consider the set of universal strict Horn formulas $\Gamma = \{\psi_k | k \in I\}$,*

$$\psi_k = \forall x \forall y \forall \bar{z}_k (\varphi_k(x, y, \bar{z}_k) \to R_{i_k}(x, y)).$$

Then there exists a frame F_Γ^+ such that

1. $F \subseteq F_\Gamma^+$
2. $F_\Gamma^+ \models \Gamma$
3. *If $G \models \Gamma$ and $f : F \to G$ is a homomorphism then the same map makes a homomorphism $g : F_\Gamma^+ \to G$.*

We can say that F_Γ^+ is the 'least' full weak extension of F satisfying Γ.

Proof We construct F_Γ^+ as the union of a sequence of frames $F_m = (W, \varrho_{m1}, \dots, \varrho_{mn})$, beginning from $F_0 = F$. Namely, let:

$$\varrho_{(m+1)j} = \varrho_{mj} \cup \bigcup_{i_k = j} \{(a, b) | F_m \models \exists \bar{z}_k \varphi_k(a, b, \bar{z}_k)\},$$

$$\varrho_j^+ = \bigcup_m \varrho_{mj}, F_\Gamma^+ = (W, \varrho_1^+, \dots, \varrho_n^+).$$

Then $F_0 \subseteq \dots F_m \subseteq F_{m+1} \dots \subseteq F_\Gamma^+$, and thus Proposition 16.73 (1) holds. Furthermore, we have:

4. if $\varphi(\bar{z})$ is a quantifier-free positive formula then the following are equivalent:
   (i) $F_\Gamma^+ \models \varphi(\bar{a})$;
   (ii) $\exists m \forall k \geq m F_k \models \varphi(\bar{a})$;
   (iii) $\exists m F_m \models \varphi(\bar{a})$.

In fact (ii)⇒(iii) is trivial. (iii)⇒(i) holds because $F_m \subseteq F_\Gamma^+$, and the truth of positive formulas is preserved by homomorphisms. (i)⇒(ii) is easily proved by induction on φ, e.g. consider the case $\varphi = \varphi_1 \wedge \varphi_2$.

Assume (i); then $F_\Gamma^+ \models \varphi_1(\bar{a}), F_\Gamma^+ \models \varphi_2(\bar{a})$. By the inductive hypothesis we have m_1, m_2 such that $F_k \models \varphi_1(\bar{a})$ for $k \geq m_1$, $F_k \models \varphi_2(\bar{a})$ for $k \geq m_2$. Then $F_k \models \varphi(\bar{a})$ provided $k \geq \max(m_1, m_2)$.

Now let us prove Proposition 16.73 (2). Assume $F_\Gamma^+ \models \varphi_k(a, b, \bar{c_k}), i_k = j$. Then by Proposition 16.73 (4), $F_m \models \varphi_k(a, b, \bar{c_k})$ for some m, and thus $(a, b) \in \varrho_{(m+1)j}$ by definition. Hence $F_\Gamma^+ \models R_k(a, b)$. Therefore $F_\Gamma^+ \models \psi_k$.

To check Proposition 16.73 (3), assume that $G \models \Gamma, f : F \to G$ is a homomorphism and show by induction that $f : F_m \to G$ is a homomorphism for any m.

To make the step, assume $a \varrho_{(m+1)j} b$. Then there are two cases.

If $a \varrho_{mj} b$ then $G \models R_j(f(a), f(b))$ by the induction hypothesis.

If $F_m \models \exists \bar{z_k} \varphi_k(a, b, \bar{z_k}), i_k = j$ then $G \models \exists \bar{z_k} \varphi_k(f(a), f(b), \bar{z_k})$ since the homomorphism $f : F_m \to G$ preserves the truth of positive formulas. Then $G \models \psi_k$ implies that $G \models R_j(f(a), f(b))$. □

By Proposition 16.70 every pseudo-transitive modal formula corresponds to a universal strict Horn formula. So Proposition 16.73 yields

Corollary 16.74. *Let Γ be a set of pseudo-transitive n-modal formulas, F be an n-modal frame. Then there exists an n-modal frame F^+_Γ such that*
(i) $F \subseteq F^+_\Gamma$;
(ii) $F^+_\Gamma \models \Gamma$;
(iii) for any n-modal frame G, if $G \models \Gamma$ and $f : F \to G$ is a homomorphism then $f : F^+_\Gamma \to G$ is a homomorphism.

Definition 16.75. The frame F^+_Γ described in Corollary 16.74 is called the *pseudo-transitive closure* of F (under Γ).

Theorem 16.76. *Every pair of PTC-logics is product-matching.*

Proof Let $\mathbf{L}_1, \mathbf{L}_2$ be PTC-logics, $A \notin [\mathbf{L}_1, \mathbf{L}_2]$. Then $F \not\models A$ some generated countable $F \in [\mathcal{V}(\mathbf{L}_1), \mathcal{V}(\mathbf{L}_2)]$, and thus by Theorem 16.61, there exists

$$f : \Phi_1 \times \Phi_2 \twoheadrightarrow F$$

for some trees Φ_1, Φ_2. Let Γ_i be the set of pseudo-transitive axioms of $\mathbf{L}_i$. Take the pseudo-transitive closures $(\Phi_i)^+_{\Gamma_i}$. Then
$\Phi_1 \times \Phi_2 \subseteq (\Phi_1)^+_{\Gamma_1} \times (\Phi_2)^+_{\Gamma_2}$, and let us show that the same map f is a p-morphism $(\Phi_1)^+_{\Gamma_1} \times (\Phi_2)^+_{\Gamma_2} \twoheadrightarrow F$.

In fact, let $\Phi_1 = (W_1, R_1, \ldots, R_n), \Phi_2 = (W_2, S_1, \ldots, S_m)$, $F_1 = (\Phi_1)^+_{\Gamma_1} = (W_1, R'_1, \ldots, R'_n)$, $F_2 = (\Phi_2)^+_{\Gamma_2} = (W_2, S'_1, \ldots, S'_m)$, $F = (V, Q_1, \ldots, Q_{n+m})$. The property (2) from Definition 16.3 holds because $f(a)Q_iy$, $i \leq n$ implies that for some $b, aR_i^\vee b, f(b) = y$. Obviously, then also $a(R'_i)^\vee b$. (If $i > n$ take $S^\vee_{i-n}$ instead of $R^\vee_i$.)

To check the monotonicity property (1), assume aR'_ib and show that for any c, $f(a,c)Q_if(b,c)$. This happens because $(V, Q_1, \ldots, Q_n) \models \Gamma_1$, the composed map $x \mapsto (x, c) \mapsto f(x, c)$ is a homomorphism $\Phi_1 \to (V, Q_1, \ldots, Q_n)$, and Corollary 16.7.11 can be applied. Similarly, it is proved that aS'_jb implies $f(c,a)Q_{n+j}f(c,b)$.

Since f is a p-morphism, we get that $F_1 \times F_2 \not\models A$.

On the other hand $F_i \models \mathbf{L}_i$. In fact $F_i \models \Gamma_i$ by our construction. As for closed axioms, they are not affected by p-morphisms (Lemma 16.65), and thus for any closed $C \in \mathbf{L}_1$ we have $F \models C, F_1 \times F_2 \models C$. Obviously, the first projection $p_1 : F_1 \times F_2 \to F_1$ is a p-morphism of n-frames (if $F_1 \times F_2$ is considered with the first n relations only). Thus $F_1 \models C$, and hence $F_1 \models \mathbf{L}_1$. $F_2 \models \mathbf{L}_2$ is checked in the same way.

Therefore $F_1 \times F_2 \in \mathcal{V}(\mathbf{L}_1) \times \mathcal{V}(\mathbf{L}_2)$, and finally we obtain $A \notin \mathbf{L}_1 \times \mathbf{L}_2$. □

16.8 Counterexamples

In this section we show that Theorem 16.76 cannot be generalised too far, and there is a continuum of non-product-matching pairs of logics.

Let us recall that *Grzegorczyk logic* is $\mathbf{Grz} = \mathbf{S4} + \Box(\Box(p \to \Box p) \to p) \to p$, and that $\mathcal{V}(\mathbf{Grz})$ consists of all *Noetherian partial orders,* i.e. of those without infinite ascending chains.

Let $\bullet$ be the reflexive singleton, $\bullet_2$ be the two-element cluster, $\updownarrow$ be the two-element reflexive chain.

Proposition 16.77.

1. $\mathbf{L}(\bullet)(= \mathbf{K} + p \equiv \Box p)$ *is the greatest consistent extension of* $\mathbf{S4}$*;*
2. *every logic strictly between* $\mathbf{S4}$ *and* $\mathbf{L}(\bullet)$ *is contained either in* $\mathbf{L}(\updownarrow)$ *or in* $\mathbf{L}(\bullet_2)$*;*
3. $\mathbf{L}(\updownarrow)$ *is the greatest logic strictly between* $\mathbf{Grz}$ *and* $\mathbf{L}(\bullet)$.

These facts are well known, cf. [Maksimova, 1979].

Theorem 16.78. *Let* $\mathbf{L}_1, \mathbf{L}_2$ *be two monomodal logics, such that* $\mathbf{Grz} \subseteq \mathbf{L}_1 \subset \mathbf{L}(\bullet)$, $\mathbf{S4} \subseteq \mathbf{L}_2 \subset \mathbf{L}(\bullet)$. *Then* $\mathbf{L}_1, \mathbf{L}_2$ *are not product-matching.*

Proof Consider the formula

$$AGrz_{21} = \Box_2\Box_1(\Box_2(p \to \Box_1 p) \to p) \to p,$$

which is obviously equivalent to

$$\sim (\sim p \wedge \Box_2\Box_1(\sim p \to \Diamond_2(p \wedge \Diamond_1 \sim p))).$$

Lemma 16.79. $AGrz_{21} \in \mathbf{L}_1 \times \mathbf{L}_2$.

Proof We have to show that if $F \models \mathbf{L}_1, G \models \mathbf{L}_2$ then $F \times G \models AGrz_{21}$. For this assume that $F = (X, R_1) \models \mathbf{Grz}, G = (Y, R_2)$ is reflexive transitive, and prove that $F \times G \models AGrz_{21}$.

Suppose the contrary. Then there exists a model $\mathbf{m}$ over $F \times G$ such that $\mathbf{m}, (x_0, y_0) \models \sim p \wedge \Box_2\Box_1(\sim p \to \Diamond_2(p \wedge \Diamond_1 \sim p))$ for some x_0, y_0. Then we can construct an infinite ascending chain in F, beginning from (x_0, y_0), such that $\mathbf{m}, (x_n, y_n) \models \sim p$ for every n. In fact, if (x_n, y_n) is constructed, we have $(x_0, y_0)(R_2^\vee \circ R_1^\vee)(x_n, y_n)$ due to the transitivity of F and G and the (1,1)-commutativity of $F \times G$. Hence $(x_n, y_n) \models (\sim p \to \Diamond_2(p \wedge \Diamond_1 \sim p))$, and thus $(x_n, y_n) \models \Diamond_2(p \wedge \Diamond_1 \sim p)$. Then we get y_{n+1} such that $(x_n, y_{n+1}) \models p \wedge \Diamond_1 \sim p$, $y_n R_2 y_{n+1}$ and furthermore, x_{n+1} such that $(x_{n+1}, y_{n+1}) \models \sim p, x_n R_1 x_{n+1}$. Obviously, $x_n \neq x_{n+1}$ because p takes different values in $(x_n, y_{n+1}), (x_{n+1}, y_{n+1})$. So we have found an infinite ascending chain in F, which is a contradiction. □

Lemma 16.80. $AGrz_{21} \notin [\mathbf{L}_1, \mathbf{L}_2]$.

Proof By Proposition 16.77, it is sufficient to prove the following two statements:
(16.80.1) $AGrz_{21}, \notin [\mathbf{L}(\updownarrow), \mathbf{L}(\updownarrow)]$;
(16.80.2) $AGrz_{21}, \notin [\mathbf{L}(\updownarrow), \mathbf{L}(\bullet_2)]$;

For (16.80.1), consider the frame $F = (\{0, 1\}, R_1, R_2)$, where $R_1 = \leq$ is the standard order, $R_2 = \geq$ is its inverse. So we have $F \models \mathbf{L}(\updownarrow) * \mathbf{L}(\updownarrow)$, and F is (1,1)-commutative, because $R_1 \circ R_2 = R_2 \circ R_1$ is the universal relation, and also $R_1^{-1} \circ R_2 = R_2 \circ R_2 = R_2 \circ R_1^{-1}$. On the other hand, $F \not\models AGrz_{21}$. In fact, consider a model $\mathbf{m} = (F, h)$ such that $h(p) = \{1\}$. Then $\mathbf{m}, 0 \models \Diamond_2(p \wedge \Diamond_1 \sim p)$, and thus $\mathbf{m}, 0 \models \sim p \wedge \Box_2\Box_1(\sim p \to \Diamond_2(p \wedge \Diamond_1 \sim p))$.

For (16.80.2), the argument is similar. Take the frame $G = (\{0,1\}, R_1, R_2)$, where $R_1 = \leq$, and R_2 is the universal relation.

Now $F \models \mathbf{L}(\updownarrow) * \mathbf{L}(\bullet_2)$, and again the commutativity holds. To refute $AGrz_{21}$ take the same h. □

Lemmas 16.79, 16.80 show that $\mathbf{L}_1 \times \mathbf{L}_2 \neq [\mathbf{L}_1, \mathbf{L}_2]$. □

16.9 Products of some Known Systems

In this section we consider products of some particular systems, and apply the previous results to them.

16.9.1 *The logic* $\mathbf{S4.3}^2 = \mathbf{S4.3} \times \mathbf{S4.3}$

Recall that $\mathbf{S4.3} = \mathbf{S4} + \Box(\Box p \to \Box q) \vee \Box(\Box q \to \Box p)$ is a Sahlqvist logic. $\mathcal{V}(\mathbf{S4.3})$ consists of all *non-branching quasi-orders*, (i.e. of all reflexive and transitive frames satisfying

$\forall x\ \forall y\ \forall z(xRy \wedge xRz \to yRz \vee zRy)$.

So Theorem 16.54 implies

Proposition 16.81. $\mathbf{S4.3}^2$ *is RE and has the c.f.p.*

The recursive axiomatisability follows also from Theorem 2.5.9 from the book [Marx and Venema, 1997] if one takes into account Proposition 16.82 below. That theorem actually describes an axiomatic system for $\mathbf{LIN}^2$, where $\mathbf{LIN}$ is the *temporal* logic of all linear orders; so $\mathbf{LIN}^2$ is a 4-modal logic. This axiomatic system contains a non-structural irreflexivity rule. $\mathbf{S4.3}^2$ itself was axiomatised recently in [Reynolds, 1997]. However, we do not know if $\mathbf{S4.3}^2 = [\mathbf{S4.3}, \mathbf{S4.3}]$.

Proposition 16.82. $\mathbf{S4.3}^2 = \mathbf{L}(\mathcal{LO} \times \mathcal{LO}) = \mathbf{L}(\{\alpha \times \alpha | \alpha \in \mathcal{LO}\}) = \mathbf{L}(\eta \times \eta)$. *(Here $\mathcal{LO}$ denotes the class of linear orders, and η denotes the linear ordering of rationals, as usual.)*

Proof First note that any generated subframe of a non-branching quasi-order is a *linear quasi-order* (i.e. a quasiorder satisfying $\forall x\ \forall y(xRy \vee yRx)$). So $GS(\mathcal{V}(\mathbf{S4.3})) = \mathcal{LQO}$, the class of all linear quasi-orders.

On the other hand,

$$\begin{aligned} \mathbf{S4.3}^2 &= \mathbf{L}(\mathcal{V}(\mathbf{S4.3})_0 \times \mathcal{V}(\mathbf{S4.3})_0) \text{ by (16.4)} \\ &= \mathbf{L}(GS(\mathcal{V}(\mathbf{S4.3})_0 \times \mathcal{V}(\mathbf{S4.3})_0)) \text{ by Lemma (16.25)} \\ &= \mathbf{L}(GS(\mathcal{V}(\mathbf{S4.3})_0) \times GS(\mathcal{V}(\mathbf{S4.3})_0)) \text{ by Lemma (16.32)} \\ &= \mathbf{L}(\mathcal{LQO}_0 \times \mathcal{LQO}_0) \text{ by the above remark.} \end{aligned}$$

Now recall that every (countable) linear quasi-order is a p-morphic image of some (countable) linear order, by the 'Bulldozer Theorem' [Segerberg, 1970], and so given $(F \times G) \in \mathcal{LQO}_0 \times \mathcal{LQO}_0$, we can take p-morphisms $f : \alpha \twoheadrightarrow F, g : \beta \twoheadrightarrow G$ from some countable linear orders α, β, and construct $h : \alpha \times \beta \twoheadrightarrow F \times G$ by Lemma 16.33. Therefore $\mathbf{L}(\mathcal{LO}_0 \times \mathcal{LO}_0) \subseteq \mathbf{L}(\alpha \times \beta) \subseteq \mathbf{L}(F \times G)$, and we get

$$\mathbf{L}(\mathcal{LO}_0 \times \mathcal{LO}_0) \subseteq \mathbf{L}(\mathcal{LQO}_0 \times \mathcal{LQO}_0) = \mathbf{S4.3}^2, \tag{16.7}$$

since F, G are arbitrary.

Lemma 16.83. *Every countable linear order is a p-morphic image of η.*

Proof Let $\alpha \in \mathcal{LO}_0$. Consider the lexicographic product $\alpha \cdot \eta$ (which is $(\alpha \times \eta, \leq)$, with $(x, y) \leq (x', y') \Leftrightarrow x \leq x'$ or $(x = x'$ and $y \leq y'))$. This is obviously a countable dense linear order without maximal and minimal elements. So $\alpha \cdot \eta$ is isomorphic to η.

Now the first projection $f : \alpha \cdot \eta \to \eta$ is a p-morphism, because it is surjective and monotonic, and both $\alpha \cdot \eta, \eta$ are linearly ordered. □

By Lemma 16.83 and Lemma 16.33 there exists an $h : \eta \times \eta \twoheadrightarrow \alpha \times \beta$. Hence $\mathbf{L}(\eta \times \eta) \subseteq \mathbf{L}(\alpha \times \beta)$, and thus

$$\mathbf{L}(\eta \times \eta) \subseteq \mathbf{L}(\mathcal{LO}_0 \times \mathcal{LO}_0). \tag{16.8}$$

Now the proposition follows from (16.7), (16.8) and the obvious inclusions

$$\mathbf{S4.3}^2 \subseteq \mathbf{L}(\mathcal{LO} \times \mathcal{LO}) \subseteq \mathbf{L}(\mathcal{LO}_0 \times \mathcal{LO}_0) \subseteq \mathbf{L}(\eta \times \eta),$$
$$\mathbf{L}(\mathcal{LO} \times \mathcal{LO}) \subseteq \mathbf{L}(\{\alpha \times \alpha | \alpha \in \mathcal{LO}\}) \subseteq \mathbf{L}(\eta \times \eta).$$

□

16.9.2 *The logic* **S4.3** $\times$ **S5**

Since both **S4.3**, **S5** are Sahlqvist logics, again by Theorem 16.54 we have:

Proposition 16.84. **S4.3** $\times$ **S5** *is RE and has the c.f.p.*

We will discuss the axiomatisation problem for this logic later on.

Recall that a *cluster* is a frame with the universal relation, i.e. that of the form $(D, D \times D)$.

Proposition 16.85. $\mathbf{S4.3} \times \mathbf{S5} = \mathbf{L}(\mathcal{LO} \times \mathcal{C}lus) = \mathbf{L}(\mathcal{LO} \times \{\bullet_\omega\}) = L(\eta \times \bullet_\omega)$.

(Here $\mathcal{C}lus$ denotes the class of all clusters; $\bullet_\omega$ denotes the frame $(\omega, \omega \times \omega)$, i.e. the countable cluster.)

Proof $\mathcal{V}(\mathbf{S5})$ is the class of all frames with the equivalence accessibility relation, and obviously, $GS(\mathcal{V}(\mathbf{S5})) = \mathcal{C}lus, GS(\mathcal{V}(\mathbf{S5})_0)$ consists of all frames isomorphic to $\bullet_\omega$. Now we use the same argument, as in the proof of 16.82:

$$\begin{aligned}\mathbf{S4.3} \times \mathbf{S5} &= \mathbf{L}(\mathcal{V}(\mathbf{S4.3})_0 \times \mathcal{V}(\mathbf{S5})_0) \text{ by (16.46)}\\ &= \mathbf{L}(GS(\mathcal{V}(\mathbf{S4.3})_0 \times \mathcal{V}(\mathbf{S5})_0)) \text{ by Lemma (16.25)}\\ &= \mathbf{L}(GS(\mathcal{V}(\mathbf{S4.3})_0) \times GS(\mathcal{V}(\mathbf{S5})_0)) \text{ by Lemma (16.32)}\\ &= \mathbf{L}(\mathcal{LQO}_0 \times \{\bullet_\omega\}),\end{aligned}$$

and hence by the bulldozer theorem and Lemma 16.33, we obtain

$$\mathbf{L}(\mathcal{LO}_0 \times \{\bullet_\omega\}) \subseteq \mathbf{L}(\mathcal{LQO}_0 \times \{\bullet_\omega\}) = \mathbf{S4.3} \times \mathbf{S5}.$$

From Lemma 16.83 and 16.33 we also have

$$\mathbf{L}(\eta \times \bullet_\omega) \subseteq \mathbf{L}(\mathcal{LO}_0 \times \{\bullet_\omega\}).$$

Now it remains to notice that

$$\mathbf{S4.3} \times \mathbf{S5} \subseteq \mathbf{L}(\mathcal{LO} \times \mathcal{C}lus) \subseteq \mathbf{L}(\mathcal{LO}_0 \times \{\bullet_\omega\}) \subseteq \mathbf{L}(\eta \times \{\bullet_\omega\}).$$

□

16.9.3 *The logics* $\mathbf{S4.1.3} \times \mathbf{S4.3}, \mathbf{S4.1.3}^2, \mathbf{S4.1.3} \times \mathbf{S5}$

We denote the logic $\mathbf{S4.1} + \mathbf{S4.3} = \mathbf{S4.3} + \Box\Diamond p \to \Diamond\Box p$ by $\mathbf{S4.1.3}$.[49]

This logic is not Sahlqvist, but is known to be Kripke-complete and elementary (cf. [van Benthem, 1983]). Namely, $\mathcal{V}(\mathbf{S4.1.3})$ is the class of all non-branching quasi-orders where every element is related to some maximal element, i.e. of those satisfying

$\forall x \exists y (xRy \wedge \forall z(yRz \to y = z))$.

So Theorem 16.54 yields

Proposition 16.86. *The logics* $\mathbf{S4.1.3} \times \mathbf{S4.3}, \mathbf{S4.1.3^2}, \mathbf{S4.1.3} \times \mathbf{S5}$ *are RE and have the c.f.p.*

Similarly to Propositions 16.82 and 16.85 we prove the following

Proposition 16.87.

- $\mathbf{S4.1.3} \times \mathbf{S4.3} = \mathbf{L}(\mathcal{LO}^{]} \times \mathcal{LO}) = \mathbf{L}((\eta + 1) \times \eta)$.
- $\mathbf{S4.1.3^2} = \mathbf{L}(\mathcal{LO}^{]} \times \mathcal{LO}^{]}) = \mathbf{L}(\{\alpha \times \alpha | \alpha \in \mathcal{LO}^{]}\}) = \mathbf{L}((\eta + 1) \times (\eta + 1))$.
- $\mathbf{S4.1.3} \times \mathbf{S5} = \mathbf{L}(\mathcal{LO}^{]} \times \mathcal{C}lus) = \mathbf{L}(\mathcal{LO}^{]} \times \{\bullet_\omega\}) = \mathbf{L}((\eta + 1) \times \bullet_\omega)$.

(Here $\mathcal{LO}^{]}$ denotes the class of all linear orders with the greatest element, $\eta + 1$ is η with the greatest element added).

Proof

1. Since $GS(\mathcal{V}(\mathbf{S4.3.1})) = \mathcal{LQO}^{]}$, the class of all linear quasi-orders with the greatest element, by 16.46, 16.25, 16.32 and the bulldozer theorem we get
   $\mathbf{S4.1.3} \times \mathbf{S4.3} \supseteq \mathbf{L}(\mathcal{LO}_0^{]} \times \mathcal{LO}_0)$.
   From Lemma 16.83 we immediately deduce that every countable linear order with the greatest element is a p-morphic image of $(\eta + 1)$. Hence it follows that
   $\mathbf{L}((\eta + 1) \times \eta) \subseteq \mathbf{L}(\mathcal{LO}_0^{]} \times \mathcal{LO}_0)$.
   Finally we notice that
   $\mathbf{S4.1.3} \times \mathbf{S4.3} \subset \mathbf{L}(\mathcal{LO}^{]} \times \mathcal{LO}) \subseteq \mathbf{L}(\mathcal{LO} \times \mathcal{LO}_0) \subseteq \mathbf{L}((\eta + 1) \times \eta)$ holds trivially.
2. We use the same method. We show that $\mathbf{S4.1.3}^2 \supseteq \mathbf{L}(\mathcal{LO}_0^{]} \times \mathcal{LO}_0^{]})$ and that $\mathbf{L}((\eta + 1) \times (\eta + 1)) \subseteq \mathbf{L}(\mathcal{LO}_0^{]} \times \mathcal{LO}_0^{]})$, and also note that
   $\mathbf{S4.1.3}^2 \subseteq \mathbf{L}(\mathcal{LO}^{]} \times \mathcal{LO}^{]}) \subseteq \mathbf{L}(\mathcal{LO}_0^{]} \times \mathcal{LO}_0^{]}) \subseteq \mathbf{L}((\eta + 1) \times (\eta + 1))$.
3. Similarly, we prove first that
   $\mathbf{L}(\mathcal{LO}_0^{]} \times \{\bullet_\omega\}) \subseteq \mathbf{L}(\mathcal{LQO}_0^{]} \times \{\bullet_\omega\}) = \mathbf{S4.1.3} \times \mathbf{S5}$,
   then
   $\mathbf{L}((\eta + 1) \times \bullet_\omega) \subseteq \mathbf{L}(\mathcal{LO}_0^{]} \times \{\bullet_\omega\})$,
   and notice that
   $\mathbf{S4.1.3} \times \mathbf{S5} \subseteq \mathbf{L}(\mathcal{LO}^{]} \times \mathcal{C}lus) \subseteq \mathbf{L}(\mathcal{LO}_0^{]} \times \{\bullet_\omega\}) \subseteq \mathbf{L}((\eta + 1) \times \bullet_\omega)$. □

[49] An older notation for this logic is $\mathbf{K}3$.

16.10 The Finite Model Property via Reduced Normal Forms

We recall that a modal logic Λ has the *finite model property* (f.m.p.) if $\Lambda = \mathbf{L}(\mathcal{C})$ for some class $\mathcal{C}$ of finite frames, or equivalently, if every Λ-consistent formula is satisfied in some finite Λ-frame. Every finitely axiomatisable logic with the f.m.p. is decidable (Harrop's theorem).

For the logics within our consideration we will prove the f.m.p. by two methods. The first is the method of normal forms introduced by Kit Fine [Fine, 1975]. Let us recall some related definitions.

Definition 16.88. The *degree* of a modal formula A (denoted by $d(A)$) is the maximal number of nested modalities in A. Or, inductively:

$$d(p_k) = d(\bot) = 0,\ d(A \to B) = \max(d(A), d(B)),\ d(\Box_i A) = d(A) + 1.$$

We introduce also modal degrees d_i w.r.t. each type of modal operators:

$$d_i(p_k) = d_i(\bot) = 0,\ d_i(A \to B) = \max(d_i(A), d_i(B)), d_i(\Box_i A) = d_i(A) + 1,$$
$$d_i(\Box_j A) = d_i(A) \text{ provided } j \neq i.$$

Definition 16.89. Consider $\mathbb{L}_n$-formulas built from the finite list of propositional letters $\{p_1, ..., p_m\}$. The set $NC_k(m)$ of *normal conjunctions* of degree k (in letters $p_1, ..., p_m$) is defined by induction.

For a formula A, let $A^1 = A, A^0 =\sim A$. We adopt the convention that $2^m = \{0,1\}^m$ is the set of all binary sequences of length m. For $\varepsilon \in 2^m$ let

- $p(m, \varepsilon) = \bigwedge\{p_i^{\varepsilon_i} | 1 \le i \le m\}$.
  Let
- $NC_0(m) = \{p(m, \varepsilon) | \varepsilon \in 2^m\}$.
  If $NC_k(m)$ is already defined, $Y \subseteq NC_k(m), 1 \le j \le n$, let
- $C_j(Y) = \bigwedge\{\Diamond_j A | A \in Y\} \wedge \bigwedge\{\sim \Diamond_j A | A \in NC_k(m) - Y\}$
  (where the empty conjunction means $\top$ as usual).
  If $X \subseteq \{1, ..., n\} \times NC_k(m), 1 \le j \le n$, let
- $X_j = \{A | (j, A) \in X\}$.
  Let finally
- $C(X) = \bigwedge\{C_j(X_j) | 1 \le j \le n\}$,
- $C(X, \varepsilon) = p(m, \varepsilon) \wedge C(X)$,
- $NC_{k+1}(m) = \{C(X, \varepsilon) | X \subseteq \{1, ..., n\} \times NC_k(m), \varepsilon \in 2^m\}$.

A *normal form of degree k* is a disjunction of some non-empty subset of $NC_k(m)$.

Definition 16.90. Let $NC(m) = \bigcup\{NC_k(m) | k \ge 0\}$. We define the relations R_1, $\dots, R_n$ in $NC(m)$ as follows: $C(X, \varepsilon) R_i A \Leftrightarrow (i, A) \in X$; also by definition $R_i(A) = \varnothing$ whenever $A \in NC_0(m)$.

The frame $NF(\mathbf{K}_n \lceil m) = (NC(m), R_1, ..., R_n)$ is called the *normal form frame* (in $p_1, ..., p_m$) for the logic $\mathbf{K}_n$.

Definition 16.91. The *normal form model* is defined as $NM(\mathbf{K}_n\lceil m) = (NF(\mathbf{K}_n\lceil m), h)$ where

$$h(p_i) = \{p(m,\varepsilon)|\varepsilon_i = 1\} \cup \{C(X,\varepsilon)|\varepsilon_i = 1, X \subseteq \{1,...,n\} \times NC_k(m), k \geq 0\}, \text{ if } i \leq m. \qquad (16.9)$$

Lemma 16.92.

1. *In $\mathbf{K}_n$ every consistent formula of degree k is equivalent to a normal form of degree $\leq k$.*
2. *For any $A \in NC$, $(NM(\mathbf{K}_n\lceil m), A) \models A$.*

Proof By induction, cf. [Fine, 1975]. See also Lemma 16.96 and Proposition 16.99 below. □

We recall that Lemma 16.92 provides a constructive proof of the well known result:

Proposition 16.93. *The logic $\mathbf{K}_n$ has the f.m.p.*

Proof **[Fine, 1975].** Let A be a $\mathbf{K}_n$-consistent formula in letters $p_1, \ldots, p_m$ written in a normal form. Take any of its disjuncts B. By Lemma 16.92 and Lemma 16.2, $(NM(\mathbf{K}_n\lceil m)^B, B) \models B$. Then obviously, $(NM(\mathbf{K}_n\lceil m)^B, B) \models A$. But the generated submodel $NM(\mathbf{K}|_n\lceil m)^B$ is finite by our construction. Thus A is satisfied in some finite Kripke model. □

After some modifications, this method can be used for other modal logics. Let us begin with the simplest case.

Definition 16.94. For a consistent n-modal logic $\mathbf{L}$ let $NC(\mathbf{L}\lceil m)$ be the set of $\mathbf{L}$-consistent normal conjunctions in $p_1, ..., p_m$, i.e. $NC(\mathbf{L}\lceil m) = \{A|A \in NC(\mathbf{K}_n\lceil m), \sim A \notin \mathbf{L}\}$. Up to an equivalence, one can also define $NC(\mathbf{L}\lceil m)$ as follows.

Let

- $NC_0(\mathbf{L}\lceil m) = NC_0(m)$;
- $C_j(Y) = \bigwedge\{\Diamond_j A|A \in Y\} \wedge \bigwedge\{\sim \Diamond_j A|A \in NC_k(\mathbf{L}\lceil m) - Y\}$ (if $NC_k(\mathbf{L}\lceil m)$ is already defined, $Y \subseteq NC_k(\mathbf{L}|m), 1 \leq j \leq n$);
- $C(X) = \bigwedge\{C_j(X_j)|1 \leq j \leq n\}$ (where $X \subseteq \{1, ..., n\} \times NC_k(\mathbf{L}\lceil m), X_j = \{A|(j, A) \in X\}$);
- $C(X,\varepsilon) = p(m,\varepsilon) \wedge C(X)$;
- $NC_{k+1}(\mathbf{L}\lceil m) = \{C(X,\varepsilon)|X \subseteq \{1,...,n\} \times NC_k(\mathbf{L}\lceil m), \varepsilon \in 2^m, C(X,\varepsilon) \text{ is } \mathbf{L}\text{-consistent}\}$;
- $NC(\mathbf{L}\lceil m) = \bigcup\{NC_k(\mathbf{L}\lceil m)|k \geq 0\}$.

The elements of $NC_k(\mathbf{L}\lceil m)$ are called $\mathbf{L}$-*reduced normal conjunctions of degree* k. Speaking loosely, they are obtained by dropping all $\mathbf{L}$-inconsistent parts from standard normal conjunctions. An $\mathbf{L}$-*reduced normal form of degree* k is a disjunction of some non-empty subset of $NC_k(\mathbf{L}\lceil m)$.

Similarly to the case of $\mathbf{K}_n$, we define the relations $R_1, ..., R_n$ in $NC(\mathbf{L}\lceil m)$:

$$C(X,\varepsilon)R_iA \Leftrightarrow (i,A) \in X;\; R_i(A) = \varnothing \text{ whenever } d(A) = 0.$$

The frame $NF(\mathbf{L}\lceil m) = (NC(\mathbf{L}\lceil m), R_1, ..., R_n)$ is called the *normal form frame* (in $p_1, ..., p_m$) for the logic $\mathbf{L}$.

The normal form model is defined as $NM(\mathbf{L}\lceil m) = (NF(\mathbf{L}\lceil m), h)$ where

$$h(p_i) = \{p(m,\varepsilon)|\varepsilon_i = 1\} \cup \{C(X,\varepsilon)|\varepsilon_i = 1, C(X,\varepsilon) \in NC(\mathbf{L}\lceil m)\} \text{ if } i \leq m. \tag{16.10}$$

Lemma 16.95. *If $A, B \in NC_k(\mathbf{L}\lceil m)$ and $A \neq B$ then $\sim (A \wedge B) \in \mathbf{K}_n$.*

Proof If $k = 0$ this is a classical case.

Suppose $k = l+1 > 0, A = C(X,\varepsilon), B = C(Y,\sigma), A \neq B$.

If $\varepsilon \neq \sigma$ then $\sim (A \wedge B) \in \mathbf{K}_n$ because $p(m,\varepsilon), p(m,\sigma)$ are classically incompatible.

If $\varepsilon = \sigma$ then for some $j, X_j \neq Y_j$, and we may assume that $Y_j \not\subseteq X_j$. If $E \in Y_j - X_j$ then $\Diamond_j E$ is a conjunct in A, whereas $\sim \Diamond_j E$ is a conjunct in B; this again implies the incompatibility of A and B. □

Lemma 16.96. *For any $A \in NC(\mathbf{L}\lceil m), (NM(\mathbf{L}\lceil m), A) \models A$.*

Proof By induction

$$p(m,\varepsilon) \models p(m,\varepsilon) \text{ since } p(m,\varepsilon) \models p_i \Leftrightarrow \varepsilon_i = 1.$$

Let $d(A) = k+1, A = C(X,\varepsilon)$. If $B \in X_j$ then $C(X,\varepsilon) \models \Diamond_j B$, since $C(X,\varepsilon)R_jB$ and $B \models B$ by the induction hypothesis.

If $B \notin X_j$ then $C(X,\varepsilon) \models \sim \Diamond_j B$, since otherwise for some $E \in NC_k(\mathbf{L}\lceil m)$ we have $C(X,\varepsilon)R_jE, E \models B$. But we have also $E \models E$ by the inductive hypothesis, and $E \neq B$ since $B \notin X_j$. Thus $E \models E \wedge B$, in contradiction with Lemma 16.95.

Also $A \models p(m,\varepsilon)$ by definition of h. Therefore $A \models A$. □

Henceforth we will sometimes simplify our notation. Having numbers n, m and an n-modal logic $\mathbf{L}$ fixed, we abbreviate $NC_k(\mathbf{L}\lceil m)$ to N_k.

Lemma 16.97. *For any $j \leq n$, for any k:*

1. $(\bigvee\{C_j(Y)|Y \subseteq N_k\}) \in \mathbf{K}_n$;
2. $(\bigvee N_k) \in \mathbf{L}$.

Proof

1. Let $N_k = \{E_1, ..., E_s\}$, and let $\varepsilon_Y \in 2^s$ be the characteristic s-tuple of the subset Y, i.e. $\varepsilon_k = 1$ iff $E_k \in Y$. Then $C_j(Y)$ is obtained from $p(s,\varepsilon_Y)$ by the substitution $(\Diamond_j E_1, ..., \Diamond_j E_s/p_1, ..., p_s)$, and thus $(\bigvee\{C_j(Y)|Y \subseteq N_k\})$ is a substitution instance of the classical tautology $(\bigvee\{p(s,\varepsilon)|\varepsilon \in 2^s\})$.
2. If $k = 0$, then $\bigvee N_k$ equals to the classical tautology $\bigvee\{p(m,\varepsilon)|\varepsilon \in 2^m\}$.

If $k = l + 1 > 0$, then $\bigvee N_k$ is equivalent in $\mathbf{L}$ to $\bigvee\{C(X,\varepsilon)|\varepsilon \in 2^m, X \subseteq \{1,...,n\} \times N_l\}$, since $\mathbf{L}$-inconsistent disjuncts can be omitted. The latter disjunction belongs to $\mathbf{K}_n$, due to the following equivalences:
$\bigvee\{C(X,\varepsilon)|\varepsilon \in 2^m, X \subseteq \{1,...,n\} \times N_l\} \equiv$
$(\bigvee\{C(X)|X \subseteq \{1,...,n\} \times N_l\}) \wedge (\bigvee\{p(m,\varepsilon)|\varepsilon \in 2^m\}) \equiv$
$\bigvee\{C(X)|X \subseteq \{1,...,n\} \times N_l\} \equiv \bigwedge_{j=1}^{n} \bigvee\{C_j(Y)|Y \subseteq N_l\};$
by (1), the latter conjunction is a theorem of $\mathbf{K}_n$. Hence (2) follows. □

Lemma 16.98. *Let $E \in N_k, j \leq n$. Then*

1. $(\Diamond_j E \equiv \bigvee\{C_j(Y)|Y \subseteq N_k, E \in Y\}) \in \mathbf{K}_n$.
2. $(\Diamond_j E \equiv \bigvee\{D|DR_jE\}) \in \mathbf{L}$.

Proof

1. By Lemma 16.97 we have (in $\mathbf{K}_n$):

$$(\Diamond_j E \equiv \Diamond_j E \wedge \bigvee\{C_j(Y)|Y \subseteq N_k\});$$

hence

$$(\Diamond_j E \equiv \bigvee\{\Diamond_j E \wedge C_j(Y)|Y \subseteq N_k\});$$

and it remains to notice that

$$\Diamond_j E \wedge C_j(Y) \equiv \begin{cases} C_j(Y) & \text{if } E \in Y; \\ \bot & \text{if } E \notin Y. \end{cases} \tag{16.11}$$

2. The following formulas are in $\mathbf{K}_n$:

$\Diamond_j E \equiv \bigvee\{C_j(Y)|Y \subseteq N_k, E \in Y\}$ (by (1)), $\bigvee\{C_i(Y)|Y \subseteq N_k\}$
for any $i \neq j$ (by Lemma 16.97),
$\bigvee\{p(m,\varepsilon)|\varepsilon \in 2^m\}$
(as a classical tautology).

Hence we have (in $\mathbf{K}_n$):

$$\Diamond_j E \equiv \bigvee\{C_1(X_1) \wedge ... \wedge C_n(X_n) \wedge p(m,\varepsilon)|X_1,...,X_n \subseteq N_k, E \in X_j, \varepsilon \in 2^m\}. \tag{16.12}$$

Finally note that

$$C_1(X_1) \wedge ... \wedge C_n(X_n) \wedge p(m,\varepsilon) = C(X,\varepsilon),$$

where

$$X = (X_1 \times \{1\}) \cup ... \cup (X_n \times \{n\});$$

and

$$E \in X_j \Leftrightarrow (j,E) \in X \Leftrightarrow C(X,\varepsilon)R_jE$$

by Definition 16.90. Now (2) follows by omitting contradictory disjuncts from (16.12). Note also that DR_jE implies $D \in N_{k+1}$. □

Proposition 16.99. *Let A be an $\mathbf{L}$-consistent modal formula in variables $p_1, ..., p_m$, $d(A) \leq k$. Then A is equivalent to some $\mathbf{L}$-reduced normal form of degree k (in variables $p_1, ..., p_m$).*

Proof By induction on the complexity of A; we may assume that A is built from propositional letters using $\vee, \sim, \Diamond_i (1 \leq i \leq n)$.

- If $A = p_l$, we notice that $(p_l \equiv p_l \wedge \bigvee N_k) \in \mathbf{L}$ due to Lemma 16.97 (b), and thus $(p_l \equiv \bigvee \{p_l \wedge B | B \in N_k\}) \in \mathbf{L}$. But $p_l \wedge B$ is equivalent either to B or to $\bot$, depending on whether p_l or $\sim p_l$ is a conjunct in B. Hence (in $\mathbf{L}$)

$$p_l \equiv \bigvee \{B | B \in N_k, p_l \text{ is a conjunct in } B\}.$$

- If $A = \Diamond_i B, d(B) \leq k - 1$, by the induction hypothesis we have that for some $C_1, ..., C_l \in N_{k-1}, (B \equiv \bigvee \{C_j | 1 \leq j \leq l\}) \in \mathbf{L}$.
Then $A \equiv \bigvee \{\Diamond_i C_j | 1 \leq j \leq l\}$. By Lemma 16.98 (2), every formula $\Diamond_i C_j$ can be written in an $\mathbf{L}$-reduced normal form of degree k, and so the same is true for A.
- If A is a disjunction, the argument is trivial.
- Let $A =\sim B$, and $B \equiv \bigvee X$ for some $X \subseteq N_k$ by the inductive hypothesis. Note that $X \neq N_k$ because otherwise $B \equiv \top$ by Lemma 16.97 (b), which implies the $\mathbf{L}$-inconsistency of A. Now we have that $A \equiv \bigvee(N_k - X)$. To show this, let $C = \bigvee(N_k - X)$. From Lemma 16.95 it follows that $\sim (C \wedge B) \in \mathbf{L}$, i.e. $(C \rightarrow \sim B) \in \mathbf{L}$. By Lemma 16.97 (b) also $(C \vee B) \in \mathbf{L}$, i.e. $(\sim B \rightarrow C) \in \mathbf{L}$. Thus $(\sim B \equiv C) \in \mathbf{L}$, and C is a normal form for A. □

Theorem 16.100. *Let $\mathbf{L}$ be a modal logic, such that $NF(\mathbf{L}\lceil m) \models \mathbf{L}$ for any m. Then $\mathbf{L}$ has the f.m.p.*

Proof (cf. Proposition 16.93). Let A be an $\mathbf{L}$-consistent formula in letters $p_1, ..., p_m$ written in normal form. Take any of its disjuncts B. By Lemma 16.96 and Lemma 16.2, $(NM(\mathbf{K}_n\lceil m)^B, B) \models B$. Then obviously, $(NM(\mathbf{K}_n\lceil m)^B, B) \models A$. Now we observe that $NF(\mathbf{L}\lceil m) \models \mathbf{L}$ implies $NF(\mathbf{L}\lceil m)^B \models \mathbf{L}$ by Lemma 16.2. Since the frame $NM(\mathbf{K}_n\lceil m)^B$ is finite, we obtain that A is satisfied in some finite $\mathbf{L}$-frame.
□

Lemma 16.101. *Let $\mathbf{L}$ be a bimodal logic containing $\Diamond_1\Diamond_2 p \rightarrow \Diamond_2\Diamond_1 p$. Then $NF(\mathbf{L}\lceil m) \models \Diamond_1\Diamond_2 p \rightarrow \Diamond_2\Diamond_1 p$.*

Proof We have to show that $R_1 \circ R_2 \subseteq R_1 \circ R_2$ in $NF(\mathbf{L}\lceil m)$. So we assume that $A(R_1 \circ R_2)B$ and prove that $A(R_2 \circ R_1)B$. By our assumption, AR_1CR_2B for some C; hence $\Diamond_1 C$ is a conjunct in A, and $\Diamond_2 B$ is a conjunct in C. Then obviously, $A \rightarrow \Diamond_1 C, C \rightarrow \Diamond_2 B \in \mathbf{L}$. Hence $(\Diamond_1 C \rightarrow \Diamond_1\Diamond_2 B) \in \mathbf{L}$, and thus $(A \rightarrow \Diamond_1\Diamond_2 B) \in \mathbf{L}$, which implies $(A \rightarrow \Diamond_2\Diamond_1 B) \in \mathbf{L}$, due to the condition of the lemma. Using the equivalence from Lemma 16.98, we obtain that $[A \rightarrow \Diamond_2(\bigvee\{E | ER_1B\})] \in \mathbf{L}$, and hence

$$(A \to \bigvee\{\Diamond_2 E | E R_1 B\}) \in \mathbf{L} \tag{16.13}$$

by the distributivity of $\Diamond_2$. Then for some $E \in R_1^{-1}(B)$ the formula $A \wedge \Diamond_2 E$ must be $\mathbf{L}$-consistent, because otherwise we have $\bigwedge\{A \to\sim \Diamond_2 E | E R_1 B\} \in \mathbf{L}$, and hence $(A \to \bigwedge\{\sim \Diamond_2 E | E R_1 B\}) \in \mathbf{L}$, which together with (16.13) implies $\sim A \in \mathbf{L}$. But this is a contradiction, because each $A \in NC(\mathbf{L}\lceil m)$ should be $\mathbf{L}$-consistent.

So take $E \in R_1^{-1}(B)$ such that $A \wedge \Diamond_2 E$ is $\mathbf{L}$-consistent. Now if $B \in N_k$, then $C \in N_{k+1}, E \in N_{k+1}, A \in N_{k+2}$. Then either $\Diamond_2 E$ or $\sim \Diamond_2 E$ is a conjunct in A, and since $A \wedge \Diamond_2 E$ is consistent, only the first possibility remains. Thus we have that $A R_2 E R_1 B$, and therefore $A(R_2 \circ R_1)B$. □

Theorem 16.102. *Let* $\mathbf{L} = \mathbf{K}_n + S, S$ *being a set of axioms of the form* $\Diamond_i \Diamond_j p \to \Diamond_j \Diamond_i p$. *Then* $\mathbf{L}$ *has the f.m.p.*

Proof Follows readily from Theorem 16.100 and Lemma 16.101. □

Theorem 16.102 can be generalised to logics axiomatised by formulas of the form $\Delta_1 \to \Delta_2$, where Δ_1, Δ_2 are sequences of $\Diamond$-modalities of equal length. On the other hand, it is well known that logics axiomatised by finite sets of axioms of this form without restrictions on the length can be undecidable, cf. e.g. [Chagrov and Shehtman, 1995].

Theorem 16.102 is not sufficient yet to prove the f.m.p. for $\mathbf{L} = \mathbf{K} \times \mathbf{K}$, because of the axiom $\Diamond_1 \Box_2 p \to \Box_2 \Diamond_1 p$. Actually one can easily see that this formula is non-valid in normal form frames, because for any $A_1, A_2 \in N_0$ there exists $B \in N_1$ such that $B R_1 A_1, B R_2 A_2$, but of course $R_2(A_1) = R_1(A_2) = \varnothing$. This happens because the normal forms considered so far lose some information about the frame $NF(\mathbf{L}\lceil m)$; they do not say that N_0 consists of endpoints. But if we change our definition and add the conjuncts $\Box_1 E \perp, \Box_2 E \perp$ to A_1 and A_2, we will no longer be able to find B, because $\Diamond_1 A_1 \wedge \Diamond_2 A_2$ becomes $\mathbf{L}$-inconsistent!

In Section 16.12 we make such a modification systematically and show that it works properly for $\mathbf{K} \times \mathbf{K}$. But first we need some other fact *(finite depth property)* to be proved in Section 16.11.

Also let us recall here the theorem by Kit Fine [Fine, 1975], saying that every modal logic above $\mathbf{D}$ axiomatised by *uniform formulas*—i.e. by those in which every two subformulas connected by $\to$ (or $\vee, \wedge$) have the same degree—has the f.m.p. This theorem can be transferred to the polymodal case, and together with Theorem 16.76, it gives us the following.

Theorem 16.103. *Let* $\mathbf{L}_1, \mathbf{L}_2$ *be an* n*- and* m*-modal logic respectively. If* $\mathbf{L}_1, \mathbf{L}_2$ *are uniform, PTC, and* $\mathbf{L}_1 \supseteq \mathbf{D}_n, \mathbf{L}_2 \supseteq \mathbf{D}_m$, *then* $\mathbf{L}_1 \times \mathbf{L}_2$ *has the f.m.p.*

16.11 The Finite Depth Property, and the Finite Model Property for $\mathbf{K}^2$

Definition 16.104. Let $F = (W, R_1, ..., R_n)$ be a frame, $x \in W$. $d_i(x)$, the R_i-depth of x is the least k such that $R_i^{k+1}(x) = \varnothing$, if such k exists; otherwise we put $d_i(x) = \infty$.

Obviously in any model over $F, x \models \Box_i^{k+1} \perp$ iff $d_i(x) \leq k$.

Lemma 16.105. *Let a frame $F = (W, R_1, .., R_{n+m})$ be (n,m)-commutative, $x, y \in W, 1 \leq i \leq n, 1 \leq j \leq m$. Then xR_jy implies $d_{n+i}(x) = d_{n+i}(y)$, and $xR_{n+i}y$ implies $d_j(x) = d_j(y)$.*

Proof Both statements are analogous, so we prove the first.

If $d_{n+i}(x) = k$ then $x \models \Box_{n+i}^{k+1} \perp \wedge \Diamond_{n+i}^k \top$. Since $(\Box_{n+i}^{k+1} \perp \rightarrow \Box_{n+i}^{k+1} \Box_j \perp) \in \mathbf{K}_{n+m}$ (from $\perp \rightarrow \Box_j \perp$ by monotonicity of $\Box_{n+i}$), we obtain that

$$(\Box_{n+i}^{k+1} \perp \rightarrow \Box_j \Box_{n+i}^{k+1} \perp) \in [\mathbf{K}_n, \mathbf{K}_m].$$

Hence $y \models \Box_{n+i}^{k+1} \perp$, i.e. $d_{n+i}(y) \leq k$.

Also $(\Diamond_{n+i}^k \top \rightarrow \Diamond_{n+i}^k \Box_j \top) \in \mathbf{K}_{n+m}$ (from $\top \rightarrow \Box_j \top$ by monotonicity of $\Diamond_{n+i}$), and thus $(\Diamond_{n+i}^k \top \rightarrow \Box_j \Diamond_{n+i}^k \top) \subset [\mathbf{K}_n, \mathbf{K}_m]$. Hence $y \models \Diamond_{n+i}^k \top$, i.e. $d_{n+i}(y) \geq k$.
$\Box$

Lemma 16.106. *Let $F = (W, R_1, R_2)$ be a (1,1)-commutative frame generated by x. Then in any model over F:*
$x \models \Box_1^{m+1} \perp \wedge \Box_2^{n+1} \perp$ *iff* $\forall y \in W (d_1(y) \leq m \wedge d_2(y) \leq n)$.

Proof 'If' is obvious. To show 'only if' note that yR_1z implies $d_2(y) = d_2(z)$ by Lemma 16.105, and also $d_1(y) > d_1(z)$ because $z \models \Box_1^k \perp$ whenever $y \models \Box_1^{k+1} \perp$. Similarly yR_2z implies $d_1(y) = d_1(z)$ and $d_2(y) > d_2(z)$.

The same argument applied l times, yields that $d_1(y) \leq d_1(x), d_2(y) \leq d_2(x)$ for any $y \in (R_1 \cup R_2)^l(x)$. Finally, notice that
$W = W^x = \bigcup_{l \geq 0} (R_1 \cup R_2)^l(x)$. $\Box$

Proposition 16.107 (Finite depth property). *Let A be a $(\mathbf{K} \times \mathbf{K})$-consistent bimodal formula, $d_1(A) = n_1, d_2(A) = n_2$. Then the formula $A' = A \wedge \Box_1^{n_1+1} \perp \wedge \Box_2^{n_2+1} \perp$ is $(\mathbf{K} \times \mathbf{K})$-consistent.*

Proof By Proposition 16.63, A is satisfied at $T_1 \times T_2, (\lambda, \lambda)$, for some standard trees T_1, T_2. Let $T_i \upharpoonright n_i$ be the restriction of T_i to sequences of length $\leq n_i$. We claim that A is satisfied at $(T_1 \upharpoonright n_1) \times (T_2 \upharpoonright n_2), (\lambda, \lambda)$. This follows from

Lemma 16.108. *Let $\mathbf{m}$ be a Kripke model over $T_1 \times T_2$, $\mathbf{m}'$ be its restriction to $(T_1 \upharpoonright n_1) \times (T_2 \upharpoonright n_2)$. Then for any $x_1 \in (T_1 \upharpoonright n_1), x_2 \in (T_2 \upharpoonright n_2)$, for any B, if $d_1(B) + l(x_1) \leq n_1, d_2(B) + l(x_2) \leq n_2$ then*

$$\mathbf{m}, (x_1, x_2) \models B \Leftrightarrow \mathbf{m}', (x_1, x_2) \models B.$$

(Recall that l denotes the length of a sequence).

Proof By induction on the complexity of B.

- If $B \in PL$, this is by definition of $\mathbf{m}'$.
- If $B = \perp$ there is nothing to prove.

- If $B = C \to D$, we have $d_i(C), d_i(D) \le d_i(B)$, and thus $d_i(C) + l(x_i) \le n_i, d_i(D) + l(x_i) \le n_i$. So the inductive hypothesis can be applied to C, D, and we get:
  $\mathbf{m}, (x_1, x_2) \models B \Leftrightarrow$ (not $\mathbf{m}, (x_1, x_2) \models C$ or $\mathbf{m}, (x_1, x_2) \models D$) $\Leftrightarrow$
  (not $\mathbf{m}', (x_1, x_2) \models C$ or $\mathbf{m}', (x_1, x_2) \models D$) $\Leftrightarrow \mathbf{m}', (x_1, x_2) \models B$.
- Let $B = \Box_1 C$. If $\mathbf{m}, (x_1, x_2) \models B$ then for some y_1 we have: $x_1 \sqsubset y_1, \mathbf{m}, (y_1, x_2) \models C$. (Recall that $\sqsubset$ denotes the successor relation in the standard tree.) Since $l(y_1) = l(x_1)+1, d_1(C) = d_1(B)-1, d_2(C) = d_2(B)$ we have: $d_1(C)+l(y_1) \le n_1, d_2(C) + l(x_2) \le n_2$, and we can apply the inductive hypothesis to C and get that $\mathbf{m}', (y_1, x_2) \models C$. Then $\mathbf{m}', (x_1, x_2) \models B$.
  The other way round, if $\mathbf{m}', (x_1, x_2) \models B$ then for some y_1 we have: $x_1 \sqsubset y_1$ and $\mathbf{m}', (y_1, x_2) \models C$. As above, this implies $\mathbf{m}, (y_1, x_2) \models C$, and thus $\mathbf{m}, (x_1, x_2) \models B$.
- The case $B = \Diamond_2 C$ is quite analogous to the previous one. □

Since $d_i(A) + l(\lambda) = n_i$, Lemma 16.108 can be applied to A, and $\mathbf{m}, (\lambda, \lambda) \models A$ implies $\mathbf{m}', (\lambda, \lambda) \models A$. But also $\mathbf{m}', (\lambda, \lambda) \models \Box_1^{n_1+1} \bot \wedge \Box_2^{n_2+1} \bot$ because lengths of coordinates of each world in $\mathbf{m}'$ are restricted by n_1 and n_2. □

Proposition 16.109. *If Λ is an $(n + m)$-modal logic, $\Lambda \supseteq [\mathbf{K}_n, \mathbf{K}_m]$, then every restricted canonical frame $F_{\Lambda\lceil k}$ is (n, m)-commutative.*

Proof By Proposition 16.29, $[\mathbf{K}_n, \mathbf{K}_m]$ is canonical. Hence $[\mathbf{K}_n, \mathbf{K}_m] \subseteq \mathbf{L}(F_{[\mathbf{K}_n, \mathbf{K}_m]\lceil k}) \subseteq \mathbf{L}(F_{\Lambda\lceil k})$ by Proposition 16.7. □

Theorem 16.110. *In a restricted canonical model for a bimodal logic $\mathbf{L} \supseteq [\mathbf{K}, \mathbf{K}]$ every set $S_n = \{x \in W_{\mathbf{L}\lceil m} | d_1(x) + d_2(x) = n\}$ is finite.*

Proof We fix $\mathbf{L}, m$ and by induction on n prove that S_n is finite. We abbreviate the notation $R_{j,\mathbf{L}\lceil m}$ just to R_j.

Obviously, $x \in S_0 \Leftrightarrow R_1(x) = R_2(x) = \varnothing$. By Lemma 16.8, we obtain that $x, y \in S_0$ coincide iff the same propositional letters from $p_1, ..., p_m$ are true at x and y. Hence it follows that $|S_0| \le 2^m$.

For the induction step, assume that $S_0, ..., S_{n-1}$ are finite, and let $c = |S_0 \cup ... \cup S_{n-1}|$. If $x \in S_n$ and xR_1z then by Lemma 16.105, $d_2(x) = d_2(z)$; also it is clear that $d_1(x) > d_1(z)$. Thus we conclude that $R_1(x) \subseteq S_0 \cup ... \cup S_{n-1}$. Similarly we have $R_2(x) \subseteq S_0 \cup ... \cup S_{n-1}$. By Lemma 16.8, every point x in the restricted canonical frame is uniquely determined by the triple $(R_1(x), R_2(x), \{i \le m | x \models p_i\})$. Therefore S_n is finite, and $|S_n| \le 2^c \cdot 2^c \cdot 2^m$. □

Theorem 16.111. $[\mathbf{K}, \mathbf{K}]$ *has the f.m.p.*

Proof Let $\mathbf{L} = [\mathbf{K}, \mathbf{K}]$, and let A be an $\mathbf{L}$-consistent $\mathcal{L}_m$-formula, $d_j(A) = n_j$ for $j = 1, 2$. By Proposition 16.107 and Theorem 16.6, $A' = A \wedge \Box_1^{n_1+1} \bot \wedge \Box_2^{n_2+1} \bot$ is satisfied at $F_{\mathbf{L}\lceil m}, x$ for some x. But then $d_1(x) \le n_1, d_2(x) \le n_2$. By Theorem 16.110 and Lemma 16.106, the generated subframe $F^x_{\mathbf{L}\lceil m}$ is finite. Since $\mathbf{L}$ is canonical (Proposition 16.29), we obtain that $F_{\mathbf{L}\lceil m}$ is an $\mathbf{L}$-frame. Hence by Lemma 16.2, $F^x_{\mathbf{L}\lceil m}$ is an $\mathbf{L}$-frame, and A is satisfied at $F^x_{\mathbf{L}\lceil m}, x$. □

16.12 Two-dimensional Normal Forms

In this section we will show how to construct suitable normal forms in the logic $\mathbf{K}^2$. We call them $\mathbf{K}^2$-*normal forms* and define by induction. So we fix a number m, a logic $\mathbf{L} \supseteq \mathbf{K}^2$, and define the finite sets of formulas $NC_{k,n}(\mathbf{L}\lceil m)$.

Definition 16.112.

- $NC_{0,0}(\mathbf{L}\lceil m) = \{p(m,\varepsilon) \wedge \Box_1 \perp \wedge \Box_2 \perp | \varepsilon \in 2^m\}$;
- if $NC_{k,n}(\mathbf{L}\lceil m)$ is already defined, $X \subseteq NC_{k,n}(\mathbf{L}\lceil m)$, $j = 1,2$, let $C_j(X) = \bigwedge\{\Diamond_j A | A \in Y\} \wedge \bigwedge\{\sim \Diamond_j A | A \in NC_{k,n}(\mathbf{L}\lceil m) - Y\}$;
- if $X \subseteq NC_{k,0}(\mathbf{L}\lceil m), \varepsilon \in 2^m$, let $C_1(X,\varepsilon) = p(m,\varepsilon) \wedge C_1(X) \wedge \Box_1^{k+2} \perp \wedge \Box_2 \perp$;
- $NC_{k+1,0}(\mathbf{L}\lceil m) = \{C_1(X,\varepsilon) | X \subseteq NC_{k,0}(\mathbf{L}\lceil m), \varepsilon \in 2^m, C_1(X,\varepsilon)$ is $\mathbf{L}$-consistent$\}$;
- if $X \subseteq NC_{0,n}(\mathbf{L}\lceil m), \varepsilon \in 2^m$, let $C_2(X,\varepsilon) = p(m,\varepsilon) \wedge C_2(X) \wedge \Box_2^{n+2} \perp \wedge \Box_1 \perp$;
- $NC_{0,n+1}(\mathbf{L}\lceil m) = \{C_2(X,\varepsilon) | X \subseteq NC_{0,n}(\mathbf{L}\lceil m), \varepsilon \in 2^m, C_2(X,\varepsilon)$ is $\mathbf{L}$-consistent$\}$;
- if $X \subseteq NC_{k,n+1}(\mathbf{L}\lceil m), Y \subseteq NC_{k+1,n}(\mathbf{L}\lceil m), \varepsilon \in 2^m$, let $C(X,Y,\varepsilon) = p(m,\varepsilon) \wedge C_1(X) \wedge C_2(Y) \wedge \Box_1^{k+2} \perp \wedge \Box_2^{n+2} \perp$;
- $NC_{k+1,n+1}(\mathbf{L}\lceil m) = \{C(X,Y,\varepsilon) | X \subseteq NC_{k,n+1}(\mathbf{L}\lceil m), Y \subseteq NC_{k+1,n}(\mathbf{L}\lceil m)$, $\varepsilon \in 2^m, C(X,Y,\varepsilon)$ is $\mathbf{L}$-consistent$\}$;
- $NC^2(\mathbf{L}\lceil m) = \bigcup\{NC_{k,n}(\mathbf{L}\lceil m) | k,n \geq 0\}$.

The elements of $NC_{k,n}(\mathbf{L}\lceil m)$ are called $\mathbf{L}$-*reduced* $\mathbf{K}^2$-*normal conjunctions of degrees* k,n. An $\mathbf{L}$-*reduced* $\mathbf{K}^2$-*normal form of degrees* k,n is a disjunction of a non-empty subset of $NC_{k,n}(\mathbf{L}\lceil m)$.

Definition 16.113. The $\mathbf{K}^2$-normal form frame for $\mathbf{L}$ $\mathbf{K}^2NF(\mathbf{L}\lceil m) = (NC^2(\mathbf{L}\lceil m), S_1, S_2)$, is defined by induction as follows.

- $S_1(A) = S_2(A) = \varnothing$, if A is of degrees $0,0$.
- If $C_1(X,\varepsilon) \in NC_{k+1,0}(\mathbf{L}\lceil m)$ then
- $C_1(X,\varepsilon) S_1 A \Leftrightarrow A \in X, \quad S_2(C_1(X,\varepsilon)) = \varnothing$.
- If $C_2(X,\varepsilon) \in NC_{0,n+1}(\mathbf{L}\lceil m)$ then $C_2(X,\varepsilon) S_2 A \Leftrightarrow A \in X, \quad S_1(C_2(X,\varepsilon)) = \varnothing$.
- If $C(X,Y,\varepsilon) \in NC_{k+1,n+1}(\mathbf{L}\lceil m)$ then $C(X,Y,\varepsilon) S_1 A \Leftrightarrow A \in X, C(X,Y,\varepsilon)$ $S_2 A \Leftrightarrow A \in Y$.

Furthermore we abbreviate $NC_{k,n}(\mathbf{L}\lceil m)$ to $NC_{k,n}$ provided $\mathbf{L}, m$ are fixed.

Definition 16.114. The $\mathbf{K}^2$-*normal form model for* $\mathbf{L}$ is defined as $\mathbf{K}^2NM(\mathbf{L}\lceil m) = (\mathbf{K}^2NF(\mathbf{L}\lceil m), h)$, where

$$\begin{aligned}
h(p_i) = &\{p(m,\varepsilon) \wedge \Box_1 \perp \wedge \Box_2 \perp | \varepsilon_i = 1\} \cup \\
&\{C_1(X,\varepsilon) | \varepsilon_i = 1, C_1(X,\varepsilon) \in NC_{k,0}; k \geq 1\} \cup \\
&\{C_2(X,\varepsilon) | \varepsilon_i = 1, C_2(X,\varepsilon) \in NC_{0,n}; n \geq 1\} \cup \\
&\{C(X,Y,\varepsilon) | \varepsilon_i = 1, C(X,Y,\varepsilon) \in NC_{k,n}; k,n \geq 1\} \text{ if } i \leq m.
\end{aligned}$$

Lemma 16.115. *If $A, B \in NC_{k,n}$ and $A \neq B$ then* $\sim (A \wedge B) \in \mathbf{K}_2$.

Proof Similar to Lemma 16.95. It is sufficient to make the following observations:

- $\sim (p(m,\varepsilon) \wedge p(m,\delta)) \in \mathbf{K}_2$ if $\varepsilon \neq \delta$;
- $\sim (C_j(X) \wedge C_j(Y)) \in \mathbf{K}_2$ if $X, Y \subseteq NC_{k,n}, X \neq Y$. □

Lemma 16.116. *For any $A \in NC^2(\mathbf{L}\lceil m), (NM^2(\mathbf{L}\lceil m), A) \models A$.*

Proof By induction, cf. Lemma 16.96.

- If $A \in NC_{0,0}$ the statement is clear.
- Let $A = C_1(X,\varepsilon) \in NC_{k+1,0}$. It is proved similarly to Lemma 16.96 that $A \models C_1(X)$, and it is clear that $A \models p(m,\varepsilon)$. Obviously we have $A \models \Box_2 \bot$ because $R_2(A) = \varnothing$. Also $A \models \Box_1^{k+2} \bot$ since AS_1B implies $B \in NC_{k,0}$, and $B \models \Box_1^{k+1} \bot$ by the induction hypothesis. Summing up, we obtain $A \models A$.
- The case $A = C_2(X,\varepsilon) \in NC_{0,n+1}$ is analogous to the previous one.
- Let $A = C(X,Y,\varepsilon) \in NC_{k+1,n+1}$. As above, one can see that $A \models C_1(X) \wedge C_2(Y) \wedge p(m,\varepsilon)$. Also we have $A \models \Box_1^{k+2} \bot$ since AS_1B implies $B \in NC_{k,n+1}$, and $B \models \Box_1^{k+1} \bot$ by the induction hypothesis. Likewise $A \models \Box_1^{n+2} \bot$, and thus $A \models A$. □

Lemma 16.117. *For any $j \in \{1,2\}$, for any $k, n \geq 0$:*

1. $(\bigvee\{C_j(X) | X \subseteq N_{k,n}\}) \in \mathbf{K}_2$;
2. $(\bigvee NC_{k,n}) \equiv \Box_1^{k+1} \bot \wedge \Box_2^{n+1} \bot \in \mathbf{L}$.

Proof

1. By the same argument as in Lemma 16.97.
2. We consider only the case when $k, n > 0$ leaving the other cases to the reader. Then we have in $\mathbf{L}$:

$$(\bigvee NC_{k,n}) \equiv (\bigvee\{C_1(X) | X \subseteq NC_{k-1,n}\}) \wedge (\bigvee\{C_2(Y) | Y \subseteq NC_{k,n-1}\})$$
$$\wedge \bigvee\{p(m,\varepsilon) | \varepsilon \in 2^m\} \wedge \Box_1^{k+1} \bot \wedge \Box_2^{n+1} \bot .$$

The first two conjuncts are $\mathbf{K}_2$-tautologies, the third is a classical tautology, so they can be eliminated. □

Lemma 16.118. *Let $E \in NC_{k,n}$. Then*

1. $(\Diamond_j E \equiv \bigvee\{C_j(X) | X \subseteq NC_{k,n}, E \in X\}) \in \mathbf{K}_2$ *(for $j = 1, 2$)*;
2. $(\Box_1^{k+2} \bot \wedge \Box_2^{n+1} \bot \wedge \Diamond_1 E \equiv \bigvee\{D | DS_1E\}) \in \mathbf{L}$;
3. $(\Box_1^{k+1} \bot \wedge \Box_2^{n+2} \bot \wedge \Diamond_2 E \equiv \bigvee\{D | DS_2E\}) \in \mathbf{L}$.

Proof

1. Similar to Lemma 16.98.

2. The proof is also similar to that of 16.98. By (1) and Lemma 16.117, we have the following equivalences in **L**:

$$\Box_1^{k+2} \bot \wedge \Box_2^{n+1} \bot \wedge \Diamond_1 E \equiv$$
$$\Box_1^{k+2} \bot \wedge \Box_2^{n+1} \bot \wedge \bigvee\{C_1(X) | X \subseteq NC_{k,n+1}, E \in X\} \wedge$$
$$\bigvee\{C_2(Y) | Y \subseteq NC_{k+1,n-1}\} \wedge \bigvee\{p(m,\varepsilon) | \varepsilon \in 2^m\} \equiv$$
$$\bigvee\{C(X,Y,\varepsilon) | C(X,Y,\varepsilon) \in NC_{k+1,n}, C(X,Y,\varepsilon) S_1 E\}.$$

3. The same as (2). □

Lemma 16.119. *The following are theorems of* $[\mathbf{K},\mathbf{K}]$:

1. $\Box_1^{k+1} \bot \wedge \Box_2^n \bot \to \Box_1(\Box_1^k \bot \wedge \Box_2^n \bot)$,
2. $\Box_1^k \bot \wedge \Box_2^{n+1} \bot \to \Box_2(\Box_1^k \bot \wedge \Box_2^n \bot)$,
3. $\Diamond_1 A \wedge \Box_1^{k+1} \bot \wedge \Box_2^n \bot \to \Diamond_1(A \wedge \Box_1^k \bot \wedge \Box_2^n \bot)$,
4. $\Diamond_2 A \wedge \Box_1^k \bot \wedge \Box_2^{n+1} \bot \to \Diamond_2(A \wedge \Box_1^k \bot \wedge \Box_2^n \bot)$.

Proof

1. follows from the observation that

$$(\Box_2^n \bot \to \Box_2^n \Box_1 \bot) \in \mathbf{K}_2, (\Box_2^n \Box_1 \bot \equiv \Box_1 \Box_2^n \bot) \in [\mathbf{K},\mathbf{K}].$$

Let A' be the premise of (3). By (1),

$$(A' \to \Diamond_1 A \wedge \Box_1(\Box_1^k \bot \wedge \Box_2^n \bot)) \in [\mathbf{K},\mathbf{K}].$$

On the other hand, $(\Diamond_1 A \wedge \Box_1 B \to \Diamond_1(A \wedge B)) \in \mathbf{K}_2$, and therefore $(A' \to \Diamond_1(A \wedge \Box_1^k \bot \wedge \Box_2^n \bot)) \in \mathbf{L}$.

2. (2) and (4) are proved similarly. □

Proposition 16.120. *Let A be a bimodal formula in variables $p_1, ..., p_m$, such that $d_1(A) \leq k, d_2(A) \leq n$. If the formula $A' = A \wedge \Box_1^{k+1} \bot \wedge \Box_2^{n+1} \bot$ is* **L**-*consistent then A' is equivalent to some* **L**-*reduced* $\mathbf{K}^2$-*normal form of degrees k, n.*

Proof Similar to Proposition 16.99.

- If $A = p_l$ then by Lemma 16.117, there are the following equivalences in **L** : $A \wedge \Box_1^{k+1} \bot \wedge \Box_2^{n+1} \bot \equiv p_l \wedge (\bigvee NC_{k,n}) \equiv \bigvee\{B | B \in NC_{k,n}, p_l$ is a conjunct in $B\}$.
- If $A = \Diamond_1 B, d_1(B) \leq k-1, d_2(B) \leq n$, then (in **L**) $A' = \Diamond_1 B \wedge \Box_1^{k+1} \bot \wedge \Box_2^{n+1} \bot \equiv$ $\Diamond_1(B \wedge \Box_1^k \bot \wedge \Box_2^{n+1} \bot) \wedge \Box_1^{k+1} \bot \wedge \Box_2^{n+1} \bot$ (due to Lemma 16.119). Then by the induction hypothesis, for some $C_1, ..., C_s \in NC_{k-1,n}$ we have in **L** :

$$A' \equiv \Diamond_1(\bigvee_{i=1}^{s} C_i) \wedge \Box_1^{k+1} \bot \wedge \Box_2^{n+1} \bot \equiv \bigvee_{i=1}^{s} (\Diamond_1 C_i \wedge \Box_1^{k+1} \bot \wedge \Box_2^{n+1} \bot).$$

By Lemma 16.118 each of the disjuncts has a normal form of degrees k, n, therefore A' also does.

- If $A = \Diamond_2 B$ the argument is similar to the above.
- If A is a negation, the proof is the same as in Proposition 16.99 (use Lemmas 16.115, 16.117).
- If A is a disjunction the proof is trivial. □

Lemma 16.121. $\mathbf{K}^2 NF(\mathbf{L}\lceil m) \models \Diamond_1\Diamond_2 p \equiv \Diamond_2\Diamond_1 p$.

Proof We will show that $\Diamond_1\Diamond_2 p \to \Diamond_2\Diamond_1 p$ holds; the converse is checked in the same way. The proof is quite similar to that of Lemma 16.101. Assume that $A(S_1 \circ S_2)B, B \in NC_{k,n}$ and prove that $A(S_2 \circ S_1)B$. As in Lemma 16.101, we get that $(A \to \Diamond_2\Diamond_1 B) \in \mathbf{L}$. Also we have that $A \in NC_{k+1,n+1}$, and thus $(A \to \Box_1^{k+2} \bot \wedge \Box_2^{n+2} \bot) \in \mathbf{L}$ by Lemma 16.117. Now by Lemma 16.119 it follows that $(A \to \Diamond_2(\Diamond_1 B \wedge \Box_1^{k+2} \bot \wedge \Box_2^{n+1} \bot)) \in \mathbf{L}$, and from Lemma 16.118 we have $[A \to \Diamond_2(\bigvee\{E | ES_1 B\})] \in \mathbf{L}$. Hence $(A \to \bigvee\{\Diamond_2 E | ES_1 B\}) \in \mathbf{L}$ by distributivity of $\Diamond_2$. Then the same argument as in Lemma 16.101 shows that for some $E \in S_1^{-1}(B)$ the formula $A \wedge \Diamond_2 E$ is $\mathbf{L}$-consistent, and finally we obtain that $AS_2 ES_1 B$. □

Lemma 16.122. $\mathbf{K}^2 NF(\mathbf{L}\lceil m) \models \Diamond_1\Box_2 p \to \Box_2\Diamond_1 p$.

Proof By Lemma 16.21, we have to show that
$S_1^{-1} \circ S_2 \subseteq S_2 \circ S_1^{-1}$.

So we assume AS_1E, AS_2B, and prove that $S_2(E) \cap S_1(B) \neq \varnothing$. From Definition 16.113 we notice that (for some k, n) $A \in NC_{k+1,n+1}, E \in NC_{k,n+1}, B \in NC_{k+1,n}$, and also that $(A \to \Diamond_1 E), (A \to \Diamond_2 B) \in \mathbf{L}$.

The formulas E and B can be presented as
$E = E_0 \wedge C_2(Y), B = B_0 \wedge C_1(X)$, for some $X, Y \subseteq NC_{k,n}$.

Now it is sufficient to show that $X \cap Y \neq \varnothing$; in fact, $X \cap Y = S_1(B) \cap S_2(E)$, according to Definition 16.113.

First note that

(16.122.1) $(B \to \Box_1(\bigvee\{Q | Q \in X\})) \in \mathbf{L}$,
(16.122.2) $(E \to \Box_2(\bigvee\{Q | Q \in Y\})) \in \mathbf{L}$.

In fact, the following formulas are in $\mathbf{L}$:
$B \to \Box_1^{k+2} \bot \wedge \Box_2^{n+1} \bot$ (by Definition 16.112),
$\Box_1^{k+2} \bot \wedge \Box_2^{n+1} \bot \to \Box_1(\Box_1^{k+1} \bot \wedge \Box_2^{n+1} \bot)$ (by Lemma 16.119),
$\Box_1(\Box_1^{k+1} \bot \wedge \Box_2^{n+1} \bot) \to \Box_1(\bigvee NC_{k,n})$ (by Lemma 16.117),
$B \to \Box_1 \sim (\bigvee\{Q | Q \in NC_{k,n} - X\})$ (from Definition 16.112). Hence we get (16.122.1); the proof of (16.122.2) is analogous.

Then since $(A \to \Diamond_1 E), (A \to \Diamond_2 B) \in \mathbf{L}$, it follows that
$(A \to \Diamond_1\Box_2(\bigvee\{P | P \in Y\}) \wedge \Diamond_2\Box_1(\bigvee\{Q | Q \in X\})) \in \mathbf{L}$.

Now suppose $X \cap Y = \varnothing$. Then by Lemma 16.115, for any $P \in Y, Q \in X$ we have $\sim (P \wedge Q) \in \mathbf{L}$. Hence $((\bigvee\{Q | Q \in X\}) \to\sim (\bigvee\{P | P \in Y\})) \in \mathbf{L}$, and furthermore, $(\Diamond_2\Box_1(\bigvee\{Q | Q \in X\}) \to \Diamond_2\Box_1 \sim (\bigvee\{P | P \in Y\})) \in \mathbf{L}$. This implies $(A \to \Diamond_1\Box_2(\bigvee\{P | P \in Y\}) \wedge \Diamond_2\Box_1 \sim (\bigvee\{P | P \in Y\})) \in \mathbf{L}$. But the conclusion of the latter implication is a contradiction in $[\mathbf{K}, \mathbf{K}]$. Therefore $\sim A \in \mathbf{L}$, which is impossible by Definition 16.112. So the assumption $X \cap Y = \varnothing$ is false, and the proof is completed. □

Theorem 16.123. *(Efficient f.m.p. for $[\mathbf{K},\mathbf{K}]$). There exists an algorithm constructing for each $[\mathbf{K},\mathbf{K}]$-consistent formula A a model over a finite $[\mathbf{K},\mathbf{K}]$-frame, where A is satisfied.*

Proof Similar to Theorem 16.100. If A is of degrees k, n then the formula $A' = A \wedge \Box_1^{k+1} \perp \wedge \Box_2^{n+1} \perp$ is $[\mathbf{K},\mathbf{K}]$-consistent by Proposition 16.107. Now we present the latter formula in a $\mathbf{K}^2$-normal form, following the proof of Proposition 16.120. One of its disjuncts generates the required submodel in the normal form model for $[\mathbf{K},\mathbf{K}]$.

The only problem is in making reductions when the normal form is being constructed. To do that, we can proceed as follows.

The sets $NC_{k,0}$ and $NC_{0,n}$ are obtained in an obvious way, and no reductions are needed. This is because $[\mathbf{K},\mathbf{K}]$ is a conservative extension of $\mathbf{K}$.

Let us construct the set $NC_{k+1,n+1}$ provided we know already all $NC_{k',n'}$ with $k'+n' \leq k+n+1$. Given $\varepsilon \in 2^m, X \subseteq NC_{k,n+1}$ and $Y \subseteq NC_{k+1,n}$, we have to decide whether the formula $B = C(X,Y,\varepsilon)$ is in $NC_{k+1,n+1}$, i.e. whether it is $\mathbf{K}^2$-consistent.

Let $\mathbf{m}^{X.Y}$ be the union of the generated submodels $\mathbf{m}^E, E \in X \cup Y$; so the worlds of $\mathbf{m}^{X,Y}$ are $\mathbf{K}^2$-normal conjunctions occurring in the construction of B, and let $F^{X,Y} = (\mathbf{m}^{X,Y}, R_1^{X,Y}, R_2^{X,Y})$ be the corresponding frame. In an obvious way, we extend the model $M^{X,Y}$ to $N = (G,\varphi)$ where

$$G = (W, R_1, R_2), W = W^{X,Y} \cup \{B\}, R_1 = R_1^{X,Y} \cup \{(B,E) \mid E \in X\},$$
$$R_2 = R_2^{X,Y} \cup \{(B,E) \mid E \in Y\}; N, B \vDash p_i \Leftrightarrow \varepsilon_i = 1.$$

Then we have that B is $\mathbf{K}^2$-consistent iff $G \vDash \mathbf{K}^2$.

Note also that to decide whether $G \vDash \mathbf{K}^2$, we need to examine only the new point B, because $F^{X,Y} \vDash \mathbf{K}^2$. This means that we have to verify that

(*) $\forall E_1 \in X \forall E_2 \in Y S_2(E_1) \cap S_1(E_2) \neq \varnothing$

and

(**) $S_2(X) = S_2(Y)$

in order to check the validity of $\Diamond_1\Box_2 p \to \Box_2\Diamond_1 p$ and of $\Box_1\Box_2 p \equiv \Box_2\Box_1 p$ respectively. $\Box$

Remark 16.124. Theorem 16.123 gives a much better deciding procedure than Theorem 16.111. One can check that the cardinality of S_n in Theorem 16.11.7 grows faster than the cardinality of $NC_{k,n}$. However, both functions are not Kalmar-elementary.

16.13 The Finite Model Property via Filtrations

This section proves the f.m.p. for products of some modal logics by the well known filtration method. We use a modification of [Gabbay, 1972] proposed in [Shehtman, 1993].

Definition 16.125. Let $\mathbf{m} = (W, R_1, ..., R_n, h)$ be an n-modal Kripke model. Let Ψ be a set of n-formulas. For $x \in W$ let $\Psi(x) = \{A \in \Psi | (\mathbf{m}, x) \models A\}$. Two worlds $x, y \in W$ are called *Ψ-equivalent in* $\mathbf{m}$ (notation: $(\mathbf{m}, x) \equiv_\Psi (\mathbf{m}, y)$, or $x \equiv_\Psi y$) iff $\Psi(x) = \Psi(y)$.

Definition 16.126. Let $\mathbf{m} = (W, R_1, ..., R_n, h)$ be an n-modal Kripke model. Let Ψ be a set of n-modal formulas which is closed under subformulas, and let $\approx$ be an equivalence relation in $\mathbf{m}$. A model $\mathbf{m}' = (W', R'_1, ..., R'_n, h')$ is called a *filtration of* $\mathbf{m}$ *through* $\Psi, \approx$ if the following holds (for any $x, y \in W, a, b \in W', i \leq n$, for any formula A):

1. $\approx \subseteq \equiv_\Psi$;
2. $W' = W/\approx$;
3. $xR_iy \Rightarrow \approx(x)R'_i \approx(y)$;
4. $aR'_ib \wedge x \in a \wedge y \in b \wedge \mathbf{m}, x \models \Box_i A \wedge \Box_i A \in \Psi \Rightarrow \mathbf{m}, y \models A$;
5. $\mathbf{m}, x \models q \wedge q \in \Psi \cap PL \Rightarrow \mathbf{m}, y \models q$.

Further on we use the notation $x^\sim$ instead of $\approx(x)$ for the equivalence class of x.

Lemma 16.127 (Filtration lemma). *Let* $\mathbf{m}, \mathbf{m}'$ *be the same as in Definition 16.126. Then for any* $x \in W$, *for any* $A \in \Psi$

$\mathbf{m}, x \models A \Leftrightarrow \mathbf{m}', x^\sim \models A.$

Proof This lemma is well known in the standard case (cf. [Gabbay, 1976], [Segerberg, 1970]. Our case brings nothing new; the proof is by induction on the complexity of A. □

Lemma 16.128. *Let* $\mathbf{m} = (W, R_1, ..., R_n, h)$ *be an* n*-modal Kripke model,* Ψ *be a set of* n*-modal formulas closed under subformulas,* $\approx$ *be an equivalence relation in* $\mathbf{m}$ *such that* $\approx \subseteq \equiv_\Psi$*; and let* $W' = W/\approx$*. Then the model* $\mathbf{m}' = (W', R'_1, ..., R'_n, h')$ *such that*

- *for any* $x, y \in W, x^\sim R'_i y^\sim \Leftrightarrow \exists x_1 \in x^\sim \exists y_1 \in y^\sim x_1 R_i y_1$;
- *for any* $q \in PL, h(q) = \{x^\sim | \mathbf{m}, x \models q\}$

is a filtration of $\mathbf{m}$ *through* $\Psi, \approx$*; it is called the least filtration.*

Proof The item (3) in Definition 16.126 holds trivially. For (4), assume $\Box_i A \in \Psi$, $x^\sim R'_i y^\sim, \mathbf{m}, x \models \Box_i A$ and show that $\mathbf{m}, y \models A$. In fact $x^\sim R'_i y^\sim$ implies that $x \approx x_1 R_i y_1 \approx y$ for some x_1, y_1. Now $\Box_i A \in \Psi, \mathbf{m}, x \models \Box_i A$ and $x \approx x_1$ imply that $\mathbf{m}, x_1 \models \Box_i A$; and thus $\mathbf{m}, y_1 \models A$ (since $x_1 R_i y_1$). Finally from $y_1 \approx y$ and $A \in \Psi$ we get that $\mathbf{m}, y \models A$.

The item (5) holds because for $q \in PL$ we have:

$\mathbf{m}', x^\sim \models q \Leftrightarrow \exists x_1 \approx x\, \mathbf{m}, x_1 \models q \Leftrightarrow \mathbf{m}, x \models q.$ □

Proposition 16.129. *Let* $F = (W, R_1, ..., R_n, S)$ *be a* $[\mathbf{K}_n, \mathbf{S5}]$*-frame,* $\mathbf{m} = (F, h)$ *be a model in* F*. Let* Ψ *be a finite set of* $(n+1)$*-modal formulas closed under subformulas. Consider the following equivalence relation in* W :

$x \approx y \Leftrightarrow x \equiv_\Psi y \wedge \{\Psi(z) | xSz\} = \{\Psi(z) | ySz\}.$

Let $\mathbf{m}' = (W', R'_1, ..., R'_n, S', h')$ *be the least filtration of* $\mathbf{m}$ *through* $\Psi, \approx$*. Then*

1. W' *is finite;*
2. $F' = (W', R'_1, ..., R'_n, S)$ *is a* $[\mathbf{K}_n, \mathbf{S5}]$*-frame.*

Proof

1. The equivalence class $x^{\sim}$ is uniquely defined by the pair of finite sets $(\Psi(x),\ \{\Psi(z)|xSz\})$. If $N = |\Psi|$ then there are $\leq 2^N$ possibilities for $\Psi(x)$, and thus $\leq 2^{2^N}$ possibilities for $\{\Psi(z)|xSz\}$. Therefore $|W'| \leq 2^{N} \cdot 2^{2^N}$.
2. We have to check that S' is the equivalence relation and that F' is $(n,1)$-commutative. First let us prove the following:

   (2.1) $(\approx) \circ S = S \circ (\approx)$.
   In fact, suppose $x \approx ySz$. Then for some $u \in S(x)$ we have $\Psi(z) = \Psi(u)$. But then also $z \approx u$, because S is an equivalence relation, and thus $\{\Psi(v)|zSv\} = \{\Psi(v)|ySv\} = \{\Psi(t)|xSt\} = \{\Psi(t)|uSt\}$. Hence $xS \circ (\approx)z$.
   So we have proved that $(\approx) \circ S \subseteq S \circ (\approx)$; by inversion we get: $(\approx) \circ S \subseteq S \circ (\approx)$ since both $\approx, S$ are equivalence relations.

   (2.2) It follows easily from the definition that S' is reflexive and symmetric. To show the transitivity, assume that $aS'bS'c$, i.e. that for some $x \in a, y \in b, y_1 \in b, z \in c$ we have xSy and y_1Sz. Then $y \approx y_1$, and from (2.1) we obtain that there exists z_1, such that $ySz_1 \approx z$. By transitivity of S we have xSz_1. Hence $aS'c$.

   (2.3) Assume $aR_i'bS'c$, and let us show that $a(S' \circ R_i')c$. By definition we have:

   $$\exists x \in a \exists y \in b \exists y_1 \in b \exists z \in c (xR_iy \wedge y_1Sz).$$

   But then $y \approx y_1Sz$, and (2.1) yields that for some t, $ySt \approx z$. Since F is $(n,1)$-commutative, there exists r such that $xSrR_it$. Hence $aS'r^{\sim}R_i'c$.

   (2.4) Similarly, assume that $aS'bR_i'c$, and show that $a(R_i' \circ S)c$. In fact, there exist $x \in a, y \in b, y_1 \in b, z \in c$ such that xSy, y_1R_iz. By (2.1) we can find x_1 such that $x \approx x_1Sy_1$. By $(n,1)$-commutativity there exists u such that x_1R_iuSz. Then we have $aR_i'u^{\sim}S'c$.

So we have proved that $F' \models C^1_{i1}$ or any $i \leq n$. Since $S' = (S')^{-1}$, the axiom C^2_{i1} also holds in F' (Lemma 16.21). □

Proposition 16.130. *The logic* $[\mathbf{K}_n, \mathbf{S5}]$ *has the f.m.p.*

Proof Let $\mathbf{L}$ be this logic. Since $\mathbf{L}$ is complete, every $\mathbf{L}$-consistent formula A is satisfied in some $\mathbf{L}$-frame. Let $\mathbf{m} = (F, h)$ be a model where A is satisfied. Let Ψ be the set of all subformulas of A. Take $\mathbf{m}' = (F', h)$ as in Proposition 16.129. Then F' is an $\mathbf{L}$-frame by Proposition 16.129, and $\mathbf{m}'$ satisfies A by the Filtration Lemma 16.127. □

The same method can be extended to many other logics. The main result is stated in Theorem 16.133 below, but now let us consider some particular cases.

Proposition 16.131. *Let* $\mathbf{L}$ *be an* n*-modal logic axiomatised by a finite set of formulas of the following kinds:*

- $\Box_i p \to \Box_i^m p$ $(m \leq 2)$;

- $\Diamond_i\Box_i p \to p$;
- *closed formulas.*

Then the logic $[\mathbf{L}, \mathbf{S5}]$ *has the f.m.p. (Recall also that* $[\mathbf{L}, \mathbf{S5}] = \mathbf{L} \times \mathbf{S5}$ *by Theorem 16.76.)*

Proof Let $\Lambda = [\mathbf{L}, \mathbf{S5}]$. By completeness, every Λ-consistent formula A is satisfied in some $\mathbf{L}$-frame. Let $\mathbf{m} = (F, h)$ be a model where A is satisfied. Let Ψ be the set of all subformulas of A and of all subformulas of closed axioms of $\mathbf{L}$. Take $\mathbf{m}' = (F', h')$ as in Proposition 16.129. Then F' is a $[\mathbf{K}_n, \mathbf{S5}]$-frame by Proposition 16.129, and $\mathbf{m}'$ satisfies A by the Filtration Lemma 16.127. Also note the following.

- Closed axioms of $\mathbf{L}$ hold in F' by the Filtration Lemma, since they are all in Ψ.
- If $(\Box_i p \to p) \in \mathbf{L}$ then $F' \models (\Box_i p \to p)$. This happens because the reflexivity of R_i implies the reflexivity of R'_i.
- If $(\Diamond_i\Box_i p \to p) \in \mathbf{L}$ then $F' \models (\Diamond_i\Box_i p \to p)$. This is because the symmetry of R_i implies the symmetry of R'_i.

However transitivity is not necessarily preserved by the least filtration. To get it, we can take transitive closures. More precisely, if $(\Box_i p \to \Box_i^2 p) \in \mathbf{L}$ we put R_i^* to be the transitive closure of R'_i; otherwise let $R_i^* = R'_i$. Let $\mathbf{m}^* = (W', R_1^*, ..., R_n^*, S', h')$, and let F^* be the frame of $\mathbf{m}^*$.

Now the following holds.

- $\mathbf{m}^*$ is a filtration of $\mathbf{m}$ through $\Psi, \approx$. This is proved in a standard way, cf. [Gabbay, 1972] or [Shehtman, 1993].
- $F^* \models \mathbf{L}$.
  In fact, symmetry and reflexivity are preserved for transitive closures. Closed axioms of $\mathbf{L}$ still hold in F^* by the Filtration Lemma.
- F^* is $(n, 1)$-commutative.
  In fact, suppose $(\Box_i p \to \Box_i^2 p) \in \mathbf{L}$. By Proposition 16.129, S' commutes with R'_i, and thus with any $(R'_i)^k$. Therefore S' commutes with R_i^* (which is the union of all $(R'_i)^k$).

So finally we obtain that F^* is a Λ-frame. □

Definition 16.132. A *QTC-logic* is a modal logic axiomatised by a finite set of formulas of the following kinds:

- $\Box_i p \to \Box_i^j p \ (j \geq 0, 1 \leq i \leq n)$;
- $\Diamond_i\Box_j p \to p$;
- closed formulas.

Evidently, QTC-logics are PTC-logics, so they are product-matching, by Theorem 16.76.

Theorem 16.133. *If* $\mathbf{L}$ *is a QTC-logic then the logic* $[\mathbf{L}, \mathbf{S5}]$ *has the f.m.p.*

The proof goes along the same lines as in Proposition 16.131, but instead of transitive closures one should consider pseudo-transitive closures. The basic idea for the monomodal case can be found in [Gabbay, 1972]. More precisely, if $(\Box_i p \to \Box_i^j p)$ for $j \in J_i$, J_i finite, we take $R_i^* = \bigcup\{(R_i')^k | k \in X_i.\}$, where

$$X_i = \{1 + \sum_{j \in J_i} (j-1)x_j | \, \forall j \in J_i, x_j \geq 0\}.$$

16.14 Applications to Predicate Logics

Some of the propositional logics studied above can be easily interpreted as one-variable fragments of modal and intermediate predicate logics. The idea of this interpretation is very well known; quantifiers are replaced by modal operators. Still let us describe it in more details.

Let us consider modal logics first. We use the first-order language $\mathbb{L}_n\mathbb{P}1$ containing

- classical propositional connectives;
- a countable set of monadic predicate letters $PL^1 = \{P_1, P_2, ...\}$;
- a single individual variable x;
- quantifiers $\forall, \exists$;
- modal operators $\Box_1, ..., \Box_n$.

For this language we use *Kripke semantics with constant domains* defined as follows.

Definition 16.134. An *n-modal predicate Kripke frame with a constant domain* is a pair (F, D) consisting of an n-modal Kripke frame $F = (W, R_1, ..., R_n)$ and a non-empty set D. Elements of W are called *worlds*, and elements of D are called *individuals*. A *Kripke model over* (F, D) is a triple $\mathbf{m} = (F, D, h)$, where $h : PL^1 \to (2^{D \times W})$ is a *valuation*.

A *D-valued formula* is a formula where all free occurrences of x are replaced by some individual $a \in D$. By induction we can define the truth of a D-valued formula at a world w of the model $\mathbf{m}$:

- $\mathbf{m}, w, \not\models \bot$;
- $\mathbf{m}, w \models P_i(a) \Leftrightarrow (w, a) \in h(P_i)$;
- $\mathbf{m}, w \models A \to B \Leftrightarrow \mathbf{m}, w, \not\models A$ and $\mathbf{m}, w \models B$;
- $\mathbf{m}, w \models \Box_i A \Leftrightarrow \forall v (w R_i v \Rightarrow \mathbf{m}, v \models A)$;
- $\mathbf{m}, w \models \forall x A(x) \Leftrightarrow \forall a \, \mathbf{m}, w \models A(a)$.

By definition a (D-valued) formula A is *true in* $\mathbf{m}$ (notation: $\mathbf{m} \models A$) iff $\forall w, \mathbf{m}, w \models A$;

A is *valid in* (F, D) (notation: $(F, D) \models A$) iff it is true in any Kripke model over (F, D).

Definition 16.135. A *modal predicate logic* (m.p.l.) (in the language $\mathbb{L}_n\mathbb{P}1$) is a set $\mathbf{L}$ of $\mathbb{L}_n\mathbb{P}1$-formulas such that

- $\mathbf{L}$ contains all substitution instances of theorems of $\mathbf{K}_n$;
- $\mathbf{L}$ contains classical axioms:
  $\forall x(A \to B) \to (\forall x A \to \forall x B)$,
  $\forall x A \to A$,

$\exists xA \equiv \sim \forall x \sim A$,
$A \to \forall xA$ (if A is a closed formula);

- $\mathbf{L}$ is closed under the rules
(modus ponens) $A, (A \to B) \in \mathbf{L} \Rightarrow B \in \mathbf{L}$;
(necessitation) $A \in \mathbf{L} \Rightarrow \Box_i A \in \mathbf{L}$;
(universalisation) $A \in \mathbf{L} \Rightarrow \forall xA \in \mathbf{L}$;
- $\mathbf{L}$ is closed under the substitution rule:
$A \in \mathbf{L} \Rightarrow (B/P_i(x))A \in \mathbf{L}$,
where $(B/P_i(x))A$ is obtained from A by replacing each occurrence of $P_i(x)$ by B.

Remark 16.136. This definition is somewhat non-standard; another way to express things is first to define full predicate logics (with countably many individual variables) and then restrict them to one-variable formulas. Full predicate logics are defined as extensions of both classical predicate logic and $\mathbf{K}_n$ and closed under the rules cited above (however we should be rather careful when defining substitutions). See e.g. [Shehtman and Skvortsov, 1990]. When the number of variables is more than one, the second way becomes much easier than the first.

The following is well known and is proved rather easily.

Proposition 16.137 (Soundness theorem). *The set of all $\mathbb{L}_n\mathbb{P}1$-formulas valid in a frame (F, D) is a modal predicate logic.*

The above logic is called the *one-variable logic of the predicate frame* (F, D) and is denoted by $\mathbf{L}(F, D)1$. The *logic of a class $\mathcal{B}$* of predicate Kripke frames is defined as

$$\mathbf{L}(\mathcal{B})1 = \bigcap\{\mathbf{L}(F, D)1 | (F, D) \in \mathcal{B}\}.$$

Logics of the latter kind are called *Kripke-complete (with constant domains)*. A special case is the following: given a class $\mathcal{C}$ of (propositional) Kripke frames, take $\mathcal{CP} = \{(F, D) | F \in \mathcal{C}\}$ (we call it a *predicate version of* $\mathcal{C}$).

Definition 16.138. Let us define a 'Wajsberg-style' translation $A \mapsto A^*$ from $\mathbb{L}_n\mathbb{P}1$-formulas to $\mathbb{L}_{n+1}$-formulas. To get A^* we replace occurrences of $P_i(x)$ by p_i, and occurrences of $\forall x$ by $\Box_{n+1}$. Obviously this translation is a bijection.

From the definitions we obtain

Proposition 16.139. *Let (F, D) be a predicate Kripke frame, $\bullet_D = (D, D \times D)$ be the cluster on the set D. Then for any $\mathbb{L}_n\mathbb{P}1$-formula A, $(F, D) \models A \Leftrightarrow F \times \bullet_D \models A^*$.*

Theorem 16.140.

1. *If a class $\mathcal{C}$ of n-modal Kripke frames is R-elementary then the logic of its predicate version $\mathbf{L}(\mathcal{CP})1$ is RE.*
2. *If a class $\mathcal{C}$ of n-modal Kripke frames is Δ-elementary then the logic of its predicate version $\mathbf{L}(\mathcal{CP})1$ has the c.f.p.*

(We should notice here that a predicate Kripke frame (F, D) is called countable if both F and D are countable.)

Proof By Proposition 16.139, $A \in \mathbf{L}(\mathcal{CP})1 \Leftrightarrow A^* \in \mathbf{L}(\mathcal{C} \times \mathcal{C}lus)$. Now (1) follows from Theorem 16.46.

The statement (2) also follows from Theorem 16.46, if we notice that (F, D) is countable whenever $F \times \bullet_D$ is countable. □

Definition 16.141. Let λ be an n-modal propositional logic. The *one-variable CD-predicate version* of λ is the predicate logic $CD(\lambda) = \mathbf{L}(\mathcal{V}(\lambda)\mathbb{P})1$ i.e. the logic of all frames (F, D) where λ is valid in F. The *one-variable predicate version* of λ is the smallest n-modal predicate logic containing λ; it is denoted by $Q\lambda 1$. The *one-variable Barcan version* of λ is the logic

$$QB(\lambda) = Q\lambda 1 + \forall x \Box P(x) \to \Box \forall x P(x),$$

i.e. the smallest extension of $Q\lambda 1$ containing the *Barcan formula*
$\forall x \Box P(x) \to \Box \forall x P(x)$.

Note that the Barcan formula corresponds exactly to the commutativity condition by the *-translation. So our results allow us in many cases to axiomatise CD-predicate versions explicitly and prove their decidability.

Theorem 16.142. *(1) If λ is a PTC-logic then $CD(\lambda) = QB(\lambda)$.*
(2) If λ is a QTC-logic described in Definition 16.132 then $CD(\lambda)$ is decidable.

Proof By Proposition 16.139, $A \in CD(\lambda) \Leftrightarrow A^* \in \mathbf{L}(\mathcal{V}(\lambda) \times \mathcal{C}lus)$. On the other hand, $A \in QB(\lambda) \Leftrightarrow A^* \in [\lambda, \mathbf{S5}]$, because, as one can see,
$[\lambda, \mathbf{S5}] = \lambda * \mathbf{S5} + \Box_{n+1}\Box_i p \to \Box_i \Box_{n+1} p$.

Finally, $[\lambda, \mathbf{S5}] = \mathbf{L}(\mathcal{V}(\lambda) \times \mathcal{C}lus)$ follows from Theorem 16.67 and Lemma 16.32. The same argument together with Definition 16.132 proves (2). □

Now let us recall briefly the definition of a superintuitionistic predicate logic. Consider the intuitionistic first-order language with one variable $\mathbb{LIP}1$. It includes

- propositional connectives $\wedge, \vee, \to, \bot$;
- a countable set of monadic predicate letters $PL^1 = \{P_1, P_2, ...\}$;
- a single individual variable x;
- quantifiers $\forall, \exists$.

A *superintuitionistic predicate logic* (in the language $\mathbb{LIP}1$) is a set $\mathbf{L}$ of $\mathbb{LIP}$-formulas such that

- $\mathbf{L}$ contains all intuitionistic predicate tautologies.
- $\mathbf{L}$ contains the axioms:
  - $\forall x A \to A, A \to \exists x A$,
  - $\forall x(A \to B) \to (\forall x A \to \forall x B)$,
  - $\forall x(A \to B) \to (\exists x A \to \exists x B)$,
  - $B \to \forall x B$ (if B is a closed formula);
  - $\exists x B \to B$ (if B is a closed formula).
- $\mathbf{L}$ is closed under the rules:
  (modus ponens), (universalisation)

- $\mathbf{L}$ is closed under the substitution rule.

Recall also that every intuitionistic formula A corresponds to a modal formula A^T by *Gödel–Tarski translation*:

- $A^T = \Box A$ for A atomic;
- $(A \wedge B)^T = A^T \wedge B^T$;
- $(A \vee B)^T = A^T \vee B^T$;
- $(A \to B)^T = \Box(A^T \to B^T)$;
- $(\exists x A)^T = \exists x A^T$;
- $(\forall x A)^T = \Box \forall x A^T$.

An intuitionistic formula A is *valid* in a predicate Kripke frame (F, D) (notation: $(F, D) \models A$) if the formula A^T is valid.

Proposition 16.143 (Soundness theorem). *If F is an* $\mathbf{S4}$*-frame, the set of all* $\mathbb{LIP}1$*-formulas valid in a frame (F, D) is a superintuitionistic predicate logic.*

We denote this logic by $\mathbf{LI}(F, D)1$. For a class of frames $\mathcal{B}$, let $\mathbf{LI}(\mathcal{B})1 = \bigcap\{\mathbf{LI}(F, D)1 | (F, D) \in \mathcal{B}\}$. By definition, we have: $A \in \mathbf{LI}(\mathcal{B})1 \Leftrightarrow A^T \in \mathbf{L}(\mathcal{B})1$. Hence using Theorem 16.142 we obtain:

Theorem 16.144. *The one-variable superintuitionistic logic of the class of all predicate* $\mathbf{S4}$*-frames with constant domains is decidable.*

It is well known (cf. [Gabbay, 1981c]) that this logic is axiomatised by intuitionistic predicate calculus **IPC** with the additional schema:

(CDP) $\forall x(A \vee B) \to \forall x A \vee B$ (where B is a closed formula).

16.15 Case Study: Transition Systems

This section gives an idea of how products of modal logics can be applied. We choose the topic of transition systems, a well-established area in theoretical computer science.

Definition 16.145. (R-based connectives; cf. Definition 16.52.) As in Section 16.5, consider a propositional language $\mathbb{L}$ with some additional connectives. Let $\#$ be an n-place connective from $\mathbb{L}$. Let $F = (S, R, a, h)$ be a Kripke frame with relation $R \subseteq S^2$. The connective $\#$ is said to be *semantically based* on R with a formula $\alpha(t, S_1, ..., S_n, R, a)$ if α is a (possibly high order) formula in the language of (S, R, a) containing the binary predicate R, the constant a and the unary predicates $S_1, ..., S_n$ ranging over subsets of S and the individual variable t ranging over S, if the following function δ defines a neighbourhood frame for $\#$ with the set of possible worlds S:

$\delta(U_1, ..., U_n) = \{t \in S | (S, R, U_1, ..., U_n) \models \alpha(t, S_1, ..., S_n, R, a)\}$.

Now let us consider some particular connectives of this kind. Let $F_i = (S_i, R_i, a_i)$, $i = 0, 1$, be two modal frames for connectives which are R_i-based respectively. Consider the product

$F_0 \times F_1 = (S_0 \times S_1, R_0^\vee, R_1^\vee, (a_0, a_1))$.

Let $\alpha_1, \beta_1 : S_1 \to S_0$ be two functions, such that

$\forall x \in S_1, \alpha_1(x) R_0 \beta_1(x)$.

Then the quadruple $(F_0, F_1, \alpha_1, \beta_1)$ is called a *T-product of* F_0 and F_1 and denoted by $[F_0, F_1, \alpha_1, \beta_1]$.

Note that each $\tau \subseteq S_1$ corresponds to a relation $R_\tau \subseteq S_0^2$ defined by
$R_\tau = \{(x,y)|\exists z \in \tau(x = \alpha_1(z) \wedge y = \beta_1(z))\}$.

Clearly, $R_\tau \subseteq R_0$, and thus the relation R_τ gives rise to a 'submodality' of R_0, which we denote by $\Box_\tau$.

Definition 16.146. A T-product $[F_0, F_1, \alpha_1, \beta_1]$ is called a *transition system* if the functions α_1, β_1 satisfy the following:
$xR_1y \Rightarrow \beta_1(x) = \alpha_1(y)$.

This condition allows us to view x as naming a transition $(\alpha_1(x), \beta_1(x))$ from the state $\alpha_1(x)$ to the state $\beta_1(x)$ and view y as a possible further transition from $\beta_1(x) = \alpha_1(y)$ to $\beta_1(y)$.

Definition 16.147. Let $\mathcal{K}_i = \{(S_n^i, R_n^i, a_n^i, h_n^i)|n = 1, 2, ...\}$, $i = 0, 1$ be two classes of Kripke models based on frames $F_n^i = (S_n^i, R_n^i)$. The *T-product* (resp. *transition product*) of $\mathcal{K}_0$ and $\mathcal{K}_1$ are all T-product (resp. transition system) models of the form $([F_0^{n_0}, F_1^{n_1}, \alpha_1, \beta_1], h)$, where h is a two-dimensional assignment (as was defined in Section 16.1.1).

Definition 16.148. Let $([F_0^{n_0}, F_1^{n_1}, \alpha_1, \beta_1], h)$ be a transition product model as in Definition 16.147. Then for every $s \in S_0, c \in S_1$ we define the value $(s,c) \models A$:

- $(s,c) \models q$ iff $(s,c) \in h(q)$
- $(s,c) \models \Box_0 A$ iff $\exists t \in R_0(s)(t,c) \models A$
- $(s,c) \models \Box_1 A$ iff $\exists d \in R_1(c)(s,d) \models A$
- $(s,c) \models \Box_\tau A$ iff $\exists t \in R_\tau(s)(t,c) \models A$
- $(s,c) \models$ source (A) iff $\exists d(\alpha_1(d) = s \wedge (s,d) \models A)$
- $(s,c) \models$ target (A) iff $\exists d(\beta_1(d) = s \wedge (s,d) \models A)$
- $(s,c) \models$ In (A) iff $(\beta_1(c), c) \models A$
- $(s,c) \models$ Out (A) iff $(\alpha_1(c), c) \models A$.

The items for classical connectives are standard.

Let A be any formula of the language of F_1. Let τ_A be the set $\{s \in S_1 | s \models A\}$. As above, this set gives rise to a relation R_A and a modality $\Box_A$.

As a practical example of T-products of modal logics we choose the case of labelled transition systems, and their associated modal logics. These involve the Henessy–Milner logic and the Dicky logic. See [Arnold, 1994] for details.

Definition 16.149. A *labelled transition system* has the form

$$\mathcal{A} = (S, T, A, \alpha, \beta, \lambda, X_1, ..., X_n, Y_1, ..., Y_m),$$

where S is the set of states, T is a set of transitions, and α, β are functions from T to S.

The pair $(\alpha(t), \beta(t))$ gives the source state and the target state of the transition t. X_i are state parameter names and Y_j are transition parameter names. With each X_i we associate a subset S_i of S, and similarly with each Y_j we associate $T_j \subseteq T$. A is a set of labels and λ is a function from T to A assigning labels to transitions. A label $a \in A$ can be thought as an action and the transitions $t \in T_a = \{t|\lambda(t) = a\}$ are perceived as the transitions which can execute the action a moving the system from the state $\alpha(t)$ to the state $\beta(t)$.

A transition system can be turned into a multimodal Kripke model as follows. The set of possible worlds is S. For each $a \in A$ we have a relation $R_a = \{(\alpha(t), \beta(t))|\lambda(t) = a\} \subseteq S^2$. The language has modalities $\Diamond_a$ for $a \in A$ and atoms $Q = \{q_1, ..., q_n\}$. The assignment h is defined for Q by $h(s, q_i) = 1$ iff $s \in S_i$.

One can define additional modalities $\Diamond_{T_j}$ with relations $R_{T_j} = \{(\alpha(t), \beta(t))|t \in T_j\}, j = 1, ..., m$. The corresponding modal logic is a variation on Hennessy-Milner logic.

The logic can be further refined by adding Boolean connectives to the set of atoms $T_1, ..., T_m$. For each Boolean expression $B(T_1, ...T_m)$ we can define a relation $R_B = \{(\alpha(x), \beta(x))|x \in B\}$. For each B let q_B be a propositional atomic letter associated with B.

Furthermore, we can consider *paths*. A path is a sequence of transitions $(t_1, ..., t_n)$ in which the target of the previous transition is the source of the next. We can concatenate two paths c and d into $c * d$, provided the source of the first transition in d is the target of the last transition in c.

We can thus assume that all transitions are paths and similarly in the definition of associated Kripke models we can define $\Diamond_c$ and R_c for paths. Truth values of formulas can be defined at pairs (s, c), with a state s and a path c as follows:

- $(s, c) \models q_B$ iff $c \in B$
- $(s, c) \models q_i$ iff $s \in S_i$
- $(s, c) \models \Diamond_B A$ iff $\exists s_1(sR_B s_1 \wedge (s_1, c) \models A)$
- $(s, c) \models$ Next A iff $c = t * d \wedge (s, d) \models A$, for some (unique) transition t
- $(s, c) \models X$ until Y iff $(s, c) \models Y$ or there exist transitions $t_1, \ldots, t_n$ and a path d such that $(s, d) \models Y$ and $(s, t_i * \ldots * t_n * d) \models X$ for any i.

The items for the classical connectives are standard.

So it is clear from the above definition that we obtain a product of the $\{\Diamond_B, q_i\}$-logic and $\{$Next, Until, $q_B\}$-logic.

Also it is possible to add the Dicky connectives to our language and connect one logic to another. So we add unary connectives *source, target, in, out* with the following truth definitions:

$(s, c) \models$ *source* A iff $\exists d(\alpha(d) = s \wedge (s, d) \models A)$,
$(s, c) \models$ *target* A iff $\exists d(\beta(d) = s \wedge (s, d) \models A)$,
$(s, c) \models$ *in* A iff $(\beta(c), c) \models A$,
$(s, c) \models$ *out* A iff $(\alpha(c), c) \models A$.

Now one can check that the formula *source* $(q_B \wedge$ *in* $A)$ is equivalent to $\Diamond_B A$ in any model.

16.16 Conclusion: Further Results and Questions

In this section we state some more results on products. Some of them were already proved by their authors, others are planned for the sequel of [Gabbay and Shehtman, 1998]. We also formulate several open questions. The reader is referred to [Gabbay and Shehtman, 1998] for a larger overview.

Theorem 16.150 (Spaan, 1993). *The logic* $\mathbf{L}((\omega, <) \times (\omega, <))$ *is* Π^1_1*-complete.*

Note that $\mathbf{L}(\omega, <)$ is a well known logic; it is finitely axiomatisable and decidable [Segerberg, 1970]. So this theorem shows that there is no general procedure for obtaining $\mathbf{L}(F_1 \times F_2)$ from $\mathbf{L}(F_1), \mathbf{L}(F_2)$. This prompts the following question.

Question 16.151. Does finite axiomatisability (decidability) transfer from logics to their products?

A trivial example of a property which transfers is tabularity. Recall that a logic is called *tabular* if it is a logic of some finite frame. So we have

Theorem 16.152. *The product of two tabular logics is tabular.*

Proof The proof follows from the observation that in this case the product is obtained as a logic of a finite set of finite frames. □

The following theorem shows that in general the f.m.p. does not transfer. Recall that **LIN** is the temporal logic of all irreflexive linear orders [Segerberg, 1970].

Theorem 16.153 (Reynolds, 1996). *The logic* **LIN** $\times$**S5** *is decidable, but lacks the f.m.p.*

Corollary 16.154. *The logics* $\mathbf{K4.3} \times \mathbf{S5}, \mathbf{S4.3} \times \mathbf{S5}$ *are decidable.*

Remembering Proposition 16.139, we obtain:

Corollary 16.155. *The one-variable modal predicate logics* $CD(\mathbf{K4.3}), CD(\mathbf{S4.3})$ *are decidable.*

Then by applying the Gödel–Tarski translation, we have

Corollary 16.156. *The one-variable intermediate predicate logic* $CD(\mathbf{LC})$ *is decidable.*

(Here $\mathbf{LC}$ is the well known intermediate propositional logic of all linear orders.)

Some further results on the f.m.p. for products can be proved by the normal forms method:

Theorem 16.157. *The product of any two logics of the types* $\mathbf{K}_n, \mathbf{D}_n, \mathbf{T}_n, \mathbf{K}_n * \mathbf{D}_m, \mathbf{K}_n * \mathbf{T}_m, \mathbf{D}_n * \mathbf{T}_m, \mathbf{K}_n * \mathbf{D}_m * \mathbf{T}_p$ *has the f.m.p.*

Let us mention also some very new results on complexity of products.

Theorem 16.158. *[Marx, 1997a, 1997b] The satisfiability problem in* $\mathbf{S5}^2$ *is NEXPTIME-complete.*

Theorem 16.159. *[Marx, 1997a, 1997b] The satisfiability problems in* $\mathbf{K}^2, \mathbf{D}^2, \mathbf{T}^2$ *are NEXPTIME-hard.*

For the latter logics, the exact bound may be much higher. Note that the upper bound we get from the normal forms method is a priori non-Kalmar elementary! It might be an interesting problem to clarify the situation here.

There are also some open questions concerning completeness and f.m.p.

Question 16.160. Consider the following classes of frames:

$\mathcal{F}$: the class of all finite 1-frames, $\mathcal{FT}$: the class of all finite 1-trees. Also for a class of frames $\mathcal{C}$ let $\mathcal{C}^2 = \{F \times F | F \in \mathcal{C}\}$. What are the following logics:

$\mathbf{L}(\mathcal{F} \times \mathcal{F}), \mathbf{L}(\mathcal{FT} \times \mathcal{FT}), \mathbf{L}(\mathcal{F}^2), \mathbf{L}(\mathcal{FT}^2)$?

Finally, some remarks about higher dimensions. The results from Section 16.5 imply recursive axiomatisability for many higher dimensional logics. But probably there are very few cases, when these logics are decidable. For example, we have the following result.

Theorem 16.161 (Maddux, 1980; Henkin *et al.*, 1985). *The logic* $\mathbf{S5} \times \mathbf{S5} \times \mathbf{S5}$ *is undecidable and non-finitely axiomatisable.*

On the other hand there are rather trivial case when the logics are much better.

Theorem 16.162 (Gabbay and Shehtman, 1998). *Let* $\mathbf{SL} = \mathbf{K} + \Diamond p \equiv \Box p$ *(tomorrow logic). Then for any* n, $\mathbf{SL}^n$ *is finitely axiomatisable and decidable; the satisfiability problem is NP-complete.*

Question 16.163. Is the logic $\mathbf{K} \times \mathbf{K} \times \mathbf{K}$ finitely axiomatisable? Is it decidable?

A related topic is the study of two-variable fragments of modal and intermediate predicate logics. We refer the reader to our paper [Gabbay and Shehtman, 1993] and to the questions posed there.

17

FIBRING INTUITIONISTIC LOGIC PROGRAMS

17.1 Introduction

The purpose of this chapter is to give a more practical example of fibring of a programming nature. We apply the technique of fibring to intuitionistic logic programs [Gabbay and Reyle, 1984], i.e., an extension of logic programming where goals can contain implications. The operational semantics of a goal $\psi \leftarrow \phi$ is that of making a hypothetical assumption: $\psi \leftarrow \phi$ succeeds from a program P if ψ succeeds under the assumption ϕ, i.e., if ψ succeeds from P augmented by an additional database D containing ϕ. Deductions in this logic programming language are sound and complete w.r.t. intuitionistic logic [Gabbay, 1985b]. Intuitionistic logic programming is well suited to express meta-level constructs in the object language, and to perform hypothetical reasoning [Gabbay and Reyle, 1984; McCarty, 1988a; McCarty, 1988b]. The particular intuitionistic logic programming language we are using is QN-PROLOG [Gabbay, 1985b] the first-order extension of N-PROLOG [Gabbay and Reyle, 1984] (see Section 17.2).

Concepts for modularising and structuring logic programming languages have been extensively studied over the last few years: algebraic approaches [O'Keefe, 1985; Mancarella and Pedreschi, 1988]; methods based on second-order logic [Nait Abdallah, 1986; Chen, 1987]; the definition of meta-predicates [Bowen and Kowalski, 1982; Brogi *et al.*, 1992]; and most recently, approaches using multi-modal logics [Farinas del Cerro and Herzig, 1992; Baldoni *et al.*, 1993], and a combination of modal logic and implication goals [Giordano and Martelli, 1994]. Extending logic programming by implication goals is a method that has often been used to introduce a module concept [Miller, 1986; Giordano *et al.*, 1988; Moscowitz and Shapiro, 1991; Giordano *et al.*, 1992]. In this context, the implicant ϕ of an implication goal $\psi \leftarrow \phi$ is not interpreted to be an assumption but to be a definition of a program part (called a *module* or a *block*) that can only be used locally. That way, by using slightly different semantics for implication goals, different module concepts can be defined (see Section 17.3.6).

Although we use a language with implication goals, the approach to structuring logic programs presented in this chapter is completely different. Instead of using the database of assumptions to model the (currently available) modules, we use it to model the information passed on to a (sub-)module. Here, modules are represented by separate programs that are combined into a single system using fibring. This flexible approach can be instantiated to resemble many of the different module concepts described in the literature.

We provide semantics for the combined (fibred) program by defining both fibred procedural and fibred model semantics. Since we are using intuitionistic logic programs, the models of the original programs are Kripke structures. The models of the fibred program are constructed by fibring these Kripke structures.

Logic programming languages are very well suited for implementing deduction calculi, because the constructs widely used in logic systems like terms and atoms do not have to be implemented; unification is provided; and non-deterministic procedures can easily be specified, which turns out to be very useful since many proof procedures are based on intrinsically non-deterministic calculi. We show that in particular *intuitionistic* logic programming is useful for implementing calculi and algorithms used in automated deduction, and that fibring can be used to combine these implementations in a useful way. As an example, we show how an implementation of semantic tableaux can be fibred with an equality reasoning module to construct a proof procedure for first-order logic with equality (Sections 17.2.6 and 17.3.5).

Finally, we discuss in Section 17.4 the possibility of applying the method of *incremental theory reasoning* [Beckert and Pape, 1996] to both fibred and non-fibred intuitionistic logic programs; this technique can be employed to re-use information that has been derived during the computations for a goal ψ when subsequently a goal of the form $\psi \leftarrow \phi$ is to be evaluated.

17.2 Intuitionistic Logic Programming

17.2.1 *Overview*

For fibring, it is important that the state a computation has reached is more explicit than it is in classical PROLOG (the state can be thought to be identical to the knowledge that the procedure has accumulated or the data it has computed up to a certain point). This allows us to associate the state of the computation with a world in the structures defining model semantics. In the models of classical PROLOG, there are no worlds and there is nothing corresponding to the states of the computation. In intuitionistic PROLOG, the state a computation has reached is more explicit: The database of assumptions that have been made represents information that the procedure has derived, and that from that point on is part of its knowledge.

If, for example, given the goal p, the rule $p \leftarrow (q \leftarrow r)$ is applied, then r becomes a part of the database and has, thus, the status of a known fact during the computations for the new goal q. This is reflected in the definition of models for intuitionistic programs where the sets of known facts (assumptions) coincide with the worlds of the model. Note that the 'knowledge' of a state w does not include the answers to goals evaluated in w (otherwise, the answer to $provable(\phi)$ could be used in the proof of ϕ).

In classical PROLOG, if a *subroutine* is called, all relevant information has to be provided as part of the goal that represents the call. Contrary to that, intuitionistic PROLOG allows one to encode into the assumption database all general knowledge that is not closely related to the subroutine call, but that nevertheless can be used by the subroutine. The advantages are similar to that encountered in theorem proving when a theorem prover is presented with a theorem ψ to be proven (the goal) and a separate set Ax of axioms (the database) instead of one big implication $Ax \rightarrow \psi$.

We do not allow negation—although it is possible to apply the techniques presented in this chapter to logic programs with negation using the procedural and model semantics from [Giordano and Olivetti, 1994]. With negation all our definitions would become much more complicated without providing more insight; in addition, the monotonicity property would be lost which is needed for applying the incremental reasoning technique (see Section 17.4). On the other hand, it may be possible to use fibring of intuitionistic programs *with negation* to handle inconsistency, since the fibring function can turn an inconsistent database into a consistent one (cf. [Gabbay *et al.*, 1995]).

17.2.2 *Syntax of QN*-PROLOG

We use the intuitionistic logic programming language QN-PROLOG [Gabbay, 1985b], which is the quantificational extension of N-PROLOG [Gabbay and Reyle, 1984].

A QN-PROLOG language $\mathbb{L}$ is built up from countable sets $\mathcal{P}$ of *predicate symbols*, $\mathcal{F}$ of *function symbols* (*constants* are functions of arity 0), and $\mathcal{V}$ of *variables*. $\mathbb{L}$-*terms* and $\mathbb{L}$-*atoms* are constructed from these in the usual manner. We use the connectives $\wedge$ (conjunction) and $\leftarrow$ (right-to-left implication).

The following definitions deviate from those given in [Gabbay, 1985b]. The differences are, however, only of a syntactic nature and do not affect the semantics itself; they turn out to be useful later for fibring programs. Instead of just adding assumptions to the program, we keep a separate database consisting of all assumptions made.

Definition 17.1. Given a QN-language $\mathbb{L}$, the set of $\mathbb{L}$-*clauses* is inductively defined as follows:

1. Positive atoms are $\mathbb{L}$-clauses.
2. If $\phi_1, \ldots, \phi_n$ are $\mathbb{L}$-clauses ($n \geq 1$) and ψ is a positive $\mathbb{L}$-atom, then

$$\psi \leftarrow \phi_1 \wedge \ldots \wedge \phi_n$$

is an $\mathbb{L}$-clause.

$\mathbb{L}$-*databases* and $\mathbb{L}$-*programs* are finite sets of $\mathbb{L}$-clauses. $\mathbb{L}$-*goals* are (finite) conjunctions of $\mathbb{L}$-clauses. An $\mathbb{L}$-*query* $D?\psi$ consists of an $\mathbb{L}$-database D and an $\mathbb{L}$-goal ψ. An $\mathbb{L}$-*substitution* is a substitution that instantiates variables from $\mathbb{L}$ with terms from $\mathbb{L}$.

The variables in $\mathbb{L}$-programs are implicitly universally quantified, they do each occur in one clause only; the variables in $\mathbb{L}$-databases, $\mathbb{L}$-goals, and $\mathbb{L}$-queries are implicitly existentially quantified. The database and the goal that are part of a query can have variables in common.

Example 17.2. If $\mathbb{L}$ consists of $\mathcal{P} = \{p, q\}$, $\mathcal{F} = \{a, f\}$, and $\mathcal{V} = \{x\}$, then

$$\begin{array}{l} p(a) \\ p(x) \leftarrow q(f(x)) \\ p(a) \leftarrow (r(a) \leftarrow q(a) \wedge q(f(a))) \end{array}$$

are $\mathbb{L}$-clauses. $(p(a) \leftarrow r(a)) \leftarrow q(a)$ is *not* an $\mathbb{L}$-clause.

17.2.3 *Procedural semantics of QN*-PROLOG

We define the notion of *computation tree*, and based on this the notions of *refutation*, *finitely terminating queries*, and *finitely failing queries*. Usually, in the literature on intuitionistic logic programs, it is only defined what it means for a query to be successful w.r.t. a program. Here, in addition, we define the notion of *answer substitutions*.

Variables in a QN-PROLOG program are always understood to be universal; they are renamed before a clause is used. Variables in the (additional) database and the goals are understood to be existential; they are not renamed when a clause from the database is used.

Definition 17.3. A *selection function* R for a QN-language $\mathbb{L}$ selects from each (finite) non-empty sequence L of $\mathbb{L}$-queries a query $R(L) \in \mathbb{L}$.

A selection function imposes a certain order on queries, in which they are evaluated. The choice of a certain selection function does not, however, affect the semantics of a program.

Definition 17.4. Given a QN-language $\mathbb{L}$, a selection function R for $\mathbb{L}$, and an $\mathbb{L}$-program P of $\mathbb{L}$, the *computation tree* T for an $\mathbb{L}$-goal ψ is defined as follows:

- T is a tree whose nodes are labelled with finite sequences of $\mathbb{L}$-queries and whose edges are labelled with $\mathbb{L}$-substitutions; if not stated otherwise, edges are labelled with the empty substitution.
- The root node of T is labelled with the singleton sequence $\langle \varnothing ? \psi \rangle$.
- Nodes labelled with the empty list of queries are leaves.
- For all nodes N in T, where N is labelled with

$$L = L' \circ \langle D?\psi \rangle \circ L'',$$

  and $R(L) = D?\psi$, the following holds:

  1. *Rule for conjunctions:* If $\psi = \rho_1 \wedge \ldots \wedge \rho_k$, $k \geq 2$, then N has one successor N' that is labelled with

  $$L' \circ \langle D?\rho_1, \ldots, D?\rho_k \rangle \circ L''.$$

  2. *Rule for implications:* If $\psi = (\phi \leftarrow \rho_1 \wedge \ldots \wedge \rho_k)$, $k \geq 1$, then N has one successor N' that is labelled with

  $$L' \circ \langle D \cup \{\rho_1, \ldots, \rho_k\}?\phi \rangle \circ L''.$$

  3. *Rule for atoms:* If ψ is an atom, then, for each clause ς that
     (a) i. is an element of D, or
        ii. is the renaming of a clause ς' in P, i.e., $\varsigma = \varsigma'\nu$ where the variable renaming ν replaces the variables in ς' by new variables not occurring in P nor in the list L of queries;
     (b) and whose head κ is unifiable with ψ with a most general unifier μ,

     the node N has a successor N' that is labelled with

  $$L'\mu \circ L''\mu$$

     in case $\varsigma = \kappa$ is an atom, and with

  $$L'\mu \circ \langle D\mu?\xi\mu \rangle \circ L''\mu$$

     in case $\varsigma = (\kappa \leftarrow \xi)$; the edge between N and N' is labelled with μ.

A maximal path in T ending in a leaf labelled with the empty list is a *refutation* for ψ.

A substitution σ is an *answer substitution* for ψ if it is the composition of all substitutions labelling edges in a refutation for ψ.

There are three kinds of paths in a computation tree:

- paths that end in a leaf labelled with the empty list (success nodes),
- paths that end in a leaf labelled with a non-empty list (failure nodes),
- infinite paths.

Additional important notions from a procedural point of view are finite termination and finite failure:

Definition 17.5. Given a QN-language $\mathbb{L}$ and an $\mathbb{L}$-program P, an $\mathbb{L}$-goal ψ *finitely terminates* for the selection function R if all paths in the computation tree for ψ constructed with R are finite.

ψ *finitely fails* for R if at least one of the leaves of the computation tree for ψ constructed with R is labelled with a non-empty list of queries.

ψ *finitely terminates* (resp. *finitely fails*) if there is a selections function R such that $D?\psi$ finitely terminates (finitely fails) for R.

Note that for a goal $\phi_1 \wedge \ldots \wedge \phi_n$ to be finitely failing, it suffices that *one* of the subgoals ϕ_i $(1 \leq i \leq n)$ is finitely *failing*, whereas for $\phi_1 \wedge \ldots \wedge \phi_n$ to be finitely *terminating*, *all* the ϕ_i have to be finitely terminating.

Proposition 17.6. *Given a QN-language $\mathbb{L}$ and an $\mathbb{L}$-program P; if an $\mathbb{L}$-goal ψ finitely terminates w.r.t. P then the set of answer substitutions for ψ is finite (where answer substitutions that are identical up to variable renaming are considered to be identical).*

17.2.4 *Model semantics of QN*-PROLOG

We define model semantics for QN-PROLOG programs using Kripke structures.

Definition 17.7. Given a QN-language $\mathbb{L}$, a (Kripke) $\mathbb{L}$-*model* is a pair $\mathbf{m} = \langle W, val\rangle$, where W is a non-empty set (of worlds) and the valuation val is a mapping from worlds to sets of *ground* $\mathbb{L}$-atoms. Thus, $val(w)$ is the set of ground atoms that are 'true' at w under the valuation val.

The reachability relation $\leq$ on the set W of worlds is for all $w, w' \in W$ defined by:

$$w \leq w' \text{ iff } val(w) \subseteq val(w').$$

The world w' is *reachable* from the world w, and w' is a *successor* of w if $w \leq w'$.

The valuation function val is extended from ground $\mathbb{L}$-atoms to ground $\mathbb{L}$-clauses and $\mathbb{L}$-goals as follows:

1. $\phi_1 \wedge \ldots \wedge \phi_n \in val(w)$ iff $\phi_i \in val(w)$ for all $i \in \{1, \ldots, n\}$.
2. $\phi \leftarrow \psi \in val(w)$ iff $\phi \in val(w')$ for all $w' \in W$ such that
   (a) $w \leq w'$ and
   (b) $\psi \in val(w')$.

If $\psi \in val(w)$, then ψ is *true* in w, denoted by $w \models \psi$. If ψ is true in all $w \in W$, then ψ is *valid* in the model $\mathbf{m} = \langle W, val \rangle$, denoted $\mathbf{m} \models \psi$.

The validity of non-ground clauses (and thus non-ground programs and databases) in a model $\mathbf{m}$ is reduced to the validity of its instances in $\mathbf{m}$:

Definition 17.8. Given a QN-language $\mathbb{L}$, a non-ground $\mathbb{L}$-goal[50] ψ is *valid* in an $\mathbb{L}$-model $\mathbf{m}$, denoted $\mathbf{m} \models \psi$, iff all ground instances of ψ are valid in $\mathbf{m}$.

In arbitrary models unexpected things may happen, because $p \leftarrow q$ is, by definition, trivially true in a world w if there is *no* successor world w' of w in which q is true. To avoid these anomalies, we define the notion of canonical models:

Definition 17.9. An $\mathbb{L}$-model $\langle W, val \rangle$ is *canonical* iff for all worlds $w \in W$ and all ground $\mathbb{L}$-atoms ψ there is a world $w' \in W$ such that $w \leq w'$ and $w' \models \psi$.

With the restriction to *canonical* models, it is still necessary to consider *all* canonical models of a program to characterise its semantics. However, using the following ordering on models, it is sufficient to only consider a single *minimal* canonical model (see Theorem 17.11).

Definition 17.10. Given $\mathbb{L}$-models $\mathbf{m} = \langle W, val \rangle$ and $\mathbf{m}' = \langle W, val' \rangle$ with the same set of worlds W; the model $\mathbf{m}$ is *smaller* than $\mathbf{m}'$, denoted $\mathbf{m} \leq \mathbf{m}'$, iff for all $w \in W$:

$$val(w) \subseteq val'(w).$$

17.2.5 *The relation between procedural and model semantics*

The following completeness theorem states the relationship between procedural and model semantics:

Theorem 17.11. *Given a QN-language $\mathbb{L}$, an $\mathbb{L}$-program P, and an $\mathbb{L}$-goal ψ, the substitution σ is an answer substitution for ψ w.r.t. P if and only if ψ is valid in some minimal canonical $\mathbb{L}$-model of P.*

An important property of QN-PROLOG is that the set of answer substitutions is (a) monotonic w.r.t. extensions of the database, and (b) stable under specialisation:

Theorem 17.12. *Given a QN-language $\mathbb{L}$, an $\mathbb{L}$-program P and an atomic $\mathbb{L}$-goal ψ, if σ is an answer substitution for ψ, then:*

1. *σ is an answer substitution for the goal $\psi \leftarrow \phi$ for all $\mathbb{L}$-clauses ϕ.*
2. *every specialisation τ of σ is an answer substitution for $\psi\tau$.*

17.2.6 *An example: semantic tableaux*

As an example we define a language $\mathbb{L}_{tab}$ and an $\mathbb{L}_{tab}$-program P_{tab} implementing first-order semantic tableaux.

[50] That includes $\mathbb{L}$-clauses, because every $\mathbb{L}$-clause is also an $\mathbb{L}$-goal.

First-order formulae are assumed to be in negation normal form (NNF) such that fewer tableau rules are necessary and the program becomes shorter. In the object language $\mathbb{L}_{fol}$, disjunction is denoted with $+$ and conjunction with $\wedge$; universal quantification is denoted with $\forall(x, \psi)$, where x is the quantified variable and ψ is the scope.

The QN-prolog language $\mathbb{L}_{tab}$ is an enrichment of the language $\mathbb{L}_{fol}$ of first-order logic such that all $\mathbb{L}_{fol}$-formulae are $\mathbb{L}_{tab}$-terms. In addition, $\mathbb{L}_{tab}$ contains list constructors $cons$ and nil[51] (lists of formulae are used to represent tableau branches) and predicate symbols *cl* ('closable'), *ob* ('on the branch'), *append*, and $replace$; the set of variables in $\mathbb{L}_{tab}$ consists of the variables in $\mathbb{L}_{fol}$ and in addition Γ, Γ', x, y, ϕ, ϕ', ψ.

Intuitively, a query $D?cl(\Gamma)$ represents a tableau branch B consisting of the (first-order) formulae in the list Γ and the formulae ψ for which $ob(\psi)$ is an element of D. If the query succeeds with an answer substitution σ, then B can be extended to a (sub-)tableau that is closed when σ is applied. In that case, the formula set

$$B\sigma = \Gamma\sigma \cup \{\psi\sigma \mid ob(\psi) \in D\}$$

has been shown to be inconsistent.

To prove a first-order formula ϕ in NNF to be inconsistent, the program is started with the goal $cl([\phi])$. The proof proceeds by considering individual branches of a tableau;[52] if the first formula in the list Γ is not a literal, the appropriate tableau expansion rule is applied. In addition, the first formula in Γ can be used for closing the branch, or it is added to the database and the next formula in Γ is considered.

If a conjunction $\phi \wedge \psi$ is to be expanded, then ϕ and ψ become separate elements of the branch (clause 1). For disjunctions we split the current branch and prove two new goals (clause 2).

Handling a universally quantified formula $\psi = \forall(x, \phi)$ requires a little more effort. First, a 'fresh' instance $\phi' = S^x_y \phi$ of the current formula is generated with $replace$ (see below).[53] The original formula ψ is put at the end of the formula list,[54] and the proof search is continued with the new instance ϕ' as the formula to be expanded next (clause 3).

The next two clauses (4 and 5) close branches; if the complement of the first formula in Γ can be unified with a formula in the database, the current branch is closed.

The last clause in P_{tab} allows us to remove the first formula from the list Γ and add it to the database. This clause is the most interesting one; it is the only one that is not a standard PROLOG clause.

[51] In the following, terms of the form $cons(\phi_1, cons(\cdots(cons(\phi_k, nil)\cdots))$ are abbreviated with $[\phi_1, \ldots, \phi_k]$.

[52] The order in which branches are considered is determined by the order in which goals are evaluated.

[53] $S^x_y \phi$ is the result of replacing all occurrences of x in ϕ with y, compare with Section 7.3.

[54] Putting it at the top of the list destroys completeness: the same universally quantified formula would be used over and over again.

$$
\begin{array}{lll}
1 & cl(cons(\phi \wedge \psi, \Gamma)) & \leftarrow cl(cons(\phi, cons(\psi, \Gamma))) \\
2 & cl(cons(\phi \vee \psi, \Gamma)) & \leftarrow cl(cons(\phi, \Gamma)) \wedge cl(cons(\psi, \Gamma)) \\
3 & cl(cons(\forall(x, \phi), \Gamma)) & \leftarrow replace(\phi, x, y, \phi') \wedge \\
 & & \quad append(\Gamma, [\forall(x, \phi)], \Gamma') \wedge \\
 & & \quad cl(cons(\phi', \Gamma')) \\
4 & cl(cons(\phi, \Gamma)) & \leftarrow ob(\neg\phi) \\
5 & cl(cons(\neg\phi, \Gamma)) & \leftarrow ob(\phi) \\
6 & cl(cons(\phi, \Gamma)) & \leftarrow (cl(\Gamma) \leftarrow ob(\phi))
\end{array}
$$

The program P_{tab} has two choice points: one is selecting between clauses, which means applying an expansion rule to a formula, using it for closing a branch, or adding it to the database. The second choice point, within the clauses 4 and 5, enumerates closing substitutions.

Besides the clauses shown, P_{tab} has to contain additional clauses that define the predicates *append* and *replace*. The meaning of $replace(\phi, x, y, \phi')$ is that ϕ' is the result of replacing all occurrences of x in ϕ with y, i.e., σ is an answer substitution for $replace(\phi, x, y, \phi')$ iff

$$\phi'\sigma = S^{x\sigma}_{y\sigma}\phi\sigma.$$

Figure 17.1 shows (part of) the computation tree for the goal $cl([\forall(x, p(x)) \wedge \neg p(a)])$, i.e., for proving the inconsistency of the formula $\forall(x, p(x)) \wedge \neg p(a)$. That part of the tree where goals of the form $replace(\cdots)$ and $append(\cdots)$ are evaluated is left out. The tree continues infinitely below the leftmost node, because the tableau rule for universal quantification can be applied infinitely often to $\forall(x, p(x))$; the subtree below that node contains infinitely many success nodes.

P_{tab} is the QN-PROLOG equivalent of the *lean* implementation of first-order semantic tableaux in standard PROLOG called lean*TAP*, that is described in [Beckert and Posegga, 1995]. Both P_{tab} and lean*TAP* are non-deterministic and rely heavily on special PROLOG features like unification. lean*TAP* is identical to P_{tab}, except that (i) it uses a limit on the number of formulae on a branch to initiate backtracking and thus resolve the non-determinism (which a deterministic implementation of QN-PROLOG would have to do as well), (ii) it separates a branch into a *current formula* and a list containing all other formulae (this is done for efficiency reasons to make use of PROLOG's indexing mechanism), and (iii) —this is the most important difference— lean*TAP* uses an additional list that 'implements' the database of QN-PROLOG and it contains an additional clause to traverse this list. In effect lean*TAP* can be seen as the transformation of P_{tab} into a classical PROLOG program. The simplicity of P_{tab}, where the predicate *cl* has only one argument, shows that QN-PROLOG is well suited for implementing proof procedures.

17.3 Fibring of QN-PROLOG Programs

17.3.1 *The intuition behind fibring QN-PROLOG programs*

In this section we describe what the semantics of the (combined) program P^* is that results from fibring QN-PROLOG programs $P_1, \ldots, P_r$. First, in Section 17.3.2, the procedural semantics of P^* is defined, which reflects our intuition of what the result

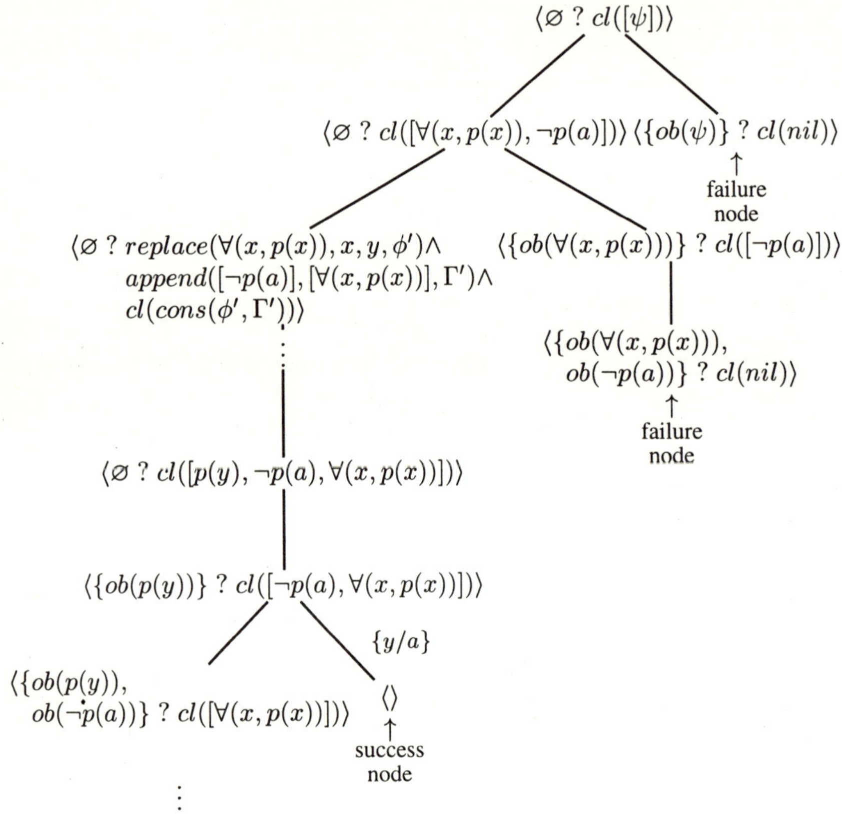

FIG. 17.1. The computation tree for proving that the first-order formula $\psi = \forall(x, p(x)) \wedge \neg p(a)$ is inconsistent.

of fibring intuitionistic logic programs should be. In Section 17.3.3, a model semantics for P^* is presented which is based on fibring models of the original programs P_i. The relationship between procedural and model semantics for fibred programs turns out to be the same as that between procedural and model semantics for non-fibred programs.

We fibre QN-PROLOG programs $P_1, \ldots, P_r$ where P_i is an $\mathbb{L}_i$-program ($1 \leq i \leq r$). The result of fibring $P_1, \ldots, P_r$ is denoted by P^*. The languages do not have to be disjoint; they may have variables, and predicate and function symbols in common. However, if the languages have any variable in common, then they must have all variables and function symbols (and therefore all terms) in common. The reason for this condition is that it has to be possible to apply an answer substitution computed by an $\mathbb{L}_i$-program to an $\mathbb{L}_j$-clause without running the risk of generating a clause that is no longer a well-formed $\mathbb{L}_j$-clause.

Definition 17.13. QN-languages $\mathbb{L}_1, \ldots, \mathbb{L}_r$ ($r \geq 1$) are *suitable for fibring* if either

- the sets $\mathcal{V}_i$ of variables of the languages $\mathbb{L}_i$ ($1 \leq i \leq r$) are all pairwise disjoint, or

- the sets $\mathcal{V}_i$ of variables and the sets $\mathcal{F}_i$ of function symbols, respectively, of the languages $\mathbb{L}_i$ $(1 \leq i \leq r)$ are all identical.

All occurring atoms are each built from symbols from a single language. Clauses, goals, queries and databases may consist of atoms from different languages but have to satisfy the conditions laid out in the following definition of the fibred language $\mathbb{L}^*$:

Definition 17.14. Given QN-languages $\mathbb{L}_1, \ldots, \mathbb{L}_r$

- ψ is an $\mathbb{L}^*$-*atom* iff it is an $\mathbb{L}_i$-atom for some $i \in \{1, \ldots, r\}$.
- ψ is an $\mathbb{L}^*$-*clause* iff it is a clause (see Definition 17.1) constructed from $\mathbb{L}^*$-atoms.
- ψ is an $\mathbb{L}^*$-*goal* iff it is of the form $\phi_1 \wedge \ldots \wedge \phi_k$ $(k \geq 1)$ where each of the ϕ_j $(1 \leq j \leq k)$ is an $\mathbb{L}_{i_j}$-clause for some $i_j \in \{1, \ldots, r\}$.
- D is an $\mathbb{L}^*$-*database* iff it is a finite set of clauses such that:
  1. The heads of the clauses in D, i.e., the atoms in the set
  $$\{\psi \mid (\psi \leftarrow \phi) \in D\} \cup \{\psi \mid \psi \in D, \psi \text{ is an atom}\},$$
  are all from the same language $\mathbb{L}_i$ for some $i \in \{1, \ldots, r\}$.
  2. The bodies of the clauses in D, i.e., the goals in the set
  $$\{\phi \mid (\psi \leftarrow \phi) \in D\},$$
  are $\mathbb{L}^*$-goals (see above).
- $D?\psi$ is an $\mathbb{L}^*$-*query* iff D is an $\mathbb{L}^*$-database and ψ is an $\mathbb{L}^*$-goal.
- σ is an $\mathbb{L}^*$-*substitution* iff it instantiates variables x from $\bigcup_{i=1}^{r} \mathcal{V}_i$ such that: if x is instantiated with t, then there is a language $\mathbb{L}_i$ such that $x \in \mathcal{V}_i$ and t is an $\mathbb{L}_i$-term.

The above definition contains two important restrictions:

1. An $\mathbb{L}^*$-goal consists of conjuncts that are each taken from a single language $\mathbb{L}_i$ (it does not have to be the same for all conjuncts).
2. The heads of the clauses in a database are all taken from a single language $\mathbb{L}_i$ (which is the same for all clauses).

The first of these restrictions implies in particular that an $\mathbb{L}^*$ goal ψ of the form $\phi \leftarrow \rho$ has to be an $\mathbb{L}_i$-goal for some $i \in \{1, \ldots, r\}$. Such non-conjunctive goals will be evaluated using a database D where all the heads of the clauses in D are from the same language $\mathbb{L}_i$. Thus, when during the evaluation of ψ the $\mathbb{L}_i$-clause ρ is added to D, then the new database D' again satisfies the second restriction and is therefore a well-formed database.

The restriction on the form of goals applies as well to bodies of clauses in a database; therefore, when an atomic goal ψ is evaluated using a clause $\psi \leftarrow (\rho_1 \wedge \ldots \wedge \rho_k)$ from D, then the new goal $\rho_1 \wedge \ldots \wedge \rho_k$ is a well-formed $\mathbb{L}^*$-goal.

In each state that is reached during the (procedural) evaluation of an $\mathbb{L}^*$-query, the fibred program P^* works in a unique mode i that is associated with the language $\mathbb{L}_i$ (and the program P_i).

Without making a real restriction, we define the program P_1 to be the 'main program'; thus, all computations initially start in mode $i = 1$. Note, that the initial database is empty and is, therefore, trivially an $\mathbb{L}_1$-database.

A problem arises if an atomic goal ψ is to be evaluated that belongs to more than one language. Then the computation can switch to different modes. There are several possible approaches to solving this problem, e.g., a completely non-deterministic choice of the next mode, or a deterministic choice that uses a preference order on the languages. We use a more flexible approach: a *mode selection function* $C(i, \psi)$ is defined that—depending on the 'current' mode i—assigns to ψ a set $\{j_1, \ldots, j_k\}$ $(k \geq 1)$ of possible 'next' modes. One of these possible next modes is chosen non-deterministically. Obviously, ψ has to be an $\mathbb{L}_j$-atom for all $j \in C(i, \psi)$.

Definition 17.15. Given QN-languages $\mathbb{L}_1, \ldots, \mathbb{L}_r$ suitable for fibring, a *mode selection function* C assigns to each pair consisting of a mode $i \in \{1, \ldots, r\}$ and an $\mathbb{L}^*$-atom ψ a set

$$\{j_1, \ldots, j_k\} \subset \{1, \ldots, r\} \qquad (k \geq 1)$$

of modes such that

1. ψ is an $\mathbb{L}_j$-atom for all $j \in \{j_1, \ldots, j_k\}$,
2. $C(i, \psi\sigma) \subset C(i, \psi)$ for all $\mathbb{L}^*$-substitutions σ.

If an $\mathbb{L}_j$-atom ψ is to be evaluated in mode j while the program is in mode i, the computation switches modes. Since the program P_j may require the knowledge contained in the database to be represented differently (e.g., using different predicates), a *procedural fibring function* $\mathbf{F}_{(i,j)}$ is applied to compute the new database used in mode j; only the database is converted, the goal remains unchanged.

To preserve the properties of (non-fibred) QN-PROLOG programs described in Theorem 17.12, both the mode selection and the fibring functions have to be stable under substitution.

Definition 17.16. Given QN-languages $\mathbb{L}_1, \ldots, \mathbb{L}_r$ suitable for fibring, (procedural) *fibring functions* $\mathbf{F}_{(i,j)}$ $(i, j \in \{1, \ldots, r\}, i \neq j)$ assign to each finite set A_i of $\mathbb{L}_i$-atoms a finite set $A_j = \mathbf{F}_{(i,j)}(A_i)$ of $\mathbb{L}_j$-atoms such that

$$\mathbf{F}_{(i,j)}(A_i\sigma) = (\mathbf{F}_{(i,j)}(A_i))\sigma$$

for all $\mathbb{L}_i$-substitutions σ.

The fibring functions are extended from sets of atoms to databases as follows: Given an $\mathbb{L}^*$-database D_i whose clause heads are $\mathbb{L}_i$-atoms, then the $\mathbb{L}^*$-database $\mathbf{F}_{(i,j)}(D_i)$ contains for each subset

$$\{\psi_1^i \leftarrow \phi_1, \ \ldots, \ \psi_l^i \leftarrow \phi_l, \ \psi_{l+1}^i, \ \ldots, \ \psi_k^i\} \qquad (0 \leq l \leq k)$$

of D_i, where $\psi_{l+1}^i, \ldots, \psi_k^i$ are atoms, the clauses in

$$\{\psi^j \leftarrow \phi_1 \wedge \ldots \wedge \phi_l \mid \psi^j \in \mathbf{F}_{(i,j)}(\{\psi_1^i, \ldots, \psi_k^i\})\}.$$

For an implementation, of course, the fibring functions as well as the mode selection function have to be reasonably easy to compute.

The following two lemmata (which are both easy to prove from the definitions of databases and fibring functions) state important properties of fibring functions:

Lemma 17.17. *Fibring functions are monotonic on databases, i.e.,*

$$\textit{if } D \subset D' \textit{ then } \mathbf{F}_{(i,j)}(D) \subset \mathbf{F}_{(i,j)}(D').$$

Lemma 17.18. *If D is a well-formed $\mathbb{L}^*$-database, then $\mathbf{F}_{(i,j)}(D)$ is a well-formed $\mathbb{L}^*$-database.*

Example 17.19. Suppose p, p', q, q' are the predicate symbols of $\mathbb{L}_1$, r, s, t, u are the predicate symbols of $\mathbb{L}_2$, and $\mathbf{F}_{(1,2)}$ is defined by:

$$\mathbf{F}_{(1,2)}(A) = \begin{cases} \{r, s\} & \text{if } A = \{p, p'\} \\ \{t\} & \text{if } A = \{p\} \\ \{u\} & \text{if } A = \varnothing \\ \varnothing & \text{else.} \end{cases}$$

If $D_1 = \{p \leftarrow q,\ p' \leftarrow q'\}$, then

$$D_2 = \mathbf{F}_{(1,2)}(D_1) = \{r \leftarrow q \wedge q',\ s \leftarrow q \wedge q',\ t \leftarrow q,\ u\}.$$

If $\mathbf{F}_{(2,1)}$ is defined by $\mathbf{F}_{(2,1)}(A) = \mathbf{F}^{-1}_{(1,2)}(A)$ in the sense that

$$\mathbf{F}_{(2,1)}(A) = \begin{cases} \{p, p'\} & \text{if } A = \{r, s\} \\ \{p\} & \text{if } A = \{t\} \\ \{p'\} & \text{if } A = \varnothing \\ \varnothing & \text{else} \end{cases}$$

then

$$D'_1 = \mathbf{F}_{(2,1)}(D_2) = \{p \leftarrow q,\ p \leftarrow q \wedge q',\ p' \leftarrow q \wedge q'\}.$$

As the above example shows, information may be lost when the database is transformed by a fibring function. It is even possible that $\mathbf{F}_{(i,j)}(D) = \varnothing$ although $D \neq \varnothing$.

17.3.2 *Fibring procedural semantics*

We now have all the ingredients needed to formally define the procedural semantics of fibred programs that has been informally described in the previous section.

If a query $q = D?(\psi \leftarrow \rho_1 \wedge \ldots \wedge \rho_k)$ is to be evaluated in the current mode i, then the computation first switches to (the next) mode $j = C(i, \psi)$. If $j \neq i$, the database D is replaced with $\mathbf{F}_{(i,j)}(D)$; then the clauses $\rho_1, \ldots, \rho_k$ are added (they are added after transforming the database). The syntactical restrictions imposed on the form of queries and the restrictions in the definition of the mode selection function C imply that $\rho_1, \ldots, \rho_k$ are $\mathbb{L}_j$-clauses, such that the new database D' is well-formed. The child nodes in the computation tree are determined for the query $D' \cup \{\rho_1, \ldots \rho_k\}?\psi$ (as before for queries with atomic goals).

Definition 17.20. Given QN-languages $\mathbb{L}_1, \ldots, \mathbb{L}_r$, QN-programs P_i (where P_i is an $\mathbb{L}_i$-program), procedural fibring functions $\mathbf{F}_{(i,j)}$ ($i, j \in \{1, \ldots, r\}$, $i \neq j$), a mode selection function C, and a selection function R for $\mathbb{L}^*$, the *fibred computation tree* T for an $\mathbb{L}^*$-goal ψ is defined as follows:

- D is a tree whose nodes are each labelled with a finite sequence of pairs $\langle i, D?\psi\rangle$, where $i \in \{1, \ldots, r\}$ and $D?\psi$ is an $\mathbb{L}^*$-query, and whose edges are labelled with $\mathbb{L}^*$-substitutions; if not stated otherwise, edges are labelled with the empty substitution.
- The root node of T is labelled with the singleton sequence $\langle\langle 1, \varnothing?\psi\rangle\rangle$.
- Nodes labelled with the empty list are leaves.
- For all nodes N in T, where N is labelled with

$$L = L' \circ \langle\langle i, D?\psi\rangle\rangle \circ L''$$

and $R(L) = \langle i, D?\psi\rangle$, the following holds:

1. *Rule for conjunctions:* If $\psi = \rho_1 \wedge \ldots \wedge \rho_k$, $k \geq 2$, then N has one successor N' that is labelled with

$$L' \circ \langle\langle i, D?\rho_1\rangle, \ldots, \langle i, D?\rho_k\rangle\rangle \circ L''.$$

2. *Combined rule for implications and atoms:* If $\psi = (\phi \leftarrow \rho_1 \wedge \ldots \wedge \rho_k)$, $k \geq 0$,[55] then for each $j \in C(i, D?\psi)$ and each clause ς that
   (a) i. is an element of

$$D' = \begin{cases} \mathbf{F}_{(i,j)}(D) \cup \{\rho_1, \ldots, \rho_k\} & \text{if } i \neq j \\ D \cup \{\rho_1, \ldots, \rho_k\} & \text{if } i = j \end{cases}$$

   ii. or is the renaming of a clause ς' in P_j, i.e., $\varsigma = \varsigma'\nu$ where the variable renaming ν replaces the variables in ς' by new variables not occurring in any of the programs $P_1, \ldots, P_r$ nor in the list L;

   (b) and whose head κ is unifiable with ϕ with a most general unifier μ,

   the node N has a successor N' where N' is labelled with

$$L'\mu \circ L''\mu$$

   in case $\varsigma = \kappa$ is an atom, and with

$$L'\mu \circ \langle\langle j, D'\mu?\xi\mu\rangle\rangle \circ L''\mu$$

   in case $\varsigma = (\kappa \leftarrow \xi)$; the edge between N and N' is labelled with μ.

The notions of refutation, answer substitution, finite failure and termination are defined as before for non-fibred programs (see Definition 17.4).

[55] That includes the case where $k = 0$ and (thus) $\psi = \phi$ is an atom.

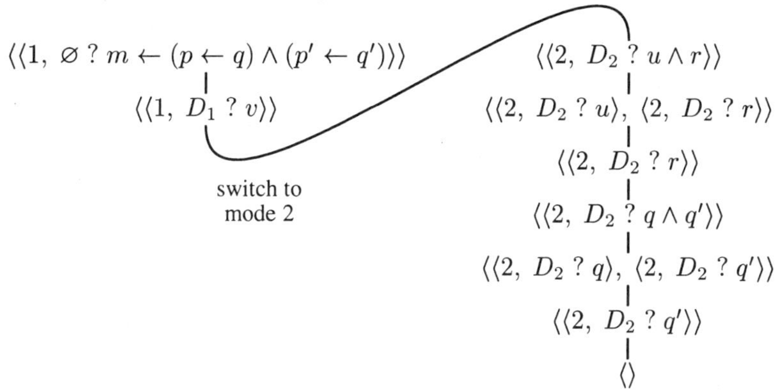

$D_1 = \{p \leftarrow q,\ p' \leftarrow q'\}$
$D_2 = \{r \leftarrow q \wedge q',\ s \leftarrow q \wedge q',\ t \leftarrow q,\ u\}$

FIG. 17.2. Computation tree for the goal $m \leftarrow (p \leftarrow q) \wedge (p' \leftarrow q')$ (see Example 17.21).

Example 17.21. We continue from Example 17.19. Suppose that m is an additional predicate symbol in $\mathbb{L}_1$, and v is an additional predicate symbol in both $\mathbb{L}_1$ and $\mathbb{L}_2$. Let

$$P_1 = \{q,\ q',\ m \leftarrow v\}$$

and

$$P_2 = \{v \leftarrow u \wedge r\}.$$

The fibring functions $\mathbf{F}_{(1,2)}$ and $\mathbf{F}_{(2,1)}$ are defined as in Example 17.19, and the mode selection function is defined by $C(i, \psi) = \{i\}$ for all atoms that belong to only one language $\mathbb{L}_i$ (which are all except v), and $C(1, v) = C(2, v) = \{2\}$, i.e., the atom v is always evaluated in mode 2. Figure 17.2 shows the computation tree (which consists of a single refutation) for the goal $m \leftarrow (p \leftarrow q) \wedge (p' \leftarrow q')$. Note that—although the mode change is not made explicit in the tree—the two goals q and q' at the end of the tree are both evaluated in mode 1.

17.3.3 *Fibring model semantics*

Model semantics for a fibred program P^* are specified by defining models for $\mathbb{L}^*$; these $\mathbb{L}^*$-models are built by fibring $\mathbb{L}_i$-models. As usual in the fibring approach, each world w in one of the $\mathbb{L}_i$-models $\mathbf{m}_i$ is assigned an $\mathbb{L}_j$-model $\mathbf{m}_j$ (using a model fibring function); and the value in w of a query q that is to be evaluated in mode j, i.e., in an $\mathbb{L}_j$-model, is then determined by the value of q in $\mathbf{m}_j$.

Note that model fibring functions $\mathbf{F}_{(i,j)}$ are different from procedural fibring functions because they assign $\mathbb{L}_j$-models to worlds in $\mathbb{L}_i$-models (instead of assigning $\mathbb{L}_j$-databases to $\mathbb{L}_i$-databases).

As in the non-fibred case we first define validity of *ground* clauses and goals in a fibred model. The validity of non-ground clauses is reduced to the validity of their instances.

Definition 17.22. Given QN-languages $\mathbb{L}_1, \ldots, \mathbb{L}_r$, a *fibred model*

$$\mathcal{M}^* = \langle (\mathcal{M}_i)_{i=1}^r,\ \mathbf{m}_1,\ (\mathbf{F}_{(i,j)})_{i,j\in\{1,\ldots,r\}, i\neq j},\ C \rangle$$

consists of

1. sets $\mathcal{M}_i$ of $\mathbb{L}_i$-models,
2. an initial model $\mathbf{m}_1 \in \mathcal{M}_1$,
3. model fibring functions $\mathbf{F}_{(i,j)}$ that assign to each world $w \in W$ of each model $\langle W, val \rangle \in \mathcal{M}_i$ a model in $\mathcal{M}_j$,
4. a mode selection function C (Definition 17.15).

An $\mathbb{L}^*$-goal ψ is *true* in a world $w \in W$ of a model $\langle W, val \rangle \in \mathcal{M}_i$, denoted $w \models^* \psi$,[56] iff either:

1. *Rule for conjunctions:* $\psi = \phi_1 \wedge \ldots \wedge \phi_k$, $k \geq 2$, and $w \models^* \phi_n$ for all $n \in \{1, \ldots, k\}$.
2. *Combined rule for implications and atoms:* $\psi = (\phi \leftarrow \rho_1 \wedge \ldots \wedge \rho_k)$, $k \geq 0$,[57] and there is a mode $j \in C(i, \psi)$ such that,
   (a) in case $j \neq i$, for all worlds w' of the $\mathbb{L}_j$-model $\mathbf{F}_{(i,j)}(w)$, and
   (b) in case $j = i$, for all worlds $w' \in W$ with $w \leq w'$,
   the following holds:

$$\text{if } w' \models^* \rho_n \text{ for all } 1 \leq n \leq k \text{ then } \phi \in val(w').$$

ψ is *valid* in $\mathcal{M}^*$, denoted $\mathcal{M}^* \models^* \psi$, if it is true in all worlds of the initial model $\mathbf{m}_1$.

Definition 17.23. A (non-ground) $\mathbb{L}^*$-goal is *valid* in a fibred model $\mathcal{M}^*$ iff all its ground instances are valid in $\mathcal{M}^*$.

A fibred program P^* is valid in a fibred model $\mathcal{M}^*$ if the programs P_i that P^* consists of are valid in the corresponding models that $\mathcal{M}^*$ consists of, i.e.: if P_i is an $\mathbb{L}_i$-program, then it has to be valid in all $\mathbb{L}_i$-models of $\mathcal{M}^*$.

Definition 17.24. Given QN-languages $\mathbb{L}_1, \ldots, \mathbb{L}_r$ and a fibred model

$$\mathcal{M}^* = \langle (\mathcal{M}_i)_{i=1}^r,\ \mathbf{m}_1,\ (\mathbf{F}_{(i,j)})_{i,j\in\{1,\ldots,r\}, i\neq j},\ C \rangle,$$

a fibred program P^* that is the result of fibring programs $P_1, \ldots, P_r$, where P_i is an $\mathbb{L}_i$-program, is *valid* in $\mathcal{M}^*$ iff the P_i $(1 \leq i \leq r)$ are valid in all models $\mathbf{m} \in \mathcal{M}_i$.

Based on the definitions for non-fibred models, the notions of canonical and minimal models can easily be adapted to fibred models.

Definition 17.25. A fibred model is *canonical* if all the (non-fibred) models it consists of are canonical (Definition 17.9).

Definition 17.26. A fibred model $\mathcal{M}_1^*$ is smaller than a fibred model $\mathcal{M}_2^*$ iff

[56] Note, that the relation $\models^*$ is different from the relation $\models$ defined earlier for the non-fibred models.

[57] That includes the case where $k = 0$ and $\psi = \phi$ is an atom.

1. $\mathcal{M}_1^*$ differs from $\mathcal{M}_2^*$ only in the evaluation functions val that are part of the (non-fibred) models that $\mathcal{M}_1^*$ and $\mathcal{M}_2^*$ consist of;
2. if $\mathbf{m}_1 = \langle W, val_1 \rangle$ and $\mathbf{m}_2 = \langle W, val_2 \rangle$ is a pair of models from $\mathcal{M}_1^*$ resp. $\mathcal{M}_2^*$ that correspond to each other, then $\mathbf{m}_1 \leq \mathbf{m}_2$, i.e., $val_1(w) \subseteq val_2(w)$ for all $w \in W$.

17.3.4 *The relation between fibred procedural and fibred model semantics*

Since both procedural and model semantics depend on the fibring functions used, they are only the same if there is a strong relationship between the procedural fibring and mode selection functions that are used in the construction of a computation tree on the one hand and the fibring and mode selection functions that are part of a model for $\mathbb{L}^*$ on the other hand. What the relationship has to be is described in the following definition.

Definition 17.27. Given QN-languages $\mathbb{L}_1, \ldots, \mathbb{L}_r$, a fibred model

$$\mathcal{M}^* = \langle (\mathcal{M}_i)_{i=1}^r,\ \mathbf{m}_1,\ (\mathbf{F}^m_{(i,j)})_{i,j \in \{1,\ldots,r\}, i \neq j},\ C^m \rangle$$

for $\mathbb{L}^*$ *satisfies* procedural fibring functions $\mathbf{F}^p_{(i,j)}$ (Definition 17.16) iff, for all $i, j \in \{1, \ldots, r\}$, $i \neq j$, all *ground* $\mathbb{L}_i$-databases D_i, all $\mathbb{L}_i$-models $\mathbf{m}_i = \langle W_i, val_i \rangle$ in $\mathcal{M}_i$, and all worlds $w_i \in W_i$, the following holds:

$$\text{if } w_i \models^* D_i \text{ , then } \mathbf{F}^m_{(i,j)}(w_i) \models^* \mathbf{F}^p_{(i,j)}(D_i).$$

$\mathcal{M}^*$ *satisfies* a mode selection function C iff $C = C^m$.

The following theorem states the relationship between fibred procedural and fibred model semantics.

Theorem 17.28. *Given QN-languages $\mathbb{L}_1, \ldots, \mathbb{L}_r$, QN-programs P_i (where P_i is an $\mathbb{L}_i$-program), procedural fibring functions $\mathbf{F}_{(i,j)}$ ($i, j \in \{1, \ldots, r\}$, $i \neq j$), and a mode selection function C. Let $\mathcal{M}^*$ be some minimal fibred $\mathbb{L}^*$-model*

1. *that is canonical,*
2. *that satisfies the fibring functions $\mathbf{F}_{(i,j)}$ and the mode selection function C, and*
3. *in which the fibred program P^* is valid.*

Then an $\mathbb{L}^*$-substitution σ is an answer substitution for an $\mathbb{L}^*$-goal ψ if and only if ψ is valid in $\mathcal{M}^*$.

Monotonicity of the set of answer substitutions w.r.t. extensions of the database and stability under instantiation is preserved when programs are fibred, i.e., the properties specified in Theorem 17.12 for non-fibred programs hold for fibred programs as well.

17.3.5 *Fibring semantic tableaux and rigid E-unification*

We continue the example from Section 17.2.6 and fibre the program P_{tab} implementing semantic tableaux with a program that implements rigid E-unification [Gallier *et al.*,

1992]. That is, given a set E of equalities and terms s and t, the program can compute (and enumerate) substitutions σ such that

$$E\sigma \models_= t\sigma = s\sigma,$$

i.e., the terms $t\sigma$ and $s\sigma$ are equal modulo the equalities in $E\sigma$, where the equalities in $E\sigma$ are held *rigid* (they are *not* implicitly universally quantified). For fibring we use the program P_{reu} that can solve the (equivalent) problem of, given a conjunction Ψ of equalities and inequalities, computing substitutions σ such that the instance $\Psi\sigma$ of Ψ is inconsistent (where again the equalities in $\Psi\sigma$ are held rigid).

We do not list such a program P_{reu} here, but only give its semantics. Suppose the language $\mathbb{L}_{reu}$ is (similar to $\mathbb{L}_{tab}$) an extension of the first-order language $\mathbb{L}_{fol}$ containing additional predicate symbols $equal$, $ineq$ and $incons$. σ is an answer substitution for the goal $incons \leftarrow \Psi$ if $\Psi\sigma$ is inconsistent, where Ψ is a conjunction of atoms of the form $equal(s,t)$ and $ineq(s,t)$.

P_{tab} and P_{reu} can be fibred such that the resulting fibred program P^* implements a proof procedure for first-order logic with equality [Beckert, 1998]. P_{tab} is the main program and uses P_{reu} as a module to check whether tableau branches can be shown to be inconsistent using equality theory and, thus, be closed.

The predicate symbol $incons$ is added to $\mathbb{L}_{tab}$, and the program P_{tab} is extended by an additional clause

$$cl(\Gamma) \leftarrow incons.$$

Thus, by checking if the set of atoms in D is inconsistent in first-order logic with inequality, P_{RUE} provides additional answer substitutions for queries $D?cl(\Gamma)$.

The appropriate mode selection function C is defined by:

$$C(i, incons) = \{reu\}$$

and $C(i,\psi) = \{i\}$ for all atoms $\psi \neq incons$.

The appropriate fibring function $\mathbf{F}_{(tab,reu)}$ is, for all finite sets A of $\mathbb{L}_{tab}$-atoms, defined by: $\mathbf{F}_{(tab,reu)}(A) =$

$$\begin{cases} \{equal(u,v)\} & \text{if } A = \{ob(u{=}v)\} \\ \{ineq(u,v)\} & \text{if } A = \{ob(\neg u{=}v)\} \\ \{ineq(\langle u_1,\ldots,u_k\rangle, \langle v_1,\ldots,v_k\rangle)\} & \text{if } A = \{ob(p(u_1,\ldots,u_k)), \\ & \qquad ob(\neg p(v_1,\ldots,v_k))\} \\ \varnothing & \text{otherwise.} \end{cases}$$

$\mathbf{F}_{(reu,tab)}$ is not needed (and can be defined arbitrarily).

Example 17.29. Figure 17.3 shows a successful refutation in the computation tree for proving $\psi = p(a) \wedge \neg p(f(f(a))) \wedge \forall(x, f(x){=}x)$ to be inconsistent, i.e., for the goal $cl([\psi])$.

In the figure, those parts of the refutation where goals of the form $replace(\cdots)$ and $append(\cdots)$ are evaluated as well as the part associated with the program P_{reu} are left out.

$$\langle\langle tab,\ \varnothing\ ?\ cl([\psi])\rangle\rangle$$

$$\langle\langle tab,\ \varnothing\ ?\ cl([p(a),\ \neg p(f(f(a)))\wedge\forall(x, f(x){=}x)])\rangle\rangle$$

$$\langle\langle tab,\ \{ob(p(a))\}\ ?\ cl([\neg p(f(f(a)))\wedge\forall(x, f(x){=}x)])\rangle\rangle$$

$$\langle\langle tab,\ \{ob(p(a))\}\ ?\ cl([\neg p(f(f(a))),\ \forall(x, f(x){=}x)])\rangle\rangle$$

$$\langle\langle tab,\ \{ob(p(a)),\ ob(\neg p(f(f(a))))\}\ ?\ cl([\forall(x, f(x){=}x)])\rangle\rangle$$

$$\begin{aligned}\langle\langle tab,\ \{ob(p(a)),\ ob(\neg p(f(f(a))))\}\ ?\ & replace(\forall(x, f(x){=}x), x, y, \phi')\wedge\\ & append(nil, [\forall(x, f(x){=}x)], \Gamma')\wedge\\ & cl(cons(\phi', \Gamma'))\rangle\rangle\end{aligned}$$

$$\vdots$$

$$\langle\langle tab,\ \{ob(p(a)),\ ob(\neg p(f(f(a))))\}\ ?\ cl([f(y){=}y,\ \forall(x, f(x){=}x)])\rangle\rangle$$

$$\langle\langle tab,\ \{ob(p(a)),\ ob(\neg p(f(f(a)))),\ ob(f(y){=}y)\}\ ?\ cl([\forall(x, f(x){=}x)])\rangle\rangle$$

$$\langle\langle reu,\ \{ineq(a, f(f(a))),\ equal(f(y){=}y)\}\ ?\ incons\rangle\rangle$$

$$\vdots$$

$$\{y/a\}$$

$$\langle\rangle$$

$\uparrow$
success node

FIG. 17.3. Refutation that proves $\psi = p(a)\wedge\neg p(f(f(a)))\wedge\forall(x, f(x){=}x)$ to be inconsistent (see Example 17.29).

17.3.6 *Instantiating the general framework*

Assume there are two programs P_1 and P_2, where P_1 is the 'main program' and P_2 is a module used by P_1; and a goal ψ is to be evaluated by P_1 using clauses from P_2. If implication goals are used to model modules, this situation is equivalent to evaluating the goal $\psi \leftarrow P_2$ in P_1, another possibility is to define a meta-predicate *demo* [Bowen and Kowalski, 1982] and evaluating the goal $demo(P_2, \psi)$ in P_1.

According to [Giordano *et al.*, 1988; Giordano and Martelli, 1994], there are three main types of semantics for modules or blocks in logic programs ('block' is synonymous with module):

Closed blocks: Only the clauses in the module P_2 can be used during the evaluation of the goal ψ, the clauses in the main program P_1 are not 'visible' and cannot be used (e.g., [Bowen and Kowalski, 1982]). Most major PROLOG systems use this kind of module concept, where only predicates are visible in other modules (or the main program) that are explicitly exported and imported.

Open blocks with static scope rules: Clauses from both P_1 and P_2 can be used for the evaluation of ψ. However, the clauses in P_2 are invisible from the outside; once the computation leaves the module P_2 and uses P_1-clauses, the clauses in P_2 cannot be used anymore (e.g., [Moscowitz and Shapiro, 1991]).

Open blocks with dynamic scope rules: Clauses from P_1 and P_2 can be used; during the evaluation, clauses from P_2 are visible in P_1 and vice versa (e.g., [Miller, 1986; Giordano *et al.*, 1992]).

Example 17.30. Let P_1 consist of the clauses $q \leftarrow r$ and $s \leftarrow p$, and let P_2 consist of $p \leftarrow q$ and r. Suppose further that the definition of p is explicitly imported from P_2 into the main program P_1; the definition of r in P_2 is not explicitly imported into P_1.

Then, if an open block semantics with dynamic scope rules is used, the goal s succeeds in P_1. Although the fact r in P_2 is not imported into P_1, it is visible in P_1 during the evaluation of p.

If, however, static scope rules are used, the goal s fails in P_1 because the fact r in P_2 is at no time visible in P_1. The goal s succeeds in the extended program $P_1' = P \cup \{r\}$.

Using closed block semantics, s fails in both P_1 and P_1' because the P_1-clause $q \leftarrow r$ is not visible in P_2.

The general fibring framework can easily be instantiated to simulate the closed block semantics for modules. A predicate p of $\mathbb{L}_i$ that is defined in module P_i, is only made a predicate symbol of other languages $\mathbb{L}_j$ if it is exported into P_j. The mode selection function is defined by

$$C(i, p(\cdots)) = \{i\} \cup \{j \mid p \text{ is imported from } P_j \text{ into } P_i\}.$$

There are different possibilities to define the fibring functions in this case. One that amounts to not passing any information stored in the database between modules is to define $\mathbf{F}_{(i,j)}(\psi) = \varnothing$ for all modes i, j and atoms ψ.

Open block semantics with static visibility rules can be simulated under the assumption that the modules form a tree structure, i.e., there is a partial order $<$ on the P_i such that module P_i imports predicates from module P_j if (and only if) $P_i < P_j$. An appropriate mode selection function is defined by

$$C(i, p(\cdots)) = \{i' \mid P_{i'} \leq P_i\} \cup \{j \mid p \text{ is imported from } P_j \text{ into } P_i\}.$$

Semantics with dynamic scope rules cannot be simulated, because the mode selection function is static in that it only depends on the current goal and the current mode, but not on the modes visited during the computation. It would, however, be easy to change the definition of the mode selection function appropriately.

Finally, it is possible to define fibring and mode selection functions such that the effect of fibring programs is that of concatenating them syntactically:

- $C(i, \psi) = \{j \mid \psi$ is an $\mathbb{L}_j$-atom$\}$ for all modes i and $\mathbb{L}_i$-atoms ψ,
- $\mathbf{F}_{(i,j)}(A) = A$ if $A = \{\psi\}$ and ψ is both an $\mathbb{L}_i$- and an $\mathbb{L}_j$-atom, and $\mathbf{F}_{(i,j)}(A) = \varnothing$ otherwise, for all modes i, j.

17.4 Incremental Databases

Besides the efficiency of two fibred programs, the interaction between them plays a critical rôle for the efficiency of the combined system if it is to be implemented. When, for a query to be successful in a (sub-)program, a certain 'amount' of assumptions (i.e. knowledge) has to be accumulated in the database, it is often a difficult problem to decide whether the database contains enough knowledge and it is useful to evaluate the query or not. In general, to give a perfect answer to this question is as difficult as the problem of computing an answer substitution itself (which might even be undecidable). Both too early and too late evaluations can lead to useless computations.

This problem can be solved (at least partially) using fibred intuitionistic PROLOG, which allows us to reuse data computed during a futile try to find an answer substitution for later evaluations of the same query with an *extended database*. Then, in case of doubt the query can be evaluated early without running the risk of doing useless computations. In addition, data can be reused multiply, if different extensions of a database have to be handled.

Such incremental methods have been successfully used in automated deduction [Beckert and Pape, 1996], where problems are typically of an incremental nature, i.e., answer substitutions for queries can only be found when enough knowledge has been accumulated.

Because of the monotonicity of (fibred and non-fibred) QN-PROLOG w.r.t. extensions of the database (Theorem 17.12), it is possible to re-use the computation tree T_1 for a query $D?\psi$ for the construction of the computation tree T_2 for a query $(D \cup D')\sigma?\psi\sigma$. This is done by first removing parts of T_1 that are below an edge labelled with a substitution that is incompatible with σ,[58] and then applying σ to the tree. The resulting tree T_1' is a subtree of T_2; only those edges and nodes of T_2 have to be added (and computed) that result from the additional clauses in $D'\sigma$. Therefore, the computations executed by an actual QN-PROLOG system to find an answer substitution for ψ with a database D and the data these computations has generated can be reused to compute an answer substitution for $\psi\sigma$ with the database $D \cup D'$.

In classical PROLOG where the database would have to be encoded into the goal, which then, for example, is of the form $\langle\psi, D\rangle$ resp. $\langle\psi\sigma, (D \circ D')\sigma\rangle$, the computation tree can—in general—not be reused, because the additional information is encoded in

[58] Substitutions σ and τ are incompatible if there is no substitution ρ that is a common specialisation of σ and τ.

the arguments (terms) of the goal, and there is no general monotonicity property w.r.t. terms in the goal.

17.5 Conclusion

In this chapter we have proposed to use the technique of fibring as a general framework for combining and structuring intuitionistic logic programs.

The flexibility that the fibring and mode selection functions provide can have advantages from a software engineering point of view, because it allows one to use different data structures for representing the same data in the programs that are fibred. The transformation between these data structures can be handled separately by (the implementation of) the fibring functions.

We have shown that monotonicity of both fibred and non-fibred intuitionistic logic programs w.r.t. extension of the database allows one to re-use data computed during the evaluation of a goal, when the same goal is to be evaluated again with an extended database. It is only possible to apply this technique if the database represents information passed to a (sub-)module and not to represent the module itself.

It should be possible to apply the methods presented in this chapter to extensions of intuitionistic programs that, for example, include disjunctions or negation as failure.

18

FIBRING SEMANTIC TABLEAUX

18.1 Introduction

In this chapter we extend the fibring approach to calculi for logical systems: we describe how to uniformly construct a sound and complete tableau calculus for the combined logic from calculi for the component logics.

Since for nearly all 'basic' logics (e.g., classical logic, modal logic, intuitionistic logic, temporal logic, and many more), tableau calculi are already known, it is possible to construct calculi for all more complex logics that can be represented by fibring basic logics, i.e., modal predicate logic, intuitionistic temporal logic, etc.

It is not possible to fibre just any proof procedures for two logics in a uniform way. First of all, 'proof procedure' can mean different things for different logics: deciding (or semi-deciding) satisfiability, validity, computing a satisfying variable instantiation, etc. Second, it is not clear where to 'plug in' the proof procedure for $\mathbf{L}_2$ into that for $\mathbf{L}_1$. Note, that a proof procedure may do something completely different from what the valuation function does that provides the truth value of a formula in a given model. For example, if the procedure P_1 is based on constructing a (counter) model, whereas the procedure P_2 uses a resolution calculus (and, therefore, does not—and cannot—compute a model), they cannot be fibred (at least uniformly).

Therefore, we only consider semantic tableau calculi that satisfy certain conditions—and are therefore known to be 'well-behaved'—such that fibring is possible. The identification and formulation of conditions that are neither too weak nor too strong is a major contribution of this paper.

If the components of a fibred calculus satisfy these conditions, then the fibred calculus is automatically sound and complete. It may only be a semi-decision procedure, i.e., only terminate for unsatisfiable input formulae, even if its components are decision procedures; this, however, is not surprising, because a fibred logic may be undecidable even if its components are decidable.

We define the notion of a *logical systems* in a very general way (Section 18.2); only indispensable properties of its syntax and semantics are part of the definition without which a useful tableau calculus for the logic cannot exist (or cannot be fibred with calculi for other logical systems).

Similarly, as few restrictions as possible are made regarding the type and form of tableau calculi. In particular, the calculus does not have to be analytical; and the tableau rules do not have to be given in the form of rule schemata, but can be described in an arbitrary way. The conditions that tableau calculi have to satisfy to be suitable for fibring are described in Section 18.3.

Related work includes [D'Agostino and Gabbay, 1996b], see Chapters 14 and 15 where tableau calculi for fibred substructural implication logics and modality has been presented, that, however, is much more restricted both w.r.t. the logics and the calculi that can be fibred.

In Section 18.4 the method of fibring logics is described in general and syntax and semantics of a fibred logic are defined, based on syntax and semantics of the component logics.

We study two examples: a calculus for first-order predicate logic is introduced in Section 18.5 and it is shown that it satisfies the conditions laid out in Section 18.3 and is therefore suitable for fibring. The fibring of tableaux for modal logic is studied in detail in Chapter 19.

In Section 18.6 a general, uniform method is presented for constructing a tableau calculus for a fibred logic from calculi for the component logics. The resulting calculus is shown to be sound and complete w.r.t. the semantics of the fibred logic. In addition, the fibred calculus satisfies the conditions ensuring that it is itself suitable for fibring with other calculi.

The most comprehensive available source of information on tableau calculi for various logics is the *Handbook of Tableau Methods* [D'Agostino *et al.*, 1998]. Its chapters cover the main variants of clausal and non-clausal, of propositional and first-order tableau systems, as well as their implementation. The most important families of non-classical logics are given treatment in specific chapters.

18.2 Logics

In this section we define the notion of a (tableaux friendly) *logical system* in a very general way; only indispensable properties of its syntax and semantics are part of the definition without which a useful tableau calculus for the logic cannot exist (or cannot be fibred with calculi for other logical systems).

The logic has to have model semantics that use Kripke-style models, i.e., models consisting of *worlds* in which formulae are true or false. That includes models consisting of just one world (for example, models of classical propositional and first-order logic);[59] and there is no restriction on what the relationship between these worlds is.

A restriction which is only made for the sake of simplicity is that only two-valued logics are considered. All notions introduced in the following could be easily extended to many-valued logics; however, no additional insight could be gained.

Definition 18.1. Associated with a *logical system* $\mathbf{L}$ (a *logic* for short) is a set *Sig* of possible signatures[60] of $\mathbf{L}$. For each signature $\Sigma \in \mathit{Sig}$ the syntax and semantics of the instance $\mathbf{L}^\Sigma$ of $\mathbf{L}$ are given by:

Syntax: A set Form^Σ of *formulae* and a set $\mathit{Atom}^\Sigma \subset \mathit{Form}^\Sigma$ of *atomic formulae* (*atoms*), where the sets Atom^Σ and Form^Σ are decidable.

[59] In fact, any kind of model can be considered to be a Kripke-style model with a single world (namely the model itself). However, since the labels of tableau formulae are interpreted as worlds of models, if there is only one world in all the models of a logic, then the interpretation of all labels is the same and they become useless for the calculus.

[60] We do not further specify what a signature is; *Sig* can be seen as a set of indices for distinguishing different instances of the logic $\mathbf{L}$ (which *usually* differ in the symbols they use).

Semantics: A set $\mathcal{M}^{\Sigma}$ of *models* where each model $\mathbf{m} \in \mathcal{M}^{\Sigma}$ contains (at least)

- a set W of *worlds*,
- an *initial world* $w^0 \in W$,
- a binary relation $\models$ between W and *Form*$^{\Sigma}$.

If $w \models \phi$ for some world $w \in W$ and some formula $\phi \in Form^{\Sigma}$, then ϕ is said to be *true* in w, else it is *false*.

A formula $\phi \in Form^{\Sigma}$ is *satisfied* by a model $\mathbf{m} \in \mathcal{M}^{\Sigma}$ if (and only if) it is true in the initial world w^0 of $\mathbf{m}$. A set $\Gamma \subset Form^{\Sigma}$ of formulae is satisfied by $\mathbf{m}$ if each of its elements is satisfied by $\mathbf{m}$.

A formula $\phi \in Form^{\Sigma}$ (a set $\Gamma \subset Form^{\Sigma}$ of formulae) is *satisfiable* if there is a model $\mathbf{m} \in \mathcal{M}$ satisfying ϕ (resp. Γ).

Although usually non-atomic formulae are constructed from atomic formulae, and their truth value is determined by the truth value of the atoms they consist of, this is *not* part of the above definition. However, the existence of a tableau calculus for a logic $\mathbf{L}$ that is well-suited for fibring, implies that the truth value of a formula ϕ is strongly related to the truth value of certain atoms (that may or may not be subformulae of ϕ).

Example 18.2. The truth value of the formula $(\forall x)(p(x))$ in a model of first-order predicate logic is not determined by the truth value of the (only) atomic formula $p(x)$ it contains, neither can it be computed from the truth value of all atomic literals (i.e. $p(t)$ for all substitutions of terms t), unless all elements of the model's domain are represented by terms. In Herbrand models, however, $(\forall x)(p(x))$ is true if and only if all atoms of the form $p(t)$ are true.

Often in tableau calculi, formulae are used that are not part of the original signature (e.g., formulae containing Skolem symbols); that is, an extended signature is used:

Definition 18.3. Given a logic $\mathbf{L}$, a signature $\Sigma' \in Sig$ is an *extension* of a signature $\Sigma \in Sig$ (and $\Sigma \in Sig$ is a *restriction* of $\Sigma' \in Sig$) if

$$Form^{\Sigma} \subset Form^{\Sigma'} \qquad \text{and} \qquad Atom^{\Sigma} \subset Atom^{\Sigma'}.$$

Definition 18.4. Given a logic $\mathbf{L}$ and signatures $\Sigma, \Sigma' \in Sig$ such that Σ is a restriction of Σ; a model $\mathbf{m} \in \mathcal{M}^{\Sigma}$ is a *restriction* of a model $\mathbf{m}' \in \mathcal{M}^{\Sigma'}$ (to the signature Σ), if there is a function f that assigns to each world of $\mathbf{m}$ a world of $\mathbf{m}'$ such that:

1. the initial world of $\mathbf{m}'$ is assigned to the initial world of $\mathbf{m}$;
2. for all formulae $\phi \in \Sigma$ and worlds w of $\mathbf{m}$:

$$w \models \phi \qquad \text{iff} \qquad f(w) \models \phi.$$

18.3 Tableau Calculi

As few restrictions as possible are made regarding the type and form of tableau calculi. In particular, the calculus does not have to be analytical; and the tableau rules do not have to be given in the form of rule schemata, but can be described in an arbitrary way.

Any function that assigns (possible) extensions to tableau branches is regarded as a tableau rule.

However, certain assumptions have to be made, the first of which ensures that tableau rule applications do not transform the whole tableau in an arbitrary way:

Condition 18.5. Tableau rule applications only have local effects, in that they extend one branch of the tableau, and do not alter or remove formulae already on the tableau.

The second assumption is that the applicability of a tableau rule to a branch B and the result of the application is solely determined by the presence of formulae on B; no other pre-conditions of rule applications are allowed, such as, for example, the absence of certain formulae, the presence of formulae on different branches, or the order of formulae on a branch:

Condition 18.6. Whether a tableau branch B can be expanded in a certain way is solely determined by the presence of certain formulae on B (the *premiss* for that expansion).

This means in particular that branches are regarded as sets, and that tableau rules are monotonic such that, when formulae are added to the branch, previous tableau rule applications are not invalidated.

Conditions 18.5 and 18.6 are syntactic restrictions that intuitively prohibit 'strange' behaviour of calculi. However, there are useful calculi that violate these conditions in certain ways and, thus, are—unfortunately—excluded: For example, (a) calculi that use free variables and apply substitutions to the whole tableau, and (b) calculi using rules with constraints demanding that certain symbols are *new* to the branch or the whole tableau. At least the latter type of rules can often be replaced by similar rules not violating Condition 18.6:

Example 18.7. In calculi for first-order predicate logic, often a tableau rule is used that allows one to derive $\phi(c)$ from formulae of the form $(\exists x)(\phi(x))$, where c is a constant symbol *new* to the tableau (or the branch); this rule violates Condition 18.6 because it demands the *absence* of formulae containing c.

If instead a special constant symbol c_ϕ is used that does not have to be new, then the rule satisfies Condition 18.6 above. Soundness is preserved provided that c_ϕ is not introduced into the tableau in any other way than by Skolemising $(\exists x)(\phi(x))$ (in particular, Skolem constants c_ϕ must not occur in the initial tableau).[61]

As we said before, we allow formulae from an extended signature Σ^* to be used in a tableau proof: the tableau formulae that are put on the tableau initially (and which are tested for satisfiability) are taken from the non-extended logic $\mathbf{L}^\Sigma$. This allows us, for example, to introduce Skolem symbols during the proof that are not elements of Σ (this possibility is important for the technique shown in Example 18.7 to work).

We now proceed to formally define our (syntactic) notions of tableaux and a tableau calculi:

Definition 18.8. Given a logic $\mathbf{L}^{\Sigma^*}$ and a set *Lab* of labels, a *tableau formula* $\mathsf{S}\phi{:}\sigma$

[61] This rule for existential formulae is an adaptation of the rule presented in [Beckert *et al.*, 1993] to the ground case [Beckert and Ahrendt, 1997].

consists of a label $\sigma \in \mathit{Lab}$, a truth value sign $\mathsf{S} \in \{\mathsf{T}, \mathsf{F}\}$, and a formula $\phi \in \mathit{Form}^{\Sigma^*}$. The set of all tableau formulae is denoted with $\mathit{TabForm}^{\Sigma^*}$.

A *tableau* is a finitely branching tree whose nodes are labelled with tableau formulae.

A *branch* of a tableau T is a maximal path in T. The set of formulae on a branch B is denoted with $\mathit{Form}(B)$.

A *branch extension* is a finite set of tableau formulae.

A tableau rule *conclusion* is a finite set of branch extensions or the symbol $\bot$ (which indicates branch closure).

A tableau calculus $\mathcal{C}$ for a logic $\mathbf{L}$ has (different) 'instances' $\mathcal{C}^{\Sigma}$ for each signature $\Sigma \in \mathit{Sig}$.

Definition 18.9. A tableau calculus $\mathcal{C}$ for a logic $\mathbf{L}$ is for each signature $\Sigma \in \mathit{Sig}$ specified by:

- an extension Σ^* of the signature Σ;
- a set Lab^{Σ} of labels and an initial label $\sigma^0 \in \mathit{Lab}^{\Sigma}$;
- a tableau (expansion and closure) rule $\mathcal{R}^{\Sigma}$, i.e., a function that assigns to each finite set $\Pi \subset \mathit{TabForm}^{\Sigma}$ of tableau formulae (each premiss)—and thus to each tableau branch B with $\Pi \subset \mathit{Form}(B)$—a set $\mathcal{R}^{\Sigma}(\Pi)$ of (possible) conclusions, satisfying the following conditions:
  1. $\mathcal{R}^{\Sigma}(\Pi)$ may be infinite but has to be enumerable;
  2. $\mathcal{R}^{\Sigma}(\Pi) \subset \mathcal{R}^{\Sigma}(\Pi \cup \Pi')$ for all $\Pi, \Pi' \subset \mathit{TabForm}^{\Sigma^*}$ (monotonicity).

In practice, tableau rules are often described by means of rule schemata[62] (see Section 18.5.2 for an example). This fits perfectly in our framework, with the exception that different rule schemata are usually considered to define (or to *be*) different rules, whereas we consider rule schemata to define different subcases of one (single) rule.

Definition 18.10. Given a tableau calculus $\mathcal{C}$ for a logic $\mathbf{L}$ and a signature Σ, the set of all tableaux for a finite set $\Gamma \subset \mathit{TabForm}^{\Sigma}$ of tableau formulae is defined inductively as follows:

1. A linear tableau whose nodes are labelled with the formulae in Γ is a tableau for Γ (an initial tableau).
2. If T is a tableau for Γ, B is a branch of T, and $C \neq \bot$ is a conclusion in $\mathcal{R}(\Pi)$ for a premiss $\Pi \subset \mathit{Form}(B)$, then a tableau T' is a tableau for Γ that is constructed from T as follows: the branch B is extended by a new subbranch for each extension in C, where the nodes in that subbranch are labelled with the tableau formulae in the extension.

A tableau is *well formed* if there is a set $\Gamma \subset \mathit{TabForm}^{\Sigma}$ such that T is a tableau for Γ.[63]

[62] Then, premises and conclusions are separated by a horizontal bar, while vertical bars in the conclusion denote different extensions.

[63] Not every tableau T is well formed, because it may be impossible to derive it from any initial tableau using tableau rule applications.

T is a tableau for a finite set $G \subset Form^{\Sigma}$ of formulae if it is a tableau for the set

$$\{\mathsf{T}\phi : \sigma^0 \mid \phi \in \Gamma\}$$

of tableau formulae.

Definition 18.11. Given a tableau calculus $\mathcal{C}$ for a logic $\mathbf{L}$ and a signature Σ, a tableau branch B is *closed* if $\bot \in \mathcal{R}(\Pi)$ for a premiss $\Pi \subset Form(B)$; a tableau is *closed* if all its branches are closed.

A tableau calculus is in general non-deterministic, since there may be any—even an infinite—number of possible conclusions that can be derived from premises on the tableau.

Conditions 18.5 and 18.6 above, which are purely syntactical, still allow calculi to behave 'strangely'. Formulae could be added to the tableau that syntactically encode certain derived knowledge, but whose semantics (i.e., truth value) has nothing to do with that of the original formulae. For example, two symbols of the signature could be used to transform the given formulae into a binary representation, and tableau rules could be employed that operate on that representation. Such calculi cannot be fibred in a uniform way (at least not without an understanding of the encoding used). To ensure a more 'conservative' behaviour one could impose additional syntactical restrictions, for example only consider tableau rules that are analytic. However, the property of tableau rules that has to be guaranteed is more of a semantic nature: the result of a rule application should be semantically related to its premiss. Before this can be formalised, the semantics of tableaux has to be defined; the first semantical condition (Condition 18.12), is already part of that definition:

Condition 18.12. The labels that are part of tableau formulae represent worlds in models and do not contain other information, and the truth value signs encode truth and falsehood of a formula.

Definition 18.13. Given a tableau calculus $\mathcal{C}$ for a logic $\mathbf{L}$ and a signature Σ, a *tableau interpretation* for $\mathcal{C}^{\Sigma}$ is a pair $\langle \mathbf{m}, I \rangle$ where

1. $\mathbf{m} \in \mathcal{M}^{\Sigma^*}$ is a model for the extended signature Σ^*.
2. I is a partial function that assigns worlds of $\mathbf{m}$ to labels $\sigma \in Lab^{\Sigma}$, such that $I(\sigma^0) = w^0$, i.e., I assigns to the initial label σ^0 the initial world w^0 of $\mathbf{m}$.

A tableau formula $\mathsf{S}\phi : \sigma \in Form^{\Sigma^*}$ is *satisfied* by $\langle \mathbf{m}, I \rangle$ if

1. $I(\sigma)$ is defined, and
2. (a) $\mathsf{S} = \mathsf{T}$ and ϕ is true in $I(\sigma)$ or (b) $\mathsf{S} = \mathsf{F}$ and ϕ is false in $I(\sigma)$.

A tableau branch B is *satisfied* by $\langle \mathbf{m}, I \rangle$ if all tableau formulae on B are satisfied by $\langle \mathbf{m}, I \rangle$. A tableau is *satisfied* by $\langle \mathbf{m}, I \rangle$, if at least one of its branches is satisfied by $\langle \mathbf{m}, I \rangle$.

Often, only tableau interpretations satisfying additional conditions are considered (for example, to define the semantics of first-order tableaux, only tableau interpretations are used whose first part is a Herbrand model are used). In the following, the set of these tableau interpretations that are actually used to define the semantics of a calculus is denoted with *TabInterp*$^{\Sigma^*}$.

The next four—semantical—conditions we impose to make calculi 'well behaved', resemble the properties that a tableau calculus is shown to have in a classical soundness and completeness proof.

Condition 18.14. *Appropriateness of the set of tableau interpretations:* If a set $G \subset$ *Form*$^{\Sigma}$ is satisfiable, then there is a tableau interpretation in *TabInterp*$^{\Sigma^*}$ that satisfies the initial tableau for G (which is important for soundness).

If a tableau interpretation $\langle \mathbf{m}', I \rangle \in$ *TabInterp*$^{\Sigma^*}$ satisfies a tableau for a set $G \subset$ *Form*$^{\Sigma}$ of formulas (and thus the initial tableau for G), then the model $\mathbf{m}'$ can be restricted to a model $\mathbf{m} \in \mathcal{M}^{\Sigma}$ that satisfies G (which is important for completeness).

Condition 18.15. *Soundness of expansion (preliminary version):* If there is a tableau interpretation $\langle \mathbf{m}, I \rangle \in$ *TabInterp*$^{\Sigma^*}$ satisfying a tableau T and T' is the result of expanding T (applying the expansion rule), then there is a tableau interpretation $\langle \mathbf{m}', I' \rangle \in$ *TabInterp*$^{\Sigma^*}$ satisfying T'.

Condition 18.16. *Soundness of closure:* If a tableau branch is closed, then it is *not* satisfied by any tableau interpretation in *TabInterp*$^{\Sigma^*}$.

Before Condition 18.18 can be formulated, which establishes completeness of a calculus, the notion of *fully expanded* branches has to be defined.[64] Intuitively a branch is fully expanded if no expansion rule application can add any new formulae to the branch.

Definition 18.17. Given a tableau calculus $\mathcal{C}$ for a logic $\mathbf{L}$ and a signature Σ, a tableau branch B is *fully expanded* if $E \subset$ *Form*(B) for all premises $\Pi \subset$ *Form*(B), all possible conclusions $C \in \mathcal{R}^{\Sigma}(\Pi)$, and all extensions $E \in C$.

Condition 18.18. *Completeness:* If a non-closed tableau branch B is fully expanded, then there is a tableau interpretation $\langle \mathbf{m}, I \rangle \in$ *TabInterp*$^{\Sigma^*}$ satisfying B.

Any calculus satisfying the above conditions is sound and complete.

Theorem 18.19. *A tableau calculus $\mathcal{C}$ for a logic* $\mathbf{L}$ *that satisfies (for all signatures $\Sigma \in$ Sig) the Conditions 18.14–18.18 is sound and complete, i.e., for all finite sets $G \subset$ Form$^{\Sigma}$ the following holds:*

There is a closed tableau for G if and only if G is not satisfiable.

Proof *Soundness:* Let G be satisfiable. We prove by contradiction that there is no closed tableau for G: Suppose that there is a sequence $T_0, \ldots, T_n$ $(n \geq 0)$ of tableaux for G such that T_0 is an initial tableau for G, T_n is closed, and T_i is constructed from T_{i-1} by an expansion rule application $(1 \leq i \leq n)$. Since T_n is closed, all its branches are closed. Thus, by Condition 18.16, none of its branches is satisfied by any tableau interpretation; therefore T_n is not satisfied by any tableau interpretation.

[64]Note, that this definition relies on the fact that tableau rules are monotonic w.r.t. the expansion of branches. Without that property of tableau rules, it is difficult to define the notion of fully expanded branches in a uniform way.

However, since G is satisfiable, by Condition 18.14 (first part), there is a tableau interpretation $\langle \mathbf{m}, I\rangle \in \mathit{TabInterp}^{\Sigma^*}$ satisfying the initial tableau T_0. Therefore, by induction on i and using Condition 18.20, all the tableaux T_i $(1 \leq i \leq n)$ are satisfied by some tableau interpretation, contradicting the fact that T_n is not satisfied by any tableau interpretation.

Thus, the assumption is wrong, and there is indeed no closed tableau for G.

Completeness: Let G be unsatisfiable. Let $(T_n)_{n \geq 0}$ be a fair sequence of tableaux for G, i.e., all possible conclusions in $\mathcal{R}(\Pi)$ for all premises Π occurring in one of the T_i is sooner or later used to expand every non-closed branch on which Π occurs (where corresponding branches in T_i and T_{i+j} are identified). Because of the monotonicity of $\mathcal{R}$ (Condition 18.6) such a sequence can be constructed for all formula sets G.

Let T_∞ be the infinite tableau that is approximated by the sequence $(T_n)_{n \geq 0}$; according to its construction, every non-closed branch in T_∞ is fully expanded (Definition 18.17). Suppose there is a non-closed (and thus fully expanded) branch B in T_∞. Then, by Condition 18.18, B is satisfied by a tableau interpretation $\langle \mathbf{m}', I\rangle \in \mathit{TabInterp}^{\Sigma^*}$. Since all the formulae in the initial tableau T_0 are on B, the initial tableau is satisfied by $\langle \mathbf{m}', I\rangle$ as well. Therefore, by Condition 18.14 (second part), $\mathbf{m}'$ can be restricted to a model $\mathbf{m} \in \mathcal{M}^\Sigma$ that satisfies G, in contradiction to the fact that G is unsatisfiable. Thus, the assumption that there is a non-closed branch in T_∞ is wrong, and indeed there is a premiss Π in each branch of T_∞ such that $\perp \in \mathcal{R}(\Pi)$. Then, because (by definition) premises are finite and the tableaux are finitely branching, König's lemma implies that there is an $n \geq 0$ such that the finite sub-tableau T_n of T_∞ is closed. This concludes the proof, because T_n is by construction a closed tableau for G. □

To be useful for fibring, however, a calculus has to satisfy two additional conditions. The first of these replaces Condition 18.15:

Condition 18.20. *Soundness of expansion:* If a tableau T is satisfied by a tableau interpretation $\langle \mathbf{m}, I\rangle \in \mathit{TabInterp}^{\Sigma^*}$ and T' is the result of expanding T (applying the expansion rule), then T' is satisfied by *the same* tableau interpretation $\langle \mathbf{m}, I\rangle$.

Example 18.21. The tableau rule for first-order logic from Example 18.7 allows one to derive $\phi(c_\phi)$ from $(\exists x)(\phi(x))$. Thus, the occurrence of the constant c_ϕ on a branch B implicitly implies $(\exists x)(\phi(x))$ to be present on B; therefore, B is unsatisfiable if it contains the formula $\neg\phi(c_\phi)$. Thus, such a tableau rule is only sound if one can be sure that the constant c_ϕ is not present on the initial tableau. This is the case if a signature Σ is used that does not contain any of the constants c_ϕ and an extended signature Σ^* that contains a constant c_ϕ for every $\phi \in \mathit{Form}^{\Sigma^*}$.

The last condition ensures that the calculus is 'analytical down to the atomic level' in the sense that the set of all the tableau interpretations satisfying a certain fully expanded branch B are already characterised by the atoms on B.[65]

[65] Tableau rules do *not* have to be analytic in the sense that all formulae in a conclusion are sub-formulae of formulae in the premiss.

Condition 18.22. If a tableau branch B is fully expanded, then every tableau interpretation satisfying the *atomic formulae* on B satisfies the whole branch B.

This (additional) condition is not vital for completeness of the calculus itself, but is needed to ensure completeness when the calculus is used for fibring.

Example 18.23. To satisfy Condition 18.18, in a tableau calculus for a modal logic it has to be possible to add to a tableau branch containing the formula $\mathsf{T}\Box p{:}\sigma$ the formulae $\mathsf{T}p{:}\tau$ for all labels τ representing a world reachable from σ.

Example 18.24. In a tableau calculus for classical propositional logic it must be possible to expand a branch containing $p \vee q$ such that there are a new subbranches containing p and q, respectively, even if one of these atoms is *pure*, i.e., occurs only positively on the branch.

Definition 18.25. A tableau calculus $\mathcal{C}$ for a logic $\mathbf{L}$ is *well-suited for fibring* if, for all signatures $\Sigma \in Sig$, there is a set $TabInterp^{\Sigma^*}$ of tableau interpretations such that Conditions 18.14–18.22 are satisfied.[66]

18.4 Fibring of Logics

To fibre two logics $\mathbf{L}_1$ and $\mathbf{L}_2$ means to consider a logic whose formulae are constructed from symbols and operators from both logics. In a first step we consider a logic $\mathbf{L}_{(1,2)}$ where $\mathbf{L}_2$-formulae can occur inside $\mathbf{L}_1$-formulae but not vice versa.

Example 18.26. Suppose $\mathbf{L}_1 = \mathbf{L}_{\mathrm{mod}}$ and $\mathbf{L}_2 = \mathbf{L}_{\mathrm{PL1}}$, then

$$\begin{array}{l} (\forall x)(p(x)) \\ \Box q \\ \Box(\forall x)(p(x)) \\ \Box(\forall x)(p(x)) \to \Diamond(\exists x)(q(x)) \end{array}$$

are in $Form^{\Sigma_{(1,2)}}$, but *not* the formula $(\forall x)(\Box p(x))$.

Logics that are a full combinations of two or more logics (where expressions from the different logics can be nested arbitrarily) can be handled by inductively repeating the construction presented here.

The semantics of the fibred logic is defined using models that consist of an $\mathbf{L}_1$-model $\mathbf{m}_1$ and $\mathbf{L}_2$-models, where each world in $\mathbf{m}_1$ is assigned an $\mathbf{L}_2$-model. Which $\mathbf{L}_2$-model can be assigned to a world is restricted (using a fibring relation $\sim_{(1,2)}$).

Definition 18.27. Given two logics $\mathbf{L}_1$ and $\mathbf{L}_2$, and a *fibring relation* $\sim_{(1,2)}$, i.e., a relation between the set of pairs $\langle \mathbf{m}_1, w_1 \rangle$, where $\mathbf{m}_1 \in \mathcal{M}_1$ and w_1 is a world in $\mathbf{m}_1$, on the one hand and $\mathcal{M}_2$ on the other hand, fibring is possible, if for all signatures $\Sigma_1 \in Sig_1$ and $\Sigma_2 \in Sig_2$ there is a signature $\Sigma_{(1,2)} \in Sig_1$ such that

- $Form_1 \cap Form_2 = \varnothing$;
- $Atom_{(1,2)} = Atom_1 \cup Atom_2$.[67]

[66] Note, that Condition 18.5–18.12 are part of the definition of tableau calculi resp. tableau interpretations.

[67] Note that $Atom_o t$ and $Form_{(1,2)}$ are defined as $Sig_{(1,2)}$ is an $\mathbf{L}_1$-signature.

An $\mathbf{L}_{(1,2)}$ model $\mathbf{m}_{(1,2)}$ consists of an $\mathbf{L}_1$-model $\mathbf{m}_1$ and a fibring function $\mathcal{F}$, that assigns to each world w in $\mathbf{m}_1$ an $\mathbf{L}_2$-model $\mathbf{m}_2$ such that $\langle \mathbf{m}_1, w\rangle \sim \mathbf{m}_2$.

The set of worlds of $\mathbf{m}_{(1,2)}$ is identical to the set of worlds of $\mathbf{m}_1$ and the initial world of $\mathbf{m}_{(1,2)}$ is the initial world of $\mathbf{m}_1$. The relation $\models_{(1,2)}$ is identical to that of $\mathbf{m}_1$

$\mathbf{m}_{(1,2)}$ is, however, only a model for $\mathbf{L}_{(1,2)}$ provided that for all worlds w and formulae $\phi \in \mathit{Form}_2$:

$$w \models_{(1,2)} \phi \text{ iff } \mathcal{F}(\langle \mathbf{m}, w\rangle) \models \phi.$$

18.5 Example: First-order Predicate Logic

18.5.1 *The logical system of first-order logic*

According to Definition 18.1, to specify the logical system $\mathbf{L}_{\mathrm{PL1}}$ of first-order logic, the set of signatures $\mathit{Sig}_{\mathrm{PL1}}$ and the syntax and semantics of $\mathbf{L}_{\mathrm{PL1}}$ have to be defined.

18.5.1.1 *The Syntax of* $\mathbf{L}_{\mathrm{PL1}}$ $\mathit{Sig}_{\mathrm{PL1}}$ is the set of all *first-order signatures* $\Sigma = \langle P_\Sigma, F_\Sigma\rangle$ consisting of a non-empty set P_Σ of predicate symbols and a set F_Σ.

For Skolemisation we do not use function symbols from F_Σ but from a special infinite set F_{sko} of *Skolem function symbols* that is disjoint from F_Σ; thus, the extended signature Σ^* is $\langle P_\Sigma, F_\Sigma \cup F_{sko}\rangle$. The symbols in P_Σ, F_Σ and F_{sko} may be used with any arity $n \geq 0$; in particular function symbols can be used as constant symbols (arity 0). In addition, there is an infinite set $\mathcal{V}$ of *object variables*.

The *logical operators* are the connectives $\vee$ (disjunction), $\wedge$ (conjunction) and $\neg$ (negation), the quantifier symbols $\forall$ and $\exists$, and the constant operators $\top$ and $\bot$.[68] Formulae that are identical up to associativity of $\vee$ and $\wedge$ are identified.

The notions of free and bound variable, literal,[69] and sentence (a formula not containing free variables) are defined as usual.

Definition 18.28. Given a signature Σ, the set Term^Σ of *terms* over Σ is defined by:

1. All variables $x \in \mathcal{V}$ and all constants $c \in \mathcal{F}^\Sigma$ are terms over Σ.
2. If $t_1, \ldots, t_n$ are terms over Σ and $f \in \mathcal{F}^\Sigma$, then $f(t_1, \ldots, t_n)$ is a term over Σ.

If $t_1, \ldots, t_n$ are terms over Σ and $p \in \mathcal{P}\Sigma$, then $p(t_1, \ldots, t_n)$ is an *atom* over Σ. Atom^Σ is the set of all ground atoms over Σ (i.e., atoms not containing variables).

The set of all *formulae* (including non-sentences) over Σ is defined by:

1. $\top$, $\bot$ and atoms over Σ are formulae over Σ.
2. If ϕ is a formula over Σ, then $\neg\phi$ is a formula over Σ.
3. If $\phi_1, \ldots, \phi_n$, $n \geq 2$, are formulae over Σ none of which is a conjunction (disjunction), then $\phi_1 \wedge \ldots \wedge \phi_n$ (resp. $\phi_1 \vee \ldots \vee \phi_n$) is a conjunction (disjunction) and a formula over Σ.
4. If ϕ is a formula over Σ and $x \in \mathcal{V}$, then $(\forall x)(\phi)$ and $(\exists x)(\phi)$ are formulae over Σ.

[68] Implication and equivalence are considered to be defined operators, i.e., $\phi \rightarrow \psi$ is the same as $\neg\phi \vee \psi$, and $\phi \leftrightarrow \psi$ is the same as $(\phi \wedge \psi) \vee (\neg\phi \vee \neg\psi)$.

[69] $\top$ and $\bot$ are considered to be literals (but they are not atoms).

The set $Form^{\Sigma}_{(1,2)}$ contains all formulae over Σ that are sentences (i.e., do not contain free variables).

Definition 18.29. A first-order *structure* $\langle D, I\rangle$ for a signature Σ consists of a domain D and an interpretation I, which gives meaning to the function and predicate symbols of Σ.

A *variable assignment* is a mapping $\mu : \mathcal{V} \to D$ from the set of variables to the domain D.

The combination of an interpretation I and an assignment μ associates (by structural recursion) with each term t over Σ an element $t^{I,\mu}$ in D.

The *evaluation function* $val_{I,\mu}$ is for all formulae $\phi \in Form^{\Sigma}$ defined by: $val_{I,\mu}(\phi) = \top$ if

- $\phi = p(t_1, \ldots, t_n)$ and $\langle t_1^{I,\mu}, \ldots, t_n^{I,\mu}\rangle \in p^I$;
- $\phi = \neg p(t_1, \ldots, t_n)$ and $\langle t_1^{I,\mu}, \ldots, t_n^{I,\mu}\rangle \notin p^I$;
- $\phi = \top$;
- $\phi = \neg\bot$;
- $\phi = \alpha$ and $val_{I,\mu}(\alpha_i) = \top$ for all α_i;
- $\phi = \beta$ and $val_{I,\mu}(\beta_i) = \top$ for some β_i;
- $\phi = \gamma$ and $val_{I,\mu[x\leftarrow d]}(\gamma_1) = \top$ for all $d \in D$;[70] or
- $\phi = \delta$ and $val_{I,\mu[x\leftarrow d]}(\delta_1) = \top$ for some $d \in D$;

and $val_{I,\mu}(\phi) = \bot$ otherwise.

$\mathcal{M}_o t^{\sigma}$ is the set of all *first-order models*, where each model $\mathbf{m}$ contains a single (initial) world $w^0 = \langle D, I\rangle$ that is a first-order structure.[71] The relation $\models_{\mathrm{PL1}}$ is defined for all first-order models defined by:

$$w^0 \models_{\mathrm{PL1}} \phi \;\text{ iff }\; val_{I,\mu}(\phi) = \top \;\text{ for all assignments } \mu$$

18.5.2 *A tableau calculus for first-order logic*

Since models of first-order logic consist of only one world (the model itself), the set Lab^{Σ} of labels contains for all signatures only the initial label $\sigma_0 = 1$, and all tableau formulae are of the form $\mathsf{S}\phi{:}1$.[72]

Following Smullyan [Smullyan, 1995], the set of formulae in $Form^{\Sigma}$ that are not literals is divided into four classes: α for formulae of conjunctive type, β for formulae of disjunctive type, γ for quantified formulae of universal type and finally δ for quantified formulae of existential type (unifying notation). This classification is motivated by the tableau expansions which are associated with each (non-literal) formula. The rules characterise the assertion of a truth value to a formula by means of asserting truth values to its direct subformulae. For example, $\phi \wedge \psi$ holds if and only if ϕ and ψ hold.

[70] $\mu' = \mu[x \leftarrow d]$ denotes the variable assignment that is defined by: $\mu'(x) = d$ and $\mu'(y) = \mu(y)$ for all other variables y.

[71] According to Definition 18.1, all models must contain a set of worlds. Since first-order models do not consist of multiple worlds, the single world w^0 is identified with the model itself.

[72] Therefore, where no confusion can arise, the label is omitted and, i.e., $\mathsf{S}{:}\phi$ is the same as $\mathsf{S}\phi{:}\sigma^0$.

Table 18.1 *Correspondence between formulae and rule types.*

α	$\alpha_1, \ldots, \alpha_n$
$\mathsf{T}\,(\phi_1 \wedge \ldots \wedge \phi_n)$	$\mathsf{T}\,\phi_1, \ldots, \mathsf{T}\,\phi_n$
$\mathsf{F}\,(\phi_1 \vee \ldots \vee \phi_n)$	$\mathsf{F}\,\phi_1, \ldots, \mathsf{F}\,\phi_n$
$\mathsf{T}\,\neg\phi$	$\mathsf{F}\,\phi$
$\mathsf{F}\,\neg\phi$	$\mathsf{T}\,\phi$

β	$\beta_1, \ldots, \beta_n$
$\mathsf{T}\,(\phi_1 \vee \ldots \vee \phi_n)$	$\mathsf{T}\,\phi_1, \ldots, \mathsf{T}\,\phi_n$
$\mathsf{F}\,(\phi_1 \wedge \ldots \wedge \phi_n)$	$\mathsf{F}\,\phi_1, \ldots, \mathsf{F}\,\phi_n$

γ	γ_1
$\mathsf{T}\,(\forall x)(\phi(x))$	$\mathsf{T}\,\phi(x)$
$\mathsf{F}\,(\exists x)(\phi(x))$	$\mathsf{F}\,\phi(x)$

δ	δ_1
$\mathsf{F}\,(\forall x)(\phi(x))$	$\mathsf{F}\,\phi(x)$
$\mathsf{T}\,(\exists x)(\phi(x))$	$\mathsf{T}\,\phi(x)$

Definition 18.30. The non-literal formulae in $Form^{\Sigma^*}$ are assigned a *type* according to Table 18.1. A formula of type $\xi \in \{\alpha, \beta, \gamma, \delta\}$ is called a *ξ-formula.*

A ξ-formula ϕ is assigned a sequence $\phi_1, \ldots, \phi_n$, $n \geq 1$, of *immediate tableau subformulae* (see Table 18.1).[73] $\Xi(\phi)$ denotes the set of all immediate tableau subformulae of ϕ. The *tableau subformula* relation is the reflexive, transitive closure of the immediate tableau subformula relation.

A ξ-formula ψ that is a tableau subformula of a formula ϕ is called a *ξ-subformula* of ϕ.

The letters α, β, γ, and δ are used to denote formulae of (and only of) the appropriate type. In the case of γ- and δ-formulae the variable x bound by the (top-most) quantifier is made explicit by writing $\gamma(x)$ and $\gamma_1(x)$ (resp. $\delta(x)$ and $\delta_1(x)$); accordingly $\gamma_1(t)$ denotes the result of replacing all occurrences of x in γ_1 by t.

To comply with Condition 18.5, which does not allow the applications of substitutions (to the whole tableau), we use the classical *ground* version of tableaux for first-order logic.[74]

In Table 18.2 the ground expansion rule schemata for the various formula types are given schematically. The formulae in an extension are implicitly conjunctively connected, and different extensions are implicitly disjunctively connected. We use n-ary α- and β-rules, i.e., when the β-rule is applied to a formula $\psi = \phi_1 \vee \ldots \vee \phi_n$, then ψ is broken up into n subformulae (instead of splitting it into two formulae $\phi_1 \vee \ldots \vee \phi_r$ and $\phi_{r+1} \vee \ldots \vee \phi_n$, $1 \leq r < n$).

Besides using n-ary rules we deviate from the classic definition of tableaux (as given by Smullyan) by using an improved δ-rule that, for the purpose of constructing the Skolem term, does not introduce a *new* Skolem function symbol. Rather, each equivalence class of δ-formulae identical up to variable renaming is assigned its own unique Skolem symbol (which can be seen as a Gödelisation of that class) [Beckert *et al.*, 1993]. This δ-rule not only complies with Condition 18.6; in addition, it is easier to

[73] Note that the index i ($1 \leq i \leq n$) of the tableau subformulae is an operator; thus, β_2 is by definition the second β-subformula.

[74] This version of tableaux is called 'ground', because universally quantified variables are replaced by *ground* terms when the γ-rule is applied.

Table 18.2 *Rule schemata for the ground version of tableaux.*

$\dfrac{\alpha}{\alpha_1}$	$\dfrac{\beta}{\beta_1 \mid \cdots \mid \beta_n}$	$\dfrac{\gamma(x)}{\gamma_1(t)}$	$\dfrac{\delta(x)}{\delta_1(c)}$
$\vdots$		where t is any ground term.	where $c = sko(\delta)$.
α_n			

implement than the classical one; and it guarantees that only a finite number of different symbols needs to be used in a tableau proof, thus restricting the search space.

Definition 18.31. Given a signature $\Sigma = \langle P^\Sigma, F^\Sigma \rangle \in Sig_{\text{PL1}}$, the function *sko* assigns to each $\delta \in Form^{\Sigma^*}$ a symbol $sko(\delta) \in F_{sko}$ such that (a) $sko(\delta) > f$ for all $f \in F_{sko}$ occurring in δ, where $>$ is an arbitrary but fixed ordering on F_{sko}, and (b) for all $\delta, \delta' \in Form^\Sigma$ the symbols $sko(\delta)$ and $sko(\delta')$ are identical if and only if δ and δ' are identical up to renaming of bound variables.

The purpose of condition (a) in the above definition of *sko* is to avoid cycles like: $sko(\delta) = f$, f occurs in δ', $sko(\delta') = g$, and g occurs in δ.

We now have everything at hand to formally define our calculus $\mathcal{C}_{\text{PL1}}$ for the logic $\mathbf{L}_{\text{PL1}}$.

Definition 18.32. The calculus $\mathcal{C}_{\text{PL1}}$ for first-order predicate logic is defined for every signature $\Sigma = \langle P^\Sigma, F^\Sigma \rangle \in Sig_{\text{PL1}}$ by:

- $\Sigma^* =$ is $P^\Sigma, F^\Sigma \cup F_{sko}\rangle$.
- $Lab^\Sigma = \{1\}$ and $\sigma^0 = 1$.
- For all premises $\Pi \subset TabForm_{\text{PL1}}$, the set $\mathcal{R}_{\text{PL1}}(\Pi)$ of possible conclusions is the smallest set such that $\mathcal{R}_{\text{PL1}}$ satisfies the monotonicity condition for expansion rules (see Definition 18.9) and the following conditions are satisfied:
  1. If $\alpha \in \Pi$, then $\{\{\alpha_1, \alpha_2\}\} \in \mathcal{R}_{\text{PL1}}(\Pi)$.
  2. If $\beta \in \Pi$, then $\{\{\beta_1\}, \{\beta_2\}\} \in \mathcal{R}_{\text{PL1}}(\Pi)$.
  3. If $\gamma \in \Pi$, then $\{\{\gamma_1(t)\}\} \in \mathcal{R}_{\text{PL1}}(\Pi)$ for all ground terms $t \in Term^\Sigma$.
  4. If $\delta \in \Pi$, then $\{\{\delta(c)\}\} \in \mathcal{R}_{\text{PL1}}(\Pi)$, where $c = sko(\delta)$.
  5. If $\mathsf{T}\bot : 1 \in \Pi$, $\mathsf{F}\top : 1 \in \Pi$, or $\mathsf{T}\phi : 1 \subset \Pi$ and $\mathsf{F}\phi : 1 \in \Pi$ for any $\phi \in Form^\Sigma$, then $\bot \in \mathcal{R}_{\text{PL1}}(\Pi)$.

To define the semantics of tableaux in PL1 we use the set $TabInterp^{\Sigma^*}$ that contain all *canonical* tableau interpretations:

Definition 18.33. A tableau interpretation is *canonical* if the first-order structure $\langle D, I \rangle$ it contains satisfies the following conditions:

1. $\langle D, I \rangle$ is a term structure, i.e., D is the set of all ground terms in $Term^\Sigma$.
2. For all variable assignments μ and all $\delta(x) \in TabForm^{\Sigma^*}$ the following holds:

$$\text{if} \quad (\mathcal{M}, \mu) \models \delta(x) \quad \text{then} \quad (\mathcal{M}, \mu) \models \delta_1(c)$$

where $c = sko(\delta)$.

Before we prove that the calculus $\mathcal{C}_{\mathrm{PL1}}$ is well suited for fibring, we formulate and prove the appropriate version of Hintikka's lemma,

Definition 18.34. A set $H \subset \mathit{TabForm}^{\Sigma^*}$ of tableau formulae is a *Hintikka* set if it satisfies the following conditions:

1. $\mathsf{T}\bot : 1 \notin H, \mathsf{F}\top : 1 \notin H$, and there are no complementary formulae $\mathsf{T}\phi : 1, \mathsf{F}\phi : 1$ in H.
2. If $\alpha \in H$, then all α_i are in H.
3. If $\beta \in H$, then some β_i is in H.
4. If $\gamma(x) \in H$, then $\gamma_1(t) \in H$ for all ground terms t in Term^{Σ^*}.
5. If $\delta(x) \in H$, then $\delta_1(t) \in H$ for some ground term t in Term^{Σ^*}.

Lemma 18.35 (Hintikka). *If H is a Hintikka set, then*

1. *H is satisfied by some tableau interpretation in $\mathit{TabForm}^{\Sigma^*}$.*
2. *Every tableau interpretation satisfying the atomic tableau formulae in H satisfies H.*

Proof The second part of the lemma is easy to prove by induction on the structure of tableau formulae in H.

A canonical tableau interpretation satisfying the atomic formulae in H can be defined by: $p^I(t_1, \ldots, t_k) = \top$ iff $p(t_1, \ldots, t_k) \in H$ for all ground atoms $p(t_1, \ldots, t_k)$ over Σ^*, which—using the second part of the lemma—proves the first part of the lemma. □

Theorem 18.36. *The tableau calculus for first-order predicate logic (Definition 18.32) is suitable for fibring.*

Proof We have to check that—using the set $\mathit{TabInterp}^{\Sigma^*}$ of tableau interpretations from Definition 18.33—the calculus $\mathcal{C}_{\mathrm{PL1}}$ satisfies Conditions 18.14–18.18.

Condition 18.14 (Part 1): If $G \subset \mathit{Form}^{\Sigma}$ is satisfiable, then, by Herbrand's theorem, there is a model consisting of a term structure $\langle D, I\rangle$ that satisfies G, where D is the set of ground terms over Σ. The Skolem function symbols in F_{sko} do not occur in G, therefore it suffices to choose their interpretation such that the resulting structure $\langle D, I^*\rangle$ is canonical, leaving the interpretation of the symbols in Σ unchanged.

The rank $rk(f)$ of the symbols $f \in F_{sko}$ is defined as follows:

- $rk(f) = 0$ if no $\delta \in \mathit{Form}^{\Sigma^*}$ exists such that $\mathit{sko}(\delta) = f$;
- $rk(f) = 1$ if $f = \mathit{sko}(\delta)$ for some $\delta \in \mathit{Form}^{\Sigma}$;
- $rk(f) = 1 + \max\{rk(f') \mid f' \in F_{sko}$ occurs in δ where $f = \mathit{sko}(\delta)\}$.

$rk(f)$ is well defined because of condition (a) in Definition 18.31.

We inductively define a sequence $(\langle D, I^n\rangle)_{n\geq 0}$ of structures that all have the domain D. $\langle D, I^n\rangle$ is a structure over the signature Σ^n that is the restriction of Σ^* to function symbols of rank not greater than n; I^{n+1} coincides with I^n on all symbols in $\Sigma^n \cup \Sigma$. I^0 is defined by $f^{I^0} = f^I$ for all $f \in F_\Sigma$, and for all $f \in F_{sko}$ of rank 0 the value of f^{I^0} is chosen arbitrarily. The function symbols

$f \in F_{sko}$ of rank $r \leq n$ have already been interpreted in $\mathcal{M}^n$. Consider $f \in F_{sko}$ of rank $n+1$ with $f = sko(\delta(x))$; for all argument tuples $b_1, \ldots, b_k \in D$ of the same length as the tuple $x_1, \ldots, x_k$ of free variables in $\delta(x)$ we define $f^{I^{n+1}}$ by: If there is a variable assignment μ with $\mu(x_i) = b_i$ (for $1 \leq i \leq n$), and $(\mathcal{M}^n, \mu) \models \delta$, choose an element $c \in D$ with $(\mathcal{M}^n, \mu[x \leftarrow c]) \models \delta_1(x)$, and set $f^{I^{n+1}}(b_1, \ldots, b_k) = c$.[75] Otherwise, if $(\mathcal{M}^n, \mu) \not\models \delta(x)$, choose $f^{I^{n+1}}(b_1, \ldots, b_k)$ to be an arbitrary element in D.

We can think of the sequence $(\langle D, I^n \rangle)_{n \geq 0}$ as an approximation to the structure $\langle D, I^* \rangle$ over Σ^*. I^* coincides with $I^n, I^{n+1}, \ldots$ on the symbols in Σ^n. The model consisting of the structure $\langle D, I^* \rangle$ is canonical by construction and satisfies the formulae in G.

Condition 18.14 (Part 2): If $\langle \mathcal{M}', I \rangle \in \mathit{TabForm}^{\Sigma^*}$ satisfies an initial tableau for G, then any restriction of $\mathcal{M}$ to Σ satisfies G, because the symbols in F_{sko} do not occur in G.

Condition 18.15: This is a weaker version of Condition 18.20.

Condition 18.16: If a branch is closed then it contains tableau formulae $\mathsf{T}\phi : 1$ and $\mathsf{F}\phi : 1$, $\mathsf{T}\bot : 1$, or $\mathsf{F}\top : 1$. Because no formula $\phi \in \mathit{Form}^{\Sigma}$ cannot be both true and false in a first-order structure that would have to be part of a tableau interpretation satisfying B, and similarly $\top$ cannot be false and $\bot$ cannot be true, there is no such tableau interpretation.

Condition 18.18: Lemma 18.35 applies, because for every non-closed, fully expanded branch B the set $\mathit{Form}(B)$ is a Hintikka set; thus, this condition is satisfied.

Condition 18.20: Let $\langle \mathbf{m}, I \rangle \in \mathit{TabInterp}^{\Sigma^*}$ be a tableau interpretation satisfying a tableau T and let T' be the result of expanding T; let B be the branch in T that is expanded, and let $\Pi \subset \mathit{Form}(B)$ be the premiss that the expansion rule is applied to.

By assumption $\langle \mathbf{m}, I \rangle$ satisfies T; thus it satisfies some branch B^0 of T. If B^0 is different from B, then B^0 is also a branch of T' and we are through.

If, on the other hand, $B^0 = B$ and therefore $\langle \mathbf{m}, I \rangle$ satisfies Π, we show that $\langle \mathbf{m}, I \rangle$ satisfies one of the branches of T' by cases according to the formula types.

β-rule ($\phi = \beta$): T' is constructed from T by adding β_i to B obtaining B^i ($1 < i \leq k$, where k is the cardinality of $\Xi(\beta)$). Since $\langle \mathbf{m}, I \rangle$ satisfies β we have, by the property of β-formulae (Definition 18.29), that $\langle \mathbf{m}, I \rangle$ satisfies β_i for some $i \in \{1, \ldots, k\}$. Therefore it satisfies B^i.

α-rule ($\phi = \alpha$): Similar to the β-rule.

γ-rule ($\phi = \gamma(x)$): T' is constructed from T by adding $\gamma_1(t)$ to B obtaining the branch B' where t is a ground term. Since $\langle \mathbf{m}, I \rangle$ satisfies $\gamma(x)$ we have, by definition of $\models_{\mathrm{PL1}}$, that $(\mathcal{M}^*, \mu[x \leftarrow d]) \models_{\mathrm{PL1}} \gamma_1(x)$ for all elements $d \in D$. Since this is in particular true for $d = I(t)$, we get that $\langle \mathbf{m}, I \rangle$ satisfies $\gamma_1(t)$ and therefore B'.

[75] Since f is of rank $n+1$, the symbols in δ are from the signature Σ^n.

δ-rule ($\phi = \delta(x)$): The tableau T' is constructed from T by adding $\delta_1(c)$ to B obtaining the branch B', where $c = sko(\delta(x))$. Since $\langle \mathbf{m}, I \rangle$ is canonical and satisfies $\delta(x)$, it satisfies $\delta_1(c)$ and thus the branch B'.

Condition 18.22: Lemma 18.35 applies, because for every non-closed, fully expanded branch B the set $Form(B)$ is a Hintikka set; thus, this condition is satisfied. □

18.6 Fibring of Tableau Calculi

A calculus for a fibred logic $\mathbf{L}_{(1,2)}$ can be constructed from two calculi $\mathcal{C}_1$ and $\mathcal{C}_2$ for $\mathbf{L}_1$ and $\mathbf{L}_2$ that are well suited for fibring (Definition 18.25), provided that the fibring relation $\sim_{(1,2)}$ when extended to tableau interpretations can be characterised using sets of tableau formulae:

Definition 18.37. Given a fibred logic $\mathbf{L}_{(1,2)}$ and calculi $\mathcal{C}_1$ and $\mathcal{C}_2$ for $\mathbf{L}_1$ resp. $\mathbf{L}_2$ that are well suited for fibring, and given a *tableau fibring function*, i.e., a function $\mathcal{F}^T$ from $Lab_1 \times TabForm^{\Sigma^*_{(1,2)}}$ to $TabForm^{\Sigma_2}$ such that:[76]

1. *Monotonicity:* if $\Pi \subset \Pi'$ then $\mathcal{F}^T(\sigma, \Pi) \subset \mathcal{F}^T(\sigma, \Pi')$;
2. *Soundness:* if $\langle \mathbf{m}_1, I_1 \rangle$ is an $\mathbf{L}_1$-tableau interpretation, w_1 is a world in $\mathbf{m}_1$, $\langle \mathbf{m}_2, I_2 \rangle$ is an $\mathbf{L}_2$-tableau interpretation, $I(\sigma) = w_1$, $\langle \mathbf{m}_1, w_1 \rangle \sim \mathbf{m}_2, I_2 \rangle$, and $\langle \mathbf{m}_1, I_1 \rangle$ satisfies Π, then $\langle \mathbf{m}_2, I_2 \rangle$ satisfies $\mathcal{F}^T(\sigma, \Pi)$;
3. *Completeness:* if $\langle \mathbf{m}_2, I_2 \rangle$ is an $\mathbf{L}_2$-tableau interpretation that satisfies $\mathcal{F}^T(\sigma, \Pi)$ and Π is $\mathbf{L}_1$-satisfiable, then there is an $\mathbf{L}_1$-tableau interpretation $\langle \mathbf{m}_1, I_1 \rangle$ such that $I(\sigma)$ is a world w_1 in $\mathbf{m}_1$ and $\langle \mathbf{m}_1, w_1 \rangle \sim \mathbf{m}_2$;

then the fibred calculus $\mathcal{C}_{(1,2)}$ is defined by:

- $Lab_{(1,2)}^{\Sigma_{(1,2)}} = Lab_1 \cup \{(\sigma_1; \sigma_2) \mid \sigma_1 \in Lab_1^{\Sigma_{(1,2)}}, \sigma_2 \in Lab_2^{\Sigma_{(1,2)}}\}$;
- for all premises $\Pi \subset TabForm_{(1,2)}$, the set $\mathcal{R}_{(1,2)}(\Pi)$ of possible conclusions is the smallest set such that $\mathcal{R}_{(1,2)}$ satisfies the monotonicity condition for expansion rules (see Definition 18.9) and the following conditions are satisfied:
  1. $\mathcal{R}_1(\Pi_1) \subset \mathcal{R}_{(1,2)}(\Pi)$, where
  $$\Pi_1 = \{\mathsf{S}\phi : \sigma_1 \mid \mathsf{S}\phi : \sigma_1 \in \Pi, \sigma_1 \in Lab_1^{\Sigma_{(1,2)}}\};$$
  2. $\mathcal{R}_2(\Pi_2^{\Sigma_1}) \subset \mathcal{R}_{(1,2)}(\Pi)$ for all $\sigma_1 \in Lab_1^{\Sigma_{(1,2)}}$, where
  $$\Pi_2^{\Sigma_1} = \{\mathsf{S}\phi : (\sigma_1; \sigma_2) \mid \mathsf{S}\phi : (\sigma_1; \sigma_2) \in \Pi, \sigma_2 \in Lab_2^{\Sigma_2}\};$$
  3. $\{\{\mathsf{S}\phi : (\sigma_1; \sigma_2) \mid \mathsf{S}\phi : \sigma_2 \in \mathcal{F}^T(\sigma_1, \Pi)\}\} \subset \mathcal{R}_{(1,2)}(\Pi)$ for all labels $\sigma_1 \in Lab_1^{\Sigma_{(1,2)}}$;
  4. $\{\{\mathsf{S}\phi : (\sigma_1; \sigma_2^0) \mid \mathsf{S}\phi : \sigma_1 \in \Pi, \sigma_1 \in Lab_1^{\Sigma_{(1,2)}}\}\} \subset \mathcal{R}_{(1,2)}(\Pi)$.

The combined calculus does not construct separate tableaux for $\mathbf{L}_1$ and $\mathbf{L}_2$ formulae, but there is only one unified structure with one unified (set of) tableau rule(s).

[76]Note, that $\Sigma^*_{(1,2)}$ is an $\mathbf{L}_1$-signature.

Table 18.3 *Tableau rules for intuitionistic propositional logics.*

$$\frac{\mathsf{T}\phi{:}\sigma}{\mathsf{T}\phi{:}\sigma.n}$$

where n is any natural number.

$$\frac{\mathsf{T}\phi \to \psi{:}\sigma}{\mathsf{F}\phi{:}\sigma \mid \mathsf{T}\psi{:}\sigma} \qquad \frac{\mathsf{F}\phi \to \psi{:}\sigma}{\begin{array}{c}\mathsf{T}\phi{:}\sigma.n\\ \mathsf{F}\psi{:}\sigma.n\end{array}}$$

where $n = goedel(\phi \to \psi)$.

$$\frac{\mathsf{T}\phi \wedge \psi{:}\sigma}{\begin{array}{c}\mathsf{T}\phi{:}\sigma\\ \mathsf{T}\psi{:}\sigma\end{array}} \qquad \frac{\mathsf{F}\phi \wedge \psi{:}\sigma}{\mathsf{F}\phi{:}\sigma \mid \mathsf{F}\psi{:}\sigma}$$

$$\frac{\mathsf{T}\neg\phi{:}\sigma}{\mathsf{F}\phi{:}\sigma} \qquad \frac{\mathsf{F}\neg\phi{:}\sigma}{\mathsf{T}\phi{:}\sigma.n}$$

where $n = goedel(\neg\phi)$.

Theorem 18.38. *The fibred calculus $\mathcal{C}_{(1,2)}$ is well suited for fibring.*

Corollary 18.39. *The fibred calculus $\mathcal{C}_{(1,2)}$ is sound and complete,*

Since the fibred calculus is itself well suited for fibring, the fibring process can be iterated.

18.6.1 *Example: fibring calculi for intuitionistic logics*

As a further example, we fibre versions of intuitionistic logic, that differ in theories which are supposed to be true in all models of that version of the logic; the theories correspond to the 'programs' in Chapter 17, where intuitionistic logic programs are fibred.

For the sake of simplicity and differing from Chapter 17, we only consider ground formulae.

All logics we consider have the same sets of signatures, of formulae, and of tableau formulae. A signature Σ is a set of propositional variables. The formulae are built in the usual way using the operators $\wedge$ (conjunction), $\to$ (intuitionistic implication). and $\neg$ (intuitionistic negation).

The calculi for the logics all have the rules shown in Table 18.3 in common. The logics differ in the theories (programs) P that are supposed to be true in all worlds. Thus, the calculi differ in an additional rule that allows us to add the formulae from P to the tableau.

We proceed to formally define the calculus for a logic $\mathbf{L}_P$ where P is the theory (program) of the logic.

Definition 18.40. The calculus $\mathcal{C}_P^\Sigma$ is defined for every signature Σ and every finite theory (program) $P \subset Form_\Sigma$ by:

- $\Sigma^* = \Sigma$.
- *Lab* is the set of all finite sequences of natural numbers that start with 1. The initial label is $\sigma^0 = 1$.
- For all premises $\Pi \subset TabForm_{\mathrm{PL1}}$, the set $\mathcal{R}_{\mathrm{PL1}}(\Pi)$ of possible conclusions is the smallest set such that $\mathcal{R}_{\mathrm{PL1}}$ satisfies the monotonicity condition for expansion rules (see Definition 18.9) and the following conditions are satisfied:
  1. $\{\{\mathsf{T}\phi{:}\sigma\}\} \in \mathcal{R}(\Pi)$ for all $\phi \in P$ and all labels $\sigma \in Lab$.
  2. If $\mathsf{T}\phi{:}\sigma \in \Pi$, then $\{\{\mathsf{T}\phi{:}\sigma.n\}\} \in \mathcal{R}(\Pi)$ for all $n \in \mathbf{N}$.
  3. If $\mathsf{T}(\phi \to \psi){:}\sigma \in \Pi$, then $\{\{\mathsf{F}\phi{:}\sigma\}, \{\mathsf{T}\phi{:}\sigma\}\} \in \mathcal{R}(\Pi)$.
  4. If $\mathsf{T}(\phi \to \psi){:}\sigma \in \Pi$, then $\{\{\mathsf{T}\phi{:}\sigma.n, \mathsf{F}\psi{:}\sigma.n\}\} \in \mathcal{R}(\Pi)$ where $n = go\dot{e}del(\phi \to \psi)$.
  5. If $\mathsf{T}(\phi \wedge \psi){:}\sigma \in \Pi$, then $\{\{\mathsf{T}\phi{:}\sigma, \mathsf{T}\psi{:}\sigma\}\} \in \mathcal{R}(\Pi)$.
  6. If $\mathsf{F}(\phi \wedge \psi){:}\sigma \in \Pi$, then $\{\{\mathsf{F}\phi{:}\sigma\}, \{\mathsf{F}\psi{:}\sigma\}\} \in \mathcal{R}(\Pi)$.
  7. If $\mathsf{T}\neg\phi{:}\sigma \in \Pi$, then $\{\{\mathsf{F}\phi{:}\sigma\}\} \in \mathcal{R}(\Pi)$.
  8. If $\mathsf{F}\neg\phi{:}\sigma \in \Pi$, then $\{\{\mathsf{T}\phi{:}\sigma.n\}\} \in \mathcal{R}(\Pi)$ where $n = goedel(\phi \to \psi)$.
  9. If $\mathsf{T}\phi{:}\sigma \in \Pi$ and $\mathsf{F}\phi{:}\sigma \in \Pi$ for any $\sigma \in Lab$ and $\phi \in Form^\Sigma$, then $\bot \in \mathcal{R}(\Pi)$.

  goedel is any bijection from $TabForm^\Sigma$ to the set $\mathbf{N}$ of natural numbers.

The calculi defined above are suitable for fibring.

In Chapter 17 a *procedural fibring function* is defined as follows: procedural fibring functions ($i, j \in \{1, \ldots, r\}$, $i \neq j$) assign to each finite set A_i of $\mathbf{L}_i$-atoms a finite set $A_j = \mathbf{F}_{(i,j)}(A_i)$ of $\mathbf{L}_j$-atoms such that

$$\mathbf{F}_{(i,j)}(A_i\sigma) = \mathbf{F}_{(i,j)}(A_i)\sigma$$

for all $\mathbf{L}_i$-substitutions σ.

The fibring functions are extended from sets of atoms to databases (i.e. complex formulae) as follows: given an $\mathbf{L}^*$-database D_i whose clause heads are $\mathbf{L}_i$-atoms, the $\mathbf{L}^*$-database $D_j = \mathbf{F}_{(i,j)}(D_i)$ contains for each subset

$$\{\phi_1^i \to \psi_1^i,\ \ldots,\ \phi_l^i \to \psi_l^i,\ \psi_{l+1}^i,\ \ldots,\ \psi_k^i\}\ \ (k \geq l \geq 0)$$

of D_i, where $\psi_{l+1}^i, \ldots, \psi_k^i$ are atoms, the clauses in

$$\{\phi_1 \wedge \ldots \wedge \phi_l \to \psi^j \mid \psi^j \in \mathbf{F}_{(i,j)}(\{\psi_1^i, \ldots, \psi_k^i\})\}.$$

It is shown in Chapter 17 that these fibring functions are monotonic. We use the same functions here for fibring. That is, we use the fibring relation that is characterised by a given procedural fibring function.

It turns out that if the tableau calculi are restricted to formulae that are QN-Prolog queries, they allow exactly those formulae to be proven, for which the corresponding QN-Prolog programs succeed. However, QN-Prolog has to be restricted to the ground case, and the switching function that is used in fibring QN-Prolog programs does not have an equivalent in the tableau case. The behaviour of tableaux is equivalent to allowing queries to be evaluated in any program where their predicates are defined.

Example 18.41. Suppose p, p', q, q' are predicate symbols in $\mathbf{L}_1$, r, s, t, u are predicate symbols in $\mathbf{L}_2$, and $F_{(1,2)}$ is defined by:

$$\begin{aligned} \mathbf{F}_{(1,2)}(\{p,p'\}) &= \{r,s\} \\ \mathbf{F}_{(1,2)}(\{p\}) &= \{t\} \\ \mathbf{F}_{(1,2)}(\{p'\}) &= \varnothing \\ \mathbf{F}_{(1,2)}(\varnothing) &= \{u\}. \end{aligned}$$

If $D_1 = \{q \to p,\ q' \to p'\}$, then

$$D_2 = \mathbf{F}_{(1,2)}(D_1) = \{q \wedge q' \to r,\ q \wedge q' \to s,\ q \to t,\ u\}.$$

If $\mathbf{F}_{(2,1)}$ is defined by $\mathbf{F}_{(2,1)}(A) = \mathbf{F}^{-1}_{(1,2)}(A)$ for all sets A of $\mathbf{L}_2$-atoms, then

$$D_1' = \mathbf{F}_{(2,1)}(D_2) = \{q \to p,\ q \wedge q' \to p,\ q \wedge q' \to p'\}.$$

Figure 18.1 shows the proof for the goal $(q \to p) \wedge (q' \to p') \to m$, i.e., the initial tableau formula $\mathsf{F}(q \to p) \wedge (q' \to p') \to m{:}1$ using the theories (programs)

$$P_1 = \{q,\ q',\ v \to m\}$$

and

$$P_2 = \{u \wedge r \to v\}.$$

This proof is equivalent to the computation tree for Examples 17.3.7 and 17.3.9 of Chapter 17.

Remark 18.42. For an example of fibring intuitionistic and modal logics through their tableau formulation see [Gabbay and Reyle, 1997].

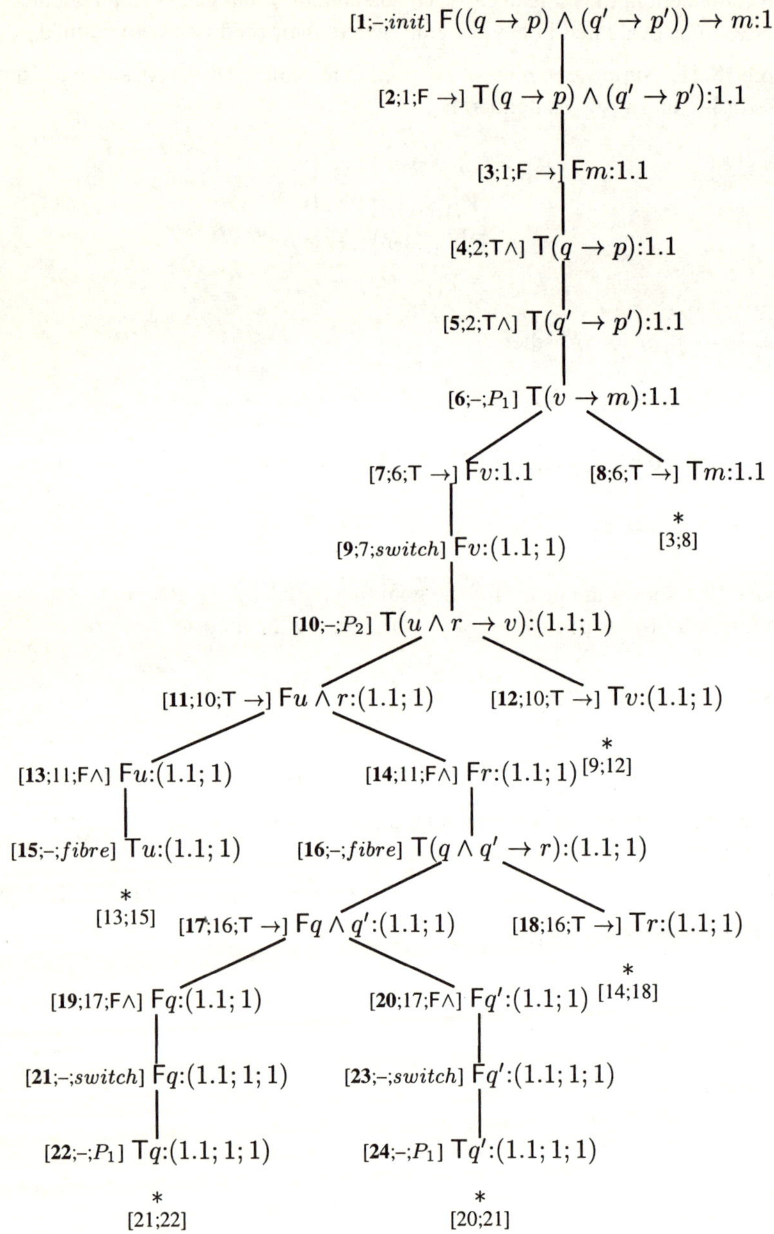

FIG. 18.1. Closed tableau proving the validity of $(q \to p) \wedge (q' \to p') \to m$.

19

FIBRING MODAL TABLEAUX

19.1 Introduction

In the previous chapter we studied under which conditions proof systems are well suited for fibring, and we argued that labelled tableaux are calculi of this kind. In Chapters 14 and 15 we have shown how D'Agostino and Mondadori's 1994 classical proof system **KE**, a combination of tableaux and natural deduction inference rules which allows for a restricted ('analytic') use of the cut rule, can be extended to deal with fibred logics, in particular for a combination of modal and substructural logics and the combination of different fragment of substructural implications. In this chapter we present another extension of **KE**, called $\mathbf{K}EM$, which has been proven to be able to cope with a wide variety of (normal) modal logics [Artosi *et al.*, 1997; Governatori, 1997]. We show how to adapt $\mathbf{K}EM$ in order to obtain a general and uniform tableau-like proof method for the fibred (dovetailed) combination of modal logics. In particular we show how the label formalism used in $\mathbf{K}EM$ matches our semantic based fibring for (multi-) modal logics. The resulting system enjoys several interesting properties: it is modular, in the sense that the proof system for each logic is developed on its own and it is reused in the combination; it is uniform, in the sense that each system has its own individual features, but the framework remains constant among the combined systems; it seems to be flexible enough to deal with other logics, as well as their combinations; finally, although we use some logical machinery in describing the system, our presentation is natural and the idea behind it is very simple and easy to grasp.

Let $\mathbf{L}_1$ and $\mathbf{L}_2$ be two modal logics for which a tableau system exists. We start a tableau for a formula A of $\mathbf{L}_1$, which has formulas of $\mathbf{L}_2$ embedded in it. In the course of processing A, when we have to process a formula of $\mathbf{L}_2$ we begin a new proof for it in the appropriate system. It turns out that we do not really need to start a separate tableau, but we graft it into the original one. The labels will have a structure which will store nicely all the information about the grafting operation. The key feature of $\mathbf{K}EM$, besides its being based neither on resolution nor on standard sequent/tableau inference techniques, is that it generates models and checks them using a label scheme during the bookkeeping for the fibred model. In [Artosi *et al.*, 1996; Governatori, 1995; Governatori, 1996; Governatori, 1997] it has been shown how this formalism can be extended to handle various systems of multi-modal logic with interaction axioms. The mechanism which $\mathbf{K}EM$ uses in manipulating labels is close to semantic fibring (dovetailing).

Let us explain the fibring idea by looking at two examples. These examples will be treated in Section 19.4 below.

Consider two logics $\mathbf{L}_1$ and $\mathbf{L}_2$ with modalities $\{\Box_1, \Diamond_1\}$ and $\{\Box_2, \Diamond_2\}$ respectively. Assume $\mathbf{L}_i$ is an extension of $\mathbf{K}$ for its modality. Consider:

$$A = \Box_1 \Diamond_2 p \to \Diamond_1 \Box_2 \Diamond_2 p$$
$$B = \Box_1 p \to \Diamond_1 \Box_2 \Diamond_2 p.$$

These formulas are in the combined language. We know how to dovetail these two logics to form a logic for the combined language. We have done dovetailing in Chapter 3. We know that if $\mathbf{L}_1$ and $\mathbf{L}_2$ are complete for classes of Kripke models $\mathcal{K}_1$ and $\mathcal{K}_2$ respectively, then we can fibre the models and get a class $\mathcal{K}$ of models of the form $\mathbf{m} = (S, R_1, R_2, w_0, h)$, with two accessibility relations. The class $\mathcal{K}$ is constructed from $\mathcal{K}_1$ and $\mathcal{K}_2$. It contains some specific models of the form above and completeness holds.

Let us see whether we can find an $\mathbf{m}$ which is a countermodel for A.

Let us check what is needed to be in $\mathbf{m}$ and then we can check whether such an $\mathbf{m}$ can be in the class $\mathcal{K}$.

We want to have

1. $w_0 \vDash \Box_1 \Diamond_2 p$
2. $w_0 \nvDash \Diamond_1 \Box_2 \Diamond_2 p.$

From (1) we get that at an arbitrary point t such that $w_0 R_1 t$ we must have $t \vDash \Diamond_2 p$. Let $t = W_1^1$ be such an arbitrary point. We use capital letters to indicate that the point t is *arbitrary* and we use the superscript '1' to indicate that $w_0 R_1 t$ holds (i.e. we are choosing t because we are evaluating a modality of $\mathbf{L}_1$). The subscript just counts that this is the first 'W^1' we are using. Hence we have

$$W_1^1 \vDash \Diamond_2 p.$$

Now we are evaluating the second modality. Therefore there exists a specific point s such that $W_1^1 R_2 s$ and $s \vDash p$. Again we use the suggestive notation $s = w_1^2$ and write $w_1^2 \vDash p$. The lower case w indicates that w_1^2 is not arbitrary but specific and the superscript '2' indicates that we are dealing with an $\mathbf{L}_2$ modality.

Similarly, since

$$w_0 \nvDash \Diamond_1 \Box_2 \Diamond_2 p$$

we must have that for an arbitrary x such that $w_0 R_1 x$ we have $x \nvDash \Box_2 \Diamond_w p$. According to our agreed notation, we represent such an arbitrary x by

$$W_2^1 \nvDash \Box_2 \Diamond_2 p.$$

The superscript '1' indicates that this point arose because of an $\mathbf{L}_1$ modality. The subscript '2' indicates that this is a second such point used so far and the capital W (as opposed to a lower case w) indicates that it is an arbitrary point (accessible to w_0).

We can continue: since

$$W_2^1 \nvDash \Box_2 \Diamond_2 p$$

there exists a specific point y, $W_2^1 R_2 y$ such that $y \nvDash \Diamond_2 p$. Using our conventions, we represent this point by

$$w_2^2 \nvDash \Diamond_2 p.$$

We can continue and represent an arbitrary point W_1^2 such that $w_2^2 R_2 W_1^2$ and $W_1^2 \nvDash p$.

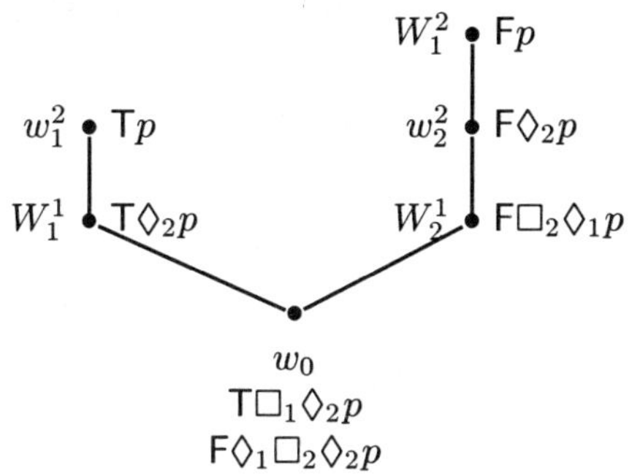

FIG. 19.1.

Point in tree	Sequence representation
w_0	w_0
W_2^1	(W_2^1, w_0)
w_2^2	$(w_2^2, (W_2^1, w_0))$
W_1^2	$(W_1^2, (w_2^2, (W_1^1, w_0)))$
W_1^1	(W_1^1, w_0)
w_1^2	$(w_1^2, (W_1^1, w_0))$.

FIG. 19.2.

We can represent what we have got so far in a tree as in Figure 19.1.

The T, F indicate whether the formula is supposed to hold or not in the node. The uppercase/lowercase distinction tells us whether the point is arbitrary or not and the superscript tells us whether the point is R_1 or R_2 accessible to the point below it. We are relying on the geometrical drawing of a tree to represent the situation. If we do not want to draw trees, we can represent each point as the *sequence* of points leading up to it. Thus we have the table in Figure 19.2.

We now address the question of whether the situation in Figure 19.1 does provide a countermodel for A. Of course, if both $\Box_1$ and $\Box_2$ are **K** modalities, then we can provide a countermodel for A, since in this case the dovetailed semantics $\mathcal{K}$ contains any model **m** with arbitrary h, R_1 and R_2. We know this fact from the completeness theorem of Chapter 3. If we do not want to rely on semantics, we can ask what would be the corresponding condition on the trees obtained (as we see in the example of Figure 19.1) that will tell us whether we can get a countermodel or not. In other words, is the tree 'closed' and therefore is there no countermodel, i.e. A is a theorem?

The following is the kind of rule we want.

In the subtree of Figure 19.3, W^i can unify with x^i. By repeated applications of the rule we try to unify points and get a contradiction. For example, if in Figure 19.3 we have W^i Tp and x^i Fp we get a contradiction.

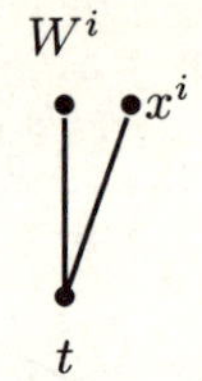

FIG. 19.3.

FIG. 19.4.

Different logics will have different unification rules. For example, if the logic $\mathbf{L}_2$ is a **T** modality, then in this case we use the following rule.

In the subtree of Figure 19.4, W^2 can unify with either x^2 or t.

A tableaux system for a dovetailed logic with $\{\Box_1, \Box_2\}$ would need the following components.

- Rules for building trees as we did in Figure 19.1 or using sequences as in Figure 19.2.
- Unification rules that unify labels in order to identify two labels; for example, one with $\mathsf{F}q$ and the other with $\mathsf{T}q$.
- Completeness theorem for the resulting tableau system for the dovetailed logic.

The next sections do all that in detail. It is a messy business and is done best by a theorem prover. Work on functional translation of this kind has been done by Ohlbach [1991].

Before we conclude this section, here is a tree (Figure 19.5) for the other example, B.

19.2 Label Formalism

In this section we introduce the label formalism we shall use in the course of the chapter. $\mathbf{K}EM$ uses *labelled signed formulas*, where a labelled signed formula is an expression of the form $SA : t$, where A is a wff of the logic, S (the truth sign) is in $\{\mathsf{T}, \mathsf{F}\}$, and t is a label. Notice that the labels we are referring to in this section are not the labels of t. In the case of modal logic we have a type of label corresponding to various modalities, and each set of atomic labels is partitioned into the set of variables and the set of constants. Formally

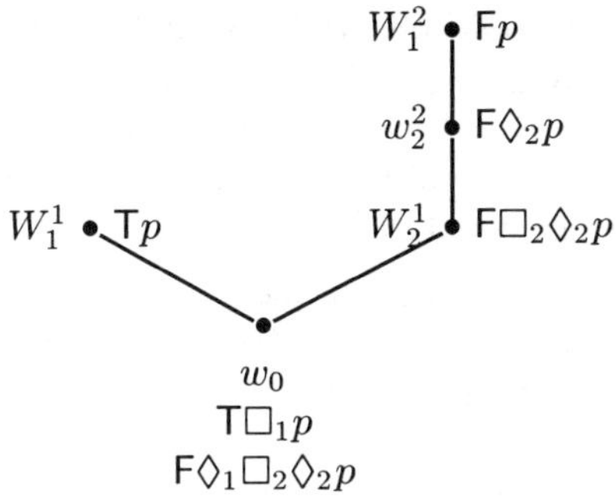

FIG. 19.5.

$$\Phi^{i\in I} = \Phi_C^{i\in I} \cup \Phi_V^{i\in I}$$

where $\Phi_C^{i\in I} = \{w_1^i, w_2^i, \ldots\}$ and $\Phi_V^{i\in I} = \{W_1^i, W_2^i, \ldots\}$. The set of constant world symbols and variable world symbols are respectively

$$\Phi_C = \bigcup_{i\in I} \Phi_C^{i\in I} \qquad \Phi_V \bigcup_{i\in I} \Phi_V^{i\in I}.$$

The set of labels $\Im$ is then defined inductively as follows:

$$\Im = \bigcup_{1\leq k} \Im_k \text{ where } \Im_k, k \in \mathbb{N} :$$

$$\begin{aligned} \Im_1 &= \Phi_C \cup \Phi_V; \\ \Im_2 &= \Im_1 \times \Phi_C; \\ \Im_{n+1} &= \Im_1 \times \Im_n, (n > 1). \end{aligned}$$

According to the above definition a label is either a (i) an element of the set Φ_C, or (ii) an element of the set Φ_V, or (iii) a path term (s', s) where (iiia) $s' \in \Phi_C \cup \Phi_V$ and (iiib) $s \in \Phi_C$ or $s = (t', t)$ where (t', t) is a label. From now on we shall use $t, s, r, \ldots$ to denote arbitrary labels.

Definition 19.1. For any label $t = (s', s)$ we shall call s' the *head* of t, s the *body* of t, and denote them by $\mathbf{h}(t)$ and $\mathbf{b}(t)$ respectively.

Notice that these notions are recursive (they correspond to projection functions): if $\mathbf{b}(t)$ denotes the body of t, then $\mathbf{b}(\mathbf{b}(t))$ will denote the body of $\mathbf{b}(t)$, $\mathbf{b}(\mathbf{b}(\mathbf{b}(t)))$ will denote the body of $\mathbf{b}(\mathbf{b}(t))$; and so on.

Definition 19.2. We call each of $\mathbf{b}(t)$, $\mathbf{b}(\mathbf{b}(t))$, etc., a *segment* of t. Let $\mathbf{b}^m(t)$, be any segment of t (obviously, by definition every segment $\mathbf{s}(t)$ of a label t is a label); then $\mathbf{h}(\mathbf{b}(t))$ will denote the head of $\mathbf{b}^m(t)$. We shall call a label t *restricted* if $\mathbf{h}(t) \in \Phi_C$, otherwise *unrestricted.*

We shall think of the $\Im_1$ labels as possible worlds in a fibred (dovetailed) model, and the path $\Im_n$ as paths leading to them starting from the actual world. For example a label such $t = (W_2^i, (w_2^j, w_0))$ denotes with respect to a fibred (dovetailed) model the set of worlds in a model for $\mathbf{L}_i$ accessible from the actual world $a^i = \mathbf{F}_i((w_2^j, w_0))$; since dovetailed models can be reduced to structures of the form $(W, R(i), h)$, the label t turns out to represent the set of worlds accessible via $R(i)$ from a world accessible through $R(j)$ from the actual world w_0.

Definition 19.3. For any label t, we define the length of t, $\ell(t)$, as the number of world-symbols in t, i.e., $\ell(t) = n \Leftrightarrow t \in \Im_n$. $\mathbf{s}^n(t)$ will denote the segment of t of length n, i.e., $\mathbf{s}^n(t) = \mathbf{s}(t)$ such that $\ell(\mathbf{s}(t)) = n$. We shall use $\mathbf{h}^n(t)$ as an abbreviations for $\mathbf{h}(\mathbf{s}^n(t))$. Notice that $\mathbf{h}(t) = \mathbf{h}^{\ell(t)}(t)$.

Definition 19.4. For any label $t, \ell(t) > n$, we define the *counter-segment-n* of t, as follows:

$$\mathbf{c}^n(t) = \mathbf{h}(t) \times (\cdots \times (\mathbf{h}^k(t) \times (\cdots \times (\mathbf{h}^{n+1}(t), w_0)))) \;\; (n < k < \ell(t))$$

where w_0 is a dummy label, i.e., a label not appearing in t (the context in which such a notion occurs will tell us what w_0 stands for. In most cases it will denote the actual world).

The counter-segment-n defines what remains of a given label after having identified the segment of length n with a 'dummy' label w_0. The appropriate dummy label will be specified in the applications where such a notion is used. However, it can be viewed also as an independent atomic label. In the context of fibring w_0 can be thought of as denoting the actual world obtained via the fibring function from the world denoted by $\mathbf{s}^n(t)$.

Example 19.5. Given the label $t = (w_4, (W_3, (w_3, (W_2, w_1))))$, according to the above definitions its length $\ell(t)$ is 5, its segment of length 3 is $\mathbf{s}^3(t) = (w_3, (W_2, w_1))$, and the relative counter-segment-3 is $\mathbf{c}^3(t) = (w_4, (W_3, w_0))$, where $w_0 = \mathbf{s}^3(t) = (w_3, (W_2, w_1))$.

To clarify the notion of counter-segment, which will be used frequently in the course of the present work, we present in the following table the list of the segments of t in the left-hand column and the relative counter-segments in the right-hand column.

$\mathbf{s}^1(t) = w_1$	$\mathbf{c}^1(t) = (w_4, (W_3, (w_3, (W_2, w_0))))$
$\mathbf{s}^2(t) = (W_2, w_1)$	$\mathbf{c}^2(t) = (w_4, (W_3, (w_3, w_0)))$
$\mathbf{s}^3(t) = (w_3, (W_2, w_1))$	$\mathbf{c}^3(t) = (w_4, (W_3, w_0))$
$\mathbf{s}^4(t) = (W_3, (w_3, (W_2, w_1)))$	$\mathbf{c}^4(t) = (w_4, w_0)$
$\mathbf{s}^5(t) = t$	$\mathbf{c}^5(t) = w_0$

So far we have provided definitions about the structure of the labels without regard to the elements they are made of. The following definitions will be concerned with the type of world symbols occurring in a label.

Let t be a label and t' an atomic label, in what follows we shall use $(t'; t)$ as a notation for the label (t', t) if $t' \neq \mathbf{h}(t)$, or for t otherwise

Definition 19.6. We say that a label t is *i-preferred* iff $\mathbf{h}(t) \in \Phi^i$.

Definition 19.7. We say that a label t is *i-pure* iff each segment of t of length $n > 1$ is i-preferred, and we shall use $\Im^i$ to denote the set of i-pure labels.

19.3 Unifications

In the course of proofs labels are manipulated in a way closely related to the semantics of the logics under analysis. Labels are confronted and matched using a specialised logic dependent unification mechanism. The notion of two labels t and s unifying means that the intersection of their denotations is not empty and that we can move to such a set of worlds, i.e., to the result of their unification.

According to the semantics each modality is evaluated in an appropriate model corresponding to a model in the class of models characterising the logic the modality corresponds to. Similarly we provide a unification for each logic, the unification characterising such a logic in $\mathbf{K}EM$ formalism, then we graft them into a single unification for the whole $\mathbf{L}_I^\delta$.

19.3.1 *Basic unifications (axiom unifications)*

We add a set of auxiliary unindexed atomic labels $\Phi^A = \{w_0, w_0', \ldots\}$, that will be used in unifications and proofs. Intuitively they stand for distinguished worlds in the various models. We define two substitutions, $\sigma^{\mathbf{D}}$ and $\sigma^{\mathbf{F}}$ respectively, dovetailing and fibring substitution, in the usual way as a mapping

$$\begin{aligned} \sigma^\delta &= \mathbf{1}_{\Phi^A \cup \Phi_C} \\ \sigma^{\mathbf{D}} &: \Phi_V^i \longrightarrow \Im^i \cup \Phi^A \\ \sigma^{\mathbf{F}} &: \Phi_V^i \longrightarrow \Im^i \end{aligned}$$

i.e., identity for constants and auxiliary labels; and a mapping of variables onto i-pure labels in the case of fibring, and either onto i-pure labels or auxiliary ones for the other. The only difference between fibring and dovetailing substitutions is that, in the former, a variable cannot be mapped onto an auxiliary label. Henceforth we use σ to mean indifferently, unless specified, either $\sigma^{\mathbf{D}}$ or $\sigma^{\mathbf{F}}$.

For two labels t and s, and a substitution σ, if σ is a unifier of t and s then we shall say that t, s are *σ-unifiable*. We shall (somewhat unconventionally) use $(t, s)\sigma$ to denote both that t and s are σ-unifiable and the result of their unification. In particular

$$\forall t, s, r \in \Im, (t, s)\sigma = r \text{ iff } \exists \sigma : \sigma(t) = \sigma(s) \text{ and } \sigma(t) = r.$$

On this basis we may define several specialised, logic-dependent notions of σ-unification. As a case study we choose the normal modal logics arising from the combination of the axioms K, D, T, 4, B, and 5. Notice that the unifications listed below mimic the conditions on the accessibility relation corresponding to the appropriate axiom (see Example 19.8 for an explanation).

$$\begin{aligned} (t, s)\sigma^K = (t, s)\sigma & \text{ if at least one of } t \text{ and } s \text{ is restricted, and} \\ & \forall n \leq \ell(t), (\mathbf{s}^n(t), \mathbf{s}^n(s))\sigma^K \end{aligned}$$

$$(t,s)\sigma^D = (t,s)\sigma$$

$$(t,s)\sigma^T = \begin{cases} (\mathbf{s}^{\ell(s)}(t), s)\sigma \text{ if } \ell(t) > \ell(s), \text{ and} \\ \qquad \forall n \geq \ell(s), (\mathbf{h}^n(t), \mathbf{h}(s))\sigma = (\mathbf{h}(t), \mathbf{h}(s))\sigma \\ (t, \mathbf{s}^{\ell(t)}(s))\sigma \text{ if } \ell(s) > \ell(t), \text{ and} \\ \qquad \forall n \geq \ell(t), (\mathbf{h}(t), \mathbf{h}^n(s))\sigma = (\mathbf{h}(t), \mathbf{h}(s))\sigma \end{cases}$$

$$(t,s)\sigma^4 = \begin{cases} \mathbf{c}^{\ell(t)}(s)\ \ell(s) > \ell(t), \mathbf{h}(t) \in \Phi_V \text{ and} \\ \qquad w_0 = (t, \mathbf{s}^{\ell(t)}(s))\sigma \\ \mathbf{c}^{\ell(s)}(t)\ \ell(t) > \ell(s), \mathbf{h}(s) \in \Phi_V \text{ and} \\ \qquad w_0 = (\mathbf{s}^{\ell(s)}(t), s)\sigma \end{cases}$$

$$(t,s)\sigma^B = \begin{cases} (\mathbf{s}^{\ell(t)-2n}(t), s)\sigma \text{ if } \mathbf{h}(t) \in \Phi_V \text{ and} \\ \qquad (\mathbf{h}(t), \mathbf{h}(s))\sigma = (\mathbf{h}^{\ell(t)-2n}(t), \mathbf{h}(s))\sigma \\ (t, \mathbf{s}^{\ell(s)-2n}(s))\sigma \text{ if } \mathbf{h}(s) \in \Phi_V \text{ and} \\ \qquad (\mathbf{h}(t), \mathbf{h}(s))\sigma = (\mathbf{h}(t), \mathbf{h}^{\ell(s)-2n}(s))\sigma \end{cases}$$

where $1 \leq n \leq V$, and $V = \ell(t) - m$, with m such that $\forall x, m \leq x \leq \ell(t), \mathbf{h}^x(t) \in \Phi_V$

$$(t,s)\sigma^5 = \begin{cases} ((\mathbf{h}(t), \mathbf{h}(s))\sigma; \mathbf{c}^1(\mathbf{s}^2(t))) \text{ if } \ell(t) > 2, \ell(s) > 1, \mathbf{h}(t) \in \Phi_V, \text{ or} \\ \qquad \mathbf{h}(t) = \mathbf{h}(s) \in \Phi_C \\ (t,s)\sigma \qquad \text{if } \ell(t) = \ell(s) = 2 \\ ((\mathbf{h}(t), \mathbf{h}(s))\sigma; \mathbf{c}^1(\mathbf{s}^2(s))) \text{ if } \ell(s) > 2, \ell(t) > 1, \mathbf{h}(s) \in \Phi_V, \text{ or} \\ \qquad \mathbf{h}(t) = \mathbf{h}(s) \in \Phi_C \end{cases}$$

where $w_0 = (\mathbf{s}^1(t), \mathbf{s}^1(s))\sigma$.

We exemplify how unifications corresponding to axioms obtained from the axioms listed above by prefixing n □'s to them can be defined.:

$$(t,s)\sigma^O = (\mathbf{c}^2(t), \mathbf{c}^2(s))\sigma^D$$

where $w_0 = (\mathbf{s}^2(t), \mathbf{s}^2(s))\sigma^K$, and $O = \Box(\Box A \rightarrow \Diamond A)$.

Example 19.8. For the notion of σ^T-unification, take for example the labels $t = (w_3, (W_1, w_1))$ and $s = (w_3, (W_2, (w_2, w_1)))$. Here $(W_2, w_3)\sigma = (w_3, w_3)\sigma$. Then t and s σ^T-unify to $(w_3, (w_2, w_1))$. This intuitively means that the world w_3, accessible from a sub-path $\mathbf{s}(s) = (W_2, (w_2, w_1))$, after the deletion of W_2 from s, is accessible from any path t which turns out to denote the same world(s) as $\mathbf{s}(s)$; in fact the step from w_2 to W_2 is irrelevant because of the reflexivity relation of the model. For the notion of σ^4-unification, take for example the labels $t = (W_3, (w_2, w_1))$ and $s = (w_5, (w_4, (w_3, (W_2, w_1))))$. Here $\mathbf{s}^{\ell(t)}(s) = (w_3, (W_2, w_1))$. Then t and s σ^4-unify to $(w_5, (w_4, (w_3, (w_2, w_1))))$ since $(t, \mathbf{s}^{\ell(t)}(s))\sigma = ((W_3, (w_2, w_1)), (w_3, (W_2, w_1)))\sigma$. This intuitively means that all the worlds accessible from a sub-path $\mathbf{s}^{\ell(t)}(s)$ of s are accessible from any path t which leads to the same world(s) denoted by $\mathbf{s}^{\ell(t)}(s)$.

19.3.2 *High unifications (combined unifications)*

We are now able to combine the above unifications corresponding to the axiom characterising a logic into a single 'high' unification which will be used for defining the unifications characterising the logic we are concerned with:

$$(t,s)\sigma^{A_1^i \cdots A_n^i} = \begin{cases} (t,s)\sigma^{A_1^i} \text{ if } C_1^i \\ \vdots \quad \vdots \\ (t,s)\sigma^{A_n^i} \text{ if } C_n^i \end{cases} \qquad i \in I \tag{19.1}$$

where $A_1^i \cdots A_n^i$ stand for the axioms characterising $\mathbf{L}_i$ and $C_i, 1 \leq i \leq n$ are conditions varying from logic to logic. For example the high unification for **T**, which is characterised by the axioms D and T, is

$$(t,s)\sigma^{DT} = \begin{cases} (t,s)\sigma^{T} \text{ if } \ell(t) \neq \ell(s) \\ (t,s)\sigma^{D} \text{ otherwise} \end{cases}$$

and for **OM** which is **K** plus $M = \Box(\Box A \to A)$, the deontic version of **T**, the corresponding high unifications is

$$(t,s)\sigma^{OM} = \begin{cases} (t,s)\sigma^{T} \quad (\mathbf{s}^2(t), \mathbf{s}^2(s))\sigma^{O} \\ (t,s)\sigma^{O}. \end{cases}$$

We then provide the definition of σ^{DT4} which is used in defining the logic unification for **S4**.

$$(t,s)\sigma^{DT4} = \begin{cases} (t,s)\sigma^{D} \text{ if } \ell(t) = \ell(s) \\ (t,s)\sigma^{T} \text{ if } \ell(t) < \ell(s), \mathbf{h}(t) \in \Phi_C \\ (t,s)\sigma^{4} \text{ if } \ell(t) < \ell(s), \mathbf{h}(t) \in \Phi_V. \end{cases} \tag{19.2}$$

19.3.3 *Low unification (logic unifications)*

Combining recursively each high unification we obtain the $\sigma_{\mathbf{L}_i}$ unification for $\mathbf{L}_i$ as follows

$$(t,s)\sigma_{\mathbf{L}_i} = \begin{cases} (\mathbf{c}^n(t), \mathbf{c}^m(s))\sigma^{A_1^i \cdots A_n^i} \\ (t,s)\sigma^{A_1^i \cdots A_n^i} \end{cases} \tag{19.3}$$

where $w_0 = (\mathbf{s}^n(t), \mathbf{s}^m(s))\sigma_{\mathbf{L}_i}$.

Although the above definition provides a general method for obtaining the unification characterising the appropriate logic, for particular systems it may be preferred to define a different unification reflecting peculiar properties; for example, for **S5** which is characterised by the class of frames where the accessibility relation is an equivalence relation, it is convenient to define $\sigma_{\mathbf{S5}}$ as follows:

$$(t,s)\sigma_{\mathbf{S5}} = \begin{cases} (\mathbf{h}(t), \mathbf{h}(s))\sigma & \min\{\ell(t), \ell(s)\} = 1 \\ ((\mathbf{h}(t), \mathbf{h}(s))\sigma, (\mathbf{s}^1(t), \mathbf{s}^1(s))\sigma) & \text{otherwise.} \end{cases}$$

The format of the unifications for the logics containing axioms obtained by prefixing n $\Box$'s to one of the five axiom listed above is slightly more complex. We exemplify it by providing the unification for **OM**:

$$(t,s)\sigma_{\mathbf{OM}} = \begin{cases} (\mathbf{c}^n(t), \mathbf{c}^m(s))\sigma^{OM} & \\ (\mathbf{c}^n(t), \mathbf{c}^m(s))\sigma^{T} & n,m > 2 \\ (t,s)\sigma^{OM} & \end{cases}$$

where $w_0 = (\mathbf{s}^n(t), \mathbf{s}^m(s))\sigma_{\mathbf{OM}}$.

Example 19.9. We provide a pair of labels t, and s that $\sigma_{\mathbf{S4}}$-unify:

$$t = (w_6,(w_5,(W_2,(w_1,(W_1,w_0))))) \quad s = (W_3,(w_4,(w_3,(w_2,(w_1,w_0)))))$$

The two labels do not σ^D-, nor σ^T-, nor σ^4-unify. We have then split them into appropriate counter-segments. First we consider $\mathbf{s}^3(t) = (w_1,(W_1,w_0))$ and $\mathbf{s}^2(s) = (w_1,w_0)$. Such labels σ^T-unify on (w_1,w_0), see Example 19.8. We identify their unification with w_0'; then $\mathbf{c}^3(t) = (w_6,(w_5,(W_2,w_0')))$ and $\mathbf{c}^2(s) = (W_3,(w_4,(w_3,(w_2,w_0'))))$. At this point we notice that $\mathbf{s}^2(\mathbf{c}^3(t)) = (W_2,w_0')$ and $\mathbf{s}^4(\mathbf{c}^2(s)) = (w_4,(w_3,(w_2,(w_1,w_0)))$ σ^4-unify: $(w_0',w_0')\sigma$ and $W_2 \in \Phi_V$. The result of their unification w_0' is $(w_4,(w_3,(w_2,(w_1,w_0))))$. Finally $\mathbf{c}^4(t) = (w_6,(w_5,w_0''))$ and $\mathbf{c}^5(s) = (W_3,w_0'')$ σ^4-unify. According to the definition of logic unification 19.3, and the previous considerations t and s $\sigma_{\mathbf{S4}}$-unify.

For a wide class of unifications and a detailed account of the labelling algebra of **K**EM see [Governatori, 1997].

19.3.4 *Fibred unification*

As $\mathbf{L}_I^\delta$ is obtained by combining the logic $\mathbf{L}_i$, $i \in I$, the corresponding unification is the combination of the $\sigma_{\mathbf{L}_i}$-unifications characterising $\mathbf{L}_i$:

$$(t,s)\sigma_{\mathbf{L}_I^\delta} = \begin{cases} \bigcup_{i\in I}(\mathbf{c}^n(t),\mathbf{c}^m(s))\sigma_{\mathbf{L}_i} \\ \bigcup_{i\in I}(t,s)\sigma_{\mathbf{L}_i} \end{cases} \tag{19.4}$$

where $w_0 = (\mathbf{s}^n(t), \mathbf{s}^m(s))\sigma_{\mathbf{L}_I^\delta}$, if $\mathbf{c}^n(t), \mathbf{c}^n(s)$ are i-pure, $i \in I$.

Example 19.10. We want to combine a deontic **D** modality $\Box_1$ with an epistemic **S5** modality $\Box_2$, let us call the resulting logic $\mathbf{ED}^\delta$. The unifications for $\mathbf{ED}^\delta$ are provided by the schema

$$(t,s)\sigma_{\mathbf{ED}^\delta} = \begin{cases} (\mathbf{c}^n(t),\mathbf{c}^m(s))\sigma_{\mathbf{S5}^2} \\ (t,s)\sigma_{\mathbf{S5}^2} \\ (\mathbf{c}^n(t),\mathbf{c}^m(s))\sigma_{\mathbf{D}^1} \\ (t,s)\sigma_{D^1} \end{cases}$$

where $w_0 = (\mathbf{s}^n(t), \mathbf{s}^m(s))\sigma_{\mathbf{ED}^\delta}$, if $\mathbf{c}^n(t)$ and $\mathbf{c}^n(s)$ are i-pure, $i \in \{1,2\}$.

According to the above definition, the labels

$$t = (W_2^2,(w_1^2,(W_1^1,w_0))) \qquad s = (w_2^2,(W_2^1,w_0))$$

$\sigma_{\mathbf{ED}^F}$-unify, and then $\sigma_{\mathbf{ED}^D}$. In fact

$$\mathbf{c}^2(t) = (W_2^2,(w_1^2,w_0') \qquad \mathbf{c}^2(s) = (w_2^2,w_0')$$

$\sigma_{\mathbf{S5}^2}$-unify, and $w_0' = ((W_1^1, w_0), (w_1^1, w_0))\sigma_{\mathbf{D}^1}$. On the contrary, if t had been $t' = (W_2^2, (w_2^1, (w_1^2, (W_1^1, w_0))))$, it would not have unified with s: $\mathbf{c}^2(t')$ is not 2-pure. Let us consider now the labels

$$t = (W_1^1, w_0) \qquad s = (W_1^2, (w_1^2, (W_2^1, w_0))).$$

t and s do not $\sigma_{\mathbf{E}D^F}$-unify, but $\sigma_{\mathbf{E}D^D}$-unify, in so far as we can map W_2^2 onto w_0', where w_0' is as before.

We prove now an interesting property of $\sigma_{\mathbf{L}_I^\delta}$-unifications.

Lemma 19.11. *$\forall t, s \in \Im$, if $(t, s)\sigma_{\mathbf{L}_I^\delta} = r$, then $(t, r)\sigma_{\mathbf{L}_I^\delta}$ and $(r, s)\sigma_{\mathbf{L}_I^\delta}$.*

Proof Since $\sigma_{\mathbf{L}_I^\delta}$ has been defined as the combination of $\sigma_{\mathbf{L}_i}$-unifications, but each of them works on i-pure labels and there are no interactions among them, we just have to prove the property for each $\sigma_{\mathbf{L}_i}$.

The proof will be by induction on the number of applications of $\sigma^{A_1^i \cdots A_n^i}$ in a $\sigma_{\mathbf{L}_i}$-unification. Let n be the number of such applications. We give the proof only for **S4** and **S5**. The proof for the others systems is similar and can be found in [Artosi *et al.*, 1997; Governatori, 1997].

If $n = 1$ then we have to prove the property for $\sigma^{A_1^i \cdots A_n^i}$.[77] We prove it by induction on the length of labels.

If $\min\{\ell(t), \ell(s)\} = 1$ then we assume that $\ell(t) = 1$ (the proof for $\ell(s) = 1$ is similar).

1. $t \in \Phi_C$. If also $\ell(s) = 1$, we apply σ^D; in every case, by obvious considerations about σ, $r = (t, s)\sigma^D = t$, but $(t, t)\sigma^D$ and $(t, s)\sigma^D$. Therefore $(t, r)\sigma^{DT4}$ and $(r, s)^{DT4}$. If $\ell(s) > 1$ then $(t, s)\sigma^{DT4} = (t, s)\sigma^T$. Therefore $r = (t, s)\sigma^T = (t, \mathbf{s}^1(s))\sigma^T = t$, hence $(t, t)\sigma^D$ and $(t, s)\sigma^T$, and so, according to 19.2, $(t, r)\sigma^{DT4}$ and $(r, s)^{DT4}$.
2. $t \in \Phi_V$ then by the definition of σ it unifies with any label, in particular $(t, s)\sigma^D = s = r$, whence $(t, s)\sigma^D$ and $(s, s)\sigma^D$, then $(t, r)\sigma^{DT4}$ and $(r, s)^{DT4}$.

Let us suppose now that $\min\{\ell(t), \ell(s)\} = n > 1$, and that the property holds up to n for $\sigma^{A_1^i \cdots A_n^i}$. Thus we have the following cases.

If $\ell(t) = \ell(s)$ then $(t, s)\sigma^D = s$; by the inductive hypothesis $(\mathbf{b}(t), \mathbf{b}(l))\sigma^D$, $(\mathbf{b}(s), \mathbf{b}(r))\sigma^D$, $(\mathbf{h}(t), \mathbf{h}(s))\sigma^D$ and $(\mathbf{h}(s), \mathbf{h}(r))\sigma^D$; therefore $(t, r)\sigma^D$ and $(s, r)\sigma^D$. Consequently $(t, r)\sigma^{DT4}$ and $(r, s)^{DT4}$.

If $\ell(t) < \ell(s)$ and $\mathbf{h}(t) \in \Phi_C$, then $(t, s)\sigma^T = r$, where, by the inductive hypothesis $(\mathbf{b}(t), \mathbf{b}(l))\sigma^D$, $(s^{\ell(\mathbf{b}(t))}(s), \mathbf{b}(r))\sigma^D$. By the definition of σ^T, we know that, for $n \leq \ell(t)$ $\mathbf{h}^n(r) = (\mathbf{h}(t), \mathbf{h}(s))\sigma = (\mathbf{h}(t), \mathbf{h}^{\ell(t)}(s))\sigma$; therefore $(t, r)\sigma^D$ and $(r, s)\sigma^T$. We can conclude that $(t, r)\sigma^{DT4}$ and $(r, s)^{DT4}$.

If $\ell(t) < \ell(s)$ and $\mathbf{h}(t) \in \Phi_V$, then $\mathbf{h}(t) \in \Phi_V$, which implies $(t, s)\sigma^4 = \mathbf{c}^{\ell(t)}(s)$ where $w_0' = (t, \mathbf{s}^{\ell(t)}(s))\sigma$. By the inductive hypothesis and the definition of σ we have

[77] Hereafter, in order to shorten proofs, when we have to consider labels of different lengths, we shall assume, unless specified, the first to be the shorter. Obviously proofs for the other cases can be carried out in the same way.

$(t, \mathbf{s}^{\ell(t)}(r))\sigma$ and $(\mathbf{s}^{\ell(t)}(t), \mathbf{s}^{\ell(t)}(l))\sigma$ and therefore $(t, s)\sigma^4$ and $(r, s)\sigma^D$, which means $(t, r)\sigma^{DT4}$ and $(r, s)^{DT4}$.

We have thus proved the inductive base for the lemma.

We can now assume that the lemma holds up to the n-th application of $\sigma^{A_1^i \cdots A_n^i}$. By the definition of $\sigma_{\mathbf{L}_i}$, $(\mathbf{s}^n(t), \mathbf{s}^m(s))\sigma_{\mathbf{L}_i} = w_0' = \mathbf{s}^r(r)$ and $(\mathbf{c}^n(t), \mathbf{c}^m(s))\sigma^{A_1^i \cdots A_n^i} = \mathbf{c}^r(r)$, but, by the inductive hypothesis, $(\mathbf{s}^n(t), \mathbf{s}^r(r))\sigma_{\mathbf{L}_i}$ and $(\mathbf{s}^m(s), \mathbf{s}^r(r))\sigma_{\mathbf{L}_i}$. By the property we have just proved for $\sigma^{A_1^i \cdots A_n^i}$ $(\mathbf{c}^n(t), \mathbf{c}^r(r))\sigma^{A_1^i \cdots A_n^i}$ and $(\mathbf{c}^m(s), \mathbf{c}^r(r))\sigma^{A_1^i \cdots A_n^i}$, which implies $(t, r)\sigma_{\mathbf{L}_i}$ and $(r, s)\sigma_{\mathbf{L}_i}$.

For **S5** if $\min\{\ell(i), \ell(s)\} = 1$, we have $(t, s)\sigma_{\mathbf{S5}}$ iff $(\mathbf{h}(t), \mathbf{h}(s))\sigma$, whence, if t is restricted, then $(t, s)\sigma_{\mathbf{S5}} = \mathbf{h}(t) = r$ and thus $(t, s)\sigma_{\mathbf{S5}}$, i.e., $(\mathbf{h}(t), \mathbf{h}(t))\sigma$, and similarly for s; otherwise $(t, s)\sigma_{\mathbf{S5}} = \mathbf{h}(s) = r$, therefore for the same reason as in the previous case $(r, s)\sigma_{\mathbf{S5}}$ and $(t, r)\sigma_{\mathbf{S5}}$. If $\min\{\ell(t), \ell(s)\} > 1$ we can repeat the same argument of the other case with the difference that $r = ((\mathbf{h}(t), \mathbf{h}(s))\sigma, (\mathbf{s}^1(t), \mathbf{s}^1(s))\sigma)$. □

Remark 19.12. It is worth noting that we have no constraints on the component logics; they may be combined logics themselves.

19.4 Inference Rules

In displaying the rules of $\mathbf{K}EM$ we shall use Smullyan–Fitting [Fitting, 1983] $\alpha, \beta, \nu_i, \pi_i, i \in I$ unifying notation as shown in the following tables:

α	α_1	α_2
$\mathsf{T}A \wedge B$	$\mathsf{T}A$	$\mathsf{T}B$
$\mathsf{F}A \vee B$	$\mathsf{F}A$	$\mathsf{F}B$
$\mathsf{F}A \to B$	$\mathsf{T}A$	$\mathsf{F}B$
$\mathsf{F}\neg A$	$\mathsf{T}A$	$\mathsf{T}A$

β	β_1	β_2
$\mathsf{F}A \wedge B$	$\mathsf{F}A$	$\mathsf{F}B$
$\mathsf{T}A \vee B$	$\mathsf{T}A$	$\mathsf{T}B$
$\mathsf{T}A \to B$	$\mathsf{F}A$	$\mathsf{T}B$
$\mathsf{T}\neg A$	$\mathsf{F}A$	$\mathsf{F}A$

Classification of α- and β-formulas

The relationships between α- and β-formula are depicted in the following table.

$$\alpha = \beta^C \qquad \alpha_1 = \beta_1^C \qquad \alpha_2 = \beta_2^C$$
$$\beta = \alpha^C \qquad \beta_1 = \alpha_1^C \qquad \beta_2 = \alpha_2^C.$$

Relationships between α- and β-formulas

Formulas of type π and ν are classified according to the following tables

ν_i	ν_0
$\mathsf{T}\Box_i A$	$\mathsf{T}A$
$\mathsf{F}\Diamond_i A$	$\mathsf{F}A$

π_i	π_0
$\mathsf{F}\Box_i$	$\mathsf{F}A$
$\mathsf{T}\Diamond_i A$	$\mathsf{T}A$

Classification of ν- and π-formulas

Similarly we provide the relationships between ν_i- and π_i-formulas.

$$\nu_i = \pi_i^C \qquad \nu_0 = \pi_0^C$$
$$\pi_i = \nu_i^C \qquad \pi_0 = \nu_0^C.$$

Relationships between ν_i- and π_i-formulas

Given a signed formula X, X^C denotes the *conjugate* of X, i.e., the result of changing the sign of X to its opposite; two LS-formulas $X : t$ and $X^C : s$ such that $(t, s)\sigma_{\mathbf{L}_I^\delta}$ and will be called $\sigma_{\mathbf{L}_I^\delta}$*-complementary*:

$$\frac{\alpha : t}{\alpha_1 : t} \qquad \frac{\alpha : t}{\alpha_2 : t} \tag{α}$$

$$\frac{\beta : t \quad \beta_2^C : s}{\beta_1 : (t, s)\sigma_{\mathbf{L}_I^\delta}}[(t, s)\sigma_{\mathbf{L}_I^\delta}] \qquad \frac{\beta : t \quad \beta_1^C : s}{\beta_2 : (t, s)\sigma_{\mathbf{L}_I^\delta}}[(t, s)\sigma_{\mathbf{L}_I^\delta}] \tag{β}$$

$$\frac{\nu_i : t}{\nu_0 : (s, t)}[s \in \Phi_V^i \text{ and new}] \tag{ν_i}$$

$$\frac{\pi_i : t}{\pi_0 : (s, t)}[s \in \Phi_C^i \text{ and new}] \tag{π_i}$$

$$\frac{}{X : t \qquad X^C : t}[t \text{ restricted}] \tag{PB}$$

$$\frac{X : t \quad X^C : s}{\times(t, s)\sigma_{\mathbf{L}_I^\delta}}[(t, s)\sigma_{\mathbf{L}_I^\delta}] \tag{PNC}$$

Here the α-rules are just the familiar linear branch-expansion rules of the tableau method, while the β-rules correspond to such common natural inference patterns such as modus ponens, modus tollens, etc. (t, s stand for arbitrary labels). The rules for the modal operators are as usual. 's new' in the proviso for the ν_i- and π_i-rule means: s must not have occurred in any label yet used. Notice that in all inferences via an α-rule the label of the premiss carries over unchanged to the conclusion, and in all inferences via a β-rule the labels of the premisses must be $\sigma_{\mathbf{L}_I^\delta}$-unifiable, so that the conclusion inherits their unification. PB (the 'principle of bivalence') represents the (LS-version of the) semantic counterpart of the cut rule of the sequent calculus (the intuitive meaning is that: a formula A is either true or false in any *given* world, whence the requirement that t should be restricted). PNC (the 'principle of non-contradiction') corresponds to the familiar branch-closure rule of the tableau method, saying that from the occurrence of a pair of $\sigma_{\mathbf{L}_I^\delta}$-complementary formulas on a branch we may infer the closure ('$\times$') of the branch. The $(t, s)\sigma_{\mathbf{L}_I^\delta}$ in the 'conclusion' of PNC means that the contradiction holds 'in the same world'. Other logics might require additional rules in order to capture the full power of their semantics. See for example [Governatori, 1996; Governatori, 1997]. As usual with refutation methods a $\mathbf{K}EM$-proof of A consists of a successful attempt to construct a countermodel for A by assuming that A is false in some arbitrary model, which means that we assume that A is false in the actual world of the model.

Let us consider the combined systems $\mathbf{ED}^\delta$ of Example 19.10 and the formulas $A = \Box_1\Diamond_2 p \to \Diamond_1\Box_2\Diamond_2 p$ and $B = \Box_1 p \to \Diamond_1\Box_2\Diamond_2 p$. It is easy to see that A is valid in $\mathbf{ED}^\delta$. On the other hand, B holds in $\mathbf{ED}^D$, but fails in $\mathbf{ED}^F$. We provide now their $\mathbf{K}EM$-proofs, and from them the fibred model where B fails can be easily seen.

1. $\mathsf{F}\Box_1\Diamond_2 p \to \Diamond_1\Box_2\Diamond_2 p$	w_0
2. $\mathsf{T}\Box_1\Diamond_2 p$	w_0
3. $\mathsf{F}\Diamond_1\Box_2\Diamond_2 p$	w_0
4. $\mathsf{T}\Diamond_2 p$	(W_1^1, w_0)
5. $\mathsf{F}\Box_2\Diamond_2 p$	(W_2^1, w_0)
6. $\mathsf{T}p$	$(w_1^2, (W_1^1, w_0))$
7. $\mathsf{F}\Diamond_2 p$	$(w_2^2, (W_2^1, w_0))$
8. $\mathsf{F}p$	$(W_1^2, (w_2^2, (W_2^1, w_0)))$

1. $\mathsf{F}\Box_1 p \to \Diamond_1\Box_2\Diamond_2 p$	w_0
2. $\mathsf{T}\Box_1$	w_0
3. $\mathsf{F}\Diamond_1\Box_2\Diamond_2 p$	w_0
4. $\mathsf{T}p$	(W_1^1, w_0)
5. $\mathsf{F}\Box_2\Diamond_2 p$	(W_2^1, w_0)
6. $\mathsf{F}\Diamond_2 p$	$(w_2^2, (W_2^1, w_0))$
7. $\mathsf{F}p$	$(W_1^2, (w_2^2, (W_2^1, w_0)))$

Steps 2, and 3 are obtained by an application of an α-rule on 1; the other steps have been obtained by straightforward applications of ν_i- and π_i-rules, until we have atomic formulas. We notice that in both trees we have complementary formulas, i.e., 6, 8 in the tree for A and 4, 7 in the tree for B. All that remains to do consists of verifying whether they are $\sigma_{\mathbf{E}D^\delta}$-complementary; i.e., we have to check if their labels $\sigma_{\mathbf{E}D^\delta}$-unify. In Example 19.10 we have seen that the labels of 6 and 8 $\sigma_{\mathbf{E}D^\delta}$-unify. However those of 4 and 7 $\sigma_{\mathbf{E}D^D}$- but not $\sigma_{\mathbf{E}D^F}$-unify.

19.5 Soundness and Completeness

In this section we prove soundness and completeness for $\mathbf{K}EM$-based fibred modal tableaux. There are very similar to the results of the previous chapter;‘ however the material presented in this section points out close connections between unifications and modal fibring.

Theorem 19.13. *If a* $\mathbf{K}EM$ *tree closes it closes atomically.*

Proof A closed $\mathbf{K}EM$ tree means that each branch is closed, i.e., it contains two $\sigma_{\mathbf{L}_I^\delta}$-complementary formulas $A : t$ and $A^C : s$.

$$\begin{array}{c} \vdots \\ A : t \\ \vdots \\ A^C : s \\ \times \end{array}$$

We prove the theorem by induction of the complexity of the complementary formulas. If they are literals then the branch closes atomically.

If they are not literals let us examine their form: if A is of type α then A^C is of type β; moreover $\alpha_1 = \beta_1^C$ and $\alpha_2 = \beta_2^C$. We apply an α-rule on $A : t$, obtaining $\alpha_1 : t$ and $\alpha_2 : t$. Because of the relations just mentioned we can apply a β-rule w.r.t. $A^C : s$ and α_n $(n = 1, 2)$, from which we derive $\beta_{3-n} : (t, s)\sigma_{\mathbf{L}_I^\delta}$. At this point the branch contains $\alpha_n : t$ and $\beta_n : (t, s)\sigma_{\mathbf{L}_I^\delta}$, which are $\sigma_{\mathbf{L}_I^\delta}$-complementary, in so far as $(t, (t, s)\sigma_{\mathbf{L}_I^\delta})\sigma_{\mathbf{L}_I^\delta}$; see Lemma 19.11. If A is of type β we repeat the above reasoning applying the α-rule on A^C instead of A.

If A is of type ν_i, then A^C is of type π_i and $\nu_0 = \pi_0^C$. We apply a ν_i-rule on $A : t$ and a π_i-rule on $A^C : s$ obtaining $\nu_0 : (W_n, t)$ and $\pi_0 : (w_m, s)$, where W_n and w_m are new in the branch. The resulting formulas are $\sigma_{\mathbf{L}_I^\delta}$-complementary due to the relationship between ν_i and π_i formulas and the fact that the labels obviously $\sigma_{\mathbf{L}_I^\delta}$-unify. If A is of type π_i, then A^C is of type ν_i and we can repeat the same argument. □

In the course of $\mathbf{K}EM$-proofs labels are used to build appropriate models. Since the structure of the labels and unifications follows closely that of dovetailed and fibred models, we can repeat the same construction 'grafting' the models for each $\mathbf{L}_i$ through $\mathbf{F}$ into fibred and dovetailed models obtaining models for $\mathbf{L}_I^\delta$.

Theorem 19.14. *Let $\mathbf{L}_i, i \in I$, be modal logics and let $\mathbf{L}_I^\delta$ be the resulting combined logic. If $\models_{K_i} A \iff \vdash_{\mathbf{K}EM(\mathbf{L}_i)} A$ then*

1. $\models_{\mathbf{L}_I^F} A \iff \vdash_{\mathbf{K}EM(\mathbf{L}_I)} A$ *using* $\sigma^{\mathbf{F}}$;
2. $\models_{\mathbf{L}_I^D} A \iff \vdash_{\mathbf{K}EM(\mathbf{L}_I)} A$ *using* $\sigma^{\mathbf{D}}$.

Proof We prove the theorem by showing (a) the set $\mathbf{T}_I^\delta$ is a subset of the set of formulas provable in $\mathbf{K}EM$ and (b) $\mathbf{K}EM$ rules are sound with respect to fibred and dovetailed models.

We start by proving (a). According to Theorems 3.10 and 3.16 a formula is a fibred (dovetailed) theorem if either it is a theorem of a component, or has been obtained by one of the axioms, or has been derived from an application of a modal fibring (dovetailing) rule or modus ponens. We have then to prove that such rules and axioms are derived in $\mathbf{K}EM$.

By hypothesis $\mathbf{T}_i$ coincides with the set of formulas provable in $\mathbf{K}EM$ for $\mathbf{L}_i$.

For axiom 1b1 (Theorem 3.10), by hypothesis $A \to \alpha_n$ is a theorem of a fibred language, therefore $\mathsf{T}A : w_0$ and $\mathsf{F}\alpha_n : w_0$ lead to a closed $\mathbf{K}EM$-tree. Let us start now a $\mathbf{K}EM$-tree for $A \to \bigvee_n \alpha_n : w_0$; we obtain

$$\begin{array}{c} \mathsf{F}A \to \bigvee_n \alpha_n : w_0 \\ \mathsf{T}A : w_0 \\ \mathsf{F}\bigvee_n \alpha_n : w_0 \\ \mathsf{F}\alpha_1 : w_0 \\ \vdots \\ \mathsf{F}\alpha_n : w_0 \end{array}$$

At this point we can graft the proof for $A \to \alpha_n$, thus closing the tree.

For axiom 1b2 (Theorem 3.10), by hypothesis $A(x_j)$ has a closed $\mathbf{K}EM$-tree, which means that each branch τ is closed; Theorem 19.13 implies that each branch is atomically closed, therefore each branch contains two $\sigma_{\mathbf{L}_i}$-complementary labelled signed literals, let us say $x_\tau : t_\tau$ and $x_\tau^C : s_\tau$. We can now replace $\Box_j\alpha_j$ with x_τ obtaining $S\Box_j\alpha_j : t_\tau$ and $S^C\Box_j\alpha_j : s_\tau$. But the last two formulas are $\sigma_{\mathbf{L}_I^F}$-complementary; then also in this case the tree is closed.

For modus ponens, by hypothesis A and $A \to B$ have closed $\mathbf{K}EM$-trees; let us call them $\mathcal{T}_1$ and $\mathcal{T}_2$:

$$
\begin{array}{ccc}
 & \mathsf{F}B : w_0 & \\
 \swarrow & & \searrow \\
\mathsf{F}A \to B : w_0 & & \mathsf{T}A \to B : w_0 \\
\mathcal{T}_2 : w_0 & & \mathsf{F}A : w_0 \\
\times & & \mathcal{T}_1 : w_0 \\
 & & \times
\end{array}
$$

For the modal and dovetailing rule, let us assume that C is a formula satisfying the conditions of the rules. Then C has a closed $\mathbf{K}EM$-tree; let us call it $\mathcal{T}$. We show now that also the tree for $\Box_j^d C$ is closed:

$$
\begin{array}{c}
\mathsf{F}\Box_j^d C : w_0 \\
\mathsf{F}\Box_j^{d-1} C : (w_1^j, w_0) \\
\vdots \\
\mathsf{F}C : (w_d^j, \ldots (w_1^j, w_0)) \\
\mathcal{T} : w_0' \\
\times
\end{array}
$$

All we have to do is to identify w_0' with $(w_d^j, \ldots (w_1^j, w_0))$ and we can repeat the proof for $C : w_0'$, in so far as $((w_d^j, \ldots (w_1^j, w_0)), (w_d^j, \ldots (w_1^j, w_0)))\sigma_{\mathbf{L}_I^s}$.

For (b) in the course of $\mathbf{K}EM$-proofs, we generate labels according to the structure of the formulas involved, but, as we have already said, they also generate (counter)-models. The labels are intended to denote possible worlds and relations among them. Remember (see Section 19.2, and Example 19.8): all the relevant information is recoded in the labels. So, to extract such information, we have to map labelled signed formulas to elements of fibred and dovetailed models. This is achieved with the help of three functions, namely g, r, and f. The function g will map labels to sets of possible worlds: a singleton for constants, a set of worlds (possibly empty) for variables, and an actual world for auxiliary labels. The accessibility relation R is assumed to be closed under specific conditions; but, we want to reconstruct it, through r, from the labels: path labels are intended to represent not only worlds, but also the chain of possible worlds leading to them. Finally, f, given an LS-formula, returns the evaluation of the formula with respect to the world(s) denoted by its label.

Let $\mathbf{m}^i = (S^i, R^i, a^i, h^i)$ be a model in $\mathcal{K}_i$ where $S^i = \Phi_C^i$; R^i is a binary relation on S^i; $a^i \in \Phi^A$ (see the beginning of Section 19.3.1) and h^i is an evaluation function.

Let g be a function from the set of i-pure labels occurring in a $\mathbf{K}EM$-proof to $\wp(S^i)$ thus defined:

$$g(t) = \begin{cases} \{\mathbf{h}(t)_t\} & \text{if } \mathbf{h}(t) \in \Phi^i_C \\ \{w_n \in S^i \mid g(\mathbf{b}(t))R^i w_n\} & \text{if } \mathbf{h}(t) \in \Phi^i_V \\ S^i & \text{if } t \in \Phi_V \\ \{a^i\} & \text{if } t \in \Phi^A. \end{cases}$$

It may be possible that two labels have the same head, but they denote different worlds; this is the way we have indexed $\mathbf{h}(t)$ with the label itsself. However we shall drop the subscript, when this is harmless.

Let r be a function from the set of i-pure labels occurring in a $\mathbf{K}EM$-proof to R^i thus defined:

$$r(t) = \begin{cases} \varnothing & \text{if } \ell(t) = 1 \\ g(t^1)R^i g(t^2), \ldots, g(t^{n-1})R^i g(\mathbf{h}(t)) & \text{if } \ell(t) = n > 1. \end{cases}$$

Let f be a function from LS-formulas occurring in a $\mathbf{K}EM$-proof to v thus defined:

$$f(SA : t) =_{def} h(w^i_n, A) = 1 \text{ if } S = \mathsf{T}$$
$$f(SA : t) =_{def} h(w^i_n, A) = 0 \text{ if } S = \mathsf{F}$$

for all $w^i_n \in g(t)$.

Until now we have examined i-pure labels. Let t be a non-i-pure label. It can be decomposed into i-pure labels as follows: let $n \in \mathbb{N}$ such that $\forall m > n$, $\mathbf{s}^m(t)$ is i-preferred. The label $\mathbf{c}^n(t)$ is i-pure. A non-i-pure label can be conceived as a recursive fibring of i-pure sublabels.

Let $\mathbf{m}$ be an SFM structure, where the fibred function $\mathbf{F}$ is defined as follows:

$$\mathbf{F}_i(w_j) = g(w_0) \tag{19.5}$$

for each $w_n \in g(\mathbf{s}^n(t))$, where $w_0 = \mathbf{h}^1(\mathbf{c}^n(t))$ and $\mathbf{c}^n(t)$ is i-pure. Moreover we require that if $w_j \neq w_k$ then $\mathbf{F}_i(w_j) \neq \mathbf{F}_i(w_k)$. In the case of dovetailing we impose $\mathbf{F}_i(w_j) = w_j$

It is easy to see that $\mathbf{m}$ is a fibred or a dovetailed model for $\mathbf{L}^\delta_I$.

Lemma 19.15. *For any $t, s \in \Im$, if $(t, s)\sigma_{\mathbf{L}^\delta_I}$ then $g(t) \cap g(s) \neq \varnothing$.*

Proof The proof is by induction on the number of applications of $\sigma_{\mathbf{L}_i}$ in $\sigma_{\mathbf{L}^\delta_I}$. First we have to prove the property for $\sigma_{\mathbf{L}_i}$ and therefore for $\sigma^{A^i_1 \cdots A^i_n}$. For a detailed proof see [Artosi *et al.*, 1997; Governatori, 1997].

The proof is by induction on the number of applications of $\sigma^{A^i_1 \cdots A^i_n}$ in $\sigma_{\mathbf{L}_i}$. We need first to prove the following:

Lemma 19.16. *For any $t, s \in \Im$, if $(t, s)\sigma^{A^i_1 \cdots A^i_n}$ then $g(t) \cap g(s) \neq \varnothing$.*

Proof We prove only the case for $DT4$; the other cases are similar and can be found in [Artosi *et al.*, 1997; Governatori, 1997]. The proof is by induction on the length of labels. If $\min\{\ell(t), \ell(s)\} = 1$, then at least one of t and s is either a constant or a variable, so that five cases will be present. By the definition of unifications t, s are either: (i) two constants, or (ii) a variable and a constant, or (iii) two variables, or (iv) a variable and a label, or (v) a constant and a label.[78]

Case (i). Two constants unify if and only if they are the same constant, and so $t = s$; therefore from the definition of g, $g(t) = g(s)$ and so $g(t) \cap g(s) \neq \varnothing$.

Case (ii). If t (respectively s) is a variable and s (respectively t) is a constant, then $g(t) = S^i$ and $g(s) \in \wp(S^i)$ therefore also in this case $g(t) \cap g(s) \neq \varnothing$.

Cases (iii) and (iv). These cases are identical to the previous ones because: (1) S^i is not empty, and (2) the variable is mapped to S^i and the label to some world(s) in it.

Case (v). This case implies that $(t, s)\sigma^T$. Let us assume, for the sake of economy, that $\ell(t) = 1$ and $\ell(s) = n > 1$. If $(t, s)\sigma^T$, then for each $\mathbf{h}(\mathbf{s}(s))$ such that $\ell(\mathbf{s}(s)) > 1$ either $\mathbf{h}(\mathbf{s}(s)) \in \Phi_V$, or $\mathbf{h}(\mathbf{s}(s)) = t$; therefore $r(s) = tR^i\mathbf{h}^2(s), \ldots, \mathbf{h}^{n-1}(s)R^i\mathbf{h}^n(s)$. If $\mathbf{h}^2(s) \in \Phi_V$, then it denotes the set of worlds accessible from t; if $\mathbf{h}^2(s) \in \Phi_C$, then $t = g(\mathbf{h}^2(s))$,[79] in any case, through reflexivity $t \subseteq g(\mathbf{h}^2(s))$, so we take t as a representative of the set denoted by $\mathbf{h}^2(s)$, which implies $tR^i\mathbf{h}^3(s)$. We repeat the same argument until we arrive at $tR^i\mathbf{h}^n(s)$: if $\mathbf{h}^n(s) \in \Phi_C$, then $t = g(\mathbf{h}^2(s))$ and so they denote the same world; if $\mathbf{h}^n(s) \in \Phi_V$, then it denotes the set of worlds accessible from t; but t belongs to such a set, therefore, in all cases $g(t) \cap g(s) \neq \varnothing$.

For the inductive step we have $\min\{\ell(t), \ell(s)\} = n > 1$. Let us assume inductively that the lemma is valid up to n; if $\ell(t) = \ell(s)$ we shall write t and s as $(\mathbf{h}(t), \mathbf{b}(t))$ and $(\mathbf{h}(s), \mathbf{b}(s))$, respectively. If $(t, s)\sigma^D$, by the definition of σ^D we get $(\mathbf{b}(t), \mathbf{b}(s))\sigma^D$, for which the lemma holds; let w_j be one of the worlds shared by $\mathbf{b}(t)$ and $\mathbf{b}(s)$, whence $w_jR^i\mathbf{h}(t)$ and $w_jR^i\mathbf{h}(s)$. We have now only to analyse what kind of labels $\mathbf{h}(t)$ and $\mathbf{h}(s)$ are, which falls under the cases (i), (ii), and (iii). Cases (i) and (ii) are the same as the inductive case. We have thus to examine case (iii). Both $\mathbf{h}(t)$ and $\mathbf{h}(s)$ denote the set of worlds accessible from w_j, but such a set is not empty because of the seriality of R.

If $\ell(t) \neq \ell(s)$, we shall assume that $\ell(t) < \ell(s)$ (the case $\ell(s) < \ell(t)$ is dealt with in the same way). If $(t, s)\sigma^T$ and $\mathbf{h}(t) \in \Phi_C$ then $(t, \mathbf{s}^{\ell(t)}(s))\sigma^D$, therefore, combining the proofs of the previous case and case (v) of the inductive base we obtain the desired result. If $\mathbf{h}(t) \in \Phi_V$, then for all $s^n, n \leq \ell(t)$, $(\mathbf{h}(t), \mathbf{h}(s))\sigma = (\mathbf{h}(t), s^n)\sigma$ which means $g(t) \cap g(\mathbf{s}^n(s)) \neq \varnothing$, and in particular $g(t) \cap g(\mathbf{s}^{\ell(t)}(s)) \neq \varnothing$.

If $(t, s)\sigma^4$ then $\mathbf{h}(t) \in \Phi_V$ and $(\mathbf{b}(t), \mathbf{s}^{\ell(t)-1}(s))\sigma^D$, for which the inductive hypothesis holds; let w_j be such a shared world. $\mathbf{h}(t)$ denotes all the worlds accessible from w_j, but, due to transitivity, the world(s) denoted by $\mathbf{h}(s)$ belong(s) to $\mathbf{h}(t)$ and so $g(t) \cap g(s) \neq \varnothing$.

This ends the proof of Lemma 19.16. □

[78] Cases (ii), (iii), and (iv) are not found in $\mathbf{K}EM$-proofs, but they are useful both for dealing with cases in the inductive step and for case (v).

[79] Due to the rules of $\mathbf{K}EM$ and the definition of the unifications, this case is possible only if $\mathbf{h}^2(s)$ has been obtained by a previous unification, and so they do denote the same world.

We return to the proof of the main lemma. If $\sigma_{\mathbf{L}_i}$ consists of a single step of $\sigma^{A_1^i \cdots A_n^i}$, then $(t,s)\sigma_{\mathbf{L}_i} = (t,s)\sigma^{A_1^i \cdots A_n^i}$; by Lemma 19.16 we obtain $g(t) \cap g(s) \neq \varnothing$.

Let us assume, inductively, that the lemma holds up to n. If $\sigma_{\mathbf{L}_i}$ consists of $n+1$ $\sigma^{A_1^i \cdots A_n^i}$-unifications, $(t,s)\sigma_{\mathbf{L}_i} = (\mathbf{c}^t(t), \mathbf{c}^s(s))\sigma^{A_1^i \cdots A_n^i}$ where $(\mathbf{s}^t(t), \mathbf{s}^s(s))\sigma_{\mathbf{L}_i}$, which contains n applications of $\sigma^{A_1^i \cdots A_n^i}$, and the lemma holds for it. We can now repeat the argument of Lemma 19.11 with respect to $(\mathbf{c}^t(t), \mathbf{c}^s(s))\sigma^{A_1^i \cdots A_n^i}$, proving thus that $g(t) \cap g(s) \neq \varnothing$. We have thus proved that the property holds also for $\sigma_{\mathbf{L}_i}$. $\sigma_{\mathbf{L}_I^\delta}$ results from the fibred combination of $\sigma_{\mathbf{L}_i}$-unifications of i-pure labels. According to the interpretation of the fibred function (19.5) we can combine recursively the same argument as before. This ends the proof of Lemma 19.15. □

Lemma 19.17. *For any $t, s \in \Im$, if $f(SA : t)$, $(t,s)\sigma_{\mathbf{L}_I^\delta}$ then $f(SA : (t,s)\sigma_{\mathbf{L}_I^\delta})$.*

Proof Let us suppose, by contradiction, that the lemma does not hold, so the proof trivially follows from Lemma 19.15, and the definition of f. □

The α-rules and PB are obviously sound rules in $\mathbf{m}$ in so far as they work locally. For the β-rules and PNC: by the hypothesis $(t,s)\sigma_{\mathbf{L}_I^\delta}$, then, by Lemmas 19.15 and 19.17, there exists a world in $g((t,s)\sigma_{\mathbf{L}_I^\delta})$, let us say w_j, where both β, β_n $(n = 1, 2)$ and β_{3-n} hold. This implies that β_n and β_n^C hold at w_j, which is a contradiction. The same argument can be applied in the case of PNC.

For the ν_i-rules, let us suppose $\nu_i = \mathsf{T}\Box_i A$; the case of $\nu_i = \mathsf{F}\Diamond_i A$ is identical. Let us suppose it does not hold. Then for all $w_i \in g(t)$, $h(w_i, \Box_i A) = 1$ and for some $w_m \in g((s,t))$, $h(A, w_m) = 0$, where (s,t) is i-preferred. If t is i-preferred, then $h^i(w_i, \Box_i A) = 1$ implies $\forall w_j (w_i R^i w_j \rightarrow h^i(w_j, A) = 1)$, By the definitions of g and r we know $w_i R^i w_m$, thus obtaining a contradiction. If t is not i-preferred, then $h^j(w_i, \Box_i A) = 1$ iff $h^i(\mathbf{F}_i(w_i), \Box_i A) = 1$. The label (s,t) is not i-pure, but (s, w_0), where $\mathbf{F}_i(w_i) = g(w_0)$, is. At this point we can repeat the same argument as before with w_0 instead of w_i. The proof for the π_i-rule is similar.

This ends the proof of Theorem 19.14. □

20

FIBRING LABELLED DEDUCTIVE SYSTEMS

20.1 Introduction

We saw in previous chapters how fibring can be done with the instrumental use of labels. This chapter applies our fibring ideas to labelled deductive systems (LDS).

LDS (see [Gabbay, 1996b]) is a general methodology for presenting and studying logical systems. It unifies many areas of logic and has extensive applications.

We are going to investigate the connection between fibring and labelled deductive systems on two levels.

1. First we view any labelled deductive system as a fibred system in itself. We regard it as the result of fibring an algebraic system $\mathcal{A}$ with a logical language $\mathbb{L}$, to form the LDS $(\mathcal{A}, \mathbb{L})$.
2. Second, we explore the options for fibring any two labelled deductive systems to form a new fibred system.
   Since a large number of diverse logical systems have an LDS formulation, fibring LDSs will provide us with a very general way of combining any two logics: we first give them an LDS form and then we combine (fibre) the LDSs.

The chapter will conclude with several case studies.

We begin by introducing a simplified version of LDS, sufficient for our fibring purposes. The reader is referred to my LDS book [Gabbay, 1996b] for full details.

An algebraic LDS is built up from two components: an algebra $\mathcal{A}$ and a logic language $\mathbb{L}$. To make things specific, let us assume that we are dealing with a particular algebraic model $\mathcal{A} = (S, \leq, f_1, \ldots, f_k)$, where S is the domain of the algebra, $\leq$ is a partial order, i.e. a reflexive, antisymmetric and transitive relation on S and $f_1, \ldots, f_k$ are function symbols on S of arities $r_1, \ldots, r_k$ respectively. The sequence $\Sigma = (\leq, f_1, \ldots, f_k)$ is called the signature of $\mathcal{A}$. It is really the language of $\mathcal{A}$ in logical terms but we use Σ to separate it from $\mathbb{L}$. We assume the functions are *isotonic*, i.e. they are either monotonic up or monotonic down in each variable, namely for each coordinate x in f we have that either

$$\forall x, y(x \leq y \to f(\ldots, x, \ldots) \leq f(\ldots, y, \ldots))$$

or

$$\forall x, y(x \leq y \to f(\ldots, y, \ldots) \leq f(\ldots, x, \ldots))$$

holds.

(If both hold in a connected model, then f does not really depend on x, because of the antisymmetry of $\leq$.)

In case $\leq$ is reflexive and transitive but not antisymmetric, we say that $\mathcal{A}$ is a *pre-algebra*. In this case the relation $x \equiv y$ defined as $x \leq y \wedge y \leq x$ is a congruence relation and so we can take the equivalence classes as the algebra. Two pre-algebras are considered equivalent if their corresponding algebras are isomorphic.

The reason for adopting this notion of algebra lies earlier in this book, in Remark 2.9, where we show that such algebras are obtained as canonical models of a fairly general class of logical systems (I-logics), where $\leq$ is the ordering arising from the consequence relation of the logic ($A \leq B$ is $B \mid\!\sim A$). It is therefore clear that for the purpose of studying connections between LDS and general fibring of logics, there is no loss of generality in restricting the algebra to the above form. The simplification that such a restriction offers in the general theory of LDS is considerable. In fact it is more convenient to work with pre-algebras, since in the case of algebras arising from a logic we manipulate formulas and not equivalence classes of formulas, which change from logic to logic.

The general theory of LDS (see [Gabbay, 1996b, Section 3.2]) requires a source of labels and a source of formulas. These together are used to form the declarative units of the form $t : A$, where t is a label and A is a formula.

The labels can be syntactical terms in some algebraic theory. The algebraic theory itself can be characterised either syntactically by giving axioms which the terms must satisfy or semantically by giving a class of models (algebras) for the language.

The formulas part of an LDS declarative unit are defined in the traditional way in some language $\mathbb{L}$.

An LDS database (theory) Δ is a set of terms and their formulas (i.e. a set of declarative units) with some relationships between the terms. In a general LDS, the terms themselves are syntactical and one always has to worry whether the required relations between the terms of Δ are possible (i.e. are consistent).

If, however, we have a semantic theory for the labels characterised by one single model (algebra), then we can take the labels to be elements of this model and consistency and relationships among the labels (elements) of Δ will always be clear—it is as dictated by the model.

We therefore present for the purpose of this chapter, a concrete definition of LDS based on a single model as an algebra of labels.[80]

Definition 20.1 (Concrete algebraic LDS).

1. A concrete algebraic LDS has the form $(\mathcal{A}, \mathbb{L})$ where:
   (a) $\mathcal{A} = (S, \leq, f_1, \ldots, f_k)$ is a concrete algebraic model. The elements of S are called labels.
   (b) $\mathbb{L}$ is a predicate language with connectives $\sharp_1, \ldots, \sharp_m$ with arities $r_1, \ldots, r_m$, and quantifiers $(Q_1 x), \ldots, (Q_{m'} x)$. The connectives can be some well-known modalities, binary conditional, etc. and the quantifiers can be some known generalised or traditional quantifiers. We assume that the traditional

[80] A general LDS will be based on a class of algebraic models, e.g. the class of all Abelian groups. In such a case the labels must be syntactical terms from the language of Abelian groups. If the LDS is based on a single algebra then the labels can be taken as elements of that algebra. We refer to such an LDS as a *concrete LDS*.

syntactical notions for $\mathbb{L}$ are defined. we assume $\mathbb{L}$ has only constants, no function symbols and that the constants of $\mathbb{L}$ are indexed by elements of the algebra $\mathcal{A}$, i.e. have the form $c_i^t, t \in S, i = 1, 2, 3, \ldots$.

2. A declarative unit has the form $t : A$, where A is a wff and $t \in S$, or it has the form $t : c_i^s$. The unit $t : A$ intuitively means 'A holds at label t' and the unit $t : c_i^s$ means 'the element c_i which was created at label s does exist in the domain of label t'.
3. A database has the form $\Delta = (D, \mathbf{f}, d, U)$ where $D \subseteq S$ is non-empty and $\mathbf{f}$ is a function associating with each $t \in D$ a set of wffs $\mathbf{f}(t) = \Delta_t$. U is a function associating with each $t \in D$ a set of terms U_t. $d \in D$ is a distinguished point in D. The theory Δ can be displayed by writing $\{t : A_1, t : A_2, s : B, r : c_3^s, \ldots, \}$, where $t : A$ indicates $A \in \mathbf{f}(t)$ and $r : c^s$ indicates $c^s \in U_r$.

Definition 20.2 (Semantics for LDS). Let $(\mathcal{A}, \mathbb{L})$ be a concrete algebraic LDS, with algebra $\mathcal{A}$ and language $\mathbb{L}$ with connectives $\{\sharp_1, \ldots, \sharp_m\}$ and quantifiers $\{Q_1, \ldots, Q_{m'}\}$ where $\sharp_i$ is an r_i place.

1. A semantical interpretation for the LDS has the form

$$\mathcal{I} = (\Psi_0(x, X), \Psi_1, \ldots, \Psi_i(x, X_1, \ldots, X_{r_i}), \ldots, \\ \Psi_m, \Psi_1'(x, Z), \ldots, \Psi_{m'}'(x, Z))$$

   where Ψ_i is a formula of the language of $\mathcal{A}$, possibly second-order, with the single free element variable x and the free set variables X as indicated, and Ψ_i' have a single free element variable x and free binary relation variable Z. We need to assume that Ψ_i and Ψ_j' have the property that if we substitute in them for the set variables sets closed under $\leq$ then the element variable coordinate is monotonic up under $\leq$. In symbols:
   - $\bigwedge_j \forall x, y(x \in X_j \wedge x \leq y \to y \in X_j) \to [x \leq y \to \Psi(x, X_j) \to \Psi(y, X_j)]$.
2. A model for the LDS has the form $\mathbf{m} = (V, h, g, d)$ where $d \in S$ is the distinguished world, V is a function associating a domain V_t with each $t \in S$, h is a function associating with each n-place atomic predicate P a subset $h(t, P) \subseteq V_t^n$.
   The following holds for all t, s and P:

$$t \leq s \text{ implies } V_t \subseteq V_s \text{ and } h(t, P) \subseteq h(s, P).$$

   g is an assignment giving each variable x an element $g(x) \in V_d$ and for each constant c_i^s an element $g(c_i^s) \in V_s$.
3. Satisfaction is defined by structural induction as follows:
   - $t \vDash P(b_1, \ldots, b_n)$ iff $(b_1, \ldots, b_n) \in h(t, P)$.
   - $t \vDash \exists x A(x)$ iff for some $b \in V_t, t \vDash A(b)$.
   - $t \vDash A \wedge B$ iff $t \vDash A$ and $t \vDash B$.

- $t \vDash \sim A$ iff $t \nvDash A$.[81]
- $t \vDash \sharp_i(A_1, \ldots, A_{r_i})$ iff $\mathcal{A} \vDash \Psi_i(t, \hat{A}_1, \ldots, \hat{A}_{r_i})$, where $\hat{A} = \{s \in S \mid s \vDash A\}$.
- $t \vDash (Q_i y)A(y)$ iff $\mathcal{A} \vDash \Psi'_i(t, \widehat{\lambda y A(y)})$, where $\widehat{\lambda y A(y)} = \{(t, y) \mid t \vDash A(y)\}$.

4. We say A holds in $\mathbf{m}$ iff $\mathcal{A} \vDash \Psi_0(d, \hat{A})$.
5. The interpretation $\mathcal{I}$ induces a translation $*$ of $\mathbb{L}$ into a two-sorted language $\mathbb{L}^*$ based on the two domains S (of the algebra $\mathcal{A}$) and $U = \bigcup_t V_t$ (of the predicates of $\mathbb{L}$) as follows:
   - Each atomic predicate $P(x_1, \ldots, x_n)$ (interpreted over the domain U) is translated into
   $$[P(x_1, \ldots, x_n)]^*_t = P^*(t, x_1, \ldots, x_n),$$
   where P^* is a two-sorted predicate with one more variable t ranging over S and $x_1, \ldots, x_n$ ranging over U.
   - $[A \wedge B]^*_t = [A]^*_t \wedge [B]^*_t$
   - $[\sim A]^*_t = \sim [A]^*_t$
   - $[\sharp_i(A_1, \ldots, A_{r_i})]^*_t = \Psi_i(t, \lambda s[A_1]^*_s, \ldots, \lambda s[A_{r_i}]^*_s)$
   - $[(Q_i y)A(y)]^*_t = \Psi'_i(t, \lambda y \lambda s[A(y)]^*_s)$
   - Let $||A||^* = \Psi_0(d, \lambda t[A]^*_t)$.
6. It is easy to show by induction that
   - $t \vDash A$ iff $[A]^*_t$ holds in the naturally defined two-sorted model.

The reader should compare this definition with [Definition 3.2.6], [Gabbay, 1996b]. Gabbay's book on LDS contains plenty of examples of such systems.

Example 20.3 (Intuitionistic canonical algebra). Let $\mathcal{A} = (S, \leq, f_{\rightarrow}, f_{\wedge}, f_n, \bot, \top)$, $n = 1, 2, 3, \ldots$ where the elements of S are all wffs of the intuitionistic fragment with $\{\wedge, \rightarrow, \bot, \forall\}$ and the following holds:

- $\alpha \leq \beta$ iff $\vdash \beta \rightarrow \alpha$ in intuitionistic logic
- $f_{\rightarrow}(\alpha, \beta) = \alpha \rightarrow \beta$
- $f_{\wedge}(\alpha, \beta) = \alpha \wedge \beta$
- $f_n(\alpha) = \forall x_n \alpha$.

$\mathcal{A}$ is the canonical algebra of the fragment; see Definition 9.8.

Let $\mathbb{L}$ be the fragment of classical logic with atomic predicates and variables only. Our LDS will have declarative units of the form $\alpha : P(x_1, \ldots, x_n)$, where α is a wff of intuitionistic logic.

We can define a model by letting

- $h(\alpha, P) = \{(x_1, \ldots, x_n) \mid \alpha \vdash P(x_1, \ldots, x_n)\}$
- $g(x) = x$, for all x

[81] Most logics do not contain this kind of classical negation but other sorts of negation, which fall under the next bullet for unary $\sharp$. Classical modal and temporal logics do contain this negation.

- $V_\alpha = V$, the set of all variables and constants of $\mathbb{L}$.

20.2 Viewing LDS as a Fibred Logic

If we examine the semantics for LDS in the previous section, we see that for a fixed label t, the model $\mathbf{m}_t = (V_t, h(t, -), g)$ is a classical model associated with t.

Consider now the equality theory of $\mathcal{A}$. When we fibre this theory with itself and write statements like $x = \varphi$, we have, according to Definition 7.13, the following:

$$t \vDash x = \varphi \text{ iff } s \vDash \varphi \text{ for all } s \in \mathbf{F}(\{g_t(x)\}, t).$$

If we consider the algebra of the previous example, where $g(x) = x$, we get:

$$t \vDash x = \varphi \text{ iff for all } s \in \mathbf{F}(\{x\}, t), s \vDash \varphi.$$

We can now let $\mathbf{F}$ be independent of t and write $\mathbf{F}(\{x\}, t) = \mathbf{m}_x$, and we therefore get:

$$t \vDash x = \varphi \text{ iff } \mathbf{m}_x \vDash \varphi.$$

Since the evaluation is now independent of t, we can present it as

$$\mathcal{A} \vDash x = \varphi \text{ iff } \mathbf{m}_x \vDash \varphi$$

or by slightly adjusting notation we can write (since $\mathcal{A}$ is fixed)

$$\vDash x = \varphi \text{ iff } \mathbf{m}_x \vDash \varphi.$$

We can thus view $x : \varphi$ as the fibring of the equality language of x with the language of φ.

To be more precise, consider the language of $\mathcal{A}$ with the available function symbols and equality. Let us self-fibre $\mathcal{A}$ with itself (as in Section 7.2) and look at wffs of the form $x = \varphi$. Then we get the fibred system above.

The above considerations suggest a general view of an LDS as a fibred system. Let $(\mathcal{A}, \mathbb{L})$ be a concrete algebraic LDS and let $\mathcal{I}$ be a semantical interpretation for it.

Assume $\mathcal{I}$ is such that all the Ψ_i and Ψ'_j involved are first order and that the models $\mathbf{m}$ involved are constant domain, i.e. $V_t = U$ for all $t \in S$. In fact, we shall further assume that $U = S$ (i.e. we assume self-reflective models[82]). In this case the translation $*$ of item 5 of Definition 16.1.2 associates with each formula A of $\mathbb{L}$ a formula $[A]^*_t$ of the first-order language of $\mathcal{A}$.

We saw above that we can have

$$\vDash x = A \text{ iff } \mathbf{m}_x \vDash \varphi.$$

We can also see that $\mathbf{m}_x \vDash A$ iff $\vDash [A]^*_x$. We therefore conclude

- $x = A$ iff $[A]^*_x$.

[82]Compare with Definitions 10.3.3 and 10.4.4 where self-reflective models are used.

We thus see that we have here a case of self-fibring of the language of $\mathcal{A}$ via a self-internal translation.

Let us check what form the above considerations take in the case of intuitionistic logic.

Example 20.4 (LDS for an intuitionistic fragment). Let $\mathcal{A}$ be the canonical algebra of Example 16.1.3. Let $\mathbb{L}$ be the language of the intuitionistic fragment with $\{\wedge, \to, \bot, \forall\}$.

Our LDS declarative units have the form $\alpha : \varphi$ where both α and φ are wffs of the intuitionistic fragment.

1. An LDS interpretation for this language has the form $\mathcal{I} = (\Psi_0, \Psi_\wedge, \Psi_\to, \Psi_\forall)$ where
$$\begin{array}{l}\Psi_0(x, X) = x \in X\\ \Psi_\wedge(x, X, Y) = (x \in X) \wedge (x \in Y)\\ \Psi_\to(x, X, Y) = \forall y(x \leq y \wedge y \in X \to y \in Y)\\ \Psi_\forall(x, Z) = \forall y \forall u(x \leq y \to Z(y, u)).\end{array}$$
We define $V_t = V =$ all variables of the language. We let g be the identity and let $h(\alpha, P) = \{(x_1, \ldots, x_n) \mid \alpha \vdash P(x_1, \ldots, x_n)\}$.
It is easy to see that the LDS model defined is the original canonical model for the fragment, i.e. we have
$$\alpha \vDash A \text{ iff } \alpha \vdash A.$$
2. We now try to present the same model as a fibred model.
Let $\mathbf{m}_t = (V, h(t, -), g)$ and let $\mathbf{F}(\{x\}, t) = \{y \mid x \leq y\}$.
Then
$$t \vDash x = \varphi \text{ iff } \forall y(x \leq y \to y \vDash \varphi).$$
It is clear from the evaluation that the value is independent of t and so we can write it as $\mathcal{A} \vDash x = \varphi$. We have therefore:
$$\mathcal{A} \vDash \alpha = \varphi \text{ iff } \alpha \vdash \varphi.$$
This is the same model again.

20.3 Fibring two algebras

An LDS is built up from an algebra $\mathcal{A}$ for the labels and a language $\mathbb{L}$ for the formulas. It is therefore clear that to fibre two LDS systems, we would need somehow to fibre their algebras and to fibre their logics. We already know how to fibre logics. We now need to find a way to fibre algebras. We assume we are dealing with pre-algebras of the form: $\mathcal{A} = (S, \leq, f_1, \ldots, f_n)$.

We assume the following:

1. $\leq$ is a reflexive and transitive relation on S.
2. f_j has arity r_j and is *isotonic*, i.e. it is either monotonic up or monotonic down (relative to $\leq$) in each argument.

3. S may or may not be freely generated from a subset $U \subseteq S$ of generators.

We now try to present, through an example, what other assumptions we need to be able to define a satisfactory fibring construction. We must keep in mind the constraint that fibring two logics must correspond to fibring their corresponding term (logic) algebras. Thus we can impose any requirements on the algebras involved in fibring as long as they hold for algebras arising from logics.

Let $\mathbb{L}_0$ and $\mathbb{L}_1$ be two languages and let $\mid\!\sim_0$ and $\mid\!\sim_1$ be two logics in the respective languages. Consider the one step fibred language $\mathbb{L}_0[\mathbb{L}_1] = \mathbb{L}_{(0,1)}$, as defined and discussed in Section 12.2. $\mathbb{L}_{(0,1)}$ is the language resulting from a one level substitution of wffs of $\mathbb{L}_1$ for atoms of formulas of $\mathbb{L}_0$.

Let $\mathcal{A}_0$ and $\mathcal{A}_1$ be the respective canonical algebras of $\mid\!\sim_0$ and $\mid\!\sim_1$. In algebraic terms the construction of $\mathbb{L}_{(0,1)}$ means that we add to $\mathcal{A}_0$ all the terms of $\mathcal{A}_1$ and get an algebra $\mathcal{A}_{(0,1)}$ by applying all functions (connectives) of $\mathcal{A}_0$ to the combined set of terms of $\mathcal{A}_0$ and $\mathcal{A}_1$. The consequence relations $\mid\!\sim_i, i = 0, 1$, of $\mathbb{L}_i$ become $\leq_i$ in the respective algebras (through $A \leq_i B$ iff $B \mid\!\sim_i A$) and so the consequence relation $\mid\!\sim_{(0,1)}$ on $\mathbb{L}_{(0,1)}$ would correspond to $\leq_{(0,1)}$ of $\mathcal{A}_{(0,1)}$ which has to be defined by using $\leq_0$ and $\leq_1$. The process of fibring must tell us how to do so.

The logic algebras are free algebras, generated by the atoms and satisfying a set of schematic axioms of the form $A(x_j) \leq_i B(y_k)$. So it is this kind of algebra that we are trying to fibre.

Let us now look at two specific algebras, not necessarily ones that arise from a logic, and see what is involved when we try to fibre them.

Consider two algebras $\mathcal{A}_0$ and $\mathcal{A}_1$ as follows: $\mathcal{A}_0 = (S_0, \leq_0, +)$, where S_0 is the domain of the algebra and it is assumed to be generated from an infinite set G_0 of generators. $+$ is a binary function symbol which is associative and commutative on S_0. $\mathcal{A}_1 = (\{0,1\}, \leq_1, \cdot)$ where $S_1 = \{0,1\}$, $\leq_1$ is the ordinary ordering and $\cdot$ is ordinary multiplication.

Let us consider one level fibring of $\mathcal{A}_1$ into $\mathcal{A}_0$. This means that we add $\{0,1\}$ to S_0 and apply the function symbols of $\mathcal{A}_0$ (i.e. $+$) to this new set of elements.

We let $S_{(0,1)}$ be the set generated by $+$ from $S_0 \cup S_1$. (This corresponds to one step fibring of logics of the form $\mathbb{L}_0[\mathbb{L}_1]$.) This set contains expressions of the form, for example, $(0+s)+1, 1+0, (s+1)+t$, for $t, s \in S_0$. The question we ask ourselves is what ordering relations do these new elements satisfy? In other words how to define the ordering $\leq_{(0,1)}$ on $S_{(0,1)}$?

Is $1+0 \leq_{(0,1)} 0+1$?

Is $(s+1)+t \leq_{(0,1)} s+(1+t)$?

The only commitment we have is to maintain the isotonicity. Since $0 \leq_1 1$ we must have $1+0 \leq_{(0,1)} 1+1$ and $(0+s)+1 \leq_{(0,1)} (1+s)+1$ but we are not committed to $t+1 \leq_{(0,1)} 1+t$, unless the commutativity of $+$ is officially part of the algebra $\mathcal{A}_0$.

If $\mathcal{A}_0$ is presented as $\mathcal{A}_0 = (S_0, \leq_0, +, \Upsilon_0)$ where Υ_0 is the set of inequalities $x+y \leq_0 y+x$ and $x+(y+z) \leq_0 (x+y)+z$, then we can require that all substitution instances of these inequalities hold in $S_{(0,1)}$.

Thus we can have, for $\alpha, \beta \in S_{(0,10)}$

$\alpha \leq^{\Upsilon}_{(0,1)} \beta$ iff by definition (α, β) is in the smallest relation containing $\leq_0 \cup \leq_1$ and is closed under isotonicity, transitivity and every substitution instance of Υ_0.

Call this definition the Υ-option for extending $\leq$.

It seems that $\leq^{\Upsilon}_{(0,1)}$ depends on the particular axiomatisation Υ_0. Υ_0 must be schematic to allow substitution. We can assume that Υ_0 is a family of general schemas of the form $t(x_j) \leq s(y_k)$ intended to hold for any substitution instance of the variables.

This is certainly the case for algebras arising as canonical algebras of logics.

In our particular algebra, we have another way of getting appropriate equations and inequalities to hold for $0, 1$ in $S_{(0,1)}$ even without an official set Υ_0. Since S_0 happens to be generated by G_0 we can map (using a function $\mathbf{F}$) the element 0 into a generator $c_0 = \mathbf{F}(0)$ and 1 into a generator $c_1 = \mathbf{F}(1)$ and let expressions like $1 + 0$ mean or be the same as $c_1 + c_0$.

In general, any expression of the form $t(x/0, y/1) \leq s(x/0, y/1)$ containing 0 and 1 will be considered to hold in $S_{(0,1)}$ if $t(x/c_0, y/c_1) \leq s(x/c_0, y/c_1)$ hold in $\mathcal{A}_0$. Let $\leq'_{(0,1)}$ be the relation obtained in this manner. It is not clear whether $\leq'_{(0,1)}$ is the same as $\leq^{\Upsilon}_{(0,1)}$. It is reasonable that it would be the same since c_0, c_1 are free generators, but we need a proof. There is a small problem, as $0 \leq_1 1$ holds but $c_0 \leq_0 c_1$ does not hold, but modulo that (i.e. if we close $\leq'_{(0,1)}$ under the consequences of this extra assumption) do we get equality?

Let $\leq^{\mathbf{F}}_{(0,1)}$ be the ordering obtained in the proper manner from the mapping $\mathbf{F}$. Let us call it the $\mathbf{F}$ option for extending $\leq$.

Is $\leq^{\mathbf{F}}_{(0,1)}$ the same as $\leq^{\Upsilon}_{(0,1)}$? We shall investigate this question later.

The advantage of this second method is that we do not need to know what Υ_0 is and the fibring is more model theoretic, independent of the presentation (axiomatisation).

Let us take another look at how $S_{(0,1)}$ is generated. It is generated from $S_0 \cup \{0, 1\}$. Since S_0 is generated from G_0, we can generate $S_{(0,1)}$ from $G_0 \cup \{0, 1\}$. Since $0 \leq_1 1$ holds, we are now generating algebras from ordered sets. It is therefore clear that this is the appropriate notion to start with.

We now summarise what concepts we need to clarify, in order to (hopefully) obtain a satisfactory definition of a fibring construction:

1. the notion of a free algebra generated from an ordered set of generators $\mathcal{P} = (U, \prec)$;
2. the notion of a family Υ of inequalities $t \leq s$ of algebraic terms and how it can contribute to generating free algebras;
3. given a free generated algebra $\mathcal{A}$ and an additional disjoint ordered set $\mathcal{P}$, we need to define a construction for $\mathcal{P}$ to be added to $\mathcal{A}$ as an additional set of generators;
4. a zigzag operation between two algebras where each can repeatedly and inductively be added to the other as a set of new generators;
5. a study of the relative properties of the Υ option as opposed to the $\mathbf{F}$ option of extending $\leq$.

We now define a series of (algebraic) constructions, to lead the way to the notion of fibring of such algebras.

Definition 20.5 (Free syntactical algebra). Let $U = \{u_1, u_2, \ldots\}$ be a set of constants (serving as specific generators) and let $V = \{x_1, x_2, \ldots\}$ be a set of variables (serving as free generators) and let $f_1, \ldots, f_k$ be function symbols. We define a notion of (*absolutely free*) terms *with variables* $x_1, \ldots, x_n$ as follows:

1. If $x \in V$ then x is (an absolutely free) term with variable x.
2. A constant u is (an absolutely free) term with no variables.
3. If f is an r-place function symbol and $t_1, \ldots, t_n$ are (absolutely free) terms with (pairwise disjoint) sets of variables $V_1, \ldots, V_r$ then $t = f(t_1, \ldots, t_r)$ is an (absolutely free) term with variables $\bigcup_{i=1}^{r} V_i$.

Lemma 20.6. *Assuming f_i are all isotonic then every absolutely free term is either monotonic up or monotonic down in each of its variables.*

Proof By induction. □

Example 20.7. If f is a two-place function symbol and $x \neq y$ are variables then $f(x, y)$ is an absolutely free term but not $f(x, x)$. If $f(x, y)$ is monotonic up in x and down in y, it is not clear whether $f(x, x)$ is monotonic at all in x.

Definition 20.8.

1. A free variable substitution θ is a function $\theta : V \mapsto V$.
2. θ is said to be *finite* if there exists a finite set $V_\theta = \{y_1, \ldots, y_n\}$ such that for all $x \notin V_\theta, \theta(x) = x$.
3. A free equation has the form $t\theta = s\theta$, where t, s are absolutely free terms and θ is a finite pure variable substitution.

Example 20.9. Let f be a unary function. The equation $f(f(x)) = x$ can be presented as the free equation

$$f(f(u))\theta = w\theta$$

where $\theta(u) = x$ and $\theta(w) = x$.

In fact any equation between terms can be presented as a free equation for suitable absolutely free terms and a finite pure variable substitution.

Consider the equation $f(f(x)) = g(x)$. This can also be presented as $x = y \rightarrow f(f(x)) = g(y)$.

In fact any expression of the form $t\theta \leq s\theta$, where t, s are absolutely free terms can be presented in the form

$$\left(\bigwedge_{\theta(x)=\theta(y)} x = y \right) \rightarrow t \leq s$$

or equivalently

$$\left(\bigwedge_{\theta(x)\leq\theta(y)} x \leq y \right) \rightarrow t \leq s.$$

Definition 20.10 (Horn rules).

1. A (universal Horn) rule has the form

$$\rho : \bigwedge_i t_i \leq s_i \to t \leq s$$

where t_i, s_i, t, s are absolutely free terms.

2. A rule is said to be a pure inequality if it has the form (x_i, y_i are variables):

$$\bigwedge_i x_i = y_i \to t \leq s.$$

3. A rule is said to be an equality if it has the form

$$\bigwedge_i x_i = y_i \to t = s.$$

4. A set of rules Υ is said to entail a rule ρ iff the universal closure of all members of Υ proves (in classical logic) the universal closure of ρ.
5. Let Υ be a set of rules in a language and let $\mathfrak{A}$ be a class of algebras for the language. We say $\mathfrak{A}$ is an Υ-class if *exactly* the universal closures of all rules of Υ are valid in all members of $\mathfrak{A}$.

Definition 20.11 (Free algebras).

1. Let $\mathcal{P} = (V, \prec)$ be an ordered set of constants with $\prec$ reflexive and transitive (intended to serve as abstract generators) and let Σ be a language with the abstract function symbols f_i of arity $r_i, i = 1, \ldots, n$. For each $j \leq r_i$ the function f_i is assumed to be *isotonic* (monotonic up or monotonic down) in its jth argument. Let Υ be a set of universal Horn rules. We now construct the free algebra $\mathcal{A}$ generated from $\mathcal{P}$ and Υ as follows:
   (a) Let T be the set of terms built up using V as a basis.
   (b) Let $\leq$ be the smallest relation on T which agrees with $\prec$ on V and is closed under Υ with the following two rules, for every function symbol f and every coordinate in f
      - $t \leq s \to f(\cdots t \cdots) \leq f(\cdots s \cdots)$ for f monotonic up in the coordinate
      - $s \leq t \to f(\cdots t \cdots) \leq f(\cdots s \cdots)$ for f monotonic down in the coordinate.

   Let $t \equiv s$ be defined as $(t \leq s) \wedge (s \leq t)$. Then clearly $\equiv$ is a congruence relation on T.

   We let $\mathbb{F}_\Sigma(\mathcal{P}, \Upsilon) = (T, \leq, f_i)$ be the pre-algebra generated by $\mathcal{P}$ and Υ in Σ. In this pre-algebra, $\leq$ is not antisymmetric but $\equiv$ is a congruence relation and we can look at the algebra $\mathbb{F}_\Sigma(\mathcal{P}, \mp)_{/\equiv}$ of equivalence classes.
2. Let $\mathcal{A}$ be an algebra of the form $\mathcal{A}_0 = \mathbb{F}_\Sigma(\mathcal{P}_0, \Upsilon)$, where $\mathcal{P}_0 = (V_0, \prec_0)$. Let $\mathcal{P}_1 = (V_1, \prec_1)$ such that $V_0 \cap V_1 = \varnothing$.

Consider $\mathcal{P}_{(0,1)} = (V_0 \cup V_1, \prec_0 \cup \prec_1)$ and construct $\mathcal{A}_{(0,1)} = \mathbb{F}_\Sigma(\mathcal{P}_{(0,1)}, \Upsilon)$. Clearly $\mathcal{A}_0 \subseteq \mathcal{A}_{(0,1)}$ (i.e. $T_0 \subseteq T_{(0,1)}$). We can proceed by induction and construct

$$\mathcal{P}_{(0,\dots,n)} = (\bigcup_{i=0}^{n} V_i, \bigcup_{i=0}^{n} \prec_i)$$

and $\mathcal{A}_{(0,\dots,n)} = \mathbb{F}_\Sigma(\mathcal{P}_{(0,\dots,n)}, \Upsilon)$. Let $\mathcal{A}_\infty = \bigcup_n \mathcal{A}_{(0,\dots,n)}$.
We have

$$\mathcal{A}_\infty = \mathbb{F}_\Sigma(\mathcal{P}, \Upsilon)$$

where

$$\mathcal{P} = (\bigcup_n V_n, \bigcup_n \prec_n).$$

3. **Zigzag construction**
   Let $\mathcal{A}_0^y = \mathbb{F}_{\Sigma_y}((V_0^y, \prec_0^y), \Upsilon_y), y = 0, 1$, be algebras in the language Σ_y respectively. We can consider $\mathcal{A}_0^{1-y}$ as an ordered sets $(V_0^{1-y}, \prec_0^{1-y})$ and construct $\mathcal{A}_{(y,1-y)}^y = \mathbb{F}_{\Sigma_y}((V_0^y \cup V_0^{1-y}, \prec_0^{1-y} \cup \prec_0^{1-y}), \Upsilon_y)$.
   Similarly we form $\mathcal{A}_{(1-y,y)}^{1-y}$.
   By induction, we can regard $\mathcal{A}_{(u_1,\dots,u_n)}^y, \mathcal{A}_{(x_1,\dots,x_n)}^{1-y}$ as ordered sets $(V_{(u_1,\dots,u_n)}^y, \prec_{(u_1,\dots,u_n)}^y)$, and $(V_{(x_1,\dots,x_n)}^{1-y}, \prec_{(x_1,\dots,x_n)}^{1-y})$ and construct $\mathcal{A}_{(y,x_1,\dots,x_n)}^y$ and similarly $\mathcal{A}_{(y-1,u_1,\dots,u_n)}^{1-y}$.
   To make sure that each $V_{\bar{x}}^y$ is a new disjoint set of constants we use a naming device. For each $a \in \mathcal{A}_{\bar{x}}^y$ we let 'a' $\in V_{\bar{x}}^y$ and similarly for $V_{\bar{x}}^{1-y}$.
   This gives us in the limit two algebras $\mathcal{A}_\infty^y$ and $\mathcal{A}_\infty^{1-y}$.

The following is a systematic way of doing this zigzag operation.

Let Σ_1 and Σ_2 be two languages and let $\mathcal{A}_i = (S_i, \leq_i, f_1^i, \dots, f_{n_i}^i, \Upsilon_i), i = 0, 1$, be two algebras in Σ_i. Assume that they are order compatible, namely that

$$\forall x, y \in S_0 \cap S_1 (x \leq_1 y \text{ iff } x \leq_0 y).$$

Let α be a finite alternating sequence $(y, x_1, \dots, x_n)$ of elements $y, x_i \in \{0, 1\}$. We define algebras $\mathcal{A}_\alpha = (S_\alpha, \leq_\alpha, f_j^y, j = 1, \dots, n_y, \Upsilon_y)$ as follows

1. $\mathcal{A}_{(0)} = \mathcal{A}_0$ and $\mathcal{A}_{(1)} = \mathcal{A}_1$.
2. Let $x \in \{0, 1\}$ and let $y = 1 - x$. Let $\alpha = (x, x_1, \dots, x_n)$ and $\gamma = (y, x, x_1, \dots, x_n)$. We define $\mathcal{A}_\gamma$ as follows:

We assume S_α and $\leq_\alpha$ are defined.

Let $c \mapsto$ 'c' be a mapping giving for each $c \in S_\alpha$ a name 'c'. Let $(V_\alpha, \prec_\alpha)$ be defined by

$$V_\alpha = \{\text{'}c\text{'} \mid c \in S_\alpha\}, \prec_\alpha = \{(\text{'}c_1\text{'}, \text{'}c_2\text{'}) \mid c_1 \leq_\alpha c_2\}.$$

Let $\mathcal{P}_\alpha = (V_\alpha \cup S_y, \prec_\alpha \cup \leq_y)$ and construct $\mathbb{F}_{\Sigma_y}(\mathcal{P}_\alpha, \Upsilon_y) = \mathcal{A}_\gamma$.

Having defined $(S_\gamma, \leq_\gamma)$ for all γ, we let $S_\infty^y = \bigcup_\gamma S_\gamma, \leq_\infty^y = \bigcup_\gamma \leq_\gamma$, for $\gamma = (y) * \alpha$. .

The elements of S^y_∞ have the form $a = t(x_1, \ldots, x_n, `c_1', \ldots, `c_k') \in S_\gamma$ for some $\gamma = (y) * \alpha$, where $x_i \in S_y$ and $c_j \in S_\alpha$.

We can now define the zigzag function $\mathbf{F}$.

Let $\mathbf{F}(a) = `a'$ be our zigzag function defined on S^y_∞. Note that $\mathbf{F}$ takes an element of $S_{(y)*\alpha}$ to an element of $S_{(1-y)*(y)*\alpha}$. Note also that $\mathbf{F}$ satisfies

$$t \leq_\infty s \to \mathbf{F}(t) \leq_\infty \mathbf{F}(s).$$

Theorem 20.12. *Let $\mathcal{A} = (S, \leq, f_i)$ be a pre-algebra with signature Σ. For each $a \in S$ let 'a' be a corresponding formal atomic constant. Let $U = \{`a' \mid a \in S\}$. Define $\prec$ on $U^2$ by letting '$a$' $\prec$ '$b$' iff (definition) $a \leq b$. Let $\mathcal{P} = (U, \prec)$ and let $\Upsilon_\mathcal{A} = \{t(`a_i') \leq s(`b_j') \mid t(a_i) \leq s(b_j)$ in $\mathcal{A}\}$.*

Then $\mathcal{B} = \mathbb{F}_\Sigma(\mathcal{P}, \Upsilon_\mathcal{A})$ is equivalent to $\mathcal{A}$, (i.e. their corresponding algebras are isomorphic.

Proof We show that $t(`a_i') \leq s(`b_j')$ holds in the algebra $\mathbb{F}_\Sigma(\mathcal{P}, \Upsilon_\mathcal{A})$ iff $t(a_i) \leq s(b_j)$ holds in $\mathcal{A}$.

One direction is clear since $\leq$ is defined as the closure under $\Upsilon_\mathcal{A}$. We show the other direction, namely that if $t(`a_i') \leq s(`b_j')$ then $t(a_i) \leq s(b_j)$. This holds because $\Upsilon_\mathcal{A}$ itself agrees with $\prec$ on U and is closed under the appropriate isotonicity rules. □

The last theorem shows that any pre-algebra $\mathcal{A}$ can be viewed as a pre-algebra on a set of generators U closed under some inequality rules $\Upsilon_\mathcal{A}$.

We can therefore present $\mathcal{A}$ as $\mathcal{A} = (S, \leq, U, \prec, f_i, \Upsilon_\mathcal{A})$, where $(S, \leq, f_i) = \mathbb{F}_\Sigma((U, \prec), \Upsilon_\mathcal{A})$, with $\Upsilon_\mathcal{A}$ inequality rules.

Definition 20.13 (Fibring algebras). Let $\mathcal{A}_i = (S_i, \leq_i, f^i_1, \ldots, f^i_{n_i}, g, \Upsilon_i)$ $i = 0, 1$, be two pre-algebras in the languages Σ_i respectively, with Υ_i pure inequality rules. We assume the two languages share some function symbols (symbolised as a single g) and that we may have $S_0 \cap S_1 = S \neq \varnothing$. We want to define a new pre-algebra $\mathbf{Fib}(\mathcal{A}_0, \mathcal{A}_1)$ which is the result of fibring the two algebras together. We need to assume that $\leq_i$ agree on S and that S is closed under the common function symbols (i.e under g). Under these conditions we are ready to begin the construction of the fibred algebra.

Let $U_i = \{`x' \mid x \in S_i\}$, where $x \mapsto `x'$ is a naming function turning the elements of S_i into formal constants. This step is to turn the algebra $\mathcal{A}_i$ into a term algebra. If $\mathcal{A}_i$ is originally a free algebra generated by a set of free generators G_i, we can take U_i as G_i. This case will be dealt with in Example 20.16 below.

Let $U = U_0 \cup U_1$. Define $\prec$ on U^2 by 'x' $\prec$ 'y' iff for some $z \in S_0 \cap S_1$ we have $x \leq_i z \leq_{i-1} y$.

The above definition makes $\prec$ the formal 'mirror' of the transitive closure of $\leq_0$ $\cup \leq_1$.[83]

[83]Note that these algebras can arise as the algebras of some logic $\mathbf{L}$. In this case U is the set of all wffs of $\mathbf{L}$ and $\leq$ is obtained from the consequence relation of the logic by $A \leq B$ iff $B \mid\!\sim A$. In this context it is important to pay attention to what happens when the algebra (logic) is decidable. If $\mathcal{A}_i$ are dedicable, this means $\leq_i$ are recursive relations. Since $\prec$ is defined as the transitive closure of $\leq_0 \cup \leq_1$, it may not be recursive. If the languages and domains of $\mathcal{A}_i$ are disjoint, $\prec$ will be recursive. Paying attention to these details will enable us to say under what conditions the fibring of two decidable logics is also decidable.

$(U, \prec)$ gives us the 'generators' of the fibred algebra. Our intention is to apply all function symbols freely to these generators (i.e. the function symbols f_j^i, g of both languages) and thus get the terms (elements) of the fibred algebra. However, we also need to extend $\prec$ in an appropriate way to these terms, and to achieve this we need a controlled inductive definition.

Let us now build inductively the terms of the fibred algebra. The induction is on the number of alternating nestings of function symbols from Σ_i over Σ_{1-i}.

Level $(i, 0)$
This case is when there is no nesting and the external function symbol is from Σ_i. The set of terms of this level, which we denote by $S_{(i,0)}$, has elements of the form $t(`x_j')$, where t is built up from Σ_i function symbols and 'x_j' are in U_i. We do not allow 'y' from $U_i - U_{1-i}$, as we consider this to be nesting.

We define the ordering $\prec_{(i,0)}$ on these terms by

$$t(`a_j') \prec_{(i,0)} s(`b_k') \text{ iff } t(a_j) \leq_i s(b_k)$$

i.e. we get a pre-algebra equivalent to the original $\mathcal{A}_i$ (as we saw in the proof of Theorem 20.12, $\prec_{(i,0)}$ is already closed under Υ_i).

Note that in case the algebras $\mathcal{A}_i$ are generated from sets of generators G_i, we need not use naming $x \mapsto `x'$ and let U_i be G_i. Thus in this case $S_{(i,0)}$ can be taken to be S_i themselves. This option will be exploited later to give a simplified construction for the case of freely generated algebras.

The next level looks at the union of $S_{(i,0)} \cup S_{(1-i,0)}$ and extends $\prec_{(i,0)}, i = 0, 1$, to this set.

Level $(i, 0) + (1 - i, 0)$
This level contains all terms in level $(i, 0)$ and level $(1 - i, 0)$. The ordering on this level is the transitive closure of $\prec_{(i,0)} \cup \prec_{(1-i,0)}$.

Level $(i, 1)$
This level has the set of terms $S_{(i,1)}$ of the form $t(`x_j', y_k)$ where 'x_j' is in U_i and where t uses function symbols of Σ_i and y_k is a level $(1 - i, 0)$ term which is not a level $(i, 0)$ term. Note that the y's in the common language are considered part of level $(i, 0)$ being the language of the external function symbols. $\prec_{(i,0)+(i,1)}$ is already defined on the set of terms of level $(i, 0) + (1 - i, 0)$. We now want to extend it to become $\prec_{(i,1)}$ defined on level $(i, 1)$ terms.

Each function symbol in each coordinate is isotonic. The idea is to extend $\prec_{(i,0)+(1-i,0)}$ to be $\prec_{(i,1)}$ by following isotonic requirements, and closing under Υ_i.

To give the construction in detail, we define $\prec_{(i,1)}$ by induction on term complexity, through a sequence of approximations $\prec^n_{(i,1)}, n = 0, 1, \ldots$.

Complexity 0
Let $\prec^0_{(i,1)}$ be $\prec_{(i,0)+(1-i,0)}$.

Complexity $\leq n + 1$
Assume $\prec^n_{(i,1)}$ was extended and is transitively closed for all terms of complexity $\leq n$.

We need to say when $\prec$ holds between terms of the form $t(\alpha_1, \ldots, \alpha_n)$, where α_i range over lower levels. We extend $\prec$ minimally by closing under Υ_i, isotonic and transitive operations and requirements.

Level $(i, 1) + (1 - i, 1)$
We now have the ordering $\prec_{(i,1)}$ extended to all $(i, 1)$ and $(1 - i, 1)$ terms by taking the transitive closure of $\prec_{(i,1)} \cup \prec_{(1-i,1)}$.

Level $(i, n + 1)$
We assume that $\prec_{(i,n)+(1-i,n)}$ is extended to all (i, n) and $(1 - i, n)$ terms. We consider $(i, n + 1)$ terms in $S_{(i,n+1)}$. These have the form $t(x_1, \ldots, y_1, \ldots)$ where x_i are atoms from U_i and $y_i, \ldots, y_k$ are $(1 - i, m)$ terms, for some $m \leq n$, which are not terms in $S_{(1-i,0)}$.

We extend the ordering to the new terms closing under Υ_i, isotonic requirements and transitivity.

Let

$$S_\infty = \bigcup_n (S_{(i,n)} \cup S_{(1-i,n)})$$
$$\prec_\infty = \bigcup_n \prec_{(i,n)+(1-i,n)} .$$

We define the fibred algebra to be

$$\mathcal{A} = \mathbf{Fib}(\mathcal{A}_0, \mathcal{A}_1) = (S_\infty, \prec_\infty, f_j^i, g).$$

Note that if $\mathcal{A}_i$ $i = 0, 1$, are algebras of recursively enumerable consequence relations (i.e. $\leq_i$ are RE) then so is $\mathbf{Fib}(\mathcal{A}_0, \mathcal{A}_1)$ and the definition shows us exactly how the fibred consequence relation is to be generated.

Example 20.14 (Fibring associative semigroups). Let $\mathcal{A}_i = (S_i, \otimes_i, e), i = 0, 1$, be two associative semigroups with multiplication $\otimes_i$ and unit e. $\leq_i$ is equality. Υ_i contains the associativity axiom and the axioms for e. The two algebras share the same unit and possibly more. We can assume $S_0 \cap S_1 = \{e\}$. Let $U = \{\text{`}x\text{'} \mid x \in S_0 \cup S_1\}$.

We check what the fibred domain looks like:

$$S_{(i,0)} = \{\text{`}x_1\text{'} \otimes_i \ldots \otimes_i \text{`}x_n\text{'} \mid x_j \in S_i\}.$$
$$S_{(i,1)} = \{(\text{`}x_{1,1}\text{'} \otimes_i \ldots \otimes_i \text{`}x_{1,n_1}\text{'}) \otimes_i (\text{`}y_{1,1}\text{'} \otimes_{1-i} \ldots \otimes_{1-i} \text{`}y_{1,m_1}\text{'}) \otimes_i \cdots \mid x_{k,j} \in S_{1-i}, y_{k,j} \in S_{i-1}\}$$

Here we took advantage of the associativity in our presentation.

The associativity allows us to use the same multiplication $(a, b) \mapsto ab$ for both semigroups and present the words in the fibred algebra in the form $a_1 a_2 a_3, \ldots, a_n$ where a_i, a_{i+1} are not from the same semigroup. The multiplication in the fibred system is defined by

$$(a_1, \ldots, a_n)(b_1, \ldots, b_m) = \begin{cases} a_1, \ldots, a_n b_1, \ldots, b_m \text{ if } a_n, b_1 \\ \quad \text{are not in the same semigroup} \\ a_1 \ldots a_{n-1}(a_n b_1) b_2 \ldots b_m \\ \quad \text{if } a_n, b_1 \text{ are in the same semigroup.} \end{cases}$$

This example is useful for fibring two substructural logics.

The reader should note that the fibring construction is very similar to free products; see [Kurosh, 1963, p. 143] on words.

Example 20.15 (Free product of groups). Let G_1 and G_2 be two subgroups of a group G. G is said to be the free product of G_1 and G_2 if these subgroups generate the group G and if every element $g \in G$ other than 1 has a unique decomposition into a product form $g = a_1 a_2, \ldots, a_n$ where $a_k \neq 1$ and $a_k \in G_1$ or $a_k \in G_2$ for $k = 1, \ldots, n$ and a_k and a_{k+1} do not belong to the same subgroup.

Example 20.16 (Fibring free algebras). The purpose of this example is to examine the fibring construction for algebras freely generated from an infinite set of generators. This is the kind of algebra arising as the canonical algebra of a logic. For such fibring we have both options, option Υ and option $\mathbf{F}$, available for the extension of $\leq$. This example will give us the opportunity to compare the two constructions.

We assume the following:

$$\mathcal{A}_i = (S_i, \leq_i, G_i, f_j^i, g_k, \Upsilon_i), j = 1, \ldots, n_i, k = 1, \ldots, n,$$

where S is freely generated from G_i using the function symbols $\{f_j^i, g_k\}$ and G_i is infinite. In fact, we let $G_i = \bigcup_m G_i^m$ where G_i^m are infinite and pairwise disjoint. Thus $\mathcal{A}_i^m = (S_i^m, \leq_i, G_i^m, f_j^i, g_k, \Upsilon_i)$ is isomorphic to $\mathcal{A}_i$. We assume that $G_0^0 = G_1^0$, i.e. that the algebras share some of the generators as well as the function symbols $g_k, k = 1, \ldots, n$, and that $\leq_0$ and $\leq_1$ agree on all terms of the common language.

Υ_i are rules of the form $t(x_1, \ldots, x_n) \leq_i s(x_1, \ldots, x_n)$ where $x_1, \ldots, x_n$ are free variables and it is understood that any substitution instances of $t\theta$ and $s\theta$ of t and s are expected to satisfy $(t\theta \leq_i s\theta)$.

$\leq_i$ is generated by Υ_i and the isotonicity of the function symbols. So Υ_i is an axiomatisation of $\leq_i$.[84]

We can simplify the construction of Definition 20.13 by using the $\mathbf{F}$-option for extending $\leq$. Recall the construction of $S_{(i,m+1)}$, the terms of level $m+1$. These have the forms $t(x_1, \ldots, y_1, \ldots)$ where x_j are atoms from U_i (i.e. from $\{`a' \mid a \in S_i\}$) and y_k are terms of lower level which are not in $S_{(i,n)}$. This construction can be viewed as adding all the y_k terms as generators to the algebra based on U_i, as done in items 2 and 3 of Definition 20.11.

Bearing this in mind, let us now construct the terms of our fibred algebras. We will fibre the algebras $\mathcal{A}_i^0$ with generators $G_i^0 = G_{1-i}^0$, leaving the rest of the generators free for future use.

1. **Level** $(i, 0)$
   Let $S_{(i,0)}^0$ be $\mathcal{A}_i^0$

[84] Two such algebras are obtained as algebras associated with logics $\mid\!\sim_i$. For example, $\mid\!\sim_0$ can be a modal logic and $\mid\!\sim_1$ can be a temporal logic. $A \leq_i B$ is $B \mid\!\sim_i A$. The shared generators G_i^0 are the atomic propositions and the shared function symbols would be the classical connectives. Note that in a logic, $\leq_i$ restricted to G_i^0 is $\varnothing$ (atoms do not logically relate to each other).

2. **Level** $(i, 0) + (1 - i, 0)$
   Take $S^0_{(i,0)} \cup S^0_{(1-i,0)}$ and extend $\leq_0 \cup \leq_1$ to this set by taking the transitive closure. Note that even though $\leq_i$ may be recursive, the transitive closure $\leq_{(i,0)+(1-i,0)}$ may not be, since there is a nonempty common language.[85]
3. **Level** $(i, 1)$
   In this level we want to add the terms of level $(i, 0) + (1 - i, 0)$ to the algebra $\mathcal{A}_i$ and apply the function symbols of $\mathcal{A}_i$ to it. Thus the elements will have the form $t(x_j, y_k)$, where x_j are in G and y_k are terms of level $(1 - i, 0)$ which are not in level $(i, 0)$, i.e. y_k begins with a function symbol from $\mathcal{A}_{1-i}$ which is not shared (compare with our construction of the corresponding level of Definition 20.13.)

This construction is fine as it is. We can, however, add to it a twist, which we will be able to appreciate later. We can have a one-to-one embedding $x \mapsto \mathbf{F}_{(1-i,0)}(x)$, which maps the elements $x \in S^0_{(1-i,0)}$ into the generators G^1_i. We can let $\prec^1_{1-i}$ be the mirror ordering of the relation $\leq_{1-i}$ exported onto G^1_i. Now we have two options to proceed.

Option 1 (Υ option)
Continue as in Definition 20.13 ignoring the possibilities offered by the mapping $\mathbf{F}_{(1-i,0)}$

Option 2 (F option)
Continue by using the mapping $\mathbf{F}$ as follows:

Let $\prec^0_i$ be $\varnothing$. We now add to $\mathcal{A}^0_i$ the new generators $(G^1_i, \prec^1_{1-i})$. We proceed to construct

$$\mathcal{A}^1_i = \mathbb{F}_{\Sigma_i}((G^0_i \cup G^1_i, \prec^0_i \cup \prec^1_i), \Upsilon_i).$$

Level $(i, n) + (1 - i, n)$ **and level** $(i, n + 1)$
We describe the two options:

Option 1
We proceed as in Definition 20.13 to construct $S^0_{(i,n)+(1-i,n)}$ and $S^0_{(i,n+1)}$ (based on the generators G^0_i).

Option 2
We assume $\mathcal{A}^n_i$ have been constructed for $i = 0, 1$ and we define embeddings $\mathbf{F}_{(1-i,n)}$ of the domain of $\mathcal{A}^n_{1-i}$ into G^{n+1}_i and construct $\mathcal{A}^{n+1}_i$ as $\mathbb{F}_{\Sigma_i}(\bigcup^{n+1}_{j=0} G^j_i, \bigcup^{n+1}_{j=0} \prec^j_i), \Upsilon_i)$.

Level ∞

Option 1
We get the algebra $\mathcal{A}_\infty$ as in Definition 20.13.

Option 2
For this option we take the union of the respective levels to get $\mathcal{A}^\infty_i$, $i = 1, 2$. This

[85] Imagine we are combining two modalities with $\Box_0$ and $\Box_1$. If both satisfy the axiom of reflexivity we get $q \leq_i \Box_i q$ and hence we get $\Diamond_i q \leq_{(i,0)+(1-i,0)} \Box_{1-i} q$. Compare with Theorem 3.10.

however is not the end of the story. We have not finished yet, as we want an algebra in the language $\Sigma_0 \cup \Sigma_1$ and $\mathcal{A}_i^\infty$, $i = 0, 1$, is in the language Σ_i.

However, we do have the mappings $\mathbf{F}_{(i,n)} : \mathcal{A}_i^n \mapsto G_{1-i}^{n+1}, n = 0, 1, \ldots$.

These mappings give rise to two mappings $\mathbf{F}_{0,1} : \mathcal{A}_0^\infty \mapsto \mathcal{A}_1^\infty$ and $\mathbf{F}_{1,0} : \mathcal{A}_1^\infty \mapsto \mathcal{A}_0^\infty$ defined as $\mathbf{F}_{i,1-i} = \bigcup_n \mathbf{F}_{(i,n)}$.

We have:

- $x \leq y \rightarrow \mathbf{F}(x) \leq \mathbf{F}(y)$.

We can now take $\mathcal{A}_i^\infty$ and allow the function symbols of Σ_{1-i} to apply to the elements of $\mathcal{A}_i$. We need to say what $f(a_1, \ldots, a_k)$ is going to be, for $f \in \Sigma_{1-i}$, and $a_j \in \mathcal{A}_i^\infty$.

Each a_j is in some minimal $\mathcal{A}_i^{m_j}$. Thus we can consider $\mathbf{F}_{i,1-i}(a_j)$. These are elements of $\mathcal{A}_{1-i}^\infty$ and so f is applicable to them. We let

$$f(a_1, \ldots, a_k) = (\text{definition})\mathbf{F}_{1-i,i}(f(\mathbf{F}_{i,1-i}(a_1), \ldots, \mathbf{F}_{i,1-i}(a_k))).$$

Since $\mathbf{F}$ is order preserving, f is isotonic in the right way.

The reader should compare **Option 2** with the independent construction in [Baader and Schulz, 1994, Section 3].

21

CONCLUSION AND DISCUSSION

We saw that the idea of fibred semantics is basically very simple. If we encounter a symbol of a foreign language while evaluating an expression in our own language, then we regard this symbol as atomic (in our language) and turn to the semantics of the foreign language to get a value for it.

The above methodology, when iterated, gives fibred semantics for the combined language. Once we have this fibred semantics, there are some natural general questions to ask about the combined system:

1. How do the properties of the combined system depend on the properties of the components (transfer theorems)?
2. What is the landscape of possible interaction axioms and their corresponding conditions on the fibred semantics?
3. Can the fibring process be mechanised?
4. Can we reformulate existing known systems as fibred systems?
5. Can we extend our methodology to quantified systems?
6. Check interpolation theorems for the combined language.

There are also specific problems to ask for each of the application areas. Here we list some currently active research problems.

Case of modal and intuitionistic logics

We saw that for this case decidability and the finite model property transfer to the combined system. We can also axiomatise the combined system in terms of the axiomatisations of the components. What remains to be done is to investigate the landscape of natural interaction axioms and their corresponding semantic conditions. For example, when dovetailing two modalities $\Box_1$ and $\Box_2$, where both are extensions of modal **K**, the basic models have the form (S, R_1, R_2, a, h). The interaction axiom $\Box_1 A \to \Box_2 A$ corresponds to the condition $R_2 \subseteq R_1$. This is easy to show. It is more difficult to find what condition corresponds to the commutativity axiom: $\Box_1\Box_2 = \Box_2\Box_1$. It is not clear whether we have completeness for models of the form $(S_1 \times S_2, R_1, R_2, (a_1, a_2))$ where R_i is a relation on S_i and where we have:

$$(x_1, x_2)R_1(y_1, y_2) \text{ iff } x_2 = y_2 \wedge x_1 R_1 y_1$$
$$(x_1, x_2)R_2(y_1, y_2) \text{ iff } x_1 y_1 \wedge x_2 R_2 y_2.$$

Of course Chapter16 has some results on products. In general, charting the landscape for two modalities means looking at some natural interaction axioms and their corresponding semantic conditions.

Case of fuzzy logics

In this case the main immediate research problem is to axiomatise the combined system in terms of the axiomatisation of its components.

Case of substructural logics

We need to analyse the various ways of adding modality to a substructural implication and compare with the current systems (linear, relevant, etc.) in the literature. We should try to explain some existing peculiarities in terms of fibring. See [D'Agostino *et al.*, 1997; D'Agostino and Gabbay, 1996a; Dörre *et al.*, 1996].

Case of self-fibring of predicate logics

There is a lot to be done here:

- Apply the fibred semantics idea to intuitionistic logic. This should be straightforward.
- Check applications of the semantics to theories of self-reference.
- Give fibred semantics to free logics and definite descriptions, and compare with existing solutions. Extend the method to intuitionistic logic.
- Investigate transfer theorems.

Case of the conditional

The connection between the metalevel non-monotonic consequence relation and the object level subjunctive conditional are well known. We know how to bring the consequence relation into the object level as a conditional, by using fibred semantics. We need to show systematically that the fibred semantics of the combined system is essentially the same as the traditional semantics of the conditional.

Case of combining metalevel with object level

Chart a methodology of bringing metalevel features into the object level in any reasonable system.

More general fibring problems

An alternative approach to combining logics and structures is in [Blackburn and de Rijke, To appear]; the theory there evolves around the idea of explicitly controlling the communication between the structures (and logics) being combined through the use of modal operators to refer to the 'communication channels' between structures. The approach has clear links with our jump operator, but the precise connection remains to be determined.

For further discussion of combining general algebraic systems see [Baader and Schulz, 1994; Baader and Schulz, 1995b].

Also there is a connection with labelled deductive systems [Gabbay, 1996b]. In view of Chapters 19 and 20, it looks like the most natural environment for performing fibring and formulating the fibred semantics and proof system.

For topological fibring see [Pfalzgraf, 1991; Pfalzgraf and Stokkermans, 1994].

REFERENCES

[Abrusci, 1991] V. M. Abrusci. Phase semantics and sequent calculus for pure non-commutative classical linear propositional logic. *Journal of Symbolic Logic*, 56:1403–1451, 1991.

[Adams, 1975] E. W. Adams. *The Logic of Conditionals*. Reidel, Dordrecht, 1975.

[Alechina, 1995] N. Alechina. *Modal Quantifiers*. PhD thesis, Institute for Logic, Language and Information, University of Amsterdam, 1995.

[Alechina and van Lambalgen, 1996] N. Alechina and M. van Lambalgen. Generalised quantification as substructural logic. *Journal of Symbolic Logic*, 61:1006–1044, 1996.

[Alexiev, 1994] V. Alexiev. Applications of linear logic to computation: An overview. *Bulletin of the IGPL*, 2(1):77–107, 1994.

[Allwein and Dunn, 1993] G. Allwein and J. M. Dunn. Kripke models for linear logic. *the Journal of Symbolic Logic*, 58:514–545, 1993.

[Amati and Pirri, 1993] G. Amati and F. Pirri. A uniform tableaux method for intuitionistic modal logics I. Manuscript, 1993.

[Anderson and Belnap Jr, 1975] A. R. Anderson and N. D. Belnap Jr. *Entailment: the Logic of Relevance and Necessity*, Vol. 1. Princeton University Press, Princeton, 1975.

[Andréka *et al.*, 1995] H. Andréka, J. van Benthem, and I. Németi. Back and forth between modal logic and classical logic. *Logic Journal of the IGPL*, 3:685–720, 1995.

[Åqvist, 1979] L. Åqvist. A conjectured axiomatization of two-dimensional reichenbachian tense logic. *Journal of Philosophical Logic*, 8:1–45, 1979.

[Apt and Turini, 1995] K. R. Apt and F. Turini, editors. *Meta-Logics and Logic Programming*. MIT Press, Cambridge, MA, 1995.

[Arnold, 1994] A. Arnold. *Finite Transition Systems*, Prentice Hall, Englewood Cliffs, NJ, 1994.

[Artosi *et al.*, 1996] A. Artosi, Guido Governatori, and Giovanni Sartor. Towards a computational treatment of deontic defeasibility. In Mark Brown and José Carmo, editors, *Deontic Logic Agency and Normative Systems*, Workshop on Computing, pages 27–46, Springer-Verlag, Berlin, 1996.

[Artosi *et al.*, 1997] A. Artosi, P. Benassi, G. Governatori and A. Rotolo. Shakespearian modal logic: a labelled treatment of modal identity. In M. Kracht, M. de Rijke, H. Wansing, and M. Zakharyaschev, editors, *Advances in Modal Logic*, SILLI. CSLI Publications, Stanford, 1997.

[Avron, 1988] A. Avron. The semantics and proof theory of linear logic. *Theoretical Computer Science*, 57:161–184, 1988.

[Baader and Schulz, 1994] F. Baader and K. Schulz. Combination of constraints solving techniques, and algebraic point of view. Technical Report CIS-Rep-94-75, University of Munich, 1994. A short version has appeared in *Proc. RTA-95*, Lecture Notes in Computer Science 914, pp. 352–366, Springer-Verlag, Berlin, 1995.

[Baader and Schulz, 1995a] F. Baader and K. Schulz. Combination of constraint solvers for free and quasi-free structures. Technical Report CIS-Rep-95-120, University of Munich, 1995.

[Baader and Schulz, 1995b] F. Baader and K. Schulz. On the combination of symbolic constraints, solution domains and constraint solvers. *Proc. CP-95*. Lecture Notes in Computer Science, 976, Springer-Verlag, Berlin, 1995.

[Baldoni *et al.*, 1993] M. Baldoni, L. Giordano and A. Martelli. A multimodal logic to define modules in logic programming. In D. Miller, editor, *Proceedings, International Logic Programming Symposium (ILPS), Vancouver, Canada*, pages 473–487. MIT Press, Cambridge, MA, 1993.

[Barwise and Cooper, 1981] J. Barwise and R. Cooper. Generalized quantifiers and natural language. *Linguistics and Philosophy*, 4:159-219, 1981.

[Beckert, 1998] B. Beckert. Semantic tableaux with equality. *Journal of Logic and Computation*, 1998. To appear.

[Beckert *et al.*, 1993] B. Beckert, R. Hähnle and P. H. Schmitt. The even more liberalized δ-rule in free variable semantic tableaux. In G. Gottlob, A. Leitsch, and D. Mundici, editors, *Proceedings, 3rd Kurt Gödel Colloquium (KGC), Brno, Czech Republic*, LNCS 713, pages 108–119. Springer-Verlag, Berlin, 1993.

[Beckert and Posegga, 1995] B. Beckert and J. Posegga. leanT^AP: Lean tableau-based deduction. *Journal of Automated Reasoning*, 15(3):339–358, 1995.

[Beckert and Pape, 1996] B. Beckert and C. Pape. Incremental theory reasoning methods for semantic tableaux. In P. Miglioli, U. Moscato, D. Mundici, and M. Ornaghi, editors, *Proceedings, 5th Workshop on Theorem Proving with Analytic Tableaux and Related Methods, Terrasini, Palermo, Italy*, LNCS 1071, pages 93–109. Springer-Verlag, Berlin, 1996.

[Beckert and Ahrendt, 1997] B. Beckert and W. Ahrendt. An improved δ-rule for ground first-order tableaux. Draft. University of Karlsruhe, Institute for Logic, Complexity and Deduction Systems, Karlsruhe, Germany, 1997.

[Bencivenga, 1986] E. Bencivenga. Free logic. In D. M. Gabbay and F. Guenthner, editors, *Handbook of Philosophical Logic*, Vol. 3. Kluwer, Dordrecht, 1986.

[Blackburn, 1993] P. Blackburn. Modal logic and attribute value structures. In M. De Rijke, editor, *Diamonds and Defaults: Studies in Pure and Applied Intensional Logic*. Kluwer Academic Publishers, Dordrecht, Synthese Library, 229:19–65, 1993.

[Blackburn and de Rijke, 1995] P. Blackburn and M. de Rijke. Why combine logics? Technical Report CS-R9533, CWI, Amsterdam, 1995. To appear in special issue of *Studia Logica*.

[Blackburn and de Rijke, To appear] P. Blackburn and M. de Rijke. Zooming in, zooming out. *Journal of Logic, Language and Information*, To appear.

[Bolc and Borowik, 1992] L. Bolc and P. Borowik. *Many-valued Logics*, Vol. 1. Springer-Verlag, Berlin, 1992.

[Božić and Došen, 1984] M. Božić and K. Došen. Models for normal intuitionistic modal logics. *Studia Logica*, 43:217–245, 1984.

[Bowen and Kowalski, 1982] K. Bowen and R. Kowalski. Amalgamating language and metalanguage in logic programming. In K. Clark, editor, *Logic Programming*. Academic Press, London, 1982.

[Broda *et al.*, 1998] K. Broda, M. D'Agostino and A. Russo. Transformation methods in LDS. In Hans Jürgen Ohlbach and Uwe Reyle, editors, *Logic, Language and Reasoning. Essays in Honor of Dov Gabbay*. Kluwer Academic Publishers, Dordrecht, 1998.

[Brogi *et al.*, 1992] A. Brogi, P. Mancarella, D. Pedreschi and F. Turini. Meta for modularizing logic programming. In A. Pettorossi, editor, *Proceedings, Third International Workshop on Meta-Programming in Logic (META), Uppsala, Sweden*, LNCS 649, pages 105–119. Springer-Verlag, Berlin, 1992.

[Büchi, 1962] J. R. Büchi. On a decision method in restricted second order arithmetic. In *Logic, Methodology, and Philosophy of Science. Proc. 1960 Intern. Congress*, pages 1–11. Stanford University Press, Stanford, CA, 1962.

[Bull and Segerberg, 1984] R. Bull and K. Segerberg. Basic modal logic. In Dov Gabbay and Franz Guenthner, editors, *Handbook of Philosophical Logic*, Vol. II, Chapter II.I, pages 1–88. Kluwer Academic Publishers, Dordrecht, 1984.

[Bull, 1965] R. A. Bull. A modal extension of intuitionist logic. *Notre Dame Journal of Formal Logic*, 6, 1965.

[Burgess and Gurevich, 1985] J. P. Burgess and Y. Gurevich. The decision problem for linear logic. *Notre Dame Journal of Formal Logic*, 26:566–582, 1985.

[Burgess, 1981] J. P. Burgess. Quick completeness proofs for some logics of conditionals. *Notre Dame Journal of Formal Logic*, 22:76–84, 1981.

[Burgess, 1982] J. P. Burgess. Axioms for tense logic i: "since" and "until". *Notre Dame Journal of Formal Logic*, 23:367–374, 1982.

[Burgess, 1984] J. P. Burgess. Basic tense logic. In D. M. Gabbay and F. Guenthner, editors, *Handbook of Philosophical Logic*, Vol. II, pages 89–133. D. Reidel Publishing Company, Dordrecht, 1984.

[Buvač and Mason, 1995] S. Buvač and I. A. Mason. Metamathematics of contexts. *Fundamenta Informatica*, 23, 1995.

[Buvač, 1996] S. Buvač. Logic of context. In *Proceedings of the 13th AAAI Conference*, 1996.

[Chagrov and Shehtman, 1995] A. Chagrov and V. Shehtman. Algorithmic aspects of tense logics In: *Proceedings of Computer Science Logic'94*, ed. by L. Pacholski and J. Tiuryn. Lecture Notes in Computer Science, Vol. 933, Springer-Verlag, Berlin, 1995.

[Chen, 1987] W. Chen. A theory of modules based on second order logic. In *Proceedings, Symposium on Logic Programming, San Francisco, USA*. IEEE Press, 1987.

[Clarke and Gabbay, 1988] M. R. B. Clarke and D. M. Gabbay. An intuitionistic basis for non monotonic reasoning. In *Non Standard Logics for Automated Reasoning*, P. Smets, A. Mamdani, D. Dubois and H. Prade, eds. pp. 163–178. Academic Press, 1988.

[Crocco *et al.*, 1995] G. Crocco, L. Fariñas del Cerro and A. Herzig. Introduction. In G. Crocco, L. Fariñas del Cerro, and A. Herzig, editors, *Conditionals: From Philosophy to Computer Science*, pages 1–12. Oxford University Press, 1995.

[D'Agostino and Gabbay, 1994] M. D'Agostino and D. M. Gabbay. A generalization of analytic deduction via labelled deductive systems.Part I: Basic substructural logics. *Journal of Automated Reasoning*, 13:243–281, 1994.

[D'Agostino and Gabbay, 1996a] M. D'Agostino and D. M. Gabbay. Fibring labelled tableaux for substructrual logics, 1996. To appear Tableaux 96 conference.

[D'Agostino and Gabbay, 1996b] M. D'Agostino and D. M. Gabbay. Fibred tableaux for multi-implication logics. In P. Miglioli, U. Moscato, D. Mundici, and M. Ornaghi, editors, *TABLEAUX '96. Proceedings of the 5th International Workshop on Theorem Proving with Analytic Tableaux and Related Methods*, Vol. 1071 of *Lecture Notes in Artificial Intelligence*, pages 1–35. Springer-Verlag, Berlin, 1996.

[D'Agostino and Mondadori, 1994] M. D'Agostino and M. Mondadori. The taming of the cut. *Journal of Logic and Computation*, 4:285–319, 1994.

[D'Agostino *et al.*, 1997] M. D'Agostino, D. M. Gabbay, and A. Russo. Grafting modalities onto substructural implication systems. *Studia Logica*, 59:65–102, 1997.

[D'Agostino *et al.*, 1998] M. D'Agostino, D. M. Gabbay, R. Hähnle and J. Posegga, editors. *Handbook of Tableau Methods*. Kluwer, Dordrecht, 1998. To appear.

[de Rijke, 1993–94] M. De Rijke. *Extending modal logic*. Ph.D.Thesis. ILLC Dissertation Series. University of Amsterdam, (1993–94).

[Delgrande, 1988] J. P. Delgrande. An approach to default reasoning based on a first order conditional logic. *Artificial Intelligence*, 36:63–90, 1988.

[Dörre *et al.*, 1996] J. Dörre, D. M. Gabbay, and E. König. Fibred semantics for feature based grammar logics. *Journal of Logic, Language and Information*, 5:387–422, 1996.

[Došen, 1985] K. Došen. Models for stronger normal intuitionistic modal logics. *Studia Logica*, 44:39–70, 1985.

[Došen, 1988] K. Došen. Sequent systems and groupoid models I. *Studia Logica*, 47:353–385, 1988.

[Došen, 1989] K. Došen. Sequent systems and groupoid models II. *Studia Logica*, 48:41–65, 1989.

[Došen, 1993] K. Došen. A historical introduction to substructural logics. In P. Schroeder Heister and Kosta Došen, editors, *Substructural Logics*, pages 1–31. Oxford University Press, 1993.

[Dunn, 1986] M. Dunn. Relevance logic and entailment. In D. M. Gabbay and F. Guenthner, editors, *Handbook of Philosophical Logic*, Vol. III, chapter 3, pages 117–224. Kluwer Academic Publishers, Dordrecht, 1986.

[Ewald, 1986] W. B. Ewald. Intuitionistic tense and modal logic. *Journal of Symbolic Logic*, 51:166–179, 1986.

[Fagin *et al.*, 1990] R. Fagin, J. Y. Halpern and M. Y. Vardi. A nonstandard approach to the logical omniscience problem. In R. Parikh, editor, *Reasoning About Knowledge, TARK 1990*, pages 41–55. Morgan Kaufmann, 1990.

[Farinas del Cerro and Herzig, 1992] L. Farinas del Cerro and A. Herzig. Metaprogramming through intensional deduction: Some examples. In A. Pettorossi, editor, *Proceedings, Third International Workshop on Meta-Programming in Logic (META), Uppsala, Sweden*, LNCS 649, pages 11–25. Springer-Verlag, Berlin, 1992.

[Fariñas del Cerro *et al.*, 1994] L. Fariñas del Cerro, A. Herzig, and J. Lang. From ordering based nonmonotonic reasoning to conditional logics. *Artificial Intelligence*, 66:375–393, 1994.

[Fariñas del Cerro and Herzig, 1996] L. Fariñas del Cerro and A. Herzig. Combining classical and intuitionistic lgoic, or: intuitionistic implication as a conditional. In *Frontiers in Combining Systems (Proc. Int. Workshop FroCos'96)*. F. Baader and K. Schulz, eds. pp. 93–102. Kluwer Academic Publishers, Dordrecht, 1996.

[Fine, 1974a] K. Fine. An ascending chain of S4 logics. *Theoria*, 40:110–116, 1974.

[Fine, 1974b] K. Fine. Logics containing K4. I. *Journal of Symbolic Logic*, 39:31–42, 1974.

[Fine, 1975] K. Fine. Normal forms in modal logic. *Notre Dame Journal of Formal Logic*, 16:(2)229–237, 1975.

[Fine and Schurz, 1991] K. Fine and G. Schurz. Transfer theorems for stratified multimodal logics, 1991. Unpublished manuscript.

[Fine and Schurz, 1997] K. Fine and G. Schurz. Transfer theorems for multimodal logics. In *Logic and Reality, Proceedings Arthur Prior Memorial Conference*, Oxford University Press, 1997, pp. 169–214.

[Finger, 1992] M. Finger. Handling database updates in two-dimensional temporal logic. *Journal of Applied Non-Classical Logic*, 2:201–224, 1992.

[Finger, 1994] M. Finger. *Changing the Past: Database Applications of Two-dimensional Temporal Logics*. PhD thesis, Imperial College, London, Department of Computing, 1994.

[Finger and Gabbay, 1992] M. Finger and D. M. Gabbay. Adding a temporal dimension to a logic. *Journal of Logic, Language and Information*, 1:203–233, 1992.

[Finger and Gabbay, 1996] M. Finger and D. M. Gabbay. Combining temporal logic systems. *Notre Dame Journal of Formal Logic*, 37:204–232, 1996.

[Fischer-Servi, 1977] G. Fischer-Servi. On modal logic with an intuitionistic base. *Studia Logica*, 36:141–149, 1977.

[Fischer-Servi, 1980] G. Fischer-Servi. Semantics for a class of intuitionistic modal calculi. In M. L. Dalla Chiara, editor, *Italian Studies in the Philosophy of Science*, pages 59–71. D. Reidel, Dordrecht, 1980.

[Fischer-Servi, 1984] G. Fischer-Servi. Axiomatizations for some intuitionistic modal logics. *Rend. Sem. Mat. Univ. Politecn. Torino*, 42:179–194, 1984.

[Fitch, 1948] F. B. Fitch. Intuitionistic modal logic with quantifiers. *Portugalia Mathematica*, 7:113–118, 1948.

[Fitting, 1969] M. Fitting. Logics with several modal operators. *Theoria*, 35:259–266, 1969.

[Fitting, 1983] M. Fitting. *Proof Methods for Modal and Intuitionistic Logics*. D. Reidel, Dordrecht, 1983.

[Fitting, 1991] M. Fitting. Many valued modal logics. *Fundamenta Informatica*, 15:235–254, 1991.

[Fitting, 1992] M. Fitting. Many valued modal logics II. *Fundamenta Informaticae*, 17:55–73, 1992.

[Fitting, 1994] M. Fitting. Tableaux for many valued modal logic. Technical Report Jan 25, 1994.

[Fitting, 1996] M. Fitting. *First-Order Logic and Automated Theorem Proving*. Springer, second edition, 1996.

[Font, 1986] J. M. Font. Modality and possibility in some intuitionistic modal logic. *Notre Dame Journal of Formal Logic*, 27:533–546, 1986.

[Gabbay, 1972] D. M Gabbay. General filtration method for modal logic. *Journal of Philos. Logic*, 1:29–34, 1972.

[Gabbay, 1976] D. M. Gabbay. *Investigations in Modal and Tense Logics*. D. Reidel, Dordrecht, 1976.

[Gabbay, 1981a] D. M. Gabbay. Expressive functional completeness in tense logic. In U. Monnich, editor, *Aspects of Philosophical Logic*, pages 91–117. D. Reidel, Dordrecht, 1981.

[Gabbay, 1981b] D. M. Gabbay. An irreflexivity lemma. In U. Monnich, editor, *Aspects of Philosophical Logic*, pages 67–89. D. Reidel, Dordrecht, 1981.

[Gabbay, 1981c] D. M. Gabbay. *Semantical Investigations in Heyting's Predicate Logic*. D. Reidel, Dordrecht, 1981.

[Gabbay, 1982] D. M. Gabbay. Intuitionistic basis for non-monotonic logic. In *Proceedings of CADE-6*, pages 260–273. LNCS 138, Springer-Verlag, Berlin, 1982.

[Gabbay, 1985a] D. M. Gabbay. Theoretical foundations for non-monotonic reasoning in expert systems. In K. R. Apt, editor, *Proceedings NATO Advanced Study Institute on Logics and Models of Concurrent Systems*, pages 439–457. Springer-Verlag, Berlin, 1985.

[Gabbay, 1985b] D. M. Gabbay. N-Prolog: An extension of Prolog with hypothetical implication. II. Logical foundations, and negation as failure. *Journal of Logic Programming*, 4:251–283, 1985.

[Gabbay, 1992] D. M. Gabbay. Metalevel in the object level: modal and temporal logic programming III. In L. Fariñas del Cerro and M. Pentonnen, eds. *Intensional Logics for Programming*, pp. 85–124. Oxford University Press, 1992.

[Gabbay, 1993a] D. M. Gabbay. Fibred semantics and the combination of logics, Part 1. Modal and intuitionistic logics, 1993. Lectures given at *Logic Colloquium 1992*, Veszprém, Hungary, August 1992. A version of the notes is published as a Technical Report, by the University of Stuttgart, Sonderforschungbereich 340, Azenbergstr 12, 70174 Stuttgart, Germany, No 36, 1993. To appear in *Journal of Symbolic Logic*.

[Gabbay, 1993b] D. M. Gabbay. Labelled deductive systems, a position paper. In *Logic Colloquium 90*, pages 66–89. Lecture Notes in Logic, vol. 2, Springer-Verlag, Berlin, 1993.

[Gabbay, 1994a] D. M. Gabbay. Classical vs non-classical logic. In *Handbook of Logic in AI*, Vol. 2, pages 349–489. Oxford University Press, 1994.

[Gabbay, 1994b] D. M. Gabbay. Labelled deductive systems, part I. Technical report, First draft 1989. Preprint, Department of Computing, Imperial College, London, SW7 2BZ, UK. 2nd intermediate draft February 1991, 265 pp, published as CIS-Bericht-90-22, Centrum für Informations und Spachverarbeitung, Universität München, Germany. Third intermediate draft, Max Planck Institute, Saarbrucken, Technical Report, MPI-

I-94-223, 460 pp, 1994.

[Gabbay, 1995a] D. M. Gabbay. Conditional implication and nonmonotonic consequence. In L. Fariñas del Cerro *et al.*, editors, *Views on Conditionals*, pages 347–369. Oxford University Press, 1995.

[Gabbay, 1995b] D. M. Gabbay. Fibred semantics and the weaving of logics, part 2. In L. Csirmak, D. Gabbay, and M. de Rijke, editors, *Logic Colloquium 92, Vezprém, Hungary*, pages 95–113. SILLI/CUP, 1995.

[Gabbay, 1995c] D. M. Gabbay. Fibred semantics and the weaving of logics, part 3: How to make your logic fuzzy. *LINZ Proceedings*, Kluwer, Dordrecht, 1998. To appear.

[Gabbay, 1996a] D. M. Gabbay. Fibred semantics and the weaving of logics, part 4: Self fibring of predicate logics. Technical report, Imperial College, London, 1996.

[Gabbay, 1996b] D. M. Gabbay. *Labelled Deductive Systems*, Volume 1—*Foundations.* Oxford University Press, 1996.

[Gabbay, 1996c] D. M. Gabbay. An overview of fibred semantics and the combination of logics. In *Frontiers of Combining Systems*, K. Schultz and F. Baader, eds. pp. 1–56. Applied Logic Series, Kluwer Academic Publishers, Dordrecht, 1996.

[Gabbay and Clarke, 1987] D. M. Gabbay and M. Clarke. An intuitionistic basis for non-monotonic reasoning. In P. Smets, editor, *Automated Reasoning for Non-standard Logic*, pages 163–179. Academic Press, 1987.

[Gabbay and Nossum, 1996] D. M. Gabbay and R. Nossum. Labelling and fibring your context. Technical report, Imperial College, London, 1996.

[Gabbay *et al.*, 1994] D. M. Gabbay, I. M. Hodkinson, and M. A. Reynolds. *Temporal Logic—Mathematical Foundations and Computational Aspects.* Oxford University Press, 1995.

[Gabbay and Reyle, 1984] D. M. Gabbay and U. Reyle. N-Prolog: An extension of Prolog with hypothetical implication. I. *Journal of Logic Programming*, 4:319–355, 1984.

[Gabbay *et al.*, 1995] D. Gabbay, L. Giordano, A. Martelli and N. Olivetti. Hypothetical updates, priority and inconsistency in a logic programming language. In V. Marek, A. Nerode, and M. Truszczynsky, editors, *Proceedings, 3rd International Conference on Logic Programming and Non-monotonic Reasoning, Lexington/KY, USA*, LNCS 928, pages 203–216. Springer-Verlag, Berlin, 1995.

[Gabbay and Shehtman, 1993] D. M. Gabbay and V. Shehtman. Undecidability of modal and intermediate first-order logics with two individual variables. *Journal of Symbolic Logic*, 58:(3)800–823, 1993.

[Gabbay and Reyle, 1997] D. M. Gabbay and U. Reyle. Labelled resolution for classical and non-classical logics. *Studia Logica*, 59:179–216, 1997.

[Gabbay and Reyle, forthcoming] D. M. Gabbay and U. Reyle. Fibred semantics for Natural Language. In preparation.

[Gabbay and Shehtman, 1998] D. M. Gabbay and V. Shehtman. Products of modal logics, Part 1. *Logic Journal of the IGPL*, 6:73–146, 1998.

[Gallier *et al.*, 1992] J. Gallier, P. Narendran, S. Raatz, and W. Snyder. Theorem proving using equational matings and rigid E-unification. *Journal of the ACM*, 39(2):377–

429, April 1992.

[Garson, 1989] J. Garson. Modularity and relevant logic. *Notre Dame Journal of Formal Logic*, 30:207–223, 1989.

[Ginsberg, 1986] M. L. Ginsberg. Counterfactuals. *Artificial Intelligence*, 30:35–79, 1986.

[Giordano and Martelli, 1994] L. Giordano and A. Martelli. Structuring logic programs: A modal approach. *Journal of Logic Programming*, 21(2):59–94, 1994.

[Giordano and Olivetti, 1994] L. Giordano and N. Olivetti. Combining negation as failure and embedded implication in logic programs. Draft, Università di Torino, Dipartimento di Informatica, 1994.

[Giordano *et al.*, 1988] L. Giordano, A. Martelli, and G. Rossi. Local definitions with static scope rules in logic programming. In *Proceedings, International Conference on Fifth Generation Computer Systems, Tokyo, Japan*, Vol. 2, pages 389–396. Springer-Verlag, Berlin, 1988.

[Giordano *et al.*, 1992] L. Giordano, A. Martelli, and G. Rossi. Extending horn clause logic with implication goals. *Theoretical Computer Science*, 95:43–74, 1992.

[Girard, 1987] J.-Y. Girard. Linear logic. *Theoretical Computer Science*, 50:1–102, 1987.

[Gargov and Goranko, 1993] G. Gargov and V. Goranko. Modal logic with names. *Journal of Philosophical Logic*, 22:607–636, 1993.

[Goldblatt, 1987] R. Goldblatt. *Logics of Time and Computation.* CSLI Lecture notes 7, CSLI, Stanford, CA, 1987.

[Goranko, 1994] V. Goranko. Temporal logic with reference pointers. In *Temporal Logic* (Bonn, 1994), pp. 133–148. Lecture Notes in Comput. Sci., 827. Springer-Verlag, Berlin, 1994.

[Goranko, 1996] V. Goranko. Hierarchies of modal and temporal logics with reference pointers. *Journal of Logic Language and Information*, 5:(1)1–24, 1996.

[Goranko and Passy, 1992] V. Goranko and S. Passy. Using the universal modality: Gains and questions. *Journal of Logic and Computation*, 2:5–30, 1992.

[Goré, 1998] R. Goré. Tableau methods for modal and temporal logics. In Marcello D'Agostino, Dov Gabbay, Reiner Hähnle, and Joachim Posegga, editors, *Handbook of Tableau Methods*, chapter 7. Kluwer, Dordrecht, 1998. In press.

[Gottwald, 1993] S. Gottwald. *Fuzzy Sets and Fuzzy Logic.* Vieweg, 1993.

[Governatori, 1995] G. Governatori. Labelled tableaux for multi-modal logics. In *Theorem Proving with Analytic Tableaux and Related Methods*, Vol. 918 of *LNAI*, pages 79–94, Springer-Verlag, Berlin, 1995.

[Governatori, 1996] G. Governatori. Labelling ideality and subideality. In Dov M. Gabbay and Hans Jurgen Ohlbach, editors, *Practical Reasoning*, Vol. 1085 of *LNAI*, pages 291–304. Springer-Verlag, Berlin, 1996.

[Governatori, 1997] G. Governatori. *Un modello formale per il ragionamento giuridico.* PhD thesis, University of Bologna, 1997.

[Halpern and Shoham, 1986] J. Y. Halpern and Y. Shoham. A propositional modal logic of time intervals. In *Proceedings of the Symposium on Logics in Computer Science – LICS86*, pages 279–292, 1986.

[Henkin *et al.*, 1985] H. Henkin, J. D. Monk and A. Tarksi. *Cylindric Algebras, Part 2*. North Holland, Amsterdam, 1985.

[Hill and Lloyd, 1994] P. Hill and J. Lloyd. *The Gödel Programming Language*. MIT Press, Cambridge, MA, 1994.

[Hughes and Cresswell, 1996] G. Hughes and M. Cresswell. *A New Introduction to Modal Logic*, Routledge, 1996.

[Jones, 1993] A. J. I. Jones. Towards a formal theory of defeasible deontic logic. *Annals of Mathematics and Artificial Intelligence*, 9:151–166, 1993.

[Jones and Pörn, 1985] A. J. I. Jones and I. Pörn. Ideality, subideality and denontic logic. *Synthese*, 65:275–290, 1985.

[Jones and Pörn, 1986] A. J. I. Jones and I. Pörn. 'Ought' and 'must'. *Synthese*, 66:89–93, 1986.

[Kamp, 1971] H. Kamp. Formal properties of now. *Theoria*, 35:227–273, 1971.

[Karlsbeek, 1995] M. Karlsbeek. *Metalogics for Logic Programming*. PhD thesis, Institute for Logic, Language and Information, University of Amsterdam, 1995.

[Katsuno and Sato, 1995] H. Katsuno and K. Sato. A unified view of consequence relation, belief revision and conditional logic. In G. Crocco, L. Fariñas del Cerro, and A. Herzig, editors, *Conditionals: From Philosophy to Computer Science*, pages 33–66. Oxford University Press, 1995.

[Kracht and Wolter, 1991] M. Kracht and F. Wolter. Properties of independently axiomatizable bimodal logics. *Journal of Symbolic Logic*, 56:1469–1485, 1991.

[Kraus *et al.*, 1990] S. Kraus, D. Lehmann, and M. Magidor. Nonmonotonic reasoning, preferential models and cumulative logics. *Artificial Intelligence*, 44:167–208, 1990.

[Kurosh, 1963] A. G. Kurosh. *Lectures on General Algebra*. Chelsea, 1963.

[Karlsbeek and Jiang, 1995] M. Karlsbeek and Y. Jiang. A vademecum of ambivalent logic. In [Apt and Turini, 1995], pp. 27–58.

[Lakemeyer, 1987] G. Lakemeyer. Tractable meta-reasoning in propositional logics of belief. In *IJCAI-87*, pages 402–408, 1987.

[Lambert and van Fraassen, 1972] K. Lambert and Bas C. van Fraassen. *Derivation and Counterexample*. Dickinson Publishing Co, 1972.

[Levesque, 1984] H. Levesque. A logic of implicit and explicit belief. In *AAI-84*, pages 198–202, 1984.

[Link, 1984] G. Link. 1984. Plural. In D, Wunderlich and A. von Stechow, editors *Semantik*. de Gruyter, Berlin, 1991

[MacCaull, 1996] W. MacCaull. Relational semantics and a relational proof system for full Lambek calculus. Technical report, Dept. Mathematics and Computing Sciences, St. Francis Xavier University, 1996.

[McDermott and Doyle, 1980] D. McDermott and J. Doyle. Non-monotonic logic I. *Artificial Intelligence*, **13**, 41–72, 1980.

[McCarthy, 1987] J. McCarthy. Generality in artificial intelligence. *Communications of the ACM*, 30:1030–1035, 1987.

[McCarthy, 1996] J. McCarthy. A logical AI approach to context. Technical report, Stanford University, 1996.

[McCarthy and Buvač, 1994] J. McCarthy and S. Buvač. Formalising context, (expanded notes). Technical Report STAN-CS-TN-94-13, Stanford University, 1994.

[Maddux, 1980] R. Maddux. The equational theory of CA3 is undecidable. *Journal of Symbolic Logic*, 45:311–315, 1980.

[Makinson, 1989] D. Makinson. General theory of cumulative inference. In M. Reinfrank, editor, *Proceedings Second International Workshop on Non-Monotonic Reasoning*. Lecture Notes in Computer Science, Springer-Verlag, Berlin, 1989.

[Maksimova, 1979] L. Maksimova. A classification of modal logics. *Algebra i Logika*, 18:(3)328–340, 385–386, 1979.

[Mancarella and Pedreschi, 1988] P. Mancarella and D. Pedreschi. An algebra of logic programs. In R. Kowalski, editor, *Proceedings, Fifth International Conference on Logic Programming, Seattle, USA*, pages 1006–1023. MIT Press, Cambridge, MA, 1988.

[McCarty, 1988a] T. McCarty. Clausal intuitionistic logic. I. Fixed-point semantics. *Journal of Logic Programming*, 5(1):1–31, 1988.

[McCarty, 1988b] T. McCarty. Clausal intuitionistic logic. II. Tableau proof procedures. *Journal of Logic Programming*, 5(2):93–132, 1988.

[Marx, 1997a] M. Marx. Complexity of modal logics of relations. University of Amsterdam, Report ML-97-02, 1997.

[Marx, 1997b] M. Marx. Complexity of products of modal logics. Imperial College, 1997.

[Marx and Venema, 1997] M. Marx and Y. Venema. *Multi-dimensional Modal Logic*, Applied Logic Series, Vol. 4. Kluwer Academic Publishers, Dordrecht, 1997.

[Meyer and Mares, 1993] R. E. Meyer and E. D. Mares. Semantics of entailment 0. In K. Došen and P. Schroeder-Heister, editors, *Substructural Logics*, pages 239–258. Oxford University Press, 1993.

[Miller, 1986] D. Miller. A theory of models for logic programming. In *Proceedings, Symposium on Logic Programming, Salt Lake City, USA*, pages 106–114. IEEE Press, 1986.

[Moscowitz and Shapiro, 1991] Y. Moscowitz and E. Shapiro. Lexical logic programs. In K. Furukawa, editor, *Proceedings, Eighth International Conference on Logic Programming, Paris, France*, pages 349–363. MIT Press, Cambridge, MA, 1991.

[Nait Abdallah, 1986] M. Nait Abdallah. Ions and local definition in logic programming. In B. Monien, editor, *Proceedings, Third Annual Symposium on Theoretical Aspects of Computer Science (STACS), Orsay, France*, LNCS 210, pages 60–72. Springer-Verlag, Berlin, 1986.

[Németi, 1995] I. Németi. Decidable versions of first-order logic and cylindric-relativised set algebras. In L. Csirmaz, D. M. Gabbay, and M. de Rijke, editors, *Logic Colloquium 92*, pages 177–241. CSLI–FOLLI publication, 1995.

[Németi, 1996] I. Németi. A fine structure analysis of first order logic. In *Arrow Logic and Multimodal Logic*, M. Marx, L. Pólos and M. Masuch, eds. pp. 221–147. FoLLi–CSLI Publications, 1996.

[Nute, 1984] D. Nute. Conditional logic. In D. Gabbay and F. Guenthner, editors, *Handbook of Philosophical Logic*, Vol. 2: Extensions of Classical Logic, pages 387–

440. Kluwer, Dordrecht, 1984.

[Ohlbach, 1991] H. J. Ohlbach. Semantic based translation methods for modal logic. *Journal of Logic and Computataion*, 1:691–746, 1991.

[O'Keefe, 1985] R. O'Keefe. Towards an algebra for constructing logic programs. In *Proceedings, Symposium on Logic Programming, Boston, USA*, pages 152–160. IEEE Press, 1985.

[Ono, 1977] H. Ono. On some intuitionistic modal logic. *Publications Research Institute of Mathematical Science*, 13:687–722, 1977.

[Ono, 1987] H. Ono. Some problems in intermediate predicate logics. *Reports on Mathematical Logic*, 21:55–68, 1987.

[Ono, 1993] H. Ono. Semantics for substructural logics. In Peter Schroeder-Heister, editor, *Substructural Logics*, pages 259–291. Oxford University Press, 1993.

[Pearl and Geffner, 1988] J. Pearl and H. Geffner. Probabilistic semantics for a subset of default reasoning. Technical report, Computer Science Department, University of California, Los Angeles, 1988.

[Pfalzgraf, 1991] J. Pfalzgraf. Logical fiberings and polycontextural systems. In Ph. Jorrand and J. Kelemen, editors, *Fundamentals of Artificial Intelligence Resarch.* LNCS 535, Springer-Verlag, 1991.

[Pfalzgraf and Stokkermans, 1994] J. Pfalzgraf and K. Stokkermans. On robotics scenarios and modeling with fibered structures. In J. Pfalzgraf and D. Wang, editors, *Automated Practical Reasoning: Algebraic Approaches.* Texts and Monographs in Symbolic Computation, Springer-Verlag, Berlin, 1994.

[Plotkin and Stirling, 1986] G. D. Plotkin and C. P. Stirling. A framework for intuitionistic modal logic. In J. Y. Halpern, editor, *Theoretical Aspects of Reasoning About Knowledge*, pages 399–406, 1986.

[Prior, 1957] A. N. Prior. *Time and Modality*. Oxford University Press, 1957.

[Rabin, 1969] M. O. Rabin. Decidability of second order theories and automata on infinite trees. *Transactions of the American Mathematical Society*, pages 595–629, 1969.

[Reiter, 1980] R. Reiter. A logic for default reasoning. *Artificial Intelligence*, 13:81–132, 1980.

[Reynolds, 1992] M. A. Reynolds. An axiomatisation for until and since over the reals without the irr rule. *Studia Logica*, 51:165–194, 1992.

[Reynolds, 1996] M. Reynolds. A decidable temporal logic of parallelism. Manuscript. December. 1996.

[Reynolds, 1997] M. Reynolds. Undecidability of the rectangular modal product. Manuscript, 1997 (submitted).

[Sahlqvist, 1975] H. Sahlqvist. Completeness and correspondence in the first and second order semantics for modal logic. *Studies in Logic and Found. Math.*, 82:110-143, 1975.

[Sambin, 1993] G. Sambin. The semantics of pretopologies. In Peter Schroeder-Heister, editor, *Substructural Logics*, pages 293–307. Oxford University Press, 1993.

[Santos, 1968] E. Santos. Maximin automata. *Information and Control*, 13:363–377, 1968.

[Segerberg, 1970] K. Segerberg. Modal logics with linear alternative relations. *Theoria*, 36:301–322, 1970.

[Segerberg, 1973] K. Segerberg. Two-dimensional modal logic. *Journal of Philosophical Logic*, 2:77–96, 1973.

[Shehtman, 1987] V. Shehtman. On some two-dimensional modal logics. In *8th International Congress on Logic, Methodology, and Philosophy of Science*, Moscow, Abstracts, 1:326–330, 1987.

[Shehtman, 1978] V. Shehtman. Two-dimensional modal logics. *Math. Zametki*, 5:759–772, (In Russian), 1978.

[Shehtman, 1993] V. Shehtman. A logic with progressive tenses. In M. De Rijke, editor, *Diamonds and Defaults: Studies in Pure and Applied Intensional Logic*. pp. 255–285, Kluwer Academic Publishers, Dordrecht, 1993.

[Shehtman and Skvortsov, 1990] V. Shehtman and D. Skvortsov. Semantics of non-classical first order predicate logics. In P. Petkov, editor, *Proceedings of Summer School and Conference on Mathematical Logic held Sept. 13–23 in Chaika*, Bulgaria, Plenum Press, pp. 105–116, 1990.

[Shoham, 1987] Y. Shoham. A semantical approach to nonmonotonic logics. In *Proceedings Logics in Computer Science*, pages 275–279. Ithaca, NY, 1987.

[Shoham, 1988] Y. Shoham. *Reasoning About Change-80*. MIT Press, Cambridge, MA, 1988.

[Simpson, 1993] A. K. Simpson. *The Proof Theory and Semantics of Intuitionistic Modal Logic*. PhD thesis, University of Edinburgh, 1993.

[Smullyan, 1995] R. M. Smullyan. *First-Order Logic*. Dover Publications, New York, second corrected edition, 1995. First published in 1968 by Springer.

[Spaan, 1993] E. Spaan. *Complexity of Modal Logics*. PhD thesis, Free University of Amsterdam, 1993.

[Suzuki, 1988] N. Y. Suzuki. An algebraic approach to intuitionistic modal logics in connection with intermediate predicate logics. *Studia Logica*, 48:141–155, 1988.

[Suzuki, 1990] N. Y. Suzuki. Kripke bundles for intermediate predicate logics and Kripke frames for intuitionistic modal logics. *Studia Logica*, 49:289–306, 1990.

[Thiele, 1993] H. Thiele. On the definition of modal operators in fuzzy logic. In *Proceedings of the 23rd International Symposium on Multiple Valued Logic*, pages 62–67. IEEE Computer Society Press, 1993.

[Thomason, 1980] S. K. Thomason. Independent propositional modal logics. *Studia Logica*, 39:143–144, 1980.

[Urquhart, 1972] A. Urquhart. Semantics for relevant logic. *Journal of Symbolic Logic*, 37:159–170, 1972.

[van Benthem, 1983] J. van Benthem. *Modal Logic and Classical Logic*. Bibliopolis, 1983.

[van Lambalgen, 1996] M. van Lambalgen. Natural deduction for generalised quantifiers. In J. van der Does and J. van Eijck, editors, *Quantifiers, Logic and Language*, pages 225–236. CSLI Lecture Notes, 1996.

[Veltman, 1986] F. Veltman. *Logics for Conditionals*. PhD thesis, Filosofisch Instituut, Universiteit van Amsterdam, 1986.

[Venema, 1990] Y. Venema. Expressiveness and completeness of an interval tense logic. *Notre Dame Journal of Formal Logic*, 31: 1990.

[Wansing, 1993] H. Wansing. *The Logic of Information Structures*. Number 681 in Lecture Notes in Artificial Intelligence. Springer-Verlag, Berlin, 1993.

[Wijesekera, 1990] D. Wijesekera. Constructive modal logic 1. *Annals of Pure and Applied Logic*, 50:271–301, 1990.

[Wolter, 1994] F. Wolter. Solution to a problem of Goranko and Passy. *Journal of Logic and Computation*, 4:21–22, 1994.

[Xu, 1988] M. Xu. On some U, S-tense logics. *Journal of Philosophical Logic*, 17:181–202, 1988.

INDEX